ANNUAL REVIEW OF PLANT PHYSIOLOGY

WINSLOW R. BRIGGS, *Editor*
Carnegie Institution of Washington, Stanford, California

PAUL B. GREEN, *Associate Editor*
Stanford University

RUSSELL L. JONES, *Associate Editor*
University of California, Berkeley

VOLUME 30

1979

ANNUAL REVIEWS INC. 4139 EL CAMINO WAY PALO ALTO, CALIFORNIA 94306

ANNUAL REVIEWS INC.
Palo Alto, California, USA

REPRINTS The conspicuous number aligned in the margin with the title of each article in this volume is a key for use in ordering reprints. Available reprints are priced at the uniform rate of $1.00 each postpaid. The minimum acceptable reprint order is 5 reprints and/or $5.00 prepaid. A quantity discount is available.

International Standard Serial Number: 0066-4294
International Standard Book Number: 0-8243-0630-9
Library of Congress Catalog Card Number: A51-1660

PRINTED AND BOUND IN THE UNITED STATES OF AMERICA

Annual Review of Plant Physiology
Volume 30, 1979

CONTENTS

ARTICLES IN OTHER *ANNUAL REVIEWS* OF INTEREST TO PLANT PHYSIOLOGISTS

From the *Annual Review of Phytopathology,* Volume 16 (1978):

Root Graft Transmission of Tree Pathogens, Abraham H. Epstein

Lectins and Their Role in Host-Pathogen Specificity, Luis Sequeira

From the *Annual Review of Microbiology,* Volume 32 (1978):

Chromatin in Eukaryotic Microbes, Paul A. Horgen and Julie C. Silver

From the *Annual Review of Genetics,* Volume 12 (1978):

The B Chromosome of Corn, Wayne R. Carlson

Transmission Genetics of Mitochondria and Chloroplasts, C. William Birkby, Jr.

From the *Annual Review of Biochemistry,* Volume 48 (1979):

RNA Processing and the Intervening Sequence Problem, John Abelson

Photoaffinity Labeling of Biological Systems, Vinay Chowdhry and F. H. Westheimer

Electron Microscopic Visualization of Nucleic Acids and of Their Complexes with Proteins, Harold W. Fisher and Robley C. Williams

Histones, Irvin Isenberg

Phosphorylation-Dephosphorylation of Enzymes, Edwin G. Krebs and Joseph A. Beavo

Structure and Function of Nitrogenase, Leonard E. Mortenson and Roger N. F. Thorneley

Regulation of Protein Synthesis in Eukaryotes, Severo Ochoa and Cesar de Haro

Lipid Asymmetry in Membranes, Jos A. F. Op den Kamp

The Isolation of Eukaryotic Messenger RNA, John M. Taylor

Mitochondrial Genes and Translation Products, Alexander Tzagoloff and Giuseppe Macino

The Assembly of Proteins into Biological Membranes: The Membrane Trigger Hypothesis, William Wickner

The Structure and Function of Eukaryotic Ribosomes, Ira G. Wool

ERRATA

Volume 29 (1978)

In "Crassulacean Acid Metabolism: A Curiosity in Context," by C. B. Osmond, the following changes should be made:

page 383, paragraph 3; please note that:

The parents of succulent orchid hybrids studied by Neales & Hew (108) and shown to have CAM are epiphytes in their natural habitats but are cultivated commercially as terrestrial plants. No orchids from natural terrestrial habitats are known to display CAM (S. C. Wong, personal communication).

page 390, Table 2; footnote b should read:

[b]Activities = μmoles/mg Chl/hr [not Chl/min]. . . .

page 406, line 4 should read:

CAM *by* short days (82), and the $\delta^{13}C$ of *K. daigremontiana* was 6 to 9‰ *less*. . . .

C. S. French

Ann. Rev. Plant Physiol. 1979. 30:1–26

FIFTY YEARS OF PHOTOSYNTHESIS

❖7662

C. Stacy French

Department of Plant Biology, Carnegie Institution of Washington, Stanford, California 94305

CONTENTS

INTRODUCTION

I first went to a scientific meeting as an undergraduate about 1928. The nature of the scientific community as well as the subject matter of plant physiology has changed so much in the past 50 years that it may be of some interest to make comparisons of then and now. I want to speak particularly about what might be called the sociology of science—the personal relations of scientists to each other that greatly affect their scientific output. The change as I see it has been essentially the transformation from a few small groups of near-amateurs with enthusiasm and interest in a wide range of

0066-4294/79/0601-0001$01.00

disciplines to a very large number of groups all intent on a small segment of science and seemingly with less interest in followers of other subjects than typical scientists used to have.

Possibly the easiest way to become a successful scientist is to cultivate the right people and to follow their way of life. By this means almost anyone, even of limited ability, may be channeled into more or less productive enterprises. Everyone's career, in science or any other field, depends largely on one's interactions with other people. Very few of us can develop a significant intellectual life from within ourselves. Therefore I will mention with appreciation some of the many people who have served as models, have been excellent guides to life in general, or who have been helpful with the details of particular scientific investigations. I hope to be forgiven for using the autobiographical format because it is by far the easiest way to string together a series of reminiscences and comments.

I was fortunate in getting a good start and continued guidance to age 25 from suitable parents. My father, Charles Ephraim French, born in Berkely, Massachusetts, in 1864 on a one-acre strawberry farm, put himself through the University of Maryland Medical School by carpentering and painting; then after European hospital experience, had a good practice of ear, eye, nose, and throat work in Lowell, Massachusetts. His comment on scientific research was "Well, every once in a while they give us something we can use." I was born in Lowell December 13, 1907. My mother, Helena Stacy, born in Colebrook, New Hampshire, in 1867, grew up in a small business and lumbering family and later lived in Bathurst, New Brunswick. She had a year in Europe, then two years at Radcliffe, followed by kindergarten teaching in Massachusetts.

My early education in Lowell public schools was not much good. Fortunately, respiratory infections made me lose so much time that a private tutor became necessary. This well-educated spinster, Flora Ewing, lightened the day's dose of Latin, English, and algebra with kitchen-scale chemical experiments. Before going away to school, I learned from my father how to do simple carpentry, wood turning, house painting, and general repair work that formed a basis for later laboratory life. I entered Loomis in Windsor, Connecticut, in 1921, failed the first year, repeated it the next and later, when the time came, was refused permission to take the Harvard entrance exam so as not to spoil the school's record. Fortunately, my mother put me in Lowell High School for a year. From there, I got into Harvard via a summer school course in botany given by Carroll W. Dodge. During the first year in college I started to prepare for engineering with math, physics, engineering drawing, and languages. In those days engineers could both draw and write. In the second year of college I discovered W. J. Crozier's course in general physiology and L. J. Henderson's introductory biochemistry, including some lectures by John Edsall, now a valued family friend 50 years later. The laboratory work in general physiology first brought contact with real science done with good apparatus and of sufficient significance to have an unpredictable outcome. From then on engineering was forgotten.

In early 1928 a few lectures on photosynthesis by Robert Emerson, who had recently returned with a PhD from Otto Warburg's laboratory in Berlin, got me interested enough to take Emerson's course on photosynthesis the following year, and I have stayed with the subject ever since.

When Harvard was somewhat smaller and under the direct personal care of President A. Lawrence Lowell, the faculty was composed largely of Boston gentlemen. Now that Harvard has become more like the large public universities, some of the local flavor that made it such a pleasant environment for quiet scholarship has been displaced by academic competitiveness. I have long been interested in observing whether academic excellence can be promoted without running into the destructive aspects of intense competition.

As an undergraduate I was lucky to have Professor George D. Birkhoff as an adviser. A prominent mathematician, he well understood the peculiarities of a creative academic career. He and his family guided me into the good life of Cambridge and Boston.

An undergraduate thesis on the temperature coefficient of catalase action under the guidance of W. J. Crozier and A. E. Navez very nearly kept me from graduating with my class in 1930 through lack of attention to the math and organic chemistry courses. My first paid scientific job was to help move the laboratory of general physiology from under the glass flowers in the Agassiz Museum to the new biology building guarded by the rhinoceroses. All went well in the new building after a pair of workmen's shoes were removed from the distilled water system.

GENERAL PHYSIOLOGY AT HARVARD

The years of graduate work and teaching assistantships in general physiology were excellent, not only because of the faculty but equally for the association with outstanding fellow graduate students and postdocs. I watched the slightly older men like Edward Castle experimenting with *Phycomyces*' response to light, Ted Stier measuring yeast respiration, Gregory Pincus being concerned with the sex life of rats, Fred Skinner studying snail behavior on sloping panels, Bob Emerson shaking manometers full of *Chlorella,* Doug Whittaker doing the same with *Fucus* eggs, and George Clarke getting equipped to measure light penetration in the ocean. Morgan Upton and Hudson Hoagland were somehow concerned about psychology —all this in one department.

My contemporaries felt the department seminars were too dominated by the professor, so after a year or two we organized the "Chlorella Club" meeting weekly in secret session to educate each other. Bill Arnold had just come to Harvard from Caltech, where he and Bob Emerson had measured the size of the chlorophyll unit. Bill and Caryl Haskins, later President of the Carnegie Institution, led most of the discussion, while Henry Kohn maintained a skeptical attitude. The Chlorella Club gave us more education than any of the biology courses. Continuing interaction with Bill Arnold and Caryl Haskins has been a lifelong pleasure. The only formal courses I remember much about during the graduate years were George Shannon Forbes' "Photochemistry" and James B. Conant's "Natural Products" and "Physical Organic." My major conclusion from all this is that exposure to great men is more valuable than the subject matter acquired from them.

Two summers at Woods Hole in the early 1930s, one to take the invertebrate course, the other to study the effect of light on respiration of a red sponge, led to a broader acquaintance with American biologists. In graduate school I eventually did some experiments on the rate of respiration of *Chlorella* at various temperatures. Having accumulated too much data, some of doubtful significance, it became nearly impossible to write a doctoral thesis. My life was saved by Pei-Sung Tang, a postdoc from China, who spent much time helping me organize the data. We also collaborated on measurements of the rate of *Chlorella* respiration as a function of oxygen pressure, which resulted in my first scientific paper appearing in the *Chinese Journal of Physiology*. It was a great pleasure to be in correspondence with Pei-Sung Tang again about 1974 after his long leadership of both academic and practical plant physiology in China. In 1934 the good times came to an end with the completion of the degree requirements. Due to the depression, few academic positions were available. My applications to teach at Williams and at Howard were turned down and there was no way to stay on at Harvard. Bob Emerson however, allowed me to go to his lab at Caltech to work on photosynthesis in purple bacteria.

WITH VAN NIEL AND EMERSON IN CALIFORNIA

On the way to Caltech I spent the summer taking Kees van Niel's famous course on microbiology at the Hopkins Marine Station at Pacific Grove, California. No one else I ever knew could lecture for four to six hours without losing the student's attention. That summer I visited the Carnegie Institution's laboratory at Stanford and met Herman A. Spoehr, James H. C. Smith, and Harold Strain. Spoehr was a most helpful and kind adviser whose guidance I relied on until his death in 1954. I became his successor in 1947.

As an undergraduate, under the stimulus of A. E. Navez, I had developed an interest in purple photosynthetic bacteria. Through van Niel's course and his personal interest, I learned to grow and work with bacteria more efficiently. At Caltech I tried to measure their photosynthetic efficiency, but the excellent skiing in the mountains near Pasadena left little time for science, so that year was not productive. Bob Emerson was justifiably disgusted with my performance and we were barely on speaking terms for the academic year. Some years later we became friends again. However, in spite of my poor performance, he arranged for me to spend the next year in Berlin with Otto Warburg, which was what saved my scientific career.

At Caltech I consulted the optics professor, Ira Bowen, later a C.I.W. colleague, about how to measure the absorption spectra of purple bacteria in the near infrared. He referred me to Theodore Dunham, who said it could be done with the astronomical spectrograph on Mount Wilson. We made a date for a particular night on the telescope with the understanding that if it was clear he would work on a star, but if it was cloudy we would photograph spectra of bacteria. The weather favored the bacteria. That was probably the only time astronomical equipment has been used for the study of microscopic objects. On another occasion I wanted to calibrate a thermopile and was told that a young assistant professor, Arnold O. Beckman, had a setup for that purpose in the chemistry department. He generously helped me out. There is an uncorroborated story that he had a market survey made when preparing to manufacture the Beckman D.U. spectrophotometer about 1940. The survey report is reputed to have been that 100 instruments would saturate the market. About 1976 production was discontinued after something like 30,000 of these spectrophotometers and a very large number of competing instruments had been sold. These figures must bear some relation to the explosion of scientific research in general as well as of biochemical plant physiology.

WITH WARBURG IN BERLIN—DAHLEM

The Cold Spring Harbor Symposium on Photosynthesis in June 1935 was my introduction to the world of professional photosynthesis investigators. The Physiological Congress in Leningrad and Moscow that summer followed by the Botanical Congress in Amsterdam provided some acquaintance with European science.

While in Leningrad I had the pleasure of a visit to Professor N. N. Lubimenko, a corresponding member of the American Society of Plant Physiologists. He was the first as far as I know to believe in the existence of different forms of chlorophyll *a* although the evidence then was not clear. His laboratory in the Botanical Garden of Leningrad was very simple—all the equipment visible was a microspectroscope and his only collaborator seemed to be one old lady. He was extremely kind and helped me revise and clarify the speech I was to give at the Physiological Congress. This was before biochemistry split off from physiology to have its own international congress, so papers on photosynthesis were a reasonable part of a physiology program.

In Stockholm I visited the biochemist, Professor H. von Euler, who treated me very well initially. After some pleasant and informative conversation, he asked about my plans. Two minutes after telling him I was on the way to work with his enemy, Professor Warburg, I was out in the street. That kind of intense personal competitiveness seems to be less common now than it was in those days—perhaps because of the greater number of people now involved in every field. By contrast with the visit to von Euler, Professor John Runnström, the zoologist, was most kind and gave me a fine country weekend complete with lessons on the details of expected behavior at Swedish dinner parties, which I still find formidable.

The year with Professor Otto Warburg at the Kaiser Wilhelm Institut, (now the Max Planck Institut) was one of intense concentration on the efficiency and action spectrum of photosynthesis in *Rhodospirillum rubrum.* Living in the laboratory building and eating at Harnack House just around the corner in Dahlem was most convenient. The laboratory was excellently supplied with optical equipment left over from Warburg's determination of the action spectrum for dissociation of the CO complex with the respiratory enzyme, which had brought him the Nobel Prize. I was treated very well by the professor and all his staff. The rigid discipline and long hours without outside distraction were just what was needed to convert an easygoing freewheeling academic type into a professional scientist.

Training in constructing apparatus specifically for particular research purposes was of great value. At Caltech I had experimented with a Christiansen filter to produce monochromatic light. The principle is that powdered optical glass suspended in an organic liquid will scatter all wavelengths of light unless the refractive index is identical for the glass and the liquid so the light can therefore pass straight through. Since the refractive index for the liquid and the glass vary differently with temperature, the wavelength transmitted can be selected by temperature control. In Dahlem I constructed several of these devices but only later found out that the optical glass powder has to be annealed to remove strains produced by the pulverizing.

The unusual intellectual climate of Dahlem under the Weimar Republic several years before my time there has been well described by Nachmanson (3). I saw little of Dahlem science outside the laboratory except for a few seminars on photosynthesis at Max Delbrück's house with Hans Gaffron and Eugene Rabinowitch. During the 1935–36 year that I was in Berlin, the good days of the Weimar Republic had largely faded away. The Hitler government was busily working up war spirit through well-planned propaganda. It was disillusioning a few years later to see the same propaganda tricks being played at home by our own government. That the well-established procedures for inciting any population to enter a war are not immediately recognized by the people being manipulated is most unfortunate.

At that time quantities of horse blood were being processed in the basement of Warburg's Institute to isolate and identify a redox substance then called coenzyme II, later TPN, and now NADP. Daily association with the small group of the professor and the assistants—Negelein, Kubowitz, Haas, Lütgens, and Gerischer—made it possible to learn many biochemical procedures by watching the experts. As far as I remember, these people, with a janitor, a secretary, two mechanics, and a courier, were all the inhabitants of the most famous biochemical laboratory of the time. (Apologies to Cambridge biochemistry where a different scale of values may prevail.) Fortunately this was before all German scientists spoke English, and I am grateful to Warburg for refusing to speak English though he could do it better than I. There one saw the advantages of a small laboratory free from the distractions of students, seminars, and committee meetings. In later years, however, that sort of isolation showed its negative effect when Warburg, no longer at the top of the field, refused to consider the advances in photosynthetic research made in other laboratories.

WITH HASTINGS AT THE HARVARD MEDICAL SCHOOL

Having lived so far on a modest inheritance, I returned to the United States in the late summer of 1936 to look for academic employment. I was rescued by Baird Hastings, who gave me an Austin Teaching Fellowship ($1000 per year) in biochemistry at Harvard Medical School for two years. This allowed half time for research, so it was possible to continue with purple bacteria photosynthesis.

One lesson I learned at the Medical School was to believe and to publish experimental results. I found that light completely stopped O_2 uptake by purple bacteria, but never having heard of the effect, I thought something must have been wrong with the experiment, so I dropped that work. Many years later the effect was discovered and taken seriously by others.

Another valuable experience was to see the difference between "big science," or at least what was big for those days, and the small scale one-man research then more typical of academic life than it is today. My objective at the time was to measure the absorption spectra of various strains of purple bacteria and to compare them with water extracts of the organisms. Attending a meeting of the Optical Society at MIT, I heard one of the older members, Frederic Ives, or one of his contemporaries make the following remark: "The time has about gone by when one can get a PhD in physics for publishing the absorption spectrum of a single substance." The only commercial spectrophotometers available then were the Koenig-Martens visual-balance monstrosities that worked only for clear solutions and in the visible spectrum. However, these bacteria have their most interesting bands in the near infrared. I wanted to build a near infrared photoelectric spectrophotometer suitable for use with scattering suspensions. The facilities at Harvard Medical School were not adequate for such a deviation from standard biochemistry, so I went over to MIT for a talk with Professor George Harrison. He had a fine laboratory for all sorts of spectroscopy with many collaborators. This was "big science" for that time. His program was to remeasure the emission lines of all the elements. He agreed that my plan was feasible, but insisted that one of his graduate students be paid $200 to follow me around to be sure I didn't spoil any of his equipment. Perhaps he had had unfortunate experiences with someone else having medical connections. However, neither I nor anyone else in my situation had $200 for research expenses. That was my first and only contact with the "big science" of that day. After a visit to Professor Theodore Lyman at the Harvard Physics Department, I was given all the equipment and help necessary. Several kind people, knowledgeable in electronics and in optics, helped me put together a workable instrument, and the measurements were completed without difficulty in a very pleasant atmosphere.

Breaking the bacteria, however, was a problem. I tried grinding them with abrasive powders between rotating glass plates and also by a cheap and simple procedure with a hypodermic syringe rotating in a lathe. The outlet of the syringe was plugged and the stationary plunger was forced in slowly. This squeezes the bacteria out between the ground glass surfaces. Also I had heard that Stuart Mudd and Alfred Loomis had broken bacteria with supersonic vibration. Luckily a magnetostrictive supersonic generator was in the physics building, the creation of Professor G. W. Pierce, who kindly let me use it. As far as I know, this was the first use of supersonic vibration to break photosynthetic organisms.

While at the Harvard Medical School, I had the good luck to watch a demonstration by O. A. Bessey of his method for titrating ascorbic acid with dichlorophenol-indophenol. Ascorbic acid had recently been isolated from cabbage by Szent-Georgy. Several years later the memory of this demonstration led me to try indophenol dyes as Hill reagents for detection of chloroplast activity. Among the people in the Hastings laboratory then were John Taylor, a former student of W. Mansfield Clark, Kenneth Fisher, and

Oliver Lowry, later to become the most often cited man in science because of his protein determination procedure. It was a stimulating group. These two years were great training in biochemistry, but for a longer period it would not have been appropriate to try to make a career of photosynthesis research in a Medical School. I am most grateful to Baird Hastings for his support while I was looking for a suitable position.

ASSISTANT TO FRANCK AT CHICAGO

In the fall of 1938, the day before the great hurricane, I left for Chicago with an appointment as instructor (research) in chemistry ($2400 per year). This was to help James Franck set up a photosynthesis laboratory with support from the Fels Foundation. The unspoken thought that seemed to be in the air was that a Nobel Prize winner in physics would easily be able to take care of the photosynthesis problem in a few years with some simple critical experiments. My first assignment was to study the time course of chlorophyll fluorescence from green leaves while the other assistant, Foster Rieke, a physicist, was to measure quantum yields of photosynthesis. In December I returned to Cambridge briefly to marry Margaret Wendell Coolidge, daughter of the mathematician, Professor Julian L. Coolidge, first Master of Lowell House at Harvard. We have had a very happy home life ever since. In addition to her work as a counselor, she has taken complete responsibility for household affairs and family life, thus leaving me free to concentrate on science. Through Margaret I again became on good terms with Bob Emerson since their fathers had been college friends.

It soon became evident that Franck and I had very different ideas about the conduct of research. Given the phenomenon of complex time courses of chlorophyll fluorescence, I wanted to explore systematically the influence of temperature, CO_2 and O_2 concentration, previous light regimes, etc on the effect. His plan was to think about the situation, develop a theory, and plan a critical experiment that would prove or disprove the theory. According to that view, my job was to do the critical experiment. I survived this unhappy situation for three years before a better position became available. After the first year, Hans Gaffron, a former Warburg collaborator, joined the group. He was better able to deal with the conditions and managed to work more independently. For many decades Gaffron remained in the small group of a dozen or so leading investigators of photosynthesis. One of his major contributions was the idea of "photosynthetic units." I learned some practical physics from Rieke to combine with the biochemistry picked up during the past three years.

At Chicago the studies of kinetics of fluorescence intensity changes with time of illumination brought the effects discovered by Kautsky in Marburg to the attention of photosynthesis workers in this country. A graduate student, Ted Puck, joined in this project. He later became famous by culturing human tissue cells. A freshman student was assigned to help in the work of the Fels group through some sort of student aid program about 1939. His first assignment was to prepare a set of neutral absorbing filters for attenuation of light beams. This set of filters is still in daily use at the Carnegie Institution and has played a significant part in experiments leading to a large number of publications by many investigators at Chicago, Minnesota, and the Carnegie Institution. This was Roderick Clayton, now a Cornell professor and member of the National Academy, well known for his studies of the photosynthetic reaction centers and his excellent books on photosynthesis. About 1940 Robert Livingston, already a distinguished photochemist, joined the group. His later work as a professor of chemistry at the University of Minnesota on spectroscopy of chlorophyll in solution free from water vapor led to some of the present ideas about chlorophyll aggregation. After I escaped from Chicago, Warren Butler did his graduate work with Franck and continues as a leader in the biophysical side of photosynthetic research at the University of California, San Diego.

My life changed greatly for the better one day when Mortimer L. Anson, of the Rockefeller Institute in Princeton, on his way home from an Arizona vacation dropped in at the Chicago chemistry department to tell James Franck about Robin Hill's discovery of oxygen evolution by isolated chloroplasts. He stayed about a month while we repeated Hill's experiments with some variations. In spite of Franck's opinion that all this had nothing to do with photosynthesis, he did allow me to continue to play with isolated chloroplasts. Tim Anson and I worked on chloroplast isolation, stabilization, and O_2 evolution kinetics. We eventually prepared a manuscript for the 1941 American Society of Plant Physiologists meeting in Texas. Neither of us could attend, so Jack Myers (2) read our paper at the meeting. Anson was trained in the customs of organic biochemistry, so he proposed to call the effect the "Hill Reaction." Following Jack Myer's presentation in Texas, the chloroplast reaction has been so known, except by Robin. Our first objective was to try all the common biochemical tricks of homogenization, freeze drying, fractionation, etc on chloroplasts.

TEACHING PLANT PHYSIOLOGY AT MINNESOTA

I continued these chloroplast projects after being brought to the University of Minnesota by George Burr to be assistant professor of botany in 1941. There the memory of Bessey's ascorbic acid titration by dichlorophenolindophenol led to use of that blue dye as a Hill reagent that could be followed photoelectrically. Fortunately, I then had at Minnesota the one and only doctoral student of my whole career, Stanley Holt. His vigor and stamina developed as a long distance runner was applied to studies of chloroplast activity and later, on his own, to chlorophyll chemistry. Together we investigated the stoichiometry of the reduction of various dyes and inorganic salts by chloroplasts. We also found by use of $^{18}O_2$ that the oxygen evolved by isolated chloroplasts comes from water, as was already known for whole cell photosynthesis.

At the Minnesota botany department I inherited a spectroscopy laboratory set up by Elmer Miller as a university-wide service department. This was intended to take care of both absorption and analytical emission spectroscopy for all comers. Fortunately only a few agronomists wanted chlorophyll determinations and some analyses of plant material for inorganic nutrient deficiencies. The service aspects were soon forgotten, so I had access to much useful equipment.

In the days before large-surface photomultiplier tubes or before Shibata's opal glass techniques for measuring the absorption of scattering material like leaves or cell suspensions, the Ulbright sphere seemed more essential than it does now. Glenn Rabideau, Stanley Holt, and I put together a large white sphere with a homemade monochromator using a large replica grating. With this we measured the absorption spectra of various leaves and algae.

For chloroplast disruption I used a piezoelectric quartz supersonic oscillator at the University of Minnesota chemistry department. This machine, imported from Germany by Freundlich, the colloid chemist, was the typical Hollywood idea of scientific equipment. It comprised a large table covered with electronics of the 1920–30 era. The two-foot high vacuum tubes were spectacular when in operation. The apparatus was made available and its use explained by a former collaborator of Freundlich. Thus chloroplasts were first intentionally broken into smaller pieces in the hope of isolating the active particles.

At Minnesota I was exposed to real botanists like Ernst Abbe, Lawrence Moyer, Orville Dahl, William S. Cooper, and Don Lawrence, the plant physiologists George Burr, Allan Brown, and Albert Frenkel, and such biochemists as Harland Wood and R. A. Gortner. One unexpected pleasure was to meet Leroy S. Palmer, whose book on carotenoids I had studied long before.

Before the second World War, I had two worries about research in photosynthesis. One was that far too little biochemical work was being done in relation to the physical thinking of the time. The other worry was the apparent shortage of young scientists interested in the subject. It is now obvious that these two concerns need not have been disturbing in view of the predominantly biochemical nature of the present work on photosynthesis. The number of people now engaged in photosynthetic research is overwhelming.

During the war years I was registered as a conscientious objector, so I had to find some occupations more acceptable to the draft board than merely teaching plant physiology. When will freedom from conscription take its rightful place in the list of basic human rights? Draft dodging led me at various times into teaching elementary physics, researching chlorophyll-containing paint for camouflage purposes, and a long project on mold selection for penicillin production in E. C. Stakman's Department of Plant Pathology. There I enjoyed an association with Clyde Christensen.

THE CARNEGIE INSTITUTION

In the summer of 1946, Herman A. Spoehr invited me to spend a month visiting the Carnegie Institution of Washington, Department of Plant Biology, at Stanford with the thought of joining the group. During the following winter he arranged a meeting for me in Chicago with Vannevar Bush, President of CIW, and with Alfred Loomis, a physicist and trustee of CIW. As a result I became Director of the Department of Plant Biology to succeed Spoehr on July 1, 1947. Spoehr remained on as a staff member while we worked closely and happily together for the rest of his life. There were two groups in the department: Biochemical Investigations (essentially limited to photosynthesis), and Experimental Taxonomy, consisting of Jens Clausen, Bill Hiesey, David Keck, and Malcolm Nobs. The latter group worked largely together on transplant studies to separate and identify the relative contributions of heredity and environment in plant development, originally under Harvey Monroe Hall, a professor at Berkeley and a research associate of the Carnegie Institution.

With the hope of producing an improved strain of range grass, the experimental taxonomy group was making crosses of different grasses that had probably not previously met in nature. This was done on a very large scale with help in testing the progeny from many experiment stations in various parts of the world. The history of that enterprise shows some of the advantages of research support by a private organization with sufficient patience. In the first place, cooperation with universities and experiment stations in various countries was easily arranged by Jens Clausen, the group leader. Secondly, the work was continued for several decades without pressure to publish prematurely. Annual reports and occasional papers on certain aspects of the results were published, but only now after 35 years is the final survey of the whole range grass project nearing completion by Hiesey and Nobs.

The photosynthesis group, mainly working individually, was made up of Spoehr, James H. C. Smith, Harold H. Strain, and Harold Milner, all basically chemists. Fergus D. H. Macdowall and Violet Koski came with me from Minnesota as graduate students. Spoehr and Milner were working on large scale algal culture. *Chlorella* culture continued with many collaborators for six more years and eventually resulted in the Institution's all time best seller, *Algal Culture from Laboratory to Pilot Plant.*

Once I asked James Smith if he was going to the next American Society of Plant Physiologists meeting. He replied, "When I was a boy I lived on the bank of the Ohio River. A small steamboat worked its way up against the current but when blowing its whistle lost steam and drifted down again. No meetings for me this year."

Until about the middle of this century, the justification for spending time and money on research in photosynthesis was that knowledge of its basic mechanism might lead to increased food production. Of course, that thought is still valid, but now the expectations are less specific and photosynthesis research itself is seen more as an essential part of the larger scientific enterprise from which practical human benefits may arise in many unpredictable ways. The *Chlorella* culture work was an attempt to force some practical value out of existing information about photosynthesis. The actual results, however, were of more value in educating scientists into the basic facts of economics than in feeding hungry people. At least we tried.

We had a small shop for wood and metal working where I usually spent more time than at the desk or in the laboratory. The shop was open to all scientists and we have had excellent mechanics, first George Schuster, who left after a few years to head the shop of the chemistry department at UCLA, then Louis Kruger, great-nephew of Oom Paul Kruger of South Africa, and now Richard Hart, who with Frank Nicholson has set high standards for construction and care of laboratory equipment. All the accounting and secretarial work of the laboratory was done when I first came by Wilbur A. Pestel, who had worked previously in the CIW administration building in Washington, at the Desert Laboratory in Tucson, Arizona, and in the department's laboratory in Carmel.

In 1902 the Carnegie Institution founded the Desert Laboratory in Tucson, Arizona, and in 1905 Dr. Daniel T. MacDougal became its director. Spoehr joined the group in 1910 after research experience in Berlin and at the University of Chicago with Ulrich Nef. Spoehr, Otto Warburg, and H. O. L. Fischer, Emil Fischer's son, all had been postdocs together in Emil Fischer's laboratory. A summer laboratory at Carmel was used by the department until early in the 1940s. The present laboratory at Stanford was constructed two years after Spoehr became chairman in 1927. Robert Emerson and Charlton Lewis had worked here for three years from 1938 and while measuring action spectra had discovered the "red drop." The later explanation of that effect and of Lawrence Blinks's chromatic transient experiments revolutionized the ideas about photosynthesis by making obvious the existence of two separate photochemical systems.

Until he retired in 1956, Dr. Vannevar Bush, President of CIW, visited the laboratory for several days once or twice each year. These visits were extraordinarily stimulating because of his intense interest in the details of our activities. His great personal warmth and enthusiasm always left us invigorated for long periods. This tradition was continued by Dr. Caryl Haskins, who succeeded Dr. Bush in 1957 as president of the Institution. It was also a pleasure to serve under Caryl, who was an old friend from graduate school times at Harvard. In addition to these two good presidents, the routine affairs of the Institution in Washington were always easily and pleasantly handled by the most tolerant and understanding executive officers, Paul A. Scherrer and later Edward A. Ackerman. Because of the small size of our department and the kindly attitude of the CIW officers in Washington, the administrative work involved in being director was simple and left most of the time for research.

RESEARCH FELLOWS AND VISITING INVESTIGATORS AT CIW

About the time I came to the Institution, Dr. Bush was starting a fellowship program largely for postdoctoral training. This program has the long-recognized values of the apprenticeship system. It also brings to each laboratory knowledge and techniques developed in other places. By careful selection of stimulating fellows it is possible to keep a small permanent staff near the advancing edge of scientific progress. To select the most promising young people for fellowships, James H. C. Smith and I traveled extensively and tried to keep in close contact with the leading laboratories in photosynthesis and related subjects. We were helped a great deal by Dr. Spoehr, who had previously spent a year as Director of Natural Sciences for the Rockefeller Foundation and knew all the photosynthesis people. In 1950 and again in 1955 Smith and I visited 50 European laboratories. After that we decided to go oftener and to cut the number of laboratories visited to a much smaller list of more specialized places. Much of our hunting later was done at meetings even though we continued to believe that laboratory visits are far more definitive in getting a picture of a man's ideas and working habits. We were continually appreciative of the wisdom of the old adage: "Never ask a professor what he is doing because he will probably tell you." This frequently takes a long time. Our systematic visiting turned up many extremely competent investigators and in looking back on the list of fellows, my major regrets concern only those we could not take for one reason or another. Of the Institution's particularly successful "graduates," I expect most of them would have done equally well regardless of their time here and that their contribution to our life exceeded whatever they may have acquired by coming. In addition to the young fellows, we also had visits of various lengths from well-established investigators on sabbatical or shorter periods of leave. The intangible residue left at a laboratory through which a number of great men have passed is significant in itself. Lists of the Institution Fellows and other associated scientists up to 1962 are given in Year Book 61, "Report of the President," p. *76* and p. *102* and were brought up to 1973 in Year Book 72, pp. 321–27. The latter article also gives an abbreviated survey of the department's history. Details of each year's work are reported in the Annual Report of the Director of the Department of Plant Biology in the CIW Year Book.

The Biophysical Research Group at Utrecht, founded by A. J. Kluyver and L. S. Ornstein with a Rockefeller grant, operated for many years under the direction of E. C. Wassink and later of J. B. Thomas. The work of that group on photosynthesis was of great interest to all researchers on the subject. James Smith and I were frequent visitors, and a succession of scientists from that group worked at the Carnegie Institution. These are Bessel Kok, L. N. M. Duysens, Joop Goedheer, C. J. P. Spruit, Cornelis Brill, and currently G. van Ginkel. Related to this group is also Jan Amesz, a "second generation Utrecht scientist" trained by Lou Duysens, now Professor of Biophysics at Leiden.

From Japan we had very productive visits from Hiroshi Tamiya, Atusi Takamiya, and Kazuo Shibata of the Tokugawa Institute, Norio Murata of Tokyo University, and Tetsuo Hiyama. All these Japanese people became close friends of our group.

From Scandinavia came various other delightful friends, now distinguished leaders in the field of plant physiology or related subjects. These were: Hemming Virgin, Göteborg; Per Halldal, Oslo; Lars Olof Björn, Lund; Axel Madsen, Diter von Wettstein, and Erik Jorgensen, Copenhagen; Hedda Nordenshiöld, Uppsala; Axel Nygren, Uppsala. Continuing friendship with Hemming and other Scandinavian colleagues lead to a very pleasant lecture tour and a Göteborg degree in 1974. Among our German Fellows or Visiting Investigators were Günter Jacobi, Alexander Müller, and Wolfgang Wiessner from Göttingen; August Ried and Eckhard W. Gauhl, Frankfurt; Friedrich Ehrendorfer, Vienna; Wilhelm Menke, Cologne; Eckhard Loos, Regensburg; Ulrich Heber, Düsseldorf; Helga Ninnemann, Tübingen; Ulrich Schreiber, Aachen via Vancouver; Carl Soeder,

Dortmund; and Wolfgang Urbach, Würzburg. Contact with German photosynthesis workers such as Karl Egle, André Pirson, and others resulted in an invitation in 1964 to speak at a meeting in Halle and become a member of Leopoldina. Kenneth Thimann was also in the group and on this occasion we celebrated our third penetration of the Iron Curtain together. With regret I omit a complete listing of our Fellows and Guest Investigators from other countries and the USA. It was a great pleasure to have Gordon Gould, a fellow biochemical sciences college classmate appear later on the Stanford faculty so that our friendship since freshman days could be resumed.

The laboratory was so small, usually about 15–20 people all together, that all visiting scientists became close personal friends of the whole group. We rarely had more than 3–4 visiting scientific workers at a time and usually all the staff followed each other's activities and those of the visitors with considerable interest. It is sometimes said that the department's Annual Reports are not very good publications because of the lack of outside reviewers. This comment is mildly amusing to those of us who lived through more than one of the Annual Report seasons when each member of a group, including visitors, independently edited each other's work in far more critical detail than would have been expended in reviewing any journal article referred by an editor.

SOME OF THE EXPERIMENTAL WORK AT CIW

Chlorophyll is clearly the most obvious organic material in the world. Nevertheless, more is known about the spectroscopy of the rare earth elements than about chlorophyll in its natural state as it occurs in leaves. This situation was mentioned in my first annual report in 1948 and much time since then has been spent on spectroscopy of in vivo chlorophyll. The revival of Lubimenko's ideas and their great expansion by Krasnovsky about 1945–1955 stimulated me to concentrate on the various forms of chlorophyll *a*. It is now clear that there are four major forms of chlorophyll *a* with absorption peaks at 662, 670, 677, and 684 nm in all green plants. Much of James Smith's work when I first came to the Institution was on chlorophyll formation and the transformation of the protochlorophyll by light. Smith, Koski, and I collaborated on the action spectrum for this photochemical effect in dark-grown seedlings. During this time Harold Strain was perfecting his chromatographic methods of pigment separation, an art only slowly being rediscovered since the pioneer work of Tswett long before. Unfortunately for plant physiology, Strain was called away to the Argonne laboratory to apply chromatography to a wider range of chemicals.

Continuing with Harold Milner and others the chloroplast fractionation experiments started in Minnesota, I was always looking for better ways to disrupt chloroplasts. My attempts to do so with detergents were never successful for lack of work with finely graded series of different concentrations. My crude experiments either did nothing with low concentrations or dissolved and ruined the whole mixture with high concentrations. Dr. Bush discussed our individual problems in considerable technical detail. He told me about some company that could make very small holes in sapphire and suggested that extruding a chloroplast suspension through such a small hole might do a good job of disruption. This sounded very promising, but I worried about getting the hole plugged up by fibers and bits of dirt. To have a small hole that could be opened up when plugged and then constricted again led me to the idea of a needle valve instead of a fixed hole. I threaded an ammonia needle valve into a steel cylinder with a hole in the center to contain the chloroplast suspension. The suspension was forced through the valve by a steel piston with a seal. This extrusion principle worked well and has since found much use for disruption of bacteria and other cells as well as for chloroplasts. When I told Dr. Bush that the device was being made commercially, he suggested that I ask the manufacturer to attach my name to the device and to give a free one to the laboratory. The first request was acceded to at once, and about 20 years later the American Instrument Company gave the department a "French Press." We still drive it with an old hydraulic jack made for automobiles.

Combining an interest in chlorophyll fluorescence from the time at Chicago with a taste for building spectroscopic equipment, I put together two homemade grating monochromators. One of these was used to irradiate the sample such as a leaf or an alga with blue light while the other monochromator had its wavelength setting swept by a synchronous motor to record the emitted fluorescence as a function of wavelength. A synchronized drum carrying a hand-drawn curve corrected the photomultiplier output to give the true emission spectrum of the sample. This was the first automatic recording fluorescence spectrophotometer ever built. For its construction and for many similar purposes I found the part-time help of students at Stanford's electrical engineering department to be indispensable.

In the summer of 1949 we enjoyed a visit from Richard H. Goodwin of Connecticut College, who used fluorescence spectroscopy to identify traces of insoluble uroporphyrin particles in the cells surrounding the guard cells of *Vicia.* These particles have a brilliant orange-red fluorescence which Goodwin duly recorded by color photomicrography and sent in for development. A package of tourist photos of the Matterhorn came back.

Shortly after I came to the laboratory, Shao-lin Chen, a research fellow, measured the action spectrum for di-chlorophenol indophenol reduction by spinach chloroplasts, confirming for this reaction the "red drop" of Emerson and Lewis. The carotenoids present did not appear to be active.

In 1953 we had a visit of a month more or less from Robin Hill. With James Smith we started to measure the absorption and fluorescence spectra of protochlorophyll. The stimulation from these experiments and the subsequent discussions were out of proportion to the results published from this brief collaboration, but they led to further related studies and to life-long friendships.

After Violet Koski and I had measured some fluorescence spectra of red algae, we found it necessary to resolve these spectra into their overlapping components. To do this we wanted to add together, in adjustable proportions, curves of the individual substances present in the mixture. With help from an engineering student, George Towner, we experimented with vertically movable tables carrying heavily inked curves. The curves were followed by a light beam and photocell assembly attached to a potentiometer. This arrangement produced a voltage proportional to the Y-axis of the curve as the X-axis was driven along. For four years I worked on this device with various engineering students. Eventually we had a so-called "Curve Analyzer and General Purpose Graphical Computer." This consisted of five curve follower tables, a recorder, two integrators, numerous adding amplifiers and servo drives. In final form it was much more versatile than the original plan. This device served us well until Stanford's IBM computer and associated plotting system displaced it about 15 years later. One of the early experimental curve followers later was converted to a curve digitizer. It is now used to tabulate numbers from plotted curves and enter them into the digital computer.

In 1953 I started to build a first derivative spectrophotometer to detect small changes in the slopes of absorption spectra. This was based on a slit vibrating over a small wavelength interval in the spectrum. The construction and subsequent modifications took several years. With this, as well as with the curve analyzer, the continued technical interest and support by Dr. Bush was a great help. At first the machine plotted the derivative of transmission against wavelength but later it was altered to plot the derivative of absorbance. We surveyed the absorption spectra of various algae. When we got to *Euglena* the strange results made us suspect trouble in the apparatus. However, the irregularity turned out to be an unexpected form of chlorophyll *a* with an absorption band at 695 nm. With Dr. Jeanette S. Brown and others, work on the different forms of in vivo chlorophyll was continued for many years.

A simple but useful technique for removal and readdition of lipid components of chloroplasts was developed in 1957 with Victoria Lynch. Extraction of dry chloroplasts with petroleum ether greatly reduced their Hill reaction capacity, but readdition of the extract or of carotene solutions in hexane to the dry material followed by evaporation of the solvent caused reactivation. The remarkable effect was that the readded substance went back to the right place. Subsequent work by others has implicated plastoquinones as well as β-carotene in the reactivation.

In that same year we enjoyed Per Halldal's first visit. He arranged a spectral projection apparatus with a vertical intensity gradient on one side of a glass vessel. On the other side uniform light from the opposite direction fell on the same vessel. Within the vessel some motile algae in suspension plotted their own action spectrum for phototaxis by swimming to one side of the vessel or to the other side.

To study the growth of algae at various temperatures and light intensities, Halldal and I fixed up a thick aluminum plate with a thin layer of inoculated agar on its surface. One side of the plate was cooled by circulating cold water; the other side was heated to produce a temperature gradient across the plate. At a right angle to this temperature gradient a light intensity gradient was established. Thus combinations of temperature and intensity were provided at different places on the plate and the resulting growth pattern gave a visual picture of the intensity and temperature effects on growth rates. Later Ruth Elliott and I used the same apparatus to study lettuce seed germination as influenced by light at various temperatures.

A new technique was brought to the group by Francis Haxo, who worked at the laboratory for several months about 1957. He showed us how to measure oxygen exchange easily. The platinum electrode arrangement originally developed by Blinks and Skow had been used by Haxo and Blinks for measuring photosynthetic action spectra of algae. About this time Jack Myers came for a sabbatical year and with this platinum electrode measured action spectra for the Emerson enhancement effect and also for Blinks' chromatic transients. Furthermore he found that the two wavelength enhancement effects persisted even though the two beams were given consecutively instead of concurrently. This established the nature of the interaction as being due to accumulation of a chemical substance produced by one light reaction and used up by another.

The rate-measuring O_2 electrode was also used in an apparatus for automatic recording of action spectra as the wavelength was continuously varied. With this device the rate of O_2 evolution as measured by the electrode output was used to control the light intensity so that the rate of O_2 evolution was kept constant as the actinic wavelength was swept through the spectrum. The reciprocal of the light intensity, so adjusted to give a constant rate of O_2 evolution, was then plotted automatically against wavelength. Later this system was improved by Per Halldal during a return visit in 1969. A variety of Blinks O_2 electrodes were used by many different investigators over a long period of years. Among this group were also: Guy McCloud, David Fork, Govindjee, Martin Gibbs, William Vidaver, Carl Soeder, Yaroslav de Kouchkovsky, August Ried, James Pickett, and Eckhard Loos. After David Fork returned from Witts laboratory in Germany and set up equipment for measuring absorbance changes, the O_2 electrode was less in demand. This was because the absorption changes can be used to tell which particular substances in the cells or chloroplasts are actually changed by a light exposure rather than merely measuring overall metabolism.

This absorption change technique was first used for photosynthesis investigations by L. N. M. Duysens at Utrecht. He came to the laboratory in 1952 and extended his previous work with absorption changes in purple bacteria to *Chlorella* and other photosynthetic species. He thus discovered the "515 change" later found to be caused by shifts in the carotenoids.

Before 1947 Harold Strain had been interested in the possibility that changes in the absorption spectrum of leaves might be caused by light. He looked at leaves with a visual spectroscope, left them in light for minutes to hours, then took them downstairs to the spectroscope to look for changes. None were observed. About that time I took a flask of *Chlorella* culture to Britton Chance's laboratory in Philadelphia where he had excellent optical equipment and techniques. We pumped the *Chlorella* suspension slowly through his apparatus. The suspension first went through one measuring chamber, then through a chamber where it was exposed to light, then through a second measuring chamber. The transmission difference for various wavelengths between the two measuring chambers was plotted automatically. No difference! One reason for Duysen's success was that in addition to the improved apparatus sensitivity, he illuminated the sample during the measurement.

In 1960, while our new house was being built in the country, I renewed a previous, very primitive interest in land surveying. Experience with simple optics applied to the illumination of plant preparations and the work on the curve analyzer led me to think about combining a range finder with a mechanical computer and alidade to partially automate plane table plotting. With help from Dick Hart, a functional but cumbersome model was built. Later Charlton Lewis, Bob Emerson's former collaborator, wrote a fine application for a patent, which was eventually granted. Part of our thinking, largely inherited from Dr. Bush, was that an occasional by-product of scientific work might be of enough use to produce some income to further facilitate research. That this idea can be productive had been very successfully demonstrated by Dr. Cottrell, a Palo Alto resident in his later years, who frequently visited the laboratory and whose dust precipitator patents had been the foundation of the Research Corporation. The patent for the automatic plotter was placed in the hands of the Research Corporation with the intent of splitting profits equally between CIW and RC. However, there weren't any. About that time laser range finding was being developed which made optical range finders obsolete. The patent never became of interest to an instrument manufacturer in spite of considerable effort. When the development was well along, I found out that a somewhat similar device had been built before and was located in the geography building at Harvard next door to the biology building where I had worked 25–30 years before. While I have no conscious knowledge of having seen that device, I often wonder if some previous but forgotten exposure to the basic idea had been behind my "invention." The main lesson learned from this adventure was to appreciate the cost in time and money of getting too interested in a sideline not directly concerned with the normal affairs of the laboratory.

About 1967 we tired of the complexities of first derivative spectroscopy and converted that instrument into a more conventional absorption spectrophotometer. The device was, however, made for convenient use at liquid N_2 temperature. We then were able to see irregularities in the absorption spectra of chloroplast preparations due to the presence of different forms of chlorophyll *a*. Through a discussion with Glenn Bailey I heard of the computer program developed at the Shell Development Laboratory by Dr. Don D. Tunnicliff, who kindly gave me a card file of the program. With the help of the Stanford computer group and of Mark Lawrence, that program was adapted to the Stanford IBM computer. Numerous modifications were made and the program has been widely distributed with permission of the original author. With this computerized curve analysis, spectra can be resolved into Gaussian or Lorentzian components. Glenn Ford has adapted a simplified version of the program to the department's Hewlett-Packard computer. Jeanette Brown and I used it to characterize various forms of chlorophyll *a*. Other specific components were similarly identified with Hemming Virgin in natural protochlorophyll spectra and with Atusi Takamiya and the Muratas in water-soluble chlorophyll protein complexes.

COMMENTS ON VISITS TO OTHER LABORATORIES AND ON LABORATORY ADMINISTRATION

Until after the 1950s, the average scientist seemed to take more interest in fellow scientists in other fields and in their work than is now common. As a biologist looking for new techniques and different approaches to photosynthesis research in the 1930–1970 period, I frequently visited and was treated hospitably by specialists in many different fields. Many of these kind people, though they may have thought of biology as butterfly collecting and naming the wild flowers, nevertheless would willingly give their time to an interested visitor. I have always valued contacts with some of the great men in such diverse and to me largely unknown fields as spectroscopy, physical chemistry, photochemistry, bacteriology, algology, mammalian physiology, photocell technology, lamp manufacture, biochemistry, servosystem theory, optics, electrical engineering and, of course, various phases of plant physiology. For arranging many of these visits, I am particularly grateful first to my major professor, W. J. Crozier, and later to Dr. Vannevar Bush, both of whom realized the potential value of cross-fertilization of fields through scientific visits without an immediate specific purpose. I hope the increase in numbers of scientists and attendant specialization will not make such contacts between men in different fields less common in the future. I fear that the increase in size of the National Academy may, or already has, led to a decrease in its potential for developing friendships or at least speaking acquaintances between practitioners of different subjects.

A particularly memorable visit was to Robert W. Wood, Professor of Physics at Hopkins, author of *Physical Optics* and of *How to Tell the Birds from the Flowers.* I showed him a curve that I had measured for the transmission of a replica grating for various wavelengths. He said he would give me a grating in exchange for the curve. This grating is still in use for fluorescence spectroscopy of photosynthetic material by Bill Hagar. I also remember having lunch at the Rockefeller Institute with Leonore Michaelis, who pointed out that mathematical analysis of kinetic experiments is of great value in very simple systems but may become nearly meaningless if the system is too complicated. Several visits to James B. Sumner at Cornell were particularly helpful when I was first trying to separate out chloroplast components by biochemical means. To see a man with only one hand pour liquid from a flask into an unsupported test tube shows the value of experience and familiarity with the tools of the trade. The take-home lesson from Sumner was to keep the protein concentration high and the liquid volume low when trying to crystallize a protein. Unfortunately my fractionation experiments never got to the point where crystallization of anything but ammonium sulfate seemed likely. George Shannon Forbes once pointed out that many famous chemists have rediscovered ammonium chloride.

Sometime during the late 1930s or early 1940s Jack Myers and I were talking about the future of photosynthesis research. The upper limit of our thinking was to establish a laboratory with a million dollar endowment. That, we thought, would enable all the interested people to get together and solve all the problems we could think of at that time. Since then many people have each run through much more than that amount of money and the same unsolved problems along with some new ones are still around. This is progress in research. Myers (2) has recently written a remarkable review of the development of the basic concepts of the process of photosynthesis.

It has been a particular pleasure to have had the stimulation of knowing some of the photosynthesis investigators of the previous generation. In addition to the long and close association with Spoehr, I appreciated several visits with Professor Harder in Göttingen and enjoyed his visit to CIW accompanied by his students, von Denfer and von Witsch. At a Physiological Congress in Switzerland Emerson introduced me to Professor Arthur Stoll, who had worked with Willstätter on photosynthesis and on chlorophyll before and during World War I. After I gave a short paper at the Congress, Stoll kindly stood up and pointed out the significance of the work which I had not made clear. He visited the CIW laboratory some years later. James Smith and I had a memorable visit with Stoll at Basel when he was President of Sandoz. Other contacts with photosynthesis pioneers were several visits with René Wurmser in Paris, with G. E. Briggs at Cambridge, W. O. James at Oxford, H. Kautsky in Marburg, J. Buder, early investigator of photosynthetic bacteria, in Halle, A. Seybold in Heidelberg, William Duggar and Farrington Daniels at Wisconsin, W. J. V. Osterhout at the Rockefeller Institute, M. G. Stolfelt at Stockholm, and various others of later dates. Having known many of the active people in the first three quarters of this century gives a sense of continuity in the field of photosynthesis research.

Throughout my time as a scientist I have been amazed at the progress that can be made by combining the techniques of different branches of science. Furthermore, a facility with the simple elements of assorted trades such as carpentry, metal machining, plumbing, painting, electrical work, and all that sort of thing can give an investigator great freedom to improvise unusual equipment. The very small amount of knowledge from other fields that is needed to significantly promote one's main line of investigation seems almost unbelievable. Applications of simple mathematics, elementary optics, primitive electronics, and basic metal working was about all that I found necessary to build various pieces of apparatus that were new to plant physiology. Using the principle of combining assorted techniques is great fun and with very little effort can easily break new ground when the problem at hand is clearly visualized.

Most everyone who is responsible for more than one or two assistants has to face the question of laboratory organization and how to develop the customs and the atmosphere that best facilitate research. It seems that the answer to this question depends primarily on whether the laboratory is basically a one-man affair with everything devoted to furthering his program or whether it is a place to develop independent scientists. Whatever attributes of a good laboratory are listed, it is easy to find a clearly successful place that drastically violates anyone's preconceived ideas of good organization. This makes the subject difficult to discuss. However, by keeping in mind two contrasting types of laboratories, some useful attributes of each type may reasonably be described.

For the one-man show, organized very likely to capture a Nobel Prize for its head, it is essential to have all the staff personally loyal to and preferably afraid of the boss. This means that independent thinkers with unconventional approaches must be suppressed. Travel of the staff to meetings and other laboratories is very dangerous and must be strictly limited. Visitors must be urged to talk about their own work and their questions evaded as much as possible. Of course, it is forbidden to talk about the work of the group to outsiders, even nonscientists. Working hours and vacation times are strictly kept. Many great discoveries have come from such groups but one hears very little from most of the workers after they leave that sort of an organization. Surprisingly enough, that kind of structure occasionally flourishes within universities as well as in more business-like research organizations. Some places of this type may be patterned on older European models or on customs in the business world. Such places are not necessarily all bad because that otherwise abominable system may provide the optimum environment to take advantage of the particular abilities of certain rare geniuses.

The contrasting kind of a laboratory would abhor all the prescriptions for success of a director-led group as just described. Perhaps its main characteristic would be the encouragement of "enthusiasm without stress" Tiselius (4). There the individual scientist usually thinks and acts independently but with frequent friendly discussions with colleagues both in his own and in outside groups. In that environment cooperation predominates over competitiveness and the workers are happier. Perhaps only in this way can true academic excellence be encouraged without danger from the almost inevitable degradation that follows from the spirit of too much competition. This good laboratory environment fortunately seems to be nearly universal in plant physiology. Strangely enough, I have learned as much or more from visiting bad laboratories as from studying good ones. It is a great help to see clear illustrations of things to avoid.

The question of the optimum size of a laboratory is entirely different from that of its form of government, i.e. dictatorship, democracy, oligarchy, or anarchy. In theory the optimum size depends on the nature of the people there, on the nature of the problems being investigated, and on the teaching responsibilities, if any, of the group. In actual fact, the size is usually determined by the available budget or space on the principle that bigger is better. The critical size below which the effectiveness of a group evaporates may be very different for trying out unusual ideas with little probability of success or for pursuing a well-defined objective. The optimum must be somewhere above the critical size but below the point where in the words of Paul Kramer (1): "Increasing numbers cover both talent and mediocrity."

THE PLEASURES OF RETIREMENT

After 26 years at CIW I retired in 1973 with the great good luck of being succeeded by Winslow Briggs. He has kindly provided me with a comfortable office and a laboratory. This is an excellent opportunity to try to refine methods of measuring action spectra of the separate steps in photosynthesis. The hope is to compare action and absorption spectra of photosynthetic material with enough precision to see if action spectra for the two photosynthetic systems can add together to match the absorption spectra accurately. It may be possible to establish definitely the presence or absence of other reactions than the well-known two photosystems.

A photostationary state, steady-deflection method for measuring DCIP oxidation has been found to give more precise measurements than direct rate measurements. The principle is that with an excess of reductant present the dye oxidation rate by light is just balanced by the chemical reduction rate. The resulting steady-state dye concentration can be measured easily because the noise level can be averaged for a long time period since nothing changes after equilibrium is established.

Strangely enough, this dye oxidation reaction seems to be driven by all the forms of chlorophyll in the preparation (*Nostoc* particles kindly provided by Arnon and Hiyama) while the dye reduction is driven, at least preferentially, by the usual system II components. I look forward to a happy old age devoted to clarifying these questions. Working alone has for me always been a pleasure and is particularly desirable as the usable length of a working day becomes shorter.

The slow decline in competence with old age is fortunately compensated by an increasing delight in the completion of simple jobs. It is a great satisfaction to be able to continue with scientific work in a renovated laboratory and with many highly skilled and modern investigators available for discussion and often needed enlightenment.

THOUGHTS ON THE FUTURE OF RESEARCH IN PHOTOSYNTHESIS

"Solving" the problems of photosynthesis really means describing the process in terms of currently understood concepts of molecular interaction. Many aspects of photosynthesis have been well described already. Probably the most comprehensive interpretation of one part of the photosynthetic problem is the path of carbon through the various intermediates from CO_2 to carbohydrate as described by Calvin and his associates in terms of organic chemistry. Also, the moderately clear pictures of the process of light absorption followed by energy transfer through various pigments to a reaction center have been widely accepted. Furthermore, the electron transport processes from the reaction center to reduced NADP and the accompanying formation of ATP are believed to be reasonably well understood and the chemical nature of the transporting substances are mostly known.

Where then are the remaining problems? In the first place many details remain to be filled in to complete and verify the current concepts of electron transport, and no doubt some major revision of the present concepts of the two-reaction scheme and associated carriers may be expected.

If past experience is any guide to the future course of development of the understanding of photosynthesis, we may confidently expect some discoveries that would make radical changes in the present concepts. It may be that some new enlightenment comparable in significance to the discovery of the two separate light reactions and to the involvement of photophosphorylation in photosynthesis may again appear.

In any case, as the relevant chemistry and physics develop further, the language in which presently known effects are described will certainly be very different in the future.

We may hope that future research on photosynthesis and related processes will show ways to increase agricultural production, possibly through influencing the path of carbon fixation.

Literature Cited

1. Kramer, P. J. 1973. Some reflections after 40 years in plant physiology. *Ann. Rev. Plant Physiol.* 24:1–24
2. Myers, J. 1974. Conceptual developments in photosynthesis, 1924–1974. *Plant Physiol.* 54:420–26
3. Nachmanson, D. 1972. Biochemistry as part of my life. *Ann. Rev. Biochem.* 41:1–28
4. Tiselius, A. 1968. Reflections from both sides of the counter. *Ann. Rev. Biochem.* 37:1–24

Ann. Rev. Plant Physiol. 1979. 30:27–40

FACULTATIVE ANOXYGENIC PHOTOSYNTHESIS IN CYANOBACTERIA[1]

❖7663

Etana Padan

Department of Microbiological Chemistry, The Hebrew University-Hadassah Medical School, Jerusalem, Israel

CONTENTS

INTRODUCTION

The notion that cyanobacteria (blue-green algae) represent a group of primitive organisms in which plant-type photosynthesis evolved is based both on their antiquity (56) and on the combination of a prokaryotic cellular organization with a plant-type oxygenic photosynthetic system (58). Recently, Cohen, Padan & Shilo (16) showed that a cyanobacterium, *Oscillatoria limnetica,* is also capable of a facultative bacterial-type anoxygenic CO_2

[1]Abbreviations used: ATP, adenosine triphosphate; DCMU, 3-(3,4-dichlorophenyl)-1,1-dimethylurea; FCCP, carbonyl cyanide-*p*-trifluoromethoxyphenylhydrazone.

0066-4294/79/0601-0027$01.00

photoassimilation with sulfide as an electron donor in a photosystem I-driven reaction. This anoxygenic reaction was thoroughly investigated (6, 13, 23, 46, 47) and later found in many other cyanobacteria (11, 12, 23).

The present review attempts to reevaluate the position that cyanobacteria occupy in the phototrophic world in the light of these results. For this purpose the gap existing between the phototrophic pattern of bacteria and that of eukaryotes will be defined at the biochemical, physiological, and ecological levels. The possibility that cyanobacteria bridge this gap will be considered. In striving for clarity and emphasis, this analysis will concentrate on several of the major factors differentiating or interlinking the three phototrophic groups.

THE GAP BETWEEN THE ANOXYGENIC PHOTOSYNTHETIC SYSTEM OF BACTERIA AND THE OXYGENIC ONE OF EUKARYOTIC PHOTOTROPHS AND CYANOBACTERIA

There is a sharp discontinuity in the phototrophic world between photosynthetic bacteria on the one hand and cyanobacteria and eukaryotic phototrophs on the other on the basis of photosynthetic patterns. Phototrophic bacteria generally contain a single photosystem, and their reducing power is provided by reduced substrates in a pathway generally linked only indirectly, if at all, with the photosynthetic system. These substrates include sulfide, sulfur, thiosulfate, and organic molecules. Light energy absorbed by chlorophyll of the photosystem causes a cyclic electron flow down a redox potential gradient established by electron and/or hydrogen carriers leading the electrons back to the chlorophyll. Through this light-initiated cyclic electron flow, energy is conserved in forms utilizable by the cell, i.e. high energy bonds of ATP and electric and osmotic gradients (for reviews see 20, 24, 30, 51, 52).

Eukaryotic phototrophs and cyanobacteria, in contrast, contain two photosystems, I and II, working in series. Light absorbed by chlorophylls of these systems maintains an electron flow from water to NADP with evolution of oxygen; energy is conserved concurrently in NADPH, ATP, chemical, and electrical gradients (for reviews see 4, 32, 38, 58).

Oxygenic photosynthesis can be turned into functional bacterial type anoxygenic photosynthesis, an observation made in the early 1960s (3). If photosystem II is inhibited by its specific inhibitor DCMU (27), or if it is not excited as when light is provided in wavelengths absorbed only by photosystem I, a cyclic electron flow is maintained around photosystem I with energy-conserving patterns similar to those in photosynthetic bacteria.

Under these conditions, these phototrophs photoassimilate organic substances (57, 58). After adaptation to hydrogen, these cells can either photoreduce CO_2 with hydrogen serving as an electron donor (21, 36) or evolve hydrogen (6, 8, 63, 66). Arnon et al (3) suggested that these photosystem I-driven reactions may represent an intermediate stage between bacterial-anoxygenic and plant-oxygenic photosynthesis. However, eukaryotic phototrophs use sulfide as an electron donor very inefficiently if at all (36, 37), in contrast to its efficient use by photosynthetic bacteria and, as recently discovered, by many cyanobacteria species (11–13, 23).

FACULTATIVE ANOXYGENIC PHOTOSYNTHESIS OF CYANOBACTERIA

Oscillatoria limnetica, a cyanobacterium isolated (16) from the sulfide-rich layers of the Solar Lake (Elat, Israel), carries out oxygenic photosynthesis inhibited by DCMU. When even a very low concentration (0.1–0.2 mM) of sulfide is added to the oxygenic system, immediate inhibition of CO_2 photoassimilation is obtained. However, if exposure to light is continued for about 2 hr in the presence of a high sulfide concentration (3 mM), photoassimilation reappears, but it is now insensitive to DCMU (16, 46). This photoassimilation is therefore anoxygenic, independent of photosystem II, and driven by photosystem I with sulfide as an electron donor.

The possible participation of photosystem II in the reaction was definitively excluded (47) by results of experiments in which photosystem II was simply not activated (rather than being inhibited) in the presence of far red light (>673 nm). Under these light conditions, oxygenic photosynthesis, requiring operation of both photosystems, was drastically inhibited ("red drop"), whereas anoxygenic photosynthesis with sulfide was fully operative. Furthermore, if both photosystems could contribute to the reaction, the enhancement in quantum yield of the photoassimilation would be predicted with respect to that obtained with only photosystem I in operation. The "enhancement phenomenon" was observed, however, only with the oxygenic photosynthesis, while none was observed with the anoxygenic one (47).

The 2 hr period of incubation in the presence of sulfide and light that is required for the anoxygenic photosynthesis may indicate that an induction process is involved. Indeed, in the presence of chloramphenicol, a specific inhibitor of protein synthesis in *O. limnetica* as well as in bacteria, anoxygenic photosynthesis is not induced. If sulfide is excluded, *O. limnetica* immediately shifts to oxygenic photosynthesis (46). Thus, the photosynthetic system of *O. limnetica* operates facultatively both oxygenically and anoxygenically.

Sulfide is oxidized to elemental sulfur by *O. limnetica* according to the following stoichiometric relationship (13):

$$2\ H_2S + CO_2 \rightarrow HCHO + 2S + H_2O$$

Elemental sulfur expelled from the cells was observed as typical refractile granules either free in the medium or adhering to the cyanobacterial filaments. The appearance of the cells under the electron microscope is the same under both anoxygenic and oxygenic conditions (A. Oren and S. Holt, unpublished observations).

In addition to photoassimilation of CO_2, sulfide donates electrons to hydrogen evolution in *O. limnetica* (S. Belkin and E. Padan, in preparation). This reaction, occurring only in sulfide-induced cells, is light- and sulfide-dependent, DCMU insensitive, and accelerated by FCCP. This acceleration indicates that hydrogenase rather than nitrogenase is involved in the reaction. The presence of CO_2 inhibits the sulfide-dependent hydrogen evolution, and FCCP relieves the inhibition. The light requirement, uncoupler, and CO_2 effects all indicate that sulfide donates electrons directly to the photosynthetic system.

Cell concentrations of light-absorbing pigments, chlorophyll *a*, phycocyanin, and carotenoids, are identical in both photosynthetic conditions (47). Quantum yield spectra of oxygenic and anoxygenic photosynthesis show (47) that whereas only a narrow sector is utilized in the oxygenic type, the entire absorbed spectrum is utilized in anoxygenic photosynthesis. The drop in efficiency of oxygenic photosynthesis at both blue and red ends of the spectrum is marked. This limited range of utilization of quantum energies in oxygenic photosynthesis of *O. limnetica* is similar to that of other cyanobacteria and eukaryotic algae with similar pigment composition (33, 40); it is markedly different, however, from that of eukaryotic algae and plants containing chlorophyll *b* as the light harvesting system. In these, almost the entire absorbed spectrum is utilized in the oxygenic photosynthesis, with the exception of the far "red drop" (28). Hence, quantum utilization in the photosynthetic system of *O. limnetica* appears to be oriented less towards oxygenic photosynthesis than in higher eukaryotic phototrophs, and the ready shift to anoxygenic photosynthesis permits efficient utilization of the total absorbed spectrum. This facultative alternation between anoxygenic and oxygenic photosynthesis by *O. limnetica* may be considered as representative of interlinking of the anoxygenic bacterial type with the oxygenic plant type of photosynthesis. There is also a possibility that the photosynthetic pattern of *O. limnetica* is an exceptional secondary specialization toward anaerobic photosynthesis. An investigation of the frequency of occurrence of this photosynthetic pattern among cyanobacteria may permit distinguishing between these alternative interpretations.

ANOXYGENIC PHOTOSYNTHESIS AMONG CYANOBACTERIA

Various cyanobacteria were tested for the capacity for anoxygenic photosynthesis, including strains from culture collections, new isolates (23), and natural populations (11, 12). All the strains—30 in number—can carry out oxygenic photosynthesis inhibited by DCMU and not operative at >700 nm light. Sixteen of them (including *O. limnetica*) are also capable of anoxygenic photosynthesis with sulfide, driven by photosystem I. In these latter cases, photoassimilation of CO_2 has been demonstrated in the presence of sulfide, DCMU, and/or >700 nm light.

Anacystis nidulans can utilize a different sulfur-containing source of electrons—oxidizing thiosulfate to sulfate (65). In both *O. limnetica* and *Aphanothece halophytica,* molecular hydrogen has been shown to be an efficient electron donor for CO_2 in an anoxygenic reaction driven by photosystem I. Hydrogen evolution was also shown in *A. halophytica* (6). Anoxygenic hydrogen metabolism is thus shared by these nonheterocystous sulfide-utilizing cyanobacteria species and heterocystous species (8, 63, 66).

The property of sulfide dependent anoxygenic photosynthesis appears in different groups of cyanobacteria—in filamentous as well as unicellular forms (12, 23, 48). Moreover, it is not confined to particular ecosystems, geographical locations, or culture conditions (11, 12, 23). Anoxygenic photosynthesis of *A. halophytica* was compared with that of *O. limnetica;* in both, an induction process is involved and the stoichiometric relations of the reaction are the same (23).

A comparison of rates of anoxygenic photosynthesis with respect to sulfide concentration was carried out among *Lyngbya* 7104, *A. halophytica,* and *O. limnetica* (23). In all three, the pattern of dependence on sulfide concentration was similar, generating an optimum curve rather than a saturation curve. The drop in photosynthetic rates at high sulfide concentrations is presumably caused by sulfide toxicity—an effect recognized both in phototrophs and heterotrophs (11)—thus the observed rates may not be the true ones. Nevertheless, the highest rate of oxygenic photosynthesis is similar to that of anoxygenic photosynthesis in both *O. limnetica* (1–2 μmole/mg protein per hr) and *A. halophytica* (0.5–1 μmole/mg protein per hr). While the trend of dependency on sulfide is similar, both the affinity toward sulfide and the tolerance for sulfide are different. Thus, each strain shows a different range of sulfide concentration at which anoxygenic photosynthesis is carried out (pH 6.8): *Lyngbya,* 0.1–0.3 mM; *A. halophytica,* 0.1–1.2 mM; and *O. limnetica,* the highest concentration range, 0.7–5 mM. This range, furthermore, is markedly affected by the medium pH (48) which governs proportions of dissociated sulfide species.

Differences in pH-dependent sulfide ranges are known to constitute a determinative factor in the ecology of photosynthetic sulfur bacteria (5, 51). At pH 7, Chlorobiaceae thrive at the highest sulfide concentrations, 4–8 mM, like *O. limnetica;* Chromatiaceae do best at 0.8–4 mM, and Rhodospirillaceae at 0.4–2 mM. The extracellular depositon of sulfur by *O. limnetica* and *A. halophytica* is carried out in a similar way by sulfide-oxidizing photosynthetic bacteria, Chlorobiaceae, and by some Chromatiaceae (52). However, the pattern of sulfide oxidation of these cyanobacteria strains seems thermodynamically inefficient; photosynthetic bacteria oxidize sulfide to sulfate (51, 52), gaining 8 electrons for each H_2S molecule, while *O. limnetica* and *A. halophytica* remove only 2 electrons per molecule. The oxidation of sulfide to sulfur may be a by-product in the sulfide oxidation process of sulfur photosynthetic bacteria (64). Finally, other sulfur-containing electron donors which are efficiently used by photosynthetic bacteria (30, 51, 52) do not seem to serve cyanobacteria photosynthesis (12).

In cyanobacteria, sulfide appears more toxic to oxygenic photosynthesis than to anoxygenic photosynthesis; in *O. limnetica* the former is inhibited by 0.1 mM sulfide (46), while the latter is only inhibited by $>$4 mM (at pH 6.8) (23, 48). These differences have also been observed in other facultative anoxygenic strains (11, 12, 23, 48). Low sulfide concentrations ($<$0.5 mM) are toxic to eukaryotic phototrophs and cyanobacterium strains which do not carry out anoxygenic photosynthesis (11, 12, 37, 48). This apparently greater sensitivity to sulfide in oxygenic photosynthesis as compared to anoxygenic photosynthesis suggests that photosystem II is more susceptible to sulfide poisoning than photosystem I.

In many photosynthetic systems, e.g. *A. halophytica, A. nidulans, Chlamydomonas* sp., and isolated tobacco chloroplasts, sulfide (0.1–0.5 mM) acts as an electron transport inhibitor at a site preceding photosystem II (A. Oren, E. Padan and S. Malkin, in preparation). Thus after exposure to sulfide in the presence of photosystem II light, the fluorescence yield of photosystem II decreased to the nonvariable value, but the full yield was restored upon addition of DCMU. This kind of inhibition is observed in the presence of hydroxylamine which acts on a site before photosystem II (43). In the dark, sulfide affected the variable fluorescence of the phototrophic systems in a way similar to that obtained by N_2 flushing—anaerobiosis—of the system, i.e. the variable fluorescence remained at high constant value (see also 50). The greater sulfide tolerance of anoxygenic photosynthesis operating with photosystem I confers a selective advantage for anoxygenic photosynthetic cyanobacteria over nonanoxygenic species in sulfide-rich habitats. This advantage is obtained even if facultative anoxygenic photosynthesizing cyanobacterium strains cannot grow under purely anaerobic conditions because they can at least maintain themselves in the presence of sulfide. However, the most successful strains are capable of growth under these conditions.

FACULTATIVE ANAEROBIC PHOTOTROPHS VERSUS OBLIGATE ANAEROBIC OR AEROBIC PHOTOTROPHS

Phototrophic bacteria differ markedly from eukaryotic algae in the pattern of phototrophic growth. Photosynthetic bacteria have an obligate anaerobic phototrophic growth physiology characteristic for obligate as well as for facultative anaerobic photosynthetic bacteria which can grow aerobically in the dark (17, 39, 51, 52). In contrast, the majority, if not all, of eukaryotic phototrophs cannot grow anaerobically either in light or dark (36); hydrogen-dependent photoreduction of CO_2 could not be shown to support anaerobic photoautotrophic growth of eukaryotic algae (21). Although aerobic photoheterotrophic growth does occur in eukaryotic algae (36, 57), anaerobic photoheterotrophic growth or anaerobic heterotrophy have not been unequivocally demonstrated (25, 36, 57); hence, eukaryotic phototrophs appear to be obligate aerobes. In the cyanobacterium *O. limnetica,* once induction to utilization of sulfide in anoxygenic CO_2 photoassimilation has taken place, it can readily carry out anaerobic photoautotrophic growth (46), and the anaerobic photoautotrophic growth rate (2 days, doubling time) is similar to the aerobic photoautotrophic rate. Once sulfide is excluded, *O. limnetica* immediately shifts to photoaerobic growth with oxygenic photosynthesis, and the anoxygenic photosynthetic capacity is diluted out by growth. Hence *O. limnetica* is a facultative anaerobic photoautotroph which readily shifts between aerobic photoautotrophic and anaerobic photoautotrophic growth patterns. It may therefore represent an intermediate photoautotrophic physiology between that of eukaryotic algae and photosynthetic bacteria.

The incapability of eukaryotic algae to photoassimilate CO_2 and grow in the presence of sulfide may be explained by the toxicity of sulfide whether to the photosynthetic system or to a different system. However, it is not self-evident why eukaryotic algae do not carry out any anaerobic growth —either heterotrophic or phototrophic growth—using H_2 as an electron donor. This difficulty occurs because there are requirements for anaerobic growth not directly linked with the capacity to use a particular electron donor. There are at least two fundamental prerequisites for anaerobic growth: (*a*) noninvolvement of O_2 in any essential enzymatic reaction; (*b*) a capacity to function under reducing conditions. Eukaryotic phototrophs possess enzymes which require O_2 such as those involved in fatty acid or sterol synthesis or in phenol oxidation (26, 31, 68). Some of these reactions may not be replaceable by alternative anaerobic pathways. Oxygen-indispensable pathways exist in *Saccharomyces,* which grows anaerobically if supplied with essential products such as sterols (2). No anaerobic growth experiments of this kind have been carried out with eukaryotic phototrophs. Reducing conditions, from –200 to –300 mV with sulfide or H_2 present, are often encountered in anaerobic habitats (5, 14, 18, 24). Growth capability in a particular environmental redox potential depends on the capacity of the cell to maintain the appropriate cell redox potential. The evolution of hydrogen in anaerobically growing photosynthetic bacteria is an example of an adaptation to reducing conditions (24). In this context, it is possible that the stringent dependence of eukaryotic algae on photoaerobic conditions for growth is a function of their inability to maintain their redox potential under anaerobic conditions (25).

Both of the requirements for anaerobic growth are met by the photosynthetic bacteria—they do not need oxygen, and they tolerate low external redox potentials. Moreover, oxygen is toxic to several metabolic steps in obligate anaerobic photosynthetic bacteria as in other obligate anaerobes (42). Negative control exerted by oxygen on the photosynthetic system of facultative anaerobic bacteria is also a well-known phenomenon (17, 39). Finally, the photosynthetic system of both facultative and obligate anaerobic bacteria appears to operate most efficiently when poised at negative redox potential (20), as expected for a cyclic system operating without change in net oxidoreduction potential.

In carrying out anaerobic growth in conditions of high sulfide concentration (3.5 mM; pH 6.8), *O. limnetica* must cope both with the low redox potential and with dispensing with oxygen in its metabolism. Recent studies of fatty acid composition in unicellular cyanobacteria have found species with polyunsaturated as well as with saturated and monounsaturated fatty acids (34), along with an oxygen-dependent desaturation mechanism (44). In almost all filamentous cyanobacteria analyzed, there was an abundance of unsaturated fatty acids (35); an oxygen-dependent mechanism for desaturation was demonstrated in *Anabaena variabilis* (7). Only several *Spirulina* spp. possibly related to the anoxygenic photosynthesizing strains were found to contain saturated and monounsaturated fatty acids (35). Recently A. Oren, A. Tiez, and E. Padan (in preparation) showed that *O. limnetica* grown either aerobically or anaerobically lacks polyunsaturated fatty acids as do anaerobic growing bacteria. In contrast, anoxygenic photosynthesizing cyanobacteria species *A. halophytica, Lyngbya* 7104, and *Lyngbya* 7004 contain polyunsaturated fatty acids (35). Thus, anoxygenic photosynthesis need not necessarily be linked with anaerobic growth and anaerobic fatty acid metabolism. This physiological spectrum, representing different degrees of anaerobism, may aid in the expansion of cyanobacteria taxonomy (35, 58).

Eukaryotic algae or photosynthetic bacteria are restricted to habitats with exclusively photoaerobic or photoanaerobic conditions, respectively. In habitats with fluctuations in conditions, a cyanobacterium with an *O. limnetica* type facultative anoxygenic photosynthesis should clearly have a selective advantage. In fact, the more frequent and marked the fluctuations, the more prominent the cyanobacteria are with respect to the other phototrophic organisms.

ECOLOGICAL IMPACT OF FACULTATIVE PHOTOAUTOTROPHIC ANAEROBIC METABOLISM OF CYANOBACTERIA

Taking into account the fluctuation in oxygen tension and sulfide concentrations in different photic aquatic systems, we may assume that there is a gap between photosynthetic bacteria and eukaryotic algae at the ecological level (48). Bacteria dominate most stable photoanaerobic situations, e.g. source waters of hot sulfur springs (11, 12), sulfide-rich layers of long duration in stratified lakes (9, 10, 51, 52), and anaerobic sediments (5, 51), while eukaryotic algae inhabit stable aerobic ecosystems such as the open sea (59) and aerobic layers of lakes (10). Moreover, both of these phototroph groups avoid ecosystems with marked and rapid fluctuations in anaerobic and aerobic conditions. Cyanobacteria, in dominating these intermediate ecosystems, seem to bridge this apparent ecological gap (48). Typical cyanobacteria habitats include: waters of hot sulfur springs with a downstream gradient in oxygen and sulfide concentrations (11, 12); eutrophic lakes with frequent anaerobic conditions in the upper layers (9, 10, 19, 54); shallow lakes of the equatorial regions with daily alternations in O_2 and occasionally also in sulfide concentrations (22, 29); also, shallow bodies of water such as mangroves, marshes, and rice paddies and their sediments (9, 19, 45, 48, 62). Littoral marine sediments are also subject to fluctuations in these conditions and are very abundant in cyanobacteria (18).

Two cases have been thoroughly studied at both the ecological and physiological levels (48). Castenholtz described the continuum of phototrophs along the downstream sulfide gradient of hot sulfur springs of New Zealand, Iceland, and Yellowstone (USA) (11, 12). Photosynthetic bacteria dominate the source waters, then a continuum of cyanobacteria dominates all intermediary sulfide concentrations downstream up to the area with photoaerobic conditions. Only two upsteam species, *Oscillatoria amphigranulata* and *Spirulina labyrinthiformis,* carry out anoxygenic photosynthesis in situ, in the area of about 1mM sulfide concentration. Strains occurring farther downstream do not utilize sulfide and are even excluded by much lower concentrations of sulfide (0.2 mM).

The Solar Lake (Israel), the type locality of *O. limnetica,* has a winter density and temperature stratification, with bottom layers reaching up to 1.5 mM sulfide (14, 15). Throughout the stagnation period, there are some eukaryotic algae and *A. halophytica* in the oxygenated upper layers; *Chromatium* sp. at the oxygen-sulfide borderline; *Prosthecochloris,* a green bacterium, in the deeper anaerobic layers; and *O. limnetica,* dominant in the deepest layer, along with other cyanobacteria species, *Oscillatoria salina* and *Microcoleus* sp. (15). Primary productivity rates (15) in this bottommost layer were very high (4.960 mg C/m^3 per day.) This photosynthetic activity appears to be anoxygenic, as it is not inhibited by DCMU (S. Garlick, unpublished data); also, the light wavelengths penetrating into the lake bottom through the photosynthetic bacterial layers, 500–600 nm (15), are more effective in driving anoxygenic photosynthesis than oxygenic photosynthesis (47). An in situ doubling time of 1–2 days was calculated (15), a very similar growth rate to that measured for *O. limnetica* under anaerobic culture conditions (46).

With the spring overturn, the Solar Lake becomes aerobic; the photosynthetic bacteria disappear; now *O. limnetica* shifts to its oxygenic metabolism and thrives throughout the water column, along with other cyanobacterium species and eukaryotic algae. In the next anaerobic period, the algae which are only capable of oxygenic photosynthesis disappear, while *O. limnetica,* which shifts to anaerobic phototrophic metabolism, continues to flourish. Hence, *O. limnetica,* by utilizing combined oxygenic and anoxygenic photosynthesis capacities, is the dominant phototroph of the Solar Lake with its fluctuating photoaerobic and photoanaerobic conditions.

CONCLUDING REMARKS

The discovery of cyanobacterial facultative anoxygenic photoautotrophic physiology clarifies certain previously obscure aspects of cyanobacterial ecology, the occurrence and predominance of cyanobacteria in habitats alternating between photoaerobic and photoanaerobic conditions. Cyanobacteria have long been known to be nitrogen-fixing organisms under aerobic conditions (60), and they can also fix nitrogen, sometimes with greater efficiency, under microaerophilic or anaerobic conditions (55, 61, 62). Considering the apparent ubiquity of cyanobacteria in photoanaerobic habitats, their contribution to the nitrogen balance in these ecosystems must be significant.

Dual capacities—oxygenic photosynthesis on the one hand and facultative anoxygenic phototrophic capacity on the other hand—permit cyanobacteria to occupy an intermediate ecological and physiological position unsuitable for organisms with a strictly anaerobic phototrophic physiology such as photosynthetic bacteria, or with a strictly aerobic physiology such as eukaryotic phototrophs. In fact, preference for low oxygen tension and for low redox potential appears to be a general characteristic of many cyanobacterium species (1, 9, 19, 62, 67). The interlinking position of cyanobacteria in the phototrophic world is compatible with the fact that cyanobacteria are prokaryotes and among the oldest organisms, with a minimum dating from the Precambrian period (56). Significantly, two of the sulfide-rich ecosystems abundant in cyanobacteria, the marine littoral sediments (18) and the hot sulfur springs (11, 12), may represent old ecosystems which may even predate the oxidized biosphere. Cyanobacteria retained the anoxygenic option with chlorophyll *a*—photosystem I—while developing plant-type oxygenic photosynthesis with photosystem II. Indeed, it is widely accepted that cyanobacteria represent the predecessors of contemporary chloroplasts (41, 58) as evidenced by the great similarity between cyanobacteria and chloroplast structure, composition, and function of the oxygenic photosynthetic system and in nucleic acid composition and function. Finally, recent works on the mechanism of energy transductions in the cyanobacterial membranes support the chloroplast cyanobacterial interrelationship (49, 53).

Acknowledgments

Many thanks to M. Shilo and A. Oren of the Department of Microbiological Chemistry, The Hebrew University-Hadassah Medical School, and M. Avron of the Department of Biochemistry, the Weitzmann Institute, Rehovot, for critically reading the manuscript, and to N. Ben-Eliahu for help in preparation of the manuscript. This project was partly supported by the Deutsche Forschungsgemeinschaft.

Literature Cited

1. Abeliovich, A., Shilo, M. 1972. Photooxidation death in blue-green algae. *J. Bacteriol.* 111:682–89
2. Andreasen, A. A., Stier, T. J. B. 1953. Anaerobic nutrition of *Saccharomyces cerevisiae. J. Cell. Comp. Physiol.* 41: 23–36
3. Arnon, D. I., Losada, M., Nozaki, M., Tagawa, K. 1961. Photoproduction of hydrogen, photofixation of nitrogen and a unified concept of photosynthesis. *Nature* 190:601–6
4. Avron, M. 1977. Energy transduction in chloroplasts. *Ann. Rev. Biochem.* 46:143–55
5. Baas Becking, L. G. M., Wood, E. J. F. 1955. Biological processes in the estuarine environment. *Proc. K. Ned. Akad. Wet., Sect. B* 58:160–81
6. Belkin, S., Padan, E. 1978. Hydrogen metabolism in the facultative anoxygenic cyanobacteria (blue-green algae) *Oscillatoria limnetica* and *Aphanothece halophytica. Arch. Microbiol.* 116: 109–11
7. Bloch, K., Barnowsky, Y., Goldfine, H., Lennarz, W. J., Light, R., Norris, A. T., Scheuerbrandt, G. 1966. Biosynthesis and metabolism of unsaturated fatty acids. *Fed. Proc.* 20:921–27
8. Bothe, H., Tennigkeit, J., Eisenberger, G., Yates, M. G. 1977. The hydrogenase-nitrogenase relationship in the blue-green alga *Anabaena cylindrica. Planta* 133:237–42
9. Brock, T. D. 1973. Evolutionary and ecological aspects of the cyanophytes. In *The Biology of Blue-Green Algae,* ed. N. G. Carr, B. A. Whitton, pp. 487–500. Oxford: Blackwell. 676 pp.
10. Caldwell, D. E. 1977. The planktonic microflora of lakes. *CRC Crit. Rev. Microbiol.* 5:305–70
11. Castenholtz, R. W. 1976. The effect of sulfide on the blue-green algae of hot springs. I. New Zealand and Iceland. *J. Phycol.* 12:54–68
12. Castenholtz, R. W. 1977. The effect of sulfide on the blue-green algae of hot springs. II. Yellowstone National Park. *Microb. Ecol.* 3:79–105
13. Cohen, Y., Jorgensen, B. B., Padan, E., Shilo, M. 1975. Sulfide-dependent anoxygenic photosynthesis in the cyanobacterium *Oscillatoria limnetica. Nature* 257:489–92
14. Cohen, Y., Krumbein, W. E., Goldberg, M., Shilo, M. 1977. Solar Lake (Sinai). 1. Physical and chemical limnology. *Limnol. Oceanogr.* 22:597–608
15. Cohen, Y., Krumbein, W. E., Shilo, M. 1977. Solar Lake (Sinai). 2. Distribution of photosynthetic microorganisms and primary production. *Limnol. Oceanogr.* 22:609–20
16. Cohen, Y., Padan, E., Shilo, M. 1975. Facultative anoxygenic photosynthesis in the cyanobacterium *Oscillatoria limnetica. J. Bacteriol.* 123:855–61
17. Cohen-Bazire, G., Sistrom, W. R., Stanier, R. Y. 1957. Kinetic studies of pigment synthesis by non-sulfur purple bacteria. *J. Cell. Comp. Physiol.* 49: 25–68
18. Fenchel, T. M., Riedl, R. J. 1970. The sulfide system: a new biotic community underneath the oxidized layer of marine sand bottoms. *Mar. Biol.* 7:255–68
19. Fogg, G. E., Stewart, W. D. P., Fay, P., Walsby, A. E. 1973. In *The Blue-Green Algae,* pp. 255–80. London: Academic. 459 pp.
20. Frenkel, A. W. 1970. Multiplicity of electron transport reactions in bacterial photosynthesis. *Biol. Rev.* 45:569–616
21. Gaffron, H. 1944. Photosynthesis, photoreduction and dark reduction of carbon dioxide in certain algae. *Biol. Rev.* 19:1–20
22. Ganf, G. G., Horne, A. J. 1975. Diurnal stratification, photosynthesis and nitrogen-fixation in a shallow, equatorial lake (Lake George, Uganda). *Freshwater Biol.* 5:13–39
23. Garlick, S., Oren, A., Padan, E. 1977. Occurrence of facultative anoxygenic photosynthesis among filamentous and unicellular cyanobacteria. *J. Bacteriol.* 129:623–29
24. Gest, H. 1972. Energy conversion and generation of reducing power in bacterial photosynthesis. *Adv. Microb. Physiol.* 7:243–82
25. Gibbs, M. 1962. Fermentation. In *Physiology and Biochemistry of Algae,* ed. R. A. Lewin, pp. 91–97. London: Academic. 929 pp.
26. Goad, L. J., Goodwin, T. W. 1972. Biosynthesis of plant sterols. *Prog. Phytochem.* 3:113–98
27. Good, N. E., Izawa, S. 1973. Inhibition of photosynthesis. In *Metabolic Inhibitors,* ed. R. M. Hochster, M. Kates, J. H. Quastel, 4:179–214. New York: Academic. 513 pp.
28. Govindjee, R., Rabinovitch, E., Govindjee. 1968. Maximum quantum yield and action spectrum of photosynthesis and fluorescence in *Chlorella. Biochim. Biophys. Acta* 162:539–44

29. Greenwood, P. H. 1976. Lake George, Uganda. *Philos. Trans. R. Soc. London, Ser. B* 274:375–91
30. Gromet-Elhanan, Z. 1977. Electron transport and photophosphorylation in photosynthetic bacteria. In *Encyclopedia of Plant Physiology,* ed. A. Trebst, M. Avron, 5:637–62. Berlin: Springer-Verlag. 730 pp.
31. Hamberg, M., Samuelsson, B., Björkhem, I., Danielsson, H. 1974. Oxygenases in fatty acid and steroid metabolism. In *Molecular Mechanisms of Oxygen Activation,* ed. O. Hayaishi, pp. 29–85. New York: Academic. 678 pp.
32. Jagendorf, A. T. 1975. Mechanisms of photophosphorylation. In *Bioenergetics of Photosynthesis,* ed. Govindjee, pp. 414–92. New York: Academic. 698 pp.
33. Jones, L. W., Myers, J. 1964. Enhancement in the blue-green alga, *Anacystis nidulans. Plant Physiol.* 39:938–46
34. Kenyon, C. N. 1972. Fatty acid composition of unicellular strains of blue-green algae. *J. Bacteriol.* 109:827–34
35. Kenyon, C. N., Rippka, R., Stanier, R. Y. 1972. Fatty acid composition and physiological properties of some filamentous blue-green algae. *Arch. Mikrobiol.* 83:216–36
36. Kessler, E. 1974. Hydrogenase, photoreduction and anaerobic growth. In *Algal Physiology and Biochemistry,* ed. W. D. P. Stewart, pp. 456–73. Oxford: Blackwell. 989 pp.
37. Knobloch, K. 1969. Sulfide oxidation via photosynthesis in green algae. In *Progress in Photosynthesis Research,* ed. H. Metzner, 2:1032–34. Tubingen: Int. Union Biol. Sci. 1127 pp.
38. Krogmann, D. W. 1977. Blue-green algae. See Ref. 30, pp. 625–36
39. Lascelles, J. 1959. Adaptation to form bacteriochlorophyll in *Rhodopseudomonas spheroides:* Change in activity of enzymes concerned in pyrrole synthesis. *Biochem. J.* 72:508–18
40. Lemasson, C., Tandeau de Marsac, N., Cohen-Bazire, G. 1973. Role of allophycocyanin as a light-harvesting pigment in cyanobacteria. *Proc. Natl. Acad. Sci. USA* 70:3130–33
41. Margulis, L. 1968. Evolutionary criteria in thallophytes: A radical alternative. *Science* 161:1020–22
42. McCord, J. M., Keele, B. B. Jr., Fridovich, I. 1971. An emzyme-based theory of obligate anaerobiosis: The physiological function of superoxide dismutase. *Proc. Natl. Acad. Sci. USA* 68:1024–27
43. Mohanty, P., Mar, T., Govindjee. 1971. Action of hydroxylamine in the red alga *Porphyridium cruentum. Biochim. Biophys. Acta* 253:213–21
44. Nichols, B. W., Harris, R. V., James, A. T. 1965. The lipid metabolism of blue-green algae. *Biochem. Biophys. Res. Commun.* 20:256–62
45. Odum, H. T. 1967. Biological circuits and the marine systems of Texas. In *Pollution and Marine Ecology,* ed. T. A. Olson, F. J. Burgess, pp. 99–157. New York: Wiley/Interscience, 364 pp.
46. Oren, A., Padan, E. 1978. Induction of anaerobic, photoautotrophic growth in the cyanobacterium, *Oscillatoria limnetica. J. Bacteriol.* 133:558–63
47. Oren, A., Padan, E., Avron, M. 1977. Quantum yields for oxygenic and anoxygenic photosynthesis in the cyanobacterium *Oscillatoria limnetica. Proc. Natl. Acad. Sci. USA* 74:2152–56
48. Padan, E. 1979. Impact of facultative anaerobic photoautotrophic metabolism on ecology of cyanobacteria (blue-green algae). *Adv. Microb. Ecol.* In press
49. Padan, E., Schuldiner, S. 1978. Energy transduction in the photosynthetic membranes of cyanobacterium (blue-green alga) *Plectonema boryanum. J. Biol. Chem.* 253:3281–86
50. Papageorgiou, G. 1975. Chlorophyll fluorescence: an intrinsic probe of photosynthesis. See Ref. 32, pp. 319–71
51. Pfennig, N. 1975. The phototrophic bacteria and their role in the sulfur cycle. *Plant & Soil* 43:1–16
52. Pfennig, N. 1977. Phototrophic green and purple bacteria: a comparative systematic survey. *Ann. Rev. Microbiol.* 31:275–90
53. Raboy, B., Padan, E. 1978. Active transport of glucose and α-methylglucoside in the cyanobacterium *Plectonema boryanum. J. Biol. Chem.* 253:3287–91
54. Reynolds, C. S., Walsby, A. E. 1975. Water-blooms. *Biol. Rev.* 50:437–81
55. Rippka, R., Waterbury, J. B. 1977. The synthesis of nitrogenase by non-heterocystous cyanobacteria. *FEMS Microbiol. Lett.* 2:83–86
56. Schopf, J. W. 1974. Paleobiology of the Precambrian: The age of blue-green algae. *Evol. Biol.* 7:1–43
57. Simonis, W., Urbach, W. 1973. Photophosphorylation in vivo. *Ann. Rev. Plant Physiol.* 24:89–114
58. Stanier, R. Y., Cohen-Bazire, G. 1977. Phototrophic prokaryotes: The cyanobacteria. *Ann. Rev. Microbiol.* 31: 225–74

59. Steemann-Nielsen, E. 1975. Marine photosynthesis with special reference on the ecological aspects. *Elsevier Oceanogr. Ser.* 13:25–31
60. Stewart, W. D. P. 1973. Nitrogen fixation. See Ref. 9, pp. 260–78
61. Stewart, W. D. P., Lex, M. 1970. Nitrogenase activity in the blue-green alga *Plectonema boryanum* Strain 594. *Arch. Microbiol.* 73:250–60
62. Stewart, W. D. P., Pearson, H. W. 1970. Effects of aerobic and anaerobic conditions on growth and metabolism of blue-green algae. *Proc. R. Soc. London Ser. B* 175:293–311
63. Tel-Or, E., Luijk, L. W., Packer, L. 1977. An inducible hydrogenase in cyanobacteria enhances N_2 fixation. *FEBS Lett.* 78:49–52
64. Trüper, H. G. 1973. The present state of knowledge of sulfur metabolism in phototrophic bacteria. *Symp. Prokaryotic Photosynth. Org., Freiburg, Germany,* pp. 160–66 (Abstr.). Deutsche Forschungsgemeinschaft, Int. Union Biol. Sci.
65. Utkilen, H. C. 1976. Thiosulfate as electron donor in the blue-green alga *Anacystis nidulans. J. Gen. Microbiol.* 95:177–80
66. Weissman, J. C., Benemann, J. R. 1977. Hydrogen production by nitrogen starved cultures of *Anabaena cylindrica. Appl. Environ. Microbiol.* 33:123–31
67. Weller, D., Doemel, W., Brock, T. D. 1975. Requirement of low oxidation-reduction potential for photosynthesis in a blue-green alga (*Phormidium* sp.). *Arch. Microbiol.* 104:7–13
68. Wood, B. J. B. 1974. Fatty acids and saponifiable lipids. See Ref. 36, pp. 236–65

Ann. Rev. Plant Physiol. 1979. 30:41–53

SULFATED POLYSACCHARIDES IN RED AND BROWN ALGAE[1]

❖7664

Esther L. McCandless

Department of Biology, McMaster University, Hamilton, Ontario, Canada L8S 4K1

James S. Craigie

Atlantic Regional Laboratory, National Research Council of Canada, Halifax, Nova Scotia, Canada B3H 3Z1

CONTENTS

Sulfated polysaccharides occur in the structural elements of both plants and animals, although in plants their distribution is limited to the algae where they may constitute up to 70% of the dry matter of some red seaweeds. Usually ester sulfate is associated with galactans in the Rhodophyceae and fucans in the Phaeophyceae. We shall discuss the general nature of these sulfated polysaccharides and what is known of their biosynthesis including physiological parameters relevant to their production.

[1]Issued as NRCC No. 17056.

0066-4294/79/0601-0041$01.00

IN RED ALGAE

It is now recognized that native agars as well as carrageenans are sulfate-containing polyanions, but the charge density is less in the agars. Aqueous solutions of agars usually form thermally reversible gels, while some carrageenans do not gel and others do so only in the presence of specific cations such as K^+, Rb^+, Cs^+. Rees and his collaborators have shown the inadequacy of studies based upon solubility properties and substituted one using a rigorous chemical definition of the polymers (89, 90). The skeleton of carrageenan is composed solely of D-galactose units alternately linked α-1→3 and β-1→4. The agar skeleton contains both D- and L-galactose residues in equimolar quantities, again with alternating α-1→3, β-1→4 linkages. The seemingly endless variations in polysaccharide composition were simplified by Anderson & Rees's introduction of the concept of masked repeating disaccharide units (4). The masking may be attributed to SO_4^{2-}, methoxyl groups, or anhydrogalactose in various combinations. Methods for isolation of these polysaccharides have been summarized (18), and excellent accounts of their chemical and physical structures are available (6, 43, 77, 78, 90, 96).

Agars

Our present understanding of the structure of agars is based largely on the chemical studies of Araki and associates in Japan (6) and the enzymatic studies of Yaphe and collaborators in Canada (34, 35). Sulfate is an important constituent of native agars, but because it appears to interfere with gelation it is generally regarded as a nuisance in agar of commerce. Another anionic constituent, pyruvic acid, was first described in agar by Hirase (46) as the ketal 4,6-0-(1-carboxyethylidine)-D-galactose and is now reported from a number of agarophytes (74, 101).

Araki (6) fractionated agar from several species into a neutral polymer "agarose" and an anionic fraction. Most recent work (34, 35, 101) indicates that the situation is more complex and that there may be a continuous spectrum of molecules ranging from neutral to charged agarose, the latter containing pyruvated but unsulfated agarose as well as sulfated galactans. We would classify as agars porphyran (4), funoran (6), and the sulfated galactans of *Laurencia pinnatifida* (12) and *Polysiphonia lanosa* (8).

Carrageenans

We group with carrageenans those galactans composed of alternating α-1→3, β-1→4 D-galactose units including the theoretical limiting case in which no sulfate or anhydrogalactose is present. Specific sulfation and subsequent desulfation of this hypothetical molecule would generate families of carrageenans that can be divided into those which are esterified with sulfate at the 4-position (the κ-family), and those which are not (the λ-family). The κ-family would encompass ι- and κ-carrageenans (2, 3) which contain 3,6-anhydro-D-galactose and gel in the presence of K^+, as well as the theoretical μ- and ν-carrageenans (90, 93). The λ-family would include λ- (89), ξ- (76), and the pyruvate containing (48) π-carrageenans (31), all of which do not gel and are very viscous. The partly methylated, sulfated D-galactans from the Grateloupiaceae (1, 47, 73) possess the repeating 1→3 and 1→4 linkages and appear to be carrageenans (1).

Biological Techniques in Polysaccharide Chemistry

Bacterial enzymes were used to degrade agar specifically and produce neoagarobiose (6, 35). In the carrageenan series a κ-carrageenase was used to prepare neocarrabiose and related oligomers (99). Information gained in this way erased any doubt about the repeating nature of the disaccharide units in these galactans. The specificities of the enzymes agarase, κ-carrageenase (99), λ-carrageenase (51), and ι-carrageenase (W. Yaphe, personal communication) provide unique structural information.

Specific antibodies may be used to study localization of polysaccharides (41) as well as to compare their structures. Because the antibody combining site accommodates a segment of the polysaccharide, probably involving several contiguous disaccharide units, we have a sensitive method for comparing common structural features of related molecules. An anti-κ-carrageenan serum appears to react specifically with a structural feature associated with 3,6-anhydro-D-galactose (50). A gamma globulin preparation from this antiserum also requires the 4-sulfated, 3-linked unit; the presence of 2-sulfate reduces the reactivity (28). This antibody was used to distinguish κ-from ι-carrageenan (29, 30). The anti-λ-carrageenan antibodies recognize sulfate ester at the 6-position of the 4-linked unit and so do not react with alkali modified λ-carrageenan, thus permitting discrimination between λ- and ξ-carrageenans (30, 39).

Sulfated Heteroglycans

The capsular polysaccharide from the unicellular *Porphyridium cruentum* is a glucuronoglucoxylogalactan in which the galactose to xylose ratio approximates unity and more than 90% of the galactose is in the L-form. The sulfate content is 7.4% and may be associated with both galactose and xylose residues (67). A similar polysaccharide containing 7.6% of ester sulfate occurs in *P. aerugineum* (84) and another sulfated, xylose-rich galactan was reported in *Rhodella maculata* (37).

IN BROWN ALGAE

Brown algae also contain major amounts of anionic polysaccharides in their cell walls. These are principally the polyuronide alginate and glycans rich in sulfated L-fucose. The complexity of the fucan sulfates varies with the algal source and probably with the extraction techniques used. Products range from the essentially pure fucan sulfate in *Fucus distichus* (49) to ascophyllan, a glucuronoxylofucan sulfate from *Ascophyllum nodosum* in which a 3-O-β-D-xylopyranosyl fucose linkage was demonstrated (54, 55). Mian & Percival (68) concluded that the fucose containing polysaccharides from brown algae comprised a spectrum of molecules from a group rich in uronic acid and poor in fucose sulfate to a relatively pure fucan sulfate containing little uronic acid. Neither *Ascophyllum nodosum* nor *Fucus vesiculosus* contain a discrete fucan sulfate (fucoidan), but the fucan backbone bears long chains composed of glucuronic, mannuronic, and guluronic acids attached by an acid labile linkage (66). Developing embryos of *Fucus distichus,* however, contain fucoidan and not glucuronoxylofucan (49, 83). The generally assumed location of the sulfate at position 4 of an α-1$\rightarrow$2 linked fucan (78) has been verified (5).

BIOLOGICAL AND ENVIRONMENTAL FACTORS AFFECTING SULFATED POLYSACCHARIDES

Unpredictable variations in proportions of the gelling κ-carrageenan and the viscous λ-carrageenan in commercial Irish moss constituted a barrier to quality control of the product until recently. Seasonal factors, habitat, and age of plants have all been invoked and investigated to explain, albeit with little success, these differences in *Chondrus crispus* (11, 40). Although the idea that differences in proportions of the two types of carrageenan might exist between gametophyte and sporophyte generations of *C. crispus* was expressed in 1949 (58), more than 20 years elapsed before the idea was revived (64) and investigated. Chen et al (15) measured 3,6-anhydrogalactose content of *Gigartina stellata* and concluded that tetrasporophytes contained little κ-carrageenan. Similar conclusions were reached on the basis of solubility of carrageenan in the presence of K^+ for *G. decipiens, G. angulata, G. atropurpurea,* and *G. lanceata* (80). Detailed fractionation and characterization of the carrageenans from *Chondrus crispus* (63) and later from *Iridaea cordata* (62) established that gametophytes produced κ-carrageenan but not λ-carrageenan, while the converse applied to tetrasporophytes. To date no exceptions to this distribution have been reported in members of the Gigartinaceae (50, 75, 80, 81, 98). Control of just one parameter, the phase of the life cycle, thus explained most of the variability observed in extracts from these algae.

Species outside the Gigartinaceae do not follow this pattern, as several *Eucheuma* spp. show the same polysaccharide components whether κ- or ι-carrageenans, in both nuclear phases (22, 23, 29, 33). A similar situation exists in *Hypnea musciformis* (50) and *Furcellaria fastigiata* (unpublished data), both of which synthesize carrageenan of the κ-type regardless of the nuclear phase of the plant.

Several studies of natural populations of carrageenophytes indicate that carbohydrate content changes with season (11, 21, 59, 60, 70), but because little attention was paid to nutrient status, temperature, irradiance regimens, growth rates, or nuclear phase, only tentative conclusions can be drawn. It appears, however, that carbohydrate levels are minimal at times when plants would be rapidly growing. This interpretation is supported by reports (71, 72) of a low carrageenan content in *C. crispus* grown in seawater enriched with nitrogen and phosphate fertilizers. In the absence of added fertilizer the percentage of carrageenan increased. The same phenomenon has been observed in experiments with *Porphyridium aerugineum* (J. Ramus, personal communication) and *Neoagardhiella* (24). When carrageenans were extracted from carposporic plants of *Chondrus crispus* from an unpolluted open coastal site, little qualitative or quantitative differences attributable to season could be discerned (61). Virtually no effect of changes in latitude or season were detected in carrageenans extracted from 12 *Gigartina* spp. in New Zealand (75, 81).

BIOSYNTHESIS

Fundamental principles in the biosynthesis of polysaccharides have been reviewed by others (42, 44, 96). Pertinent to our discussion is the mechanism of linking sugar units, the substitution reactions involving sulfate, methyl, and pyruvate groups, and modification of the hexose units to anhydrogalactose in the red algae. Our assumption that modification occurs by sequential steps is prompted by the observations (*a*) that SO_4^{2-} is more rapidly incorporated into polymers than is radiocarbon (10, 16, 25); (*b*) the demonstration that sulfate is incorporated into fucoidan of *Fucus distichus* embryos at a time when there is no net increase in fucan (82); (*c*) the fact that no activated sugar sulfates have been isolated from algae; and (*d*) the formation of anhydrogalactose by a sulfohydrolase acting on a sulfated polymeric precursor (56, 100).

The synthesis of the polysaccharide chain is believed to occur in the Golgi apparatus (14, 36). Several glycosyltransferases from Golgi preparations have been described in mammalian systems (91), but only in the past year have two UDP-galactosyltransferases been demonstrated in dictyosome preparations of the brown alga *Fucus serratus* (17). One of these, UDP-galactose: N-acetylglucosamine β-4-galactosyltransferase (EC 2.4.1.38), is considered to be a Golgi marker enzyme in animals and functions in glycosylation of glycoproteins and oligosaccharides. The second Golgi transferase, UDP-galactose: L-fucose galactosyltransferase, is presumably involved in the linkage of galactose to fucan. This demonstration strengthens the earlier hypothesis, based on evidence from histochemistry and electron microscopy, that the Golgi apparatus is an important site of polysaccharide synthesis in algae (79, 84).

Polysaccharide sulfation is believed to occur in the dictyosomes as originally postulated on the basis of evidence from electron microscopy (86). Histochemical and autoradiographic procedures show radiosulfur-labeling of polysaccharides in the Golgi-rich perinuclear region of cells in both *Pelvetia canaliculata* and *Laminaria* spp. (38), and in the unicellular red alga *Rhodella maculata* (37). Although the weight of evidence favors the Golgi as the main site of polysaccharide sulfation, the inner cell wall may also be involved (53). It is interesting that UDP-acetylglucosamine sulfate in hen oviducts does not appear to participate in sulfate incorporation into mucopolysaccharides (94), and that chondroitin sulfate is synthesized by direct sulfation of a preexisting polysaccharide chain (27).

Sulfate is activated by ATP sulfurylase (ATP: sulfate adenylyl-transferase, EC 2.7.74) to produce adenosine 5' phosphosulfate (APS) which in turn may be phosphorylated by APS kinase (ATP: adenylyl-sulfate 3'-phosphotransferase, EC 2.7.1.25) to adenosine 3'-phosphate 5'-phosphosulfate (PAPS) (92). The latter was demonstrated in *Porphyridium* and its formation was suppressed by molybdate, an inhibitor of ATP sulfurylase (85, 97). Recently both SO_4^{2-} activating enzymes were demonstrated in *Rhodella* (69). Radiosulfate rapidly enters activated intermediates in intact *Fucus serratus,* but this uptake is not affected by molybdate (16). Tsang & Schiff (95) postulate that PAPS is unnecessary for sulfate reduction in a number of algal autotrophs, but no direct evidence exists as to which form of activated sulfate is involved in esterification of polysaccharides. De Lestang & Quillet (25) implicate cytidine diphosphosulfate, a homologue of PAPS, in the biosynthesis of fucoidan.

The entry of exogenous $^{35}SO_4^{2-}$ into fucoidan or its complexes with glucuronic acid, xylose, or galactose has been studied in *Fucus vesiculosus* (10), *F. distichus* (82), *F. serratus* (16), *Pelvetia canaliculata* (26, 38), and *Laminaria* spp. (38). Quatrano and collaborators used developing zygotes of *Fucus* in an interesting study of embryo development (20, 82). They found no SO_4^{2-} incorporation until 10 hr after fertilization and no sulfated polysaccharides before this time, in good accord with earlier histological studies on maturing oogonia of *F. edentatus* (65). The replacement of SO_4^{2-} by methionine suppressed sulfation of the fucan with little effect on rhizoid formation; thus, the synthesis of fucoidan is independent of the establishment of polarity and of rhizoid formation. Two fucans are actively synthesized in early zygote development (83). One is closely associated with xylose and cellulose in the developing cell walls; another becomes heavily sulfated about 10 hr after fertilization and moves into the developing rhizoid. Further work is required to ascertain the relationship between this sulfated fucan and the sulfated glucuronoxylofucans characteristic of mature thalli of *Fucus* and other brown algae (66, 68).

The rates of $^{35}SO_4^{2-}$ and $H^{14}CO_3^-$ incorporation into sulfated polysaccharides are not at all similar (10, 16, 25). One interpretation of the results is that turnover of the SO_4^{2-} is very rapid with the hydrolysis of a sulfate ester bond, activation of SO_4^{2-}, and resynthesis of a new ester. An alternative explanation is that $^{35}SO_4^{2-}$ may be derived rather directly from the external medium, while the fucan-carbon arises from endogenous, chiefly nonradioactive sources (10, 16).

Few biochemical studies have been carried out on sulfation of red algal polysaccharides. Ramus & Groves (86, 87) investigated precursor product relationships of intracellular SO_4^{2-}, capsular, and extracellular polysaccharides in *Porphyridium aerugineum.* Half of the free SO_4^{2-} pool was subsequently incorporated into the capsular polysaccharide, thus establishing the major importance of polysaccharide esterification in the sulfur economy of the cell. The dynamic nature of the cell wall may be adduced from the appearance of ^{35}S-labeled polysaccharide in the medium within 15 min. The preliminary investigation of $^{35}SO_4^{2-}$ incorporation into carrageenans of *Chondrus crispus* (57) was carried out before it was known that κ- and λ-carrageenans were synthesized by separate plants, and is now being reexamined (S. G. Jackson & E. L. McCandless, unpublished).

Biosynthetic mechanisms for O-methylation and for the addition of pyruvate residues to the algal polysaccharides have not been elucidated. Work with higher plants suggests that the methyl group of S-adenosylmethionine provides the methoxyl group in pectins and hemicellulose (52). The activation and incorporation of pyruvate residues could involve the readily available intermediate phosphoenolpyruvate.

An important alteration in the basic structure of some red algal polysaccharides is the reaction yielding anhydrogalactose. Rees (88) isolated an enzyme from *Porphyra umbilicalis* which removed ester sulfate from C-6 of the 4-linked L-galactose unit of porphyran with the concomitant formation of 3,6-anhydro-L-galactose. Subsequently Lawson & Rees (56) demonstrated enzyme activity in extracts of *Gigartina stellata* which converted the 6-sulfated D-galactose unit of carrageenan to the corresponding 3,6-anhydrogalactose. These workers raised the question of whether the appropriate precursor unit for this reaction was the galactose-6-SO_4 or the galactose-2,6-disulfate. Fractionation of radioactively labeled soluble carrageenans of gametophytic *C. crispus* with increasing KCl concentrations yielded a preparation of intermediate solubility and high specific activity, and with an enhanced infrared absorption at 805 cm^{-1} suggesting a higher 3,6-anhydrogalactose-2-SO_4 content (S. G. Jackson & E. L. McCandless, unpublished). Evidence (32) that the KCl-soluble carrageenan in gametophytes of *C. crispus* and several other carrageenophytes is enriched in galactose-2,6-disulfate and that all of the units modified by alkaline borohydride are of this nature raises the possibility that κ-carrageenan is derived from ι-carrageenan by desulfation of 3,6-anhydrogalactose-2-SO_4. Wong & Craigie (100) reinvestigated the sulfohydrolase of *Gigartina stellata* and extended the study to *C. crispus.* Enzyme activity present in both haploid and diploid plants of the latter modified a precursor polysaccharide sulfate to form anhydrogalactose-containing carrageenan. Inhibition of the sulfohydrolase by λ-carrageenan, as well as the lack of a suitable substrate, may explain the absence of 3,6-anhydrogalactose in diploid plants. The enzyme from *Gigartina stellata* was of higher activity but otherwise similar to that from *Chondrus.*

An ι-like carrageenan fraction was obtained (19) by degrading κ-carrageenan from *C. crispus* with the bacterial enzyme κ-carrageenase. Approximately one-half of the original κ-carrageenan remained as an enzyme resistant fraction enriched in 3,6-anhydrogalactose-2-SO_4 and 4-linked galactose-2,6-disulfate. ^{13}C-nmr evidence (9) confirms that the repeating disaccharide unit of ι-carrageenan exists in κ-carrageenan from several carrageenophytes. The fact that pure carrageenans with the postulated theoretical structures have not been isolated may be due to the extraction techniques, or because dissimilar units actually exist as blocks within the same polymer. Such a block-copolymer structure could result from the sequential modification of the nascent polysaccharide.

Evidence suggests that the carrageenans may be grouped according to biosynthetically related sequences (19) to yield the λ- or κ-families. It is perhaps significant that the λ-family of carrageenans, in which a 4-sulfate is absent, has been found only in tetrasporophytes of the Gigartinaceae. Carrageenans of the κ-family are more widely distributed ranging from furcellaran to ι-carrageenan as sulfation on C-2 of the 4-linked unit is increased.

Only preliminary information is available on biosynthesis and deposition of sulfated polysaccharides in algal cells, and virtually nothing is known of their function or whether they are further metabolized by the algae. Research is hampered by unsophisticated techniques for isolation, fractionation, and specific degradation of the polymers. The nature of the covalent linkages in the cell wall structures is little known although it is likely that glycoproteins (13) are involved. Preparations from red algae generally contain low levels of "contaminant" sugar residues, e.g. xylose (7), which could be involved in binding to serine or threonine of protein as recently established for *Porphyridium* (45). Clearly, few questions about algal sulfated polysaccharides have been resolved and fascinating problems await research.

Literature Cited

1. Allsobrook, A. J. R., Nunn, J. R., Parolis, H. 1971. Sulphated polysaccharides of the Grateloupiaceae family. V. A polysaccharide from *Aeodes ulvoidea*. *Carbohydr. Res.* 16:71–78
2. Anderson, N. S., Dolan, T. C. S., Penman, A., Rees, D. A., Mueller, G. P., Stancioff, D. J., Stanley, N. F. 1968. Carrageenans. IV. Variations in the structure and gel properties of κ-carrageenan, and the characterisation of sulphate esters by infrared spectroscopy. *J. Chem. Soc. C*:602–06
3. Anderson, N. S., Dolan, T. C. S., Rees, D. A. 1973. Carrageenans. VII. Polysaccharides from *Eucheuma spinosum* and *Eucheuma cottonii*. The covalent structure of ι-carrageenan. *J. Chem. Soc. Perkin Trans. 1*, pp. 2173–76
4. Anderson, N. S., Rees, D. A. 1965. Porphyran: A polysaccharide with a masked repeating structure. *J. Chem. Soc.* 5880–87
5. Anno, K., Seno, N., Ota, M. 1969. Structural studies on fucoidan from *Pelvetia wrightii*. *Proc. Int. Seaweed Symp.* 6:421–26
6. Araki, C. 1966. Some recent studies on the polysaccharides of agarophytes. *Proc. Int. Seaweed Symp.* 5:3–17
7. Araki, C., Arai, K., Hirase, S. 1967. Studies on the chemical constitution of agar-agar. XXIII. Isolation of D-xylose, 6-O-methyl-D-galactose, 4-O-methyl-L-galactose and O-methylpentose. *Bull. Chem. Soc. Jpn.* 40:959–62
8. Batey, J. F., Turvey, J. R. 1975. The galactan sulphate of the red alga *Polysiphonia lanosa*. *Carbohydr. Res.* 43: 133–43
9. Bhattacharjee, S. S., Yaphe, W., Hamer, G. K. 1978. A study of agar and carrageenan by ^{13}C nmr spectroscopy. *Proc. Int. Seaweed Symp.* 9. In press
10. Bidwell, R. G. S., Ghosh, N. R. 1963. Photosynthesis and metabolism in marine algae. VI. The uptake and incorporation of S^{35}-sulphate in *Fucus vesiculosus*. *Can. J. Bot.* 41:209–20
11. Black, W. A. P., Blakemore, W. R., Colquhoun, J. A., Dewar, E. T. 1965. The evaluation of some red marine algae as a source of carrageenan and of its κ- and λ-components. *J. Sci. Food Agric.* 16:573–85
12. Bowker, D. M., Turvey, J. R. 1968. Water-soluble polysaccharides of the red alga *Laurencia pinnatifida*. I. Constituent units. *J. Chem. Soc. C*:983–88
13. Brown, R. G., Kimmins, W. C. 1977. Glycoproteins. *Int. Rev. Biochem* 13: 183–209
14. Brown, R. M. Jr., Herth, W., Franke, W. W., Romanovicz, D. 1973. The role of the Golgi apparatus in the biosynthesis and secretion of a cellulosic glycoprotein in *Pleurochrysis*: A model system for the synthesis of structural polysaccharides. In *Biogenesis of Plant Cell Wall Polysaccharides*, ed. F. Loewus, pp. 207–57. New York: Academic. 379 pp.
15. Chen, L. C-M., McLachlan, J., Neish, A. C., Shacklock, P. F. 1973. The ratio of kappa- to lambda-carrageenan in nuclear phases of the rhodophycean algae, *Chondrus crispus* and *Gigartina stellata*. *J. Mar. Biol. Assoc. UK* 53:11–16
16. Coughlan, S. 1977. Sulphate uptake in *Fucus serratus*. *J. Exp. Bot.* 28:1207–15

17. Coughlan, S., Evans, L. V. 1978. Isolation and characterization of Golgi bodies from vegetative tissue of the brown alga *Fucus serratus*. *J. Exp. Bot.* 29:55–68
18. Craigie, J. S., Leigh, C. 1978. Carrageenans and agars. In *Handbook of Phycological Methods. Physiological and Biochemical Methods*, ed. J. A. Hellebust, J. S. Craigie, pp. 109–31. New York: Cambridge. 512 pp.
19. Craigie, J. S., Wong, K. F. 1978. Carrageenan biosynthesis. *Proc. Int. Seaweed Symp.* 9. In press
20. Crayton, M. A., Wilson, E., Quatrano, R. S. 1974. Sulfation of fucoidan in *Fucus* embryos. II. Separation from initiation of polar growth. *Dev. Biol.* 39:134–37
21. Dawes, C. J., Lawrence, J. M., Cheney, D. P., Mathieson, A. C. 1974. Ecological studies of Floridian *Eucheuma* (Rhodophyta, Gigartinales). III. Seasonal variation of carrageenan, total carbohydrate, protein, and lipid. *Bull. Mar. Sci.* 24:286–99
22. Dawes, C. J., Stanley, N. F., Moon, R. E. 1977. Physiological and biochemical studies on the ι-carrageenan producing red alga *Eucheuma uncinatum* Setchell and Gardner from the gulf of California. *Bot. Mar.* 20:437–42
23. Dawes, C. J., Stanley, N. F., Stancioff, D. J. 1977. Seasonal and reproductive aspects of plant chemistry, and ι-carrageenan from Floridian *Eucheuma* (Rhodophyta, Gigartinales). *Bot. Mar.* 20:137–47
24. DeBoer, J. A. 1978. Effects of nitrogen enrichment on growth rate and phycocolloid content in *Gracilaria foliifera* and *Neoagardhiella baileyi*. *Proc. Int. Seaweed Symp.* 9. In press
25. de Lestang, G., Quillet, M. 1974. Comportement du fucoïdane sulfurylé de *Pelvetia canaliculata* (Dcne & Thur.) vis à vis des cations de la mer: propriétés d'échange, renouvellement des radicaux sulfuriques, coenzyme d'activation des sulfates. Intérêt fonctionnel. *Physiol. Vég.* 12:199–227
26. de Lestang-Brémond, G., Quillet, M. 1976. Étude, à l'aide de $^{35}SO_4$, du turnover des sulfates estérifiant le fucoïdane de *Pelvetia canaliculata* (Dcne et Thur.). *Physiol. Vég.* 14:259–69
27. DeLuca, S., Richmond, M. E., Silbert, J. E. 1973. Biosynthesis of chondroitin sulfate. Sulfation of the polysaccharide chain. *Biochemistry* 12:3911–15
28. DiNinno, V., McCandless, E. L. 1978. Anti-carrageenans. *Immunochemistry* 15:273–74
29. DiNinno, V., McCandless, E. L. 1978. The chemistry and immunochemistry of carrageenans from *Eucheuma* and related algal species. *Carbohydr. Res.* 66:85–93
30. DiNinno, V., McCandless, E. L. 1978. The immunochemistry of λ-type carrageenans from certain red algae. *Carbohydr. Res.* 67:235–41
31. DiNinno, V. L. 1978. *Chemistry and immunochemistry of carrageenans.* PhD thesis. McMaster University, Hamilton, Ontario, Canada
32. DiNinno, V. L., McCandless, E. L. 1979. Immunochemistry of κ-type carrageenans from certain red algae. *Carbohydr. Res.* In press
33. Doty, M. S., Santos, G. A. 1978. Carrageenans from tetrasporic and cystocarpic *Eucheuma* species. *Aquat. Bot.* 4:143–49
34. Duckworth, M., Yaphe, W. 1971. The structure of agar. I. Fractionation of a complex mixture of polysaccharides. *Carbohydr. Res.* 16:189–97
35. Duckworth, M., Yaphe, W. 1971. The structure of agar. II. The use of a bacterial agarase to elucidate structural features of the charged polysaccharides in agar. *Carbohydr. Res.* 16:435–45
36. Evans, L. V., Callow, M. E. 1976. Secretory processes in seaweeds. In *Perspectives in Experimental Biology, Vol. 2 Botany*, ed. N. Sunderland, pp. 487–99. Oxford: Pergamon. 523 pp.
37. Evans, L. V., Callow, M. E., Percival, E., Fareed, V. 1974. Studies on the synthesis and composition of extracellular mucilage in the unicellular red alga *Rhodella*. *J. Cell Sci.* 16:1–21
38. Evans, L. V., Simpson, M., Callow, M. E. 1973. Sulphated polysaccharide synthesis in brown algae. *Planta* 110: 237–52
39. Evelegh, M. J., Vollmer, C. M., McCandless, E. L. 1978. Differentiation of lambda carrageenans from Rhodophyta with immunological and spectroscopic techniques. *J. Phycol.* 14:89–91
40. Fuller, S. W., Mathieson, A. C. 1972. Ecological studies of economic red algae. IV. Variations of carrageenan concentration and properties in *Chondrus crispus* Stackhouse. *J. Exp. Mar. Biol. Ecol.* 10:49–58
41. Gordon-Mills, E. M., McCandless, E. L. 1975. Carrageenans in the cell walls of *Chondrus crispus* Stack. (Rhodophyceae, Gigartinales). I. Localization

with fluorescent antibody. *Phycologia* 14:275–81
42. Hassid, W. Z. 1970. Biosynthesis of sugars and polysaccharides. In *The Carbohydrates: Chemistry* and *Biochemistry,* ed. W. Pigman, D. Horton, A. Herp, 2A:301–73. London: Academic. 469 pp. 2nd ed.
43. Haug, A. 1974. Chemistry and biochemistry of algal cell-wall polysaccharides. In *Plant Biochemistry,* ed. D. H. Northcote. *MTP Int. Rev. Sci.: Biochem. Ser.* I, 11:51–88. London: Butterworths. 287 pp.
44. Haug, A., Larsen, B. 1974. Biosynthesis of algal polysaccharides. In *Plant Carbohydrate Biochemistry,* ed. J. B. Pridham. *Ann. Proc. Phytochem. Soc.* No. 10, pp. 207–18. London: Academic. 269 pp.
45. Heaney-Kieras, J., Rodén, L., Chapman, D. J. 1977. The covalent linkage of protein to carbohydrate in the extracellular protein-polysaccharide from the red alga *Porphyridium cruentum. Biochem. J.* 165:1–9
46. Hirase, S. 1957. Studies on the chemical constitution of agar-agar. XIX. Pyruvic acid as a constituent of agar-agar. *Bull. Chem. Soc. Jpn.* 30:68–79
47. Hirase, S., Araki, C., Watanabe, K. 1967. Component sugars of the polysaccharide of the red seaweed *Grateloupia elliptica. Bull. Chem. Soc. Jpn.* 40: 1445–48
48. Hirase, S., Watanabe, K. 1972. The presence of pyruvate residues in λ-carrageenan and a similar polysaccharide. *Bull. Inst. Chem. Res., Kyoto Univ.* 50:332–36
49. Hogsett, W. E., Quatrano, R. S. 1975. Isolation of polysaccharides sulfated during early embryogenesis in *Fucus. Plant Physiol.* 55:25–29
50. Hosford, S. P. C., McCandless, E. L. 1975. Immunochemistry of carrageenans from gametophytes and sporophytes of certain red algae. *Can. J. Bot.* 53:2835–41
51. Johnston, K. H., McCandless, E. L. 1973. Enzymic hydrolysis of the potassium chloride soluble fraction of carrageenan: properties of "λ-carrageenases" from *Pseudomonas carrageenovora. Can. J. Microbiol.* 19:779–88
52. Kauss, H., Hassid, W. Z. 1967. Biosynthesis of the 4-O-methyl-D-glucuronic acid unit of hemicellulose B by transmethylation from S-adenosyl-L-methionine. *J. Biol. Chem.* 242:1680–84
53. LaClaire, J. W. II, Dawes, C. J. 1976. An autoradiographic and histochemical localization of sulfated polysaccharides in *Eucheuma nudum* (Rhodophyta). *J. Phycol.* 12:368–75
54. Larsen, B. 1967. Sulphated polysaccharides in brown algae. II. Isolation of 3-O-β-D-xylopyranosyl-L-fucose from ascophyllan. *Acta Chem. Scand.* 21: 1395–96
55. Larsen, B., Haug, A., Painter, T. J. 1966. Sulphated polysaccharides in brown algae. I. Isolation and preliminary characterisation of three sulphated polysaccharides from *Ascophyllum nodosum* (L.) Le Jol. *Acta Chem. Scand.* 20:219–30
56. Lawson, C. J., Rees, D. A. 1970. An enzyme for the metabolic control of polysaccharide conformation and function. *Nature* 227:392–93
57. Loewus, F., Wagner, G., Schiff, J. A., Weistrop, J. 1971. The incorporation of ^{35}S-labeled sulfate into carrageenan in *Chondrus crispus. Plant Physiol.* 48: 373–75
58. Marshall, S. M., Newton, L., Orr, A. P. 1949. *A Study of Certain British Seaweeds and Their Utilisation in the Preparation of Agar.* London: HMSO. 184 pp.
59. Mathieson, A. C., Tveter, E. 1975. Carrageenan ecology of *Chondrus crispus* Stackhouse. *Aquat. Bot.* 1:25–43
60. Mathieson, A. C., Tveter, E. 1976. Carrageenan ecology of *Gigartina stellata,* (Stackhouse) Batters. *Aquat. Bot.* 2:353–61
61. McCandless, E. L., Craigie, J. S. 1974. Reevaluation of seasonal factors involved in carrageenan production by *Chondrus crispus:* Carrageenans of carposporic plants. *Bot. Mar.* 17:125–29
62. McCandless, E. L., Craigie, J. S., Hansen, J. E. 1975. Carrageenans of gametangial and tetrasporangial stages of *Iridaea cordata* (Gigartinaceae). *Can. J. Bot.* 53:2315–18
63. McCandless, E. L., Craigie, J. S., Walter, J. A. 1973. Carrageenans in the gametophytic and sporophytic stages of *Chondrus crispus. Planta* 112:201–12
64. McCandless, E. L., Richer, S. M. 1972. ^{14}C studies of carrageenan synthesis. *Proc. Int. Seaweed Symp.* 7:477–84
65. McCully, M. E. 1968. Histological studies on the genus *Fucus.* II. Histology of the reproductive tissues. *Protoplasma* 66:205–30
66. Medcalf, D. G., Larsen, B. 1977. Fucose-containing polysaccharides in the brown algae *Ascophyllum nodosum* and *Fucus vesiculosis. Carbohydr. Res.* 59:531–37

67. Medcalf, D. G., Scott, J. R., Brannon, J. H., Hemerick, G. A., Cunningham, R. L., Chessen, J. H., Shah, J. 1975. Some structural features and viscometric properties of the extracellular polysaccharide from *Porphyridium cruentum*. *Carbohydr. Res.* 44:87–96
68. Mian, A. J., Percival, E. 1973. Carbohydrates of the brown seaweeds *Himanthalia lorea* and *Bifurcaria bifurcata*. II. Structural studies of the "fucans." *Carbohydr. Res.* 26:147–61
69. Møller, M. E., Evans, L. V. 1976. Sulphate activation in the unicellular red alga *Rhodella*. *Phytochemistry* 15: 1623–26
70. Mollion, J. 1978. Seasonal variations in the carrageenan from *Hypnea musciformis*. *Proc. Int. Seaweed Symp.* 9. In press
71. Neish, A. C., Shacklock, P. F. 1971. Greenhouse experiments (1971) on the propagation of strain T4 of Irish moss. *Tech. Rep. 14.* Atlantic Reg. Lab., Natl. Res. Counc. Canada, No. 12253. 25 pp.
72. Neish, A. C., Shacklock, P. F., Fox, C. H., Simpson, F. J. 1977. The cultivation of *Chondrus crispus*. Factors affecting growth under greenhouse conditions. *Can. J. Bot.* 55:2263–71
73. Nunn, J. R., Parolis, H. 1968. A polysaccharide from *Aeodes orbitosa*. *Carbohydr. Res.* 6:1–11
74. Nunn, J. R., Parolis, H., Russell, I. 1972. A polysaccharide from *Anatheca dentata*. *Proc. Int. Seaweed Symp.* 7:436–38
75. Parsons, M. J., Pickmere, S. E., Bailey, R. W. 1977. Carrageenan composition in New Zealand species of *Gigartina* (Rhodophyta): Geographic variation and interspecific differences. *N. Z. J. Bot.* 15:589–95
76. Penman, A., Rees, D. A. 1973. Carrageenans. IX. Methylation analysis of galactan sulphates from *Furcellaria fastigiata, Gigartina canaliculata, Gigartina chamissoi, Gigartina atropurpurea, Ahnfeltia durvillaei, Gymnogongrus furcellatus, Eucheuma isiforme, Eucheuma uncinatum, Aghardhiella tenera, Pachymenia hymantophora,* and *Gloiopeltis cervicornis*. Structure of ξ-carrageenan. *J. Chem. Soc. Perkin Trans. 1*:2182–87
77. Percival, E. 1970. Algal polysaccharides. In *The Carbohydrates. Chemistry and Biochemistry,* ed. W. Pigman, D. Horton, A. Herp, 2B:537–68. London: Academic. 853 pp.
78. Percival, E., McDowell, R. H. 1967. *Chemistry and Enzymology of Marine Algal Polysaccharides*. London: Academic. 219 pp.
79. Peyrière, M. 1970. Evolution de l'appareil de Golgi au cours de la tétrasporogenèse de *Griffithsia flosculosa* (Rhodophycée). *C. R. Acad. Sci.* 270:2071–74
80. Pickmere, S. E., Parsons, M. J., Bailey, R. W. 1973. Composition of *Gigartina* carrageenan in relation to sporophyte and gametophyte stages of the life cycle. *Phytochemistry* 12:2441–44
81. Pickmere, S. E., Parsons, M. J., Bailey, R. W. 1975. Variations in carrageenan levels and composition in three New Zealand species of Gigartina. *N. Z. J. Sci.* 18:585–90
82. Quatrano, R. S., Crayton, M. A. 1973. Sulfation of fucoidan in *Fucus* embryos. I. Possible role in localization. *Dev. Biol.* 30:29–41
83. Quatrano, R. S., Stevens, P. T. 1976. Cell wall assembly in *Fucus* zygotes. I. Characterization of the polysaccharide components. *Plant Physiol.* 58:224–31
84. Ramus, J. 1972. The production of extracellular polysaccharide by the unicellular red alga *Porphyridium aerugineum*. *J. Phycol.* 8:97–111
85. Ramus, J. 1974. *In vivo* molybdate inhibition of sulfate transfer to *Porphyridium* capsular polysaccharide. *Plant Physiol.* 54:945–49
86. Ramus, J., Groves, S. T. 1972. Incorporation of sulfate into the capsular polysaccharide of the red alga *Porphyridium*. *J. Cell Biol.* 54:399–407
87. Ramus, J., Groves, S. T. 1974. Precursor-product relationships during sulfate incorporation into *Porphyridium* capsular polysaccharide. *Plant Physiol.* 53:434–39
88. Rees, D. A. 1961. Enzymic synthesis of 3:6-anhydro-L-galactose within porphyran from L-galactose 6-sulphate units. *Biochem. J.* 81:347–52
89. Rees, D. A. 1963. The carrageenan system of polysaccharides. 1. The relation between the κ- and λ-components. *J. Chem. Soc.:* 1821–32
90. Rees, D. A. 1969. Structure, conformation, and mechanism in the formation of polysaccharide gels and networks. *Adv. Carbohydr. Chem. Biochem.* 24:267–332
91. Rodén, L., Schwartz, N. B. 1975. Biosynthesis of connective tissue proteoglycans. In *Biochemistry of Carbohydrates,* ed. W. J. Whelan, pp. 95–152. London: Butterworths. 441 pp.
92. Schiff, J. A., Hodson, R. C. 1973. The

metabolism of sulfate. *Ann. Rev. Plant Physiol.* 24:381–414
93. Stancioff, D. J., Stanley, N. F. 1969. Infrared and chemical studies of algal polysaccharides. *Proc. Int. Seaweed Symp.* 6:595–609
94. Suzuki, S., Strominger, J. L. 1960. Enzymatic sulfation of mucopolysaccharides in hen oviduct. I. Transfer of sulfate from 3'-phosphoadenosine 5'-phosphosulfate to mucopolysaccharides. *J. Biol. Chem.* 235:257–66
95. Tsang, M. L-S., Schiff, J. A. 1975. Studies of sulfate utilization by algae. 14. Distribution of adenosine-3'-phosphate-5'-phosphosulfate (PAPS) and adenosine-5'-phosphosulfate (APS) sulfotransferases in assimilatory sulfate reducers. *Plant Sci. Lett.* 4:301–7
96. Turvey, J. R. 1978. Biochemistry of algal polysaccharides. In *Biochemistry of Carbohydrates II,* ed. D. J. Manners, pp. 151–77. Baltimore: University Park. 259 pp.
97. Varma, A. K., Nicholas, D. J. D. 1971. Purification and properties of ATP-sulphurylase from *Nitrobacter agilis. Biochim. Biophys. Acta* 227:373–89
98. Waaland, J. R. 1975. Differences in carrageenan in gametophytes and tetrasporophytes of red algae. *Phytochemistry* 14:1359–62
99. Weigl, J., Yaphe, W. 1966. The enzymatic hydrolysis of carrageenan by *Pseudomonas carrageenovora:* purification of κ-carrageenase. *Can. J. Microbiol.* 12:939–47
100. Wong, K. F., Craigie, J. S. 1978. Sulfohydrolase activity and carrageenan biosynthesis in *Chrondrus crispus* (Rhodophyceae). *Plant Physiol.* 61: 663–66
101. Young, K., Duckworth, M., Yaphe, W. 1971. The structure of agar. III. Pyruvic acid, a common feature of agars from different agarophytes. *Carbohydr. Res.* 16:446–48

Ann. Rev. Plant Physiol. 1979. 30:55–78

POLYSACCHARIDE CONFORMATION AND CELL WALL FUNCTION

❖7665

R. D. Preston

Emeritus Professor of Plant Biophysics and recent Head of the Astbury Department of Biophysics, University of Leeds; Visiting Professor, Department of Botany and Plant Technology, Imperial College of Science and Technology, University of London.

CONTENTS

INTRODUCTION

The time has long gone since the wall of a living cell could be regarded merely as an outer, inactive, secreted envelope. It is now recognized instead as a delicately balanced entity with specific functions, open to the metabolic machinery of the plant, essential to the continuing existence of the cell, and therefore, in a tissue, to the continuation of the plant itself. The structures of many of the components of the wall are now well documented and the general architecture to which they contribute is known in outline. In order to understand the derivations of the functions of the wall, however, and perhaps thereby achieve some control, we need to know the conformations of the constituent molecules as they occur in the fresh untreated wall, the

0066-4294/79/0601-0055$01.00

relative dispositions of these molecules, and the nature and distribution of the bonds between them. It is the purpose of this article to examine how far the conformations of the polysaccharide constituents particularly, such as have been derived during recent years, taken with physiological and biophysical findings with regard to function, can lead to a model of the wall upon which understanding of function can be founded.

Two types of wall function can be defined which seem at first sight to demand irreconcilable differences in the construction of the wall. It has come to be realized, for instance, that the wall carries recognition sites whereby a parasite can recognize its host (54, 74a) and, though it has now been claimed that lectins with strong carbohydrate binding activity may be extracted from the walls of growing cells (37), the recognition envisaged is a carbohydrate—carbohydrate recognition. This would imply the presence in the wall of species-specific polysaccharides. In principle this does not present difficulties. Polysaccharides may contain one, a few, or all of the residues of glucose, mannose, galactose, arabinose, xylose, rhamnose, glucuronic acid, and galacturonic acid, to take only the commoner saccharides. If any one of the residues in a chain of 100 in, therefore, only a low molecular weight polysaccharide can be any one of these eight, then there are possible 8^{100} permutations and an enormous number of different polysaccharides. Remembering that some polysaccharides carry branches of various length, some are sulphated (in the algae) and some are acetylated (in higher plants) and so forth, then clearly polysaccharides specific to species can be envisaged. This would make cell walls very variable whereas the second type of function, exemplified by the necessity of all primary walls to extend in much the same way as conditioned by turgor forces and growth substances, seems to demand the uniformity of structure built into some models which will be discussed below.

These two needs are not, however, mutually exclusive. It has already been pointed out (43) that polysaccharides with very different chemistries can present to the outer world very much the same bonding pattern giving, therefore, aggregates with very much the same mechanical properties. It is with the function of the wall as expressed through its mechanical properties that this article deals. These are reflected both in the primary wall of the growing cell and, in a different way and through different qualities, in the secondary walls of mature or dead cells. We shall here be concerned only with the former of these two; the latter has been dealt with briefly elsewhere (58). In this context, cell wall polysaccharides can be grouped into a limited number of conformational types, and these will be examined first with particular regard to the intermolecular bonding which each conformation allows and the corresponding variable strength of the bonding to other specified types of polysaccharides. The possible consequences of the presence of protein in growing walls to the bonding pattern will also be considered. This will leave the way open for a consideration of the biophysical qualities of the wall involved in the growth process in terms of these bonding possibilities, with particular regard for the relevant models of wall structure which have been proposed during recent years.

POLYSACCHARIDE CONFORMATION

Polysaccharide conformation—the shape which the chain molecule adopts —is determined by the type of linkage between the sugar units of the chain, by modifications arising from the presence of branches, and by adjustments dictated by close packing with neighboring molecules. Very few wall polysaccharides are crystalline in the native state, but several can be extracted and crystallized, and in all these cases the method of X-ray diffraction analysis is available for structure determination. In no case, however, is the information thus available sufficient for a full determination of conformation. The situation has improved out of recognition during recent years through the application of so-called conformational analysis guided by the X-ray diagram. Brief descriptions of the method will be found eleswhere (58, 63), and conformational analysis itself will not be discussed here; only the results will be used.

In the polysaccharides, constituent sugars are all cyclic, usually with six atoms in a ring, one of them always being oxygen. This is a conformation found in pyran and the sugars are accordingly called pyranose sugars. D-glucopyranose is illustrated in Figure 1, parts a and b in two forms dependent on the configuration at carbon 1. These are both called chair forms, from the shape of the molecule, and symbolized C1. There is another chair form, 1C, which occurs more rarely because it is energetically rather less probable and which is found only occasionally in plants (Figure 1c). Carbon 1 is an important atom in the ring since it is the only carbon connected to two oxygens; accordingly, when two sugars are joined one of the carbons in the bond is always carbon 1. The carbon of the other sugar unit may be carbon 2, 3, 4, or 6 (with hexose sugars). In pentoses,

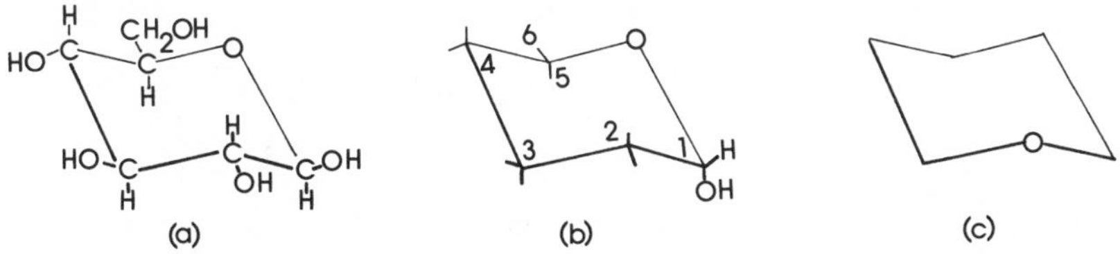

Figure 1 Structural formulae for D-glucose; rings seen from the side. (a) β-D-glucose; the carbon atoms are numbered as in (b). (b) α-D-glucose; carbons omitted, groups at 2, 3, 4, 5, and 6 as in (a). (c) The two previous rings are in the C1 form; this shows the skeleton of the ring in the 1C form.

the $-CH_2OH$ attached to carbon 5 is replaced by $-H$, and β-D-glucopyranose, for instance, becomes β-D-xylopyranose. Similarly, replacement of $-CH_2OH$ by $-COOH$ produces glucuronic acid. Sugars other than glucose can formally be derived by switching the $-OH$ groups on other carbons from one side of the ring to the other. Mannose, for example, can be figured by switching the $-OH$ on carbon 2 only. Sugars also occur in plants in a 5-atom ring configuration. The similarity of these to furan causes these to be named furanose sugars; some will be met with here.

Skeletal Polysaccharides

The great bulk of plant cell walls contains one polysaccharide which is crystalline in the native state and has visible form in the electron microscope. Three are known: cellulose (β-(1→4)-linked-D-glucopyranose), β-(1→3)-linked-D-xylopyranose, and β-(1→4)-linked-D-mannopyranose.

When two molecules of β-glucose are joined through carbons 1 and 4, the resulting disaccharide, cellobiose, is exceptionally stable, in part because a hydrogen bond forms between carbon 3 of one sugar residue and the ring oxygen of the other, giving the conformation shown in Figure 2. Further condensation of sugar units on each end of this molecule, until thousands are joined in a chain, accordingly gives a straight, flat, ribbon-like molecule, cellulose (Figure 3). It will be noted in Figure 3 that neighboring glucose residues are turned through 180° about the chain axis, so that if the chain is turned over around its axis, it will coincide with itself only when moved upward or downward by 5.15 A, the length of a glucose residue. The chain

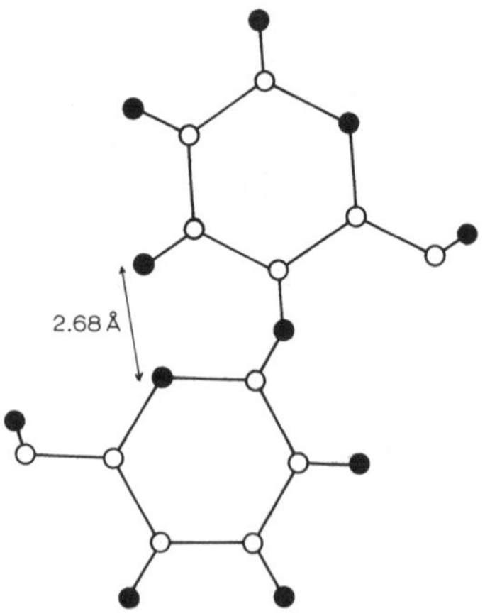

Figure 2 Face view of the two glucose rings in the disaccharide cellobiose; open rings represent carbon, solid rings oxygen, hydrogen omitted. Note that one glucose ring is turned 180° with respect to the other; note also the hydrogen bond between carbon 3 of one ring and the ring oxygen of the other.

has a twofold screw axis. The chain bristles along each edge of the ribbon with the potentially hydrogen bonding groups –OH; when two such chains lie side by side, all the hydroxyl groups facing each other can come into register, hydrogen bonds are formed in enormous numbers, and the association is stable. Correspondingly, the molecular chains are found aggregated in bundles (though this is not how cellulose is synthesized) which again are flattish ribbons, estimated in higher plants to be some 5 nm by 3 nm in section (about some 17 nm by 10 nm in some algae) (58). These sizes have been further confirmed in the most recent electron microscope findings (30, 72).

It will be clear from Figure 3 that each chain has an up and a down sense. The chains of the crystalline aggregates of cellulose were long considered to lie alternately up and down in an antiparallel arrangement. More recently, using conformational analysis, there has been reason to think that the arrangement is, on the contrary, parallel (29). The latest tendency has been to confirm the parallel structure through electron diffraction studies

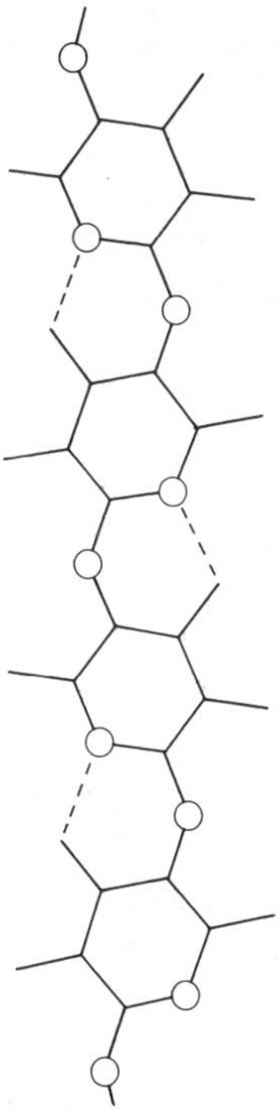

Figure 3 Four glucose residues in a chain molecule of cellulose. The sugars are all β-D-glucose in the C1 conformation and some 10,000 are linked in the chain. The open rings represent oxygen; carbon, hydrogen, and hydroxyl groups omitted. Note the intramolecular hydrogen bonds (broken lines) and the twofold screw axis.

of cellulose itself (16) and by detailed analysis of cellulose triacetate (75). It is therefore odd that mercerized cellulose, which can be produced from native cellulose without dissolution, is still considered to be antiparallel (63). The change to a parallel arrangement is of material consequence because the mechanical properties of the crystallite are vital for the quantitative understanding of the mechanical properties of the cell wall, and these have up to now been calculated on the basis of antiparallel chains. A parallel chain arrangement will give a different hydrogen bonding sequence and therefore different mechanical properties. In fact, the structure of cellulose varies somewhat in detail among both lower and higher plants (58), traceable perhaps to the finding (3) that in single free chains the $-CH_2OH$ on each carbon 5 is free to rotate about the C5-C6 bond (59).

The surface of the crystallite is covered closely by –OH groups numbering of the order of 5 nm^{-2}, and this means that other compatible polysaccharides in the wall will bond strongly to the crystallite by hydrogen bonds. The nature of the polysaccharides expected, and found, to do this will be examined later; collectively they bring up the size of the aggregate to about 8–10 nm in higher plants giving structures which are very long and thin called microfibrils. These can be seen in some walls without any treatment (e.g. in algae such as *Valonia* and the Cladophorales where the microfibrils are much broader), but extraction of the surrounding matrix polysaccharides is often needed for clarity, so that the cross-sectional size may depend to some extent upon the method of extraction.

In a few plants, all members of the old algal order, the Siphonales, cellulose is replaced by β-(1→4)-linked mannan. The structure is similar to that of cellulose though the switching of the –OH on each sugar ring to the other side of the ring makes for a different hydrogen bonding pattern, presumably a somewhat different set of mechanical properties and certainly a lower stability than cellulose. The structure has been worked out (57) under the customary assumption that the size and shape of each sugar ring is fixed and rigid; it has recently been modified and refined (83) by allowing some flexibility in the ring. This polysaccharide, like cellulose, is microfibrillar, but the microfibrils are much more sensitive to chemical extractive agents than is cellulose and care is needed to preserve them intact.

The third type of skeletal polysaccharide also occurs in some members of the Siphonales and, like mannan, appears in other plants including higher plants as a matrix polysaccharide. The chain molecules of this xylan, being uniformly 1→3 linked, are curved. In the crystallite they adopt a helical configuration with three chains coiled together in a triple helix (4, 5; see also 58, 59). This is stabilized by strong hydrogen bonding between the three chains and by the presence of water in the lattice, a difference from both cellulose and mannan, neither of which permits the entry of water. The triple helices are stacked in hexagonal close packing into microfibrils closely resembling those of cellulose. Again, this polysaccharide is less stable than cellulose. Moreover, the frequency of –OH groups on the surface is much lower than in cellulose or mannan, of the order of 1 nm^{-2} so that the whole wall is probably weaker.

Matrix Polysaccharides

The skeletal polysaccharides are homopolymers, or at least there are long stretches of the chain molecules with only one sugar unit, with the same interunit bond, which can pack together in crystalline array. The matrix polysaccharides are much more diverse. Each chain molecule may contain one, a few, or several different sugar residues with the same or different bonds. They may be branched to varying degrees and the conformation may be like a ribbon, a twisted ribbon, an open helix, or the molecule may be completely disordered. With some, the succession of sugars in the backbone chain or the branches may be such that parts of a chain have a specific conformation while other parts are disordered. There exists among these molecular species, therefore, a varying compatibility with cellulose and with each other. Moreover, the constituent sugars may carry O-acetyl groups (higher plants only) conferring some solubility, or half ester sulphate groups (absent in higher plants) by replacement of $-CH_2OH$ on carbon 5 by $-CH_2SO_3$.

RIBBON-LIKE CHAINS These include β-(1→4)-linked mannan, β-(1→4)-linked xylan, α-(1→4)-linked rhamnogalacturonan (pectic acid) and, of course, noncellulosic β-(1→4)-linked glucan. All these are of very wide occurrence.

Mannose is commonly copolymerized with glucose or galactose, forming glucomannan or galactomannan with the nonmannose sugar residues as side chains on a mannose backbone. These polysaccharides are copiously present in many plants, particularly gymnosperms. The side chains occur sometimes in blocks (21) so that long lengths of the chain are pure mannan. Such a chain would be compatible with cellulose though the hydrogen bonding of a mannan chain to a cellulose chain would appear to be only about two thirds as strong as the bonding between cellulose chains on account of the configuration at carbon 2 of mannose. Nevertheless, mannan is held firmly to the cellulose of some plants. It may then be removed only by treatments which cause the microfibril to degenerate into short rods which are probably cellulose crystallites (24), and the material so removed from the microfibrils yields glucose as well as mannose on hydrolysis so that it might be a glucomannan. It seems unlikely, however, that a branched chain would be held so firmly and the "adsorbed" chains are probably glucan and mannan separately.

β-(1→4)-linked xylan, characteristic particularly of angiosperms, is conceptually more complex. This is again a linear molecule with only occasional side groups of L-arabinose or with 4-O-methyl-D-glucuronopyranosyl side groups. The unbranched segments are long enough that on dissolution they may be persuaded to come together in ordered chain bundles to give X-ray diagrams as good as any found in polysaccharides. The unbranched segments of the chain have a threefold axis instead of the twofold found with cellulose and mannan, with a repeat distance along the chain of 1.485 nm (56) instead of the 1.03 of cellulose and mannan. Yet conformational analysis shows that a twofold screw axis here is more stable than a threefold screw by 4–5 kcal/mol residue. The explanation is that, in the crystal, water enters the lattice and stabilizes it with the chains in the threefold screw. On the other hand, when the xylan is acetylated the chain becomes a twofold screw axis with a repeat distance along the chain of 1.03 nm like cellulose. Thus it could be that in the wall, ranges of xylan with no side groups could possess a threefold screw axis while other lengths of the same chain are twofold. The degree of branching varies widely among xylans. Birch xylan has side chains of 4-O-methyl-glucuronic acid residues irregularly disposed (70) so that long lengths might be smooth. In slash pine, branches of this kind, together with L-arabinofuranose and D-xylopyranose, are frequent (66). In grasses, however, the xylan is linear (55).

Only a twofold screw axis will make the xylan compatible with either cellulose or mannan as far as hydrogen bonding is concerned, and even then the absence of the long reach of $-CH_2OH$ from carbon 5 reduces the hydrogen bonding intensity. At best only two thirds of the hydroxyls of cellulose can be accepted for hydrogen bonding, as with mannan. Nevertheless, xylan is known to be strongly bonded to cellulose for exactly the reasons already given for mannan. The material thus firmly held by cellulose could be a xyloglucan, as suggested for the walls of cultured cells of sycamore callus (7) or a glucoxylan. Neither of these, however, would be expected to form copious hydrogen bonds with cellulose and, failing a covalent linkage, it seems most likely that the chains held by cellulose are pure xylan. In that case the chains should possess a twofold axis induced perhaps by the neighboring microfibril surface. The possible association of xylan chains with a threefold axis, such as may occur in a wet wall, will be considered later.

Pectic acid, α-(1→4)-linked-polygalacturonan, occurs very widely as the Na or Ca ester or fully or partially methylated. It is sometimes crystalline in the wall and can in any case be extracted and partially crystallized. In this way it was shown more than 30 years ago that the chain, owing to the α link, is kinked and has a threefold screw axis with a repeating unit 1.305 nm long. More recent conformational analysis has confirmed this view (65).

The analysis shows that the chain is held in as extended a form as possible and is exceptionally stiff. Until about 1969 it was thought that pectic acid was a homopolymer with only galacturonic acid residues in the chain. Since then it has been shown that few pectins, if any, are free of neutral sugars, notably L-rhamnose, as insertions in the chain (1, 2, 62, 64; Figure 4). The effect of such insertions cannot be examined by X-ray diffraction, but conformational analysis has shown that whether the glycosidic configuration of the rhamnose residue is α or β an insertion puts a marked kink in the chain. Figure 4 shows the extent of the kink if the configuration is β.

With an axial rise of 1.3 nm and a threefold screw axis, pectin cannot be expected to hydrogen bond strongly to any other polysaccharide in the wall though it may be compatible to a reduced extent with xylan when this has a threefold axis. Perhaps this is in part why pectic compounds are readily extractable in hot water alone. On the other hand, pectic acid can be expected to form Ca or Mg cross bridges with itself or with other polysaccharides with acidic groups. Single calcium links could be of little significance though they have figured widely in the past in some models of wall bonding, particularly in terms of the effects of auxin on growth. The buckled conformation of the polygalacturonic chain, however, leaves spaces for the insertion of a series of cations (Figure 6a—the egg-box model—derived for polyguluronate), all of which may be filled since the binding of one ion

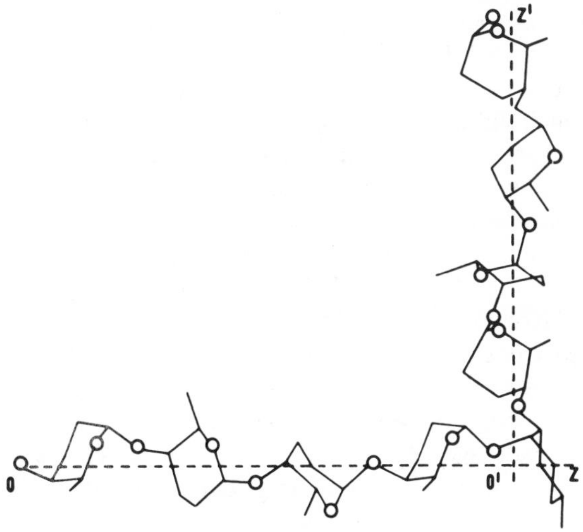

Figure 4 The molecular chain of rhamnogalacturonan (pectic acid) as a computer drawing; open rings represent oxygen, carbon hydrogen and hydroxyl groups omitted. The rhamnose residue lies in the angle at the bend. Note the threefold screw axis of each arm. (By courtesy of Prof. D. A. Rees.)

causes chain alignment which facilitates insertion of the next. Such cooperation of weak bonds would make in total for a strong bond such as seems to be required even in growing walls.

Polysaccharides based on other sugars such as galactose and arabinose found widely in cell walls may also be ribbon-like in chain form but the conformations are not yet known. β-(1→4)-linked galactan is known, for instance, in garlic (22) to contain only galactose residues and should certainly be ribbon-like; even this, however, has single galactose residues as side branches on every third residue in the backbone. Most galactans, like most arabinans, are highly branched. On the other hand, galactomannans have been isolated from *Ceratomia siliqua* (locust), *Caesalpina spinosa,* and *Cyamopsis tetragonolobulus* seeds which are based on a β-(1→4)-linked mannan backbone, of which the first two possess long unbranched lengths separated by regions of about 25 residues which carry sidechains of (1→6)-linked galactosyl residues (23). The smooth regions of these are therefore flat ribbons which could associate with cellulose and other ribbon-like molecules. The third glucomannan, on the other hand, has the side branches along one side only, giving a smooth side which could associate and a rough side which could not.

HELICAL CHAINS The demonstration that the skeletal polysaccharide β-(1→3)-linked xylan is helical (4, 5, 27) prompted the speculation that all (1→3)-linked polysaccharides should be helical (27). More recent observations have supported this view. This is important because polysaccharides of this kind are widespread in plants and have strong gel-forming properties. Those for which the conformation has been worked out all come, however, from fungi; all are glucans and have a fiber repeat (the length of the repeating unit measured along the helix axis) of about 0.6 nm (34, 45), like the xylan. Lentinan, from *Lentinus elodes,* takes the form of a triple helix (9, 71), again like xylan, and curdlan, from *Alcaligenes faecalis,* gives an X-ray diagram so similar that, though it is not definitive enough to allow a derivation of structure, it must represent a very similar structure. Pachyman, derived from the tree root fungus *Poria cocos,* is also thought to constitute a triple helix (8, 10). The triple helices are bulky throughout, of the order of 1.5 nm in diameter, and have an open center usually crossed by hydrogen bonds, again like xylan. Some of them have side chains, as have most of the (1→3)-linked glucans of fungi, but these are probably lost during extraction.

It is now to be anticipated that the smooth regions of any (1→3)-linked glycan will tend to adopt a helical conformation. These have been found in the galactan of root tissue (50), the arabinan of a number of plants (1) including young stems of roses (36) (though this is highly branched), and a glucan in cotton fibers (49). It is to be expected that almost all cell walls will contain β-(1→3)-linked glycans and they could be important because, though they may not be very effective in hydrogen bonding to other polysaccharides, their tendency to curl would provide a physical entanglement holding other polysaccharides bound. (1→3) links in the backbone of a chain would be expected to induce a helical conformation even if interrupted by (1→4) links. Glucans with both links in the backbone have been found as homopolymers in a wide range of grasses, including bamboo, and a possible role in wall elongation has been claimed (81).

The epitome of helical polysaccharides is probably carrageenan, an important polysaccharide not for its wide distribution for it is confined to the red algae, but for the light it throws on the structure and the behavior of helical molecules in solution and on the specificity of polysaccharides in plant cell walls. With carrageenan may be classed also agar, porphyran, furcellaran, and furonan because, though they show superficial differences, the backbone of each contains only galactose residues joined alternately by β-(1→3) and α-(1→4) links. The (1→3) linked units usually have the D configuration whereas the (1→4) linked units may have D or L, often as the 3,6 anhydride. These polysaccharides are very strongly gel-forming and their gelation properties have been studied intensively (63). The conformation of agarose (from agar) and carrageenan have been worked out (see 59 and references therein); both are double helices, agarose with a pitch of 1.9 nm and carrageenan of 1.3 nm. Carrageenan is nowadays regarded as a family of closely related polysaccharides and six idealized types have been defined. The double helical conformation has been proved for only two of these, called κ and ι [for details of these see (59, 63)]. The specificity mentioned above concerns the types κ and λ which differ only in that the former contains anhydrogalactose residues. In a variety of red algae κ occurs only (or at least to the extent of 95%) in the gametophyte and λ is confined (perhaps again only to the extent of 95%) to the sporophyte (46, 47, 80). This recalls the situation of *Halicystis* which is the gametophytic stage of the sporophyte *Derbesia,* since the skeletal polysaccharide of the former is cellulose [though the wall also contains β-(1→3)-linked xylan] and that of the latter is mannan only (58). These differences have been found only because all the polysaccharides concerned carry markers which are easily recognizable. The implication is clear that specificity of polysaccharides to particular cell walls may be widespread, though at present unrecognized because the differences between them are sophisticated.

The sequence of repeating units in carrageenan is interrupted in various places by the insertion of galactose 2,6 units replacing a proportion of the 3,6 anhydrogalactose -2-sulphate. This introduces a kink in the chain, leading to regions of the chain which can partake in a double helix and parts which cannot. In solution, therefore, it is conceived that the ordered regions of one chain can form a double helix with the ordered region of another, with the unordered regions serving as links between helices, and this is why a gel is formed (63; Figure 6b). If this can be taken over to cell walls, the potential importance of helical polysaccharides is obvious.

BRANCHED CHAINS It will now be clear that many wall polysaccharides are branched to varying degrees. Heavily branched chains were identified long ago in xylem, such as arabinogalactan with a backbone of (1→3)-linked galactose residues all carrying a galactosyl or arabinosyl side branch two residues long. More recently, similar polysaccharides including xyloglucan (13, 40, 76) have been described in a number of herbaceous plants. Arabinogalactan is a constituent of cultured sycamore cells (76) and possibly occurs in monocotyledons (13) which also contain an arabinoxylan as does also the aleurone layer of barley (48). Arabino-4-O-methylglucuronoxylan is common in grasses and has recently been found in the reed *Arundo donax* (35). A complex xylan, galactoarabinoxylan, was detected sometime ago in *Avena* (12).

Branched polysaccharides present difficulties because they do not crystallize easily and therefore there is no guide to conformation. They are mostly thought to have branches that stand out at right angles to the backbone, for which there is no evidence, and appear in that form in a number of models of the wall including a recent one (38) which will be discussed later. Two branched polysaccharides have now been defined, however, and it is immaterial that both come from bacteria. A capsular polysaccharide from *Escherichia coli* occurs in the form of chains with a repeating unit of [-2-α-mannopyranose-(1→3)-β-D-glucopyranose-(1→3)-β-D-glucuronic acid-(1→3)-α-D-galactopyranose-(1→] with a disaccharide [β-D-glucopyranose-(1→2)-α-D-mannopyranose] attached by a (1→4) link from the mannopyranose to the glucuronic acid. In the crystallized condition, for which the conformation has been determined, the side chains do not stand out but are instead folded back against the backbone chain (Figure 5, 6) (53). The other branched polysaccharide is based on a backbone identical with that of cellulose but with a substituent [(3→1)-α-D-mannopyranose-6-O-acetate-(2→1)-β D-glucuronic acid-(4→1)-β-D-mannopyranose] on every other glucose residue. No crystallographic data are given for this one, but the authors claim that the side chains are again folded back (54). Of greater significance here is the demonstration that the same conformation holds in solution. There is nothing exceptional about either of these two polysaccharides, and the implication is clear that all branched polysaccharides may have to be envisaged as with the side chains folded back along the main chain, both in solution and in the solid state, unless there is some overriding energetic factor in the environment that dictates otherwise.

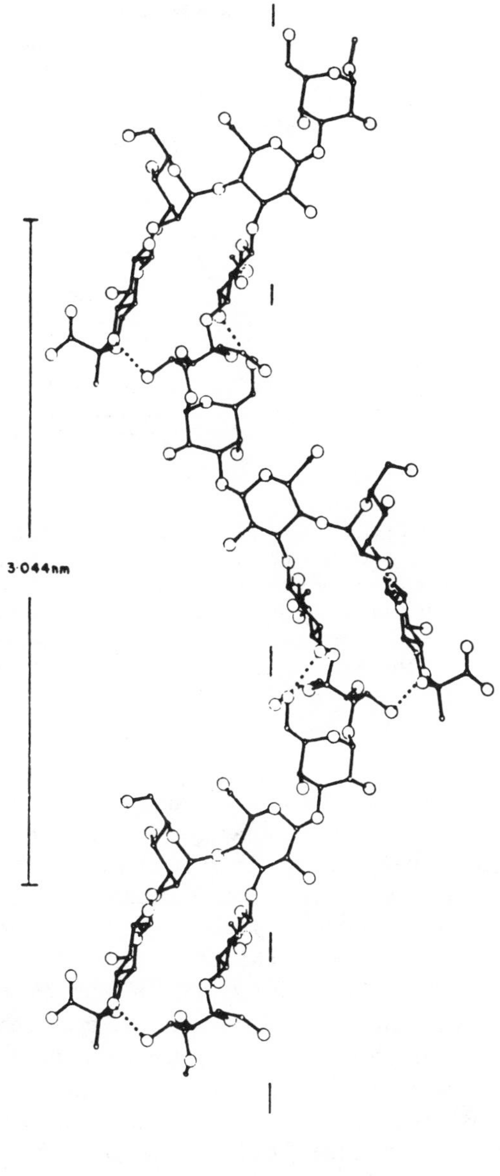

Figure 5 Molecular conformation of the branched polysaccharide from the M41 mutant of *Escherichia coli;* open circles represent oxygen, hydrogen omitted. Broken lines represent hydrogen bonds. Note that all three side chains visible in this section of the main chain are folded back along the main chain and hydrogen bonded to it. (By courtesy of Dr. Struther Arnott.)

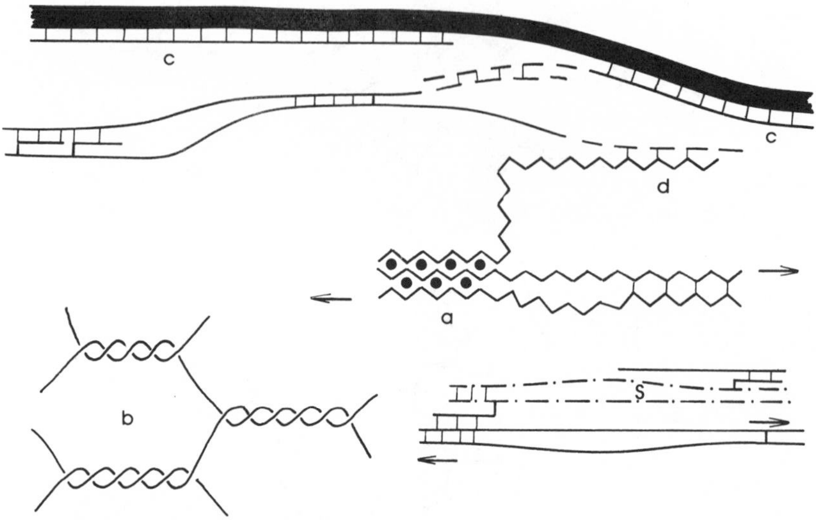

Figure 6 Diagrammatic representations of some possible bonding schemes in primary walls. Solid, straight, or smoothly curved lines—ribbon-like polysaccharides; broken lines—twisted ribbon-like polysaccharides, e.g. xylan with a threefold screw axis; zig-zag lines—rhamnogalacturonan (pectic acid); broken and dotted lines—protein; helices—helical chain molecules such as 1–3 linked xylan. Full circles—calcium ions in the "egg box." Short thinner lines represent hydrogen bonds. Arrows show possible shear stresses allowing yield and creep. S represents a disulphide bond. (Not to scale.)

WALL PROTEIN

There can no longer be any dissent from the view that primary walls contain glycoprotein which is in some way attached to the polysaccharides in the wall and is of structural significance. There is as yet, however, no complete agreement as to the sugar covalently bound in the glycoprotein, as to the polysaccharides to which the glycoprotein is bound, or indeed whether there is a single protein or several. The protein mostly, but not invariably, contains hydroxyproline often in a sequence of four together along the chain, the residues on either side of this sequence being serine. The first suggestion was that the hydroxyprolines carry side chains of arabinose residues of unknown length while the serine carries galactose (41, 42), and on the whole this has tended to stand. The claim has been made very recently, however, that with the glycoprotein extracted from parenchyma of *Phaseolus vulgaris* by alkali (which may therefore have lost some carbohydrates attached to serine by β-elimination), arabinose and glucose are attached to serine as well as galactose (11). Moreover, though some hydroxyproline residues carry only arabinose, others are said to carry galactose and glucose as well. As another variant, an arabinogalactan-peptide from wheat endosperm with a molecular weight of 22,000 is said to carry an oligogalactan attached to hydroxyproline as a chain with arabinose side chains (25). In this glycoprotein the arabinose adopts the furanose configuration as found for some other wall glycoproteins (11, 39). This leads to the concept that there are at least two proteins in cell walls, one with most of the carbohydrate moiety attached through serine and the other through hydroxyproline. It had been considered from the first by many authors that more than one protein is present. This view has more recently been supported using the protein extracted by delignification of *Phaseolus coccineus* tissue by sodium chlorite and acetic acid (74). In this case one protein is relatively rich in hydroxyproline and the other relatively poor.

The finding that the glycoproteins of some plants are released by alkali (39, 51) while others are not (74) and that still others are solubilized by delignification suggests that the association between glycoprotein and polysaccharides in the wall is variable. Views recently expressed range from the idea that at least part of the glycoprotein chains are not linked covalently together nor to any wall constituent (6) to the belief that the protein is firmly linked to all wall components (76). An intermediate situation is adopted by most workers. An early view that the attachment is to the hemicelluloses (42) has received support (76) though through the galactose side chains of the glycoprotein, not the arabinose, and sometimes the support is hesitant (15). It has also been claimed that if two proteins are present, only the protein poor in hydroxyproline is bound to the hemicellulose (74). Equally, it has been asserted that there may be no link to hemicellulose (52). With other tissues a firm bond has been claimed to the polyuronides (38), though by some physical association not necessarily a covalent link (39), a possibility which has equally been denied (52, 74). The view that the bonding is direct to cellulose microfibrils is not without strong support either (52), though this may be through galactan (51). It may be only the hydroxyproline-rich protein which is linked directly to cellulose (74).

There can be no doubt that the glycoprotein is somehow firmly bound in the wall if only because it can be extracted only with strong reagents. In this instance the conformation is of no consequence in assessing the bonding since it is not known. The oligogalactan attached to serine (but less likely the oligoarabinan chain because the constituent sugars are in the furanose form) might provide hydrogen bonding to wall polysaccharides, but it seems unlikely that this could induce sufficiently strong bonding since the degree of polymerization of this oligosaccharide has been estimated at between 1 and 5. Even if these side chains are much longer than this, the situation would still be uncertain since it is not known if the side chains are folded back along the peptide main chain as they would be in a branched polysaccharide. Only a general view seems acceptable at present, that the protein in some way is linked to a glycan network (42) and that the nature and distribution of the linkage may vary from plant to plant.

WALL EXTENSION

During the growth of a cell the wall increases considerably, often enormously, in area while retaining about the same thickness because new wall lamellae are continually deposited over the inner wall surface, though recent work has confirmed earlier views that the wall sometimes becomes thinner (77). The mutual disposition of the polysaccharides and proteins in the wall, and the bonding between them, must be such as to allow the morphological changes involved and to conform with a controlled yielding to the mechanical stresses which must be associated with extension.

Growing walls are clearly under tension and wall extension must in some way reflect the resultant strain. Polarization microscopy shows that the *net* cellulose chain orientation in the walls of most elongating cells is transverse to the direction of extension, and the fact that this apparent transverse orientation is maintained during growth seemed for a long time to be satisfactorily explained by the multinet growth hypothesis (67). In this, every wall lamella is progressively elongated during growth so that, since new lamellae are continually deposited on the inner face, lamellae have been the more extended the further they are from this face. If the orientation of microfibrils in the lamella just being deposited at any time is constant, then there must be a progressive reorientation through the wall. The approval and the criticisms of this hypothesis have been considered elsewhere (58) sometime ago, and attention will here be given only to more recent investigations.

The multinet growth hypothesis unfortunately confuses two issues, the yielding of the wall to stress and the consequence for reorientation if the structure of the innermost lamella always remains constant. If the structure of the innermost lamella varies from time to time, then clearly there will be no observable reorientation through the wall, but each lamella may nevertheless be individually yielding to stress and undergoing reorientation —and this alone is the fundamental principle at issue. The observation that some growing parenchyma cell walls show crossed fibrillar structure [i.e. with lamellae with microfibrils lying alternately "longitudinal" and "transverse" (14, 68, 69)] does not of necessity negate this fundamental principle though it does call into question the validity of the name of the hypothesis as a whole. Cases are known in which each lamella of a growing wall with crossed microfibrillar structure is demonstrably extending and reorienting during growth (26). Issue must be taken, however, with the concepts expressed (68, 69) that normal extraction and shadowcasting of wall lamellae which have been used in supporting the multinet growth hypothesis have grossly distorted wall structure and that changes in microfibril orientation can easily be observed in wall section. When carried out with care, shadow-cast surface views of wall lamellae in some instances fully support the whole hypothesis as evidenced by recent work on cotton hairs (82). Indeed, in some very delicate walls, extraction processes have been found to improve orientation, not destroy it (27, 28). Changes with reorientation of the length of the microfibrils in a packed array visible in transverse wall section will scarcely be noticeable unless the extension has been very great, and it seems unfortunate that a claim for the lack of reorientation judged by this means should be based on cells at the maximum *rate* of growth rather than after *extensive* growth (68, 69). There seems no alternative for a critical assessment of reorientation to observation of lamellae in surface view. Moreover, the rate at which reorientation is to be expected through a wall depends on the rate of wall deposition as well as rate of extension, and this should always be looked at. This applies not only to observation of parenchyma cells mentioned here but also to much of the earlier work on collenchyma cells and epidermal cells, for which it has recently been shown that the crossed fibrillar structure demands, not a negation of the multinet growth hypothesis, but a modification (73). The hypothesis has recently received further support from careful polarization microscopical observation of growing walls (29a).

How these morphological reactions to stress are accommodated by the wall architecture and what bonds are involved in the displacement of the constituent chain molecules which must occur have been the subject of some investigation. Some of this has been dealt with elsewhere (58) and will be treated only briefly here. The findings need to be interpreted in the light of the observation that turgor pressure in a cell must exceed a certain threshold value before growth can occur (17, 31). Several workers over the years have observed a close correlation between rate of growth in cells and the elastic mudulus of the constituent walls. It seems likely that this is a chaemera, since continued growth would need a continually increasing turgor pressure, which is not what is observed. The mechanical property which seems better conceived to control the yielding of the wall is something analogous to creep (60, 61), i.e. a slow yielding of the wall under constant stress. With *Nitella* internodal cells the rate of growth of the cell parallels precisely the rate of creep in longitudinal strips of the wall taken from it. The rate of creep under the stress exerted on it when it was part of the living cell is, however, more than three times the rate of growth so that some other factor, certainly a metabolic factor, must intervene to reduce the rate of creep in the living cell; the controlling wall phenomenon may thus be called "biochemical creep." Observations have been made on oat coleoptiles which lead to the same view (18). Wall strips do not creep, however, unless the stress on them exceeds a certain minimal value. This value is, as nearly as one can tell, equivalent to the threshold turgor pressure (58), lending further support to the view that creep is the critical factor in growth.

Creep experiments say nothing, however, about the bonds which must be broken during extension and growth, except that strong bonds (giving the critical turgor pressure requirement) and weak bonds which are readily remade (allowing creep) are essentials. Accordingly, note has been taken of the converse phenomenon, stress relaxation. When a body is held under a fixed, constant strain, the stress declines with time and the activation energy of the intermolecular bonds involved can be calculated. The basis of the appropriate theory is not very secure as applied to wall strips so that only a rough estimate is possible. This is equivalent to the breaking of only a few hydrogen bonds in *Nitella, Penicillus,* and *Acetabularia* (with walls of very different chemistry) (32, 33) possibly between 5 and 20. Similar considerations apply to oat coleoptiles (19, 20). If this is of wide application, it means that the carbohydrates and proteins of the wall must be so associated that strong bonds must be broken before the chains can slide past each other and that the sliding involves flow units with only a few hydrogen bonds joining them. The circumstance that Ca^+ decreases, and K^+ increases, the rate of creep means that the flow units must be in the matrix (60, 61) and upon this there is general agreement.

This concept of a few strong bonds and many groups of weak bonds refers strictly to those lamellae in the wall which bear the bulk of the load induced by turgor pressure, and these cannot at present be defined. Although in pure multinet growth neither the innermost nor the outermost lamellae are load bearing [the latter because it is torn (26)] so that the stress must fall on intermediate lamellae, the distribution of stress cannot be calculated at present. In those walls with crossed fibrillar structure it might be expected that lamellae with longitudinally oriented microfibrils would bear most of the load, but the distribution of stress will depend materially on the relative thickness of the two kinds of lamella. In either case, however, the critical lamellae are those which have already been subject to strain, and the distribution and type of bonding looked for in the understanding of a growing wall are not necessarily those applying to a lamella as it is first laid down at the plasmalemma. At any rate, the direction of creep is clearly neither parallel nor perpendicular to the overall microfibril direction in the load-bearing lamellae.

WALL MODELS

In attempting to model the effective parts of the wall, therefore, we are looking for a few strong bonds and many small groups of weak bonds. There are a few guiding considerations. 1. Although the chemical details will vary from plant to plant, we may expect any primary wall to contain representatives of each of the types of polysaccharide listed above, together with protein. 2. Since treatment successively with hot water, chlorination, and 4N KOH removes in sequence polyuronides and hemicelluloses, leaving so-called α-cellulose without the wall falling apart, there must be links between polyuronides (weak), between polyuronides and hemicelluloses (weak), between hemicelluloses (stronger), and between hemicelluloses and cellulose (strongest), and directly between the microfibrils themselves (unless these links are formed de novo as other compounds are extracted). 3. There is reason to believe that unlike chains react to each other in such a way as to exclude each type of chain from the domain of the other, examples being dextran and galactomannan (23). 4. When like chains form a gel, laser light scattering shows that the density fluctuates widely along any line in the gel so that there are large interstitial spaces between clusters of chains; with calcium alginate and agarose gels, if the structure is maintained by junction between chain molecules, the junction must consist of some hundreds of chains (44). It is to be expected that in places the microfibrils are in close contact—and electron micrographs show indeed that they are in places twisted around each other—so that the separation between them is variable and that in the spaces between them the chains of the matrix lie in clusters of like kind. Models of the wall depending upon uniform separation of microfibrils such as recently proposed (38), and even a model which is commendably presented as a partial model only (54), could be misleading in terms of function.

The strong bonds could be one or more of a variety of bonds expected to be present.

(*a*) It could be a covalent bond. This is a feature of one of the models proposed (38) in which all the matrix polysaccharides and the glycoprotein are covalently linked though connected to the microfibrils by xyloglucan which is only hydrogen bonded to them. This model depends upon the introduction of a powerful tool in cell wall research, a combination of degradation by specific enzymes and the detection of covalent bonds by methylation and hydrolysis. This is a tool of great promise but currently suffers certain drawbacks. Individual polysaccharides are not isolated, so their nature and conformation can only be implied. Moreover, unless it can be guaranteed that methylation is complete, some of the so-called covalent bonds are spurious. Even further, it has been alleged (81) that during methylation only a fraction of the hemicelluloses is represented in the final product, and it is not safe to transfer the significance of the methylated product to the substance before methylation. In any case, the idea that a growing wall is a single macromolecule (38) is surely unacceptable since it is incompatible with the known swelling and shrinkage properties of primary walls and with extension during growth. We need a situation with much fewer covalent bonds than proposed in this tightly designed model.

(*b*) It might be a bond between glycoprotein chains. It was claimed sometime ago, on the basis of the lack of cohesion in a wall after treatment with dithiothreitol, that the protein chains might be linked by disulphide bonds which are important in wall architecture, with some caution because dithiothreitol may have effects other than the breakage of these bonds (78). This has received more recent support from the observation that analogs of uracil (which itself has no effect on cell elongation) stimulate growth only if they contain thiol groups, interpreted as meaning that such compounds delay the formation of disulphide bonds (79). None of this evidence is conclusive, but the possibility of a protein-protein bond still seems open.

(*c*) It might be the result of a sequence of Ca bonds between polyuronides. Something like 20 of these bonds, working in collusion, might be necessary (63).

(*d*) It might represent a situation in which two or more polysaccharide chains are held together by numerous closely spaced hydrogen bonds such as would occur between ribbon-like molecules of like kind. It is, for example, such a conspiracy of hydrogen bonds which holds a cellulose crystallite together and a ribbon-like glycan to its surface.

(*e*) Finally, it might be a simulated bond induced by an entanglement between chains which tend to twist as figured for carrageenan (63) and possibly with all (1→3) linked polysaccharides and the protein which, at least over the stretches of hydroxyproline, will lie in the α-helix. This possibility has already been mooted (52).

Though outside the expressed scope of this article, it is relevant to note also that during recent years increasing weight is being laid on another class of compounds as important in maintaining coherence in the wall—the phenolics (31a). Walls in the leaf laminae of rye grass and wheat contain phenolic compounds, mainly *trans*-ferulic acid though with some *cis*-ferulic acid, *trans-p*-coumaric acid and *trans trans*-diferulic acid, ester linked at -COOH to wall polysaccharides. These are released as phenolic esters of carbohydrates by treatment of the wall with carbohydrases of which cellulase is one. Cell walls richer in these esters (as judged by the depth of stain with *p*-nitrobenzene diazonium tetrafluroborate) are not degraded by the cellulase whereas walls showing only a weak reaction are badly affected. Here, therefore, are compounds attached to carbohydrates —and perhaps even to cellulose—in walls which, incidentally, are not lignified. They might function by cross-linking polysaccharides and, if widespread, could be important.

Whatever the bond, it must in general lie in parallel with a group of hydrogen bonds so that when the strong bond is severed the system can creep. These groups of bonds could lie between polyuronides, between twisted ribbon-like chains, between disordered lengths of otherwise helical chains, between long branches in branched chains, or between any two of them. It is not possible at the moment to construct a meaningful model and only partial models of parts of the wall can be supported. Taking all the evidence presented in this article, however, certain fixed points emerge.

The cellulose crystallite must be clothed with flat, ribbon-like molecules, strongly hydrogen bonded because they are fully compatible with cellulose (Figure 6c). Such molecules are β-(1→4)-linked glucans, xylans, or mannans; xyloglucans proposed in one model (38) seem unlikely owing to the extensive branching of the molecule. These in turn would preferentially hydrogen bond to lengths of a chain compatible with cellulose, such as xylan, of which neighboring lengths might have a threefold axis not so compatible. These lengths would, however, form suitable links to a polyuronide aggregate (Figure 6d) linked together by Ca bridges with hydrogen bonds in series. It could be relevant in this context to note that the kinking induced in a polygalcturonan chain by the presence of L-rhamnose might be a means of cross-linking many polysaccharide chains. Elsewhere in the wall, other polysaccharides might also present strong bonds with hydrogen bonds in series as in Figure 6, but if these are present, we have no way of knowing how they are mutually organized.

The prospect of improving the situation by defining the mutual disposition of the chains, or even the chain types, within a wall does not seem bright. It might help if the bonding could be more clearly defined by some direct method. Magnetic resonance spectroscopy comes immediately to mind. This has already been mentioned elsewhere with respect to the strength properties of secondary walls, but application to growing walls does not seem to have been aired. Perhaps this omission should now be rectified.

Literature Cited

1. Aspinall, G. O. 1970. In *The Carbohydrates,* ed. W. Pigman, D. Horton, 116:515–36. New York: Academic
2. Aspinall, G. O., Cottrell, I. W., Molloy, J. A., Uddin, M. 1970. Lemon peel pectin III fractionation of acids from lemon peel and lucerne. *Can. J. Chem.* 48:1290–95
3. Atkins, E. D. T., Hopper, E. D. A., Nieduszinski, I. A. 1973. Polysaccharide conformation: Effect of side-group geometry on 4 diequatorially (1–4)-linked polysaccharides. *Carbohydr. Res.* 27:29–37
4. Atkins, E. D. T., Parker, K. D. 1969. The helical structure of a β-D-1,3-xylan. *J. Polym. Sci. Part C* 28:69–77
5. Atkins, E. D. T., Parker, K. D., Preston, R. D. 1969. The helical structure of the β-1,3-linked xylan in some siphoneous green algae. *Proc. R. Soc. Ser. B* 173:209–21
6. Bailey, R. W., Kauss, H. 1974. Extraction of hydroxyproline-containing proteins and pectic substances from cell walls of growing and non-growing mung bean hypocotyl segments. *Planta* 119:233–45
7. Bauer, W. D., Talmadge, K. W., Keegstra, K., Albersheim, P. 1973. The structure of plant cell walls II. The

hemicelluloses of the walls of suspension-cultured sycamore cells. *Plant Physiol.* 51:174–87
8. Bluhm, T. L., Sarko, A. 1975. Crystal structure of Pachyman triacetate. A preliminary report. *Biopolymers* 14: 2639–43
9. Bluhm, T. L., Sarko, A. 1977. The triple helical structure of lentinan, a linear β-(1→3)-D-glucan. *Can. J. Chem.* 55: 293–99
10. Bluhm, T. L., Sarko, A. 1977. Packing analysis of carbohydrates and polysaccharides V. Structure of two polymorphs of Pachyman triacetate. *Biopolymers* 16:2067–89
11. Brown, R. G., Kimmins, W. C. 1978. Protein-polysaccharide linkages in glycoproteins from *Phaseolus vulgaris*. *Phytochemistry* 17:29–33
12. Buchala, A. J., Fraser, C. G., Wilkie, K. C. B. 1972. An acidic galactoarabinoxylan from the stem of *Avena sativa*. *Phytochemistry* 12:1373–76
13. Burke, D., Kaufman, P., McNeil, M., Albersheim, P. 1974. The structure of plant cell walls VI. A survey of the walls of suspension-cultured monocotyledons. *Plant Physiol.* 54:109–15
14. Chafe, S. C., Chauret, G. 1974. Cell wall structure in the xylem parenchyma of trembling aspen. *Protoplasma* 80: 129–47
15. Cho, Y-P., Chrispeels, M. J. 1976. Serine-O-glactosyl linkages in glycopeptides from carrot cell walls. *Phytochemistry* 15:165–68
16. Claffey, W., Blackwell, J. 1976. Electron diffraction of *Valonia* cellulose. A quantitative interpretation. *Biopolymers* 15:1903–18
17. Cleland, R. 1967. Extensibility of isolated cell walls: measurement and change during cell elongation. *Planta* 74:197–209
18. Cleland, R. 1971. The mechanical behaviour of isolated *Avena* coleoptile walls subjected to constant stress. *Plant Physiol.* 47:805–11
19. Cleland, R., Haughton, P. M. 1971. The effect of auxin on stress relaxation in isolated *Avena* coleoptiles. *Plant Physiol.* 47:812–15
20. Cleland, R., Thompson, W. F., Haughton, P. M., Rayle, D. L. 1972. In *Plant Growth Substances 1970,* ed. D. J. Carr. Berlin: Springer
21. Courtois, J. E., Le Dizet, P. 1970. Récherches sur les Galactomannanes VI. Action de quelques mannanases. *Bull. Soc. Chim. Biol.* 52:15–22
22. Das, W. W., Das, A. 1977. Structure of the D-galactan isolated from garlic (*Allium sativum*) bulbs. *Carbohydr. Res.* 56:337–49
23. Dea, I. C. M., Morris, E. R., Rees, D. A., Welsh, E. J., Barnes, H. A., Price, J. 1977. Association of like and unlike polysaccharides: Mechanism and specificity in galactomannans, interacting polysaccharides and related systems. *Carbohydr. Res.* 57:249–72
24. Dennis, D. T., Preston, R. D. 1961. Constitution of cellulose microfibrils. *Nature* 191:667–68
25. Fincher, G. B., Sawyer, W. H., Stone, B. A. 1974. Chemical and physical properties of an arabinogalactan-peptide from wheat endosperm. *Biochem. J.* 139:535–45
26. Frei, E., Preston, R. D. 1961. Cell wall organisation and cell growth in the filamentous green algae *Cladophora* and *Chaetomorpha* I. The basic structure and its formation. *Proc. R. Soc. Ser. B* 154:70–94; II. Spiral structure and spiral growth. *Proc. R. Soc. Ser. B* 155:55–77
27. Frei, E., Preston, R. D. 1964. Non-cellulosic structural polysaccharides in algal cell walls I. Xylan in siphoneous green algae. *Proc. R. Soc. Ser. B* 160:293–327
28. Frei, E., Preston, R. D. 1968. Non-cellulosic structural polysaccharides in algal cell walls III. Mannan in siphoneous green algae. *Proc. R. Soc. Ser. B* 169:127–45
29. Gardner, K. H., Blackwell, J. 1974. The structure of native cellulose. *Biopolymers* 13:1975–81; The hydrogen bonding in native cellulose. *Biochim. Biophys. Acta* 343:232–38
29a. Gertel, E. T., Green, P. B. 1977. Cell growth pattern and wall microfibrillar arrangement. Experiments with *Nitella*. *Plant Physiol.* 60:247–54
30. Goto, T., Harada, H. 1975. Cross-sectional view of microfibrils in gelatinous layer of poplar tension wood (*Populus euramericana*). *Mokuzai Gakkaishi* 21:537–42
31. Green, P. B., Erickson, R. O., Buggy, J. 1971. Metabolic and physical control of cell elongation rate. *Plant Physiol.* 47:423–30
31a. Hartley, R. D., Harris, P. J. 1978. Degradability and phenolic components of cell walls of wheat in relation to susceptibility to *Puccinia struiformis*. *Ann. Appl. Biol.* 88:153–58
32. Haughton, P. M., Sellen, D. B. 1969. Dynamic mechanical properties of the

cell walls of some green algae. *J. Exp. Bot.* 20:516–35
33. Haughton, P. M., Sellen, P. B., Preston, R. D. 1968. Dynamic mechanical properties of the cell wall of *Nitella opaca*. *J. Exp. Bot.* 19:1–12
34. Jelsma, J., Kreger, D. R. 1975. Ultrastructural observations on (1→3)-β-D-glucan from fungal cell walls. *Carbohydr. Res.* 43:200–3
35. Joseleau, J. P., Barnoud, F. 1975. Hemicelluloses of *Arundo donax* at different stages of maturity. *Phytochemistry* 14:71–75
36. Joseleau, J. P., Chambat, G., Vignon, M., Barnoud, F. 1977. Chemical and ^{13}C NMR studies in two arabinans from inner bark of young stems of *Rosa glauca*. *Carbohydr. Res.* 58:165–75
37. Kauss, H., Bowles, D. J. 1976. Some properties of carbohydrate-binding proteins (lectins) solubilised from cell walls of *Phaseolus aureus*. *Planta* 130:169–74
38. Keegstra, K., Talmadge, K. W., Bauer, W. D., Albersheim, P. 1973. The structure of plant cell walls III. A model of the walls of suspension-cultured sycamore cells based on the interconnections of the macromolecular components. *Plant Physiol.* 51:188–97
39. Knee, M. 1975. Soluble and wall-bound glycoproteins of apple fruit tissue. *Phytochemistry* 14:2181–88
40. Labavitch, J., Ray, P. 1974. Relationships between promotion of xyloglucan metabolism and induction of elongation by indoleacetic acid. *Plant Physiol.* 54:499–502; Turnover of cell wall polysaccharides in elongating stem segments. *Plant Physiol.* 53:669–73
41. Lamport, D. T. A. 1970. Cell wall metabolism. *Ann. Rev. Plant Physiol.* 21:235–70
42. Lamport, D. T. A. 1973. In *Biogenesis of Plant Cell Wall Polysaccharides,* ed. F. A. Loewus, p. 149. New York: Academic. See also *The Structure, Biosynthesis and Degradation of Wood,* ed. F. A. Loewus, V. C. Runeckles. 1977. *Recent Adv. Phytochem.* 11:79–116
43. Mackie, W., Preston, R. D. 1974. In *Algal Physiology and Biochemistry,* ed. W. P. D. Stewart, pp. 40–85. London: Blackwell
44. Mackie, W., Sellen, D. B., Sutcliffe, J. 1978. Spectral broadening of light scattered from polysaccharide gels. *Polymer* 19:9–16
45. Marchessault, R. H., Deslandes, Y., Ogawa, K., Sundararajan, P. R. 1977. X-ray diffraction data for β-(1–3)-D-glucan. *Can. J. Chem.* 53:300–3
46. McCandless, E. L., Craigie, J. S., Hansen, J. E. 1975. Carrageenans of gametangial and tetrasporangial stages of *Iridaea cordata* (Gigartinaceae). *Can. J. Bot.* 53:2315–18
47. McCandless, E. L., Craigie, J. S., Walter, J. A. 1973. Carrageenans in the gametophytic and sporophytic stages of *Chondrus crispus*. *Planta* 112:201–2
48. McNeil, M., Albersheim, P., Taiz, L., Jones, R. L. 1975. The structure of plant cell walls VII. Barley aleurone cells. *Plant Physiol.* 55:14–18
49. Meinert, M., Delmer, D. P. 1977. Changes in biochemical composition of the cell walls of the cotton fiber during development. *Plant Physiol.* 58: 1088–97
50. Mollard, A., Barnoud, F. 1975. Caractérisation d'une D-galactanne linéaire a liaison β-(1→3) dans les cellules de rosier cultivées *in vitro*. *Carbohydr. Res.* 39:C16–C17
51. Monro, J. A., Bailey, P. W., Penny, D. 1974. Cell wall hydroxyproline-polysaccharide associations in *Lupinus* hypocotyls. *Phytochemistry* 13:375–82
52. Monro, J. A., Penny, D., Raymond, W. B. 1976. The organisation and growth of primary cell walls of lupin hypocotyls. *Phytochemistry* 15:1193–98
53. Moorhouse, R., Winter, W. T., Struther, A. 1977. Conformation and molecular organisation in fibres of the capsular polysaccharide from *Escherichia coli* M41 mutant. *J. Mol. Biol.* 109:373–91
54. Morris, E. R., Rees, D. A., Young, G., Walkinshaw, M. D., Darke, A. 1977. Order-disorder transitions for a bacterial polysaccharide in solution. A role for polysaccharide conformation in recognition between *Xanthomonas* pathogen and its plant host. *J. Mol. Biol.* 110:1–16
55. Morrison, I. 1974. Changes in the hemicellulose polysaccharides of rye grass with increasing maturity. *Carbohydr. Res.* 34:45–51
56. Nieduszinski, I. A., Marchessault, R. H. 1972. Structure of β-D-(1→4')-xylan hydrate. *Biopolymers* 11:1335–44
57. Nieduszinski, I. A., Marchessault, R. H. 1972. The crystalline structure of poly- β-D-(1→4') mannan; Mannan I. *Can. J. Chem.* 50:2130–37
58. Preston, R. D. 1974. *The Physical Biology of Plant Cell Walls.* London: Chapman & Hall
59. Preston, R. D. 1975. X-ray analysis and the structure of the components of plant cell walls. *Phys. Rep.* 216:184–226

60. Probine, M. C., Preston, R. D. 1961. Cell growth and the structure and mechanical properties of the wall in internodal cells of *Nitella opaca.* I. Wall structure and growth. *J. Exp. Bot.* 12:261–82
61. Probine, M. C., Preston, R. D. 1962. Cell growth and the structure and mechanical properties of the wall in internodal cells of *Nitella opaca,* II. Mechanical properties of the walls. *J. Exp. Bot.* 13:111–27
62. Rees, D. A. 1969. Conformational analysis of polysaccharides II. Alternating copolymers of the agar-carrageenan-chondroitin type by model building in the computer with calculation of helical parameters. *J. Chem. Soc. B* 217–26
63. Rees, D. A. 1977. *Polysaccharide Shapes:* Outline Studies in Botany series. London: Chapman & Hall; New York: Wiley
64. Rees, D. A., Wight, A. W. 1969. Molecular cohesion in plant cell walls. Methylation analysis of pectic polysaccharides from the cotyledons of white mustard. *Biochem. J.* 115:431–39
65. Rees, D. A., Wight, A. W. 1971. Polysaccharide conformation VII. Model building computations for α-1,4 galacturonan and the kinking function of L-rhamnose residues in pectic substances. *J. Chem Soc. B* 1366–72
66. Richards, G. N., Whistler, R. L. 1973. Isolation of two pure polysaccharides from the hemicelluloses of slash pine. *Carbohydr. Res.* 31:47–55
67. Roelofsen, P. A. 1959. *The Plant Cell Wall* Berlin: Borntraeger
68. Roland, J. C., Vian, B., Reis, D. 1975. Observations with cytochemistry and ultramicotomy on the fine structure of the expanding walls in actively elongating plant cells. *J. Cell Sci.* 19:239–59
69. Roland, J. C., Vian, B., Reis, D. 1977. Further observations on cell wall morphogenesis and polysaccharide arrangement during plant growth. *Protoplasma* 91:125–41
70. Rosell, K. G., Svensson, S. 1975. Studies of the distribution of the 4-O-methyl-D-glucuronic acid residues in birch xylan. *Carbohydr. Res.* 42:297–304
71. Saito, H., Ohki, O., Takasuka, N., Sasaki, T. 1977. A ^{13}C-NMR spectral study of a gel-forming branched (1→3)-β-D-glucan (lentinan) from *Lentinus elodes. Carbohydr. Res.* 58:293–305
72. Saka, S., Goto, T., Harada, H., Saiki, H. 1976. The width of cellulose microfibrils in the pit membrane of softwood tracheid. *Bull. Kyoto Univ. Forests* No. 48
73. Sargent, C. 1978. Differentiation of the crossed-fibrillar outer epidermal wall during extension growth in *Heracleum vulgaris* L. *Protoplasma.* In press
74. Selvendran, R. R. 1975. Cell wall glycoproteins and polysaccharides of parenchyma of *Phaseolus coccineus. Phytochemistry* 14:2175–80. See also Selvendran, R. R., Davis, M. C., Tidder, E. 1975. Cell wall glycoproteins and polysaccharides of mature runner beans. *Phytochemistry* 14:2169–74
74a. Smith, H. 1978. Recognition and defense in plants. *Nature* 273:266–68
75. Stepanovic, A. J., Sarko, A. 1978. Molecular and crystal structure of cellulose triacetate I. A parallel chain structure. *Polymer* 19:3–8
76. Talmadge, K. W., Keegstra, K., Bauer, W. D., Albersheim, P. 1975. Plant cell walls I. The macromolecular components of the walls of suspended-culture sycamore cells with a detailed analysis of the pectic polysaccharides. *Plant Physiol.* 51:158–73
77. Thomas, R. J., Doyle, W. F. 1976. Changes in the carbohydrate constituents of elongating *Lophocolea heterophylla* setae (Hepaticae). *Am. J. Bot.* 63(8):1054–59
78. Thompson, E. W., Preston, R. D. 1968. Evidence for a structural role of protein in algal cell walls. *J. Exp. Bot.* 19:690–97
79. Vaughan, D., Cusens, E. 1975. Some effects of analogues of uracil on cell wall elongation and wall metabolism in excised pea root systems. *Planta* 122:227–38
80. Waaland, J. R. 1975. Differences in carrageenan in gametophytes and soporophytes in red algae. *Phytochemistry* 14:1359–62
81. Wilkie, K. C. B., Woo, S. L. 1976. Non-cellulosic β-D-glucans from bamboo and interpretative problems in the study of all hemicelluloses. *Carbohydr. Res.* 49:399–409
82. Willison, J. H. M., Brown, J. M. Jr. 1977. An examination of the developing cotton fiber; wall and plasmalemma. *Protoplasma* 92:21–41
83. Zugenmaier, P. 1974. Conformation and packing analysis of polysaccharides and derivatives I. Mannan. *Biopolymers B*:1112–39

Ann. Rev. Plant Physiol. 1979. 30:79–104

ROLES OF A COUPLING FACTOR FOR PHOTOPHOSPHORYLATION IN CHLOROPLASTS[1]

❖7666

Richard E. McCarty

Section of Biochemistry, Molecular and Cell Biology, Cornell University, Ithaca, New York 14853

CONTENTS

INTRODUCTION

In the quarter of a century since Arnon et al (5) reported the discovery of photosynthetic phosphorylation, the light-dependent synthesis of ATP from ADP and P_i, in isolated chloroplast preparations, an understanding

[1]Abbreviations: CF_1, chloroplast coupling factor 1; DCCD, N,N'-dicyclohexyl carbodiimide; F_1, mitochondrial coupling factor 1; F_o, membrane components of the ATPase complex; NBD, 7-chloro-4-nitrobenzo-2-oxa-1,3-diazole.

0066-4294/79/0601-0079$01.00

of at least the basic underlying mechanism of this process has been reached. Moreover, we now appreciate that photophosphorylation and oxidative phosphorylation in mitochondria or in bacteria are very closely related. Peter Mitchell's chemiosmotic theory of oxidative and photosynthetic phosphorylation (63, 64), first set forth in 1961, has stood the test of extensive experimentation (see 38, 39, 54, for reviews). Although some of the details are controversial, it is now clear that electron flow from reduced substrates to oxygen in mitochondria and bacteria, or light-dependent electron transfer in chloroplasts and photosynthetic bacteria, generates an electrochemical proton gradient across the membranes in which the components of the electron transfer chains are present. The flow of protons down the electrochemical gradient through a special device provides the energy needed for the synthesis of ATP. This device thus catalyzes proton translocation linked to ATP synthesis. As is often the case in biochemistry, this enzyme complex is named for the reverse reaction of ATP formation, ATP hydrolysis. Mitchell named the complex the reversible, proton translocating ATPase which is often shortened to the ATPase complex. The coupling factor discussed in this article is part of the ATPase complex of chloroplast thylakoid membranes.

What is a Coupling Factor?

Photophosphorylation is linked to or coupled to electron flow via a common intermediate, the electrochemical proton gradient. A number of reagents or treatments enhance the rate of electron flow while they inhibit ATP formation. Clearly, phosphorylation is uncoupled from electron flow by these reagents or treatments. This uncoupling may be elicited by treatments which cause the release of a protein(s) from the membrane. Under appropriate conditions, the protein may rebind to the membrane, resulting in a stimulation of photophosphorylation without a corresponding stimulation of electron flow. Thus, the protein has resulted in the recoupling of ATP synthesis to electron transport and is termed a "coupling factor." Under some circumstances reagents not present in chloroplast membranes enhance phosphorylation in uncoupled preparations. For example, N,N'-dicyclohexyl carbodiimide (DCCD) stimulates photophosphorylation in EDTA-treated chloroplast thylakoids (61). Since a coupling factor must be of chloroplast membrane origin, the carbodiimide is not classified as a coupling factor even though it mimics its action.

Discovery of Coupling Factor 1

Resolution and reconstruction analysis, a classical technique in biochemistiry, has been applied only comparatively recently to membrane processes. Pullman et al (87) solubilized a Mg^{2+}-dependent ATPase from mitochondria, and Penefsky et al (81) showed that this ATPase partially restored oxidative phosphorylation to ATPase-deficient submitochondrial particles. The ATPase was called F_1, for factor 1; other protein coupling factors were subsequently discovered.

Jagendorf & Smith (40) first found that photophosphorylation is uncoupled by treatment of thylakoids with dilute EDTA solutions. Shortly thereafter, Avron (6) showed that a heat-labile factor was released by the EDTA treatment which, when added to the EDTA-extracted thylakoids in the presence of Mg^{2+}, partially restored phosphorylation to the uncoupled membranes.

Although chloroplast thylakoids have only very low ATPase activity, Vambutas & Racker (112) demonstrated that trypsin treatment of thylakoids markedly activates a Ca^{2+}-dependent ATPase. A latent, Ca^{2+}-dependent ATPase was solubilized by buffer extraction of acetone-precipitated thylakoids and was partially purified. These preparations stimulated photophosphorylation in sonicated thylakoid preparations exposed to a low ionic strength wash. Since this preparation seemed analogous to F_1, it was called CF_1, for chloroplast coupling factor 1.

The coupling factor released by EDTA treatment was found by McCarty & Racker (60) to be identical to CF_1. Homogeneous preparations of CF_1 were both a latent, Ca^{2+}-dependent ATPase and a coupling factor. Moreover, monospecific antisera against CF_1 strongly inhibited ATPase activity and photophosphorylation.

Thus, by 1966 it was evident that CF_1 plays an essential role in photophosphorylation. In this article the structure of CF_1 and its functions in photophosphorylation will be discussed. The properties of F_1 molecules from other coupling membranes are similar and some studies on these molecules will be presented. Since CF_1 is but part of a functional ATPase complex, it is also appropriate to discuss the non-CF_1 components of the complex. Because CF_1 research has been reviewed rather recently (39, 56, 71), I will emphasize those aspects of the field which have changed most dramatically over the past few years.

SOME PROPERTIES OF PURIFIED COUPLING FACTOR 1

CF_1 comprises about 10% of the thylakoid membrane protein and can be readily removed from the membranes since it is a peripheral protein. Extraction of washed thylakoids with either EDTA (6, 60) or pyrophosphate (108) causes the nearly specific release of CF_1 from the membranes. Recently a convenient procedure for the large scale preparation of CF_1 was described (10). Using this method, one can prepare over 50 mg of pure

CF_1 from only 1 kg of spinach leaves. A chloroform extraction procedure for the isolation of CF_1 in a modified form has also been described (121). Although spinach or romaine lettuce leaves are used most commonly as the source of CF_1, it has been prepared from the leaves of other higher plants and from some algae.

Structure

CF_1 is a colorless, water soluble protein. In contains tightly, but not covalently bound nucleotides. Although Harris & Slater (33) reported that CF_1 contains slightly less than 1 mole of ADP and ATP per mole of CF_1, others find mostly ADP bound, in about a 1 : 1 molar ratio (86; N. Shavit, personal communication). Electrophoretically purified CF_1 has also been reported to contain about 3 moles carbohydrate per 100 moles of amino acid (4). Ribose, galactose, and glucose were the major sugars detected.

Negatively stained CF_1 preparations appear as 90 to 95 Å spheres in the electron microscope (26, 67). Similar particles present on the outer surfaces of thylakoids have been shown (26, 67) to be CF_1.

The molecular weight of CF_1, determined by sedimentation equilibrium, is 325,000 (23). This value is somewhat lower than that of coupling factor-ATPases from other coupling membranes. For example, the molecular weight of mitochondrial F_1 is reported to be 360,000 (98).

As would be expected for such a large protein, CF_1 is composed of subunits. The subunit structure of CF_1 was evident in electron micrographs and from average molecular weights of the enzyme dissociated in guanidine hydrochloride (23). With the advent of sodium dodecylsulfate gel electrophoresis, it was shown (90) that CF_1 contains five polypeptide components. This feature seems to be a general property of coupling factor-ATPases (98). The CF_1 subunits are labeled alpha to epsilon in order of decreasing molecular weight, and some of their properties are given in Table 1. Pyridine treatment, followed by ammonium sulfate fractionation and chromatographic techniques, was used to separate CF_1 subunits in denatured form (72, 78). Since the polypeptides have different amino acid compositions and antigenic properties, they are clearly distinct from one another. More recently, an improved procedure for the separation of the subunits was reported (10). This procedure involves dissociation of the enzyme by sodium dodecylsulfate, followed by chromatography on hydroxylapatite and Bio-Gel P300. Another approach to the isolation of F_1 polypeptides has been very successful with the *E. coli* enzyme. Like mitochondrial (87) and chloroplast F_1 (60), bacterial F_1 is cold-labile and dissociates when exposed to the cold (116). Subunits have been purified from cold-dissociated *E. coli* F_1 in native form (25). Although CF_1 dissociates in the cold (45), there are no reports on using cold inactivation as a means for isolating the subunits.

Table 1 Some properties of the subunits of coupling factor 1 from chloroplasts

CF_1 subunit	Molecular weight × 10^{-3}	Probable stoichiometry	Cysteine content per subunit	Suggested functions
Alpha	59[a, b], 61[c]	2	2[c], 2[a]	Regulatory
Beta	54[b], 56[a], 57[c]	2	2, 2	Active site
Gamma	34.5[c], 37[a], 39[b]	1	3, 6	Transmits protons to active site
Delta	17.5[a], 20[b], 20.8[c]	1	1, —	Membrane attachment
Epsilon	13[a], 13.5[b], 15.7[c]	2	1, 1	Inhibitor of ATPase activity; membrane attachment

The molecular weight of CF_1 is 325,000 (23) and the total SH content is about 13/mol (24). Using average molecular weights of the subunits and the stoichiometry shown, a molecular weight of 316,000 is calculated for CF_1. The calculated SH content is 14/mol CF_1 using the data of Binder et al (10) and 18/mol using the data of Nelson et al (72).
[a] Data of Nelson et al (72).
[b] Data of Baird & Hammes (7).
[c] Data of Binder et al (10).

The amino acid composition of the subunits has been determined by two groups. The two independent analyses were similar except for methionine, tyrosine, and, most notably, cysteine. Nelson et al (72) determined the cysteic acid content of performic acid-oxidized CF_1 subunits to be α,2; β,3; γ,6; δ,0 and ϵ,1. In contrast, Binder et al (10) found α,2; β,2; γ,3, δ,1 and ϵ,1 by both cysteic acid determination and titration of the isolated subunits with 2,2'-dithiopyridine. The presence of a sulfhydryl on the δ subunit is somewhat surprising in view of the fact that it does not appear to react at all with maleimides under conditions where 12–13 maleimides reacted per CF_1 (M. A. Weiss and R. E. McCarty, unpublished). Moreover, cross-linking of the ϵ subunit to other CF_1 components by copper 1,10-phenathroline treatment can still occur when the ϵ subunit has been reacted with N-ethylmaleimide (7). Unless the copper-phenanthroline reagent causes cross-links other than disulfide bonds to form, this result could indicate that the epsilon subunit contains more than 1 SH.

Although there is general agreement that coupling factor-ATPases from all sources contain five kinds of subunits, the stoichiometry of these subunits is controversial. A number of proposals have been made including: $\alpha_3\beta_3\gamma\delta\epsilon$, $\alpha_2\beta_2\gamma_2\delta\epsilon$, $\alpha_2\beta_2\gamma_2\delta_{1-2}\epsilon_2$ and $\alpha_2\beta_2\gamma\delta\epsilon_2$ (98). Four independent methods have been used to estimate the stoichiometry of CF_1 subunits. Baird & Hammes (7) used five cross-linking reagents with CF_1 and deduced a minimum stoichiometry of $\alpha_2\beta_2\gamma\delta\epsilon_2$. Pea plants were grown in $^{14}CO_2$ and the radioactive CF_1 precipitated from EDTA extracts by CF_1 antibody. The relative radioactivity in the CF_1 subunits after sodium dodecyl sulfate gel electrophoresis was 2.1:2.0:0.79:0.93:1.2 for the $\alpha,\beta,\gamma,\delta$, and ϵ subunits respectively (71). The low stoichiometry of the ϵ subunit may be explained by the unpublished finding of Nelson's group that the ϵ subunit of pea chloroplast CF_1 dissociates more readily than that in CF_1 complexes from other sources. Binder et al (10) determined the amounts of Coomassie blue dye which bound to known amounts of purified subunits of spinach chloroplast CF_1 in slab gels. Using these values, they calculated the subunit stoichiometry from the amount of dye absorbed by the subunits in dissociated CF_1. Relative to the γ subunit, which was assumed to be 1.0 subunit/mole, they found an average of 2.1 α, 2.1 β, 0.4 δ, and 1.5 ϵ. The δ subunit is readily lost from spinach CF_1 (121).

The sulfhydryl content of CF_1 and its subunits seems most consistent with a 2:2:1:1:2 stoichiometry, if the values of Binder et al (10) for the SH content of the subunits are used. Farron & Racker (24) detected 12–13 cysteines per mole CF_1 and Binder et al (10) found 13 cysteic acids per mole of performic acid-oxidized CF_1. Assuming a subunit stoichiometry of 2:2:1:1:2, CF_1 would have 14 cysteines per mole if the SH/subunit figures of Binder et al are used, and 18 if the figures of Nelson et al are used. It seems likely that the cysteic acid determinations of Nelson et al are too high. Using radioactive N-ethylmaleimide, we found (M. A. Weiss and R. E. McCarty, unpublished) about 8 maleimide-reactive groups for the α and β subunits together, 3 in γ and 1 in ϵ.

Finally, when the molecular weights of the subunits are summed taking a 2:2:1:1:2 stoichiometry into account, the calculated molecular weight is only about 3% different from the measured molecular weight of CF_1. A 2:2 stoichiometry for the two larger subunits also makes sense in view of the finding that CF_1 has two high affinity nucleotide binding sites (13, 28).

The cross-linking data (7) have been combined with the results of extensive studies of intersubunit distances established by fluorescence energy transfer techniques (14–16, 36). A model of CF_1 based on these studies is shown in Figure 1.

Recently, CF_1 (G. Hammes and S. J. Edelstein, personal communication) as well as F_1 from rat liver mitochondria (2) and a thermophilic bacterium (Y. Kagawa, personal communications) have been crystallized. Image reconstruction analysis of electron micrographs of thin sections of fixed CF_1 crystals and of F_1 from a thermophilic bacterium are in progress. X-ray diffraction analysis of mitochondrial F_1 has also been initiated. These structural studies should resolve the controversy about the stoichiometry of subunits in coupling factors from different sources. In view of the many similarities between the coupling factors, it might seem most likely that the coupling factors have similar subunit stoichiometries. However, the possibility that they differ cannot be ruled out at present. Moreover, the ease of dissociation of subunits from various F_1 complexes complicates the determination of subunit stoichiometries.

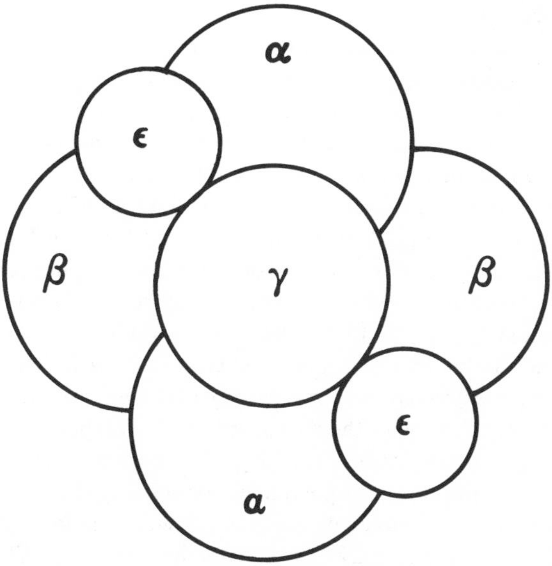

Figure 1 A view of the structure of CF_1. This working model is based primarily on cross-linking studies (7) and on distances between subunits estimated by fluorescence energy transfer techniques (14–16, 36). The view is from above the plane of the thylakoid membrane, so that the delta subunit, which is required for attachment of CF_1 to the membrane, is not visible. Adapted from Holowka & Hammes (36).

Enzymatic Activity and Ligand Binding

Purified CF_1 preparations have low ATPase activity which is markedly activated by a number of treatments, including trypsin (112), heat (24, 112), and incubation with high concentrations of SH compounds (62). In the absence of high concentrations of dicarboxylates or of bicarbonate, the ATPase activity is Ca^{2+}-dependent and is inhibited by Mg^{2+} (112). A more detailed analysis (35) of the cation requirements revealed, however, that low concentrations of divalent cations enhanced ATPase activity of heat-treated CF_1 in the following order: $Mn^{2+} > Mg^{2+} > Ca^{2+} > Co^{2+}$. High concentrations of free ATP or of the cations inhibited ATPase activity, and cation-ATP complexes are the true substrates for the reaction. The K_i values for free cation were 4, 20, and 7000 μM for Mn^{2+}, Mg^{2+} and Ca^{2+}, respectively. Thus, Ca^{2+} supports faster rates of ATPase activity at higher cation concentrations because free Ca^{2+} is a poor inhibitor. The Mg^{2+} inhibition of ATPase activity of CF_1 is apparently partially relieved by dicarboxylates or bicarbonate since high rates of Mg^{2+}-dependent ATPase were observed when these reagents were present (77).

The apparent K_m values for ATP at 37° and pH 8.0 in the presence of excess Ca^{2+} range from 0.8 to 1.3 mM. Hochman et al (35) calculated a

K_m for $(Ca\text{-}ATP)^{2-}$ complex of 2.5 mM. V_{max} values vary with the mode of activation and assay conditions but should be in the range of 35–45 μ mole P_i formed min^{-1} mg^{-1} CF_1 at 37°. ATP is the preferred substrate, although GTP and ITP are hydrolyzed at good rates (112). ADP is an allosteric inhibitor of the Ca^{2+}-ATPase of CF_1 (77). That is, ADP (0.6 mM) changes the apparent reaction order from 1 to close to 2.

Even after activation, CF_1 does not catalyze ADP-ATP exchange (112). A conversion of added ADP to bound ATP and less tightly bound AMP is catalyzed by CF_1 (68, 93). The rate of this transphosphorylation process is, however, exceedingly slow. Cantley & Hammes (13) report a rate of AMP formation by CF_1 of 15 pmoles min^{-1} mg CF_1^{-1}.

The nucleotide-binding properties of CF_1 have received considerable attention and have been studied by a variety of different methods (see Table 2 for a summary). Although the data from different laboratories conflict in some respects, it is very clear that CF_1, like coupling factors from other membranes (98), contains multiple nucleotide binding sites (13, 28, 93, 109, 113). Approximately 2 moles of [γ^{32}P]ATP were found in early experiments to bind to 1 mole of CF_1 (47). After incubation of CF_1 with [^{14}C] ADP for 2 hr and passage of the mixture through a Sephadex column, Roy & Moudrianakis (93) and Tiefert et al (109) detected about 1.5 to 2 moles

Table 2 Nucleotide binding to purified CF_1

Authors	Conditions	Method	Ligand	Number of sites/CF_1	Dissociation constants – μM
Roy & Moudrianakis, 1971 (93)	10 mM Tris-SO_4 (pH 8.0); 1 mg/ml CF_1 (DTT[a]-activated); 2 hr incubation	Gel filtration to remove unbound nucleotide	ADP	2	2, 35[b]
Tiefert et al 1977 (109)	As above	As above	ADP	2	2.4, 28[b]
			AMP-PNP	2	5.5, 35[b]
			PP_i	1.25	7.3, 21[b]
Cantley & Hammes, 1975 (13)	50 mM Tris-HCl (pH 8.0); 2mM $MgCl_2$, 0.1M $MgCl_2$, 2–10 mg/ml CF_1, passed through two Sephadex G-25 columns to remove exchangeable nucleotides	Equilibrium (forced dialysis)	ADP	3	1.8 (2 sites); about 100 (1 site)
			AMP-PNP	2	1.4
			ϵ-ADP[c]	2	1.8
Girault & Galmiche, 1977 (27)	20 mM Tricine buffer, 20 mM $MgCl_2$ (pH 8.4); CF_1 (prepared in absence of ATP) 0.19 mg/ml	Equilibrium (dialysis)	ϵ-ADP,	2.5	1, ?
			ϵ-ATP	2.5	1, ?
VanderMeulen & Govindjee, 1977 (114)	30 mM Tricine, 5 mM $MgCl_2$ (pH 8); CF_1 (lettuce) 1.4 – 1.5 mg/ml	Equilibrium (polarization of ϵ-ADP fluorescence)	ϵ-ADP	2	0.5, 2.0

[a] DTT stands for dithiothreitol.
[b] Concentrations which give half-maximal binding.
[c] ϵ-ADP and ϵ-ATP stand for the 1, N^6-derivatives of ADP and ATP, respectively.

of ^{14}C nucleotide bound to CF_1. The saturation curve was biphasic; 50% saturation was achieved at 2 μM and about 28 to 35 μM ADP. Using a forced dialysis technique to determine equilibrium binding, Cantley & Hammes (13) found that CF_1 contains two apparently identical sites which bind ADP, adenylylimidodiphosphate and 1,N^6-etheno-ADP (but not AMP) with dissociation constants of from 1.35 to 1.8 μM. A number of much weaker ADP binding sites were also evident.

Interestingly, another adenylylimidodiphosphate binding site was exposed by heat activation of the ATPase. Binding to this site was characterized by a dissociation constant of 7.6 μM a value close to the K_I for adenylylimidodiphosphate as a competitive inhibitor of the heat-activated ATPase (13). Other investigators have employed either the polarization of fluorescence (113, 114) of 1,N^6-etheno-ADP and the triphosphate derivative or the decrease in fluorescence (27) which occurs when these nucleotides bind. They find that the tight nucleotide sites are not equivalent. In the presence of 5 mM $MgCl_2$, VanderMeulen & Govindjee (114) detected two tight 1,N^6-etheno-ADP binding sites characterized by dissociation constants of about 0.5 and 2 μM. Under similar conditions two sites for the etheno-ATP derivative with slightly higher dissociation constants were detected. P_i (0.5 mM) increased the dissociation constants for both sites to 7 μM. Mg^{2+}, Ca^{2+}, and especially ADP enhanced the polarization of intrinsic fluorescence of CF_1. P_i decreased the effects of Mg^{2+} and ADP. These results indicate that cations, P_i and ADP cause changes in the conformation of CF_1. Girault & Galmiche (27) also found two tight 1,N^6-etheno ADP sites with dissociation constants of 1.25 and 2.50 μM. In equilibrium experiments, they found a maximum of 2.5 to 2.6 mole of 1,N^6-etheno-ADP or ATP bound per mole CF_1. Changes in the intrinsic fluorescence of CF_1, as well as that of a probe, 4-nitrobenzofuran, were related to the binding of nucleotide to the lower affinity site (27). Moreover, hydrogen exchange studies shown that CF_1 undergoes conformational changes when it binds ADP or ATP (70). A model in which the high affinity sites are nonequivalent was proposed (27). Binding to the second site was thought to be dependent on conformational changes in the enzyme elicited by nucleotide binding to the first site.

It is not clear why Cantley & Hammes (13) found the high affinity sites to be apparently equivalent or why Moudrianakis's group obtains quite different results. However, variations in the conditions under which the assays were performed as well as in the methods used could explain the discrepancies. Tiefert et al (109) point out that since some conversion of ADP to ATP and AMP takes place, ADP binding should not be treated as an equilibrium process. However, this objection should not apply to AMP-PNP binding, and Cantley & Hammes (13) found ADP binding and

AMP-PNP binding to the tight sites to be very similar. Moreover, Girault & Galmiche (27) mention that the extent of 1,N^6-etheno ADP binding to CF_1 assayed by gel filtration to remove unbound nucleotides was lower than that assayed by equilibrium dialysis. Some ADP may be lost from the enzyme during gel filtration as well. The presence of bound nucleotides in isolated CF_1 complicates the measurement of nucleotide binding to CF_1. Some of the binding could in fact be exchange. Moreover, if there was a slow dissociation of bound nucleotide from the enzyme, the radioactive or fluorescent nucleotides would be diluted.

In view of the slow rate of binding of ADP or ATP and their analogs to the high affinity sites, it seems unlikely that they are catalytic sites. A model in which these tight sites are control sites for the ATPase activity of CF_1 has been proposed (13). More recently it was found (20) that the rate of binding formycin triphosphate, a fluorescent ATP analog, to a CF_1 preparation which contained only four subunits (121), was increased fortyfold or more by activation of the Ca^{2+}-ATPase by dithiothreitol. Whether this increased rate of binding is related to activation of ATPase activity is uncertain. Further characterization of nucleotide binding to CF_1 is required before the physiological functions of the multiple sites are elucidated. Moreover, it may be dangerous to extrapolate the results of nucleotide-soluble CF_1 interactions to the membrane-bound enzyme since the properties of the soluble enzyme are likely to be very different from its membrane counterpart. For example, much faster rates of ATP hydrolysis by the membrane bound enzyme are obtained in the presence of high levels of Mg^{2+} than in the presence of Ca^{2+} (62). The sensitivity of the ATPase activity of thylakoids to adenylylimidodiphosphate is also much less than that of purified CF_1 (R. P. Magnusson, R. Wong, and R. E. McCarty, unpublished).

Purified CF_1 also binds Mn^{2+} to about five sites characterized by dissociation constants of 9 to 53 μM (35). The higher affinity Mn^{2+} sites, which apparently bind Mg^{2+} and Ca^{2+} more weakly, are probably involved in the inhibition of ATPase activity by free cations.

PROPERTIES OF MEMBRANE-BOUND COUPLING FACTOR 1

CF_1 is only a part of the enzymatic apparatus which couples proton efflux from the thylakoid interior to the synthesis of ATP. In view of its capacity to bind nucleotides and to catalyze ATP hydrolysis, it is very likely that CF_1 is the enzyme responsible for ATP synthesis. Other more hydrophobic factors, however, are probably required to stabilize the binding of CF_1 to thylakoid membranes as well as to allow the flow of protons across the membrane to the CF_1. CF_1 together with these membrane factors is termed the ATPase complex. The hydrophobic component(s) are collectively called F_0 so that the complex is also denoted CF_1-F_0. It should be noted that the "o" in F_0 stands for oligomycin since F_0 was originally isolated as a factor which confers oligomycin sensitivity to soluble F_1 ATPase activity (88).

The ATPase Complex and its Activities

In the early phase of research on coupling factors, the isolation and purification of individual components of the phosphorylation apparatus were attempted. More recently, however, isolation and purification of the ATPase complex in active form has dominated research in this area. This approach was made possible by the pioneering work of Kagawa & Racker (42). A detergent-solubilized preparation from mitochondrial membranes which contained ATPase activity was reconstituted into phospholipid vesicles. The reconstituted vesicles catalyzed $^{32}P_i$-ATP exchange, and similar preparations were later shown to catalyze ATP synthesis provided a generator of an electrochemical proton gradient such as segments of the respiratory chain (99) or bacteriorhodopsin (91) was also incorporated into the vesicles. Since then, ATPase complexes in active form have been isolated from a variety of coupling membranes, including chloroplasts.

Carmeli & Racker (19) used cholate treatment in the presence of $(NH_4)_2SO_4$ to solubilize proteins from thylakoid membranes. The cholate extract contained little chlorophyll and cytochromes, but was rich in ATPase activity and lipids. Vesicles with a low $^{32}P_i$-ATP exchange activity spontaneously formed when the cholate was removed from the extract by dialysis. Winget et al (118) obtained some purification of the complex and showed that phospholipid vesicles containing the complex catalyzed ATP synthesis powered by an electrochemical proton gradient generated by the light-driven bacteriorhodopsin proton pump. This preparation contained approximately eight different polypeptide chains in addition to the CF_1 subunits. More recently, Pick and Racker (in preparation) have achieved substantial further purification of the complex. Octylglucoside was included in the extraction medium with cholate and an ammonium sulfate precipitation step was further refined. Sucrose density gradient centrifugation in the presence of Triton X-100 and phospholipids yielded fractions which catalyzed rapid rates of $^{32}P_i$-ATP exchange when incorporated into phospholipid vesicles. These fractions showed only 3 to 4 polypeptides in addition to the CF_1 subunits.

The functions of the hydrophobic components of the complex are beginning to be elucidated. For over 10 years it has been known that partial removal of CF_1 from thylakoid membranes by EDTA extraction increases the permeability of the membranes to protons (60, 61), and this finding has been confirmed (29, 48, 96). Interestingly, this increased permeability appears to be specific for protons (79). Low proton permeability can be partially restored by the rebinding of CF_1 to the depleted membranes or by dicyclohexylcarbodiimide (DCCD). DCCD inhibits the ATPase activity of membrane-bound F_1 molecules from all sources, but has little if any effect on the ATPase of the soluble F_1 molecules. It also inhibits ATP synthesis. Removal of CF_1 may expose a proton channel in the membrane, and the reaction of a component(s) with DCCD would then block the channel. If a proton-conducting component were present in the F_o part of the complex, it would be expected that F_o would contain a DCCD-reactive protein. Winget et al (118) showed that DCCD became bound to their partially purified preparation of the complex. More recently, Nelson et al (75) extracted and purified a hydrophobic protein that bound DCCD. Light-dependent proton uptake by bacteriorhodopsin chloroplast lipid vesicles was decreased by this protein. In fact, one population of vesicles obtained by ficoll density gradient centrifugation showed no light-dependent proton uptake at all unless DCCD was present. Thus, this hydrophobic protein, which has a molecular weight of 8000, behaves as expected if it were the proton ionophore of the ATPase complex. This protein is also present in the ATPase preparation of Pick and Racker (personal communication). It seems, therefore, that one of the hydrophobic components of the membrane part of the complex functions as a proton channel. The other component(s) is likely to be involved in the binding of CF_1, but this remains to be proved.

Although illuminated chloroplasts catalyze rapid rates of ATP formation from ADP and P_i, only very low rates of ATPase activity are detected in the dark. Thus, the ATPase activity of CF_1 in situ is also not expressed. However, illumination of thylakoids in the presence of SH compounds markedly enhances ATPase activity, but does not inhibit photophosphorylation (53, 82, 83). Once activated, Mg^{2+}-dependent ATP hydrolysis continues in the dark for many minutes. Although the mechanism of activation has not been elucidated, it probably involves energy-dependent conformational changes in CF_1. These changes could result in the exposure of a disulfide bond to reduction by SH compounds as well as dislocation of an inhibitory subunit. Uncouplers of photophosphorylation strongly inhibit ATPase activation (62, 82) and light-induced conformational changes of CF_1 in thylakoids (95). Activation may also involve changes in the nucleotide binding properties of the membrane bound CF_1. Permanent as well as more transitory changes in CF_1 occur during the activation process. CF_1 isolated from thylakoids illuminated in the presence of dithiothreitol is active as a Ca^{2+}-ATPase (62). The ability of CF_1 in thylakoids to hydrolyze Mg^{2+}-ATP after activation decays in the dark, but the Ca^{2+}-ATPase activity of CF_1 solubilized from these membranes is stable after a dark period.

However, ATP hydrolysis can be restored to the thylakoids by brief illumination in the absence of SH compounds (8). Thus, CF_1 in thylakoids may have to assume an activated state or conformation before it catalyzes ATP hydrolysis.

There are many indications that the light- and SH-activated, Mg^{2+}-dependent ATPase is an expression of photophosphorylation operating in reverse. ATP hydrolysis is coupled to inward proton translocation (17), and by dissipating the resulting proton gradient, uncouplers stimulate ATPase activity (8) in previously activated thylakoids. Concentrations of uncouplers which more fully collapse the gradient inhibit the ATPase, probably because the maintenance of the activated form of CF_1 requires energy (8). $^{32}P_i$-ATP (18) exchange activity is activated with the ATPase activity. Moreover, ATPase and $^{32}P_i$-ATP exchange are sensitive to photophosphorylation inhibitors including DCCD and antisera to CF_1 (62). ATP can even drive reverse electron flow in isolated chloroplasts (92).

Conformational Changes and Nucleotide Binding

Definitive proof that CF_1 undergoes conformational changes upon establishment of a proton gradient across thylakoid membranes was provided by the hydrogen exchange experiments of Ryrie & Jagendorf (95). CF_1 purified from thylakoids illuminated in the presence of ^3H_2O contained as much as 100 moles of ^3H per mole of CF_1, whereas that from thylakoids incubated with ^3H_2O in the dark had essentially none. Thus, groups in CF_1 that bear exchangeable hydrogens are exposed to the medium upon illumination and exchange with the ^3H. These groups become buried once again when the thylakoids are returned to darkness. Hydrogen exchange was rapid, sensitive to uncouplers, and could also be elicited in the dark by the formation of artificial pH gradients across the membranes. The simultaneous presence of ADP and P_i reduced the extent of exchange by about one-half (95). Since inhibition of photophosphorylation by phlorizin did not reverse this effect, phosphorylation cannot be the cause of the inhibition. Instead, ADP and P_i binding must alter the structure of CF_1.

The conditional reaction of CF_1 in thylakoids with modifying reagents has also been exploited as a means for studying conformational changes in CF_1. N-Ethylmaleimide inhibited photophosphorylation only after the thylakoids were illuminated in the presence of this SH-alkylating reagent (59). This light-dependent, uncoupler-sensitive inhibition was correlated to a reaction of 0.5 to 0.7 mole of N-ethylmaleimide per mole of CF_1. Nearly all of the light-dependent incorporation is into the γ subunit (57). ADP and ATP (10–50 μM) partially prevent the development of the N-ethylmaleimide inhibition. Their effectiveness is, however, enhanced by P_i which by itself has little effect (49). Bis-dinitropyridine (3) and *o*-iodosobenzoate

(110) were subsequently found to inhibit photophosphorylation in a light-dependent manner. These reagents may act to form new disulfide bonds in CF_1. The lysine-modifying reagent trinitrobenzene sulfonate (80) inhibits ATPase activity of thylakoid membranes previously treated with methyl-acetamidate in the dark to protect exposed lysine residues. The inhibition is light-dependent, and incorporation of about 4 moles of trinitrobenzene sulfonate per mole of CF_1 was found. Since trinitrobenzene sulfonate was incorporated into the α and β subunits as well as the γ, it is apparent that illumination causes conformational changes in the larger subunits in addition to the γ subunit.

The fluorescence of amine-specific fluorophore, fluorescamine, attached to CF_1 in reconstituted EDTA-treated thylakoids underwent a rapid (half time 45 ms) quenching and blue shifting upon illumination. These effects, which were sensitive to uncouplers and a phosphorylation inhibitor, are consistent with the notion that conformational changes in CF_1 cause some lysine groups to become buried (43).

Tightly bound nucleotides in membrane-associated CF_1 exchange with ADP or ATP in the medium only very slowly in the dark. However, Harris & Slater (33) demonstrated that illumination promotes an uncoupler-sensitive exchange between the medium and bound nucleotides. Most, if not all, of the exchangeable nucleotide is associated with CF_1 (33, 51, 105), and this nucleotide site seems to be present on the α and/or β subunits (51). ATP and ADP exchange into this site equally well with apparent dissociation constants of 2–5 μM (52, 105). The energy-dependent step in the exchange is the release of bound nucleotide, which is predominately ADP (107). Thylakoids illuminated in the absence of added ADP or ATP retain much of their capacity to bind these nucleotides in the dark long after the proton gradient has decayed (52, 105). It seems likely that energy-dependent nucleotide exchange is a reflection of conformational changes in CF_1 which result in exposure of a nucleotide site to the medium.

Nucleotide exchange and phosphorylation respond in a similar manner to a number of factors, including light intensity and uncouplers (9, 52). Moreover, the fact that ATP is converted to mostly bound ADP during or within seconds after the exchange (51) suggests that this nucleotide exchange site has at least some catalytic activity. However, there are a number of indications that the nucleotide exchange site is not a catalytic site for phosphorylation. Bound ADP is phosphorylated more slowly than medium ADP (100). The rate of nucleotide exchange is also 50– to 100-fold slower than phosphorylation (105), although Gräber et al (30) have reported an initial faster phase of binding. The nucleotide specificity for exchange differs from that for phosphorylation. GDP is phosphorylated well by chloroplasts, but exchanges very poorly (52). Rapid phosphorylation of GDP (1–2 mM) reduces the extent of ADP (10 μM) binding by only 30–50% (S. R. Ketcham and R. E. McCarthy, unpublished). An extensive comparison of the nucleotide specificity of exchange and phosphorylation revealed that the two processes respond differently to base-modified adenine nucleotide analogs (106). At the present time, it is not possible to relate the nucleotide exchange site to nucleotide binding sites of soluble CF_1. However, the exchangeable site bears some resemblance to the site(s) involved in the stimulation of the extent of proton uptake (58) and protection of membrane-bound CF_1 from modifying reagents (see 55 for a review).

Acid, heat, or urea cause the dissociation of nucleotides bound to the exchangeable site of CF_1 in thylakoids. When acid (pH less than 2) was used to dissociate nucleotides, ATP and ADP were released in about a 1 : 3 ratio. Heat or urea caused the release of ADP almost exclusively (50). Acid either preserves bound ATP or causes its formation from bound ADP and P_i during rapid denaturation. The former possibility is probably correct since plunging thylakoids into hot 12M urea causes the release of about the same amount of ATP as does treatment with perchloric acid (115). Moreover, no $^{32}P_i$ is incorporated into the ATP released by trichloroacetic acid treatment of thylakoids (S. R. Ketcham and R. E. McCarty, unpublished).

The role(s) of tightly bound nucleotides remains to be established. However, they may be required for the assembly of F_1 molecules, since the reconstitution of ATPase activity from isolated *E. coli* F_1 subunits required ATP (25). Moreover, some function for bound nucleotides in oxidative phosphorylation is indicated by the observation (44) that their removal strongly inhibited ATP-linked reactions and ATP synthesis.

The conformation of CF_1 can also apparently affect electron flow rates and proton permeability. ATP or ADP (half-maximal effect about 2 μM) inhibits electron flow but stimulates proton uptake (58). Other nucleotides are effective only at much higher concentrations, although they have more effect in the presence of P_i (119). Early on it was found (58) that the enhancement of the extent of proton uptake is sensitive to a CF_1 antiserum. Conformational changes in bound CF_1 were proposed to regulate proton flux through the membrane. In agreement with this notion, ATP and ADP protect phosphorylation from inhibition by maleimides (49). Moreover, the development of the inhibition of phosphorylation by N-ethylmaleimide shows a similar dependence on ΔpH as does the light-induced increase in proton flux (85). It appears that in the absence of added nucleotides, bound CF_1 undergoes a change at high ΔpH values (greater than 2.8 to 2.9) which allows a faster rate of proton efflux through the ATPase complex. This faster rate of proton efflux induces an increased rate of electron flow since under nonphosphorylating conditions, the rate of electron transport is proportional to the proton efflux rate (85). By partially preventing a conformational change, ATP offsets the induced proton permeability and enhanced electron flow. DCCD also prevents these effects by blocking the proton channel in F_o.

The significance of the CF_1 structural changes is not clear. Those changes which take place in the absence of added nucleotides are not physiologically significant because the adenine nucleotide content of intact chloroplasts is too high. Nonetheless, energy-dependent conformational changes in CF_1, as revealed by the hydrogen exchange studies (95), occur in the presence of substrate levels of ADP plus P_i. CF_1 may have to assume an active conformation before it catalyzes photophosphorylation (32). As suggested by Boyer (11) and Slater (101), conformation changes in coupling factor-ATPases could be an integral part of the mechanism. On the other hand, these alterations could also be coincidental to the functioning of the ATPase complex.

Role of AMP

In 1971, a scheme for photophosphorylation was proposed in which AMP was suggested to be the primary phosphate acceptor (94). Phosphorylation of AMP to produce ADP bound to CF_1 and the transfer of the β-phosphate of the bound ADP to medium ADP to form ATP was proposed. This scheme is based on the findings that soluble CF_1 catalyzes a slow transphosphorylation of ADP (93) and that illumination of chloroplast thylakoids in the presence of AMP plus P_i causes ADP to become tightly bound to CF_1 (94), probably to the nucleotide exchange site. If AMP were the primary phosphate acceptor, ADP would be labeled more rapidly than ATP in the early phase of the phosphorylation process. Although conflicting reports have appeared (12, 115), it is evident that the phosphorylation of ATP driven either by light (115) or by an artificial pH gradient (102) can occur without significant AMP phosphorylation. Moreover, the observation that ADP becomes bound to CF_1 in thylakoids illuminated in the presence of AMP plus P_i (94) does not necessarily mean that AMP is phosphorylated directly by CF_1. Phosphorylation of ADP, either present as a minute contaminant or dissociated from CF_1 in the light, combined with the action of adenylate kinase, results in a net conversion of AMP to ADP and ATP (56a). These nucleotides then bind to the nucleotide exchange site. Moreover, even though GDP is phosphorylated well by thylakoids, GMP does not support $^{32}P_i$ incorporation into CF_1, whereas AMP does (74). This observation is in accord with the fact that GTP and probably GDP also bind poorly to the nucleotide exchange site (52).

Although AMP binds to neither soluble CF_1 nor to the exchange site of bound CF_1, there are other indications that it can interact in some manner with the membrane-bound enzyme. AMP promotes a $^{32}P_i$-ATP exchange in illuminated thylakoids (111) which cannot be explained by adenylate kinase activity. In addition, AMP plus arsenate overcome the inhibition of electron transport by ATP (69). Clearly, added AMP can affect processes which require CF_1 action. Whether AMP is more directly involved in ATP synthesis is uncertain at best.

Roles of Coupling Factor 1 Subunits

A variety of approaches have been used to attempt to elucidate the functions of the individual subunits of coupling factors from various sources. These include immunological, resolution and reconstruction analysis, and inhibitor studies. Nelson et al (72) used antisera against CF_1 subunits and found that antibodies to the α and γ subunits strongly inhibited photophosphorylation. The anti-α serum also prevented the stimulation of proton uptake by ATP. The ATPase activity of soluble CF_1 was resistent to each antiserum tested alone, but was sensitive to anti-α and anti-γ sera together.

Prolonged digestion of CF_1 with trypsin free of chymotrypic activity yielded an active ATPase preparation which, under dissociating conditions, gave precipitin lines with anti-α and β sera, but not with sera to the three smaller subunits (22). Thus, the active site(s) of CF_1 are present on the α and/or β subunits. Although it is clear that the γ, δ, and ϵ subunits are digested by the trypsin, the possibility that fragments of these subunits are resistant to trypsin and remain associated with the α and β components is difficult to rule out. The observation that anti-γ serum together with anti-α serum inhibit ATPase activity (72) could suggest a role for the γ subunit in this activity. Alternately, the combination of these two sera could simply induce unfavorable conformational changes in CF_1. Interestingly, a γ fragment as well as the α and β subunits survive trypsin digestion of the *E. coli* coupling factor (103).

The ATPase activity of CF_1 is sensitive to the inhibitor, 7-chloro-4-nitrobenzo-2-oxa-1,3-diazole (NBD). NBD is largely bound to the β subunit of CF_1, most likely to a tyrosine residue (22). Recently, it was shown (36) that NBD modification induces major conformational changes in CF_1. For example, the reactivity of a group in the ϵ subunit was decreased by NBD treatment, even though the NBD site(s) on the subunit is more than 44 Å from this group. Thus, the inhibition of ATPase activity by NBD could be the result of these conformational changes rather than of the modification of an essential tyrosine residue.

Remarkably, the treatment of *Rhodospirillum rubrum* chromatophores with 2M LiCl in the presence of ATP results in the extraction of the β

component of the coupling factor of this organism (84) and the loss of ATP hydrolysis and synthesis. Purified β subunit preparations restored both activities to the treated membranes.

The subunits of bacterial ATPases have been purified in reconstitutively active form (25, 120). Reconstitution of the ATPase activity of the *E. coli* enzyme requires the α, β, and γ components as well as ATP. The α subunit binds ATP (25). Thus, on the whole, although it is clear that δ and ϵ play no direct role in the ATPase activity of coupling factors, a function for γ or a fragment of γ cannot be entirely ruled out. However, the need for γ for reconstitution of ATPase activity may reflect a requirement for proper assembly.

A role for the γ subunit in phosphorylation is indicated not only by the sensitivity of phosphorylation to antisera directed against the subunit (72), but also by the observation that maleimides inhibit phosphorylation by specifically reacting with the γ subunit (57, 59). The bifunctional maleimide, *o*-phenylene bismaleimide, cross-links two groups in the γ component of CF_1 in illuminated thylakoids (117). This cross-linking apparently modifies the structure of the enzyme so as to increase the proton permeability of the membrane and, thus, uncouples phosphorylation. Recently, dithiobisethylmaleimide was synthesized. This compound also uncouples photophosphorylation and inhibitis proton uptake in thylakoids illuminated in its presence. However, high concentrations of SH compounds partly restore phosphorylation and proton uptake, probably because they reduce the disulfide bond in this bifunctional maleimide and break the cross-link (J. V. Moroney and R. E. McCarty, unpublished). In view of these results, it is tempting to conclude that the γ subunit functions to transmit protons to the active site(s) of ATP synthesis. The fact that the γ component, together with δ and ϵ, are required to block the proton channel formed in phospholipid vesicles by F_0 from a thermophilic bacterium (120) is in line with this concept. Moreover, the proton permeability of *R. rubrum* chromatophores is not enhanced by removal of the β subunit (84).

The δ subunit is clearly involved in binding CF_1 to F_0. The enzyme lacking the δ component fails to bind to depleted membranes or to recouple phosphorylation. Addition of δ restores these functions (76, 121).

The smallest CF_1 subunit, ϵ, is an inhibitor of the ATPase activity of CF_1 (78). It dissociates along with δ from CF_1 when the enzyme is heated in the presence of digitonin. In view of this finding, and of the trypsin sensitivity of ϵ, the mechanism of the activation of ATPase by heat or trypsin is explained. However, thiol activation of the ATPase does not appear to cause the permanent inactivation or dissociation of ϵ, so that more subtle changes may be involved. There are indications that the ϵ subunit interacts with γ (72). It is interesting to note that the mitochondrial ATPase inhibitor is not the ϵ component of F_1 (98). A clear role for ϵ in the *E. Coli* coupling factor in binding of the factor to the membrane was recently shown (104). If this finding also holds for CF_1, ϵ and δ are likely to be in contact with F_0. Extrapolating from the results found for the bacterial systems, γ interacts both with α and β and with δ and/or ϵ. Thus, γ could be at the center of the molecule, surrounded by the α and β subunits. This interpretation of the arrangement of CF_1 subunits is largely in accord with those proposed by others (36, 71). Based on cross-linking studies (7), δ and ϵ were suggested to be on opposite sides of the enzyme. However, this interpretation was considered inconclusive because it is based on negative results. The finding that the anti-γ serum inhibits phosphorylation, but does not agglutinate thylakoid membranes (72), suggests that the antigenic sites on γ are close to the membrane. On the other hand, cross-linking studies suggest that the γ subunit is distal to the δ and, therefore, away from the membrane. Further work is clearly needed on the structure of CF_1, both in solution and bound to the membrane.

Problems with Reconstitution Studies

The fact that CF_1 stimulates phosphorylation in deficient thylakoids does not prove that the added CF_1 is functioning in a catalytic sense. Removal of only 30–50% of the CF_1 renders the membranes proton leaky (61), and as a result, phosphorylation is totally uncoupled. Added CF_1, like DCCD (61), blocks the leak (60, 96), and the residual CF_1, rather than the added CF_1, could be active in phosphorylation. Recently it was shown (97) through the use of tentoxin, a phytotoxin which acts as a species-specific phosphorylation inhibitor, that only about 25% of the added CF_1 functions catalytically. In principle, then, inactive CF_1 could stimulate phosphorylation in EDTA-extracted thylakoids nearly as well as native CF_1 provided it could still bind and prevent the leak. This notion is supported by our unpublished observations that cold-inactivated CF_1 that was devoid of ATPase activity was still a coupling factor.

For the assay of coupling factor activity, thylakoid membrane preparations devoid of CF_1 would be most useful. Unfortunately, irreversible damage to the membranes occurs when CF_1 is removed (34). It is difficult to resolve all of the CF_1 from thylakoids by EDTA extraction. Why a fraction of the CF_1 is so difficult to remove is unclear, but it could be related to changes in the membrane. Treatment of subchloroplast particles with silicontungstic acid removes most of the CF_1, but severely damages the particles since phosphorylation is restored to only a low level by added CF_1 (46). An abstract recently appeared (73) in which it is stated that CF_1 restores phosphorylation by NaBr-treated thylakoids to rates similar to those of intact thylakoids. Since the NaBr treatment effectively removes CF_1, the search for a suitable preparation for the assay of coupling activity may thus be over.

SUGGESTED MECHANISMS

Although it is generally agreed that ATP synthesis in mitochondira, chloroplasts, and bacteria is driven by proton flux through the ATPase complex, the mechanism of this coupling is obscure. Several distinct hypotheses have been formulated and may be classified as either *direct* or *indirect* on the basis of the way in which the protons are used. Since these hypotheses have been summarized recently in some detail, only their essential features will be considered here.

In the direct mechanism formulated by Mitchell (65, 66), protons are conducted through the F_0 channel to the active site region of F_1. P_i in a specific state of protonation is attacked by two protons resulting in the formation of water and a positively charged phosphorous center (a phosphorylium). Nucleophilic attack of this center by ADP-O^- produces ATP. Mitchell has written his mechanisms so that two protons are translocated for each ATP formed or hydrolyzed. However, the proton stoichiometry of the chloroplast ATPase complex is probably three per ATP (56). Mitchell points out that his formulation can accomodate higher proton/ATP ratios without major alterations.

Boyer (11) and Slater (101) have proposed mechanisms in which conformational changes in coupling factor-ATPases are involved. Boyer's suggestions were originally based on the finding (21) that the exchange of ^{18}O between water and P_i which accompanies ATP hydrolysis by mitochondria is relatively uncoupler insensitive. If this exchange is related to hydrolysis and reformation of bound ATP, this result would mean that the formation of the phosphate anhydride bond of ATP is not a major energy-requiring step in oxidative phosphorylation. The release of newly formed ATP from F_1, rather than its synthesis, was suggested to be the major energy-dependent step. Conformational changes in F_1, linked in some way to proton translocation, are thought to provide the needed energy for dissociation of bound ATP. Detailed analysis of ^{18}O exchange data and of the participation of bound ADP in photophosphorylation led to the suggestion of an alternating site model for F_1 (11). In this model, the binding of ADP and P_i to one site is thought to promote the release of ATP at another site. Interactions between active sites were also proposed for a bacterial coupling factor-ATPase (1).

Slater's proposal (101) that energy input in oxidative and photosynthetic phosphorylation was required primarily for the dissociation of ATP from the ATPase complex was based on the findings that coupling factors contain tightly bound nucleotides and that exchange of the nucleotide bound to CF_1 is energy-dependent (33).

By analogy with other ion translocating ATPases, Racker (89) suggested that a phosphoenzyme intermediate may be involved in the mechanism of proton-translocating ATPases. The binding of Mg^{2+} was proposed to elicit conformational changes which drive the formation of a phosphorylated intermediate. Protons translocated across the membrane down the proton gradient displace the tightly bound Mg^{2+}, ATP is formed, and the coupling factor is now in a form which can rebind Mg^{2+}.

A role for lipoic acid (or a derivative) in oxidative phosphorylation has been suggested on the basis of results from Griffiths' laboratory (31). However, a lipoic acid-deficient mutant of *E. coli* carried out normal respiration and ATP-driven proton translocation (41). The DCCD-sensitive synthesis of ATP using oleoyl phosphate as the phosphoryl donor has been demonstrated in all types of coupling membranes, including chloroplasts (37). Moreover, soluble preparations derived from these membranes by chloroform extraction catalyzed rather active oleoyl phosphate-dependent ATP synthesis. Purified mitochondrial F_1, however, was inactive. This reaction appears to be specific for oleoyl phosphate. The chloroform-solubilized proteins of heart submitochondrial particles carried out oleoyl phosphate-dependent $^{32}P_i$-ATP exchange and oleoyl phosphatase activity. These intriguing findings may suggest that coupling factor-ATPases can act as oleoyl phosphokinases. Whether this activity is central to the mechanism at the ATPases is not yet established. Even if oleoyl phosphate is shown to be a "high energy" phosphorylated intermediate in ATP synthesis, the question remains as to the mechanism of its formation. Mitchell a number of years ago (64) derived schemes by which the synthesis of an anhydride bond may be linked to proton translocation.

The facts that CF_1 in thylakoids undergoes energy-dependent changes in its conformation and nucleotide affinity are consistent with the indirect mechanisms, but at the same time are not inconsistent with other mechanisms. Conformational changes of enzymes during catalysis are hardly exceptional. Moreover, the site which is responsible for energy-dependent nucleotide exchange is probably not a catalytic site. The presence of alternating, interacting sites in coupling factors is not inconsistent with the direct hypothesis.

CONCLUDING REMARKS

There has been a very noticeable shift in the emphasis of research in the biochemistry of oxidative and photosynthetic phosphorylation in the past few years. Previously the central question in this area was what couples electron flow to ADP phosphorylation. As it became increasingly more evident that Mitchell's revolutionary ideas were correct, more attention was directed by experimentalists to the mechanisms of proton-translocating ATPases and of electron transport-linked proton translocation. A great deal of progress has been made in defining the structure of CF_1 and possible roles for its individual subunits. Similar progress will undoubtedly be made with the components of F_0 in the near future. Despite these advances, the mechanism of the ATPase complex remains elusive. A more thorough characterization of soluble CF_1, as well as the ATPase complex of which it is a part, will certainly be required before this mechanism is revealed.

Literature Cited

1. Adolfsen, R., Moudrianakis, E. N. 1976. Binding of adenine nucleotides to purified 13S coupling factor of bacterial oxidative phosphorylation. *Arch. Biochem. Biophys.* 172:425–33
2. Amzel, L. M., Pedersen, P. L. 1978. Adenosine triphosphatase from rat liver mitochondria: Crystallization and x-ray diffraction studies of the F_1-component of the enzyme. *J. Biol. Chem.* 253: 2067–69
3. Andreo, C. S., Vallejos, R. H. 1976. Light-dependent inhibition of photophosphorylation by the sulphydryl reagent 2,2'-dithio bis(5-nitropyridine) *Biochim. Biophys. Acta* 423:590–601
4. Andreu, J. M., Warth, R., Muñoz, E. 1978. Glycoprotein nature of energy-transducing ATPases. *FEBS Lett.* 86: 1–5
5. Arnon, D. I., Allen, M. B., Whatley, F. R. 1954. Photosynthesis by isolated chloroplasts. *Nature* 24:394–96
6. Avron, M. 1963. A coupling factor in photophosphorylation. *Biochim. Biophys. Acta* 77:699–702
7. Baird, B. A., Hammes, G. G. 1976. Chemical cross-linking studies of chloroplast coupling factor 1. *J. Biol. Chem.* 251:6953–62
8. Bakker-Grunwald, T. 1977. ATPase. In *Encyclopedia of Plant Physiology New Series,* ed. A. Trebst, M. Avron, 5:369–73. Berlin: Springer-Verlag
9. Bickel-Sandkötter, S., Strotmann, H. 1976. Effects of external factors on photophosphorylation and exchange of CF_1-bound adenine nucleotides. *FEBS Lett.* 65:102–6
10. Binder, A., Jagendorf, A. T., Ngo, E. 1978. Isolation and composition of the subunits of spinach coupling factor protein. *J. Biol. Chem.* 253:3094–3100
11. Boyer, P. D. 1977. Coupling mechanisms in capture, transmission and use of energy. *Ann. Rev. Biochem.* 46: 957–66
12. Boyer, P. D., Stokes, B. O., Wolcott, R. G., Degani, C. 1975. Coupling of "high-energy" phosphate bonds to energy transductions. *Fed. Proc.* 34:1711–17
13. Cantley, L. C., Hammes, G. G. 1975. Characterization of nucleotide binding sites on chloroplast coupling factor 1. *Biochemistry* 14:2968–75
14. Cantley, L. C., Hammes, G. G. 1975. Fluorescence energy transfer between ligand binding sites on chloroplast coupling factor 1. *Biochemistry* 14:2976–81
15. Cantley, L. C., Hammes, G. G. 1976. Investigation of quercitin binding sites on chloroplast coupling factor 1. *Biochemistry* 15:1–8
16. Cantley, L. C., Hammes, G. G. 1976. Characterization of sulfhydryl groups on chloroplast coupling factor 1 exposed by heat activation. *Biochemistry* 15:9–14
17. Carmeli, C. 1970. Proton translocation induced by ATPase activity in chloroplasts. *FEBS Lett.* 7:297–300
18. Carmeli, C. 1977. Exchange reactions. See Ref. 8, pp. 492–500
19. Carmeli, C., Racker, E. 1973. Reconstitution of chlorophyll-deficient vesicles catalyzing phosphate-adenosine triphosphate exchange. *J. Biol. Chem.* 248:8281–87
20. Chipman, D. M., Shoshan, V., Shavit, N. 1977. Kinetics of binding of fluorescent nucleotide analogs to CF_1. *Int. Congr. Photosynth. Abstr. 4th, Reading,* p. 68
21. Cross, R. L., Boyer, P. D. 1975. The rapid labeling of adenosine triphosphate by ^{32}P-labeled inorganic phosphate and the exchange of phosphate oxygens as related to conformational coupling in oxidative phosphorylation. *Biochemistry* 14:392–98
22. Deters, D. W., Racker, E., Nelson, N., Nelson, H. 1975. Approaches to the active site of coupling factor 1. *J. Biol. Chem.* 250:1041–47
23. Farron, F. 1970. Isolation and properties of a chloroplast coupling factor and heat-activated adenosine triphosphatase. *Biochemistry* 9:3823–28
24. Farron, F., Racker, E. 1970. Studies on the mechanism of the conversion of coupling factor 1 from chloroplasts to an active adenosine triphosphatase. *Biochemistry* 9:3829–36

25. Futai, M. 1977. Reconstitution of ATPase activity from the isolated α, β, and γ subunits of the coupling factor, F_1, of *Escherichia coli. Biochem. Biophys. Res. Commun.* 79:1231–37
26. Garber, M. P., Steponkis, P. L. 1974. Identification of chloroplast coupling factor by freeze-etching and negative staining techniques. *J. Cell Biol.* 63: 24–34
27. Girault, G., Galmiche, J.-M. 1977. Further study of nucleotide-binding site on chloroplast coupling factor 1. *Eur. J. Biochem.* 77:501–10
28. Girault, G., Galmiche, J.-M., Michel-Villaz, M., Thiéry, J. 1973. Comparative study of photophosphorylation coupling factor ligand complex by circular dichroism and chemical isolation. *Eur. J. Biochem.* 38:473–78
29. Girault, G., Galmiche, J.-M., Vermeglio, A. 1974. CF_1 reconstitution of EDTA treated membrane fragments observed by means of photophosphorylation and 515 mm absorbance change. *Proc. Int. Congr. Photosynth., 3rd, Rehovot,* pp. 839–47
30. Gräber, P., Schlodder, E., Witt, H. T. 1977. Conformational change of the chloroplast ATPase induced by a transmembrane electric field and its correlation to phosphorylation. *Biochim. Biophys. Acta* 461:426–40
31. Griffiths, D. E. 1976. Net synthesis of adenosine triphosphate by isolated adenosine triphosphate synthase preparations. A role for lipic acid and unsaturated fatty acids. *Biochem. J.* 160:809–12
32. Harris, D. A., Crofts, A. R. 1978. The initial steps of photophosphorylation. Studies using excitation by saturating, short flashes of light. *Biochim. Biophys. Acta* 502:87–102
33. Harris, D. A., Slater, E. C. 1975. Tightly bound nucleotides of the energy-transducing ATPase of chloroplasts and their role in photophosphorylation. *Biochim. Biophys. Acta* 387:335–48
34. Hesse, H., Jank-Ladwig, R., Strotmann, H. 1976. On the reconstitution of photophosphorylation in CF_1-extracted chloroplasts. *Z. Naturforsch.* 31c:445–51
35. Hochman, Y., Lanir, A., Carmeli, C. 1976. Relations between divalent cation binding and ATPase activity in coupling factor from chloroplast. *FEBS Lett.* 61:255–59
36. Holowka, D. A., Hammes, G. G. 1977. Chemical modification and fluorescence studies of chloroplast coupling factor. *Biochemistry* 16:5538–45
37. Hyams, R. L., Carver, M. A., Parks, M. D., Griffiths, D. E. 1977. Studies of energy-linked reactions: oleoyl phosphate-dependent ATP synthesis (oleoyl phosphokinase) activity of membrane ATPases and soluble ATPases from mitochondria, chloroplasts, chromatophores, and *Escherichia coli* plasma membrane. *FEBS Lett.* 82:307–13
38. Jagendorf, A. T. 1975. The mechanism of photophosphorylation. In *Bioenergetics of Photosynthesis,* ed. Govindjee, pp. 413–92. New York: Academic
39. Jagendorf, A. T. 1977. Photophosphorylation. See Ref. 8, pp. 307–37
40. Jagendorf, A. T., Smith, M. 1962. Uncoupling phosphorylation in spinach chloroplasts by absence of cations. *Plant Physiol.* 37:135–41
41. Jones, R. W., Haddock, B. A., Garland, P. B. 1978. Vectorial organization of proton-translocating oxidoreductions of *Escherichia coli.* In *The Proton and Calcium Pumps,* ed. G. F. Azzone, M. Avron, J. C. Metcalfe, E. Quagliarello, N. Silliprandi, pp. 71–80. Amsterdam: Elsevier
42. Kagawa, Y., Racker, E. 1971. Reconstitution of vesicles catalyzing $^{32}P_i$-adenosine triphosphate exchange. *J. Biol. Chem.* 246:5477–87
43. Kraayenhof, R., Slater, E. C. 1974. Studies of chloroplast energy conservation with electrostatic and covalent fluorophores. *Proc. Int. Congr. Photosynth., 3rd, Rehovot,* pp. 985–96
44. Leimgruber, R. M., Senior, A. E. 1976. Removal of "tightly bound" nucleotides from phosphorylating submitochondrial particles. *J. Biol. Chem.* 251: 7110–13
45. Lein, S., Berzborn, R. J., Racker, E. 1972. Studies on the subunit structure of coupling factor 1 from chloroplasts. *J. Biol. Chem.* 247:3520–24
46. Lein, S., Racker, E. 1971. Properties of silicotungstate-treated subchloroplast particles. *J. Biol. Chem.* 246: 4298–4307
47. Livne, A., Racker, E. 1969. Interaction of coupling factor 1 from chloroplasts with ribonucleic acid and lipids. *J. Biol. Chem.* 244:1332–38
48. Lynn, W. S., Straub, K. D. 1968. Isolation and properties of a protein from chloroplasts required for phosphorylation and H^+ uptake. *Biochemistry* 8: 4789–93
49. Magnusson, R. P., McCarty, R. E. 1975. Influence of adenine nucleotides

on the inhibition of photophosphorylation by N-ethylmaleimide. *J. Biol. Chem.* 250:2593–98
50. Magnusson, R. P., McCarty, R. E. 1976. Acid-induced phosphorylation of adenosine 5'-diphosphate bound to coupling factor 1 in spinach chloroplasts thylakoids. *J. Biol. Chem.* 251:6874–77
51. Magnusson, R. P., McCarty, R. E. 1976. Illumination of chloroplast thylakoids in the presence of ATP causes the binding of ADP to one of the large subunits of coupling factor 1. *Biochem. Biophys. Res. Commun.* 70:1283–89
52. Magnusson, R. P., McCarty, R. E. 1976. Light-induced exchange of nucleotides into coupling factor 1 in spinach chloroplast thylakoids. *J. Biol. Chem.* 251:7417–22
53. Marchant, R. H., Packer, L. 1963. Light and dark stages in the hydrolysis of adenosine triphosphate by chloroplasts. *Biochim. Biophys. Acta* 75: 458–66
54. McCarty, R. E. 1976. Ion transport and energy conservation in chloroplasts. In *Encyclopedia of Plant Physiology New Series,* ed. C. R. Stocking, U. Heber, 3:347–76. Berlin: Springer-Verlag
55. McCarty, R. E. 1979. Interactions between nucleotides and coupling factor 1 in chloroplasts. *Trends in Biochemical Science.* In press
56. McCarty, R. E. 1978. The ATPase complex of chloroplasts and chromatophores. *Curr. Top. Bioenerg.* 7:245–78
56a. McCarty, R. E. 1978. *FEBS Lett.* 95: 299–302
57. McCarty, R. E., Fagan, J. 1973. Light-stimulated incorporation of N-ethylmaleimide into coupling factor 1 in spinach chloroplasts. *Biochemistry* 12: 1503–7
58. McCarty, R. E., Fuhrman, J. S., Tsuchiya, Y. 1971. Effects of adenine nucleotides on hydrogen ion transport in chloroplasts. *Proc. Natl. Acad. Sci. USA* 68:2522–26
59. McCarty, R. E., Pittman, P. R., Tsuchiya, Y. 1972. Light-dependent inhibition of photophosphorylation by N-ethylmaleimide. *J. Biol. Chem.* 247: 3048–51
60. McCarty, R. E., Racker, E. 1966. Effect of a coupling factor and its antiserum on photophosphorylation and hydrogen ion transport. *Brookhaven Symp. Biol.* 19:202–14
61. McCarty, R. E., Racker, E. 1967. The inhibition and stimulation of photophosphorylation by N,N'-dicyclohexylcarbodiimide. *J. Biol. Chem.* 242: 3435–39
62. McCarty, R. E., Racker, E. 1968. Activation of adenosine triphosphatase and ^{32}P-labeled orthophosphate-adenosine triphosphate exchange in chloroplasts. *J. Biol. Chem.* 243:129–37
63. Mitchell, P. 1961. Coupling of phosphorylation to electron and hydrogen transfer by a chemiosmotic type of mechanism. *Nature* 191:144–48
64. Mitchell, P. 1966. Chemiosmotic coupling in oxidative and photosynthetic phosphorylation. *Biol. Rev. Cambridge Philos. Soc.* 41:445–502
65. Mitchell, P. 1974. A chemiosmotic molecular mechanism for proton-translocating adenosine triphosphatases. *FEBS Lett.* 43:189–94
66. Mitchell, P. 1977. Vectorial chemiosmotic processes. *Ann. Rev. Biochem.* 46:996–1005
67. Moudrianakis, E. N. 1968. Structural and functional aspects of photosynthetic lamellae. *Fed. Proc.* 27:1180–85
68. Moudrianakis, E. N., Tiefert, M. A. 1976. Synthesis of bound adenosine triphosphate from bound adenosine diphosphate by the purfied coupling factor 1 of chloroplasts: Evidence for direct involvement of the coupling factor in this "adenylate kinase-like" reaction. *J. Biol. Chem.* 251:7796–7801
69. Mukohata, Y., Yagi, T. 1974. Electron transport coupled to quasi-arsenylation in isolated chloroplasts. *Bioenergetics* 7:111–20
70. Nabedryk-Viala, E., Calvet, P., Thiéry, J. M., Galmiche, J.-M., Girault, G. 1977. Interaction of adenine nucleotides with the coupling factor of spinach chloroplasts: A hydrogen-deuterium exchange study. *FEBS Lett.* 79:139–43
71. Nelson, N. 1977. Structure and function of chloroplast ATPase. *Biochim. Biophys. Acta* 456:314–38
72. Nelson, N., Deters, D. W., Nelson, H., Racker, E. 1973. Properties of isolated subunits of coupling factor 1 from spinach chloroplasts. *J. Biol. Chem.* 248: 2049–55
73. Nelson, N., Eytan, E. 1978. Preparation of chloroplasts highly depleted of CF_1 with a high degree of reconstitution. *Plant Physiol.* 61s:89
74. Nelson, N., Eytan, E., Julian, C. 1977. The function of individual polypeptides in photosynthetic energy transduction. *Proc. Int. Congr. Photosynth., 4th, Reading,* pp. 559–70
75. Nelson, N., Eytan, E., Notsani, B.-E., Sigrist, H., Sigrist-Nelson, K., Gittler,

C. 1977. Isolation of a chloroplast N,N'-dicyclohexylcarbodiimide-binding proteolipid, active in proton translocation. *Proc. Natl. Acad. Sci. USA* 74:2375–78
76. Nelson, N., Karny, O. 1976. The role of δ subunit in the coupling activity of coupling factor 1. *FEBS Lett.* 70: 249–53
77. Nelson, N., Nelson, H., Racker, E. 1972. Magnesium-adenosine triphosphatase properties of heat-activated coupling factor 1 from chloroplasts. *J. Biol. Chem.* 247:6506–10
78. Nelson, N., Nelson, H., Racker, E. 1972. Purification and properties of an inhibitor isolated from chloroplast coupling factor 1. *J. Biol. Chem.* 247: 7657–62
79. O'Keefe, D. P., Dilley, R. A. 1977. The effect of chloroplast coupling factor removal on thylakoid membrane permeability. *Biochim. Biophys. Acta* 461: 48–60
80. Oliver, D., Jagendorf, A. T. 1976. Exposure of free amino groups in the coupling factor of energized spinach chloroplasts. *J. Biol. Chem.* 251:7168–75
81. Penefsky, H. S., Pullman, M. E., Datta, A., Racker, E. 1960. Participation of a soluble adenosine triphosphatase in oxidative phosphorylation. *J. Biol. Chem.* 235:3330–36
82. Petrack, B., Cranston, A., Sheppy, F., Farron, F. 1965. Studies on the hydrolysis of adenosine triphosphate by chloroplasts. *J. Biol. Chem.* 240:906–12
83. Petrack, B., Lipmann, F. 1961. Photophosphorylation and photohydrolysis in cell-free preparations of blue-green algae. In *Light and Life,* ed. W. D. McElroy, H. B. Glass, pp. 621–24. Baltimore: Johns Hopkins
84. Philosoph, S., Binder, A., Gromet-Elhanan, Z. 1977. Coupling factor ATPase complex of *Rhodospirillum rubrum.* Purification and properties of a reconstitutively active single subunit. *J. Biol. Chem.* 252:8747–52
85. Portis, A. R. Jr., Magnusson, R. P., McCarty, R. E. 1975. Conformational changes in coupling factor 1 may control the rate of electron flow in spinach chloroplast. *Biochem. Biophys. Res. Commun.* 64:877–84
86. Posorske, L., Jagendorf, A. T. 1976. Nucleotide and metal interactions affecting inactivation of spinach chloroplast coupling factor by NaCl in the cold. *Arch. Biochem. Biophys.* 177: 276–83
87. Pullman, M. E., Penefsky, H. S., Datta, A., Racker, E. 1960. Purification and properties of soluble, dinitrophenol stimulated adenosine triphosphatase. *J. Biol. Chem.* 235:3322–29
88. Racker, E. 1967. Resolution and reconstitution of the inner mitochondrial membrane. *Fed. Proc.* 26:1335–40
89. Racker, E. 1977. Mechanism of energy transformations. *Ann. Rev. Biochem.* 46:1006–14
90. Racker, E., Hauska, G. A., Lein, S., Berzborn, R. J., Nelson, N. 1971. Resolution and reconstitution of the system of photophosphorylation. *Proc. Int. Congr. Photosynth. 2nd, Stresa,* pp. 1097–1113
91. Racker, E., Stoeckenius, W. 1974. Reconstitution of purple membrane vesicles catalyzing light-driven proton uptake and adenosine triphosphate formation. *J. Biol. Chem.* 249:662–63
92. Rienits, K. G., Hardt, H., Avron, M. 1974. Energy-dependent reverse electron flow in chloroplasts. *Eur. J. Biochem.* 43:291–98
93. Roy, H., Moudrianakis, E. N. 1971. Interactions between ADP and the coupling factor of photophosphorylation. *Proc. Natl. Acad. Sci. USA* 68:464–68
94. Roy, H., Moudrianakis, E. N. 1971. Synthesis and discharge of the coupling factor-adenosine diphosphate complex in spinach chloroplast lamellae. *Proc. Natl. Acad. Sci. USA* 68:2720–24
95. Ryrie, I. J., Jagendorf, A. T. 1972. Correlation between a conformational change in the coupling factor protein and the high energy state in chloroplasts. *J. Biol. Chem.* 247:4453–59
96. Schmid, R., Junge, W. 1974. On the influence of extraction and recondensation of CF_1 on the electric properties of the thylakoid membrane. See ref. 29, pp. 821–30
97. Selman, B. R., Durbin, R. D. 1978. Evidence for a catalytic function of the coupling factor protein reconstituted with chloroplast thylakoid membranes. *Biochim. Biophys. Acta* 502:29–37
98. Senior, A. E. 1978. The Mitochondrial ATPase. In *Membrane Proteins in Energy Transduction,* ed. R. A. Capaldi, M. Dekker. In press
99. Serrano, R., Kanner, B. I., Racker, E. 1976. Purification and properties of the proton-translocating adenosine triphosphatase of bovine heart mitochondria. *J. Biol. Chem.* 251:2453–61
100. Shavit, N., Lein, S., San Pietro, A. 1977. On the role of membrane-bound ADP and ATP in photophosphorylation in

chloroplast membranes. *FEBS Lett.* 73:55–58

101. Slater, E. C. 1977. Mechanism of oxidative phosphorylation. *Ann. Rev. Biochem.* 46:1015–26
102. Smith, D. J., Stokes, B. O., Boyer, P. D. 1976. Probes of initial events in ATP synthesis by chloroplasts. *J. Biol. Chem.* 251:4165–71
103. Smith, J. B., Wilkonski, C. 1978. A fragment of subunit γ remains tightly bound to the *E. coli* F_1-ATPase after trypsinization. *Fed. Proc.* (Abstr.) 37: 1521
104. Sternweis, P. C. 1978. The ϵ subunit of *Escherichia coli* coupling factor 1 is required for its binding to the cytoplasmic membrane. *J. Biol. Chem.* 253:3123–28
105. Strotmann, H., Bickel-Sandkötter, S. 1977. Energy-dependent exchange of adenine nucleotides on chloroplast coupling factor (CF_1). *Biochim. Biophys. Acta* 400:126–35
106. Strotmann, H., Bickel-Sandkötter, S., Edelman, N. K., Schlimme, E., Boos, K. S., Lüstoff, J. 1977. Studies on the tight adenine nucleotide binding sites of chloroplast coupling factor (CF_1). In *Structure and Function of Energy Transducing Membranes,* ed. K. Van Dam, B. F. Van Gelder, pp. 307–17. Amsterdam: Elsevier
107. Strotmann, H., Bickel, S., Huchzermeyer, B. 1976. Energy-dependent release of adenine nucleotides tightly bound to chloroplast coupling factor CF_1. *FEBS Lett.* 61:194–98
108. Strotmann, H., Hesse, H., Edelmann, K. 1973. Quantitative determination of coupling factor CF_1 of chloroplasts. *Biochim. Biophys. Acta* 314:202–10
109. Tiefert, M. A., Roy, H., Moudrianakis, E. N. 1977. Binding of adenine nucleotides and pyrophosphate by the purified coupling factor of photophosphorylation. *Biochemistry* 16:2396–2404
110. Vallejos, R. H., Andreo, C. S. 1976. Sulphydryl groups in photosynthetic energy conservation: Further evidence of vicinal dithiols involvement as shown by light-dependent effects of o-iodosobenzoate. *FEBS Lett.* 61:95–99
111. Vambutas, V. K., Bertsch, W. 1976. Does AMP participate in photosynthetic phosphorylation? *Biochem. Biophys. Res. Commun.* 73:686–93
112. Vambutas, V. K., Racker, E. 1965. Stimulation of photophosphorylation by a preparation of a latent, Ca^{++}-dependent adenosine triphosphatase from chloroplasts. *J. Biol. Chem.* 240:2660–67
113. VanderMeulen, D. L., Govindjee. 1975. Interactions of fluorescent analogs of adenine nucleotides with coupling factor protein isolated from spinach chloroplasts. *FEBS Lett.* 57:272–74
114. VanderMeulen, D. L., Govindjee. 1977. Binding of modified adenine nucleotides to isolated coupling factor from chloroplasts as measured by polarization of fluorescence. *Eur. J. Biochem.* 78: 585–98
115. Vinkler, C., Rosen, G., Boyer, P. D. 1978. Light-driven ATP formation from $^{32}P_i$ by chloroplast thylakoids without detectable labeling of ADP, as measured by rapid mixing and acid quench techniques. *J. Biol. Chem.* 253:2507–10
116. Vogel, G., Steinhart, R. 1976. ATPase of *Escherichia coli:* Purification, dissociation and reconstitution of the active complex from the isolated subunits. *Biochemistry* 15:208–16
117. Weiss, M. A., McCarty, R. E. 1977. Cross-linking within a subunit of coupling factor 1 increases the proton permeability of spinach chloroplasts thylakoids. *J. Biol. Chem.* 252:8007–12
118. Winget, G. D., Kanner, N., Racker, E. 1977. Formation of ATP by the adenosine triphosphatase complex from spinach chloroplasts reconstituted together with bacteriorhodopsin. *Biochim. Biophys. Acta* 460:490–502
119. Yagi, T., Mukohata, Y. 1977. Effects of purine nucleotides on photosynthetic electron transport in isolated chloroplasts. *J. Bioenerg. Biomemb.* 9:31–40
120. Yoshida, M., Okamoto, H., Sone, N., Hirata, H., Kagawa, Y. 1977. Reconstitution of thermostable ATPase capable of energy coupling from its purified subunits. *Proc. Natl. Acad. Sci. USA* 74:936–40
121. Younis, H., Winget, G. D., Racker, E. 1977. Requirement of the δ subunit of chloroplast coupling factor 1 for photophosphorylation. *J. Biol. Chem.* 252: 1814–18

Ann. Rev. Plant Physiol. 1979. 30:105–30

ENZYMIC CONTROLS IN THE BIOSYNTHESIS OF LIGNIN AND FLAVONOIDS[1]

❖7667

Klaus Hahlbrock and Hans Grisebach

Biologisches Institut II der Universität, D-7800 Freiburg im Breisgau, West Germany

CONTENTS

INTRODUCTION

Phenylpropanoid units derived from the shikimate pathway are common structural elements for both lignin and flavonoids. Two of the three aromatic amino acids originating from the shikimate pathway, phenylalanine

[1]Abbreviations: PAL, phenylalanine ammonia-lyase (EC 4.3.1.5); FS, flavanone synthase; SAM, S-adenosyl-L-methionine.

0066-4294/79/0601-0105$01.00

and, to a much lesser extent, tyrosine, were found in tracer studies to serve as substrates for the phenylpropanoid building units of lignin and flavonoids (4, 12, 61). Large variations of the substitution patterns and rearrangements of the carbon atoms of the side chain of the phenylpropanoid residue were used for defining various types or subgroups of lignin and flavonoids. These include syringyl-type and guaiacyl-type lignins, and chalcones, flavanones, flavones, flavonols, isoflavones, anthocyanins etc. For the present purpose, we do not intend to deal comprehensively with all of these classes of compounds. We shall rather confine ourselves to those biosynthetic pathways whose enzymology and regulation have been studied in sufficient detail to allow us to draw some general conclusions concerning mechanisms of enzymic control.

The common origin of the phenylpropanoid building units, the identity of some of the early intermediates derived from phenylalanine, and the frequent co-occurrence of lignin and flavonoids in the same plant are ample reason for treating both types of compounds together in the same review. General schemes for the biosynthesis of lignin and flavonoids were originally proposed on the basis of results from tracer studies with radioactive substrates. The rapid progression of work at the enzymic level during the past one or two decades largely confirmed the original, hypothetic schemes for the sequences of biosynthetic steps and supplemented them with further details. The results have been summarized in several more or less comprehensive reviews (4, 12, 15, 16, 18, 21, 22, 27, 61, 66). The aim of our present review is to discuss some of the most recent work from our own and other laboratories, with particular emphasis placed on results obtained with cell suspension cultures. Since their first use for studies of enzymes of phenylpropanoid metabolism in 1966 (14), plant cell cultures have proved to be an excellent tool for research in this field (21). A large number of papers describing such studies have since appeared and have added greatly to our present knowledge of enzymic controls in the biosynthesis of lignin and flavonoids. Because of the limited amount of space, and in order to concentrate on recent developments within the areas selected for the following account, we shall refer to the pertinent literature mostly by citing appropriate review articles or recent original publications from which earlier references can be taken.

A third class of phenylpropanoid compounds which occurs as widely in higher plants as lignin and flavonoids comprises the esters of various cinnamic acids. Since the phenylpropanoid moieties of these substances have the same origin as those of lignin and flavonoids, the interplay of mechanisms for enzymic control of all three classes of phenylpropanoid compounds will be mentioned where appropriate data are available.

BIOSYNTHETIC PATHWAYS

General Phenylpropanoid Metabolism

The three early steps in the conversion of phenylalanine to derivatives of cinnamic acid are common to all major phenylpropanoid pathways (Figure 1). This sequence of reactions was therefore termed "general phenylpropanoid metabolism" (11). The enzymes catalyzing the individual steps are phenylalanine ammonia-lyase (PAL[1]), cinnamate 4-hydroxylase, and 4-coumarate:CoA ligase[2] (Table 1). We discuss below some of the regulatory and kinetic properties of these enzymes from several plants, including the occurrence of isoenzymes of 4-coumarate:CoA ligase.

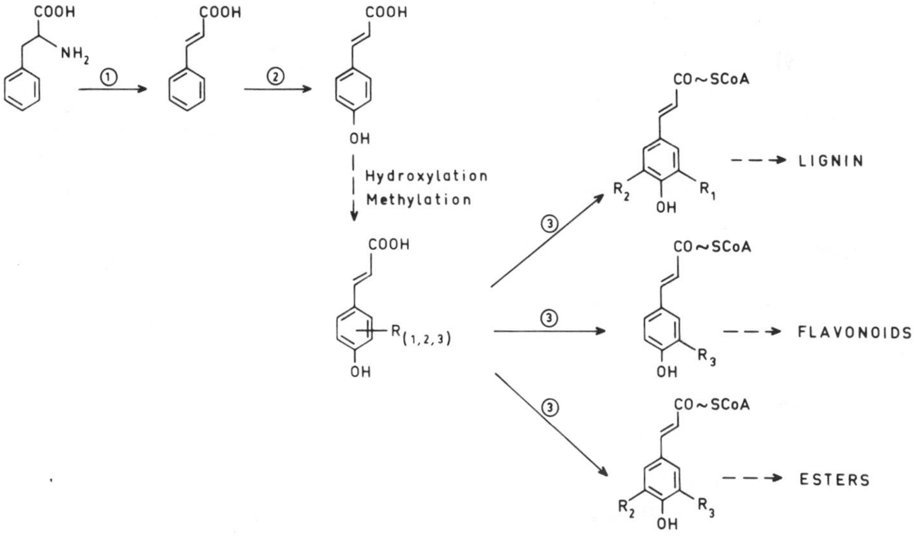

Figure 1 Scheme illustrating the sequence of reactions of general phenylpropanoid metabolism. The enzymes marked by numbers are listed in Table 1. Enzyme No. 3 occurs in several tissues in the form of two or three isoenzymes with properties suggesting their preferential involvement in the lignin, flavonoid glycoside, and cinnamate ester pathways, respectively (see text for details). The dashed arrow indicates (a) hypothetical 3- and 5-hydroxylase(s) as well as a methyltransferase which is not a typical enzyme of this sequence. R_1, R_2 = H or OCH_3; R_3 = H or OH.

[2]Some authors have designated the ligase as "hydroxycinnamate:CoA ligase." Because 4-coumarate was one of the most efficient substrates for nearly all ligases reported so far to be involved in phenylpropanoid metabolism (18, 35), and the E. C. number was assigned to 4-coumarate:CoA ligase, this latter term is used here throughout.

Table 1 List of enzymes indicated in figures

Enzyme	E.C. number	Key to figure
General phenylpropanoid metabolism		
Phenylalanine ammonia-lyase (PAL[1])	4.3.1.5	Fig. 1, No. 1
Cinnamate 4-hydroxylase	1.14.13.11	Fig. 1, No. 2
4-Coumarate:CoA ligase[2]	6.2.1.12	Fig. 1, No. 3
Additional reaction preceding the CoA ligase step		
SAM:Caffeate 3-O-methyltransferase	2.1.1.–	Fig. 1
Lignin pathway		
Cinnamoyl-CoA:NADPH oxidoreductase	1.1.1.–	Fig. 2, No. 1
Cinnamyl alcohol:NADP oxidoreductase	1.1.1.–	Fig. 2, No. 2
Flavone and flavonol glycoside pathways		
Acetyl-CoA carboxylase	6.4.1.2	Fig. 3, No. 1
Flavanone synthase (FS[1])		Fig. 3, No. 2
Chalcone isomerase	5.5.1.6	Fig. 3, No. 3
"Flavanone oxidase"		Fig. 3, No. 4
SAM:Flavone/flavonol 3′-O-methyltransferase	2.1.1.42	Fig. 3, No. 5
UDP-Glucose:flavone/flavonol 7-O-glucosyltransferase	2.4.1.82	Fig. 3, No. 6
UDP-Glucose:flavonol 3-O-glucosyltransferase	2.4.1.–	Fig. 3, No. 7
UDP-Apiose synthase		Fig. 3, No. 8
UDP-Apiose:flavone-7-O-glucoside 2″-O-apiosyltransferase	2.4.2.25	Fig. 3, No. 9
Malonyl-CoA:flavone/flavonol-7-O-glycoside malonyltransferase	2.3.1.–	Fig. 3, No. 10
Malonyl-CoA:flavonol-3-O-glucoside malonyltransferase	2.3.1.–	Fig. 3, No. 11

Some of the specific pathways into which general phenylpropanoid metabolism diverges, in particular the lignin pathway, require in addition to 4-hydroxylation further substitution(s) of cinnamate prior to activation by the ligase. The most frequently occurring additional reactions are hydroxylation and subsequent methylation in the 3- and 5-positions of the aromatic ring. Of the enzymes involved, only the methyltransferases have so far been characterized to any extent. The lignin-specific methyltransferase in soybean cell cultures was found to be regulated independently from the enzymes of general phenylpropanoid metabolism (see below). Since flavonoid-specific methyltransferases also exist, it is unlikely that general phenylpropanoid metabolism includes a "general" methyltransferase. We indicate therefore the methylation reaction in Figure 1 by a broken arrow. It is not known at present whether one or even two "general" 3- and 5-hydroxylase(s) exist in addition to the three enzymes represented by solid arrows.

Lignin Pathway

The various types of lignin were classified according to their relative content of phenylpropanoid building units with the characteristic 4(OH)-, 3(OCH_3), 4(OH)-, and 3,5(OCH_3)$_2$, 4(OH)-substitution patterns (61). Their respective substrates are the CoA esters of 4-coumaric, ferulic, and sinapic acids. These activated acids formed in the last step of the general pathway are first reduced in a two-step reaction to give the corresponding alcohols which are then polymerized through the action of a peroxidase. This last step in the biosynthesis of lignin may or may not be preceded by the intermediate formation of glucosides of the phenylpropanoid alcohols, e.g. coniferin (15). The sequence of reactions leading from 4-coumaroyl-CoA, feruloyl-CoA, and sinapoyl-CoA to the polymeric structure of lignin is depicted in Figure 2.

Flavonoid Pathways

The scheme in Figure 3 illustrates our present knowledge of the biosynthetic pathways of two of the most important subgroups of flavonoid compounds, flavone and flavonol glycosides. For comparison we give the chemical structures of an anthocyanin and an isoflavonoid whose biosynthetic pathways are not known in detail.

The function of chalcone isomerase, the first enzyme of flavonoid metabolism which was studied in some detail (48), is still unclear. The occurrence of this enzyme has been reported for a large number of flavonoid-producing plants and plant tissues; its absence from defined mutants was correlated with an abnormally high rate of chalcone accumulation and a concomitant block in the synthesis of other flavonoids (39). A major point of uncertainty is the question whether a flavanone or the isomeric chalcone is the true substrate for the enzymes catalyzing the formation of flavones, flavonols, isoflavones, anthocyanins etc. Although it has been demonstrated in *Petroselinum hortense* that the first product of the flavonoid pathway formed by flavanone synthase (FS[1]), is the flavanone naringenin (37), it is possible that the isomerase is involved in one of the following steps of

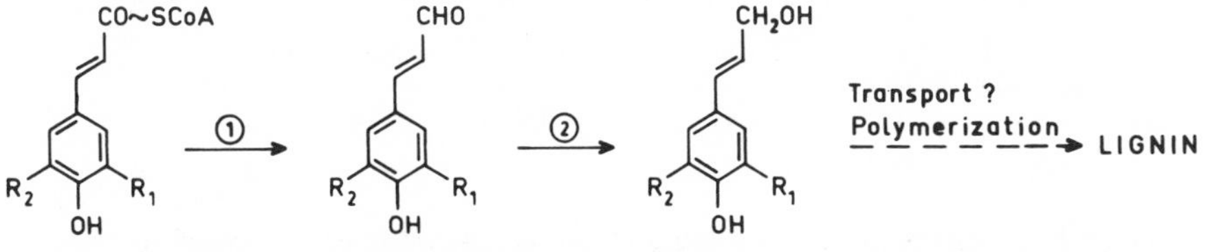

Figure 2 Scheme illustrating the sequence of reactions of the lignin pathway. The enzymes marked by numbers are listed in Table 1. R_1, R_2 = H or OCH_3.

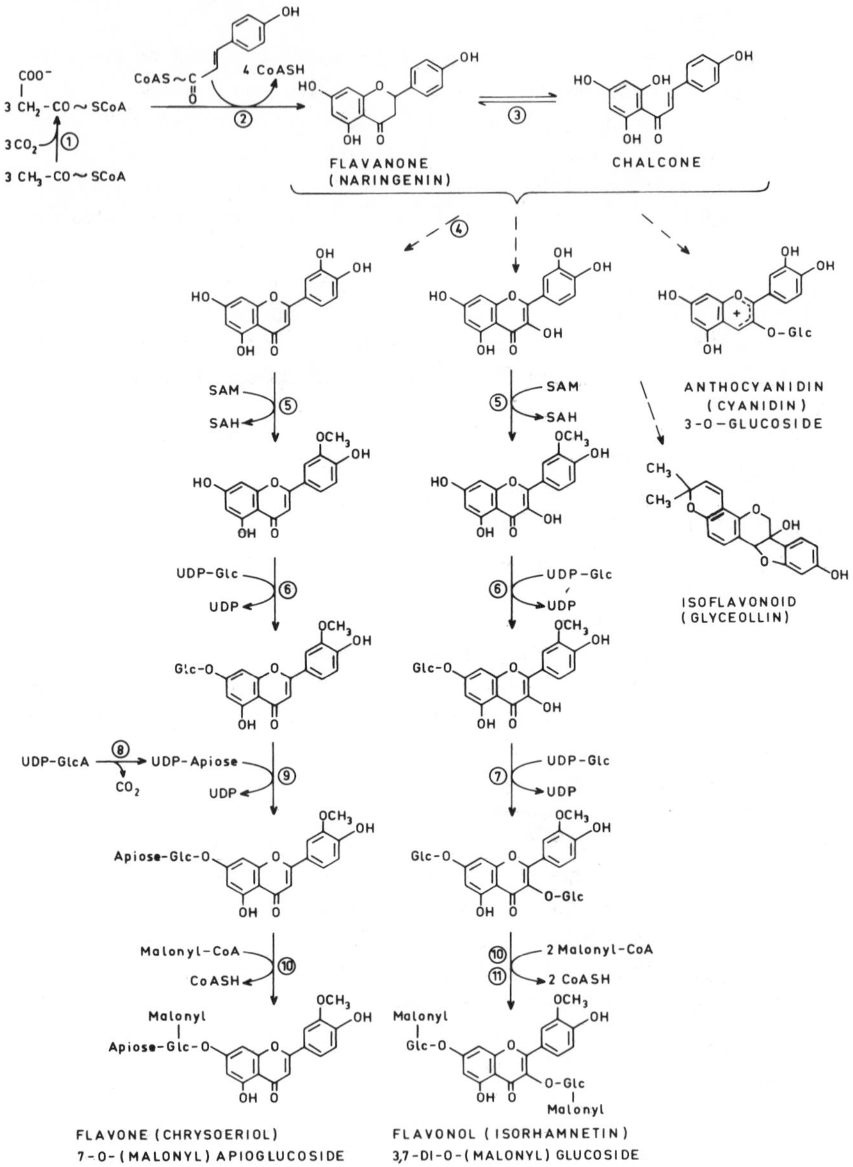

Figure 3 Scheme illustrating the sequence of reactions of the flavone and flavonol glycoside pathways. The enzymes marked by numbers are listed in Table 1. Dashed arrows indicate unknown reactions. For comparison, the structures of an anthocyanidin and an isoflavonoid are given.

flavone or flavonol glycoside formation. Because of this unsolved question, which is open also for flavonoid pathways in other plants, the roles of flavanones and chalcones are not defined precisely in Figure 3.

Biosynthesis of Cinnamate Esters

Substituted cinnamic acids, predominantly 4-coumaric, caffeic, ferulic, and sinapic acids, occur frequently as esters of glucose, quinic acid, shikimic acid, choline, hydroxysubstituted carboxylic acids, and other compounds containing hydroxy groups. Typical representatives of this class of substances are chlorogenic acid (3-caffeoyl quinic acid) or sinapin (sinapoyl choline). Chlorogenic acid, for example, is formed from caffeoyl-CoA and quinic acid by a caffeoyltransferase (57, 68). It should be noted, however, that CoA esters of substituted cinnamic acids are not in all cases the activated substrates for such reactions. For instance, 1-(4-coumaroyl) glucose was reported to be synthesized from 4-coumarate and UDP-glucose (42).

CONTROL OF BIOSYNTHESIS

The possible control mechanisms for biosynthetic pathways in multicellular eukaryotic organisms are manifold. Various lines of evidence suggest that the following mechanisms of enzymic control play an important role in the biosynthesis of phenylpropanoid compounds in higher plants: product inhibition and substrate specificity of enzymes; rates of enzyme synthesis; rates of enzyme degradation; subcellular and cellular compartmentation of enzymes, substrates, and products.

SYSTEMS USED FOR STUDIES OF ENZYMIC CONTROL

Cell suspension cultures from various plants are extremely useful systems for studying enzymic control in the biosynthesis of lignin and flavonoids. Although it is obvious that some mechanisms of control can only be studied with intact plants, many of the basic principles of an interdependent regulation of lignin and flavonoid biosynthesis are far more conveniently investigated with cell cultures. We shall therefore first discuss the results obtained with cell cultures and then proceed to the more complex systems of intact plants and plant tissues.

Cultured Soybean Cells

Cells of soybean (*Glycine max* L.) in suspension culture are capable of producing lignin (47, 50) as well as several types of flavonoid compounds, including the flavone apigenin (19) and the isoflavonoid glyceollin (9). While the formation of lignin and most of the flavonoids is regulated by unknown endogenous mechanisms, the phytoalexin glyceollin is accumulated only after treatment of the cells with a polysaccharide fraction isolated from a plant pathogen or from certain other sources (9). The fact that lignin and flavonoids are synthesized simultaneously by the same cell culture is a particular advantage of this system which is enabling us to compare directly properties and regulation of the enzymes involved in the synthesis of the two classes of compounds.

PROPERTIES OF ENZYMES

General phenylpropanoid metabolism The first two enzymes of this sequence, phenylalanine ammonia-lyase and cinnamate 4-hydroxylase, were not studied in detail in soybean cells with respect to their catalytic properties. They are assumed to have similar, narrow substrate specificities as reported for these enzymes from other sources (7, 26, 59) and to provide the substrates for both the lignin and flavonoid pathways. On the other hand, we have extensively investigated the 4-coumarate:CoA ligase activity of soybean cell cultures. The activity could be separated into two isoenzymes, each with characteristic properties (34). The first species, designated as isoenzyme 1, has relatively low K_m and high V/K_m values only for the three typical substrates of the lignin pathway, 4-coumarate, ferulate, and sinapate, indicating that these acids are substrates of the enzyme in vivo. The other species, isoenzyme 2, has relatively high affinities for 4-coumarate and caffeate and does not activate sinapate. Hence in soybean cells general phenylpropanoid metabolism branches out at its last step into two 4-coumarate:CoA ligase isoenzymes with substrate specificities fitting the needs of the lignin and flavonoid pathways (see Figure 1). The assignment of an essential function in the biosynthesis of lignin to isoenzyme 1 is obvious from its substrate specificity. A similar, clear-cut relation between isoenzyme 2 and the formation of flavonoids cannot be deduced with certainty from the substrate specificity of the enzyme alone. Both enzymes could provide the substrate(s) for the flavonoid pathway, 4-coumaroyl-CoA, and perhaps caffeoyl-CoA (30, 60), with similar efficiency (34). The involvement of isoenzyme 2 in the flavonoid pathway was indirectly concluded from the great similarity of its substrate specificity with the substrate specificity of a 4-coumarate:CoA ligase from parsley cells (35), whose specific role in flavonoid biosynthesis is evident (see below). However, this conclusion does not rule out the possibility that isoenzyme 2 of 4-coumarate:CoA ligase in soybean cells, which does not activate highly methoxylated cinnamic acids (34), is also involved in lignin formation. It was shown that the relative amounts of building units of lignin bearing methoxy groups are low in young tissues and cell cultures even from plants producing the typical syringyl-type angiosperm lignin in fully differentiated, older tissue (18, 49).

Apart from differing considerably with respect to substrate specificity, the two 4-coumarate: CoA ligase isoenzymes from soybean cells provide a further possible mechanism for differential regulation of CoA ester synthesis. The activity of both ligases is inhibited efficiently, but to a different degree, by AMP, one of the reaction products. The degree of inhibition depends on the concentration of ATP, that is, on the energy charge of the system (34). Thus, the occurrence of isoenzymes differing in both catalytic and regulatory properties might indicate that 4-coumarate: CoA ligase represents a sensitive point of control for the rates at which the products of the general pathway are distributed among the subsequent, specific pathways of phenylpropanoid metabolism (21).

Lignin pathway The two enzymes of the lignin pathway catalyzing the reduction of CoA esters to the corresponding alcohols (see Figure 2) exhibit substrate specificities which strikingly match the pattern of products produced by the preceding enzyme, the lignin-specific ligase isoenzyme 1 (21). Both reductases, a cinnamoyl-CoA: NADPH oxidoreductase (74) and a cinnamoyl alcohol:NADP oxidoreductase (75), preferentially act on substrates bearing the same 4(OH)-, 3(OCH_3), 4 (OH)-, or 3,5(OCH_3)$_2$, 4(OH)-residues of the aromatic ring as provided by 4-coumaroyl-CoA, feruloyl-CoA, and sinapoyl-CoA. The substrate specificity of enzymes involved in the lignin-specific reactions includes the methyltransferase catalyzing the formation of ferulate and sinapate. This enzyme, SAM: caffeate O-methyltransferase, has a particularly high affinity for caffeate and 5-hydroxyferulate, the respective precursors of ferulate and sinapate (53).

Flavonoid pathways Much less information is available on the flavonoid pathways in soybean cell cultures. Attempts to demonstrate activity of the key enzyme FS (see Figure 3) in cell cultures have failed so far, although this enzyme activity is easily detected in soybean cotyledons (76). However, one important observation with respect to the differentiation between specific reactions of the flavonoid and other pathways of phenylpropanoid metabolism has been made with the cell cultures. We have demonstrated the existence of a flavonoid-specific SAM: *ortho*-dihydric phenol O-methyltransferase catalyzing the formation of 3'-O-methylated flavones and flavonols (54). The occurrence of this enzyme suggests that in soybean, as in parsley (see below), the introduction of methyl groups into flavonoid compounds takes place after formation of the flavonoid ring structure and not at the cinnamic acid stage, as in the case of methylated lignin precursors.

Thus, at least the initial sequence of reactions shown in Figure 3 for the flavonoid pathways in parsley might be applicable also to the formation of flavones and flavonols in soybean. However, it is unknown whether malonylation, the last step of flavonoid biosynthesis in parsley, occurs at all in soybean cell cultures.

It seems justified to conclude that the relative flow rates of substrates through the lignin pathway are controlled at least in part by the substrate specificities of all enzymes involved, beginning with the last enzyme of general phenylpropanoid metabolism. In addition to the resulting "quality control" in product formation by the catalytic properties of the enzymes, there is good evidence for a control of phenylpropanoid metabolism at the level of enzyme quantity, as suggested by the following observations.

CHANGES IN ENZYME ACTIVITIES Large changes in most of the enzyme activities related to phenylpropanoid metabolism occur during the growth cycle of soybean cell suspension cultures. In particular, the three enzymes of the first general pathway (including both isoenzymes of 4-coumarate:CoA ligase) increase simultaneously at least about tenfold within a period of only about 20 hr at the end of the linear growth phase of the culture (11). A subsequent rapid decline of all three enzyme activities occurs with the same high degree of coordination as observed for the increase, indicating that the expression of this sequence of enzymes is regulated by a common mechanism. The behavior of these enzymes as a regulatory unit was also demonstrated by their identical responses to light. The initial endogenously induced increase of all three enzyme activities was further stimulated almost twofold when cell cultures of the appropriate growth stage were irradiated for 15 hr (21).

A similar, but nevertheless clearly distinguishable pattern was observed for changes in the activities of the two known enzymes of the lignin pathway, cinnamoyl-CoA:NADPH oxidoreductase and cinnamyl alcohol:NADP oxidoreductase (17), and of the flavonoid-specific methyltransferase (52). Although the specific activities of these enzymes increased only about three- to fourfold, they exhibited characteristic maxima at the same growth stage as shown for the enzymes of general phenylpropanoid metabolism. But no significant changes during the growth cycle of a soybean cell culture were observed for the lignin-specific methyltransferase (11, 52).

When cultured soybean cells are treated with a glucan elicitor isolated from the fungal pathogen *Phytophthora megasperma* var. *sojae,* rapid accumulation is observed of the pterocarpanoid phytoalexin glyceollin, an isoflavonoid compound (see Figure 3) (9). Accumulation of glyceollin is preceded by a large increase in the enzyme activities of general phenylpropanoid metabolism (9; J. Ebel, unpublished results). Thus, these enzymes can be induced in soybean cells in several ways: by an unknown, endogenous mechanism operating at the end of the linear growth phase, by light, or by the action of an elicitor of microbial origin. It is unknown whether the different modes of induction lead to the formation of different phenylpropanoid compounds, except that glyceollin is produced only in the presence of an appropriate polysaccharide (9).

Cultured Parsley Cells

Parsley cells (*Petroselinum hortense* Hoffm.) in suspension culture have been used most extensively as a model system for studies of the enzymology and regulation of flavonoid biosynthesis. The formation of various flavone and flavonol glycosides in these cells is dependent on induction of the enzymes involved in their synthesis. A highly selective and efficient induction of these enzymes by light allowed various possible control mechanisms of flavonoid biosynthesis to be investigated in this system. All but two, or perhaps three, of about 16 enzymic steps leading to the formation of various flavone and flavonol glycosides from phenylalanine and several other compounds of intermediary metabolism (Figures 1 and 3) have been demonstrated in parsley cells. Most of these enzymes have been purified and characterized with respect to both catalytic properties and patterns of light-induced activity changes. The results strongly indicate that the enzymes of general phenylpropanoid metabolism and of the flavonoid glycoside pathways proper are controlled as separate, albeit closely related, regulatory units. The two sequences of enzymes were therefore designated as group I and group II, respectively. This terminology will be used throughout the following discussion.

PROPERTIES OF ENZYMES

General phenylpropanoid metabolism Of the three enzymes of group I, those at the beginning and the end of the sequence, PAL and 4-coumarate:CoA ligase, which seemed likely to have regulatory properties, were studied in detail. PAL is a tetrameric enzyme composed of four large subunits ($M_r = 83{,}000$). At least under certain conditions, this enzyme seems to be rate-limiting for the formation of flavonoid glycosides (see below). Its catalytic properties are very similar to those of phenylalanine ammonia-lyases from various other sources (77).

Only one species of 4-coumarate:CoA ligase could be demonstrated in parsley cells (35). The properties of this enzyme are similar to those of isoenzyme 2 from soybean cells (see above). These two ligases have very similar substrate specificities, including their common inability to activate sinapate (35). This lack of activation of sinapate, one of the most efficient precursors of lignin, might be an especially useful feature for differentiating these ligases from those which are preferentially involved in lignin biosynthesis. The involvement of the enzyme from irradiated parsley cells in flavonoid biosynthesis is evident from its coordinated induction by light with all other enzymes of the general and the flavonoid glycoside pathways (see below). It seems possible, therefore, that many ligases from other species possessing similar substrate specificities have an important regulatory function in directing metabolites from general phenylpropanoid metabolism towards flavonoid biosynthesis. Another common feature of the two similar 4-coumarate:CoA ligases from parsley and soybean cells is their strong inhibition by AMP, which acts as a competitive inhibitor with respect to ATP. Thus the activity of both enzymes might be regulated in vivo by the energy charge of the cells (34, 35).

Enzymes of the flavone and flavonol glycoside pathways The properties of several of the enzymes of group II have been summarized (16, 22). More recent reports on FS (30, 37, 38, 60) and UDP-apiose synthase (46) have revealed that these enzymes, which catalyze rather complex reactions, fall into the same molecular weight range (M_r = ca 50,000–80,000) and are "soluble enzymes" as are all other enzymes of this pathway. The same applies to chalcone isomerase and to two malonyltransferases which are specific for flavonoid glycosides as substrates but have not been investigated in detail in parsley cells. While the role of the isomerase in flavonoid biosynthesis in parsley is not clear (see Figure 3), the two malonyltransferases are specific for the malonylation of flavone and flavonol glycosides in the glucose moiety at C-7 and of flavonol glycosides at C-3, respectively (U. Matern, unpublished results; cf Figure 3).

The enzymes of group II have generally rather narrow substrate specificities (21, 22) which are in close agreement with the substitution pattern of the flavonoids isolated from irradiated parsley cells (36). An interesting observation was made with acetyl-CoA carboxylase. This enzyme which is also related to fatty acid biosynthesis is co-induced with the enzymes of the flavonoid glycoside pathways. Acetyl-CoA carboxylase is therefore a true member of group II, in contrast to several other enzymes which also provide compounds of intermediary metabolism as substrates for flavonoid glycoside biosynthesis (e.g. *S*-adenosylmethionine, UDP-glucose, UDP-glucoronic acid) but are not co-induced with the enzymes of group II (10). Another interesting phenomenon was observed with the second enzyme of group II, FS. This enzyme is specific for 4-coumaroyl-CoA as substrate at the pH optimum around pH 8 (30), whereas equal amounts of 4-coumaroyl-CoA and caffeoyl-CoA (see Figure 1) are converted to the corresponding flavanone at pH 6.5 (60). It was speculated that, besides the substrate concentration, the pH at the site of the synthase reaction could play a role in determining the relative rates of incorporation of 4-coumarate and caffeate into flavonoids (60). Feruloyl-CoA (see Figure 1) was not an efficient substrate for flavanone synthesis (30, 60), suggesting that methylation takes place after this stage. This conclusion is in line with the occurrence of a flavonoid-specific 3'-0-methyltransferase with a preference for the flavone luteolin and the flavonol quercetin as substrates (10, 22). The only so far unexplained exception to the otherwise observed specificity of the enzymes of group II for their respective substrates and products related to flavonoid biosynthesis is UDP-apiose synthase. In vitro, this enzyme catalyzes the formation of two products, UDP-apiose and UDP-xylose, at a constant ratio throughout a purification to apparent homogeneity (46). While apiose is a constituent of several flavone glycosides in parsley cells, xylose was not detected in these compounds (36).

CHANGES IN ENZYME ACTIVITIES Upon dilution of a parsley cell suspension culture either into fresh medium or into water, all three enzymes of group I increase greatly (5- to 50-fold) in activity for about 12–15 hr and then return to the original low level (24, 25). The changes in activity occur in a highly coordinated manner, including an initial apparent lag of about 2 hr. The lag preceding detectable increases in enzyme activities is not a general phenomenon in the cell cultures applying to all enzymes and, for example, was not observed for nitrate reductase (J. Vieregge, unpublished results). The enzymes of group II are not induced upon dilution of the cultures, unless the cells are irradiated simultaneously (see below). Flavonoids are therefore not produced by diluted parsley cell cultures in the dark. Since lignin is not formed under these conditions either (28), it can only be speculated that the increased activities of the enzymes of group I lead to the formation of yet another class of compounds such as esters of (substituted) cinnamic acid(s). This would be in agreement with the isolation from the cultures of a caffeoyl ester whose complete structure has not been elucidated (H. Kühnl, unpublished results). The induction in diluted cultures of the enzymes of the general pathway independent of the enzymes of the flavone and flavonol glycoside pathways is used as one important criterion for the classification for the two sequences of enzymes into group I and group II, respectively.

Both groups of enzymes are induced simultaneously by irradiation of parsley cells with UV light. Although each enzyme reaches a peak in activity at a slightly different time after the onset of induction, the general shapes of the curves for activity changes are similar for all of the enzymes under continuous irradiation. The peak positions vary between about 15–40 hr (10, 23) and depend greatly on the growth stage of the cell culture (20).

At a given growth stage, the enzymes of group I reach their maximal activities several hours earlier than the enzymes of group II (10, 22, 23). Another characteristic difference between the two groups is the length of the lag period preceding detectable increases in the enzyme activities (about 2 hr for group I and somewhat longer for group II; see below).

Not only the peak position, but also the extent to which each individual enzyme activity can be induced by light varies greatly with the growth stage of the parsley cell culture. While at least two of the three enzymes of group I, PAL (20, 77) and 4-coumarate : CoA ligase, are maximally induced at the end of the linear growth phase, all five measured enzymes of group II—acetyl-CoA carboxylase, FS, chalcone isomerase, UDP-apiose synthase, and malonyltransferase—exhibit maximal inducibility much earlier during the log phase of growth of the culture (J. Ebel, B. Egin-Bühler, S. E. Gardiner, W. Heller, K. H. Knobloch, U. Matern, and J. Vieregge, unpublished results). This additional striking difference between the two groups, besides differences in the response to culture dilution and in the lengths of lag periods in irradiated cells, is taken as a third criterion indicating that the activities of the enzymes of groups I and II are separately regulated. The enzymes of the lignin pathway are not induced by irradiation of parsley cells (28).

CHANGES IN mRNA ACTIVITIES For three of the light-induced enzymes, PAL, FS, and UDP-apiose synthase, the changes in activity were shown to be caused by corresponding changes in mRNA activity (63, 65; S. E. Gardiner, J. Schröder, U. Matern, unpublished results). Essentially the same result was obtained for the changes in PAL activity in diluted cell cultures (64). A detailed analysis of the light-induced changes in PAL and FS mRNA activities (65) allows the following conclusions to be drawn regarding the cause of the differences in the lengths of the lag periods and the times required to reach maximal activities for each of these two representatives of groups I and II, respectively.

Studies at the mRNA level and labeling experiments in vivo revealed that the lag observed for detectable changes in enzyme activity was not a true lag, but rather a matter of assay sensitivity. Although the rate of PAL synthesis in vivo (3), and probably also the mRNA activity in vitro (65), increases with no detectable lag in irradiated cells, the increase is exponential during the initial period of induction and, therefore, causes only a very slight increase in enzyme activity. Until about 2 hr after the onset of irradiation, this increase is within the limits of experimental error (3). FS mRNA activity also increases slowly, perhaps exponentially, at the beginning, but starts from a much lower level relative to the highest value than PAL mRNA activity (65). A lag of about 4 hr preceding detectable increases in FS activity could be explained by these results.

Subsequent to the period of exponential increase in mRNA activity and to the resulting apparent lag for the increase in enzyme activity, both mRNA and catalytic activities of these two enzymes increase rapidly during a period of several hours. However, the points of maximal mRNA activity differ. This difference is not only observed under continuous irradiation, but also after short-term irradiation for 2.5 hr, when the increases in PAL and FS mRNA activities continue for a few hours even in the dark period following irradiation. The conclusion from these results is that the lengths of the periods for which PAL and FS mRNAs are induced by light are different, and perhaps this difference holds true for all of the enzymes of groups I and II (65).

RELATION TO QUALITY AND QUANTITY OF LIGHT In contrast to the usual mechanism of action of red light via phytochrome in the induction of the enzymes of phenylpropanoid metabolism (62), parsley cells require an "activating" irradiation with UV light (70, 72). Although the action spectrum for the UV effect was determined (70), the UV receptor has not been identified. After (or in addition to) irradiation with UV light, parsley cells show the typical phytochrome-mediated induction of the enzymes of groups I and II, as demonstrated by induction/reversion experiments with red and far-red light (70, 72). The dose/response relationship for enzyme induction with UV light is linear, at least in the case of PAL and chalcone isomerase, up to about one third of the maximal effect (N. Duell, unpublished results).

THEORETICAL CALCULATIONS A particularly useful method for estimating the significance of the various possible control mechanisms in the phenylpropanoid pathways was the comparison of experimentally derived data with the results of simple mathematical calculations. In the case of flavonoid biosynthesis in parsley cells, such calculations were carried out at two different stages, at the stage of mRNA translation and at the stage of flavonoid glycoside accumulation.

Rates of enzyme synthesis If the amount of enzyme activity is exclusively determined by the rates of synthesis and degradation of the enzyme, the simple equation

$$dE(t)/dt = {}^{o}k_s(t) - {}^{1}k_d \cdot E(t) \qquad 1.$$

can be applied, provided that synthesis and degradation follow zero-order and first-order kinetics, respectively, and that the rate of degradation is constant. Using a proper transformation of Equation 1 (3), the expected light-induced changes of enzyme activity (E) with time (t) were calculated from the measured changes in the rate of enzyme synthesis (${}^{o}k_s$) and the rate of enzyme degradation ($^{1}k_d$). The results were in close agreement with the corresponding experimental data and therefore indicated that at least in the two cases of PAL and FS the changes in enzyme activity were due to corresponding changes in mRNA activity (3, 63, 65). This conclusion implies that no parameter influencing enzyme activity, other than mRNA activity, varied significantly during enzyme induction. The apparent half-lives of the two enzymes which were used to calculate the $^{1}k_d$ values were of the order of about 5–10 hr and remained approximately constant throughout a given period of enzyme induction (3, 20, 65).

Product accumulation A second simple mathematical approach was to calculate the amounts of products formed by the various light-induced enzymes under the presupposition that product inhibition (or substrate activation) did not play a significant role. For this purpose, the curves obtained for changes in the enzyme activities were integrated using the equation

$$P(t) = \int_{o}^{t} E(\tau)\mathrm{d}\tau \qquad 2.$$

where P is the amount of product, changing with time (t, τ) as enzyme activity (E) changes in response to irradiation of the cells. The results, obtained with parsley cells at the beginning of the stationary phase of a culture, suggest that PAL is the rate-limiting enzyme in flavonoid glycoside biosynthesis under these conditions (23). Furthermore, the close correspondence between the curve for the integrated values of PAL activity and the accumulation curve for flavonoid glycosides suggests that these compounds do not exert end-product inhibition under the experimental conditions used. However, two important aspects concerning these results must be borne in mind. First, a prerequisite for such calculations is the availability of reliable data on the rate(s) of degradation of the product(s). Under the growth conditions mentioned above, the rate of flavonoid degradation in parsley cells is negligible (2). Thus, no correction of the calculated data is necessary. Second, PAL is by no means the rate-limiting enzyme of flavonoid biosynthesis under all conditions of induction. This is best illustrated by the fact that no flavonoids at all are produced when PAL and the other enzymes of group I are induced upon dilution of parsley cell cultures (25).

A schematic representation of the known steps of control in the biosynthesis of flavonoid glycosides in irradiated parsley cells is depicted in Figure 4. The scheme shows the normalized curves for light-induced changes in mRNA and catalytic activities of PAL and for flavonoid accumulation in an early stationary-phase cell culture. For comparison, the curve for changes in FS activity under the same conditions is included. Integration

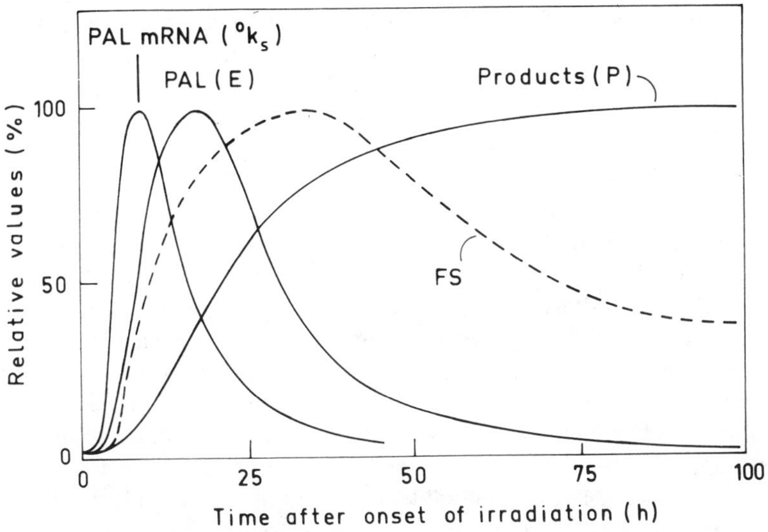

Figure 4 Schematic representation of changes in PAL mRNA, PAL and FS activities, and of the curve for flavonoid glycoside accumulation in continuously irradiated parsley cells from a late stage of culture growth [from (23)]. See text for methods of calculating solid curves from one another.

of this curve and of all other curves for changes in the enzyme activities of group II does not result in a similar coincidence with the accumulation curve for flavonoids as observed for PAL (23).

Cultured Haplopappus Cells

Cultured cells of *Haplopappus gracilis* produce cyanidin 3-glucosides in short wavelength light (69). Work on the biosynthesis of cyanidin (Figure 3) with such cells has been covered in an earlier review (22). In callus cultures grown on agar, anthocyanin formation can be induced by blue and near UV light, and the action spectrum was found to have two peaks, one at 438 nm and the other at 372 nm (40). With cell suspension cultures, only UV light below 345 nm stimulated anthocyanin synthesis under the conditions used (73). A linear relationship between UV dose and flavonoid accumulation, as found with parsley cell cultures (72), was not observed with *H. gracilis* cells as only continuous irradiation with high doses of UV was effective. Cells transferred to fresh medium prior to irradiation gave a much larger UV response. Large and concomitant increases in the activities of PAL, FS, and chalcone isomerase were observed under continuous UV light (73). FS from *Haplopappus* cell cultures was partially purified and shown to have different pH optima for the synthesis of naringenin (pH 8) and eriodictyol (pH 6.5–7) (60). It is possible that 4-coumaroyl-CoA and caffeoyl-CoA are both substrates for the synthase and that the pH at the site of the synthase reaction could play a role in determining how much naringenin or eriodictyol is formed (see above for FS from parsley cells).

Intact Plants and Plant Tissues

Disks isolated from swede root (*Brassica napo-brassica*) and aged for 24 hr develop a lignin-like substance. Formation of this material can be stimulated by treatment of the disks with ethylene during aging, and this system has been used to study the enzymes involved in the lignin pathway (56). The enzymes catalyzing the activation and reduction of ferulic acid to coniferyl alcohol have very similar properties to the corresponding enzymes from *Forsythia, Salix alba,* and cell cultures of soybean (58). Coniferyl alcohol: NADP oxidoreductase was separated on a DEAE-cellulose column into two active fractions. One fraction had much lower K_m values for coniferyl alcohol (2 μmol/1) and coniferaldehyde (2 μmol/1) than the other fraction (550 and 140 μmol/1, respectively). Two separate activity peaks were also found on a DEAE-cellulose column for feruloyl-CoA : NADPH oxidoreductase but they had identical K_m values for feruloyl-CoA. In both cases it has not been established with certainty that these fractions belong to separate enzyme entities.

The activity changes of nine enzymes involved in the formation of lignin were followed during aging of the disks in the presence and absence of ethylene (58). Peroxidase, cinnamyl alcohol dehydrogenase, and phenylalanine transaminase showed very little change in activity. PAL, cinnamic acid 4-hydroxylase and 4-coumarate : CoA ligase increased about 20- to 30-fold on aging and were strongly stimulated by ethylene. In contrast, methyltransferase, shikimate dehydrogenase, and feruloyl-CoA reductase increased two- to threefold on aging and were relatively insensitive to ethylene.

Young xylem tissue from *Forsythia suspensa* was used for an extensive purification of 4-coumarate : CoA ligase, cinnamoyl-CoA : NADPH oxidoreductase, and cinnamyl alcohol : NADP oxidoreductase (18). The substrate specificities of the enzymes (18) are similar to the substrate specificities of 4-coumarate : CoA ligase isoenzyme 2 and the two oxidoreductases, respectively, from soybean cells (see above).

Experiments with labeled coniferin (coniferyl alcohol β-D-glucoside) have shown that this compound can act as a lignin precursor in a variety of species (4). However, the role of coniferin in lignification remains uncertain. In 100-day-old spruce (*Picea abies*) seedlings containing approximately 50–60 μg coniferin/g fresh weight of tissue, a turnover of coniferin with a half-life of 60–120 hr was determined by pulse labeling experiments

(44). β-Glucosidase activity which was held responsible for the hydrolysis of coniferin, a necessary prerequisite for polymerization, has been detected adjacent to the cambial zone of *Araucaria excelsa* by a histochemical method using indican as substrate (13). Spruce seeds contain a soluble β-glucosidase which is active with 4-nitrophenyl β-D-glucoside but does not hydrolyze coniferin. However, immediately after germination a glucosidase with high activity toward coniferin is detectable in cell wall fractions of roots and hypocotyls (45). The glucosidase activity from the hypocotyl cell wall fraction could be solubilized with 0.6 mol/1 NaCl, and two glucosidases with respectively a high and a low activity ratio for coniferin/4-nitrophenyl-glucoside were present in this extract. The enzyme with high activity toward coniferin (glucosidase 1) also catalyzes the hydrolysis of syringin (sinapyl alcohol β-D-glucoside) with good efficiency. A histochemical localization of this glucosidase with the indican method was not possible because the enzyme has only very weak activity with indican as substrate. Similar properties were reported for β-glucosidase isoenzymes from chick-pea (*Cicer arietinum*) cell cultures (29). The relative activity with indican was only about 1% of that found with coniferin. Immunofluorescent labeling of the glucosidase in situ using transsections of spruce hypocotyls indicate that the enzyme is localized at the inner layer of the secondary wall (43). The presence in spruce seedlings of cell-wall–bound glucosidase capable of hydrolyzing coniferin would be consistent with its participation in lignification.

Enzymes for the synthesis of coniferin from coniferyl alcohol and UDP-glucose have also been found in cambial sap of spruce (B. Egin, unpublished results), in cell cultures of Paul's scarlet rose (32), and in *Forsythia suspensa* (31). However, quantitative assessment of the role of coniferin in lignin biosynthesis is not yet possible. Large increases in the activities of cinnamoyl-CoA : NADPH oxidoreductase and cinnamyl alcohol : NADP oxidoreductase were observed during the development of spruce seedlings. Maximal enzyme activities were reached approximately at day 12 after sowing, and then declined to a fairly constant level. The appearance of the enzyme activities in the seedlings can be correlated with the beginning of lignification in the vascular bundles at the transition zone of root and hypocotyl (T. Lüderitz, unpublished results).

Resistance of soybean (*Glycine max* L.) seedlings to the fungal pathogen *Phytophtora megasperma* var. *sojae* is due in part to the accumulation in infected tissue of the pterocarpanoid phytoalexin glyceollin (5) (Figure 3) and two of its isomers (41). The accumulation of glyceollin in soybean seedlings can be induced not only by the pathogen but also by a glucan (elicitor) purified from the mycelial walls of *Phytophtera megasperma* (1). The stimulation of PAL, chalcone isomerase, and peroxidase activities in resistant or susceptible soybean cultivars after wounding and inoculation of hypocotyls with the pathogen and a similar response in wounded controls with the pathogen omitted has been reported (51). This led the authors to conclude that the activity increases were caused primarily by wounding and that they have no causal role in the stimulation of glyceollin biosynthesis in inoculated resistant plants. In contrast to these results, large increases and subsequent decreases in the activities of PAL, cinnamic acid 4-hydroxylase, 4-coumarate: CoA ligase, and FS in elicitor-induced wounded soybean cotyledons were observed, whereas wounding alone did not yield significant changes in the activities of these enzymes (76; J. Ebel, unpublished results). PAL activity and glyceollin were induced with similar relative efficiency at progressively increasing rates when the amount of elicitor applied to wounded tissue was raised from 1 μg/ml to 1 mg/ml (J. Ebel, unpublished results).

Anthers of *Tulipa* cv. "Apeldoorn" were shown to be a particularly useful system for studies of the correlation between developmental changes in enzyme activities and in the rates of accumulation and interconversion of various phenylpropanoids. PAL and 4-coumarate: CoA ligase reached their maximal activities early during microsporogenesis (67). The activities of FS (R. Wiermann, unpublished results), chalcone isomerase, and methyltransferase (67) reached their peaks at a later stage. The peaks in the enzyme activities could be correlated with defined stages of cinnamate ester, chalcone, and flavonol accumulation. The tapetum is thought to play an important role in the supply of enzymes necessary for the biosynthesis of flavonoids in the anther (67).

Another system of particular interest is that of leaves and flowers from *Petunia hybrida.* The leaves produce all three types of phenylpropanoid compounds, lignin, flavonoids (flavonol glycosides), and cinnamate esters. The most abundant cinnamate esters are caffeoyl quinic (chlorogenic) and feruloyl quinic acids (55). Enzymology and regulation of the biosynthesis of the various phenylpropanoids were studied, including the separation and partial purification of three isoenzymes of 4-coumarate: CoA ligase and of a hydroxycinnamoyl-CoA: quinate hydroxycinnamoyltransferase. The involvement of the three 4-coumarate: CoA ligases in the biosynthesis of lignin, flavonoids and cinnamate esters, respectively, was concluded from their different properties with respect to feedback inhibition (55). In illuminated buckwheat (*Fagopyrum esculentum*) seedlings, the hydroxycinnamoyl-CoA: quinate (shikimate) hydroxycinnamoyltransferase activity was shown to be regulated concomitantly with the enzymes of general phenylpropanoid metabolism (68).

A second important aspect of research with *Petunia hybrida* is the use of genetically defined mutants for studies of flavonoid biosynthesis (33).

Callistephus chinensis (39) and *Zea mays* (8) have been used for the same purpose. To mention only one interesting result, a correlation between the accumulation of a chalcone in the blossoms from several recessive genotypes of *Callistephus chinensis* and a deficiency of chalcone isomerase activity seemed to indicate that the chalcone is the first flavonoid intermediate in this plant (39). This would be in contrast to the results obtained with parsley cell cultures (see above).

A final very important point concerning phenylpropanoid biosynthesis needs to be emphasized. The synthesis not only of lignin (15, 18) but also of flavonoids and cinnamate esters is strongly related to both the developmental stage of the whole organism and compartmentation between and within cells of a particular organ. Space does not permit dealing with this aspect comprehensively in this article. Only two recent examples should be mentioned. In mustard cotyledons, (*Sinapis alba* L.), light-induced PAL is regulated differently in the upper and lower epidermis and is involved in anthocyanin production in one case and in flavonol production in the other (71). Butt & Wilkinson (6; and unpublished results) observed a tissue-specific induction of two 4-coumarate:CoA ligases upon illumination of etiolated pea seedlings (*Pisum sativum*). One enzyme was assumed to be involved in lignin formation in the stems, and a different form of the enzyme was ascribed to flavonoid biosynthesis in the buds. The light-induced changes in all three enzyme activities of general phenylpropanoid metabolism occurred in a strictly parallel manner (6). The possibility of intracellular compartmentation in phenylpropanoid metabolism has recently been discussed (66).

CONCLUSIONS AND OUTLOOK

Nearly all of the enzymes involved in the biosynthesis of lignin, flavone, and flavonol glycosides, and the biosynthetically related class of cinnamate esters, have been purified and characterized to a considerable extent. Their properties in vitro, particularly their rather narrow substrate specificities, are generally in close agreement with the pattern of phenylpropanoid compounds formed by the specific organism or cell type. With respect to regulation at the level of substrate specificity, the enzyme at the branching point between general phenylpropanoid metabolism and the specific pathways, 4-coumarate:CoA ligase, seems to play a pivotal role. This enzyme occurs frequently in the form of isoenzymes with substrate specificities strikingly fitting the requirements of the respective subsequent pathways.

The occurrence of all three classes of phenylpropanoid compounds, lignin, flavonoids, and cinnamate esters, is closely related to the developmental stage and environmental conditions of the producing organism or cell.

Accordingly, the enzymes of the individual pathways change greatly in activity under appropriate conditions and can be induced up to over 100-fold by various different stimuli such as light, hormones, plant pathogens etc. Drastic changes in the rates of enzyme synthesis are therefore among the most important control mechanisms in the biosynthesis of lignin and flavonoids. Only scarce information is available about possible enzymic controls at the level of product or end-product inhibition in vivo. Although such inhibitory effects were observed in some cases in vitro, these might be due partly to a general inhibition of enzymes by phenolic substances. Furthermore, intracellular compartmentation is likely to separate completely most enzymes and end-products of lignin and flavonoid biosynthesis in vivo.

Much of the work on enzymology and regulation of lignin and flavonoid biosynthesis has been done with plant cell suspension cultures. This material offers great advantages over intact plants or whole plant tissues, and some of the remaining problems, such as the elucidation of the enzymic reactions leading to the formation of anthocyanins and isoflavonoids, perhaps even the mechanism of polymerization involved in lignin biosynthesis, might advantageously be tackled by using appropriate cell cultures. On the other hand, since developmental and environmental conditions of whole plant tissues have important regulatory functions, the entire multitude of endogenous and exogenous factors controlling the biosynthesis of lignin and flavonoids can only be studied with intact plants. However, this task seems to be a long way off, and cell cultures will probably render good service in the near future for further studies of enzymic controls in the biosynthesis of lignin and flavonoids.

Acknowledgments

We thank Drs. Jürgen Ebel and Susan E. Gardiner for their helpful comments on the manuscript. Research cited from our laboratories was supported by Deutsche Forschungsgemeinschaft (partly through SFB 46) and Fonds der Chemischen Industrie. This support is gratefully acknowledged.

Literature Cited

1. Ayers, A. R., Ebel, J., Valent, B., Albersheim, P. 1976. Host-pathogen interactions. X. Fractionation and biological activity of an elicitor isolated from the mycelial walls of *Phytophthora megasperma* var. *sojae*. *Plant Physiol.* 57: 760–65
2. Barz, W. 1977. Catabolism of endogenous and exogenous compounds by plant cell cultures. In *Plant Tissue Culture and Its Bio-technological Application,* ed. W. Barz, E. Reinhard, M. H. Zenk, pp. 153–77. Berlin: Springer. 419 pp.
3. Betz, B., Schäfer, E., Hahlbrock, K. 1978. Light-induced phenylalanine ammonia-lyase in cell suspension cultures of *Petroselinum hortense.* Quantitative comparison of rates of synthesis and degradation. *Arch. Biochem. Biophys.* 190:126–35
4. Brown, S. A. 1966. Lignins. *Ann. Rev. Plant Physiol.* 17:223–44
5. Burden, R. S., Bailey, J. A. 1975. Structure of the phytoalexin from soybean. *Phytochemistry* 14:1389–90
6. Butt, V. S., Wilkinson, E. M. 1978. Enzyme changes accompanying the induction of lignin and flavonoid synthesis in pea shoots by light. *Abstr. 12th FEBS Meet.,* Dresden, No. 0307
7. Camm, E. L., Towers, G. H. N. 1973. Phenylalanine ammonia-lyase. *Phytochemistry* 12:961–73
8. Dooner, H. K., Nelson, O. E. 1977. Genetic control of UDPglucose: flavonol 3-O-glucosyltransferase in the endosperm of maize. *Biochem. Genet.* 15: 509–19
9. Ebel, J., Ayers, A. R., Albersheim, P. 1976. Host-pathogen interactions. XII. Response of suspension-cultured soybean cells to the elicitor isolated from *Phytophthora megasperma* var. *sojae,* a fungal pathogen of soybeans. *Plant Physiol.* 57:775–79
10. Ebel, J., Hahlbrock, K. 1977. Enzymes of flavone and flavonol glycoside biosynthesis. Coordinated and selective induction in cell suspension cultures of *Petroselinum hortense. Eur. J. Biochem.* 75:201–9
11. Ebel, J., Schaller-Hekeler, B., Knobloch, K.-H., Wellmann, E., Grisebach, H., Hahlbrock, K. 1974. Coordinated changes in enzyme activities of phenylpropanoid metabolism during the growth of soybean cell suspension cultures. *Biochim. Biophys. Acta* 362: 417–24
12. Freudenberg, K., Neish, A. C. 1968. *Constitution and Biosynthesis of Lignin.* Berlin: Springer. 129 pp.
13. Freudenberg, K., Reznik, H., Boesenberg, H., Rasenack, D. 1952. Das an der Verholzung beteiligte Fermentsystem. *Chem. Ber.* 85:641–47
14. Gamborg, O. L. 1966. Aromatic metabolism in plants. II. Enzymes of the shikimate pathway in suspension cultures of plant cells. *Can. J. Biochem.* 44:791–99
15. Grisebach, H. 1977. Biochemistry of lignification. *Naturwissenschaften* 64: 619–25
16. Grisebach, H., Hahlbrock, K. 1974. Enzymology and regulation of flavonoid and lignin biosynthesis in plants and plant cell suspension cultures. *Recent Adv. Phytochem.* 8:21–52
17. Grisebach, H., Wengenmayer, H., Wyrambik, D. 1977. Cinnamoyl-CoA: NADPH oxidoreductase and cinnamyl alcohol dehydrogenase: Two enzymes of lignin monomer biosynthesis. In *Pyridine Nucleotide-Dependent Dehydrogenases,* ed. H. Sund, pp. 458–71. Berlin: de Gruyter. 513 pp.
18. Gross, G. G. 1977. The structure, biosynthesis, and degradation of wood. *Recent Adv. Phytochem.* 11:141–84
19. Hahlbrock, K. 1972. Isolation of apigenin from illuminated cell suspension cultures of soybean, *Glycine max. Phytochemistry* 11:165–66
20. Hahlbrock, K. 1976. Regulation of phenylalanine ammonia-lyase activity in cell suspension cultures of *Petroselinum hortense.* Apparent rates of enzyme synthesis and degradation. *Eur. J. Biochem.* 63:137–45
21. Hahlbrock, K. 1977. Regulatory aspects of phenylpropanoid biosynthesis in cell cultures. See Ref. 2, pp. 95–111
22. Hahlbrock, K., Grisebach, H. 1975. Biosynthesis (Chapter 9). In *The Flavonoids,* ed. J. B. Harborne, T. J. Mabry, H. Mabry, pp. 866–915. London: Chapman & Hall. 1204 pp.
23. Hahlbrock, K., Knobloch, K.-H., Kreuzaler, F., Potts, J. R. M., Wellmann, E. 1976. Coordinated induction and subsequent activity changes in two groups of metabolically interrelated enzymes. Light-induced synthesis of flavonoid glycosides in cell suspension cultures of *Petroselinum hortense. Eur. J. Biochem.* 61:199–206
24. Hahlbrock, K., Schröder, J. 1975. Specific effects on enzyme activities upon dilution of *Petroselinum hortense* cell

cultures into water. *Arch. Biochem. Biophys.* 171:500–6

25. Hahlbrock, K., Wellmann, E. 1973. Light-independent induction of enzymes related to phenylpropanoid metabolism in cell suspension cultures of parsley. *Biochim. Biophys. Acta* 304: 702–6
26. Hanson, K. R., Havir, E. A. 1972. The enzymic elimination of ammonia. In *The Enzymes,* ed. P. D. Boyer, 7:75–166. New York: Academic. 959 pp. 3rd ed.
27. Higuchi, T. 1971. Formation and biological degradation of lignins. *Adv. Enzymol.* 34:207–83
28. Hösel, W., Borgmann, E. 1978. Development of lignin biosynthetic enzymes during lignification of cell suspension cultures of *Petroselinum hortense. Abstr. 12th FEBS Meet.,* Dresden, No. 0308
29. Hösel, W., Gurholt, E., Borgmann, E. 1978. Characterization of β-glucosidase isoenzymes possibly involved in lignification from chick pea (*Cicer arietum* L.) cell suspension cultures. *Eur. J. Biochem.* 84:487–92
30. Hrazdina, G., Kreuzaler, F., Hahlbrock, K., Grisebach, H. 1976. Substrate specificity of flavanone synthase from cell suspension cultures of parsley and structure of release products in vitro. *Arch. Biochem. Biophys.* 175: 392–99
31. Ibrahim, R. K. 1977. Glucosylation of lignin precursors by uridine diphosphate glucose : coniferyl alcohol glucosyltransferase in higher plants. *Z. Pflanzenphysiol.* 85:253–62
32. Ibrahim, R. K., Grisebach, H. 1976. Purification and properties of UDP-glucose : coniferyl alcohol glucosyltransferase from suspension cultures of *Paul's Scarlet Rose. Arch. Biochem. Biophys.* 176:700–8
33. Kho, K. F. F., Bolsman-Louwen, A. C., Vuik, J. C., Bennink, G. J. H. 1977. Anthocyanin synthesis in a white flowering mutant of *Petunia hybrida.* II. Accumulation of dihydroflavonol intermediates in white flowering mutants; uptake of intermediates in isolated corollas and conversion into anthocyanins. *Planta* 135:109–18
34. Knobloch, K.-H., Hahlbrock, K. 1975. Isoenzymes of *p*-coumarate : CoA ligase from cell suspension cultures of *Glycine max. Eur. J. Biochem.* 52:311–20
35. Knobloch, K.-H., Hahlbrock, K. 1977. 4-Coumarate : CoA ligase from cell suspension cultures of *Petroselinum hortense* Hoffm. Partial purification, substrate specificity, and further properties. *Arch. Biochem. Biophys.* 184:237–48
36. Kreuzaler, F., Hahlbrock, K. 1973. Flavonoid glycosides from illuminated cell suspension cultures of *Petroselinum hortense. Phytochemistry* 12:1149–52
37. Kreuzaler, F., Hahlbrock, K. 1975. Enzymic synthesis of an aromatic ring from acetate units. Partial purification and some properties of flavanone synthase from cell suspension cultures of *Petroselinum hortense. Eur. J. Biochem.* 56:205–13
38. Kreuzaler, F., Hahlbrock, K. 1975. Enzymatic synthesis of aromatic compounds in higher plants. Formation of Bisnoryangonin (4-hydroxy-6-[4-hydroxystyryl]2-pyrone) from *p*-coumaroyl-CoA and malonyl-CoA. *Arch. Biochem. Biophys.* 169:84–90
39. Kühn, B., Forkmann, G., Seyffert, W. 1978. Genetic control of chalcone-flavanone isomerase activity in *Callistephus chinensis. Planta* 138:199–203
40. Lackmann, I. 1971. Wirkungsspektren der Anthocyansynthese in Gewebekulturen und Keimlingen von *Haplopappus gracilis. Planta* 98:258–69
41. Lyne, R. L., Mulheim, L. J., Leworthy, D. P. 1976. New pterocarpinoid phytoalexins of soybean. *J. Chem. Soc. Chem. Commun.* 497–98
42. Macheix, J. J. 1974. *Les esters hydroxycinnamiques de la pomme: identification, variations au cours de la croissance du fruit et métabolisme.* PhD thesis. Univ. Paris, Paris, France
43. Marcinowski, S., Falk, H., Hammer, D. W., Hoyer, B., Grisebach, H. 1979. Appearance and localization of a β-glucosidase hydrolyzing coniferin in spruce (*Picea abies*) seedlings. *Planta* 144:161–65
44. Marcinowski, S., Grisebach, H. 1977. Turnover of coniferin in spruce seedlings. *Phytochemistry* 16:1665–67
45. Marcinowski, S., Grisebach, H. 1978. Enzymology of lignification. Cell-wall-bound β-glucosidase for coniferin from spruce (*Picea abies*) seedlings. *Eur. J. Biochem.* 87:37–44
46. Matern, U., Grisebach, H. 1977. UDP-Apiose/UDP-xylose synthase. Subunit composition and binding studies. *Eur. J. Biochem.* 74:303–12
47. Moore, T. S. Jr. 1973. An extracellular macromolecular complex from the surface of soybean suspension cultures. *Plant Physiol.* 51:529–36
48. Moustafa, E., Wong, E. 1967. Purification and properties of chalcone-flava-

none isomerase from soya bean seed. *Phytochemistry* 6:625–32
49. Nakamura, Y., Fushiki, H., Higuchi, T. 1974. Metabolic differences between gymnosperms and angiosperms in the formation of syringyl lignin. *Phytochemistry* 13:1777–84
50. Nimz, H., Ebel, J., Grisebach, H. 1975. On the structure of lignin from soybean cell suspension cultures. *Z. Naturforsch.* 30c:442–44
51. Partridge, J. E., Keen, N. T. 1977. Soybean phytoalexins: Rates of synthesis are not regulated by activation of initial enzymes in flavonoid biosynthesis. *Phytopathology* 67:50–55
52. Poulton, J., Grisebach, H., Ebel, J., Schaller-Hekeler, B., Hahlbrock, K. 1976. Two distinct S-adenosyl-L-methionine: 3,4-dihydric phenol 3-O-methyltransferases of phenylpropanoid metabolism in soybean cell suspension cultures. *Arch. Biochem. Biophys.* 173: 301–5
53. Poulton, J., Hahlbrock, K., Grisebach, H. 1976. Enzymic synthesis of lignin precursors. Purification and properties of the S-adenosyl-L-methionine: caffeic acid 3-O-methyltransferase from soybean cell suspension cultures. *Arch. Biochem. Biophys.* 176:449–56
54. Poulton, J., Hahlbrock, K., Grisebach, H. 1977. O-Methylation of flavonoid substrates by partially purified enzyme from soybean cell suspension cultures. *Arch. Biochem. Biophys.* 180:543–49
55. Ranjeva, R. 1978. *La biosynthèse des derivés hydroxycinnamiques chez Petunia hybrida (Vilmour): Un exemple de diversification biochimique.* PhD thesis. Univ. Paul Sabatier, Toulouse, France. 129 pp.
56. Rhodes, M. J. C., Hill, A. C. R., Wooltorton, L. S. C. 1976. Activity of enzymes involved in lignin biosynthesis in swede root disks. *Phytochemistry* 15: 707–10
57. Rhodes, M. J. C., Wooltorton, L. S. C. 1976. The enzymic conversion of hydroxycinnamic acids to *p*-coumaroylquinic and chlorogenic acids in tomato fruits. *Phytochemistry* 15:947–51
58. Rhodes, M. J. C., Wooltorton, L. S. C. 1975. Enzymes involved in the reduction of ferulic acid to coniferyl alcohol during aging of disks of swede root tissue. *Phytochemistry* 14:1235–40
59. Russell, D. W. 1971. The metabolism of aromatic compounds in higher plants. X. Properties of the cinnamic acid 4-hydroxylase of pea seedlings and some aspects of its metabolic and developmental control. *J. Biol. Chem.* 246: 3870–78
60. Saleh, N. A. M., Fritsch, H., Kreuzaler, F., Grisebach, H. 1978. Flavanone synthase from cell suspension cultures of *Haplopappus gracilis* and comparison with the synthase from parsley. *Phytochemistry* 17:183–86
61. Sarkanen, K. V., Ludwig, C. H. 1971. *Lignins: Occurrence, Formation, Structure and Reactions.* New York: Wiley Interscience. 916 pp.
62. Schopfer, P. 1977. Phytochrome control of enzymes. *Ann. Rev. Plant Physiol.* 28:223–52
63. Schröder, J. 1977. Light-induced increase of messenger RNA for phenylalanine ammonia-lyase in cell suspension cultures of *Petroselinum hortense. Arch. Biochem. Biophys.* 182:488–96
64. Schröder, J., Betz, B., Hahlbrock, K. 1977. Messenger RNA-controlled increase of phenylalanine ammonia-lyase activity in parsley. Light-independent induction by dilution of cell suspension cultures into water. *Plant Physiol.* 60:440–45
65. Schröder, J., Kreuzaler, F., Schäfer, E., Hahlbrock, K. 1979. Concomitant induction of phenylalanine ammonia-lyase and flavnone synthase mRNAs in irradiated plant cells. *J. Biol. Chem.* In press
66. Stafford, H. A. 1974. The metabolism of aromatic compounds. *Ann. Rev. Plant Physiol.* 25:459–86
67. Sütfeld, R., Wiermann, R. 1974. Über die Bedeutung des Antherentapetums für die Akkumulation phenylpropanoider Verbindungen am Pollen. *Ber. Dtsch. Bot. Ges.* 87:167–74
68. Ulbrich, B., Stöckigt, J., Zenk, M. H. 1976. Induction by light of hydroxycinnamoyl-CoA-quinate-transferase activity in buckwheat hypocotyls. *Naturwissenschaften* 63:484
69. Von Ardenne, R. 1965. Bestimmung der Natur der Anthocyane in Gewebekulturen von *Haplopappus gracilis. Z. Naturforsch.* 20b:186–87
70. Wellmann, E. 1974. Regulation der Flavonoidbiosynthese durch ultraviolettes Licht und Phytochrom in Zellkulturen und Keimlingen von Petersilie (*Petroselinum hortense* Hoffm.). *Ber. Dtsch. Bot. Ges.* 87:267–73
71. Wellmann, E. 1974. Gewebespezifische Kontrolle von Enzymen des Flavonoidstoffwechsels durch Phytochrom in Kotyledonen des Senfkeimlings (*Sinapis alba* L.). *Ber. Dtsch. Bot. Ges.* 87:275–79

72. Wellmann, E. 1975. UV Dose-dependent induction of enzymes related to flavonoid biosynthesis in cell suspension cultures of parsley. *FEBS Lett.* 51: 105–7
73. Wellmann, E., Hrazdina, G., Grisebach, H. 1976. Induction of anthocyanin formation and of enzymes related to its biosynthesis by UV light in cell cultures of *Haplopappus gracilis. Phytochemistry* 15:913–15
74. Wengenmayer, H., Ebel, J., Grisebach, H. 1976. Enzymic synthesis of lignin precursors. Purification and properties of a cinnamoyl-CoA:NADPH reductase from cell suspension cultures of soybean (*Glycine max*). *Eur. J. Biochem.* 65:529–36
75. Wyrambik, D., Grisebach, H. 1975. Purification and properties of isoenzymes of cinnamyl alcohol dehydrogenase from soybean cell suspension cultures. *Eur. J. Biochem.* 59:9–15
76. Zähringer, U., Ebel, J., Grisebach, H. 1978. Induction of phytoalexin synthesis in soybean. Elicitor-induced increase in enzyme activities of flavonoid biosynthesis and incorporation of mevalonate into glyceollin. *Arch. Biochem. Biophys.* 188:450–55
77. Zimmermann, A., Hahlbrock, K. 1975. Light-induced changes of enzyme activities in parsley cell suspension cultures. Purification and some properties of phenylalanine ammonia-lyase (EC 4.3. 1.5). *Arch. Biochem. Biophys.* 166: 54–62

Ann. Rev. Plant Physiol. 1979. 30:131–58

THE CENTRAL ROLE OF PHOSPHOENOLPYRUVATE IN PLANT METABOLISM

❖7668

D. D. Davies[1]

School of Biological Sciences, University of East Anglia, Norwich, NR4 7TJ, Norfolk, United Kingdom

CONTENTS

[1]The preparation of this review was supported by SRC grant B/RG 90891.

0066-4294/79/0601-0131$01.00

INTRODUCTION

The importance of phosphoenolpyruvate in metabolism stems largely from the fact, implicit in the abbreviation PEP, that it is a high energy compound. In 1941 Kalckar (78) and Lipman (92) independently introduced the concept of high and low energy linkages. The hydrolysis of PEP has a particularly large ΔG^1 of −14.8 kcal compared with −7.3 kcal for the terminal phosphate of ATP and −3.3 kcal for the low energy linkage of glucose-6-phosphate. Following the objections of Gillespie et al (53), we avoid the use of the term high energy *bond* and question why ATP and PEP are high energy *compounds.* This has been the subject of a number of theoretical considerations, including opposing resonance (114), electrostatic repulsion (69), electron distribution based on simple π electron Hückel calculations (49, 56, 125), electron distribution using an extended Hückel method (18), and the effect of solvation, given the fascinating title "Squiggle-H_2O" (50). As far as PEP is concerned, a simplistic approach is to note that the equilibrium between the keto and enol forms of pyruvate greatly favors the keto form; thus the enol form must be at a much higher energy level, and it is convenient, if not precise, to think of PEP as a high energy compound due to stabilization of the enol form. The importance of PEP in metabolism is then seen to be a consequence of the large amount of energy available for biosynthesis and to the range of reactions made possible by enzyme cleavage on either side of the enol oxygen atom; C-O cleavage producing an active pyruvyl group and O-P cleavage producing an active phosphoryl group (Figure 1).

METABOLIC BRANCH POINTS

Common Ancestry of Enzymes Utilizing PEP

The central role of PEP as a branch point of metabolism is reflected in the evolutionary conservation of the stereochemical binding of PEP to a num-

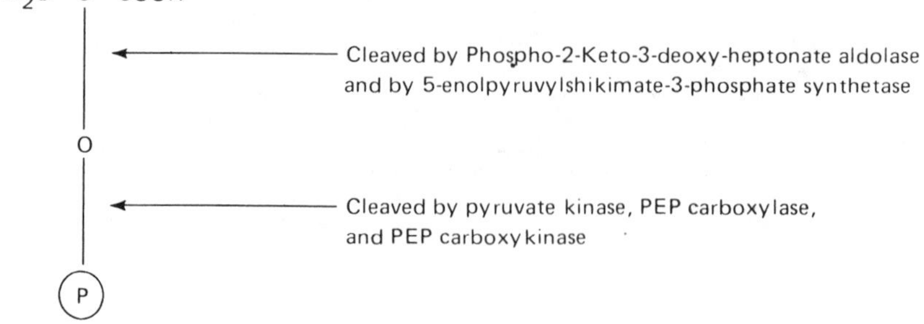

Figure 1 Points of cleavage of PEP.

ber of enzymes which catalyze the allylic replacement of phosphate (142). The addition of CO_2 catalyzed by PEP carboxylase and PEP carboxykinase (143), the addition of hydrogen catalyzed by pyruvate kinase (14, 141), and the addition of erythrose-4-phosphate catalyzed by DAHP synthetase (115) all occur on C-3 of PEP from the *si* face. The carboxylation of PEP is shown in Figure 2, where the stereochemistry of oxaloacetate is determined by the attack of CO_2 on the *si* face of PEP.

There appears to be no advantage in orientating the *re* or *si* face of PEP toward the electrophilic reagent. Thus the fact that in all four cases addition is from the *si* face suggests that once the binding site for PEP had evolved, the basic structure was retained when branched metabolic pathways evolved. The validity of this conclusion is analogous to the demonstration of bias in a coin-tossing experiment—each determination has a 50% significance level and the number of tests is limited by the number of reactions involving PEP!

Metabolic Reactions Involving PEP

The central role of PEP in plant metabolism is illustrated by Figure 3.

Carboxylase or Carboxykinase?

Because of the basic similarity between PEP carboxylase, carboxykinase, and carboxytransphosphorylase, it is not always easy to decide which enzyme is responsible for CO_2 fixation. For example, CO_2 fixation attributed to PEP carboxylase could be due to the combined action of PEP carboxytransphosphorylase and a pyrophosphatase.

$$\begin{array}{ll} & \text{PEP} + P_i + HCO_3^- \rightarrow \text{Oxaloacetate} + \text{PP} \\ & \text{PP} \rightarrow 2\,P_i \\ \hline \text{Sum} & \text{PEP} + HCO_3^- \rightarrow \text{Oxaloacetate} + P_i \end{array}$$

This mechanism predicts that the decarboxylation would be stimulated by P_i, and a number of apparent PEP carboxylases are stimulated by P_i (161,

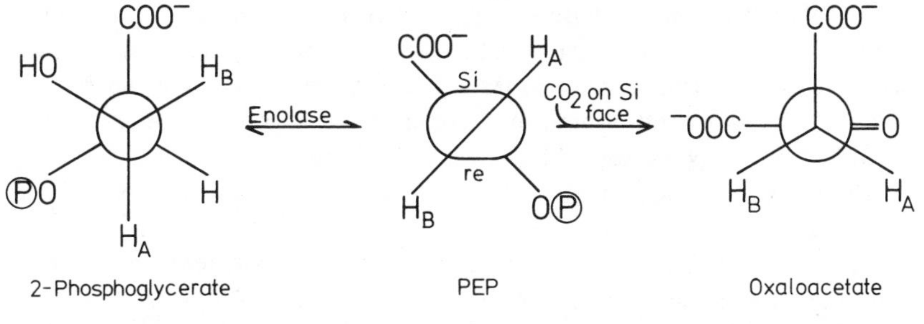

Figure 2 Stereospecificity of carboxylation of PEP. The *si* and *re* faces (58) refer to C-2 since the ligands on C-3 are identical.

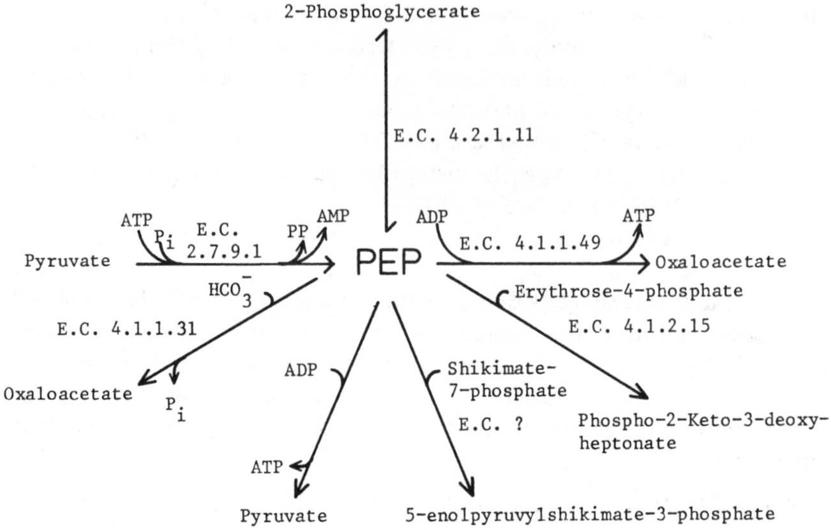

Figure 3 Reaction involving PEP in plants.

E.C. Number	*Name*
2.7.9.1.	Pyruvate phosphate dikinase
4.1.1.31	PEP carboxylase
2.7.1.40	Pyruvate kinase
–	Pyruvylshikimate phosphate synthetase
4.1.2.15	Phospho-2-keto-3-deoxy-heptonate aldolase
4.1.1.49	PEP carboxykinase (ATP)

173, 188). However, the possible involvement of a carboxytransphosphorylase in an apparent spinach PEP carboxylase has been eliminated by showing that the preparation does not catalyze the incorporation of ^{32}P into PP and that the preparation does not contain pyrophosphatase (149). There is no evidence for the presence of PEP carboxytransphosphorylase in plants (189).

Stimulation of carboxylation by ADP was noted by Mazelis & Vennesland (99) in a particulate preparation of PEP carboxylase from spinach which was free from carboxykinase. The authors were surprised by this finding, but after conditioning by the concept of allosteric regulation, few would express surprise today! However, the stimulation of carboxylation by ADP led Weidner & Küppers (181) to classify the enzyme they isolated from *Laminaria* as a PEP carboxykinase without demonstrating the production of ATP. Stimulation of PEP carboxylase by nucleotides has been reported with enzymes from several organisms (113, 144, 160, 188). Consequently, stimulation by ADP is insufficient to allow the enzyme from *Lami-*

naria to be classified as a carboxykinase, and the criteria formulated by Utter & Kolenbrander (170) should be adopted.

The carboxylating enzymes from marine phytoplankton were originally classified as PEP carboxylases. The enzyme from the diatom *Phaeodactylum tricornutum* is now classified as a carboxykinase since ATP is a product (72). However, the enzyme differs from other PEP carboxykinases in apparently favoring carboxylation and using HCO_3^- rather than CO_2. The carboxylating enzyme from the dinoflagellate *Amphidinuum carterae* resembles an ADP-stimulated PEP carboxylase, but Holdsworth (71) does not consider it to be a carboxylase.

Two reports of heat-stable PEP carboxylases with quite unusual properties have appeared (91, 119), but Coombs et al (25) have argued that these apparent activities are artifacts of the assay system.

HCO_3^- or CO_2?

Maruyama et al (97) used $HC^{18}O_3^-$ to study the carboxylation of PEP and concluded that despite the rapid exchange between $HC^{18}O_3^-$ and H_2O, their data permitted the conclusion that the active species for peanut PEP carboxylase was HCO_3^-. Cooper et al (29) reinvestigated the carboxylation reactions using $H^{14}CO_3^-$ or $^{14}CO_2$ and concluded that HCO_3^- was the active species for the carboxylase while CO_2 was the active species for carboxykinase and carboxytransphosphorylase. Waygood et al (179) used a kinetic method similar to that employed by Dalziel & Londesborough (32a) and concluded that CO_2 was the active species for maize PEP carboxylase. The validity of this kinetic method requires (*a*) that the buffering capacity should prevent any significant changes of pH, and (*b*) the concentrations of CO_2 or HCO_3^- should be small relative to their apparent K_m values. Waygood et al (179) failed to meet either requirement.

Coombs et al (26) recognized the need to maintain the pH constant and the advantages of measuring the reaction at pH 8, but failed to recognize the need for low concentrations of HCO_3^-. They used 10 mM HCO_3^-, which is two orders of magnitude greater than the apparent K_m, and similar conditions were used by Reibach & Benedict (139). Thus both sets of kinetic experiments join those of Waygood et al in failing to meet the theoretical conditions necessary for valid conclusions. It happens that these results (26, 139) contradict the results of Waygood et al, but valid conclusions cannot be drawn from an invalid method. However, Coombs et al (26) also used a modification of the isotope method (29), and other similar experiments (15, 102) have confirmed that HCO_3^- is the active species for PEP carboxylase. The enzyme prepared from *Mesembryanthemum crystallinum,* when assayed at pH 5.5, has 50% of the activity at pH 8 (57). This property is consistent with the plant's capacity to form large amounts of malic acid during darkness, but it raises interesting questions concerning the nature of the carboxylation since at pH 5.5 the HCO_3^- concentration is only 20% of the CO_2 concentration. The carboxykinase of phytoplankton may use HCO_3^- (73), but there is uncertainty about the classification of this enzyme.

CONTROL OF BRANCHED REACTIONS

Variations in Substrate Concentrations

When two or more compounds are formed from a single substrate, the partitioning of that substrate between the reactions may be affected by changes in substrate concentration. In the case of nonenzymic simultaneous side reactions, partitioning is independent of substrate concentration (180). However, in the case of irreversible enzyme reactions with Michaelian kinetics, it can be shown that

$$V_1/V_2 = V_{max_1}(K_{m_2} + S)/V_{max_2}(K_{m_1} + S) \qquad \text{[Davies (33)].}$$

This simple treatment has been applied to the competition for PEP between pyruvate kinase and alkaline phosphatase in mammalian tissues (146). Although alkaline phosphatase has a high affinity for PEP at neutral pH, the role of this enzyme in the metabolism of PEP is not clear. Most plants contain phosphatases which are very active against PEP (42) and a *specific* PEP phosphatase which is stimulated by NaCl conceivably could be present in *Suaeda fructicosa* (2). However, this phosphatase was not characterized, and my personal prejudice is against a role for phosphatases in the metabolism of PEP. I accept the viewpoint of Matile (98) that unspecific phosphatases are located in the lysosomal system. A report that acid phosphatase is distributed equally between the vacuole and the cytosol (20) appears to be due to contamination from the cell wall-degrading enzymes used to prepare the protoplasts, and reinvestigation suggests that acid phosphatase is located in the vacuole (13).

The flux through branched pathways will depend on the kinetic constants of all enzymes in each pathway (77), and if the initiating reactions are reversible, the partitioning will show reciprocal interactions (177). However, the initiating reactions at the PEP branch point are irreversible, and the simplified kinetics can be applied, provided the reactions occur in the same compartment. If both initiating reactions obey Michaelian kinetics, large differences in K_m will produce large changes in partitioning in response to changes in substrate concentration (Figure 4). If one of the enzymes shows sigmoid kinetics, low concentrations of substrate will favor the enzyme with Michaelian kinetics. At substrate concentrations close to S(0.5), small increases in concentration will greatly favor the reaction with sigmoid kinetics.

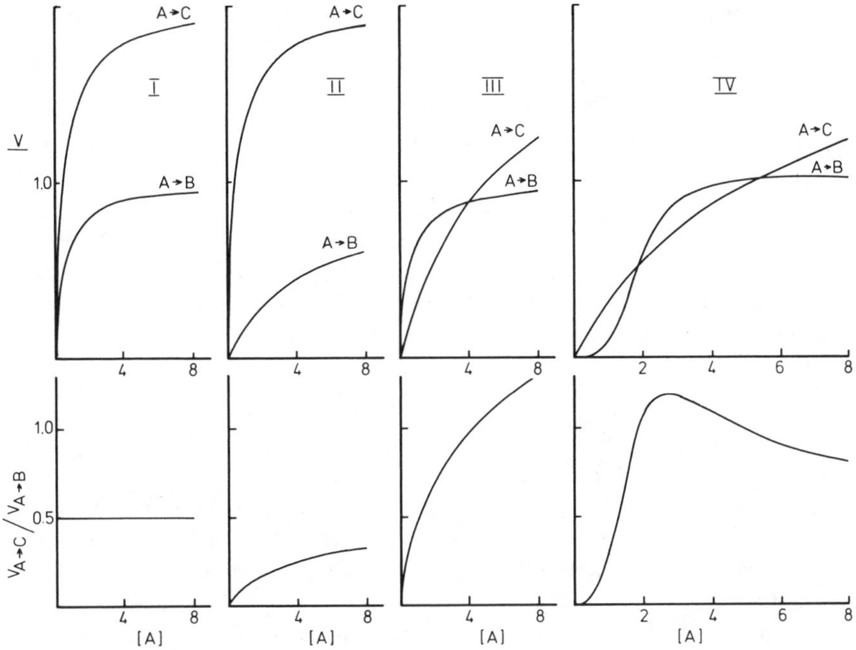

Figure 4 Effect of substrate concentration on the rate of simultaneous side reactions. The curves are calculated assuming Michaelian kinetics, except for IV A → B which is calculated from $v = V_{A\to B} \times (A)^4/K_{mA\to B} + (A)^4$

I. $V_{A\to B} = 1$ $V_{A\to C} = 2$, $K_{mA\to B} = K_{mA\to C} = 0.5$
II. $V_{A\to B} = 1$ $V_{A\to C} = 2$, $K_{mA\to B} = 5$, $K_{mA\to C} = 0.5$
III. $V_{A\to B} = 1$ $V_{A\to C} = 2$, $K_{mA\to B} = 0.5$, $K_{mA\to C} = 5$
IV. $V_{A\to B} = 1$ $V_{A\to C} = 2$, $K_{mA\to B} = 10$ $K_{mA\to C} = 5$

In applying these considerations to data from the literature, it should be noted that most authors use the *total* concentration of PEP, but some use the *free* or Mg or Mn *bound* PEP. The values can be interconverted using the dissociation constant of Mg–PEP = 5.56 mM or Mn–PEP = 179 μM (186). Pyruvate kinase from nine plants examined (167) obeyed Michaelian kinetics with K_m's for PEP from 10 to 50 μM (pH 7.4, Mg^{2+} 8 mM). The enzyme from cotton seed (42) has a K_m for PEP of 85 μM (pH 7.5, Mg^{2+} 10 mM), and from castor bean (106) the K_m is 58 μM (pH 6.5, Mg^{2+} 10mM). PEP carboxylases frequently show deviations from Michaelian kinetics when examined over a wide range of concentrations (103). Sigmoid kinetics may be observed at one pH but not at another (169) or at one temperature but not at another (188). However, in most cases the kinetics approximate to Michaelian and in ten species of C_4 plants the mean

K_m for PEP was 0.59 ± 0.35 mM and for eight species of C_3 plants the mean K_m was 0.14 ± 0.07 mM at pH 7.8, all with saturating Mg^{2+} (166). It appears that the affinity of pyruvate kinase for PEP is an order of magnitude greater than that of PEP carboxylase, and Figure 4 suggests that if the concentration of PEP rises, it will partition in favor of oxaloacetate production. In the case of pea seeds, Bonugli (15) has measured the kinetic constants of pyruvate kinase and PEP carboxylase and calculated the partitioning of PEP between the two reactions. The calculations predict a sharp increase in the relative activity of PEP carboxylase when PEP is increased in the range 0 to 250 μM, and direct measurements were in good agreement with values calculated from the kinetic data.

Feedback Control at the PEP Crossroads

The partitioning of PEP at the metabolic crossroads is complicated by an intricate network of feedback controls (Figure 5). The metabolic pathways are common to the majority of plants, but the controls vary from species to species. For example, starch phosphorylase from a CAM plant is strongly inhibited by PEP, whereas the enzyme from a C_4 plant *Atriplex spongiosa* is not (157). PEP carboxylase from leaves of monocotyledonous C_4 plants is activated by glycine, while the enzyme from leaves of dicotyledonous C_4 plants and mono- and dicotyledonous C_3 plants is not activated (111).

The enzymologist investigating an enzyme showing allosteric responses has a daunting task—a response to a metabolite may be observed at one pH but not at another (188), or the kinetic behavior may depend on whether Mg^{2+} or Mn^{2+} is present (102). The magnitude of the task stems from the need to undertake 1,679,616 assays in order to establish the complete rate law for a regulatory enzyme with eight reactants and effectors and at one pH (145). The physiologist faces difficulties in interpreting the kinetic data. For example, are in vitro effects likely to occur at in vivo concentrations, or are allosteric effects (or their absence) a result of isolating or purifying the enzyme? Clearly the student of feedback control needs patience and judgment!

Futile Cycles

If a nonequilibrium reaction A → B is opposed by another nonequilibrium reaction converting B → A by a different chemical reaction, then the simultaneous operation of the two reactions will produce a futile cycle.

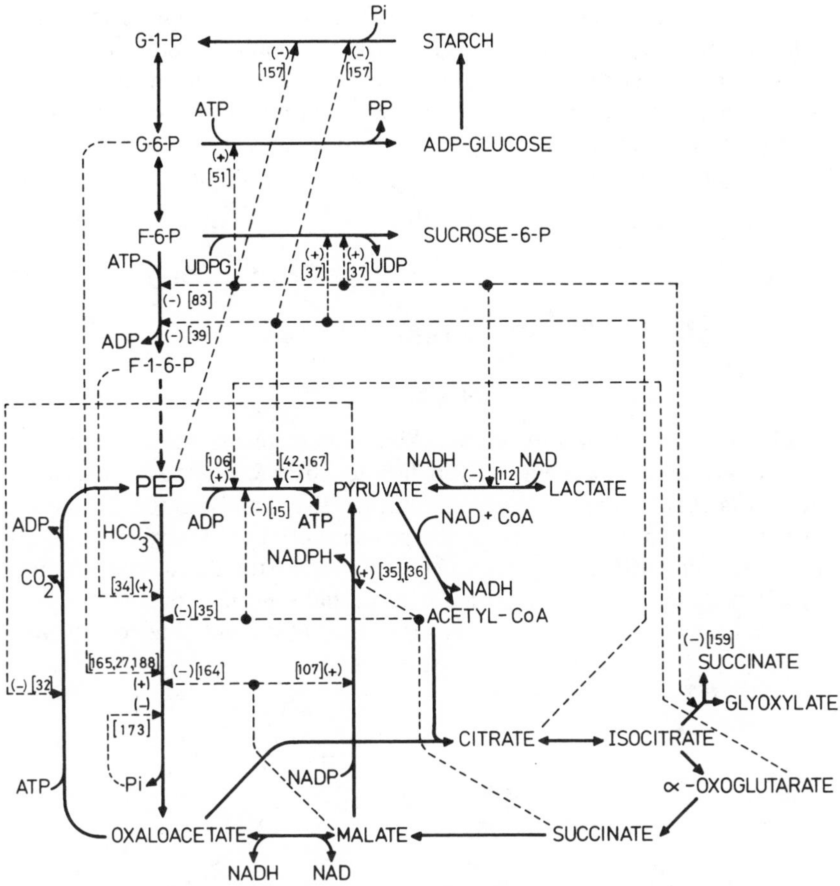

Figure 5 Metabolic interactions involving PEP. Metabolic pathways indicated by solid bars. Control interactions by dashed line. (+) is activation, (–) is inhibition and [N] gives the literature citation.

Futile cycles are energetically wasteful, but in some cases they may have a metabolic role when the term substrate cycle would be more appropriate (109). A number of the reactions involving PEP could interact to produce futile or possibly substrate cycles as shown in Figure 6.

I. OXALOACETATE DECARBOXYLASE An α-decarboxylating peroxidase forming malonate from oxaloacetate has been isolated from plants (147), but oxaloacetate decarboxylase (E.C. 4.1.1.3) is a bacterial enzyme. However, a number of other enzymes can catalyze the β decarboxylation of oxaloacetate. Thus oxaloacetate decarboxylase activity appears to be an inherent property of pyruvate kinase (30), and malic enzyme can decarboxylate oxaloacetate at low pH (41). A futile cycle involving malic enzyme and malate dehydrogenase could operate if both enzymes with the same specificity for the nicotinamide adenine dinucleotide occurred in the same cell compartment.

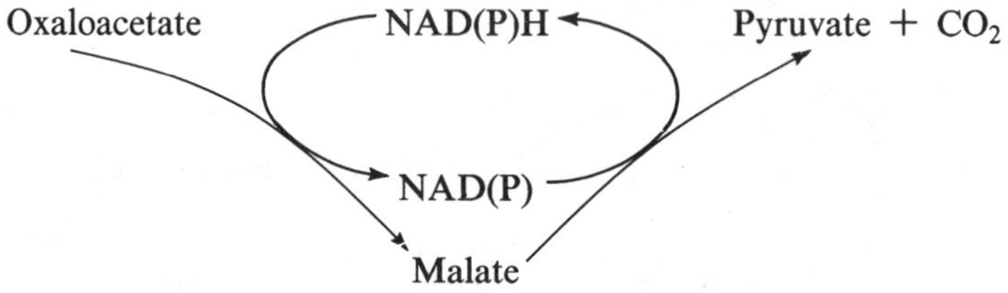

The possibility that PEP carboxykinase from plants has oxaloacetate decarboxylase activity is currently in dispute. Utter & Kurahashi (171) first noted that PEP carboxykinase from chicken liver catalyzes the formation of pyruvate from oxaloacetate in the presence of IDP. A similar reaction has been demonstrated in yeast (21), and ADP-stimulated oxaloacetate decarboxylation has been reported with partially purified preparations of PEP carboxykinase from *Panicum maximum* (135) and pineapple leaves

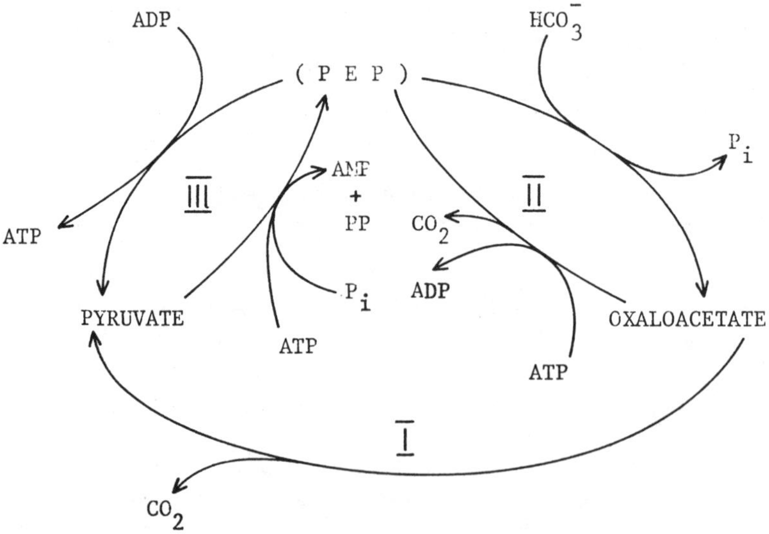

Figure 6 Possible futile cycles associated with the metabolism of PEP. Roman numerals refer to reactions discussed in the text.

(32). However, the absence of ADP-stimulated oxaloacetate decarboxylation in preparations of PEP carboxykinase from *Chloris gayana* has been reported (62).

The ADP-stimulated decarboxylation could be an inherent property of plant PEP carboxykinases—in which case this property has been lost during purification from *Chloris gayana.* Alternatively, the observed activity could be due to a futile cycle involving pyruvate kinase.

$$\begin{array}{c} \text{Oxaloacetate} + \text{ATP} \rightleftharpoons \text{ADP} + \text{PEP} + CO_2 \\ \text{PEP} + \text{ADP} \rightarrow \text{Pyruvate} + \text{ATP} \\ \hline \text{Sum} \quad \text{Oxaloacetate} \rightarrow \text{Pyruvate} + CO_2 \end{array}$$

Most preparations of ADP contain some ATP, so this reaction could occur in the absence of adenylate kinase. The possibility that this futile cycle could operate in fat-bearing seeds is discussed later.

II. FUTILE CYCLE INVOLVING PEP CARBOXYLASE AND PEP CARBOXYKINASE

$$\begin{array}{c} \text{PEP} + CO_2 \rightarrow \text{Oxaloacetate} + P_i \\ \text{Oxaloacetate} + \text{ATP} \rightleftharpoons \text{ADP} + \text{PEP} + CO_2 \\ \hline \text{Sum} \quad \text{ATP} \rightarrow \text{ADP} + P_i \end{array}$$

Since both enzymes have been reported in the cytosol of castor beans (28), futile cycling to produce ATPase activity could occur in vivo. There is currently no information on factors which might operate to control this futile cycle.

III. FUTILE CYCLE BETWEEN PYRUVATE KINASE AND PYRUVATE DIPHOSPHOKINASE

$$\begin{array}{c} \text{PEP} + \text{ADP} \rightarrow \text{Pyruvate} + \text{ATP} \\ \text{Pyruvate} + \text{ATP} + P_i \rightarrow \text{PEP} + \text{AMP} + \text{PP} \\ \hline \text{Sum} \quad \text{ADP} + P_i \rightarrow \text{AMP} + \text{PP} \end{array}$$

Pyruvate diphosphokinase is only found in some C_4 and CAM plants, and in these plants the enzyme is in the chloroplasts (63) so that cycling can be partly controlled by compartmentation and partly by light, because the diphosphokinase is a light-activated enzyme.

The implicit assumption in this discussion has been that substrate cycles are futile cycles which the plant must avoid or at least control. There is no clear evidence that these cycles operate in plants although the methodology for studying them is available (81). Possibly physiological roles for substrate cycles include thermogenesis, flux control by amplification (108, 109), and stabilization of ATP levels (131).

PEP AND CARBOHYDRATE METABOLISM

The analysis of the control of a metabolic pathway is difficult, even in cases such as erythrocytes, where one pathway (glycolysis) is dominant (130). The analysis is much more complex when glycolysis, gluconeogenesis, the pentose phosphate pathway, and the citric acid cycle are operating simultaneously. These pathways are doubtless coordinated, but the difficulties of comprehension make it necessary to consider the pathways separately. Fortunately this area has been thoroughly reviewed recently (168) so that a brief discussion can suffice.

Control Points of Glycolysis

The control of glycolysis is most frequently studied by stimulating the process by anoxia and correlating the increased flux with changes in concentration of metabolites. Most workers recognize that all the reactions of glycolysis contribute to the control (67, 77), but in practice, analysis is based on two propositions: 1. that control is likely to be associated with irreversible reactions occurring early in the sequence, and 2. that if the flux increases, the substrate concentration of the controlled step falls (88). Data are analyzed by crossover plots which emphasize changes in the concentration of the *product* of the controlled reaction, and since irreversible reactions are not affected by their products (140), there is a certain illogicality in this graphical approach. However, the large increase in concentration of fructose diphosphate, which occurs when the glycolytic flux is increased, makes a visually dramatic crossover plot, even though the decrease in substrate may be very small (46, 54, 86). A number of workers have used crossover plots in the study of developmental processes (1, 150, 151), but it should be emphasized that there is no theoretical basis for the application of the crossover plot in a time-dependent process.

The identification of phosphofructokinase as a control point is in accord with the linear correlation between respiration and the concentration of fructose diphosphate and triose phosphates (45). Triose phosphates were included because of the analytical methods used, and the correlation can be compared with that between glucose utilization and fructose diphosphate concentration in *E. coli* (40). A number of workers (1, 86) have, on the basis of crossover plots, identified pyruvate kinase as a control point. In view of the difficulty of interpreting a second crossover point (67), this identification should be treated with caution.

The complex kinetics and allosteric properties shown by certain enzymes have convinced many workers that these enzymes are pacemakers. Thus the complex kinetics of phosphofructokinase have, no doubt, contributed to its wide acceptance as a control point of glycolysis. A number of metabolites have been shown to affect pyruvate kinase, but Tomlinson & Turner (167) are somewhat skeptical about the regulatory role of this enzyme, although Duggleby & Dennis (42) argue for a regulatory role in tissues which simultaneously carry out glycolysis and gluconeogenesis.

Crossover plots do not usually include data for starch, sucrose, glucose, or fructose, but the reactions which lead from these compounds into the glycolytic sequence may well play an important part in the control of glycolysis.

Coordinated Control of Glycolysis

Atkinson (6) has developed the concept of energy charge.

$$\text{Energy charge} = \frac{\text{ATP} + 0.5\ \text{ADP}}{\text{ATP} + \text{ADP} + \text{AMP}}$$

The energy charge is maintained constant at ca 0.8 even in anoxia and exerts a central controlling influence in metabolism. Some enzymes from plants respond to energy charge in the manner proposed by Atkinson (6) but significant differences exist. Thus anoxic plants—with the exception of rice—do not maintain an energy charge of 0.8 but seem adapted to survive by maintaining a low energy charge of ca 0.2 and by a reduced metabolic activity (124). The control of phosphofructokinase in plants and animals appears fundamentally different. The current theory of the control of glycolysis in animals (110) proposes that phosphofructokinase is normally inhibited by ATP. When the level of ATP drops slightly, the adenylate kinase equilibrium ensures that there will be a large percent increase in AMP, and it is the AMP which activates phosphofructokinase. This mechanism cannot apply to plants since all plant phosphofructokinases are inhibited by AMP. The proposal (38) that the enzyme is directly controlled by the level of ATP is unlikely, unless the concentration of ATP in the cytoplasm of carrots is at least 50-fold higher than the concentration expressed per unit fresh weight (1). The alternative proposal (83) that phosphofructokinase is controlled by PEP seems more likely and offers a basis for the integration of control.

3-Phosphoglycerate and PEP are readily interconverted, and the net gain of citrate requires a corresponding fixation of CO_2 by PEP. For the purpose of this discussion we will consider that PEP, 3-phosphoglycerate, and citrate are equivalent and examine the sequence of interactions which occur when the metabolic system shown in Figure 5 is perturbed by the uptake of excess cations. The increased pH stimulates the carboxylation of PEP which has two effects: (*a*) the concentration of PEP falls and (*b*) the concentration of OAA rises. The fall in PEP produces three effects: (*a*) the activity of adenosine diphosphate glucose pyrophosphorylase is reduced, (*b*) the activity of sucrose phosphate synthetase is reduced, and (*c*) phosphofructokinase activity is increased. The combined result of these three effects is to direct hexosephosphates away from starch and sucrose synthesis and toward an increased rate of glycolysis. Counterbalancing this series of events are the consequences of the enhanced fixation of CO_2 which produces more oxaloacetate. The extra oxaloacetate combines with the extra acetyl CoA produced by the enhanced glycolysis and so tends to increase the concentration of citrate, which then opposes the effects initiated by the decline in PEP. The extent to which this description represents reality requires quantitative information.

The Role of PEP Carboxykinase in Gluconeogenesis

Meyerhof et al (100) found the pyruvate kinase reaction to be irreversible. However, Kalckar (79) observed the oxidation of malate with the formation of PEP and stated with great foresight "without doubt the formation of PEP from malic acid is related to the synthesis of sugars." The observation (7) that the formation of PEP from pyruvate required HCO_3^- led Krebs (87) to formulate the pathway of gluconeogenesis, in which he identified three kinases which are involved in glycolysis and where ΔG^1 is so large and negative that the reactions could not be reversed.

Glucose + ATP → glucose-6-phosphate + ADP (ΔG^1 – 4.6 kcal)

Fructose-6-phosphate + ATP → Fructose diphosphate + ADP (ΔG^1 – 4.4 kcal)

PEP + ADP → Pyruvate + ATP (ΔG^1 – 5.2 kcal)

Krebs proposed alternative bypass reactions for each of these steps but did not mention the other reaction of gluconeogenesis with a large ΔG

1,3-Diphosphoglycerate + ADP → 3-Phosphoglycerate + ATP (ΔG^1 – 4.7 kcal).

Kelly et al (82) stated that "phosphoglycerate kinase catalyses a freely reversible reaction." Despite this statement the equilibrium position of this reaction makes it the second least reversible reaction of glycolysis, and possible bypass reactions have been discussed (70). Bypass reactions have not been found in plants (55), and presumably the maintenance of high ratios of ATP/ADP and NADPH/NADP in photosynthesis are adequate to drive the reaction in the direction of glucose synthesis (118).

The bypass of the pyruvate kinase reaction was proposed largely on thermodynamic grounds, but also for kinetic reasons. Thus, although Lardy & Ziegler (90) demonstrated the reversibility of pyruvate kinase and others demonstrated the feasibility of this reaction participating in gluconeogenesis

(43, 89), most workers consider the reaction too slow to be of physiological significance, except possibly in muscle. The bypass of the pyruvate kinase reaction is achieved in C_4 plants, in some CAM plants, and in the propionic acid bacteria, via pyruvate phosphate dikinase. In gluconeogenesis from fats, the pyruvate kinase bypass is not necessary since PEP is formed directly from oxaloacetate arising from the glyoxylate cycle (8). The enzyme PEP carboxykinase has been detected in crude extracts of castor beans (10) and recently has been characterized (91a). Interaction with pyruvate kinase will produce a futile cycle, and some control is necessary.

$$\begin{array}{l} \text{PEP} + \text{ADP} \rightarrow \text{Pyruvate} + \text{ATP} \\ \underline{\text{Oxaloacetate} + \text{ATP} \rightleftharpoons \text{PEP} + CO_2 + \text{ADP}} \\ \text{Sum} \quad \text{Oxaloacetate} \rightarrow \text{Pyruvate} + CO_2 \end{array}$$

Compartmentation is one possibility, and Thomas & ap Rees (162) cite Kobr & Beevers (86) for the view that gluconeogenesis occurs in proplastids. However, PEP carboxykinase is not associated with organelles (28) and neither is fructose diphosphatase (117), so it is probable that gluconeogenesis occurs in the cytosol, where 85% of the pyruvate kinase is also found (106) and some mechanism to control the potential futile cycle is necessary. If PEP carboxykinase in castor bean is inhibited by pyruvate as shown for the pineapple leaf enzyme (32), then a possible control is

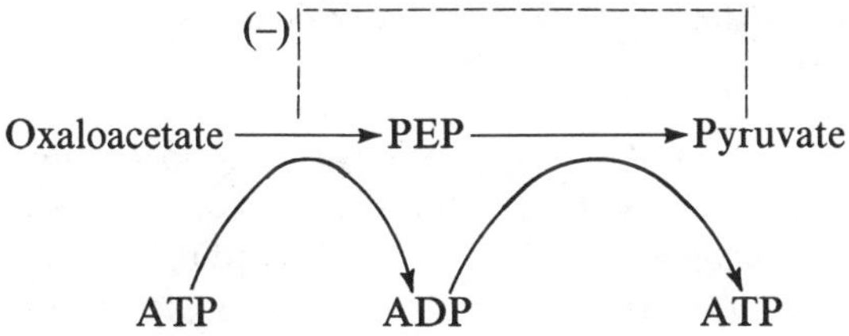

where pyruvate would need to be a strong inhibitor for effective control.

Effect of NH_4^+ on Carbohydrate Metabolism

There is strong evidence that respiration increases when NH_4^+ is being assimilated. It has been suggested (158, 191) that NH_4^+ stimulates the synthesis of glutamine, and the utilization of ATP during this reaction is responsible for the increased respiration. Kanazawa et al (80) confirmed the rise in glutamine when NH_4^+ is supplied to *Chlorella* and reported a transient fall in ATP, a fall in PEP, and a rise in alanine. They consider the primary effects of NH_4^+ to be a partial uncoupling of oxidative phosphorylation and an activation of pyruvate kinase. More recently, workers using photosynthesizing leaf discs (123) and mesophyll cells (120) have concluded that NH_4^+ activates pyruvate kinase and PEP carboxylase. The suggested explanation is that uncoupling of electron transport lowers the energy charge and the enzymes are activated by the rise in AMP and fall in ATP. In the case of root nodules where CO_2 and N_2 fixation are correlated (22), Peterson & Evans (121) suggest NH_4^+ *inhibits* pyruvate kinase by replacing K^+ so that PEP is directed towards the synthesis of oxaloacetate and aspartate.

The Role of PEP in the Control of pH

The effect of NH_4^+ noted in the previous section may be an example of the general response to a change in ion flux, which involves carboxylation and decarboxylation (11, 75, 134). The movement of ions can produce large changes in the acid-base balance; however, the cytoplasm needs to be maintained at a constant pH, and fine control could be achieved by a balance between carboxylation and decarboxylation,

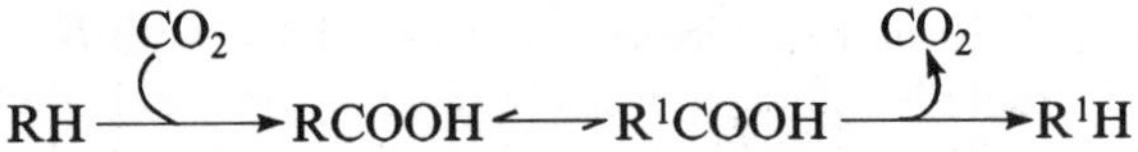

I have suggested (16, 34, 35) that PEP carboxylase, in association with malic dehydrogenase and malic enzyme, could function as a pH-stat in which the sensitivity is determined by the slopes of the activity-versus-pH curves and is enhanced by effectors which act in opposing ways on the two enzymes. The pH-stat is discussed in this volume by Smith and Raven, but two aspects will be reviewed briefly here.

STOMATAL MOVEMENT CO_2 fixation leading to malate synthesis has been shown to accompany stomatal movements (3, 132, 184). Bowling (17) has proposed a malate-switch hypothesis, according to which malate and potassium move into the guard cells when stomata are open. Travis & Mansfield (167a) reject this hypothesis on the grounds that guard cells can accumulate malate in the absence of metabolically active surrounding cells. They also noted that increasing the concentration of external CO_2 did not increase the accumulation of malate and questioned the proposition (133) that PEP carboxylase produces the malate which accumulates in guard cells. It should be noted that the stomatal aperture was not affected by CO_2 and it is possible that the treatment (pH 4.5 for 2.5 hr) used to kill cells surrounding the guard cells modified the activity of the guard cells.

IAA-STIMULATED GROWTH According to the acid-growth theory of IAA stimulated growth, H^+ ions are secreted into the cell wall solution, thereby increasing the extensibility either by activating enzymes or by breaking acid-labile links (23, 136). There is also extensive evidence that the phytotoxin fusicoccin produces similar effects (96). The secretion of H^+ will tend to raise the pH of the cytoplasm unless some form of pH-stat is involved. Evidence consistent with this proposal is that IAA stimulates CO_2 fixation producing an accumulation of malate which is stoichiometrically correlated with K^+ uptake (60, 61) and the demonstration that the PEP carboxylase of *Avena* coleoptiles has kinetic properties consistent with its proposed role (68). There is no evidence of a direct effect of IAA or fusicoccin on PEP carboxylase (152). Fusicoccin and IAA probably act on the system which catalyzes the energy-dependent electrogenic extrusion of H^+ coupled to the uptake of K^+ (95), and the consequential rise in pH activates the PEP carboxylase and so leads to a stoichiometric increase in malate.

PEP and the Synthesis of Phenylpropanoid Compounds

In view of the quantitative importance of the synthesis of phenylpropanoids, it is sad to note the limited research effort in this area. PEP provides the carbon skeleton of the side chain of phenylpropanoid compounds by combining with shikimate-3-phosphate, but the enzyme responsible has not been purified from plants. PEP contributes to the carbon skeleton of the aromatic ring by condensing with erythrose-4-phosphate to yield 3-deoxy-arabino-heptulosonic acid-7-phosphate. The synthetase involved has been purified from cauliflower (74) and shown to exhibit non-Michaelian kinetics, but not to respond to the compounds which produce allosteric responses in microorganisms. One interesting possibility to fulfill the need for control of aromatic biosynthesis (9, 52) stems from the observation (19) that in long days, leaves of *Kalanchoë blossfeldiana* produce a semispecific inhibitor of PEP carboxylase whose concentration varies daily. The inhibitor appears to be a phenolic-type substance of ca 500 mol wt and may regulate the balance between PEP carboxylation and entry into the shikimate pathway (128).

PEP AND THE SYNTHESIS OF LIPIDS

Isolated chloroplasts readily form long chain fatty acids from acetyl CoA (154), but there is uncertainty about the origin of acetyl CoA. Sherratt & Givan (148) reported the lack of incorporation of $^{14}CO_2$ into lipids and proposed an extrachloroplastic system to provide acetyl CoA. Murphy & Leech (104), using intact chloroplasts, observed $^{14}CO_2$ incorporation into lipids and supported the sequence (190)

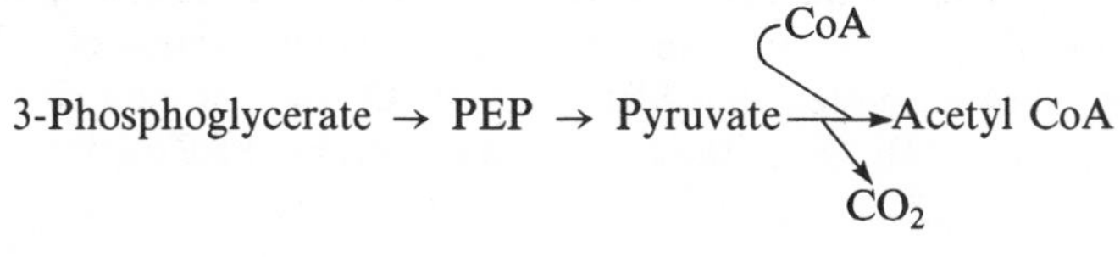

The presence of the necessary enzymes, phosphoglyceromutase, enolase (105), pyruvate kinase (12), and pyruvate dehydrogenase (153) in chloroplasts adds support to this pathway. However, other routes are possible; Wolpert & Ernst-Fonberg (187) have isolated a multienzyme complex from *Euglena gracilis,* which consists of acetyl CoA carboxylase, PEP carboxylase, and malate dehydrogenase. They propose the sequence

$$\text{PEP} \xrightarrow{CO_2} \text{Oxaloacetate} \rightarrow \text{Malate} \xrightarrow[\searrow CO_2]{} \text{Pyruvate} \xrightarrow[\searrow CO_2]{} \text{Acetyl}$$

$$\text{CoA} \xrightarrow[CO_2]{} \text{Malonyl CoA}$$

and suggest that CO_2 captured by PEP carboxylase is channeled specifically to acetyl CoA carboxylase. They propose that the complex may be important in plants and bacteria which do not have acetyl CoA carboxylases that are responsive to metabolites.

Jones (76) proposes a new pathway for fatty acid synthesis in foetal liver:

$$\text{PEP} \xrightarrow{CO_2} \text{Oxaloacetate} \rightarrow \text{Succinyl CoA} \xrightarrow{\nearrow CO_2} \text{Propionyl CoA} \rightarrow \text{fatty acids}$$

This pathway is highly speculative, particularly the reactions from propionyl CoA to fatty acids. Succinyl CoA synthetase is probably not present in chloroplasts (48).

THE ROLE OF PEP CARBOXYLASE IN FRUIT DEVELOPMENT

Net photosynthesis in the reproductive organs of cereals contributes up to 50% of the photosynthate to the grain (163). In many plants, e.g. beans (31) and soybeans (126), the reproductive organs are capable of little or no *net* photosynthesis. In the case of the pea pod, the outer layers are adapted for assimilating CO_2 from the external atmosphere and produce a net daytime gain of CO_2 (47). In high light intensities the inner layers can assimilate most of the CO_2 released by the seeds, but there is negligible CO_2 fixation in the dark (5). Harvey et al (59) concluded that a significant amount of CO_2 fixation occurs in the dark because of an apparent substantial increase in loss of CO_2 by the fruit when the external concentration of CO_2 was reduced. They suggested that high levels of CO_2 in the pod cavity maintain high levels of HCO_3^- in the cytoplasm and sustain PEP carboxylation. Hole (73) pointed out that these results would not be expected on the basis of Fick's law, and he indicated an error in the method used for measuring CO_2. It should also be noted that a high concentration of CO_2 will not necessarily lead to an enhanced rate of CO_2 fixation by PEP carboxylase, since high HCO_3^- concentration in the cytoplasm would saturate PEP carboxylase and the kinetics would be zero order with respect to HCO_3^-. Furthermore, a high concentration of CO_2 would lower the pH of the cytoplasm and so reduce the activity of PEP carboxylase. Thus Wager (176) has shown that when developing seeds are transferred from air to 10% CO_2, the concentration of malate *decreases* and on return to air malate *increases*—a result which he interprets in terms of the effect of pH changes on PEP carboxylase.

CRASSULACEAN ACID METABOLISM AND C_4 METABOLISM

Since CAM and C_4 metabolism have been reviewed recently, we will consider only two aspects.

Changes in Activity and/or Capacity of PEP Carboxylase

In CAM plants, acids accumulate at night and are removed during the day, but this circadian rhythm can occur in constant darkness (178, 182). Explanations for the switch-off of acid production include: (*a*) Changes in *activity* of PEP carboxylase brought about by feedback inhibition (24, 85, 165). (*b*) A change of net malic acid influx into the vacuole to a net efflux, brought about by turgor pressure (93, 94). (*c*) Changes in the *capacity* of PEP carboxylase (127, 128).

These explanations are not mutually exclusive, and feedback regulation may be more effective when applied to an already oscillating clock-controlled enzyme system (129). A philosophical trap is to assume a causal relationship between two rhythms. Thus the statement that "rhythmic changes in PEP carboxylase are causally involved in the endogenous rhythm of CO_2 output" (183) may mean little more than that both reactions share the same "clock."

Osmond (116) states: "it is difficult to assess the regulatory implications of a diurnal change in enzyme capacity, usually about ± 30% of the mean, when the capacity of the extracted enzyme is one or more orders of magnitude greater than the rate of the process it catalyses in vivo." The requisite theory has been provided by the concept of "sensitivity coefficient" (77) or "control strength" (67). As for practical difficulties, mixing experiments (129) and the use of polyethylene-glycol to remove small molecules (128) make it reasonable to equate extractable enzyme activity with enzyme capacity and to support the existence of diurnal changes in PEP carboxylase capacity, as well as the persistence of this rhythm in darkness (183).

Pierre & Queiroz (122) have shown that when *Kalanchoë blossfeldiana* is showing strong CAM, the intermediates and the enzymes of glycolysis show changes in concentration. In these experiments the extracts were not treated with Sephadex so that the possibility of small molecules acting as activators or inhibitors cannot be eliminated. However, if these findings are confirmed, then taken in conjunction with changes in metabolites, they offer exciting possibilities for understanding metabolic control.

When *Mesembryanthemum crystallinum* is subjected to a water stress (300 – 400 mM NaCl), the pattern of metabolism changes from C_3 to CAM (185a). This change is restricted to the older leaves and involves a marked increase in PEP carboxylase activity and the formation of a new form of the enzyme (175). More recently, von Willert (174) has shown an in vivo light inactivation of PEP carboxylase in *Premia sladeniana* and *Sceletium joubertii,* which occurs in leaves which have accumulated high concentrations of malate and is fully reversible, provided the malate concentration has been reduced to the level which occurs at the end of the light period. In *Mesembryanthemum crystallinum* diurnal changes in PEP carboxylase are apparent when the enzyme is assayed at pH 7 within 1 – 2 min of extracting the enzyme, but not when assayed at pH 8 or when the extracts are stored before assay (57). These results were interpreted to mean that PEP carboxylase occurs in two forms: a night form with high activity at pH 7–8 and a day form with low activity at pH 7. The day form of the enzyme rapidly changes into a form resembling the night form when stored at 0°.

The molecular basis of the changes in PEP carboxylase in CAM or during the greening of C_3 or C_4 plants remains to be established. During greening of sugarcane, there was no indication of turnover or de novo synthesis of PEP carboxylase (54a). However, the incorporation of ^{3}H-leucine during the greening of maize and the parallel relationship between enzyme activity and enzyme protein measured with anti-PEP carboxylase suggests de novo synthesis (155). In sorghum, greening produced a new molecular form of PEP carboxylase and the disappearance of the form found in the dark (172). Similarly the treatment of *Mesembryanthemum* with NaCl produces a new form of PEP carboxylase which disappears when NaCl is removed from the soil, and this was interpreted to mean that the enzyme turns over rapidly (175). The NaCl-induced enzyme is unstable when extracted and another enzyme, not necessarily a protease, may be involved in the inactivation (185). Clearly more work is necessary for an understanding of the molecular changes in PEP carboxylase and this is particularly true for the diurnal variations.

Pyruvate Phosphate Dikinase

Pyruvate phosphate dikinase was discovered independently in C_4 plants (64, 65) in *Entamoeba histolytica* (137) and in *Pseudomonas sp.* (44) and subsequently in some CAM plants (84, 156). The stoichiometry of the reaction is

$$\text{Pyruvate} + \text{ATP} + P_1 \rightleftharpoons \text{PEP} + \text{AMP} + \text{PP}$$

but there is disagreement about K equilibrium.

$$K(\text{equil}) = (\text{ATP})\ (P_i)\ (\text{Pyruvate})/(\text{AMP})\ (\text{PP})\ (\text{PEP}) = 2$$

at pH 8.4, according to Reeves et al (138), but according to Hatch & Slack (65) K(equil) = 5×10^3 at pH 8.3. The mechanism of the reaction proposed for the enzyme from *Pseudomonas shermanii* is

$$\text{Enzyme} + \text{ATP} \rightleftharpoons \text{Enz. PP} + \text{AMP}$$
$$\text{Enz. PP} + P_i \rightleftharpoons \text{Enz. P} + \text{PP}$$
$$\text{Enz. P} + \text{Pyruvate} \rightleftharpoons \text{PEP} + \text{Enz.}$$

This mechanism is supported by the isolation of Enzyme PP (101).

However, Andrews & Hatch (4) have proposed the mechanism

$$\text{Enzyme} + \text{ATP} + P_i \rightleftharpoons \text{Enz. P} + \text{AMP} + \text{PP}$$
$$\text{Enzyme P} + \text{Pyruvate} \rightleftharpoons \text{Enz.} + \text{PEP}$$

for the enzyme from sugarcane.

The function of the enzyme in *Entamoeba* is in the formation of pyruvate; in the other cases the enzyme provides a mechanism for the synthesis of PEP. In CAM plants with high levels of malic enzyme, the product of deacidification is pyruvate and such plants possess the dikinase. CAM plants with high levels of PEP carboxykinase produce PEP by decarboxylation and lack the dikinase.

Control of the dikinase involves a rapid light activation and dark inactivation, with a heat-labile factor catalyzing the P_i dependent activation in the presence of a thiol (66). The active and inactive forms of the dikinase are indistinguishable in size, charge, and amino acid composition (155a). The activating factor has been purified by affinity chromatography and appears to catalyze the reaction

$$\underset{(\text{Inactive})}{\text{Dikinase-X}} + P_1 \rightarrow \underset{(\text{Active})}{\text{Dikinase}} + \text{XP}$$

XP is tentatively identified as ADP (155). This suggests that some form of adenylation, deadenylation may be involved and future work will no doubt resolve this problem.

CONCLUSIONS

As befits an expanding research area, there are no clear conclusions, no broad generalizations; instead there are problems galore, loose ends to be taken up, and central problems to be attacked. What more could a research scientist ask for?

Literature Cited

1. Adam, P. B., Rowan, K. S. 1970. Glycolytic control of respiration during ageing of carrot root tissue. *Plant Physiol.* 45:490–94
2. Ahmad, R., Hewitt, E. J. 1971. Studies on the growth and phosphatase activities in *Suaeda fructicosa. Plant Soil* 34:691–96
3. Allaway, W. G. 1973. Accumulation of malate in guard cells of *Vicia faba* during stomatal opening. *Planta* 110:63–70
4. Andrews, T. J., Hatch, M. D. 1969. Properties and mechanisms of action of pyruvate, phosphate dikinase from leaves. *Biochem. J.* 114:117–25
5. Atkins, C. A., Kuo, J., Pate, J. S., Flinn, A. M., Steele, T. W. 1977. Photosynthetic pod wall of pea (*Pisum sativum* L). *Plant Physiol.* 60:779–86
6. Atkinson, D. E. 1968. The energy charge of the adenylate pool as a regulatory parameter. Interaction with feedback modifiers. *Biochemistry* 7:4030–34
7. Bartley, W. 1954. The formation of PEP by washed suspensions of sheep kidney particles. *Biochem. J.* 56:387–90
8. Beevers, H. 1969. Glyoxosomes of castor bean endosperm and their relation to gluconeogenesis. *Ann. NY Acad. Sci.* 168:313–24
9. Belser, W. L., Murphy, J. B., Delmer, D. P., Mills, S. E. 1971. End product control of tryptophan biosynthesis in extracts and intact cells of the higher plant *Nictotiana tabacum* Wisconsin 38. *Biochim. Biophys. Acta* 237:1–10
10. Benedict, C. R., Beevers, H. 1961. Formation of sucrose from malate in germinating castor beans. *Plant Physiol.* 36:540–44
11. Ben-Zioni, A., Vaadia, Y., Lips, S. H. 1970. Correlation between nitrate reduction, protein synthesis and malate accumulation. *Plant Physiol.* 23:1037–47
12. Bird, I. F., Cornelius, M. J., Dyer, T. A., Keys, A. J. 1973. The purity of chloroplasts isolated in non-aqueous media. *J. Exp. Bot.* 24:211–15
13. Boller, T., Kendle, A. 1978. Vacuolar enzymes from cultured tobacco cells. Personal communication
14. Bondinell, W. E., Sprinson, D. B. 1970. The stereochemistry of pyruvate kinase. *Biochem. Biophys. Res. Commun.* 40:1464–67
15. Bonugli, K. J. 1976. *The metabolic role of PEP carboxylase in plants.* PhD thesis. Univ. East Anglia, Norwich, U.K.
16. Bonugli, K. J., Davies, D. D. 1977. The regulation of potato PEP carboxylase in relation to a metabolic pH stat. *Planta* 133:281–87
17. Bowling, D. J. F. 1976. Malate-switch hypothesis to explain the action of stomata. *Nature* 262:393–94
18. Boyd, D. B., Lipscomb, W. N. 1969. Electronic structures for energy-rich phosphates. *J. Theor. Biol.* 25:403–20
19. Brulfert, J., Guerrier, D., Queiroz, O. 1973. Photoperiodism and enzyme activity. Balance between inhibition and induction of the crassulacean acid metabolism. *Plant Physiol.* 51:220–22
20. Butcher, H. C., Wagner, G. J., Siegelman, H. W. 1977. Localization of acid hydrolases in protoplasts. *Plant Physiol.* 59:1098–1103
21. Cannata, J. J. B., De Flombaum, M. A. C. 1974. PEP carboxykinase from bakers' yeast. *J. Biol. Chem.* 249:3356–65
22. Christeller, J. T., Laing, W. A., Sutton, W. D. 1977. Carbon dioxide fixation by lupin root nodules. *Plant Physiol.* 60:47–50
23. Cleland, R. E. 1975. Auxin-induced hydrogen ion excretion; correlation with growth and control by external pH and water stress. *Planta* 127:233–42
24. Cockburn, W., McAulay, A. 1977. Changes in metabolite levels in *Kalanchoë daigremontiana* and the regulation of malic acid accumulation in crassulacean acid metabolism. *Plant Physiol.* 59:455–58
25. Coombs, J., Baldry, C. W., Yuill-Higgs, A. 1977. PEP carboxylase-assays and artefacts. *Abstr. 4th Int. Congr. Photosynth.* p. 74
26. Coombs, J., Maw, S. L., Baldry, C. W. 1975. Metabolic regulation in C_4 photo-

synthesis: the inorganic carbon substrate for PEP carboxylase. *Plant Sci. Lett.* 4:97–102

27. Coombs, J., Baldry, C. W., Bucke, C. 1973. The C_4 pathway in *Pennisetum purpureum*. 1. The allosteric nature of PEP carboxylase. *Planta* 110:95–107
28. Cooper, T. G., Beevers, H. 1969. Mitochondria and glyoxosomes from castor bean endosperm. *J. Biol. Chem.* 244:3507–13
29. Cooper, T. G., Tchen, T. T., Wood, H. G., Benedict, C. R. 1968. The carboxylation of PEP and pyruvate. *J. Biol. Chem.* 243:3857–63
30. Creighton, D. J., Rose, I. A. 1976. Oxaloacetate decarboxylase activity in muscle is due to pyruvate kinase. *J. Biol. Chem.* 251:69–79
31. Crookston, R. K., O'Toole, J., Ozbun, J. L. 1974. Characterization of the bean pod as a photosynthetic organ. *Crop. Sci.* 14:708–12
32. Daley, L. S., Ray, T. B., Vines, H. M., Black, C. C. 1977. Characterization of PEP carboxykinase from pineapple leaves. *Plant Physiol.* 59:618–22

32a. Dalziel, K., Londesborough, J. C. 1968. The mechanisms of reductive carboxylation reactions. Carbon dioxide or bicarbonate as substrate of nicotinamide-adenine dinucleotide phosphate-linked isocitrate dehydrogenase and 'malic' enzyme. *Biochem. J.* 110: 223–30

33. Davies, D. D. 1961. *Intermediary Metabolism in Plants.* Cambridge Univ. Press
34. Davies, D. D. 1973. Control of and by pH. *Symp. Soc. Exp. Biol.* 27:513–30
35. Davies, D. D. 1977. Control of pH and glycolysis. *Phytochem. Soc. Symp.* No. 14, ed. H. Smith, pp. 41–62. New York: Academic
36. Davies, D. D., Patil, K. D. 1974. Regulation of malic enzyme of *Solanum tuberosum* by metabolites. *Biochem. J.* 137:45–53
37. de Fekete, M. A. R. 1971. The regulative properties of UDP glucose: D-fructose-6-phosphate 2 glucosyl transferase (sucrose phosphate synthetase) from *Vicia faba* cotyledons. *Eur. J. Biochem.* 19:73–80
38. Dennis, D. T., Coultate, T. P. 1966. Phosphofructokinase, a regulatory enzyme in plants. *Biochem. Biophys. Res. Commun.* 25:187–91
39. Dennis, D. T., Coultate, T. P. 1967. The regulatory properties of a plant phosphofructokinase during leaf development. *Biochim. Biophys. Acta* 146: 129–37
40. Dietzler, D. N., Leckie, M. P., Bergstein, P. E., Sughrue, M. J. J. 1975. Evidence for the coordinate control of glycogen synthesis, glucose utilisation and glycolysis in *E. coli. J. Biol. Chem.* 256:7188–93
41. Dilley, D. R. 1966. Purification and properties of apple fruit malic enzyme. *Plant Physiol.* 41:214–20
42. Duggleby, R. G., Dennis, D. T. 1973. Pyruvate kinase, a possible regulatory enzyme in higher plants. *Plant Physiol.* 52:312–17
43. Dyson, R. D., Cardenas, J. M., Barsolti, R. J. 1975. The reversibility of skeletal muscle pyruvate kinase and an assessment of its capacity to support glyconeogenesis. *J. Biol. Chem.* 256: 3316–21
44. Evans, H. J., Wood, H. G. 1968. The mechanism of the pyruvate phosphate dikinase reaction. *Proc. Natl. Acad. Sci. USA* 61:1448–53
45. Everson, R. G., Rowan, K. 1965. Phosphate metabolism and induced respiration in washed carrot slices. *Plant Physiol.* 40:1247–50
46. Faiz-ur-Rahman, A. T. M., Trewavas, A. J., Davies, D. D. 1974. The Pasteur effect in carrot root tissue. *Planta* 118:195–210
47. Flinn, A. M., Atkins, C. A., Pate, J. S. 1977. Significance of photosynthetic and respiratory exchanges in the carbon economy of the developing pea fruit. *Plant Physiol.* 60:412–18
48. Fluhr, R., Harel, E. 1975. Succinyl-CoA synthetase in greening maize leaves. *Phytochemistry* 14:2157–60
49. Fukui, K., Morokuma, K., Nagata, C. 1960. A molecular orbital treatment of phosphate bonds of biochemical interest. *Bull. Chem. Soc. Jpn.* 33:1214–19
50. George, P., Witonsky, R. J., Trachtman, M., Wu, C., Dorivart, W., Richman, L., Richman, W., Shutayh, F., Lentz, B. 1970. "Squiggle-H_2O." *Biochim. Biophys. Acta.* 223:1–15
51. Ghosh, H. P., Preiss, J. 1960. ADP glucose pyrophosphorylase—a regulatory enzyme in the biosynthesis of starch in spinach leaf chloroplasts. *J. Biol. Chem.* 241:4491–4504
52. Gilchrist, D. G., Kosuge, T. 1974. Regulation of aromatic amino acid biosynthesis in higher plants. *Arch. Biochem. Biophys.* 164:95–105
53. Gillespie, R. J., Maw, G. A., Vernon, C. A. 1953. The concept of phosphate bond energy. *Nature* 171:1147–49

54. Givan, C. V. 1968. Short-term changes in hexose phosphates and ATP in intact cells of *Acer pseudoplatanus,* subject to anoxia. *Plant Physiol.* 43:948–52
54a. Goatly, M. B., Coombs, J., Smith, H. 1975. Development of C_4 photosynthesis in sugar cane: changes in properties of PEP carboxylase during greening. *Planta* 125:15–24
55. Goodwin, T. W., Mercer, E. I. 1972. *Introduction to Plant Biochemistry.* Oxford: Pergamon
56. Grabe, B. 1958. Electron distribution in some high-energy phosphates and transfer of energy from catabolism to anabolism. *Biochim. Biophys. Acta* 30:560–69
57. Greenway, H., Winter, K., Lüttge, U. 1978. PEP carboxylase during development of crassulacean acid metabolism and during a diurnal cycle in *Mesembryanthemum crystallinum. J. Exp. Bot.* 29. In press
58. Hanson, K. R. 1966. Applications of the sequence rule 1. Naming the paired ligands g,g at a tetrahedral atom Xggij. *J. Am. Chem. Soc.* 88:2731–42
59. Harvey, D. M., Hedley, C. L., Keely, R. 1976. Photosynthetic and respiratory studies during pod and seed development of *Pisum sativum* L. *Ann. Bot.* 40:993–1001
60. Haschke, H. P., Lüttge, U. 1975. Stoichiometric correlation of malate accumulation with auxin-dependent K^+-H^+ exchange and growth in *Avena* coleoptile segments. *Plant Physiol.* 56:696–98
61. Haschke, H. P., Lüttge, U. 1977. Action of auxin on CO_2 dark fixation in *Avena* coleoptile segments as related to elongation growth. *Plant Sci. Lett.* 8:53–58
62. Hatch, M. D., Mau, S. 1977. Properties of PEP carboxykinase operative in C_4 pathway photosynthesis. *Aust. J. Plant Physiol.* 4:207–16
63. Hatch, M. D., Osmond, C. B. 1976. Compartmentation and transport in C_4 photosynthesis. In *Encyclopaedia of Plant Physiology,* (New ser.), ed. C. R. Stocking, U. Heber, pp. 144–84
64. Hatch, M. D., Slack, C. R. 1967. The participation of PEP synthetase in photosynthetic CO_2 fixation of tropical grasses. *Arch. Biochem. Biophys.* 120: 224–25
65. Hatch, M. D., Slack, C. R. 1968. A new enzyme for the interconversion of pyruvate and PEP and its role in the C_4 dicarboxylic acid pathway of photosynthesis. *Biochem. J.* 106:141–46
66. Hatch, M. D., Slack, C. R. 1969. Studies on the mechanism of activation and inactivation of pyruvate phosphate dikinase. *Biochem. J.* 112:549–58
67. Heinrich, R., Rapoport, S. M., Rapoport, T. A. 1977. Metabolic regulation and mathematical modelling. *Prog. Biophys. Mol. Biol.* 32:1–82
68. Hill, B. C., Brown, A. W. 1978. PEP carboxylase activity from *Avena* coleoptile tissue. Regulation by H^+ and malate. *Can. J. Bot.* 56:404–7
69. Hill, T. L., Morales, M. F. 1951. On "high energy phosphate bonds" of biochemical interest. *J. Am. Chem. Soc.* 73:1656–60
70. Hochachka, P. W., Somero, G. N. 1973. *Strategies of Biochemical Adaptation.* Philadelphia: Saunders
71. Holdsworth, E. S. 1978. Private communication
72. Holdsworth, E. S., Bruck, K. 1977. Enzymes concerned with β-carboxylation in marine phytoplankter. *Arch. Biochem. Biophys.* 182:87–94
73. Hole, C. C. 1977. The effect of a reduction in carbon dioxide concentration on the loss of carbon dioxide from pea fruits. *Ann. Bot.* 41:1367–70
74. Huisman, D. D., Kosuge, T. 1974. Regulation of aromatic amino acid biosynthesis in higher plants. *J. Biol. Chem.* 249:6842–48
75. Jacoby, B., Laties, G. G. 1971. Bicarbonate fixation and malate compartmentation in relation to salt-induced stoichiometric synthesis of organic acid. *Plant Physiol.* 47:525–31
76. Jones, C. T. 1976. A new pathway for the synthesis of fatty acids: A role for PEP carboxylase in lipogenesis. *FEBS Lett.* 63:77–81
77. Kacser, H., Burns, J. A. 1973. The control of flux. *Symp. Soc. Exp. Biol.* 27:65–104
78. Kalckar, H. 1941. The nature of energetic coupling in biological synthesis. *Chem. Rev.* 28:71–178
79. Kalckar, H. 1939. The nature of phosphoric esters formed in kidney extracts. *Biochem. J.* 33:631–41
80. Kanazawa, T., Kanazawa, K., Kirk, M. R., Bassham, J. A. 1972. Regulatory effects of ammonia on carbon metabolism in *Chlorella pyrenoidosa* during photosynthesis and respiration. *Biochim. Biophys. Acta* 256:656–69
81. Katz, J., Rognstad, R. 1975. Futile cycles in the metabolism of glucose. *Curr. Top. Cell. Regul.* 10:237–89
82. Kelly, G. J., Latzko, E., Gibbs, M. 1976. Regulatory aspects of photosyn-

thetic carbon metabolism. *Ann. Rev. Plant Physiol.* 27:181–205
83. Kelly, G. J., Turner, J. F. 1969. The regulation of pea-seed phosphofructokinase by PEP. *Biochem. J.* 115:481–87
84. Kluge, M., Osmond, C. B. 1971. Pyruvate P_i dikinase in CAM. *Naturwissenschaften* 58:414–15
85. Kluge, M., Osmond, C. B. 1972. Studies on PEP carboxylase and other enzymes of crassulacean acid metabolism of *Bryophyllum tubiflorum* and *Sedum praealatum. Z. Pflanzenphysiol.* 66:97–105
86. Kobr, M. J., Beevers, H. 1971. Gluconeogenesis in the castor bean endosperm 1. Changes in glycolytic intermediates. *Plant Physiol.* 47:48–52
87. Krebs, H. A. 1954. Considerations concerning the pathways of synthesis in living matter. Synthesis of glycogen from noncarbohydrate precursors. *Johns Hopkins Hosp. Bull.* 95:19–33
88. Krebs, H. A. 1957. Control of metabolic processes. *Endeavour* 16:125–32
89. Krimsky, I. 1959. Phosphorylation of pyruvate by the pyruvate kinase reaction and reversal of glycolysis in a reconstructed system. *J. Biol. Chem.* 234:232–36
90. Lardy, H. A., Ziegler, J. 1945. Enzymatic synthesis of PEP from pyruvate. *J. Biol. Chem.* 159:343–51
91. Leblova, S., Mares, J. 1975. Thermally stable PEP carboxylase from pea, tobacco and maize green leaves. *Photosynthetica* 9:177–84
91a. Leegood, R. C., apRees, T. 1978. Phosphoenolpyruvate carboxykinase in cotyledons of *Cucurbita pepo. Biochim. Biophys. Acta* 524:207–18
92. Lipmann, F. 1941. Metabolic generation and utilization of phosphate bond energy. *Adv. Enzymol.* 1:99–162
93. Lüttge, U., Ball, E., Greenway, H. 1977. Effects of water and turgor potential on malate efflux from leaf slices of *Kalanchoë diagremontiana. Plant Physiol.* 60:521–23
94. Lüttge, U., Kluge, M., Ball, E. 1975. Effects of osmotic gradients on vacuolar malic acid storage. *Plant Physiol.* 56:613–16
95. Marré, E. 1977. Effects of fusicoccin and hormones on plant cell membrane activities; observations and hypothesis. In *Regulation of Cell Membrane Activities in Plants,* ed. E. Marré, O. Ciferri. Amsterdam: Elsevier
96. Marré, E., Lado, P., Rasi Caldogno, F., Colombo, R. 1973. Correlation between cell enlargement in pea internode segments and decrease in the pH of the medium of incubation. 1. Effects of fusicoccin, natural and synthetic auxins and mannitol. *Plant Sci. Lett.* 1:179–84
97. Maruyama, H., Easterday, R. L., Chang, H., Lane, M. D. 1966. The enzymatic carboxylation of PEP. *J. Biol. Chem.* 241:2408–12
98. Matile, P. 1978. Biochemistry and function of vacuoles. *Ann. Rev. Plant Physiol.* 29:193–213
99. Mazelis, M., Vennesland, B. 1957. Carbon dioxide fixation into oxaloacetate in higher plants. *Plant Physiol.* 32:591–600
100. Meyerhof, O., Ohlmeyer, P., Gentrer, W., Maier-Leibniz, H. 1938. Studies on the intermediate reactions of glucolysis with the aid of radioactive phosphorus. *Biochem. Z.* 298:396–411
101. Milner, Y., Wood, H. G. 1976. Steady-state and exchange kinetics of pyruvate, phosphate dikinase from *Propionibacterium shermanii. J. Biol. Chem.* 251: 7920–28
102. Mukerji, S. K. 1977. Corn leaf PEP carboxylases: the effect of divalent cations on activity. *Arch. Biochem. Biophys.* 182:343–51
103. Mukerji, S. K., Ting, I. P. 1971. PEP carboxylase isoenzymes: separation and properties of three forms from cotton leaf tissue. *Arch. Biochem. Biophys.* 142:297–317
104. Murphy, D. J., Leech, R. M. 1977. Lipid biosynthesis from ^{14}C bicarbonate, 2–^{14}C pyruvate and 1–^{14}C acetate during photosynthesis by isolated spinach chloroplasts. *FEBS Lett.* 77:164–67
105. Murphy, D. J., Leech, R. M. 1978. The pathway of ^{14}C bicarbonate incorporation into lipids in isolated photosynthesising spinach chloroplasts. *FEBS Lett.* 88:192–96
106. Nakayama, H., Fujii, M., Miura, K. 1976. Partial purification and some regulatory properties of pyruvate kinase from germinating castor bean endosperm. *Plant Cell Physiol.* 17:653–60
107. Nascimento, K. H., Davies, D. D., Patil, K. D. 1975. Unidirectional inhibition and activation of malic enzyme of *Solanum tuberosum* by meso-tartrate. *Biochem. J.* 149:349–35
108. Newsholme, E. A., Crabtree, B. 1973. Metabolic aspects of enzyme regulation. *Symp. Soc. Exp. Biol.* 27:429–60
109. Newsholme, E. A., Crabtree, B. 1976. Substrate cycles in metabolic regulation and in heat generation. *Biochem. Soc. Symp.* 41:61–109

110. Newsholme, E. A., Start, C. 1973. *Regulation in Metabolism*. London: Wiley
111. Nishikido, T., Takanashi, H. 1973. Glycine activation of PEP carboxylase from monocotyldeneous C_4 plants. *Biochem. Biophys. Res. Commun.* 53:126–33
112. Oba, K., Murakami, S., Uritani, I. 1977. Partial purification and characterisation of L-lactate dehydrogenase isozymes from sweet potato roots. *J. Biochem.* 81:1193–1201
113. O'Brien, R. W., Chuang, D. T., Taylor, B. L., Utter, M. F. 1977. Novel enzymic machinery for the metabolism of oxaloacetate, PEP and pyruvate in *Pseudomonas citronellolis. J. Biol. Chem.* 252:1257–63
114. Oesper, P. 1950. Sources of the high energy content in energy-rich phosphates. *Arch. Biochem. Biophys.* 27: 255–70
115. Onderka, D. K., Floss, H. G. 1969. Stereospecificity of the DAHP synthetase reaction. *Biochem. Biophys. Res. Commun.* 35:801–4
116. Osmond, C. B. 1978. Crassulacean acid metabolism: A curiosity in context. *Ann. Rev. Plant Physiol.* 29:379–414
117. Osmond, C. B., Akazawa, T., Beevers, H. 1975. Localization and properties of ribulose diphosphate carboxylase from castor bean endosperm. *Plant Physiol.* 55:226–30
118. Pacold, I., Anderson, L. 1973. Energy charge control of the Calvin cycle enzyme 3-phosphoglyceric acid kinase. *Biochem. Biophys. Res. Commun.* 57:139–43
119. Pan, D., Waygood, E. R. 1971. A fundamental thermostable cyanide sensitive PEP carboxylase in photosynthetic and other organisms. *Can. J. Bot.* 49:631–43
120. Paul, J. S., Cornwell, K. L., Bassham, J. A. 1978. Ammonia effects on carbon metabolism in photosynthesizing leaf-free mesophyll cells from *Papaver somniferum. Planta.* In press
121. Peterson, J. B., Evans, H. J. 1977. The role of soybean cytosol pyruvate kinase in N_2 fixation. Ann. Meet. Suppl. *Plant Physiol.* 59 Abst. 286
122. Pierre, J. N., Queiroz, O. 1978. Regulation of glycolysis and level of the crassulacean acid metabolism. *Planta.* In press
123. Platt, S. G., Plant, Z., Bassham, J. A. 1977. Ammonia regulation of carbon metabolism in photosynthesizing leaf discs. *Plant Physiol.* 60:739–42
124. Pradet, A. 1969. Etude des Adénosine 5[1]-mono, di et tri-phosphates dans les tissues végétaux. *V* Effet, *in vivo,* sur le niveau de la charge énergétique d'un déséquilibre induit entre fourniture et utilisation de l'énergie dans les semences de Laitue. *Physiol. Veg.* 7:261–75
125. Pullman, B., Pullman, A. 1963. *Quantum Biochemistry.* New York: Interscience
126. Quebedeaux, B., Chollet, R. 1975. Growth and development of soybean pods. CO_2 exchange and enzyme studies. *Plant Physiol.* 55:745–48
127. Queiroz, O. 1974. Circadian rhythms and metabolic patterns. *Ann. Rev. Plant Physiol.* 25:115–34
128. Queiroz, O. 1978. CAM: rhythms of enzyme capacity and activity as adaptive mechanisms in photosynthesis. VII. In *Encyclopedia of Plant Physiology* (New ser.), ed. M. Gibbs, E. Latzko. Berlin: Springer Verlag
129. Queiroz, O., Morel, C. 1974. Photoperiodism and enzyme activity: towards a model for the control of circadian metabolic rhythms in the crassulacean acid metabolism. *Plant Physiol.* 53:596–602
130. Rapoport, T. A., Heinrich, R. 1975. Mathematical analysis of multienzyme systems. 1. Modelling of the glycolysis of human erythrocytes. *Biosystems* 7:120–29
131. Rapoport, T. A., Heinrich, R., Rapoport, S. 1976. The regulatory principles of glycolysis in erythrocytes *in vivo* and *in vitro. Biochem. J.* 154:449–69
132. Raschke, K. 1975. Simultaneous requirement of carbon dioxide and abscisic acid for stomatal closing in *Xanthium strumarium* L. *Planta* 125: 243–49
133. Raschke, K. 1975. Stomatal action. *Ann. Rev. Plant Physiol.* 26:309–40
134. Raven, J. A., Smith, F. A. 1976. Nitrogen assimilation and transport in vascular land plants in relation to intracellular pH regulation. *New Phytol.* 76: 415–31
135. Ray, T. B., Black, C. C. 1976. Characterization of PEP carboxykinase from *Panicum maximum. Plant Physiol.* 58:603–7
136. Rayle, D. L. 1973. Auxin-induced hydrogen-ion secretion in *Avena* coleoptiles and its implications. *Planta* 114: 68–73
137. Reeves, R. E. 1968. A new enzyme with the glycolytic function of pyruvate kinase. *J. Biol. Chem.* 243:3202–4
138. Reeves, R. E., Menzies, R. A., Hsu, D. S. 1968. The pyruvate-phosphate diki-

nase reaction. *J. Biol. Chem.* 243:5486–91
139. Reibach, P. H., Benedict, C. R. 1977. Fractionation of stable carbon isotopes by PEP carboxylase from C_4 plants. *Plant Physiol.* 59:564–68
140. Rolleston, F. S. 1972. A theoretical background to the use of measured concentrations of intermediates in study of the control of intermediary metabolism. *Curr. Top. Cell. Regul.* 5:47–75
141. Rose, I. A. 1970. Stereochemistry of pyruvate kinase, pyruvate carboxylase and malate enzyme reactions. *J. Biol. Chem.* 245:6052–56
142. Rose, I. A. 1972. Enzyme reaction stereospecificity. *CRC Crit. Rev. Biochem.* 1:33–57
143. Rose, I. A., O'Connell, E. L., Noce, P., Utter, M. F., Wood, H. G., Willard, J. M., Cooper, T. G., Benziman, M. 1969. Stereochemistry of the enzymatic decarboxylation of PEP. *J. Biol. Chem.* 244:6130–33
144. Sanwal, B. D., Maeba, P. 1966. PEP carboxylase: activation by nucleotides as a possible compensatory feedback effect. *J. Biol. Chem.* 241:4557–62
145. Savageau, M. A. 1972. The behaviour of intact biochemical control systems. *Curr. Top. Cell. Regul.* 7:63–103
146. Seibert, G., Pfaender, P., Kesselring, K. 1968. Competition of several enzymes for a common substrate. *Adv. Metab. Regul.* 7:131–48
147. Shannon, L. M., De Vellis, J., Lew, J. Y. 1963. Malonic acid biosynthesis in bush bean roots. II. Purification properties of enzyme catalyzing oxidative decarboxylation of oxaloacetate. *Plant Physiol.* 38:691–97
148. Sherratt, D., Givan, C. V. 1973. The apparent absence of a pathway for synthesis of acetyl coenzyme A in pea chloroplasts. *Planta* 113:47–52
149. Siu, P. 1962. Carbon dioxide fixation by PEP carboxylase from spinach. *Biochim. Biophys. Acta* 63:520–22
150. Solomos, T., Laties, G. G. 1974. Similarities between the actions of ethylene and eganide in initiating the climacterine and ripening of avocados. *Plant Physiol.* 54:506–11
151. Solomos, T., Laties, G. G. 1975. The mechanism of ethylene and cyanide action in triggering the rise in respiration in potato tubers. *Plant Physiol.* 55:73–78
152. Stout, R. G., Cleland, R. E. 1978. Effects of fusicoccin on the activity of a key pH-stat enzyme, PEP carboxylase. *Planta* 139:43–45
153. Stumpf, P. K. 1975. Biosynthesis of fatty acids in chloroplasts. In *Recent Advances in the Chemistry and Biochemistry of Plant Lipids,* ed. T. Galliard, E. I. Mercer, pp. 95–113. London: Academic
154. Stumpf, P. K., Brooks, J., Galliard, T., Hawke, J. C., Simoni, R. 1967. Biosynthesis of fatty acids by photosynthetic tissues of higher plants. In *Biochemistry of Chloroplasts,* ed. T. W. Goodwin, 2:213–239. London: Academic
155. Sugiyama, T. 1978. Private communication
155a. Sugiyama, T., Iwaki, H. 1977. Purification and partial characterization of inactive pyruvate orthophosphate dikinase from dark-treated maize leaves. *Agric. Biol. Chem.* 41:1239–44
156. Sugiyama, T., Laetsch, W. M. 1975. Occurrence of pyruvate orthophosphate dikinase in the succulent plant *Kalanchoë daigremontiana. Plant Physiol.* 56:605–7
157. Sutton, B. G. 1975. Kinetic properties of phosphorylase and 6-phosphofructokinase of *Kalanchoë daigremontiana* and *Atriplex spongiosa. Aust. J. Plant Physiol.* 2:403–11
158. Syrett, P. J. 1959. The assimilation of ammonia by nitrogen-starved cells of *chlorella vulgaris. Ann. Bot. (N.S.)* 17:1–19
159. Syrett, P. J., John, P. C. L. 1968. Isocitrate lyase: Determination of K_m values and inhibition by PEP. *Biochim. Biophys. Acta* 151:295–97
160. Taguchi, M., Izui, K., Katsuki, H. 1977. Activation of *E. coli* PEP carboxylase by guanosine-5^1-diphosphate-3^1-diphosphate. *FEBS Lett.* 77:270–72
161. Tchen, T. T., Vennesland, B. 1955. Enzymatic carbon dioxide fixation into oxaloacetate in wheatgerm. *J. Biol. Chem.* 213:533–46
162. Thomas, S. M., ap Rees, T. 1972. Glycolysis during gluconeogenesis in cotyledons of *Cucurbita pepo. Phytochemistry* 11:2187–94
163. Thorne, G. N. 1965. Photosynthesis of ears and flag leaves of wheat and barley. *Ann. Bot.* 29:317–29
164. Ting, I. P. 1968. CO_2 metabolism in corn roots. III. Inhibition of PEP carboxylase by L-malate. *Plant Physiol.* 43:1919–24
165. Ting, I. P., Osmond, C. B. 1973. Activation of plant PEP carboxylase by glucose-6-phosphate: a particular role in crassulacean acid metabolism. *Plant Sci. Lett.* 1:123–28

166. Ting, I. P., Osmond, C. B. 1973. Photosynthetic PEP carboxylases: characteristics of allozymes from leaves of C_3 and C_4 plants. *Plant Physiol.* 51:439–47
167. Tomlinson, J. D., Turner, J. F. 1973. Pyruvate kinase of higher plants. *Biochim. Biophys. Acta* 329:128–39
167a. Travis, A. J., Mansfield, T. A. 1977. Studies of malate formation in isolated guard cells. *New Phytol.* 78:451–46
168. Turner, J. F., Turner, D. H. 1975. The regulation of carbohydrate metabolism. *Ann. Rev. Plant Physiol.* 26:159–86
169. Uedan, K., Sugiyama, T. 1976. Purification and characterisation of PEP carboxylase from maize leaves. *Plant Physiol.* 57:906–10
170. Utter, M. F., Kolenbrander, H. M. 1972. Formation of oxaloacetate by CO_2 fixation on PEP. In *Enzymes* 6: 117–68
171. Utter, M. F., Kurahashi, K. 1954. Purification of oxaloacetic carboxylase from chicken liver. *J. Biol. Chem.* 207:787–841
172. Vidal, J., Cavalie, G., Gadal, P. 1976. Etude de la PEP carboxylase du haricot et du sorgho par electrophorese sur gel de polyacrylamide. *Plant Sci. Lett.* 7:265–70
173. von Willert, D. J. 1975. Die Bedeutung des anorganischen phosphats für die regulation der PEP carboxylase von *Mesembryanthemum crystallinum* L. *Planta* 122:273–80
174. von Willert, D. J. 1978. Private communication
175. von Willert, D. J., Trachel, S., Kirst, G. O., Curdts, E. 1976. Environmentally controlled changes of PEP carboxylase in *Mesembryanthemum. Phytochemistry* 15:1435–36
176. Wager, H. G. 1974. The effect of subjecting peas to air enriched with carbon dioxide. *J. Exp. Bot.* 25:338–51
177. Waley, S. G. 1962. A note on the kinetics of multi-enzyme systems. *Biochem. J.* 91:514–17
178. Warren, D. M., Wilkins, M. B. 1961. An endogenous rhythm in the rate of dark-fixation of carbon dioxide in leaves of *Bryophyllum fedtschenkoi. Nature* 191:686–88
179. Waygood, E. R., Macke, R., Tan, C. K. 1969. Carbon dioxide, the substrate for PEP carboxylase from leaves of maize. *Can. J. Bot.* 47:1455–58
180. Wegscheider, R. 1899. *Z. Phys. Chem.* 30:593
181. Weidner, M., Küppers, U. 1973. PEP carboxykinase und ribulose-1, 5 diphosphat-carboxylase von *Laminaria hyperborea. Planta* 114:365–72
182. Wilkins, M. B. 1967. An endogenous rhythm in the rate of carbon dioxide output in *Bryophyllum.* V. The dependence rhythmicity upon aerobic metabolism. *Planta* 72:66–77
183. Wilkinson, M. J., Smith, H. 1976. Properties of PEP carboxylase from *Bryophyllum fedtschenkoi* leaves and fluctuations in carboxylase activity during the endogenous rhythm of carbon dioxide output. *Plant Sci. Lett.* 6:319–24
184. Willmer, C. M., Pallas, J. E., Black, C. C. 1973. Carbon dioxide metabolism in leaf epidermal tissue. *Plant Physiol.* 52:448–52
185. Winter, K., Greenway, H. 1978. PEP carboxylase from *Mesembryanthemum crystallinum:* Its isolation and inactivation *in vitro. J. Exp. Bot.* 29. In press
185a. Winter, K., von Willert, D. J. 1972. NaCl-induzierter crassulaceen-Saurestoffwechsel bei *Mesembryanthemum crystallinum. Z. Pflanzenphysiol.* 67: 166–70
186. Wold, F., Ballou, C. E. 1957. Studies on the enzyme enolase: 1. Equilibrium studies. *J. Biol. Chem.* 227:301–12
187. Wolpert, J. S., Ernst-Fonberg, M. L. 1975. A multienzyme complex for CO_2 fixation. *Biochemistry* 14:1095–1102
188. Wong, K. F., Davies, D. D. 1973. Regulation of PEP carboxylase of *Zea mays* by metabolites. *Biochem. J.* 131:451–58
189. Wood, H. G., O'Brien, W. E., Michaels, G. 1977. Properties of carboxytransphosphorylase; pyruvate, phosphate dikinase; pyrophosphate-phosphofructokinase and pyrophosphate-acetate kinase and their roles in the metabolism of inorganic pyrophosphate. *Adv. Enzymol.* 45:85–155
190. Yamada, M., Nakamura, Y. 1975. Fatty acid synthesis by spinach chloroplasts. II The path from PGA to fatty acids. *Plant Cell Physiol.* 16:151–62
191. Yemm, E. W., Willis, A. J. 1956. The respiration of barley plants, IX. *IX* The metabolism of roots during the assimilation of nitrogen. *New Phytol.* 55:229–52

Ann. Rev. Plant Physiol. 1979. 30:159–93

MICROBODIES IN HIGHER PLANTS[1]

❖7669

Harry Beevers[2]

Thimann Laboratories, University of California, Santa Cruz, California 95064

CONTENTS

INTRODUCTION

This review describes developments in the field since 1971, when the subject was first covered comprehensively in these pages by Tolbert (173). At that time, only 4 years had elapsed since the original isolation of microbodies from fatty seedling tissues (19, 20), and shortly afterwards from leaves

[1]Abbreviations used: ER, endoplasmic reticulum; MDH, malate dehydrogenase; P-lipid, phospholipid; PC, phosphatidyl choline; PE, phosphatidyl ethanolamine; PG, phosphatidyl glycerol; PI, phosphatidyl inositol; PS, phosphatidyl serine.

[2]Supported by DOE Contract EY-76-5-03-0034 and NSF Grant PCM 75-23566.

0066-4294/79/0601-0159$01.00

(175), but the major features of enzyme constitution and function had already been established. In the succeeding years, a wealth of papers and some reviews (4, 5, 7, 47, 183) have appeared; a recent book on the subject by Gerhardt (53) lists over 700 references. Clearly then, this short review must be selective. Only higher plants will be considered, and only certain aspects will be covered in depth.

Microbodies are now recognized as constituents of plant cells generally. They are seen in electron micrographs as organelles 0.2–1.5 μm in diameter with a single bounding membrane. They usually appear to be spherical or oblate, but a range of shapes, including invaginated forms, has been described. The matrix is amorphous but occasionally striking crystalline and other inclusions are present. Since the first morphological description of these organelles by Mollenhauer et al in 1966 (129) they have been examined extensively by electron microscopy. The structural aspects are dealt with admirably in reviews by Vigil (182, 183) and by Newcomb's group (47).

The distinguishing general feature of microbodies that can be recognized in electron micrographs of suitably fixed and treated preparations is the possession of catalase. Procedures and problems in the successful application of the diamino benzoate staining procedure (which detects the peroxidative activity of the dissociated catalase subunits) have been described (8, 12, 48, 109, 181).

The isolation of microbodies from plant tissues is normally achieved by first disrupting the cells in a suitably buffered sucrose solution and centrifuging the homogenate (or a resuspended pellet obtained by first centrifuging at 10,000 *g*) on a sucrose gradient (6, 172). On assaying for catalase activity across the gradient, a particulate catalase component at equilibrium density ~1.25 g/cm^3 is obtained, and this corresponds to unbroken microbodies, which usually contain uricase and glycolate oxidase in addition (7, 76).

Three types of plant microbodies have been recognized (Figure 1). The first to be isolated were those from the endosperm of young castor bean seedlings where they account for ~20% of the organelle protein sedimenting at 10,000 *g* (19, 20, 28). In addition to the above three characteristic enzymes, and more significantly for their function, these microbodies contain each of the enzymes of the glyoxylate cycle and a complete sequence for the β-oxidation of fatty acids (20, 27, 28, 83). These highly specialized microbodies were named glyoxysomes (19), and they play a major role in the conversion of fat to sucrose in fatty seedling tissues (4).

A second specialized form of plant microbodies is that from leaves (175, 176). These contain high levels of glycolate oxidase, hydroxypyruvate reductase, and certain transaminases (Figure 1) which fit them for a role in

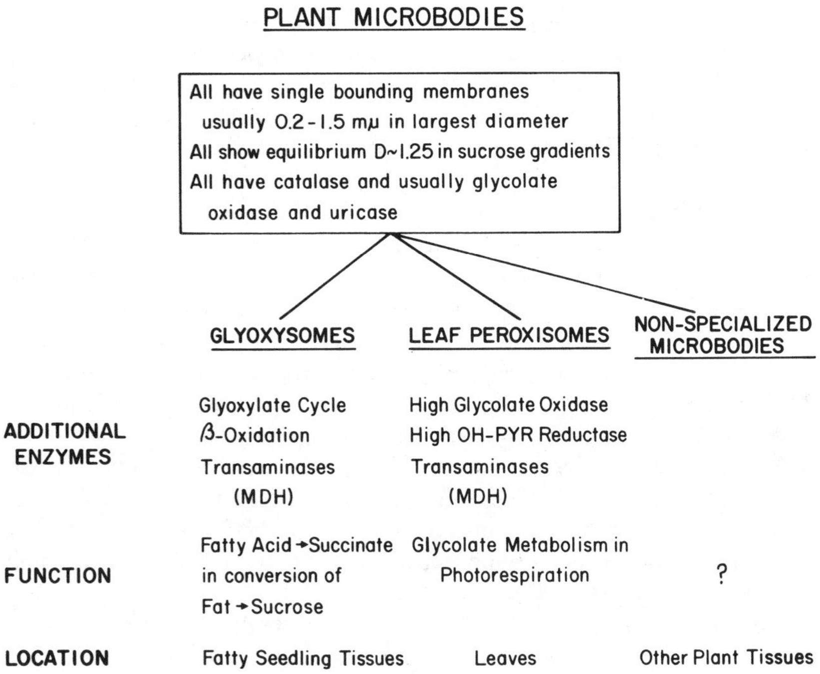

Figure 1 Properties and nomenclature of plant microbodies.

photorespiration (173). They account for only a few percent of the particulate protein and they contain catalase (173) and uricase (76). These microbodies are known as leaf peroxisomes (173).

Microbodies with apparently only catalase, uricase, and glycolate oxidase as major enzyme constituents have been isolated from a variety of other plant tissues (76). Electron microscopy of the parent tissues shows that they are few in number and comprise less than 1% of the particulate protein. Apart from a possible function in detoxication of H_2O_2 and purine catabolism, their function is unknown (Figure 1). We classify these organelles as nonspecialized microbodies (76).

Each of the three types is structurally similar, they share the three microbody enzymes, and they show an equilibrium density of ~1.25 g/cm^3 on sucrose gradients. They are also extremely fragile and osmotically sensitive. Evidence from electron microscopy suggests that the microbodies contain essentially all of the cellular catalase, but when the cells are broken by conventional grinding or blending, only a fraction of the total catalase is usually recovered in the 10,000 *g* pellet, and further losses from the microbodies occur when the pellet is resuspended and centrifuged on the sucrose gradient. Using gentler grinding or chopping with razor blades increases the yield of particulate catalase and other enzymes of microbodies, and if the initial pelleting and resuspension is avoided by layering the homogenate directly onto the gradient, improved yields of intact microbodies are obtained. In castor bean endosperm more than 90% of the cellular catalase (and other glyoxysomal enzymes) is recovered in the microbody fraction prepared in this way (118, 119, 136). This tissue is a particularly favorable one for organelle isolation, and the same methods applied to other tissues do not necessarily eliminate breakage. In most conventional preparations less than 20% of the microbodies are recovered from gradients, even when the yield of intact mitochondria is high. Nishimura et al (141) have made what appears to be the most complete separation of organelles from leaves by first making protoplasts from spinach leaves, breaking them by passage through a hypodermic needle, and applying directly to a sucrose gradient. An essentially complete yield of intact chloroplasts was obtained and the intact leaf peroxisomes were cleanly separated, but, as shown by the activities of marker enzymes at the top of the gradient, more than 60% of the leaf peroxisomes had been broken. Virtually no catalase was associated with the chloroplasts (141).

Thus while it is possible to isolate some microbodies from plant tissues, their fragility poses problems and each particular tissue, by virtue of its idiosyncracies of toughness and noxious constituents, presents its own challenges to the investigator. It will be recognized from the foregoing that preparations of "mitochondria" or "chloroplasts" made by centrifuging a homogenate at 5–10,000 *g* will inevitably be contaminated by microbodies. At the same time it should not be assumed without further evidence that the gradient fractions with catalase activity are microbodies uncontaminated with other cell constituents.

SOURCES OF MICROBODIES: MICROBODY CONSTITUENTS

Glyoxysomes

These have now been isolated from several fatty seedling tissues (26, 75, 88, 110, 113, 178), from aleurone (33, 86, 140) and scutella (109–111) of cereal seedlings and from the Jojoba bean in which wax is the major reserve (137). Huang (75) compared glyoxysomes from five fatty tissues and showed that there was a remarkable similarity in the relative amounts of enzymes concerned with β-oxidation, the glyoxylate cycle, and in catalase and glycolate oxidase. An alkaline "lipase," hydrolyzing monoglycerides, was found in glyoxysomes from castor bean endosperm (138), and this enzyme was shown to be present to varying degrees in the glyoxysomes from other tissues (75). Uricase was shown to be a general glyoxysome component (168), and allantoinase is also present in these organelles from some tissues (1, 168).

From previous work it was thought that glyoxylate reductase was absent from glyoxysomes (20, 28). With the recognition that this microbody enzyme functions primarily with hydroxypyruvate as the preferred substrate (173), this question was reexamined. Hydroxypyruvate reductase (NAD-linked) is indeed present in glyoxysomes, and glyoxylate is reduced by this enzyme when very high substrate concentrations are supplied (116). The K_m for glyoxylate is almost 10^3-fold greater than that for hydroxypyruvate, and since the affinity of malate synthetase for glyoxylate is high, it is most unlikely that the glyoxylate-glycolate conversion accounts for any significant NADH oxidation in glyoxysomes (116). Schnarrenberger et al (157) also drew attention to the hydroxypyruvate reductase activity in glyoxysomes from both castor bean and sunflower seedlings and showed, furthermore, that transaminases not recognized earlier (28) were present. Serine-glyoxylate transaminase, and particularly glutamate-glyoxylate transaminase, showed high activities (157).

In the first report describing glyoxysomes from the fatty gametophyte of pine seedlings, it was concluded that the glyoxysomes had autonomous protein-synthesizing ability (26). However, ribosomes have never been observed to be intrinsic components of microbodies (47, 183), and it was later shown that no specific DNA component was associated with glyoxysomes from castor bean (38). Glyoxysomes purified by flotation do contain a low level of RNA in their membranes (54), which probably reflects their origin from the ER (see below). It now seems clear that glyoxysomes (and other microbodies) depend completely for the synthesis of their proteins on cytoplasmic ribosomes.

Some of the glyoxysomal enzymes have now been purified to homogeneity and their kinetic properties determined. Isocitrate lyase from flax showed a K_m for isocitrate of 0.29 mM and a mol wt of 264,000 (93), while in two different reports, that from cucumber was reported to be 325,000 (99) and 220,000 (94). Malate synthetase from maize had a mol wt of 500,000 (161) and from castor bean 575,000 (17). The K_m values for acetyl CoA were respectively 20 μM and 10 μM, close to that reported in an earlier partial purification (192). Catalase from cucumber has a mol wt of 225,000 (99).

Occasional reports of low levels of isocitrate lyase activity, unaccompanied by malate synthetase, in nonfatty tissues have appeared (58, 82, 142). Hunt et al (82) showed that from suspension cultures of rose cells the isocitrate lyase activity (less than 1% of that in castor bean endosperm) is not associated with microbodies, but is mitochondrial. They suggested that the enzyme might play a role in the provision of glyoxylate for glycine and serine synthesis and showed that the activity was adequate for this. It seems that whenever malate synthetase and isocitrate lyase are present together in higher plant cells they are found in glyoxysomes.

Leaf Peroxisomes

These organelles have been observed in all samples examined by electron microscopy (37, 47, 49, 64, 70, 183), including achlorophyllous leaves (63), those with and without overt photorespiration (49, 70), and those of CAM plants (92). Rather more microbody profiles were observed in sections of bundle sheath cells than in those of mesophyll cells of C_4 plants (49, 70). Since ribulose diphosphate carboxylase is present only in the bundle sheath cells of C_4 species and there is reason to believe that the glycolate substrate for photorespiration arises from the oxygenase function of this enzyme, the possibility that only the bundle sheath microbodies were true leaf peroxisomes was raised. This question has not been examined exhaustively, but in the C_4 species *Atriplex rosea,* it was shown that microbodies from both cell types had the enzyme constitution typical of spinach leaf peroxisomes (77). In sorghum, the bundle sheath microbodies proved to be leaf peroxisomes with low total activities, and the mesophyll microbodies had relatively low glycolate oxidase activity (77). A detailed examination of the transaminases in spinach leaf peroxisomes has been made (146). Uricase, not detected earlier (173), was shown to be present in leaf peroxisomes from spinach (76, 168). Typical leaf peroxisomes have recently been isolated from several CAM plants (69).

Microbodies from Other Tissues

Rocha & Ting (148), using spinach hypocotyl and roots, and Ruis (152a), using potato tuber, were among the first to isolate such microbodies. The isolation of microbodies from nine plant materials, including, among others, roots, potato tubers, petals, and apple fruit, was described by Huang & Beevers (76), and the enzyme constituents compared to those of spinach leaf peroxisomes and castor bean glyoxysomes. From every tissue a particulate fraction containing catalase, clearly separated from mitochondria and lower in the sucrose gradient (density 1.20–1.25 g/cm^3), was obtained. Coinciding with this peak was glycolate oxidase and, in all but two examples, uricase. Only from the spinach leaf, castor bean endosperm, and potato tuber was there a clearly visible band of microbody protein. The potato microbodies did not contain the typical specific enzymes of leaf peroxisomes or glyoxysomes or ten other enzymes for which they were tested. Nonspecialized microbodies of this same general type (Figure 1) have been described from a range of other tissues, including soybean (135) and rose (82) cells in suspension culture, aroid appendices (8, 144), and roots (76, 143, 148). It should be emphasized that relatively few microbodies are present in tissues other than leaves and fatty tissues and the yield from sucrose gradients is small. In spite of the present evidence to the contrary, it is still conceivable that the nonspecialized microbodies from such tissues will be shown to be the *major* intracellular site of some other enzyme activity. For the moment their function seems to be a limited one.

In this regard it should be noted that for microbodies from several sources, a small fraction of the total cellular activity of a variety of enzymes such as those concerned with the metabolism of aromatic acids (10, 60, 95, 96, 153, 154) has been shown to be associated with these organelles. The possible contribution to the overall metabolism of these compounds in vivo is not established. In one instance, the apparent association of a shikimate dehydrogenase with microbodies from pea (152) was shown to be caused by plastid contamination (44).

In a series of papers, Lips and his colleagues (102, 103, 150, 151) have maintained that significant amounts of enzymes concerned with nitrate reduction, and more recently the reductive pentose phosphate pathway (87), are associated with leaf peroxisomes. They emphasize that such enzymes, as well as glycolate oxidase, an accepted marker enzyme, are only loosely attached to the microbodies they have studied, and that this association is lost under a variety of preparative and centrifugal procedures, while catalase activity is retained. Further, they find that a variety of pretreatments of the tissue with hormones and light affect the distribution of enzymes in the gradients (103a). Additional experiments with other systems are needed to establish the generality and significance of these observations. Although not directly accounting for the contrary observations of Lips, there seems now to be a consensus that nitrate reductase is a truly cytosolic enzyme and that nitrite reductase and other enzymes of ammonia metabolism are associated with plastids in leaves and roots (30, 31, 127, 128).

DISTRIBUTION OF ENZYMES WITHIN MICROBODIES

When castor bean glyoxysomes are recovered from a sucrose gradient (ca 52% sucrose) and diluted, most of the protein is immediately solubilized. However, the membranes are not completely disrupted by this treatment and are recovered as ghosts with the original dimensions (78). Many enzymes are solubilized, but more than half of the activities of MDH, fatty acyl CoA dehydrogenase and crotonase, and 90% of the malate synthetase and citrate synthetase activities are recovered with the ghosts. However, subsequent treatment with 0.05 M KCl removes the first three enzymes from the membranes, and the two synthetases are solubilized by 0.2 M KCl while the ghost remains intact (23, 78). Thus the enzymes do not appear to be intrinsic membrane components, but the ionic association of the synthetases with the membrane has been exploited in studies on biogenesis (below). Ruis and his colleagues independently and at about the same time reported similar observations on the association of the same enzyme activities with the glyoxysomal membrane (9, 10), and they have since made further important investigations on the membrane proteins (11). The association of enzyme activities with membranes appears to be somewhat different in glyoxysomes from maize scutella. Although citrate synthetase and part of the MDH was ionically associated with the membrane, malate synthetase was readily removed by osmotic shock (107). It was also observed that considerable loss of enzymes occurred during resuspension and during rapid centrifugation (107), recalling Lips' experience with glycolate oxidase (150, 151).

Confirmation that the enzymes loosely associated with glyoxysomal membranes from castor bean are not integral proteins has been provided by Breidenbach and his colleagues (21, 184). When examined by ESR the membranes showed phase changes in response to temperature, but the enzymes did not respond similarly. However, one enzyme component, the alkaline lipase, is firmly bound to the membrane as an integral component in castor bean (138) and other (75) glyoxysomes.

In experiments with leaf peroxisomes, none of the enzymes was found to be even loosely associated with the membrane (78).

ISOENZYMES IN GLYOXYSOMES AND LEAF PEROXISOMES

With the isolation and characterization of the two specialized forms of microbodies came the recognition that they contain enzyme activities that are also present in other cellular compartments. Thus leaf peroxisomes contain MDH and transaminases, and glyoxysomes contain MDH, citrate synthetase, aconitase, and transaminases that are in this category.

Isoenzymes of MDH

The MDH enzymes have been the most intensively studied, particularly by the groups of Ting, Hock, and Huang [for a review see Ting et al (171)]. It is now established that in leaves, including those of C_3 species (148, 149, 193, 195, 196), C_4 species (197), and CAM plants (164), and in fatty cotyledons (71, 72, 104, 185, 186, 188, 190), endosperm tissue (18, 19, 28, 80), and scutella (108), there are distinctive cytosolic, mitochondrial, and microbody forms of MDH that show clearly different electrophoretic behavior on gels and/or in DEAE cellulose chromatography, and thus are isoenzymes. Like other microbody enzymes the MDH shows low electrophoretic mobility on starch gels. The microbody MDH has been completely purified and characterized from spinach leaf peroxisomes and showed a mol wt of 70,000 (196). Walk & Hock (186) used affinity chromatography on 5'-AMP sepharose to purify the glyoxysomal and mitochondrial forms of MDH from watermelon to homogeneity. The mol wts were respectively 67,000 and 70,000 (190). The kinetic properties were established (188). At high enzyme concentration the microbody MDH tends to form aggregates (18, 108, 190). The microbody forms of MDH are considerably less stable to heat than the other MDH isoenzymes (80, 149, 190, 196), but in their kinetic properties they are not greatly different from the mitochondrial isoenzyme (80, 149, 185, 188, 193, 195, 196).

Antibodies have been raised against the MDH in microbody extracts (80) and in purified preparations (185, 187, 196) which again clearly distinguish the microbody MDH from other isoenzymes by selective inhibition and precipitation. Antibodies to microbody MDH show greater cross-reactivity with MDH in microbodies of other species (both glyoxysomes and leaf peroxisomes) than to other isoenzymes in the parent tissue. The antibodies to microbody MDH have been used as a sensitive means to follow specific MDH synthesis (185, 187, 189) and microbody turnover (185) during functional changes (see below) and to detect microbody MDH in other tissues (164).

Isoenzymes of Citrate Synthetase

The citrate synthetases of glyoxysomes and mitochondria have been compared in castor bean endosperm and corn scutella. In castor bean the kinetic properties of the two enzymes are not greatly different, and both forms of the enzyme are inhibited in parallel when challenged with increasing amounts of antibody prepared against glyoxysomal enzyme (80). A differential response to ATP reported for mitochondrial and glyoxysomal citrate synthetases using organelle suspensions (2) was not observed when solubilized enzymes were used (80). However, the two enzymes can be clearly separated chromatographically (T. Kagawa and E. Gonzalez, unpublished; C. Schnarrenberger, unpublished). The two citrate synthetases from corn scutella are also distinct isoenzymes (3). Their kinetic properties are somewhat different; they were separated by gel filtration and the glyoxysomal enzyme was more sensitive to heat. Both forms were inhibited by ATP, but the mitochondrial enzyme was more sensitive, and whereas the inhibition of the mitochondrial enzyme was competitive with acetyl CoA, that of the glyoxysomal enzyme was noncompetitive (3).

Isoenzymes of Aspartate-α-Ketoglutarate Transaminase

In addition to the enzymes that irreversibly transfer amino groups from serine and glutamate to glyoxylate, leaf peroxisomes also contain the classical aspartate-α-ketoglutarate transaminase (146). Rehfeld & Tolbert (146) showed that the leaf peroxisomes they isolated from spinach contained two distinctive aspartate-α-ketoglutarate isoenzymes that were not detected in other organelles, in addition to one isoenzyme present in mitochondria, chloroplasts, and microbodies, but they recognized that the presence of this isoenzyme in microbodies might be due to cross contamination. Huang et al (81) prepared purer organelle fractions and showed by starch gel electrophoresis that four distinct isoenzymes were present, one each in mitochondria, chloroplasts, leaf peroxisomes, and cytosol. Rehfeld & Tolbert (146) showed that the two distinctive aspartate-α-ketoglutarate transaminases they observed in leaf peroxisomes were separable on polyacrylamide gels and triethylamino ethyl cellulose but not on starch gels, so it is still possible that these organelles contain two isoenzymes.

A more complex pattern, with six isoenzymes separable on starch gels, was observed for the asparate-α-ketoglutarate transaminases in cucumber seedlings (105). Two of the isoenzymes were shown to be glyoxysomal and were apparently lost as the glyoxysomes disappeared during greening. The plastids contained three isoenzymes, but in contrast to the castor bean, where both mitochondria and glyoxysomes have high activities of the aspartate-α-ketoglutarate transaminase (28), no activity of this enzyme was found in the mitochondria from cucumber.

FUNCTIONS OF GLYOXYSOMES AND LEAF PEROXISOMES

The major functions of these two specialized forms of microbodies, deducible from their distinctive enzyme constituents, were established before 1971 and little needs to be added to the outline provided by Tolbert (173).

Investigations with glyoxysomes from a variety of tissues have established that they all have a similar enzymatic complement and that they play a key role in the conversion of fat to sugars that is the dominant event in their metabolism. Thus the whole sequence of reactions whereby long chain fatty acids are converted to succinate by way of β-oxidation and the glyoxylate cycle (29) takes place within these organelles [reviewed by Beevers et al (4, 5, 7)]. Catalase in the glyoxysomes destroys the H_2O_2 generated by the oxidation of flavin reduced in the fatty acyl CoA dehydrogenase reaction. The NADH generated in the second oxidative step in β-oxidation and in malate oxidation in the glyoxylate cycle is not oxidized by the glyoxysomes and NAD must be regenerated by mitochondrial oxidations. The mitochondria can oxidize NADH effectively, but it is still not established whether there is a shuttle of NAD(H) between mitochondria and glyoxysomes or whether a more complex shuttle operates, perhaps involving products of the transaminases present in both organelles (28, 157). The succinate generated in glyoxysomes is oxidized in the mitochondria, and the oxalacetate resulting from this oxidation is decarboxylated in the cytosol by the phosphoenolpyruvate-carboxykinase and the resulting phosphopyruvate quantitatively converted to sucrose (4, 5).

It should be emphasized that in the castor bean endosperm these reactions constitute the respiration of the tissue; all of the acetyl CoA is channeled into the glyoxylate cycle and none is oxidized to CO_2 in the mitochondria (4). The glyoxysomes themselves can convert the long chain fatty acids to their CoA esters (27), and their membranes contain an esterase that hydrolyzes monoglycerides (138). Presumably triglyceride lipase elsewhere in the cell initiates the breakdown of the stored fat, which is almost exclusively triricinolein (34). Hutton & Stumpf (84) showed that castor bean glyoxysomes have the enzymatic equipment to account for the additional reactions needed in the β-oxidation of this unusual fatty acid.

Apart from some clarification of the transaminase constituents (146) there appear to be no new enzymes from leaf peroxisomes that would augment their important role in photorespiration established earlier (173). For more recent assessments of the role of leaf peroxisomes in photorespiration the reviews of Schnarrenberger & Fock (156) and Tolbert (174) should be consulted. The leaf peroxisomes are the unique intracellular location of reactions oxidizing glycolate generated in the chloroplasts as a by-product of photosynthesis and converting it to glycine. The conversion of two moles of glycine to serine, CO_2 and NH_3 in the glycolate pathway occurs outside the leaf peroxisome, but this organelle has the transaminase and hydroxypyruvate reductase that would result in the conversion of serine to glycerate.

What still has not been established is the stoichiometry of the overall conversion of glycolate to glycerate in vivo and how closely the theoretical 75% recovery of glycolate carbon as glycerate is approached. Attention has been drawn to reactions that consume glyoxylate in other ways, which, if operating in the green cell, would result in additional photorespiratory losses of carbon from the glycolate pathway and diminish the yield of glycerate.

Since the glyoxylate is generated in the leaf peroxisome, and active transaminases are present there, it is unlikely that it escapes to the chloroplast, where oxidations driven by light-generated oxidants (40, 194) might overtake it. However, Butt and his colleagues (61, 62, 66, 67, 101) have emphasized that glyoxylate oxidation can readily occur within the leaf peroxisomes themselves, in competition with the conversion to glycine, and that some features of this oxidation correlate well with known responses of the photorespiratory release of CO_2. Grodzinksi & Butt (61, 62) showed that sufficient H_2O_2 escapes catalase action to induce glyoxylate decarboxylation. They suggest that the striking increase in photorespiration when the temperature is raised from 25°C to 35°C can be accounted for by their observations with isolated leaf peroxisomes, where, due to the different temperature coefficients of catalase and glycolate oxidase, much more H_2O_2 becomes available at the higher temperature and correspondingly more glyoxylate is oxidized.

DEVELOPMENT OF MICROBODIES

It is in this area of research that most progress has been made since the last review. The microbodies do not contain their own genetic machinery (38), and they must depend for their proteins on synthesis by cytoplasmic ribosomes directed by nuclear events. Microbodies are frequently seen in electron micrographs to be close to ER profiles and occasionally fusing with them (113, 124, 130, 139, 182, 183). The suggestion has thus been made that microbodies are derived from ER, but on this evidence alone the reverse might just as well be true.

Glyoxysome Biogenesis

Most of the biochemical investigations have been made on microbody development in fatty seedling tissues where, within a few days, there is usually a rapid and more or less synchronous development of glyoxysomal enzymes (55, 74, 79, 110, 170) from essentially zero levels in the dry seed, and there is a corresponding increase in the numbers of glyoxysomes during this period (55, 88). Between days 2 and 4 in the endosperm of the castor bean seedling growing at 30°C the amount of glyoxysomal protein (equivalent to number of glyoxysomes) increases sevenfold (55), and electron micrographs clearly show this increase in the glyoxysomal population (182). As the reserve fat is depleted the glyoxysomes disappear (55), as they do in other fatty seedling tissues (23, 24, 50, 65, 88, 111, 159).

It was established early that the increases in isocitrate lyase (57, 106) and malate synthetase (106) activity occurred by de novo enzyme synthesis. This has now been shown also for the specific glyoxysomal MDH (187). The responses of the glyoxysomal enzymes malate synthetase and isocitrate lyase suggested that it was only during the early stages of germination that DNA transcription for these enzymes occurred in watermelon seedlings (74). Although it appeared from the first experiments with cotton seedlings that messenger RNA preformed in the ripening seed was used for isocitrate lyase synthesis (85), more recent experiments have shown that here, too, there is transcription very early in germination but that thereafter isocitrate lyase development is not affected by actinomycin D (145, 163). Recently new techniques have made it possible to investigate glyoxysomal enzyme synthesis at the subcellular level (see below).

In the meantime, substantial progress has been made on glyoxysomal biogenesis in castor bean endosperm by studying first the origin of the P-lipid constituents of their membranes (reviewed in 4, 5). The resolution of whole homogenates on sucrose gradients (118, 119, 136) greatly facilitated this work. On such gradients essentially complete recovery of intact glyoxysomes and mitochondria is obtained and, in addition, lighter components are simultaneously resolved into three fractions (118, 136). These are components which would be present in the heterogeneous "microsomal" pellet obtained by centrifuging the 10,000 *g* supernatant at 100,000 *g*. One of these components, a clear band at density 1.12 g/cm^3, was shown to be comprised of vesicles derived from the ER network that fragmented during homogenization (119). The ribosomes, most of which are seen to be associated with the ER in vivo (186), are recovered separately at density 1.11 g/cm^3 after the 2–3 hr centrifugation period (119). When Mg^{2+} is included in the gradients, the association of ER and ribosomes is maintained, and the ER vesicles (with their marker enzymes) are then recovered as a broad band lower in the gradient and appear in electron micrographs as rough ER (119).

The glyoxysomal membranes have PC, PI, and PE as their major constituents (35, 36), and when ^{14}C-choline was supplied to the endosperm tissue and the organelles separated in the standard gradient (without Mg^{2+}), it was found that the PC in glyoxysomes (and mitochondria) had become labeled (89). However, kinetic analysis showed that the first membranes to become labeled from choline were those of the ER, and glyoxysomes were labeled secondarily. As the percent of the total labeled PC declined in the ER, it increased progressively in glyoxysomes and mitochondria. Chase experiments confirmed that PC in the glyoxysomes was derived from that in ER, although the ER remained the most heavily labeled component throughout (89).

The biochemical basis for this observation was provided when the distribution of CDP-choline-diglyceride transferase, the final enzyme in the PC synthesis pathway, was examined. This enzyme was present exclusively in the ER (118). The alternative pathway of PC biosynthesis, by methylation of PE, was subsequently shown to be present at about 5% of the level of the CDP-choline transferase. Most of this activity was again present in the ER and the glyoxysomes showed no activity (134).

It was then found that the ER contains, in addition, the enzymes bringing about the following reactions of P-lipid synthesis: (*a*) synthesis of PI from inositol and CDP-diglyceride (136); (*b*) synthesis of PE from CDP-ethanolamine and diglyceride (113a); (*c*) exchange reaction between PE and PS (136); (*d*) synthesis of PG, also present in mitochondria (132); (*e*) esterification of glycerol-phosphate by fatty acids from acyl CoA precursors (180); and (*f*) the hydrolysis of the resulting phosphatidic acid to give the diglyceride precursor of the P-lipids (136). Moore (132–134) has made detailed investigations of the properties of several of these enzymes.

Thus, starting with glycerol-P, the CoA derivatives of the fatty acids [generated from acetyl CoA in the plastids (180a)], the activated bases and inositol, the synthesis of each of the P-lipids of the glyoxysomal membrane is brought about in the ER. Since the glyoxysomes themselves do not have these activities, it is clear that they must depend on the ER for the synthesis of the P-lipids in their membranes. It should be noted that in other tissues the Golgi vesicles are an additional site of PC and PE synthesis (131). In the castor bean endosperm Golgi vesicles are seen to be a very minor component (182), and no enzymes of P-lipid synthesis are found in the sucrose gradients in regions where these organelles would be expected (118, 136).

Additional evidence for the role of the ER in the biogenesis of P-lipids destined for glyoxysomal membranes has come from P-lipid analyses. The relative amounts of the individual P-lipids in the glyoxysomal membranes are quite similar to those in the ER and different from those in the mitochondria (35, 36). Further, within each of the P-lipids of the glyoxysome the percent composition of the substituent fatty acids is very close to that in the corresponding P-lipid from the ER (35). When labeled acetate and glycerol were provided to endosperm tissue, the P-lipids in the ER were again the first to become labeled and the glyoxysomes only secondarily (34). The role of the ER in biogenesis of glyoxysomes is further supported by the observation that several of the enzymes of P-lipid synthesis have been shown to be present early in germination, before the main surge of glyoxysomal development (13, 180).

On the basis of these observations, we proposed (4, 119) that the membrane of the glyoxysomes was derived rather directly, by a process of vesiculation and excision, from specific regions of the ER in vivo (Figure 2). The other possibility, that is not completely excluded by the evidence, is that the P-lipids are transferred individually from the ER and assembled in the cytosol to form the developing membrane of the glyoxysome. Such transport must occur into developing mitochondria and plastids, but these organelles are partially autonomous and their precursors are present in the dry seed, providing targets for the reception of newly synthesized P-lipids required for their expansion and multiplication. There is no evidence that in the castor bean endosperm such precursor structures are present for glyoxysomes or that they undergo fission. Nevertheless, transport of P-

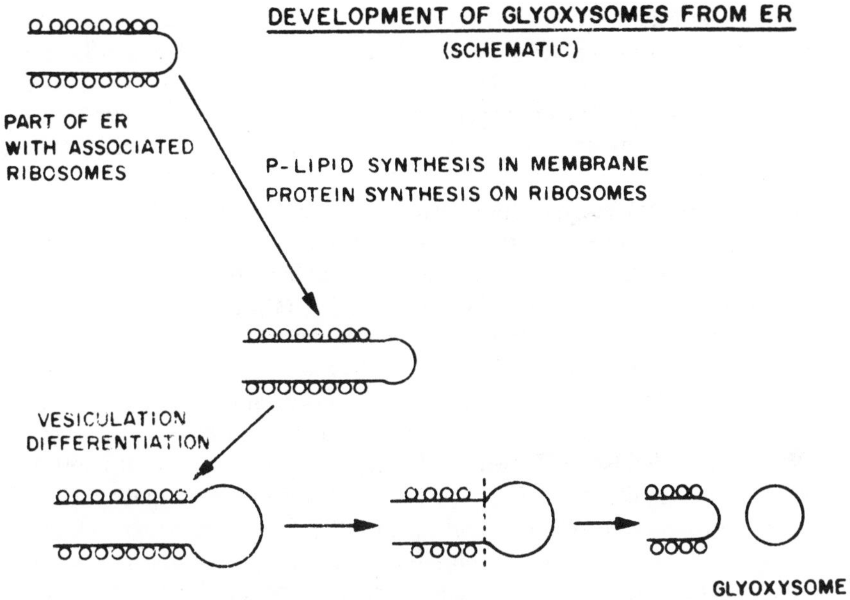

Figure 2 Biogenesis of the glyoxysome from the ER.

lipids from ER to existing glyoxysomes may occur, and Mazliak (124a) has provided some evidence that proteins responsible for this transfer are present. In experiments in which CDP-choline and CDP-ethanolamine were provided to crude homogenates (in which the ER is disrupted and the proposed vesiculation to form glyoxysomes would not be occurring), the PC and PE in ER were heavily labeled and only a small amount of transfer (exchange) to glyoxysomes was observed (114).

Evidence for the origin of the glyoxysomes from the ER has also come from examination of the protein constituents of membrane and matrix, which suggests that these glyoxysomal constituents are synthesized in vivo by the ribosomes investing the ER and introduced into the expanding vesicle, representing the developing glyoxysome. For matrix enzymes the fact that the ER network is disrupted and soluble constituents presumably lost from the lumen during vesicle formation makes an investigation of their possible origin by the model proposed difficult. However, as pointed out earlier, there are two enzymes, malate synthetase and citrate synthetase, which in the mature glyoxysome are associated with their membranes (78). After 4 days of growth more than 90% of the malate synthetase is recovered in the glyoxysomes. However, at 2 days, when only a fraction of the total enzyme activities have developed, Gonzalez & Beevers (59) showed that the glyoxysomes contained only ~50% of the total malate synthetase activity and that the rest was present in the ER fraction. This malate synthetase could be removed from the ER by salt treatment as it is from the mature glyoxysomes. During further development it was shown that the percentage of total malate synthetase in the glyoxysomes rose as that in the ER fell to very low levels.

A similar picture emerged for citrate synthetase, complicated only by the fact that this enzyme activity is found both in glyoxysomes and mitochondria (59). In these experiments, as expected from the argument above, catalase, which shows no membrane association in the mature glyoxysome, was not recovered in the ER at any stage but was present higher in the gradient at a position expected from soluble catalase with its known high molecular weight. We attribute the origin of this catalase to the ER lumen or preglyoxysomal vesicles originally attached to the ER (59). In a very recent paper, Choinski & Trelease (26a) describe experiments with cotton seed in which (in distinction from castor bean) some glyoxysomal enzymes develop during seed maturation and ripening. They showed cytochemically that catalase was present within dilations of the rough ER and that some activity was recovered in "microsomal" fractions obtained by centrifuging at 150,000 *g*. Köller & Kindl (97, 98) have recently investigated the association of malate synthetase with ER in germinating cucumber seeds. They find that under certain centrifugation conditions, malate synthetase in the ER can be separated into two components, and that part of the malate synthetase is separated from the main peak of NADH-cytochrome *c* reductase, the ER marker. They suggest that this component may be enriched in preglyoxysomal vesicles (98).

Gonzalez & Beevers (59) also showed, using mixed antibodies prepared against extracts of mature glyoxysomes (80) from castor bean, that glyoxysomal proteins other than malate synthetase and citrate synthetase were present in 2-day ER and its membranes, and the alkaline lipase, an integral component of the glyoxysomal membrane, is also found in the ER (138). Our model of glyoxysomal biogenesis (Figure 2) has received additional strong support from experiments by Bowden & Lord. They showed first that the malate synthetases in ER and glyoxysomes were identical proteins in every respect examined, including immunological properties (17). They then used antibody prepared against purified glyoxysomal malate synthetase to follow early events in malate synthetase production, supplying labeled methionine to endosperm tissue and specifically precipitating the newly synthesized enzyme in the separated organelles (117). At the earlier stages most of the malate synthetase was present in the ER, and the enzyme in this component was clearly labeled before that in the glyoxysomes (15, 17). They then showed in elegant chase experiments that malate synthetase was lost from the ER and quantitatively recovered in the glyoxysomes (117). Lord has also demonstrated that proliferation of the ER precedes the development of glyoxysomes in the endosperm tissue (115).

The proteins of glyoxysomal membranes have also been examined by Lord (14, 16, 22, 23), Ruis (11), Ludwig & Kindl (120, 121), and Hock (73). It has been found that proteins from ER and glyoxysomes separated on SDS gels and by isoelectric focusing are remarkably similar, and that when radioactive amino acids are supplied to the tissue, the membrane proteins of the ER become labeled before those in the glyoxysomes.

To reiterate, the foregoing evidence taken together supports a model of glyoxysome biogenesis (Figure 2) in which proteins synthesized on ribosomes attached to ER membranes in vivo are introduced into a vesicle differentiating from the ER and whose phospholipids are concurrently synthesized in the growing membrane (5). The model is consistent with the EM evidence and does not preclude the subsequent addition of components to the primary glyoxysome. Its further testing in the castor bean and other systems is awaited.

The detailed investigation of molecular events in glyoxysomal enzyme synthesis and packaging is presently being made in several different laboratories, including those of Lord, Hock, Becker, and Gonzalez. Antibodies to purified glyoxysomal enzymes as sensitive detection devices for labeled proteins and separations of messenger RNAs and bound polysomes are suitably in vogue (15–17, 73, 100, 117, 187, 189). The signal hypothesis of Blobel & Dobberstein (12a), in which enzyme segregation and vectorial insertion through membranes is achieved by way of a higher molecular weight precursor, has attracted attention (15, 115, 117, 189). Walk & Hock (189) have recently achieved the cell-free synthesis of glyoxysomal MDH directed by poly A-rich mRNA from germinating watermelon cotyledons. Interestingly enough, they found that in addition to the authentic glyoxysomal MDH (subunit mol wt 33,000), the major product was an immunologically identical MDH with subunit size of 38,000 (189). Again, glycosylation may be important in enzyme segregation, and the recent observations that several of the glyoxysomal proteins are glycosylated (125, 147) and that glycosylation occurred during or immediately after synthesis in the ER (126) are clearly of great interest.

Glyoxysomes disappear from fatty seedling tissues as fat utilization is completed. In seeds such as castor bean this is the prelude to complete breakdown of the endosperm tissue itself and absorption of the products by the embryo. During the phase of declining activities of the glyoxysomal enzymes in castor bean, it was found that the amount of glyoxysomal protein declined in parallel (55) as it does in pumpkin (23). However, the glyoxysomal enzymes were not recovered in the supernatant fractions, as would be expected if the organelles lost their integrity and the enzymes were released into the cytosol. In fact, the specific activities of isocitrate lyase and malate synthetase in the surviving glyoxysomes remained remarkably constant from the time that peak activity was reached until tissue senescence. This suggested that the glyoxysomes were segregated as whole units into a cellular compartment and rapidly digested there (55). The autophagic vacuole represents such a compartment and these are in fact observed with glyoxysomes inside them in electronmicrographs of older cells from castor bean endosperm (182). A different fate was deduced for the glyoxysomes in maize scutellum, which remains functional as an absorptive organ after fat utilization is complete and the glyoxysomes have disappeared. In this tissue, where part of the β-oxidation is thought to be mitochondrial (112), the disappearance of the glyoxysomes was apparently accompanied by the release of glyoxysomal protein into the cytosol (111). Theimer (165) found that in sunflower cotyledons at the time of glyoxysomal disappearance, a proteinaceous inactivator specific for isocitrate lyase was present. He suggested that in particular tissues there may be a special breakdown system for the individual glyoxysomal enzymes.

Biogenesis of Leaf Peroxisomes

Relatively little work has been done on the development of these organelles in true leaves where they persist as functional units for much longer during the life of the leaf than the glyoxysomes in seedlings. Feierabend and his colleagues have made the most intensive investigations, in developing cereal leaves (41–43, 46). Catalase and hydroxypyruvate reductase are present in low activities in the shoot of ungerminated wheat seeds, and the activity of these enzymes as well as that of another leaf peroxisomal enzyme, glycolate oxidase, develop to some extent in complete darkness (see also 64). However, the provision of continuous light greatly stimulates enzyme development, particularly that of glycolate oxidase and hydroxypyruvate reductase, and chlorophyll development occurs concommitantly. The development of the leaf peroxisomal enzymes was not synchronous, with the light-induced stimulation of catalase showing a distinct lag. When senescence was induced by excision, each of the leaf peroxisomal enzymes declined, but the loss of catalase was delayed (42). In fully greened leaves the activities of the leaf peroxisomal enzymes were recovered together in a sharp band at a density of 1.25 g/cm^3 on sucrose gradients. During development in darkness the marker enzymes were broadly distributed at lower densities and not completely coinciding on the gradients (43). Only after exposure to light, as the enzyme activity increased, was a shift to the final equilibrium density of 1.25 observed. However, it was shown that although light was required for the production both of leaf peroxisomes and chloroplasts, the two are regulated independently. Thus, leaf peroxisomes developed normally when the leaves were given short light treatments and inhibitors which did not allow chlorophyll development to occur (43).

It thus appears that in darkness a leaf peroxisomal precursor particle with low levels of enzyme activities is produced and that light is required to generate the complete leaf peroxisome. In their ultrastructural investigation of development of greening bean leaves, Gruber et al (64) provided evidence for small and enlarging leaf peroxisomes, and concluded that the life time of the leaf peroxisome was probably at least 4 days.

In further investigations, Feierabend showed, by short exposures to red and far-red light, that the phytochrome system had a controlling influence over the development of leaf peroxisomes. However, the response to light of other wavelengths could not fully be accounted for by this pigment system acting alone (41). The independence of microbody and chloroplast development was again neatly demonstrated in rye and other cereal leaves exposed to slightly elevated temperatures (45, 46). Under these conditions a lesion is induced that leads to a complete lack of development of chloroplastic ribosomes and thus of completed chloroplasts. The development of leaf peroxisomes and cytosolic enzymes was essentially unaffected by this heat treatment (45, 46).

Microbody Transition in Fatty Cotyledons

In many seeds, the reserve fat is present in cotyledons which in the normal course of events emerge from the ground and become functional leaves. During the early stages of development when fat is being consumed, glyoxysomes are functioning in the cotyledons, but later, leaf peroxisomes with their typical enzymes are present. This changeover occurs normally when the cotyledons gain exposure to light; the glyoxysomal enzymes decline sharply at the time that the leaf peroxisomal enzymes are produced. This is an interesting transition, and no aspect of microbody development has attracted more attention from more investigators. What makes the investigation troublesome is that the glyoxysomes and leaf peroxisomes are morphologically identical and cannot be clearly separated from each other on sucrose gradients.

It should be emphasized at the outset, however, that this process is restricted to those fatty cotyledons that become (briefly) functional leaves; glyoxysomes are never present in true leaves and the leaf peroxisomes arise de novo. To account for this transition two basic models have been suggested. One, proposed initially by Newcomb and his students and referred to as the one-population or repackaging hypothesis (Figure 3), suggests that the transition is accomplished by a drastic change in the enzymatic complement of existing microbodies (178). The other, with which I and (at least

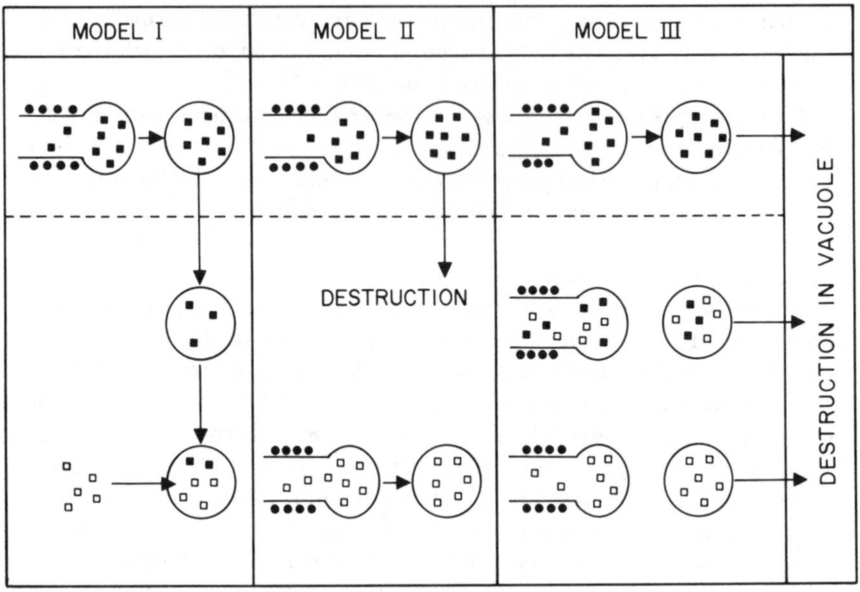

Figure 3 Models proposed for the transition of glyoxysomes to leaf peroxisomes in fatty cotyledons. Phase I, above the dashed line indicating the onset of light, shows glyoxysome formation. Vesiculation of a portion of the ER is shown, with ● representing ribosomes and ■ glyoxysomal enzymes. Model I, the one-population or repackaging model, shows the loss of some glyoxysomal enzymes and the insertion of leaf peroxisomal enzymes □. Model II, the two population model, shows the destruction of glyoxysomes and the de novo biogenesis of leaf peroxisomes initiated by light. Model III is that of Schopfer et al (see text).

some of) my former colleagues are identified, proposes that the glyoxysomes themselves disappear during the transition and are replaced by newly generated leaf peroxisomes—the two-population hypothesis (88, 90, 91). We attach significance to the fact (90) that in other organs precedents are already established for the complete destruction of glyoxysomes after fat utilization is complete (the castor bean endosperm) and for the de novo synthesis of leaf peroxisomes (true leaves), and suppose, as Occam might, that in the fatty cotyledons these same two processes also occur and normally overlap in time (Figure 3).

In approaching this question, several different kinds of fatty cotyledons have been used, including those of watermelon (88, 90, 91), cucumber (25, 104, 105, 167, 177, 178, 185), pumpkin (24), sunflower (50–52, 56, 157, 167), and mustard (39, 158, 159, 179), and the assumption is implicit that whatever the mechanism of transition is, it will be the same in all. Certain general features are established, although the precise timing of the changes is different in different species. If the seedlings are maintained in darkness, the distinctive enzymes of the glyoxysomes, which increase early in growth, decline to very low levels in a few days as the fat is depleted. The provision of light accelerates this decline and at the same time elicits the rapid development of glycolate oxidase and hydroxypyruvate reductase. The production of glycolate oxidase in sunflower and mustard is known to occur by de novo synthesis (51). Short light treatments induce the development of the two leaf peroxisomal enzymes in the absence of chlorophyll synthesis (91), and it has been clearly shown for mustard that the enzyme induction is under the control of the phytochrome system (39, 159). In both of these respects the development of leaf peroxisomes in the cotyledons is identical to the development of these organelles in true leaves (see above). Furthermore, in the watermelon [and in pumpkin (24)], this process of leaf peroxisomal development can be experimentally dissociated in time from the decline in glyoxysomal activity with which it normally overlaps (88, 91). Leaf peroxisomes and indeed the whole functional photosynthetic apparatus was developed (although not to normal levels) when light was supplied to watermelon seedlings held in darkness for 10 days, by which time the activities of glyoxysomal enzymes had declined to almost zero. The decline of enzymes in darkness was accompanied by a fall in glyoxysomal protein that was not reversed by light exposures, and we interpret this as representing a net loss of glyoxysomes (see also 23, 24). It is our contention that the numbers of leaf peroxisomes in the cotyledons after greening must be considerably less than the numbers of glyoxysomes at the peak of their development, and thus, if the one-population hypothesis is valid at all, only a part of the glyoxysome population can undergo conversion to leaf peroxisomes (88, 90, 91).

It is important in assessing the one-population hypothesis (Figure 3) that its full implications are recognized. The glyoxysomal membrane is supposed to persist and to become the leaf peroxisomal membrane during the transition. The distinctive enzymes of the glyoxylate cycle and of β-oxidation [and apparently some transaminase isoenzymes (105)] must be destroyed. However, part of the catalase must remain (see below), and it appears that this applies also to the MDH, since glyoxysomal and leaf peroxisomal isoenzymes are immunologically indistinguishable (104, 185). However, Liu & Huang (104) point out that it is still possible that new MDH is necessary, and certainly newly synthesized glycolate oxidase and hydroxypyruvate reductase must be introduced through the surviving membrane in the one-population model (Figure 3).

It also appears that a distinctive component of the glyoxysomal membrane, the alkaline lipase, must also be destroyed, since this enzyme is not present later in the leaf peroxisomes (A. Huang, personal communication). It is clear from the experiments in which leaf peroxisomes are induced after prolonged dark treatments that functional glyoxysomes are not required for the generation of leaf peroxisomes in the cotyledons, since under these conditions the glyoxysomal enzymes virtually (88, 91) or completely (24) disappear. It might still be argued that under these conditions the glyoxysomal membrane persists and is then repackaged, but there is no experimental evidence for this. Indeed, evidence has been presented to show that at the time of transition there is a loss of glyoxysomal membrane as judged from the decline in previously labeled PC (90). Using double labeling experiments with ^{14}C- and ^{3}H-labeled choline, Kagawa et al (90) showed that in spite of this net loss, a distinctive change occurred in the isotope ratio in the microbody membrane, specifically induced by light. It was concluded that this represents the synthesis of new leaf peroxisomal membrane (90). Neither the light induced loss of microbody PC or the specific increase in light are expected in the one-population model (Figure 2).

A major piece of evidence that has been advanced against the two-population hypothesis is that gained from electron microscopy of sections of cotyledons (178). In passing it should be noted that however elegant the pictures, they sample only a minute fraction of the total organ, and it is exceedingly difficult to estimate the numbers of organelles per cell, particularly when the cells are large and their geometry is changing. Nevertheless, in a careful investigation of cucumber cotyledons at different stages of development, Trelease et al (178) could find no ultrastructural evidence either for glyoxysomal destruction or for the genesis of leaf peroxisomes during the transition. However, it is known from biochemical evidence that during this time there is also a sharp decline in mitochondria, and this also was not apparently observed in the electron micrographs. Furthermore, even in true leaves, where the generation of new leaf peroxisomes is not in question, electron micrographs by the same group did not reveal in any abundance structures that could be definitively recognized as leaf peroxisomes in the actual process of biogenesis (64).

After counting glyoxysomal profiles in many sections of cotyledons from cucumber seedlings maintained in darkness for 4, 7, and 10 days, Trelease et al (178) concluded that there was no marked decline in the numbers of these organelles in the cells. However, it should be pointed out that the enzyme changes in this material (although strangely erratic) showed that malate synthetase was still present at 10 days, with at least one-third of its peak value, and storage lipid was still visible in the electron micrographs from 10 day cotyledons (178). No estimates were made of the numbers of microbodies per cell in the cotyledon after it had become green, or of any changes in microbody numbers during the crucial period of transition. Clearly we do not feel that the present ultrastructural evidence is decisive (88, 90).

Trelease et al (178) drew attention to the fact that at the time of transition, and primarily at this time, many of the microbodies showed evidence of invaginations with cytoplasmic inclusions, and this has also been observed by Theimer's group (167) working with sunflower and Schopfer working with mustard (159). It was suggested that this may be related to events in the microbody transition (178). Attention was also drawn to the changes observed in the spatial relationships between microbodies and other organelles at different developmental stages. During fat utilization the glyoxysomes were seen to be interspersed among the spherosomes (lipid bodies or oleosomes), but after greening they became preferentially associated with chloroplasts (178). Since there is only a limited cytoplasmic space available, and the electron micrographs give only a transiently static representation of what must be a dynamic situation in vivo, one may question whether any functional significance can be attached to these associations; but the authors were persuaded by the frequency with which they were observed that they did reflect functional interactions. And, it must be emphasized, similar associations, though much more intimate, are seen in mustard (158, 159; see below). An in vitro interaction between chloroplasts and leaf peroxisomes induced by phosphate has been reported (155).

The behavior of catalase during the microbody transition is of particular interest, since this is an enzyme common to both organelles. Catalase, like the other glyoxysomal enzymes, declines in darkness as expected if the glyoxysomes were themselves destroyed. The provision of light at any time after 4 days actually leads to an accelerated decline in catalase activity in watermelon (88, 91) but in these, as in other materials, a fraction of the total activity remains—that associated with glycolate oxidase and hydroxypyruvate reductase in the leaf peroxisomes. However, Drumm & Schopfer (39) showed that in mustard, far-red light treatments that induce the synthesis of glycolate oxidase and hydroxypyruvate reductase (179) also lead to an increase in catalase activity, with all three enzymes increasing coordinately. Furthermore, although previous work using other methods had not revealed the existence of multiple forms of catalase, several isoenzymes were resolved on starch gels with some of these arising during the transition. On its face, this evidence seems to support the notion that there is a de novo synthesis of leaf peroxisomes at this time, but the authors (39) claim that since the original isoenzymes remain (as would indeed be expected from surviving glyoxysomes), the new isoenzymes supplement rather than replace the existing ones, and they conclude that their evidence is more consistent with the one-population hypothesis. In sunflower, as in other tissues, there is a net loss of catalase in the light during the transition, and no evidence of new isoenzymes was found, even by starch gel electrophoresis (50, 56).

Of course the two-population model (Figure 3) would require that there is synthesis of new catalase for the leaf peroxisomes during the transition. The fact that total catalase is declining rapidly at this time makes an investigation of this question difficult, but Gerhardt, fully recognizing the difficulties, has made a detailed investigation of this question using inhibitor treatments, radioactive precursors, and density labeling (50, 52, 56). He concludes that although there is catalase synthesis at the time of transition, the rate is very small in relation to that during glyoxysomal biogenesis and specifically no greater than that shown later when the leaf peroxisomes are fully developed and deduced to be due simply to catalase turnover (half life ~2 days). This evidence is clearly difficult to reconcile with the two-population hypothesis.

Experiments have also been made with D_2O to investigate whether there is a change in the specific density of whole microbodies during the transition. It is argued that if there were a de novo synthesis of leaf peroxisomes at this time, they should have a greater density than leaf peroxisomes produced in H_2O. Although it would appear difficult to predict the changes anticipated in the densities of whole organelles, as opposed to single enzyme proteins (see also 158), Theimer et al (167) found that the chloroplasts and thylakoids produced in D_2O during the transition in sunflower were heavier by ca 0.03 g/cm^3 than those in water. The change in density of the microbodies was judged not to be significant, although in D_2O the peak of hydroxypyruvate reductase appears to have increased in density by 0.007 g/cm^3 and the catalase peak by a similar amount. The authors conclude that their experiments constitute direct evidence against the two-population hypothesis. A precisely opposite conclusion was drawn by Brown & Merrett (24) from their experiments on microbody development in D_2O in pumpkin cotyledons, although the shifts in density of the leaf peroxisomal markers were essentially no greater than those observed by Theimer et al (167). Brown & Merrett (24) argue that the changes are significant and calculate, though somewhat precariously, that they are those expected if a major de novo synthesis had occurred.

A different approach to the problem was made by Burke & Trelease (25). They reasoned that if the two-population hypothesis was correct, it should be possible during the transition to demonstrate by cytochemical tests on the organelles in situ that newly synthesized leaf peroxisomes were present. Leaf peroxisomes produced by the one-population model (and incidentally surviving glyoxysomes) would, by contrast, contain both enzymes. This is an attractive idea, and having developed methods for localizing glycolate oxidase and malate synthetase (177) in organelles examined by electron microscopy, they applied them to cucumber cotyledons (25). To be completely convincing this method should, if possible, be applied to successive serial sections of the tissue, stained alternately for malate synthetase and glycolate oxidase so that individual microbodies can be monitored. This was not the method employed. The actual counts of stained microbodies were made on sections of pellets from isolated organelle fractions individually stained for the two enzymes. Estimates of the percentage of the microbodies containing glycolate oxidase and the percentage containing malate synthetase were made by counting profiles. Comparable sections were stained for catalase to arrive at a correction factor for profiles not staining. The conclusion was reached that 94–97% of the microbodies contained both enzymes (25). On its face this appears to be evidence for the one-population hypothesis. However, when the figures showing the actual sections are examined (presumably these are the best ones available), the picture is not so clear. That showing staining for catalase [Figure 4 in (25)] is quite convincing. The section stained for malate synthetase (Figure 5) is less clear, and that for glycolate oxidase (Figure 7) is hardly recognizable as being a similar fraction to that shown in Figure 4; there is much more amorphous and assorted stained and unstained material, and clear microbody profiles are difficult to distinguish. A trained observer may be able to make objective microbody counts on such a section, and the numbers arrived at are certainly impressive. Nevertheless, a reader biased against the one-population hypothesis for other reasons would need more rigorous cytochemical evidence than that now available to be persuaded completely against it.

Other pertinent reservations concerning this and other work are outlined by Schopfer et al (158). They introduced yet another approach, again from electron microscopy, using mustard cotyledons. These workers had previously drawn attention (159) to the fact that in this tissue glyoxysomes are frequently in the most intimate contact with spherosomes, whereas after the transition they were preferentially appressed to plastids. They therefore proposed to decide whether a microbody in a given electron micrograph was a functional glyoxysome or leaf peroxisome by which of spherosome or plastid it was in surface contact. They made extensive counts on sections of cotyledons removed at various times during the rigidly controlled 2 day transitional period. Somewhat surprisingly (at least to me), given the complexity of the structures and the changing organelle population, some general and regular trends were nevertheless observed. Thus the percentage of microbody profiles attached only to spherosomes declined from about 80% at the time of light provision to less than 20% in 2 days. The corresponding figures for microbodies attached only to plastids rose from 5% to 25% in the final 12 hours. Particular significance was attached to an increase, from 5% to over 40%, in the microbody profiles attached both to spherosomes and to plastids. From this observation alone it was deduced that these organelles must be playing a dual function, and indeed they were dignified by a new name, the "glyoxyperoxisome." It was claimed that the dual association with spherosomes and plastids is "clearly incompatible" with the two-population model, and the one-population model was summarily dismissed.

To account for their results, Schopfer et al (158) proposed yet another model (Figure 3). They adopted the model that we have described above for glyoxysome biogenesis on the ER and the same model for synthesis of leaf peroxisomes after the transition, and introduced an intervening stage. They proposed that there is a continuous generation of microbodies from the ER and continuous destruction in the vacuole. During the transition it is supposed that there is a gradual decrease in the rate of synthesis of glyoxysomal enzymes and a concommitant increase in the rate of synthesis of leaf peroxisomal enzymes, hence the hypothetical "glyoxyperoxisome."

It is the basis for the proposal of this new organelle, the assumed link between proximity and function, that is likely to raise the most questions. The biochemists will certainly need more evidence for its existence, but until methods are available for separating or distinguishing unequivocally between glyoxysomes and peroxisomes in a single tissue, this question is likely to remain unresolved. It might reasonably be asked how the proposed model applies in other tissues where the microbody contacts with other organelles are less obvious. It could be argued of course that Phase II of the Schopfer model may be very much briefer and less dominant than it is supposed to be in mustard, but it should be especially noted that it is demonstrably completely absent in cucurbit cotyledons when they are kept in darkness until the glyoxysomal enzymes have disappeared before the development of leaf peroxisomes is induced by light.

Phases I and III of the Schopfer model are, of course, Phases I and II of our two-population model (Figure 3). And in fact the only real difference between the two is that we suppose that when light is supplied, the induced synthesis of the distinctive leaf peroxisomal enzymes is initiated at new points on the ER to give rise to a new population, the leaf peroxisomes, and the biogenesis of glyoxysomes is simultaneously shut down. In fact, far from the continuing synthesis of glyoxysomal enzymes shown in Schopfer's Stage II, it is the fact that provision of light actually leads to an accelerated decline in the amounts of these enzymes in most tissues (88, 91). This, it seems, could not be accommodated in Schopfer's model without invoking a second and independent effect of light on glyoxysome destruction.

Of course the Schopfer model is subject to the same questions that have been raised against the two-population model. But, we maintain, these questions largely concern predictions of the model that are not found to be realized, particularly the synthesis of new catalase when light is supplied, and the ultrastructural evidence for the destruction of whole glyoxysomes and the light-induced synthesis of leaf peroxisomes. In fact, rather than providing positive evidence in favor of the one-population hypothesis, the proponents of this model [with the possible exception of Burke & Trelease (25)] have sought only evidence against the other. Evidence for the actual repackaging of the glyoxysome with new enzymes at the time that some others within the organelles are being selectively or partially destroyed, the key postulate of the one-population model, is notably lacking.

As Gerhardt has pointed out in a thoughtful assessment (56), there are still no data on which everyone can agree about the actual numbers of microbodies per cell before, during, and after the transition. A full understanding of the transition and of microbody biology in general will also require information on the dynamics of turnover of microbodies and their enzymes during the life of the cell.

Hormonal Influences

The question of hormonal influences on the development of microbodies and their enzymes has been considered by several investigators. Even in those instances where stimulation is seen, it is not clear that specific effects on microbody development only are being observed, or only a more general influence on cell development or cell differentiation.

The development of glyoxysomes and other organelles in the endosperm of castor bean is independent of the embryo, since it occurs at the same rate even when the embryo is removed (79, 122). Nevertheless, the addition of GA_3 somewhat stimulates the rate of fat utilization (122) and the development of enzymes of the glyoxylate cycle (123, 191), although the quantity of glyoxysomal protein that was produced was not affected (191). In one report (191) the synthesis of PC did not respond to GA_3, but in a more extended investigation, Gonzalez (58a) has shown that GA treatment resulted in increased levels of all ER marker enzymes and the associated malate synthetase.

It is clear that, as for the synthesis of other enzymes in the cereal aleurone layer, the embryo is required here for the production of enzymes of the glyoxylate cycle and that GA can substitute for this requirement (33, 140). Stimulatory effects have been observed in other seeds (68 and references therein). De Boer & Feierabend (32) observed that derooted rye seedlings in darkness responded to cytokinins by an enhanced synthesis of chloroplast and microbody enzymes, while other, cytosolic, enzymes were not affected, but the levels attained were considerably less than those achieved in normal development in the light. The removal of roots from etiolated sunflower seedlings resulted in an enhanced rate of loss of glyoxysomal enzymes from the cotyledons (166). This premature loss was prevented by kinetin and the development of leaf peroxisomal enzymes was enhanced (166). Benzyladenine showed similar effects in excised cotyledons (160). Nitrogenous salts also affected the development of microbody enzymes, as had been demonstrated earlier for uricase in *Phaseolus* (169).

An effect of IAA on the numbers of microbodies in subapical segments of oat coleptiles has been reported (162).

CONCLUSIONS

In the time since microbodies were first recognized and isolated from plants in the late 1960s, considerable progress has been made in establishing their generality and enzyme constitution. The function of the two specialized forms, glyoxysomes in fatty seedlings and leaf peroxisomes, are now firmly established, but the function of the nonspecialized microbodies in other tissues seems to be a limited one. Glyoxysomes, which have a relatively brief existence, have been much more intensively studied, and many features of their biogenesis are now clear. The specific molecular events in the early phases of enzyme synthesis are now being actively examined. With this information in hand it is to be expected that a fuller understanding of glyoxysome development and the microbody transition in fatty cotyledons will be forthcoming. Further studies are needed on the development and biology of microbodies in true leaves.

Literature Cited

1. Angelo, A. J. S., Ory, R. L. 1971. Localization of allantoinase in glyoxysomes of germinating castor beans. *Biochem. Biophys. Res. Commun.* 40:290–96
2. Axelrod, B., Beevers, H. 1972. Differential responses of mitochondrial and glyoxysomal citrate synthase to ATP. *Biochim. Biophys. Acta* 256:175–78
3. Barbareschi, E., Longo, G. P., Servettaz, O., Zulian, T., Longo, C. P. 1974. Citrate synthetase in mitochondria and glyoxysomes of maize scutellum. *Plant Physiol.* 53:802–7
4. Beevers, H. 1975. Organelles from castor bean seedlings: Biochemical roles in gluconeogenesis and phospholipid synthesis. In *Recent Advances in the Chemistry and Biochemistry of Plant Lipids,* ed. T. Galliard, E. I. Mercer, pp. 287–99. New York: Academic
5. Beevers, H. 1978. Microbodies in plants: biogenesis of glyoxysomes. In *Regulation of Developmental Processes in Plants,* ed. H. R. Schütte, D. Gross, pp. 159–72. Jena: Fischer
6. Beevers, H., Breidenbach, R. W. 1974. Glyoxysomes. *Methods Enzymol.* 31: 565–71
7. Beevers, H., Theimer, R. R., Feierabend, J. 1974. Microbodies (glyoxysomes, peroxisomes). In *Biochemische Zytologie der Pflanzenzelle,* ed. G. Jacobi, pp. 127–46. Stuttgart: Thieme
8. Berger, C., Gerhardt, B. 1971. Charakterisierung der Microbodies aus Spadix-Appendices von Arum maculatum und Sauromatum guttatum. *Planta* 96: 326–38
9. Bieglmayer, C., Graf, J., Ruis, H. 1973. Membranes of glyoxysomes from castor bean endosperm. Enzymes bound to purified membrane preparations. *Eur. J. Biochem.* 37:379–89
10. Bieglmayer, C., Nahler, G., Ruis, H. 1974. Membranen von Glyoxysomen aus Ricinus Endosperm. *Z. Physiol. Chem.* 355:1121–28
11. Bieglmayer, C., Ruis, H. 1974. Protein composition of the glyoxysomal membrane. *FEBS Lett.* 47:53–55
12. Bieglmayer, C., Ruis, H., Graf, J. 1974. Cytochemical localization of catalase activity in glyoxysomes from castor bean endosperm. *Plant Physiol.* 53: 276–78

13. Bowden, L., Lord, J. M. 1975. Development of phospholipid synthesizing enzymes in castor bean endosperm. *FEBS Lett.* 49:369–71
14. Bowden, L., Lord, J. M. 1976. Similarities in the polypeptide composition of glyoxysomal and endoplasmic reticulum membranes from castor bean endosperm. *Biochem. J.* 154:491–99
15. Bowden, L., Lord, J. M. 1976. The cellular origin of glyoxysomal proteins in germinating castor bean endosperm. *Biochem. J.* 154:501–6
16. Bowden, L., Lord, J. M. 1977. Serological and developmental relationships between endoplasmic reticulum and glyoxysomal proteins of castor bean endosperm. *Planta* 134:267–72
17. Bowden, L., Lord, J. M. 1978. Purification and comparative properties of microsomal and glyoxysomal malate synthetase from castor bean endosperm. *Plant Physiol.* 61:237–45
18. Breidenbach, R. W. 1969. Characterization of some glyoxysomal proteins. *Ann. NY Acad. Sci.* 168:342–47
19. Breidenbach, R. W., Beevers, H. 1967. Association of the glyoxylate cycle enzymes in a novel subcellular particle from castor bean endosperm. *Biochem. Biophys. Res. Commun.* 27:462–69
20. Breidenbach, R. W., Kahn, A., Beevers, H. 1968. Characterization of glyoxysomes from castor bean endosperm. *Plant Physiol.* 43:705–13
21. Breidenbach, R. W., Wade, N. L., Lyons, J. M. 1974. Effect of chilling temperatures on the activities of glyoxysomal and mitochondrial enzymes from castor bean seedlings. *Plant Physiol.* 54:324–27
22. Brown, R. H., Bowden, L., Lord, J. M. 1976. Isoelectric focusing of polypeptides from endoplasmic reticulum and glyoxysomal membranes of castor bean endosperm. *Planta* 130:95–96
23. Brown, R. H., Lord, J. M., Merrett, M. J. 1974. Fractionation of the proteins of plant microbodies. *Biochem. J.* 144: 559–66
24. Brown, R. H., Merrett, M. J. 1977. Density labelling during microbody development in cotyledons. *New Phytol.* 79:73–81
25. Burke, J. J., Trelease, R. N. 1975. Cytochemical demonstration of malate synthetase and glycolate oxidase in microbodies of cucumber cotyledons. *Plant Physiol.* 56:710–17
26. Ching, T. M. 1970. Glyoxysomes in megagametophyte of germinating ponderosa pine seeds. *Plant Physiol.* 46:475–82
26a. Choinski, J. S., Trelease, R. N. 1978. Control of enzyme activities in cotton cotyledons during maturation and germination. *Plant Physiol.* 62:141–45
27. Cooper, T. G. 1971. The activation of fatty acids in castor bean endosperm. *J. Biol. Chem.* 246:3451–55
28. Cooper, T. G., Beevers, H. 1969. Mitochondria and glyoxysomes from castor bean endosperm. Enzyme constituents and catalytic capacity. *J. Biol. Chem.* 244:3507–13
29. Cooper, T. G., Beevers, H. 1969. β-oxidation in glyoxysomes from castor bean endosperm. *J. Biol. Chem.* 244:3514–20
30. Dalling, M. J., Tolbert, N. E., Hageman, R. H. 1972. Intracellular location of nitrate reductase and nitrite reductase. I. Spinach and tobacco leaves. *Biochim. Biophys. Acta* 283:505–12
31. Dalling, M. J., Tolbert, N. E., Hageman, R. H. 1972. Intracellular location of nitrate reductase and nitrite reductase. II. Wheat roots. *Biochim. Biophys. Acta* 283:513–19
32. De Boer, J., Feierabend, J. 1974. Comparison of the effects of cytokinins on enzyme development in different cell compartments of the shoot organs of rye seedlings. *Z. Pflanzenphysiol.* 71: 261–70
33. Doig, R. I., Colborne, A. J., Morris, G., Laidman, D. L. 1975. The induction of glyoxysomal enzyme activities in the aleurone cells of germinating wheat. *J. Exp. Bot.* 26:387–98
34. Donaldson, R. P. 1976. Membrane lipid metabolism in germinating castor bean endosperm. *Plant Physiol.* 57:510–15
35. Donaldson, R. P., Beevers, H. 1977. Lipid composition of glyoxysomal membranes. *Plant Physiol.* 59:259–63
36. Donaldson, R. P., Tolbert, N. E., Schnarrenberger, C. 1972. A comparison of microbody membranes with microsomes and mitochondria from plant and animal tissue. *Arch. Biochem. Biophys.* 152:199–215
37. Doohan, M. E., Newcomb, E. H. 1976. Leaf ultrastructure and δ C^{13} of three seagrasses from the Great Barrier Reef. *Aust. J. Plant Physiol.* 3:9–23
38. Douglass, S. A., Criddle, R. S., Breidenbach, R. W. 1973. Characterization of deoxyribonucleic acid species from castor bean endosperm. Inability to detect a unique deoxyribonucleic acid species associated with glyoxysomes. *Plant Physiol.* 51:902–6

39. Drumm, H., Schopfer, P. 1974. Effect of phytochrome on development of catalase activity and isoenzyme pattern in mustard seedlings. A reinvestigation. *Planta* 120:13–30
40. Elstner, E. F., Heupel, A. 1973. On the decarboxylation of α-keto acids by isolated chloroplasts. *Biochim. Biophys. Acta* 325:182–88
41. Feierabend, J. 1975. Developmental studies on microbodies in wheat leaves. III. On the photocontrol of microbody development. *Planta* 123:63–77
42. Feierabend, J., Beevers, H. 1972. Developmental studies on microbodies in wheat leaves. I. Conditions influencing enzyme development. *Plant Physiol.* 49:28–32
43. Feierabend, J., Beevers, H. 1972. Developmental studies on microbodies in wheat leaves. II. Ontogeny of particulate enzyme associations. *Plant Physiol.* 49:33–39
44. Feierabend, J., Brassel, D. 1977. Subcellular localization of shikimate dehydrogenase in higher plants. *Z. Pflanzenphysiol.* 82:334–46
45. Feierabend, J., Mikus, M. 1977. Occurrence of a high temperature sensitivity of chloroplast ribosome formation in several higher plants. *Plant Physiol.* 59:863–67
46. Feierabend, J., Schrader-Reichardt, V. 1976. Biochemical differentiation of plastids and other organelles in rye leaves with a high-temperature-induced deficiency of plastid ribosomes. *Planta* 129:133–45
47. Frederick, S. E., Gruber, P. J., Newcomb, E. H. 1975. Plant microbodies. *Protoplasma* 84:1–29
48. Frederick, S. E., Newcomb, E. H. 1969. Cytochemical localization of catalase in leaf microbodies (peroxisomes). *J. Cell Biol.* 43:343–53
49. Frederick, S. E., Newcomb, E. H. 1971. Structure and distribution of microbodies in leaves of grasses with and without CO_2 photorespiration. *Planta* 96: 152–76
50. Gerhardt, B. 1973. Untersuchungen zur Funktionsänderung der Microbodies in den Keimblättern von Helianthus annuus. *Planta* 110:15–28
51. Gerhardt, B. 1974. Studies on the formation of glycolate oxidase in developing cotyledons of *Helianthus annuus* and *Sinapsis alba. Z. Pflanzenphysiol.* 74:14–21
52. Gerhardt, B. 1979. Apparent catalase synthesis in sunflower cotyledons during the change in microbody function. A mathematical approach for the quantitative evaluation of density-labeling data. *Plant Physiol.* 63. In press
53. Gerhardt, B. 1978. Microbodies/Peroxisomen pflanzlicher Zellen. *Cell Biology Monographs* Volume 5. Wien: Springer. 283 pp.
54. Gerhardt, B., Beevers, H. 1969. Occurrence of RNA in glyoxysome from castor bean endosperm. *Plant Physiol.* 44:1475–77
55. Gerhardt, B., Beevers, H. 1970. Developmental studies on glyoxysomes from castor bean endosperm. *J. Cell Biol.* 40:94–102
56. Gerhardt, B., Betsche, T. 1976. The change of microbodies from glyoxysomal to peroxisomal function within fatty, greening cotyledons: hypotheses, results, problems. *Ber. Dtsch. Bot. Ges.* 89:321–34
57. Gientka-Rychter, A., Cherry, J. H. 1968. De novo synthesis of isocitratase in peanut cotyledons. *Plant Physiol.* 43:653–59
58. Godovari, H. R., Badour, S. S., Waygood, E. R. 1973. Isocitrate lyase in green leaves. *Plant Physiol.* 51:863–67
58a. Gonzalez, E. 1978. The effect of GA_3 on the endoplasmic reticulum and formation of glyoxysomes in the endosperm of germinating castor bean. *Plant Physiol.* 62:449–53
59. Gonzalez, E., Beevers, H. 1976. Role of the endoplasmic reticulum in glyoxysome formation in castor bean endosperm. *Plant Physiol.* 57:406–9
60. Gregor, H. D. 1976. Studies on phenylalanine ammonia lyase from castor bean endosperm. I. Subcellular localization of the enzyme. *Z. Pflanzenphysiol.* 77:454–63
61. Grodzinski, B., Butt, V. S. 1976. Hydrogen peroxide production and the release of carbon dioxide during glycolate oxidation in leaf peroxisomes. *Planta* 128:225–31
62. Grodzinski, B., Butt, V. S. 1977. The effect of temperature on glycollate decarboxylation in leaf peroxisomes. *Planta* 133:261–66
63. Gruber, P. J., Becker, W. M., Newcomb, E. H. 1972. The occurrence of microbodies and peroxisomal enzymes in achlorophyllous leaves. *Planta* 105:114–38
64. Gruber, P. J., Becker, W. M., Newcomb, E. H. 1973. The development of microbodies and peroxisomal enzymes in greening bean leaves. *J. Cell Biol.* 56:500–18

65. Gruber, P. J., Trelease, R. N., Becker, W. M., Newcomb, E. H. 1970. A correlative ultrastructural and enzymatic study of cotyledonary microbodies following germination of fat-storing seeds. *Planta* 93:262–88
66. Halliwell, B. 1974. Oxidation of formate by peroxisomes and mitochondria from spinach leaves. *Biochem. J.* 138:77–85
67. Halliwell, B., Butt, V. S. 1974. Oxidative decarboxylation of glycolate and glyoxylate by leaf peroxisomes. *Biochem. J.* 138:217–24
68. Hawker, J. S., Bungey, D. M. 1976. Isocitrate lyase in germinating seeds of *Prunus dulcis. Phytochemistry* 15:79–81
69. Herbert, M., Burkhard, C., Schnarrenberger, C. 1978. Cell organelles from CAM plants. I. Enzymes in isolated peroxisomes. *Planta.* 143:279–84
70. Hilliard, J. H., Gracen, V. E., West, S. H. 1971. Leaf microbodies (peroxisomes) and catalase localization in plants differing in their photosynthetic carbon pathways. *Planta* 97:93–105
71. Hock, B. 1973. Isoenzyme der MDH aus Watermelonenkeimlingen. Mikroheterogenität und deren Aufhebung bei der Samenkeimung. *Planta* 110: 329–44
72. Hock, B. 1973. Kompartimentierung und Eigenschaften der MDH-Isoenzyme aus Watermelonenkeimblättern. *Planta* 112:137–48
73. Hock, B. 1974. Antikorper gegen glyoxysomenmembranen. *Planta* 115:271–80
74. Hock, B., Beevers, H. 1966. Development and decline of glyoxylate cycle enzymes in watermelon seedlings—effects of dactinomycin and cycloheximide. *Z. Pflanzenphysiol.* 55:405–14
75. Huang, A. 1975. Comparative studies of glyoxysomes from various fatty seedlings. *Plant Physiol.* 55:870–74
76. Huang, A., Beevers, H. 1971. Isolation of microbodies from plant tissues. *Plant Physiol.* 48:637–41
77. Huang, A., Beevers, H. 1972. Microbody enzymes and carboxylases in sequential extracts from C_4 and C_3 leaves. *Plant Physiol.* 50:242–48
78. Huang, A., Beevers, H. 1973. Localization of enzymes within microbodies. *J. Cell Biol.* 58:379–89
79. Huang, A., Beevers, H. 1974. Developmental changes in endosperm of germinating castor bean independent of embryonic axis. *Plant Physiol.* 54: 277–79
80. Huang, A., Bowman, P. D., Beevers, H. 1974. Immunological and biochemical studies on isozymes of malate dehydrogenase and citrate synthetase in castor bean glyoxysomes. *Plant Physiol.* 54:364–68
81. Huang, A., Liu, K. D. F., Youle, R. J. 1976. Organelle-specific isozymes of aspartate-α-ketoglutarate transaminase in spinach leaves. *Plant Physiol.* 58:110–13
82. Hunt, L., Skvarla, J. J., Fletcher, J. S. 1978. Subcellular localization of isocitrate lyase in non-green tissue culture cells. *Plant Physiol.* 61:1010–13
83. Hutton, D., Stumpf, P. K. 1969. Fat metabolism in higher plants. XXXVII. Characterization of the β-oxidation systems from maturing and germinating castor bean seeds. *Plant Physiol.* 44:508–16
84. Hutton, D., Stumpf, P. K. 1971. The pathway of ricinoleic acid catabolism in the germinating castor bean and pea. *Arch. Biochem. Biophys.* 142:48–60
85. Ihle, J. N., Dure, L. 1972. The developmental biochemistry of cottonseed embryogenesis and germination. *J. Biol. Chem.* 247:5048–55
86. Jones, R. L. 1972. Fractionation of the enzymes of the barley aleurone layer: evidence for a soluble mode of enzyme release. *Planta* 103:95–103
87. Kagan-Zur, V., Lips, S. H. 1975. Studies on the intracellular location of enzymes of the photosynthetic carbon-reduction cycle. *Eur. J. Biochem.* 59:17–23
88. Kagawa, T., Beevers, H. 1975. The development of microbodies (glyoxysomes and leaf peroxisomes) in cotyledons of germinating seedlings. *Plant Physiol.* 55:258–64
89. Kagawa, T., Lord, J. M., Beevers, H. 1973. The origin and turnover of organelle membranes in castor bean. *Plant Physiol.* 51:61–65
90. Kagawa, T., Lord, J. M., Beevers, H. 1975. Lecithin synthesis during microbody biogenesis in watermelon cotyledons. *Arch. Biochem. Biophys* 167: 45–53
91. Kagawa, T., McGregor, D. J., Beevers, H. 1973. Development of enzymes in the cotyledons of watermelon seedlings. *Plant Physiol.* 51:66–71
92. Kapil, R. N., Pugh, T. D., Newcomb, E. H. 1975. Microbodies and an anomalous "microcylinder" in the ultrastructure of plants with crassulacean acid metabolism. *Planta* 124:231–44
93. Khan, F. R., Saleemuddin, M., Siddiqi, M., McFadden, B. H. 1971. Purification and properties of isocitrate lyase from

flax seedlings. *Arch. Biochem. Biophys.* 183:13–23

94. Kindl, H., Frevert, J. 1977. Glyoxysomal matrix enzymes: isocitrate lyase, enoyl-CoA hydratase, 3-hydroxyacyl-CoA dehydrogenase, acetyl-CoA acetyltransferase. *Z. Physiol. Chem.* 358: 1202
95. Kindl, H., Ruis, H. 1971. Subcellular distribution of p-hydroxybenzoic acid formation in castor bean endosperm. *Z. Naturforsch.* 266:1379–80
96. Kindl, H., Ruis, H. 1971. Metabolism of aromatic amino acids in glyoxysomes. *Phytochemistry* 10:2633–36
97. Köller, W., Kindl, H. 1977. Dynamics and properties of two malate synthetases bound to different types of membranes of cucumber cotyledons. *Z. Physiol. Chem.* 358:1233
98. Köller, W., Kindl, H. 1978. The appearance of several malate synthetase-containing cell structures during the stage of glyoxysome biosynthesis. *FEBS Lett.* 88:83–86
99. Lamb, J. E., Riezman, H., Leaver, C. J., Becker, W. M. 1978. Regulation of glyoxysomal enzymes during germination of cucumber. *Plant Physiol.* 62: 754–60
100. Leaver, C. J., Weir, E. M., Walden, R., Becker, W. M. 1977. In *Translation of Natural and Synthetic Polynucleotides,* ed. A. B. Legock, pp. 232–37. Poznan, Poland: Poznan Agric. Univ. Publ.
101. Leek, A. E., Halliwell, B., Butt, V. S. 1972. Oxidation of formate and oxalate in peroxisomal preparation from leaves of spinach beet. *Biochim. Biophys. Acta* 286:209–11
102. Lips, S. H. 1975. Enzyme content of plant microbodies as affected by experimental procedures. *Plant Physiol.* 55:598–601
103. Lips, S. H., Avissar, Y. 1972. Plant leaf microbodies as the intracellular site of nitrate reductase and nitrite reductase. *Eur. J. Biochem.* 29:20–26

103a. Lips, S. H., Roth-Bejerano, N. 1973. Plant hormones and the organization of multi-enzyme systems in plant microbodies. In *Plant Growth Substances,* ed. Y. Sumiki, pp. 719–24. Tokyo: Kirokawa

104. Liu, K. D. F., Huang, A. 1976. Developmental studies on NAD linked MDH isozymes in cucumber cotyledons in light and darkness. *Planta* 131:279–84
105. Liu, K. D. F., Huang, A. 1977. Subcellular localization and developmental changes of aspartate-α-ketoglutarate transaminase isozymes in the cotyledons of cucumber seedlings. *Plant Physiol.* 59:777–82
106. Longo, C. P. 1968. Evidence of de novo synthesis of isocitratase and malate synthetase in germinating peanut cotyledons. *Plant Physiol.* 43:660–66
107. Longo, C. P., Bernasconi, E., Longo, P. G. 1975. Solubilization of enzymes from glyoxysomes of maize scutellum. *Plant Physiol.* 55:1115–19
108. Longo, G. P., Bracci, C., Bucceri, C., Pedretti, M., Longo, C. P. 1977. Malate dehydrogenase from maize scutellum glyoxysomes. 1. Localization within the organelle. *Plant Sci. Lett.* 9:381–90
109. Longo, G. P., Dragonetti, C., Longo, C. P. 1972. Cytochemical localization of catalase in glyoxysomes isolated from maize scutella. *Plant Physiol.* 50:463–68
110. Longo, C. P., Longo, G. P. 1970. The development of glyoxysomes in peanut cotyledons and maize scutella. *Plant Physiol.* 45:249–54
111. Longo, G. P., Longo, C. P. 1970. Development of glyoxysomes in maize scutella. *Plant Physiol.* 46:599–604
112. Longo, G. P., Longo, C. P. 1975. Development of mitochondrial enzyme activities in germinating maize scutellum. *Plant Sci. Lett.* 5:339–46
113. Lopez-Perez, M. J., Gimenez-Solves, A., Calonge, F. D., Santos-Ruiz, A. 1974. Evidence of glyoxysomes in germinating pine seeds. *Plant Sci. Lett.* 2:377–86

113a. Lord, J. M. 1975. Evidence that phosphatidylcholine and phosphatidylethanolamine are synthesized by a single enzyme present in the endoplasmic reticulum of castor bean endosperm. *Biochem. J.* 151:451–53

114. Lord, J. M. 1976. Phospholipid synthesis and exchange in castor bean endosperm homogenates. *Plant Physiol.* 57:218–23
115. Lord, J. M. 1978. Evidence that proliferation of the ER precedes formation of glyoxysomes and mitochondria in germinating castor bean endosperm. *J. Exp. Bot.* 29:13–23
116. Lord, J. M., Beevers, H. 1972. The problem of NADH oxidation in glyoxysomes. *Plant Physiol.* 49:249–51
117. Lord, J. M., Bowden, L. 1978. Evidence that glyoxysomal malate synthetase is segregated by ER in castor bean endosperm. *Plant Physiol.* 61:266–70
118. Lord, J. M., Kagawa, T., Beevers, H. 1972. Intracellular distribution of enzymes of the CDP-choline pathway in castor bean endosperm. *Proc. Natl. Acad. Sci. USA* 69:2429–32

119. Lord, J. M., Kagawa, T., Moore, T. S., Beevers, H. 1973. Endoplasmic reticulum as the site of lecithin formation in castor bean endosperm. *J. Cell Biol.* 57:659–67
120. Ludwig, B., Kindl, H. 1976. Plant microbody proteins. II. Purification and characterization of the major protein component (SP-63) of peroxisome membranes. *Z. Physiol. Chem.* 357: 177–86
121. Ludwig, B., Kindl, H. 1976. Plant microbody proteins. III. Labeling of the peroxisomal membrane protein SP-63 in vitro and in vivo. *Z. Physiol. Chem.* 357:393–99
122. Marriott, K. M., Northcote, D. H. 1975. The breakdown of lipid reserves in the endosperm of germinating castor beans. *Biochem. J.* 148:139–44
123. Marriott, K. M., Northcote, D. H. 1975. The induction of enzyme activity in the endosperm of germinating castor bean. *Biochem. J.* 152:65–70
124. Matsushima, H. 1972. The microbody with a crystalloid core in tobacco cultured cell clone XD-6s. III. Developmental studies on the microbody. *J. Electron Microsc.* 21:793–99
124a. Mazliak, P., Douady, D., Demandre, C., Kader, J. C. 1975. Exchange processes between organelles involved in membrane lipid biosynthesis. In *Recent Advances in the Chemistry and Biochemistry of Plant Lipids,* ed. T. Galliard, E. I. Mercer, pp. 301–18. New York: Academic
125. Mellor, R. B., Bowden, L., Lord, J. M. 1978. Glycoproteins of the glyoxysomal matrix. *FEBS Lett.* 90:275–78
126. Mellor, R. B., Lord, J. M. 1978. Incorporation of D-(^{14}C) galactose into organelle glycoprotein in castor bean endosperm. *Planta* 141:329–32
127. Miflin, B. 1974. The location of nitrite reductase and other enzymes related to amino acid biosynthesis in the plastids of roots and leaves. *Plant Physiol.* 54:550–55
128. Miflin, B., Beevers, H. 1974. Isolation of intact plastids from a range of plant tissues. *Plant Physiol.* 53:870–74
129. Mollenhauer, H. H., Morré, D. J., Kelley, A. G. 1966. The widespread occurrence of plant cytosomes resembling animal microbodies. *Protoplasma* 62: 44–52
130. Mollenhauer, H. H., Totten, C. 1970. Studies on seeds. V. Microbodies, glyoxysomes and ricinosomes of castor bean endosperm. *Plant Physiol.* 46:794–99
131. Montague, M. J., Ray, P. M. 1977. Phospholipid-synthesizing enzymes associated with Golgi dictyosomes from pea tissue. *Plant Physiol.* 59:225–30
132. Moore, T. S. 1974. Phosphatidylglycerol synthesis in castor bean endosperm. Kinetics, requirements, and intracellular localization. *Plant Physiol.* 54:164–68
133. Moore, T. S. 1975. Phosphatidylserine synthesis in castor bean endosperm. *Plant Physiol.* 56:177–80
134. Moore, T. S. 1976. Phosphatidyl choline synthesis in castor bean endosperm. *Plant Physiol.* 57:383–86
135. Moore, T. S., Beevers, H. 1974. Isolation and characterization of organelles from soybean suspension cultures. *Plant Physiol.* 53:261–65
136. Moore, T. S., Lord, J. M., Kagawa, T., Beevers, H. 1973. Enzymes of phospholipid metabolism in the endoplasmic reticulum of castor bean endosperm. *Plant Physiol.* 52:50–53
137. Moreau, R. A., Huang, A. 1977. Gluconeogenesis from storage wax in the cotyledons of *Jojoba* seedlings. *Plant Physiol.* 60:329–33
138. Muto, S., Beevers, H. 1974. Lipase activities in castor bean endosperm. *Plant Physiol.* 54:23–28
139. Newcomb, E. H., Frederick, S. E. 1971. In *Photosynthesis and Photorespiration,* ed. M. D. Hatch, B. C. Osmond, R. O. Slatyer, pp. 442–57. New York: Wiley
140. Newman, J. C., Briggs, D. E. 1976. Glyceride metabolism and gluconeogenesis in barley endosperm. *Phytochemistry* 15:1453–58
141. Nishimura, M., Graham, D., Akazawa, T. 1976. Isolation of intact chloroplasts and other cell organelles from spinach leaf protoplasts. *Plant Physiol.* 58: 309–14
142. Osmond, C. B., Avadhani, P. N. 1968. Acid metabolism in *Atriplex.* II. Oxalate synthesis during acid metabolism in the dark. *Aust. J. Biol. Sci.* 21:917–27
143. Parish, R. W. 1972. Urate oxidase in peroxisomes from maize root tips. *Planta* 104:247–51
144. Parish, R. W. 1972. Peroxisomes from the *Arum italicum* appendix. *Z. Pflanzenphysiol.* 67:430–42
145. Radin, J. W., Trelease, R. N. 1976. Control of enzyme activities in cotton cotyledons during maturation and germination. *Plant Physiol.* 57:902–5
146. Rehfeld, D. W., Tolbert, N. E. 1972. Aminotransferases in peroxisomes from spinach leaves. *J. Biol. Chem.* 247: 4803–11

147. Riezman, H., Mersey, B. G., Lamb, J. E., Becker, W. M. 1978. Glyoxysomal glycoproteins in membranes of glyoxysomes. *Plant Physiol.* 61:S105
148. Rocha, V., Ting, I. P. 1970. Tissue distribution of microbody, mitochondria and soluble MDH isoenzymes. *Plant Physiol.* 46:754–56
149. Rocha, V., Ting, I. P. 1971. Malate dehydrogenases of leaf tissue from *Spinacia oleracea.* Properties of three isoenzymes. *Arch. Biochem. Biophys.* 147:114–22
150. Roth-Bejerano, R. N., Lips, H. S. 1975. Glycolate oxidase content of microbodies as affected by nitrate. *Plant Physiol.* 55:270–72
151. Roth-Bejerano, N., Lips, S. H. 1978. Binding of glycolate oxidase to peroxisomal membrane as affected by light. *Photochem. Photobiol.* 27:171–75
152. Rothe, G. 1974. Intracellular compartmentation and regulation of two shikimate dehydrogenase isoenzymes in *Pisum sativum. Z. Pflanzenphysiol.* 74:152–59
152a. Ruis, H. 1971. Isolation and characterization of peroxisomes from potato tubers. *Z. Physiol. Chem.* 352:1105–12
153. Ruis, H., Kindl, H. 1970. Distribution of ammonia-lyases in organelles of castor bean endosperm. *Z. Physiol. Chem.* 351:1425–27
154. Ruis, H., Kindl, H. 1971. Formation of α,β-unsaturated carboxylic acids from amino acids in plant peroxisomes. *Phytochemistry* 10:2627–31
155. Schnarrenberger, C., Burkhard, C. 1977. In vitro interaction between chloroplasts and peroxisomes as controlled by inorganic phosphate. *Planta* 134:109–14
156. Schnarrenberger, C., Fock, H. 1976. Interactions among organelles involved in photorespiration. *Encyclopedia of Plant Physiology, New Series* 3:185–234
157. Schnarrenberger, C., Oeser, A., Tolbert, N. E. 1971. Development of microbodies in sunflower cotyledons and castor bean endosperm during germination. *Plant Physiol.* 48:566–74
158. Schopfer, P., Bajracharya, D., Bergfeld, R., Falk, A. 1977. Phytochrome mediated transformation of glyoxysomes to leaf peroxisomes in mustard seedlings. *Planta* 133:73–80
159. Schopfer, P., Bajracharya, D., Falk, H., Thien, W. 1975. Phytochrom-gestuerte Entwicklung von Zellorganellen (Plastiden, microbodies, mitochondrien). *Ber. Dtsch. Bot. Ges.* 88:245–68
160. Servettaz, O., Cortesi, F., Longo, C. P. 1976. Effect of benzyladenine on some enzymes of mitochondria and microbodies in excised sunflower cotyledons. *Plant Physiol.* 58:569–72
161. Servattaz, O., Filippini, M., Longo, C. P. 1973. Purification and properties of malate synthetase from maize scutella. *Plant Sci. Lett.* 1:71–80
162. Shen-Miller, J., Gawlik, S. R. 1977. Effects of indoleacetic acid on the quantity of mitochondria, microbodies, and plastids in the apical and expanding cells of dark-grown oat coleoptiles. *Plant Physiol.* 60:323–28
163. Smith, R. H., Schubert, A. M., Benedict, C. R. 1974. The development of isocitric lyase activity in germinating cotton seed. *Plant Physiol.* 54:197–200
164. Sternberg, L., Ting, I. P., Hanscom, Z. 1977. Polymorphism of microbody malate dehydrogenase in *Opuntia basilaris. Plant Physiol.* 59:329–30
165. Theimer, R. R. 1976. A specific inactivator of glyoxysomal isocitrate lyase from sunflower cotyledons. *FEBS Lett.* 62:297–300
166. Theimer, R. R., Anding, G., Matzner, P. 1976. Kinetin action on the development of microbody enzymes in sunflower cotyledons in the dark. *Planta* 128:41–47
167. Theimer, R. R., Anding, G., Schmid-Neuhaus, B. 1975. Density labeling evidence against a de novo formation of peroxisomes during greening of fat-storing cotyledons. *FEBS Lett.* 57:89–92
168. Theimer, R. R., Beevers, H. 1971. Uricase and allantoinase in glyoxysomes. *Plant Physiol.* 47:246–51
169. Theimer, R. R., Heidinger, P. 1974. Control of particulate urate oxidase activity in bean roots by external nitrogen supply. *Z. Pflanzenphysiol.* 73:360–70
170. Theimer, R. R., Theimer, E. 1975. Studies on the development and localization of catalase and H_2O_2 generating oxidases in the endosperm of germinating castor beans. *Plant Physiol.* 56:100–4
171. Ting, I. P., Fuhr, I., Curry, R., Zschoche, W. C. 1975. Malate dehydrogenase isozymes in plants: preparation, properties, and biological significance. In *Isozymes,* ed. C. L. Markert, 2:369–86. New York: Academic
172. Tolbert, N. E. 1971. Isolation of leaf peroxisomes. *Methods Enzymol.* 23:665–87
173. Tolbert, N. E. 1971. Microbodies—peroxisomes and glyoxysomes. *Ann. Rev. Plant Physiol.* 22:45–74

174. Tolbert, N. E. 1974. Photorespiration. In *Algal Physiology*, ed. W. D. P. Stewart, pp. 474–504. Oxford: Blackwells
175. Tolbert, N. E., Oeser, A., Kisaki, T., Hageman, R. H., Yamazaki, R. K. 1968. Peroxisomes from spinach leaves containing enzymes related to glycolate metabolism. *J. Biol. Chem.* 243: 5179–84
176. Tolbert, N. E., Oeser, A., Yamazaki, R. K., Hageman, R. H., Kisaki, T. 1969. A survey of plants for leaf peroxisomes. *Plant Physiol.* 44:135–47
177. Trelease, R. N., Becker, W. M., Burke, J. J. 1974. Cytochemical localization of malate synthetase in glyoxysomes. *J. Cell. Biol.* 60:483–95
178. Trelease, R. N., Becker, W. M., Gruber, P. J., Newcomb, E. H. 1971. Microbodies (glyoxysomes and peroxisomes) in cucumber cotyledons. Correlative biochemical and ultrastructural study in light and dark grown seedlings. *Plant Physiol.* 48:461–75
179. Van Poucke, M., Cerff, R., Barthe, F., Mohr, H. 1970. Simultaneous induction of glycolate oxidase and glyoxylate reductase in white mustard seedlings by phytochrome. *Naturwissenschaften* 57: 132–33
180. Vick, B., Beevers, H. 1977. Phosphatidic acid synthesis in castor bean endosperm. *Plant Physiol.* 59:459–63
180a. Vick, B., Beevers, H. 1978. Fatty acid synthesis in young castor bean seedlings. *Plant Physiol.* 62:173–78
181. Vigil, E. L. 1969. Intracellular localization of catalase (peroxidatic) activity in plant microbodies. *J. Histochem. Cytochem.* 17:425–28
182. Vigil, E. L. 1970. Cytochemical and developmental changes in microbodies (glyoxysomes) and related organelles of castor bean endosperm. *J. Cell Biol.* 46:453–54
183. Vigil, E. L. 1973. Structure and function of plant microbodies. *Sub-Cell. Biochem.* 2:237–85
184. Wade, N. L., Breidenbach, R. W., Lyons, J. M., Keith, A. D. 1974. Temperature-induced phase changes in the membranes of glyoxysomes, mitochondria and proplastids from germinating castor bean endosperm. *Plant Physiol.* 54:320–23
185. Wainwright, I. M., Ting, I. P. 1976. Microbody malate dehydrogenase isozymes in cotyledons of *Cucumis sativa* L. during development. *Plant Physiol.* 58:447–52
186. Walk, R. A., Hock, B. 1976. Separation of malate dehydrogenase isoenzymes by affinity chromatography on 5-AMP-sepharose. *Eur. J. Biochem.* 71:25–32
187. Walk, R. A., Hock, B. 1977. Glyoxysomal malate dehydrogenase of watermelon cotyledons: de novo synthesis on cytoplasmic ribosomes. *Planta* 134: 277–85
188. Walk, R. A., Hock, B. 1977. Glyoxysomal and mitochondrial malate dehydrogenase of watermelon (*Citrullus vulgaris*) cotyledons. II. Kinetic properties of the purified isoenzymes. *Planta* 136:221–28
189. Walk, R. A., Hock, B. 1978. Cell-free synthesis of glyoxysomal malate dehydrogenase. *Biochem. Biophys. Res. Commun.* 81:636–43
190. Walk, R. A., Michaeli, S., Hock, B. 1977. Glyoxysomal and mitochondrial malate dehydrogenase of watermelon (*Citrullus vulgaris*) cotyledons. I. Molecular properties of the purified isoenzymes. *Planta* 136:211–20
191. Wrigley, A., Lord, J. M. 1977. Effects of GA on organelle biogenesis in castor bean endosperm. *J. Exp. Bot.* 28:345–53
192. Yamamoto, Y., Beevers, H. 1961. Purification and properties of malate synthetase from castor bean endosperm. *Biochim. Biophys. Acta* 48:20–25
193. Yamazaki, R., Tolbert, N. E. 1969. Malate dehydrogenase in leaf peroxisomes. *Biochim. Biophys. Acta* 178:11–20
194. Zelitch, I. 1972. The photooxidation of glyoxylate by envelope-free spinach chloroplasts and its relation to photorespiration. *Arch. Biochem. Biophys.* 150:698–707
195. Zschoche, W. C., Ting, I. P. 1973. Malate dehydrogenase of *Pisum sativum*. Tissue distribution and properties of the particulate forms. *Plant Physiol.* 51:1076–81
196. Zschoche, W. C., Ting, I. P. 1973. Purification and properties of microbody malate dehydrogenase from *Spinacia oleracea* leaf tissue. *Arch. Biochem. Biophys.* 159:767–76
197. Zschoche, W. C., Ting, I. P. 1977. Microbody MDH in plants with C_4 photosynthesis. *Plant Sci. Lett.* 9:103–6

Ann. Rev. Plant Physiol. 1979. 30:195–238

PHYSIOLOGICAL ASPECTS OF DESICCATION TOLERANCE

❖7670

J. Derek Bewley

Department of Biology, University of Calgary, Calgary,
Alberta T2N 1N4, Canada

CONTENTS

INTRODUCTION

By far the greatest number of studies on the responses of plants to water deficits have been on the effects of relatively mild stress on species which can survive only limited drought. This is not surprising when one considers that agronomically important plants are almost exclusively of this type, and an understanding of how water stress affects their growth, development, metabolism, and yield is of obvious practical value. Far less attention has been paid to plants which have a high degree of drought tolerance.

Plants growing under conditions of frequent and often severe water stress may be adapted to survive in one of three ways (54). They may evade drought, avoid drought, or tolerate drought. Drought evaders, such as the

0066-4294/79/0601-0195$01.00

ephemeral annuals, have an abbreviated life cycle; they complete the vegetative and reproductive phases of the cycle while moisture is available, and survive dry periods as the seed. Drought avoiders are adapted to retard water loss and/or to increase water absorption; these plants also may be regarded as being drought resisters since they resist losing water. But it is upon the third category of plants, the drought tolerators, that this review is concentrated and, moreover, upon those plants which can survive extreme drought, i.e. desiccation.

To clarify some of the terminology to be found in this review the following comments are pertinent. A plant which can survive desiccation is one which can revive from the air-dry state (the air being of low relative humidity), and a plant which is desiccated is one from which all available water has been lost to the surrounding dry atmosphere. The hydrature of desiccation-tolerant plants invariably is determined by the dryness of the air. Alternate terms that have been used to describe plants that fall into this category are poikilohydrous (278) and poikiloxerophytic (133). Unfortunately, some workers have referred to drought-intolerant species as being "desiccated" when in effect they have been subjected to only relatively mild, and often reversible, water stress. This confusion of terms should be avoided. A plant that can survive desiccation and can suspend its metabolism in the dry state ["anabiosis" (134)] will be termed desiccation tolerant, and one that cannot do so will be termed desiccation intolerant (or desiccation sensitive). Plants that can survive severe water deficits only by avoiding or resisting becoming desiccated are desiccation-resistant plants, and these fall into the desiccation-intolerant category.

Some 60–70 monocots, dicots, ferns, and fern allies have been reported as being desiccation tolerant (97, 100), and there are larger but unknown numbers of mosses, lichens, and algae which can withstand air-drying. The physiological responses of such plants to desiccation have not received extensive review in recent times; a brief review appeared in 1975 (166) and Parker (211), Iljin (146), and Henckel (128) outlined some early work on the subject. While available space does not permit this review to be comprehensive, the author hopes to achieve an overview of the nature of cellular responses to severe water deficits and to present some general conclusions and speculations. Although seeds may be regarded as specialized types of desiccation-tolerant organisms, being adapted propagating structures, their responses to drying and rehydration are included because they tell us much about what might be expected to occur in desiccation-tolerant vegetative tissues. Omitted are the responses of plants to desiccation due to freezing, for while such studies give us a valuable insight into how plants survive cold temperatures, they are less informative on how plants survive water loss.

ALGAE

That algae can survive desiccation has been on record in the scientific literature for over a century. A useful list containing close to 420 species of desiccation-tolerant algae (mainly greens and blue-greens, with fewer reds, browns, and diatoms) has been compiled by Davis (66), largely from pre-1970 publications. Structures which are tolerant include zygotes, akinetes, cysts, propagules, and spores, as well as unmodified vegetative cells. Not all structures of the same species can withstand desiccation equally well and, as might be expected, the resting stages of the life cycle are generally the most tolerant. Examples are known of algae surviving in herbaria and in stored soil samples for many decades (10, 35, 210), although the surviving structure is not always known. Blue-green algae seem to have an overall better survival rate than do green algae, and the longevity record of 107 years is held by the nonspore-forming blue-green *Nostoc commune* (43). It is worth noting, however, that some viability records of algae in stored soils might be exaggerated. Some samples might have become contaminated by fresh spores from outside during storage, or the soil might not have been sufficiently dry to prevent the production of offspring.

Desiccation-tolerant algae have been found in an extremely diverse range of habitats. For example, they occur in such uncompromising locations as rock surfaces in the Antarctic (140), in soil crusts in Death Valley, California (44, 79), and as slimes on the roads around Calcutta (27). They exist in freshwater and on the margins of ponds as they dry out (81), as well as in intertidal regions where they may be (littorally!) left high and dry by the receding tide. Despite the many species available for study, remarkably little work has been done to determine the physiological responses of algae to drying and rehydration.

Several workers have studied the effects of desiccation upon nitrogen fixation by free-living species of the blue-green algae. This process appears to occur optimally at less than maximum moisture contents (153, 214) but is low at much reduced moisture contents and does not occur in dry algae (152, 153, 214, 265). Resumption of nitrogen fixation has been claimed to occur within minutes of rewetting (214), although supporting evidence is not presented. It has been shown to occur within an hour, reaching control values within 5 hours (153). Whether or not the nitrogenase enzyme system is stable during desiccation remains to be elucidated.

A variety of factors are involved in determining the vertical distribution of algae in the eulittoral (intertidal) regions of the seashore (12, 243, 287). Not least of these are the abilities to tolerate or to resist desiccation. In the upper eulittoral zone are found species capable of withstanding more or less complete loss of water (e.g. *Porphyra perforata, P. tenera, Pelvetia canaliculata, Fucus vesiculosus, Chondrus crispus,* and *Ulva lactuca*), whereas those occurring in the lower zones often avoid water loss through morphological adaptations and reduce evaporation by growing in dense mats or under the cover of more resistant species.

There have been several studies to show that during drying of eulittoral algae photosynthetic capacity is gradually reduced, although most work has been done with nontolerant species which do not recover after severe water stress (47). Photosynthesis in a variety of seaweeds in the middle and upper zones is maximal after some water loss (about 25 percent loss from *F. vesiculosus* and *Ascophyllum nodosum*), but then declines with declining fresh weight (34, 150). Initially water loss from the surface of the fronds may enhance the uptake of CO_2, hence an increase in photosynthesis, but as water loss progresses the photosynthetic mechanism is presumably disturbed. The kinetics of recovery of carbon fixation after drying of desiccation-tolerant marine algae do not appear to have been studied. Nevertheless, it is evident that the ability to tolerate desiccation and to resume photosynthesis and growth when resubmerged is greatest in species found highest on the shore and is progressively less in species inhabiting successively lower levels (244). In the red alga *Porphyra* the primary photochemical mechanism of photosynthesis is stable in the dry state and on rehydration is reactivated instantaneously (93). Recently, using the technique of chlorophyll fluorescence induction, it has been shown that during extensive water loss from desiccation-tolerant marine algae there is complete inactivation of photosynthetic partial processes. The first sensitive site affected by desiccation is electron transport between photosystems II and I, possibly between plastoquinone and P700. On rehydration, recovery is rapid and complete. In contrast, desiccation-intolerant algae do not recover when rehydrated following even limited water loss (281), suggesting irreversible chloroplast damage. Respiration in *A. nodosum, C. crispus, U. lactuca, F. vesiculosus,* and *Gigartina stellata* declines during drying (following a brief rise after some water loss in the last two species) (34, 154, 189), and recovery on rehydration can occur after at least 80 percent water loss (154) (with the exception of *A. nodosum*). Similar observations have been made on other marine and freshwater species (204). Desiccation-tolerant terrestrial algae gradually cease to respire and photosynthesize during drying, recovering these faculties on subsequent rehydration (15, 36, 37, 94).

It has been suggested that the accumulation of fats or oils in the protoplasm of some algae (maybe during water loss) increases their capacity to withstand desiccation (58, 81, 83). Evidence for this is lacking, however, and in some tolerant species such accumulation does not occur (115, 193). Another suggestion, which has been reviewed by Davis (66), is that algae produce pigments (e.g. carotenoids) for photoprotection, i.e. to act as filters to reduce the level of irradiation impinging on sensitive structures within the cell. The merits of this suggestion are debatable. Some algae survive slow desiccation better than rapid desiccation (82, 141). Therefore the observation that some desiccation-tolerant species possess thick cell walls and can produce mucilage may have some significance, for both could retard water loss from cells in a drying environment.

Because of the paucity of available information, few general conclusions can be drawn about the cellular responses or mechanisms of adaptation of desiccation-tolerant algae to water loss. Key metabolic events like photosynthesis and respiration resume after desiccation of tolerant species, and in blue-green algae nitrogen fixation recommences. The topic of desiccation tolerance in algae is a fertile field for research. Information on the kinetics of metabolic changes occurring during drying and rehydration would be valuable, as would be studies on variously tolerant species to determine the stability or lability of subcellular components during drying and the role of synthetic/replacement systems during rehydration.

LICHENS

There have been a number of interesting studies on the metabolic responses of lichens to water deficits under natural (field) conditions. On the other hand, relatively little work has been done at the cellular level of organization to determine the consequences of water loss. The responses to drying of the whole lichen thallus are, of course, a combination of those of the algal (phycobiont) and of the fungal (mycobiont) component. Desiccation-induced changes in photosynthesis, for example, reflect on the metabolic state of the algal component which, although it comprises only some 3–10 percent of the thallus by weight, is the fixer of carbon which subsequently is transported as carbohydrate to the more massive fungal compoment. This latter is the major site of respiration. Separation of the two components of the lichen association has been achieved, and the effects of drying them when apart has been studied. While this approach can yield useful information on the responses to desiccation of the individual components, some caution must be expressed, for their responses might not be the same as when they are in the lichenized state.

By far the most studied metabolic processes in lichens are photosynthesis and respiration, and exchange of carbon dioxide has been the most popular technique for assessing these phenomena. A number of comprehensive and critical reviews have been written on the effects of environmental stresses on lichen survival and carbon assimilation (86, 87, 125, 155, 256), and the reader should consult these for further details and for a more extensive list of references than can be presented here.

Lichens lose water rapidly when placed in a drying environment, although the rate at which some species lose water may be related to their morphological characteristics. For example, water may evaporate from finely branched filamentous thalli more rapidly than from flat, leathery, and dense thalli (173). Whether lichens can effectively conserve water or can control their rate of water loss is a debatable point, however (155). Little is known about the metabolic changes which lichens undergo during drying itself, although Lange (169) has demonstrated that water loss from *Ramalina maciformis* is paralleled by a continuous decline in CO_2 assimilation and dark respiration.

The extent to which lichens can tolerate being dried out is, not surprisingly, related to the moisture conditions to which they are adapted in their natural habitat. In general, xeric species (and ecotypes) recover more quickly, and from longer periods of drying, than do mesic species. Aquatic lichens are irreversibly damaged by drying (168, 232, 233, 235). The longevity of collected and stored desiccation-tolerant species is variable, but those adapted to desert conditions survive for from several weeks to many months (155, 169, 236) in excess of the time to which they are subjectd to dryness under natural conditions. Tolerance of a particular period of desiccation, as well as tolerance of desiccation itself, is undoubtedly a factor limiting species distribution. It is not known what changes occur in lichens in the dry state to cause them eventually to lose their viability. Loss of chlorophyll has been suggested (280), but the evidence is scant.

The degree of "dryness" at which a lichen is maintained is important for its survival (169). *R. maciformis* survives considerably longer at very low water contents than at higher ones. For example, after 34 weeks at 1.1 percent water content, control levels of CO_2 assimilation are regained within 15 days of spraying the thallus with water, but reaching control levels takes 26 days after keeping the lichen at 2.5 percent water content for 34 weeks. If the lichen is maintained at 15 percent water content for this time, the algal component dies and no CO_2 assimilation is possible. Thus the drier the lichen during storage, the less severe are the aftereffects on rewetting. We can only speculate on why the maintenance of *R. maciformis* at certain low water contents is deleterious. Perhaps in the partially hydrated state cells are capable of limited metabolic activities which, with increasing time of storage, lead to oxidation of key substrates and to partial completion of linked reactions, and hence to permanent disturbances in cellular functions. A state of completely suspended metabolism (anabiosis) appears to be preferable.

Recovery from desiccation has been studied under both field and laboratory conditions. In their natural habitat some desert lichens are rarely rewetted by rainfall, but rather resume their metabolism in atmospheres of high relative humidity and during early morning dew formation. Dry *Evernia* and *Ramalina* species can absorb sufficient water from moist air to resume CO_2 exchange; in fact, at relative humidities above 30 percent their respiratory and photosynthetic activities increase nonlinearly with increasing moisture content of the ambient air (14, 169). These lichens can achieve at least 90 percent of their maximum activities in water-saturated air, and yet under these conditions (where care is taken to prevent water condensing into droplets) the water content of *R. maciformis,* for example, only reaches 35 percent. The compensation point of this species was found to be reached at a relative humidity of 80 percent (10,000 lux, 10°C), corresponding to a water content of about 20 percent and a tissue water potential close to –300 bars. Fixation of radioactive CO_2 by photosynthesis in *Lepraria membranaceae* has been observed at water potentials as low as –450 bars (37).

When *R. maciformis* is allowed to absorb water from moist air (to a water content of 37 percent), on subsequent wetting of the thallus with liquid water (to 60 percent water content) there is an increase in CO_2 assimilation by only 20 percent. On the other hand, dark respiration more than doubles (169). The reasons for this are not known, but these observations are of interest in relation to the environmental conditions to which this lichen is subjected in its natural habitat—the central Negev desert. Here the intervals between winter rainfalls can exceed 12 months, but this and other lichens are often moistened at night by humid air and by early morning dew. Before dawn the water content of the *R. maciformis* thallus may reach 31 percent, but the night temperatures are too low to allow for much respiration. As dawn breaks and the light intensity increases, net assimilation of CO_2 occurs, but photosynthesis then declines within 2–3 hours as the thallus loses water to the now drier and warmer air. Respiration also resumes, although less rapidly, for by the time optimal temperature conditions are reached the thallus has dried out again (172). It has been calculated that, on a 24-hour basis, the mg CO_2 yield per gram dry weight by assimilation is 1.32 and loss of respiration is 0.78, to give a net yield of 0.54 mg per gram dry weight, or an equivalent of 146 μg of organic carbon fixed [summarized in (156, 171)]. Further calculations (156, 171), based on the fact that *R. maciformis* will be subjected normally to 198 dew events per annum in the central Negev, suggest a yearly increment of thallus growth of 8.4 percent. Not all desert-inhabiting lichens resume photosynthesis in response to increasing air humidities (102); *Chondropsis semiviridis,* a native of Australian deserts, probably requires liquid water (dew or rain) (236).

Although a slight digression from the main topic of this review, it is of interest to note that photosynthesis in a number of lichen species occurs maximally at some water content less than saturation (63, 124, 158, 234). On the other hand, while respiration may occur maximally in a thallus which is not fully hydrated, it does not decline at higher water contents. Hence the net assimilation rate (NAR) of some lichens is greatest in thalli which are not fully saturated. For a particular species or ecotype, the water content at which NAR is greatest may be correlated with its habitat (124, 233, 234). As an example, *Xantheria fallax,* a lichen of widespread occurrence on isolated trees, shows the highest NAR values at 45–65 percent saturation in relatively mesic microsites (in drainage tracks), whereas xeric forms in dry and very dry sites exhibit maximum rates of 35–50 and 30–45 percent saturation respectively (158). Why maximal NAR are achieved at different water saturation levels in different lichen species and ecotypes is not known, but both morphological and physiological features of the thallus have been implicated.

Most information on the metabolic responses of lichens to rehydration following desiccation has come from laboratory studies in which dried thalli have been rewetted with and maintained on (or in) liquid water—conditions which might not be always ideal. Immediately on introduction of water there is an intense but brief (1–2 min) nonmetabolic release of previously adsorbed gases (89, 257). This "wetting burst" is a purely physical process also exhibited by nonliving tissues and inert substances, and is of no significance to desiccation tolerance. It is followed by a period of increased oxygen consumption and CO_2 output over and above basal levels of the undesiccated thallus, and this is known as "resaturation respiration," a phenomenon which may last for from one to many hours or even days (89, 232, 235, 257). Even when lichens are not fully dried, they experience resaturation respiration on rehydration, e.g. *Peltigera polydactyla* after drying to only 40 percent of full saturation (257). The length and magnitude of this resaturation respiration varies with species and probably also with conditions of drying and period of desiccation. *C. semiviridis* dried in atmospheres of high relative humidity (so that the rate of drying probably was slow) showed very little resaturation respiration on rehydration, basal respiration rates being achieved within an hour (236). If lichens become hydrated by absorbing moisture from the air before being saturated with liquid water, then no resaturation respiration occurs (14, 169). The nature of the resaturation respiration is unknown, but it is claimed that in *Hypogymnia physodes* and *P. polydactyla,* unlike basal respiration, it is cyanide-sensitive (89, 257). Whether or not this implicates mitochondria as the site of resaturation respiration has not been resolved. Resaturation respiration is also azide-sensitive in *P. polydactyla* and DNP-sensitive in *H. physodes.* But the significance of these observations is obscure because it is known that these inhibitors have different effects upon metabolism at different concentrations, and yet the effect of only one concentration was reported. Thus further studies are necessary to determine the modes of action of various respiratory inhibitors on rehydrated lichens. Also, information on the relationship between oxygen consumption, ATP production, and ATP requirements of rehydrated thalli would be useful.

The major substrates for respiration (including resaturation respiration) are polyols (e.g. mannitol and arabitol in the fungus and ribitol in the alga), plus the nonpolyol sucrose (89, 257). The highest concentration of polyols is in the fungus, where they are formed by modification of carbon skeletons provided by photosynthetic activity of the alga (256). The net loss of polyols from the thallus of *H. physodes* on rehydration exceeds that which can be accounted for by resaturation respiration alone. The balance is lost due to extensive leakage (from alga and fungus) into the surrounding medium during the first few minutes of rehydration (89). Such leakage is indicative of membrane damage, the implications of which are discussed in more detail in a later section. Discussion of the possible reasons for resaturation respiration is to be found in the section "Photosynthesis, Respiration, and Dark Fixation of CO_2."

Many mesic and xeric lichens cannot survive in a continuously wet habitat and thrive only under conditions of alternating wet and dry periods. When lichens are rewetted by water vapor there is no leakage of polyols nor any resaturation respiration, and hence (presumably, though it has not been established) no depletion of the pools of respirable substrate. But in lichens from temperate climates, where wetting by rain is a common occurrence, there is a potential for depletion of respirable substrates. Farrar (88) has hypothesized that the establishment of a sizeable and readily accessible pool of respirable substrate is an important "physiological buffer," so that in times of metabolic stress, imposed by alternate wetting and drying cycles, substrate may be drawn from the pool, and hence degradation of insoluble cell components (e.g. carbohydrate and protein polymers) does not occur. In time, the pool is replenished by photosynthesis when favorable conditions prevail. If the polyol pool is depleted below a critical level, then recovery from desiccation is not possible (88, 257). Although this concept of physiological buffering is an interesting one, its acceptance as being important to lichens in their natural habitat can come only after further experimentation.

The presence of polyols in cells might be important during desiccation to protect macromolecules because their effective concentration will increase as water of hydration is lost. Reducing sugars are present also in dry lichens (229).

Very few ultrastructural studies have been carried out on desiccated and rehydrated lichens, and hence little information is available on stress-induced cellular changes. The algal component in some tolerant species may change little in structure during 4 years of dryness (225), although in others some loss of plastoglobuli (lipid-containing bodies in the chloroplast), starch and mitochondria, and changes in chloroplast structure have been reported as a result of drying (6, 39, 224). Further studies are needed to determine if such observations are typical of drought-tolerant lichens and to eliminate the possibility that some of the observed changes are not artifacts arising during "fixing" of dry thalli.

Only one other metabolic process has been studied during desiccation of lichens: the fixation of nitrogen by nitrogenase, as assayed by the acetylene reduction reaction. In a study of nitrogenase activity in *Lichnia pygmaea* from the seashore of Morocco, it was found that this enzyme exhibits a daily rhythm of activity. It is low in the dampness of early morning, reaches a peak during submersion at high tide (around 1 P.M.), and then falls to zero as the thallus dries out in the afternoon sun when the tide recedes (231). It is a general observation that nitrogenase is inactive in dry lichens, and for a number of species its lower limit of activity is at a thallus water content of 75–90 percent oven dry weight (138). Saturation of the thallus does not usually lead to a reduction in nitrogenase activity (63) (cf thallus saturation and net CO_2 exchange). Enzyme activity recovers quickly on rehydration of the dried thallus, and may return to control levels within a few hours (179), although the longer the period of dryness the slower is the recovery (136, 145, 159). Even so, nitrogenase activity in *Stereocaulon paschale* fully recovers within 36 hours of rehydration after 75 weeks storage of the thallus (145). In *P. polydactyla* nitrogenase activity increases more quickly on rewetting after air-drying than after drying over $CaCl_2$ (159); whether this is because the lichens dried at different rates or to different extents is not known.

To summarize this section: there is a considerable body of knowledge on desiccation-induced changes in CO_2-exchange by, and nitrogenase activity of, lichens in their natural habitats. How and why these changes occur still is largely unknown. Biochemical (metabolic) and ultrastructural studies are needed, with emphasis (where possible) on the effects of desiccation and rehydration on the individual components of the lichen association.

FUNGI

As noted in the previous section, fungi in the lichenized state can withstand desiccation, but little work has been done on the physiological aspects of their desiccation tolerance. Little more seems to have been done on the effects of water loss on the cellular activities of free-living fungi. There is a considerable body of work on the metabolism of fungal spores during germination, and several prominent experimenters in this field have contributed to a recent review on this topic (107, 177, 230, 259, 274). Much of the published work on germination, however, has been with spores which were not dry before harvesting (or were harvested by flooding colonies with water) and hence is not relevant to this review. Not all spores can withstand desiccation, but many can, and a popular form of storage of fungal spores is in the dry (and frozen) state. Very few studies have been carried out on desiccated and rehydrated nongerminated spores; some are mentioned below to illustrate the types of cellular changes that they undergo.

Dry conidia of *Aspergillus nidulans* contain a single nucleus, mitochondria, abundant ribosomal material, small vacuoles, endoplasmic reticulum, and concentric membranous structures. Within 30 min of hydration the cell wall changes from a single layer, or zone, found in the dry spore to one of three layers. More mitochondria appear to be present, the concentric membrane structures disappear, and electron-transparent vacuoles increase in size and number (92). These electron microscope studies have yet to be correlated with biochemical ones. Dry spores of *Neurospora crassa* and *Rhizopus stolonifer* contain single ribosomes and almost no polysomes (197, 274). The ribosomes in dry spores of *Aspergillus oryzae* appear to be identical to those of hydrated spores (161). In *N. crassa* polysomes increase tenfold within an hour of dry spores (conidia) being placed on minimal media (197). This suggests that polysome integrity might be lost during spore drying (although carbon source deprivation could be another contributing factor) and that they can reform rapidly on rehydration. It has not been established if this reformation is due to the recombination of conserved messenger RNA and ribosomes, or if their synthesis is required. Dry (11–16 percent water content) uredospores of *Puccinia graminis* resume respiration within minutes of hydration (180). Hence there is some evidence that desiccation-tolerant propagating structures of fungi resume metabolism soon after rehydration. Details of the changes they undergo on rehydration are few, and little (if any) work has been carried out on the cellular changes associated with the drying process.

BRYOPHYTES, PTERIDOPHYTES, AND SPERMATOPHYTES

Most studies on the effects of desiccation at the cellular level have been carried out using species from these plant divisions, so for convenience they will be considered together. Several different metabolic processes have been studied, in particular RNA and protein synthesis, respiration and photosynthesis, and membrane stability and synthesis; in a few instances their interrelationships have been considered.

RNA and Protein Synthesis

There is evidence from a number of publications that very mild to moderate water stress reduces the level of protein synthesis in drought-sensitive vegetative tissues (143). For example, in coleoptilar node segments of *Zea mays* seedlings, a decline in water potential of as little as 5 or 6 bars results in a marked reduction in polysome levels, which increase on subsequent removal of stress (142). Similarly, when buds of black locust seedlings are stressed by a xylem pressure potential of some –25 bars, polysomes decline; but they reform upon full hydration (31). For both tissues severe water loss is irreversibly destructive to cellular functions.

Our understanding of the effects of desiccation on RNA and protein synthesis has come largely from studies on the gametophytes of mosses and on the developing and germinating seeds of higher plants. Moss gametophytes lose water in an apparently uncontrolled manner whenever the ambient is less than water-saturated, whereas water loss from seeds is a normal, "preprogrammed" part of maturation, Despite this and other obvious differences between mosses and seeds, their responses to drying, at least as far as the protein synthesizing complex is concerned, are quite similar.

MOSSES In desiccation-tolerant mosses the capacity for protein synthesis declines as a consequence of water loss and is regained on subsequent rehydration (16, 17, 135, 245). In *Tortula ruralis* the effect of desiccation on the stability of polysomes varies with the speed at which water loss occurs (114). Rapidly dried moss has about half the polysome content of the undesiccated control, but slowly dried moss has no polysomes. Loss of polysomes is not due to production or activation of ribonuclease during drying (71), for while activity of this enzyme increases during both rapid and slow drying, it does so only after polysome levels have declined to their minimum. Furthermore, during drying (at both speeds) extracted ribosomes show a decreasing ability to utilize puromycin in vitro to form peptidyl-puromycin (71), evidence that nonpolysomal ribosomes in dry moss are not complexed with mRNA. Thus the primary cause of polysome loss during desiccation appears to be runoff of ribosomes from mRNA, coupled with failure to reform the initiation complex. During slow drying, runoff is complete, but presumably during rapid drying, critical water loss occurs before runoff can be completed. Why reinitiation is restricted by water stress is unknown. Inhibition due to increased cellular salts concentration is possible [initiation complex formation in *E. coli,* for example, is labilized at high salt concentrations (85)], although this could not explain why polysomes decline in mildly stressed plants, e.g. *Zea mays* (142) or perhaps moderately stressed *T. ruralis* (72). Studies have not been carried out on the factors associated with initiation complex formation in *T. ruralis,* and so it is not known if they change or become limiting during desiccation. Dry wheat embryos, however, contain all the supernatant factors essential for in vitro protein synthesis, including initiation factors. That protein synthesis is restricted by declining ATP availability is unlikely, for there is a lack of temporal coincidence and quantitative correlation between ATP levels and those of polysomes during drying (22). Levels of GTP during drying have not been studied. That protein synthesis might be reduced because of stress-induced changes to membrane attachment sites for "bound" polysomes (75) has been suggested (see also the concluding section). It is worth noting here, however, that because of technical difficulties it has not been possible to gain an accurate estimation of the ratio of free to membrane-bound polysomes in mosses.

Rapidly and slowly dried *T. ruralis* contain ribosomes which are active in vitro, the latter containing more ribosomes than the former (113). Cytoplasmic ribosomal RNA (17S and 24S) and low molecular weight RNA (4–5S) are not degraded during drying (271, 272); the fate of organelle RNA species has not been studied. That mRNA also is conserved in desiccated moss has been shown indirectly by inhibitor studies (20, 74, 114) and by studies on the qualitative aspects of protein synthesis using the the double-label ratio technique (74). Direct evidence has come from studies in which poly(A)$^+$RNA has been extracted from rapidly and slowly dried *T. ruralis* and made to catalyze the synthesis of polypeptides in vitro. In slowly dried moss, therefore, ribosomes, mRNA, and low molecular weight RNA species are conserved as separated components of the protein synthesizing complex. Poly(A$^+$)-containing RNA has been extracted also from hydrated and desiccated gametophytes of *Polytrichum commune* (245). It is not known if the poly(A)$^+$ RNA in this species in the dry state is free or is associated with polysomes, nor has it been demonstrated that it has messenger-like properties by in vitro assay.

Following desiccation of *T. ruralis, P. commune,* and *Neckera crispa,* protein synthesis recommences within minutes of their reintroduction to water (17, 114, 135, 245). In *T. ruralis* control levels of polysomes are restored within 2 hours (71, 114). Protein synthesis resumes at a faster rate after slow than after rapid desiccation (114). The reason for this is not readily apparent, because polysomes are already present in rapidly dried moss [and these resume activity on rehydration (17)], whereas for slowly dried moss to resume protein synthesis, a recombination of separated mRNA and ribosomes must occur. Rapid drying is harsher on cellular integrity and metabolism than is slow drying (see next two sections), and the lower rate of protein synthesis may be another manifestation of this.

RNA synthesis in *T. ruralis* also recommences rapidly upon rehydration after both speeds of drying (25, 74, 272), although the rates of synthesis have not been compared. Newly synthesized ribosomal RNA does not become incorporated into ribosomes until after about 2 hours from rehydration, and even then these ribosomes are not active in protein synthesis (272). Hence protein synthesis in *T. ruralis* can recommence following desiccation without a requirement for mRNA or rRNA synthesis. It is claimed that RNA synthesis in rehydrated *P. commune* is delayed some 45 min (245), although low levels of incorporation of ^{3}H-uridine were recorded before this time. Limited uptake of the precursor could account for the initial low incorporation.

Similar comparative studies have been carried out on desiccation-intolerant (semi-aquatic) moss species. Rapidly dried *Bryum pseudotriquetrum* and *Cratoneuron filicinum* [originally misidentified as *Hygrohypnum luridum* in several publications (18, 21, 22, 73, 113)] retain few or no polysomes (18, 25); polysome reformation and protein synthesis do not resume on subsequent rehydration. Polysomes are lost during slow and very slow desiccation of *C. filicinum,* and only limited protein synthesis occurs on rehydration (167), which is indicative of the survival of some cells. Thus there appears to be a relationship between dryness of habitat and capacity to retain protein synthesizing ability in the few moss species that have been studied.

SEEDS No attempt will be made to present an overview of seed development and germination. Only those events associated with water loss in the final stages of seed maturation and during early stages of germination during and following imbibition will be considered.

Synthesis of proteins is an integral part of the development of seed tissues and of germination. Interpolated between these two events, in many seeds at least, is a period of desiccation. In fact, it is known that some seeds fail to germinate and that metabolic processes associated with germination are impaired if they do not dry out first. The fate of the protein synthesizing complex of cereal embryos and dicot seed axes during drying has received little attention, although it has been inferred from electron microscope studies that polysomes are lost from lima bean (*Phaseolus lunatus*) axes as they lose water (163). This probably is a general phenomenon, for there are a number of reports that dried embryos and axes do not contain polysomes (216); the claims that they do are inadequately supported. They do, however, conserve active ribosomes and mRNA (see later).

The effects of drying on protein synthesis during maturation of persistent storage tissues has received some attention. In *Pisum sativum* cotyledons, drying is accompanied by a loss of polysomes and an increase in ribosomes

(11). Ribosomes are present in dry cotyledons of other legumes, and generally it has been observed that there is no major decline in RNA content during drying (195, 218, 258). The fate of mRNA is unknown, but it is not unreasonable to assume that at least those involved in storage material deposition are degraded. In castor bean endosperms both polysomes and ribosomes decrease during the final drying stage of seed maturation, and this is accompanied by a decline in total RNA, although this may not be caused by desiccation per se (55). There is a need for more information on the fate of the components of the protein synthesizing complex in seeds during the final stages of development. Moreover, it is necessary to know if the changes occurring during the drying phase are due ultimately to water loss from the seed tissues or if they are a "preprogrammed" part of development which would occur even if the seed were not allowed to dry out.

Our understanding of the mechanism of protein synthesis in plants has come from studies on in vitro systems derived from dry seeds, especially those by Marcus and coworkers, who have used isolated wheat embryos and germ. While protein synthesis is not possible in the dry embryo, there is ample evidence that components of the protein synthesizing complex (including ribosomes, mRNA, tRNA, initiation and transfer factors) are conserved therein and that they are active when placed in an appropriate in vitro assay system (104, 185, 186). The presence of active ribosomes and/or supernatant components in the dry seeds of a few other species has been confirmed, e.g. embryos of rice (5), rye (46, 223), oats (T. Akalehiywot, unpublished), *Pinus thunbergii* (285), *P. resinosa* (241), and *P. lambertiana* (9), in peanut axes (183), in the cotyledons of peanuts (147, 183) and *P. sativum* (279), and in whole lettuce seeds (80). Evidence for the presence of conserved (long-lived) mRNA in dry seeds has received excellent review by Payne (216), and little more than a brief mention is necessary here. A poly(A)$^+$ mRNA fraction has been extracted from dry peas, rapeseed, and rye embryos, and its capacity to act as template for a number of discrete proteins has been demonstrated (106). Dry wheat embryos contain an mRNA fraction that does not require polyadenylation to be activated [i.e. it already contains poly(A)$^+$ sequences] (261), but only 25–40 percent of this mRNA becomes incorporated into polysomes during and following imbibition. Yet the incorporated mRNA is not uniquely different from that which is not (38). RNA with poly(A)$^+$ sequences has been extracted from mature cotton seeds [although not translated in vitro to prove it is mRNA (121)] and here, as in air-dry rape seeds (220), it appears to be located within the nucleus. This may be a means of protection for conserved mRNAs.

Protein synthesis in vivo has been detected within 30–60 min of the start of imbibition in some species, e.g. excised axes of lima beans (162), *Phaseolus vulgaris* (105), and isolated embryos of rye (247), rice (26), wheat (135,

184), and oats (T. Akalehiywot, unpublished), but only after several hours in others. In the latter reports, however, failure to detect protein synthesis earlier is due invariably to the use of inadequate techniques. In wheat embryos polysomes are formed within 10–15 min of the start of imbibition without a requirement for ribosomes or mRNA which have been synthesized de novo. Likewise in intact cotton seeds, neither new RNA synthesis nor processing of conserved mRNA is essential for initial protein synthesis on imbibition (121). Even so, RNA synthesis does recommence rapidly following the start of imbibition (and as with protein synthesis, failure to detect it can be blamed upon technical inadequacies). Synthesis of mRNA and rRNA in wheat embryos occurs within the first hour of imbibition (262), but it is not known when and if they are utilized prior to the completion of germination. Thus, although rRNA synthesis is linear between 0.5 and 3.5 hours after imbibition starts, the incorporation of newly synthesized RNA into ribosomes is very limited over this time period—the delay may be the time required to process rRNA precursors before assembly. In rye embryos, however, maturation of rRNA occurs quite rapidly, for within 60 min of imbibition the 31S RNA precursor has begun to mature into the 25S rRNA of the large ribosomal subunit and the 18S rRNA of the small one (247). It has not been established if these subunits participate in protein synthesis at this time. Early resumption of synthesis of poly(A)$^+$-rich RNA and tRNA in rye embryos (217, 247), and of several RNA species in rape seeds (219), in *Agrostemma githago* embryos (126), and in embryonic axes of radish (68) has been reported also.

It is apparent, therefore, that embryos and their axes are capable of synthesizing proteins and RNA soon after tissue hydration. Synthesis of RNA is not an essential prelude to the synthesis of certain proteins, thus reinforcing the concept of conserved RNA components of the protein synthesizing complex in the dry seed. Whether or not RNA synthesis, and particularly mRNA synthesis, is essential for successful completion of germination is unresolved, and discussion of this topic is beyond the intended scope of this review. It has been suggested, however, that the conserved components are eventually degraded and that their functions are assumed by new synthesized RNAs (68, 69).

Few studies have been carried out on the recommencement of RNA and protein synthesis early on rehydration of dry seed storage organs. It is not unlikely, however, that RNA and supernatant components are conserved in dry cotyledons [as in peanut cotyledons (147)] and become reactivated upon rehydration. In castor bean endosperm (which loses RNA during the final stages of maturation), there is massive synthesis of ribosomal and soluble (transfer) RNA on rehydration, which is claimed to begin within one hour (188). Autoradiographic studies on slices of onion endosperm suggest that supplied uridine becomes associated with the nucleolus after only 15 min of imbibition (215), indicating early rRNA synthesis. It is not known if ribosomes are conserved in dry onion endosperms.

Desiccation of Germinating Seeds and Seedlings

Desiccation of seeds between early stages of imbibition and the time of cell division and vacuolation of the developing seedling usually has no permanent deleterious effects on subsequent germination and/or growth. In fact, there are several claims that postimbibitional desiccation actually advances the onset of germination compared with nondried controls (7, 13, 62, 122, 157). This is not so, however, for while radicle emergence is faster from treated seeds (i.e. those hydrated-desiccated-rehydrated) after rehydration than from controls (i.e. hydrated only once), the total time of hydration received by both seeds is the same. This is discussed in greater detail elsewhere (28). Since treated seeds complete germination in the same total time as controls, it is likely that events occurring during the initial hydration period prior to desiccation are "remembered" and do not have to be repeated for radicle protrusion to occur. In effect, the consequences of metabolism on initial imbibition are not reversed by subsequent drying. Hydrated seeds eventually pass through this desiccation-insensitive phase and become sensitive (196) about the time of duplication of the genome (48, 70, 246). Desiccation of seedlings of cereal grains often results in permanent injury to the primary root, and further root growth proceeds from secondary initials.

Desiccation of oat grains (T. Akalehiywot, unpublished) and wheat embryos (48) during the desiccation-insensitive stage results in a decline in polysomes which reform on subsequent rehydration. The desiccated grains of both cereals contain active ribosomes, and desiccated oat grains contain poly(A)$^+$-rich RNA which catalyzes in vitro protein synthesis (T. Akalehiywot, unpublished); desiccated wheat embryos also may contain mRNA (48). RNA, as well as protein synthesis, resumes rapidly on rehydration of desiccated rye embryos and oat grains (246; T. Akalehiywot, unpublished), but the dependence of the latter on the former is unknown. Maize and wheat embryos desiccated during their sensitive stage of seedling development irreversibly lose their capacity to conduct protein, RNA, and DNA synthesis on rehydration; chromatin remains condensed and DNA may be damaged (48, 62). It is claimed that in wheat embryos, desiccation in the sensitive stage results in the destruction of endogenous mRNA and the inability to produce intact mRNAs on rehydration (48). While this is a reasonable expectation, the published evidence is equivocal. In the endosperm of germinated castor bean seed (which does not undergo cell division and hence probably does not have a desiccation-sensitive phase), loss of water results in loss of polysomes and an increase in single ribosomes which retain their in vitro activity (268). On reintroduction to water, polysomes reform.

Ultrastructural studies of the shoot apex and of the radicle tip of wheat and maize embryos during the desiccation-insensitive phase of seedling growth have shown condensation of chromatin and formation of granular aggregations within the nucleus, although the nuclear membrane remains well defined (62, 187). In wheat embryo shoot apical cells, bundles of fibrils form during drying, the ER tends to become associated with vacuoles and lipid droplets, and mitochondria and plastids lose internal definition. On rehydration these changes are reversed, possibly due to efficient synthetic or repair mechanisms. Similar changes occur during desiccation of wheat embryo shoot apices and maize embryo radicle tips in the sensitive phase, but on rehydration increased disorganization occurs, including fragmentation of nuclear, mitochondrial, and plasmalemma membranes (62, 187). In isolated young maize roots, a similar complete disorganization of cell ultrastructure occurs following lethal desiccation (201).

To summarize: germinating seeds are more or less insensitive to desiccation stress. As seedlings develop they become desiccation sensitive and irreversible changes occur on drying. This susceptibility to water loss may not be due directly to any metabolic changes, however, for increased vacuolation of developing cells would make them more sensitive to drying, and decreased cellular activity could be a consequence of increased cytoplasmic disruption.

"PRESOWING DROUGHT HARDENING" In 1883 Will (282) reported that repeated soaking and drying of seeds resulted in increased drought and frost resistance by the vegetative plant. Similar claims have been made by other investigators, and in particular by the Russian scientist P. A. Henckel and his coworkers (127–129, 132); a number of their early studies have been reviewed (191). They not only claim increased resistance to environmental stresses of the vegetative plant due to this "presowing drought hardening" treatment, but also cite increases in yield, particularly under conditions of moisture deficiency. Unfortunately, supporting statistical data invariably are missing. Other workers have failed to find any substantial beneficial effects of this treatment or have found inconsistencies in the responses of different species or cultivars following treatment (e.g. 84, 103, 144, 148, 149, 237, 276, 284; see also 28 for details). Henckel (127, 128) has suggested that "hardening" resulting from presowing treatments is due to a number of physicochemical changes to cell cytoplasm, including (*a*) greater hydration of colloids, (*b*) higher viscosity and elasticity of the cytoplasm, (*c*) increase in bound water, (*d*) increase in hydrophilic and decrease in lipophilic colloids, and (*e*) increase in the temperature required for protein coagulation. It is interesting to speculate how these changes, which presumably must be induced within the cytoplasm of cells of the germinating seed or growing seedling during hardening, can persist in all of the newly formed cells and tissues of the developing "hardened" plant. Some consequences of these cellular changes are claimed to include a more xeromorphic structure, more intense transpiration, lower water deficits, the ability to retain a greater quantity of water, and a more efficient root system. Most of these suggestions and claims have been challenged by various workers, for the evidence in their favor is not substantial. There is limited support for the possibility that "hardened" plants develop a more extensive root system, thus enabling them to survive better under field drought conditions (e.g. 45, 67, 144, 284). It is possible that earlier radicle emergence and seedling establishment on planting in the field following dehydration-rehydration treatments simply gives plants a better start than non-"hardened" controls and hence enables them to survive adverse environmental stresses more easily because of their more advanced state of development. It is apparent that far more extensive and rigorous studies are required before the phenomenon of "hardening" and the physiological explanations for its occurrence are acceptable.

Photosynthesis, Respiration, and Dark Fixation of CO_2

BRYOPHYTES Stomata are absent from the gametophytes of bryophytes, and gaseous exchange occurs freely at the hydrated cell surfaces. It is evident from a number of studies (23, 76, 78, 110, 111, 130, 133, 137, 174, 192, 227, 255, 283) that decline in water content is accompanied by a decline in ability to photosynthesize and respire, that gas exchange does not occur in the desiccated plant, and that in desiccation-tolerant species this resumes rapidly on reintroduction of water. Desiccation-intolerant species, on the other hand, do not recover on subsequent rehydration. Deviations from this general pattern of events have been claimed from some species, however, and while some are inexplicable (and may be artifacts of technique), the causes for others will be reviewed here. Since different workers have used different conditions of desiccation and rehydration and different techniques for measuring gas exchange, it is hardly surprising that there are discrepancies in the results obtained. For example, speed of desiccation has a profound effect upon the rate of gas exchange on subsequent rehydration.

Oxygen evolution (indicative of photosynthesis) declines steadily during drying of *T. ruralis* and *C. filicinum,* but falls off sharply as dryness is approached (23). In some species there may be an initial elevation of photosynthesis as water is lost (111, 130; T. J. K. Dilks and M. C. F. Proctor, to be published), which may result from evaporation of excess surface moisture, thus obviating CO_2 diffusion. It has been suggested (133) that desiccation-tolerant species can be distinguished from intolerant because in the former, respiration (oxygen uptake) intensity falls continually with water loss, whereas in the latter, it rises during the period when most water is lost and then declines sharply. More species need to be tested, however.

On rehydration of *T. ruralis* after rapid drying, there is a doubling of oxygen consumption over control levels (cf "resaturation respiration" in lichens), which are not regained until after about 24 hours (23). Evolution of CO_2 in darkness on rehydration is not elevated by more than 10 percent, however (W. E. Winner, unpublished IRGA data). On rehydration after slow drying, a moderate increase in oxygen consumption occurs for some 10–12 hours, and after very slow drying the elevation is small and brief (23). It is interesting to note that it is the speed of water loss during the latter part of drying which is most important in determining the subsequent pattern of oxygen consumption on rehydration (23). Like basal respiration, resaturation respiration in *T. ruralis* is cyanide-sensitive (J. E. Krochko, unpublished). Photosynthesis on rehydration, as measured by oxygen evolution and CO_2 uptake (23; W. E. Winner, unpublished IRGA data), is less affected by the prior speed of desiccation, and control levels are regained within 15 min to 2 hours, depending upon light intensity. In *C. filicinum* oxygen consumption shows a minor and relatively brief elevation above control levels on rehydration after slow drying, but after rapid drying consumption is small (23). This is not surprising, for ultrastructural studies (24, 167) have shown that after rapid drying there is extensive disruption of cellular integrity, although some cells retain their integrity after very slow drying.

A source of confusion over the response of intolerant species to desiccation appears to have been the failure of some workers to maintain rehydrated moss in sterile conditions. Rehydration of intolerant species results in considerable cellular damage and leakage of materials into the surrounding medium, which can serve as a nutrient source for contaminating bacteria. In time, these proliferate and their respiration makes the major contribution to the measured oxygen consumption (76, 110, 167), leading to the false conclusion that the moss is recovering from desiccation.

Another factor which may affect the pattern of gaseous exchange on rehydration is the duration that the plant has been in the dry state. In their natural environments most desiccation-tolerant bryophytes may remain dry for several weeks or months between rainfalls, although not all species need to take up liquid water to resume metabolism; some mosses recommence photosynthesis after absorbing water vapor from a moist atmosphere (170). A dry period of several days to several weeks probably has little effect on the metabolic recovery of tolerant mosses (78, 137). For example, in *Anomodon viticulosus* and *Porella platyphylla,* recovery of net assimilation is complete within 3–4 hours of rewetting after maintenance in the dry state for 22 and 60 days, respectively. But after longer periods, net assimilation is initially negative and may never recover. Other factors to be considered are possible seasonal variations in desiccation tolerance (77, 174) and variations between wet and dry races of the same species (174).

Chlorophyll content of desiccation-tolerant species is largely unaffected by drying and rehydration (25), although considerable chlorophyll loss occurs from intolerant species (112; J. D. Bewley, unpublished observations). This presumably is a reflection of general cellular degradation. In *T. ruralis* RUDP carboxylase activity declines by about 30 percent during desiccation [and recovers on rehydration (248)], but this is unlikely to be the cause of the observed changes in photosynthesis during water loss. Desiccation of the moss *Acrocladium cuspidatum* has little effect on the in vitro activities of nine photosynthetic enzymes (including RUDP carboxylase) even though photosynthesis is completely (and to a considerable extent irreversibly) suppressed (174, 264). The enzyme, $NADP^+$-dependent glyceraldehyde-3-phosphate dehydrogenase, however, shows reduced in vitro activity when extracted from dried moss, and this can be restored by the addition of GSH (reduced glutathione). Perhaps desiccation brings about the oxidation of sulfydryl residues of the $NADP^+$-GAPDH enzyme, thus accounting, at least in part, for loss of photosynthetic capacity (264). In species where photosynthesis does not resume fully on rehydration, this oxidation presumably cannot be reversed in vivo; in tolerant species the enzyme might be protected from oxidation, or oxidations might be reversible. Loss of enzyme activity from intolerant species is unlikely to be the sole cause of irreversible changes occurring during desiccation (and/or rehydration), for the resultant disorganization of chloroplast integrity (24) could be equally or more important.

Changes in ATP levels during desiccation of *T. ruralis* and *C. filicinum* are very similar (23). During very slow drying there is a gradual loss of ATP until in the dried moss negligible amounts remain. After slow drying there is a reduction by about 66 percent below undesiccated control levels, but after rapid drying ATP levels are more or less undiminished. Rehydration of *T. ruralis* after all speeds of drying results in a rapid return to control levels, usually within 5 min of rewetting (22, 23). ATP levels in *C. filicinum* increase in rewetting following slow and very slow drying, but not after rapid drying; these observations are consistent with those on recovery of respiration. The link between oxygen consumption and ATP synthesis appears to be tenuous, for the amounts of oxygen consumed after rapid and very slow desiccation are very different, and for many hours, and yet ATP levels are similar. It could be that after rapid desiccation ATP production is largely uncoupled from oxygen consumption, but still sufficiently coupled to allow for the maintenance of a constant pool. Certainly mitochondria are considerably swollen after rapid desiccation (24, 273), which could indicate that their function is impaired. But mitochondria are equally swollen after very slow desiccation, when oxygen consumption is elevated very little. An alternative is that after rapid desiccation ATP production and utilization are both higher and to an equal extent.

It has been postulated (133) that the maintenance of control levels of "acid-soluble organic phosphorus with energy rich bonds" (presumably ATP) during drying is an important feature of desiccation tolerance, to allow their "energy" to be transferred to organelle/membrane structures to facilitate retention of their structural integrity. An apparent decline in ATP levels in desiccation-intolerant *Atrichum undulatum* has been correlated with a rise in ATPase activity, while in the tolerant moss *Neckera crispa* ATP levels are unaffected by drying (133, 228). This hypothesis is questionable. Not only do the results obtained for *T. ruralis* clearly contradict it, but the evidence in its favor comes from experiments in which only one speed of desiccation (rapid) was used, in which nucleoside triphosphates were not assayed directly, and in which the involvement of ATPase in controlling ATP levels was not established. It is more likely that ATP depletion during desiccation is inconsequential if it can be resynthesized on rehydration.

The major soluble carbohydrate in *Tortula* spp. and in *C. filicinum* is sucrose (23, 283). Increased oxygen consumption on rehydration (in both light and darkness) after rapid desiccation of *T. ruralis* does not detectably deplete the sucrose pool (23). Amino acid pools (including proline), organic acid pools, and starch content are unaffected by desiccation and rehydration (23, 248). Considerable amounts of sucrose, amino acids, and other organic compounds are lost from intolerant moss and liverwort species on rehydration, but in large part this is due to leakage (23, 108, 109).

A third aspect of CO_2 exchange, that of dark- or nonautotrophic-fixation of CO_2 has been studied in *T. ruralis* and *C. filicinum* (248). During drying of *T. ruralis,* dark-fixation of CO_2 declines at a slower speed than does photosynthesis. Whereas it accounts for only 5 percent of the total CO_2 fixed in hydrated control moss, in nearly dry moss it accounts for 60–100 percent. Desiccated moss does not fix CO_2, but on rehydration this capacity is regained fully within 5 min, irrespective of prior desiccation speed. Hence, initially upon rehydration dark-fixation may account for a considerable percentage of total CO_2 assimilation, falling steadily as photosynthetic capacity increases. In *C. filicinum,* the intolerant species, dark-fixation of CO_2 is very sensitive to water loss, even more so than is photosynthesis.

After rapid desiccation the moss is incapable of light- or dark-fixation of CO_2, and does so only poorly after slower speeds of desiccation. In vitro activities of PEP carboxylase, glutamic-oxaloacetic transaminase (GOT), and malate dehydrogenase (MDH) do not decline during drying and subsequent rehydration of either *T. ruralis* or *C. filicinum,* so loss of dark-fixation ability is not the result of enzyme destruction within the latter. An alternate, but entirely speculative, possibility is that increasing water deficits cause a decline in acceptor molecules (e.g. PEP) or cofactors.

VEGETATIVE TISSUES OF VASCULAR PLANTS The effects of mild to moderate water stress on CO_2 exchange by desiccation-intolerant higher plants has received considerable attention, and several extensive reviews (in particular on the effects of drought on photosynthesis) are available (29, 60, 61, 143). Desiccation of intolerant species results in irreversible disruption of cellular organization and metabolism, although photosynthetic reactions have been claimed to occur at severe water deficits (238). In the desiccation-tolerant fern *Polypodium polypodiodes,* both respiration and photosynthesis are reduced as water content declines, and below 35–40 percent water content they cease (267). Whether reduced photosynthesis due to water loss is mediated through effects on stomata or by changes in the chloroplasts, or both, has not been investigated thoroughly, although the photosynthetic pigment content of *P. polypodiodes* leaves is unaffected by desiccation (267). On rehydration of this fern and of *Ceterach officinarum* (207), respiration resumes at a higher level than in undesiccated controls. In *C. officinarum* it takes several days for respiration to return to control levels; data are not available for *P. polypodiodes.* Photosynthesis resumes soon after rehydration in this latter species (267).

Myrothamnus flabellifolia is a desiccation-tolerant angiosperm which can resume photosynthesis and respiration after long periods in the dry state (134, 139, 164). A number of important soluble enzymes involved in these processes remain potentially active in the desiccated tissue [Gündel in (164)], and the chlorophyll and carotenoid content of the plant is unchanged during desiccation and rehydration (139). It has been implied (164) that desiccation-induced changes in membrane components of mitochondria and chloroplasts are responsible for the decline in gas exchange, and that their reconstitution on rehydration is essential for resumption of photosynthesis and respiration. Some suggestive evidence has been provided in a brief electron microscope study (139), but more extensive and critical work is required.

Not all desiccation-tolerant plants retain their photosynthetic pigments within more or less intact chloroplasts during desiccation. Some angiosperms lose both, and reconstitute or resynthesize them on rehydration—

such plants are "poikilochlorophyllous" [see (97) for examples]. Changes in gas exchange during drying and rehydration unfortunately have not been followed in these plants, although electron microscopy studies on organelle integrity have been carried out (see section on Ultrastructure, Permeability, and Membranes, etc). It is interesting to note that *Borya nitida* only survives desiccation if it is dried slowly, during which time chlorophyll is lost and the dried plant is yellow. Rapid drying, which does not allow for destruction of chlorophyll, is fatal (98).

SEEDS Oxygen consumption by seeds decreases as they mature and lose water, although the extent of this decrease varies greatly between seeds and seed parts, depending upon their degree of dryness and physiological state at time of maturity (1, 2, 151, 165, 205, 249). Gas exchange has been reported to occur in dry onion seeds (40), but the extent of dryness was not ascertained, nor was possible bacterial contamination eliminated. Many "dry" seeds, in fact, have a water content of some 10–15 percent. ATP content of seed from two cultivars of rape declines to a very low level as maturity is reached, as does the soluble carbohydrate content (53). Whether or not this decline in soluble substrate plays some role in limiting ATP synthesis is a matter for speculation. Very low levels of ATP in mature dry seeds of other species have been recorded also (4, 51, 52, 198, 203, 226). Mitochondria lose internal definition during drying and those in dry seeds lack cristae (e.g. 8, 117, 163, 254, 263, 286), although some loss of organelle integrity might be attributable to improper fixation of dry tissues (see next section). In pea cotyledons during drying there is a decrease in mitochondrial succinic dehydrogenase, succinic oxidase, malate oxidase, and α-ketoglutarate oxidase activities, though MDH activity is unaffected (165). Phosphorylation efficiency and respiratory control ratio also decrease, which may be indicative of disturbance of the electron transport chain as water is lost; this presumably correlates with loss of internal integrity of the mitochondria.

There is a considerable volume of research on respiration in germinating seeds, and summaries of this work can be found elsewhere (19, 50). Introduction of dry seeds to water results in an immediate and rapid evolution of gas which may last for several minutes (116, 213). This is not due to respiration but to release of adsorbed gases as water is imbibed. Keto acids important for respiratory pathways (e.g. α-ketoglutarate and pyruvate) are chemically unstable and may be absent from dry seeds, being stored therein as the appropriate amino acids. Within 15 min of the onset of imbibition, keto acids may be reformed from amino acids by deamination and transamination, indicating that the citric acid cycle is functioning (56, 57). Oxygen consumption rises rapidly on imbibition, but no "resaturation respiration" is evident. The structurally deficient mitochondria undergo several changes upon imbibition, e.g. levels of respiratory enzymes, including succinic dehydrogenase, malate dehydrogenase, and cytochrome oxidase, increase rapidly (176, 199, 260, 277, 283a). Initially, oxidative phosphorylation is only loosely coupled, but an increase in respiratory control occurs with time (199, 260, 283a). Also, at early times of hydration an alternate terminal oxidase may operate until the deficient cytochrome *c*/cytochrome oxidase pathway becomes fully operational. Which pathway operates at different times after imbibition, however, appears to be a matter for debate (41).

Whether the recovery of normal respiration and tight respiratory control is due primarily to the synthesis of new mitochondria (which take over from those damaged by drying and rehydration), or whether there is effective repair or reconstitution of existing mitochondria, is unresolved. It has been suggested that in peanut axes increased respiration and respiratory efficiency is associated with synthesis of new mitochondria (283a), but no evidence has been provided. Ultrastructural studies of the root primordium of germinating rye embryos suggest that early increases in respiration are accompanied by an increase in the number of mitochondria per cell (120). Moreover, the number of cristae per mitochondrion increases. In fact, increasing regularity of mitochondrial structure after imbibition appears to be a feature common to many embryos, but because of a lack of definitive biochemical data, the reasons for this are unknown. In imbibed cotyledons two distinct patterns of mitochondrial development have been proposed. In pea cotyledons the evidence for reconstitution of mitochondria is compelling. In conditions where synthesis of mitochondrial proteins is prevented, respiratory enzyme levels rise and membrane development within mitochondria is even enhanced (182, 200). Thus preformed structural and enzymic proteins appear to be transferred into preformed, immature mitochondria. That controlled disassembly of mitochondria occurs during the drying phase of seed maturation must now be demonstrated. In peanut cotyledons, cytochromes *b* and *c*, cytochrome oxidase, succinic dehydrogenase, succinoxidase, and mitochondrial proteins and DNA increase with time after imbibition (32, 33). While such results are more consistent with organelle biogenesis than with activation, studies of the type conducted on pea cotyledons are desirable.

Synthesis of ATP has been studied on imbibition of only a few species of seeds but, unfortunately, not in conjuction with changes in mitochondrial structure and activity. In a number of embryos and seeds (4, 49, 51, 198, 203, 226), there is a rapid rise in ATP and other nucleoside triphosphates following hydration until a plateau is reached after one to several hours. In lettuce and crimson clover seeds and in soybean axes, the increase in ATP is accompanied by an increase in the total pool of adenine nucleotides (3,

51, 226). This suggests an early requirement for their synthesis which might be via the nucleotide salvage pathway (4). Adenine nucleotides increase by about 25 percent in wheat embryos from the first to the sixth hour after imbibition starts (203), although their content in the dry embryos is not known. Energy charge also increases from 0.6 to 0.8 in imbibed wheat embryos over the same time period, when there is little net production of ATP. Hence the wheat embryo appears first to rapidly establish a high ATP level (within 60 min of hydration) and then adjusts the energy charge to favor optimum cellular activity. In crimson clover, on the other hand, the increase in energy charge parallels increased ATP levels (51).

Thus, a general observation on the effects of desiccation on tolerant tissues is that there is a perturbation of metabolism associated with water loss and/or rehydration and eventual recovery. Changes in organelle structure, substrate quality and quantity, enzyme levels, and ATP pools may be manifestations of desiccation stress, and the ability to reverse these changes may be an essential feature of desiccation tolerance.

Ultrastructure, Permeability, and Membranes, and the Possible Bases of Desiccation Tolerance

As pointed out above, mitochondria of desiccation-tolerant seeds, mosses, and vascular plants undergo structural changes during desiccation and rehydration. Essentially they lose internal definition as they dry out, swell on rehydration, and eventually recover their normal morphology. Chloroplasts of mosses, ferns, and some angiosperms also undergo reversible swelling on rehydration. The extent to which these changes occur varies between species and desiccation conditions.

First, a brief word of caution is necessary concerning the interpretation of some ultrastructural studies on dry tissues, particularly those on dry seeds, because many have involved the use of aqueous fixatives. Cells imbibe water rapidly from such fixatives, resulting in partial to complete hydration before adequate fixation occurs (42). Cellular inclusions (e.g. protein bodies, lipid droplets, mitochondria) of seed tissues fixed in the absence of water, either by freeze-etching or osmium vapor, have a more irregular shape than do those treated with aqueous fixatives (221, 269). Now that nonaqueous fixatives are available (118), it may be appropriate to repeat some of the earlier dry seed studies.

In poikilochlorophyllous *Borya nitida* and *Xerophyta villosa,* extreme ultrastructural changes occur during desiccation and rehydration, although the possibility that some are artifacts of technique cannot be eliminated (99, 101, 119). For example, dry leaves of *B. nitida* show extensively degraded protoplasm; bounding membranes of mitochondria and chloroplasts are indistinct; starch grains and grana are absent; cristae and thylakoids are reduced in number; and, of course, chlorophyll is absent. On the other hand, the nucleus and nucleolus retain their integrity, as does the tonoplast. After many hours of rehydration, normal chloroplasts are evident again. Dry *X. villosa* also shows disorganization of the protoplasm and loss of cristae, grana, and thylakoids, but nuclei and plastid membranes remain intact. Cytoplasmic reorganization occurs within 1–2 hours from rehydration, mitochondria develop cristae within 8 hours, and by 48 hours chloroplast structure is normal. One might consider what is the minimum that must be retained within the dried cells of these plants to allow metabolism and restitution processes to commence on rehydration. It is highly unlikely that restitution can occur without a ready supply of ATP and without the synthesis of appropriate enzymes and perhaps structural proteins. Also, for reformation of organelles their DNA must be conserved, which argues against there being too extensive a disruption of their bounding membranes. That nuclear integrity, and hence the integrity of the cell's genetic material, is retained during desiccation may be an important aspect of desiccation tolerance. No biochemical studies have been conducted on poikilochlorophyllous plants, so this is a promising field for investigation.

Nonlethal water stress applied to desiccation-intolerant higher plants results in changes in cell ultrastructure (90, 275), e.g. disruption of internal lamellar structures of chloroplasts, rounding or swelling of mitochondria, and loss of definition of cristae. On removal of water deficits, structures return to normal within a few hours. Plants stressed to lethal water deficits are unable to reverse these changes, and increasing fragmentation of organelles and membrane structures occurs.

In the desiccation-intolerant moss *Cratoneuron filicinum,* speed of water loss has a determining effect upon ultrastructural changes (24, 167). Forty-five minutes after rehydration of the gametophyte of *C. filicinum* following rapid desiccation, the cytoplasm of all cells of the phyllidia is disorganized, mitochondria are swollen and have an ill-defined internal structure, the outer chloroplast membrane is lost, and lamellae are dispersed. At the same time of rehydration after very slow desiccation, however, only about 20 percent of the cells are similarly disrupted. The rest are fairly normal in appearance. Twenty-four hours after rehydration, all cells of the rapidly desiccated moss are extensively disrupted, but some 50 percent of the cells of the very slowly dried moss retain their integrity. We can only speculate on the advantages of very slow drying. Perhaps gradual water loss allows for controlled changes in the configuration of macromolecules which favor their stability or which facilitate limited (bio)chemical reactions and hence increase resistance to disruption by dehydration. Presumably even in very slowly dried *C. filicinum,* some cells dry out more slowly than others, and these are the ones which survive. It would be of interest to determine if very slow water loss (coupled with subsequent slow water uptake) allows for survival of cells of other desiccation-intolerant plants. As mentioned previously, even poikilochlorophyllous plants cannot survive desiccation unless water loss is slow, possibly allowing for controlled "dismantling" of cell structure. On rehydration following rapid desiccation of the tolerant moss *Tortula ruralis,* both mitochondria and chloroplasts swell and lose internal definition, but regain their normal morphology within 24 hours (273). Similar changes occur in other tolerant and intolerant mosses (181, 202).

The ultrastructure of the radicle tip of growing maize embryos after desiccation (over $CaCl_2$; probably rapid) and rehydration has been studied 24 and 72 hours after the start of imbibition (62). After 24 hours, when the seedling is still desiccation insensitive, drying results in condensation of chromatin (as in the initially dry embryo), in some loss of internal definition of mitochondria, and in reorganization of RER. Desiccation of shoot apices of wheat in the insensitive phase also results in chromatin condensation (187), although apparently it is not condensed in the dry embryo. After 72 hours hydration, maize embryos are sensitive and desiccation causes aggregation of chromatin into large masses dispersed throughout the nucleoplasm; nucleoli are difficult to distinguish, as are mitochondria and plastids (62). On rehydration of these embryos in the desiccation-insensitive phase of development, cell fine structure returns to normal. In the sensitive phase, however, chromatin remains aggregated (suggesting irreversible disruption of genetic material) and membrane structures increase in their degree of disruption. Why cells cease to be insensitive as seedling development progresses is open to speculation. Increased disruption of cytoplasm on desiccation of cells of older seedlings could be a consequence of increased vacuolation, and hence a natural manifestation of development.

Little is known about the changes which the plasmalemma and tonoplast undergo during drying and rehydration, although leakage studies indicate that some do occur. These studies have been reviewed exhaustively by Simon (250, 251), who summarizes the evidence that dry viable seeds, lichens, yeasts, pollen, fungi, and their spores all leak solutes when placed in liquid medium. Leaked substances include amino acids, mono-, di-, and trisaccharides, sugar alcohols, organic acids, hormones, phenolics, phosphates, and various fluorescent materials and electrolytes. A common feature appears to be that the major leakage of solutes is a transient phenomenon, often lasting only minutes after the addition of water. Prolonged loss of solutes is characteristic of irreversibly damaged tissues. Examples of leakage during imbibition will be drawn from work on seeds and bryophytes. Other organisms are more difficult to review in terms of desiccation alone: work on lichens is scant; leakage occurs from undesiccated pollen grains, and some substances leak from the walls and not the cytoplasm; leakage from dry fungal spores may not occur until some secondary treatment (e.g. heat shock) has been administered.

Leakage from dry, isolated pea embryos (190, 252) and from soybean cotyledons (213) occurs most rapidly within the first minute of imbibition, falling to some 50 percent of this rate within 5–10 min. Initial leakage is probably from peripheral cells as they become rapidly hydrated, and in some isolated embryos (i.e. in the absence of the testa) the inrush of water into these cells may be so violent as to cause irreversible damage (222). As the water boundary moves inward, cells become hydrated more slowly, and leakage becomes slower and eventually ceases as the length of the pathway for diffusion of solutes to the outside becomes longer. Pea seeds removed from the pod before the drying phase of maturation do not leak when placed in water, nor do those with a water content greater than 30 percent (253). However, dry seeds which have been allowed to imbibe and leak for 30 min, then redried and returned to water, exhibit similar leakage patterns to the original dry seeds (252).

The extent of leakage from the one cell-thick phyllia of the desiccation-tolerant moss *Tortula ruralis* on rehydration is determined by its prior speed of desiccation (73). After slow drying, leakage is less than half that from rapidly desiccated moss and is similar to that from undried controls. After 1 hour the leaked material in the medium is taken up by the moss again. Hence the rapid inrush of water is not lethal. Loss of material from both slowly and rapidly desiccated intolerant *Cratoneuron filicinum* [misnamed *Hygrohypnum luridum* (73)] is more extensive than from *T. ruralis* and is not reversed with time. Even so, it is less after slow than after rapid desiccation. The implication from these studies is that membranes of *T. ruralis* undergo certain reversible conformational changes during desiccation to retain their integrity. During slow drying these can be accomplished fully, but after rapid drying some repair mechanism must be put into effect quickly on rehydration. On the other hand, damage to *C. filicinum* membranes during desiccation and subsequent rehydration is apparently too extensive for repair, and/or repair mechanisms themselves are damaged (also see below). Other desiccation-tolerant and intolerant bryophytes exhibit similar patterns of leakage following water loss (108, 109).

Leakage from tissues has been explained in terms of changes which take place in the structure of membranes during desiccation (250, 251). While corroborative evidence is lacking, the explanation nevertheless is interesting. Membranes are composed principally of proteins and phospholipids arranged in a fluid bilayer. Individual phospholipid molecules are lined up side by side with their polar head groups facing the aqueous phase on either side of the membrane and with their hydrocarbon chains forming the hydrophobic central region. This molecular organization is stabilized by the relationship established between membrane components and water. Low-angle X-ray diffraction patterns of isolated membranes have shown that 20–30 percent hydration (above fully dried weight) is essential for the maintenance of the lipoprotein association in membranes (91, 178). At lower water content the membrane bilayer becomes rearranged into a hexagonal phase, which is largely hydrophobic but is pierced by long water-filled channels lined by the polar heads of the phospholipids (178). The drier the preparations, the narrower these channels become. The structure of membrane proteins appears to be unaffected by dehydration (242), but they may be displaced from, or within, the membrane as phospholipid molecules become rearranged. Rapid imbibition of dry tissues results in the phospholipid and protein components spontaneously reconstituting the bilayer configuration once the critical degree of hydration is reached. Membrane components may become displaced (or further displaced if already so) by the rapid influx of water. In cells of desiccation-intolerant plants, this displacement may be irreversible or too slow to prevent the loss of critical amounts of solute. Displacement also may be greater in some cells of desiccation-tolerant plants (e.g. those rapidly dried or very dry) than in others (e.g. after slow drying or with a membrane water content close to the critical 20–30 percent minimum). Differences in the degree of disruption of dry membranes could be a consequence of their inherent properties and/or be due to changes occurring during desiccation itself. Even in desiccation-tolerant cells there may be a delay before all displacements are reversed, allowing for solute leakage. The longer it takes for membrane reconstitution to occur [e.g. by imbibition at low temperatures (30)], the greater the amount of leakage.

Even when its components are reassembled and leakage is halted, the membrane still may not have regained its functionality. Additional "repair mechanisms" may be necessary to add back essential lipid or protein components. For example, during desiccation mitochondrial membranes may have lost certain enzymes, enzyme complexes, or key structural components, which subsequently are replaced over several hours of rehydration. Until such replacement is complete (and the harsher the drying regime the longer this could take), it might be expected that the organelle will show symptoms of its incomplete nature. The respiratory deficiencies outlined in the section on "Photosynthesis, Respiration, and Dark Fixation of CO_2" might be such symptoms.

Alterations to membrane properties by lipid peroxidation could account for some of the observed cellular changes associated with desiccation and rehydration (250). A number of biological oxidations, both enzymatic and spontaneous, generate the free superoxide radical ($O_2^{\cdot -}$) which is cytotoxic, and in turn can react with H_2O_2 to produce singlet oxygen and the hydroxyl radical (OH·)—highly potent oxidants (96). These can induce considerable destruction, particularly to large nucleic acids, proteins and polysaccharides, and to membrane lipids (194). In hydrated tissues, free radical production is normally controlled by free radical absorbents or scavenging reactions. One such scavenger is the enzyme superoxide dismutase (SOD) (95, 96), which converts ($O_2{}^{\cdot -}$) to H_2O_2, and this in turn can be removed by catalase. It is possible that water loss from cells of desiccation-intolerant plants can upset the balance between free radical producing and scavenging reactions in favor of the former, whereas in desiccation-tolerant plants the balance is maintained. There is some preliminary evidence that this is so (R. S. Dhindsa, W. Matowe, and J. D. Bewley, unpublished). In the desiccation-tolerant moss *Tortula ruralis,* during slow desiccation SOD and catalase activities increase and lipid peroxidation decreases. On rehydration lipid peroxidation rises again, but only to control levels, and enzyme levels decline. During rapid desiccation there are no changes, but on rehydration there is a sharp rise in lipid peroxidation, but SOD and catalase levels rise also, sufficiently, presumably, to limit membrane damage and to allow the moss to resume normal metabolism (perhaps including new membrane synthesis). In desiccation-intolerant *Cratoneuron filicinum,* catalase and SOD decline during drying at either speed, and their activities decline precipitously on rehydration. Lipid peroxidation is greatest on rehydration after slow desiccation, and is four times greater after 24 hours than in rehydrated rapidly desiccated moss. This is probably related to the fact that the latter consumes very little oxygen compared with the former. Studies on leakage (73) show that this is greatest when lipid peroxidation is highest, and vice versa.

Free radical formation in dry stored seeds could result in progressive inactivation of enzymes, denaturation of other proteins, disruption of integrity of DNA and RNA, and of membranes. The evidence for free radical formation is not substantial, and studies on seeds of various ages using ESR techniques have failed to find free radical concentrations at sufficient levels to account, for example, for associated genetic damage (59). On the other hand, some observations have been made which are consistent with (but by no means proof of) the free radical hypothesis (123, 209). Further studies on membrane stability and synthesis and free radical formation and sequestering in dry and in dry-stored organisms could provide insight into the mechanisms of tolerance of desiccated tissues.

When enzymes, structural proteins, nucleic acids, macromolecular complexes, etc are desiccated in their native state, the integrity of the molecules can be retained if some water remains associated with them to prevent the formation of unfavorable conformations (for reviews see 143, 175, 212, 270) or fragmentation (65). Hence the production or availability of substances (e.g. sugar, polyols, amino acids, anions) to maintain bound water content (and, incidentally, reduce free radical activity) could be an important feature of desiccation tolerance. Little work has been done to support or reject the possibility. Increased production of polyols by lichens (89) may have some importance in this respect, although some desiccation-tolerant plants [e.g. mosses (23; and unpublished data)] do not exhibit changes in the pool sizes of low molecular weight molecules. Protection of some fungal enzymes may depend upon their conjugation with carbohydrate as glycosylated proteins (64). The universality of this feature of proteins in tolerant plants, compared to intolerant, is worthy of determination. Protection of chloroplast membrane structures against desiccation damage by sugars has been proposed, their action being to stabilize proteins (239, 240). Even so, it is evident that chloroplast membranes of desiccation-tolerant plants do suffer some damage due to water loss. Thus, while protection against extensive damage might be attributed to some extent to sugars, ions, etc, the capacity to repair what damage is elicited is important also. Desiccation naturally leads to the concentration of ions within cells, and the possibility that this results in some damage cannot be ignored, e.g. K^+ may displace Ca^{2+} from membranes, thus weakening its structure (208). Changes were once thought to occur in the composition of macromolecules, particularly nucleic acids, snythesized under stress conditions to yield more stable configurations (160). This now seems unlikely (143).

The above observations and comments suggest that the ability of plants to tolerate desiccation is related to some inherent properties of their cellular contents, i.e. tolerance is "protoplasmic." It is evident that many desiccation-tolerant plants do not undergo severe water deficits without exhibiting cellular changes, some of which may be regarded as quite extensive damage. What may be critical features of desiccation tolerance are the abilities (*a*) to limit this damage during desiccation; (*b*) to maintain physiological integrity in the dry state so that metabolism can be reactivated quickly on rehydration; and (*c*) to put a repair mechanism into effect on rehydration, in particular to regain integrity of membranes and membrane-bound organelles. Some workers have argued that it is the mechanical properties of cells which are more important (131, 146; reviewed in 175, 181, 206). Claims have been made that cell size and shape, wall structure, and vacuole size are determining factors in desiccation tolerance. Additionally there are claims that cells of tolerant plants lack plasmodesmata, that protoplasm undergoes gelification during desiccation, that vacuoles become filled in to limit shrinkage, and that cell wall collapse occurs as turgor is lost in order to prevent protoplasmic tearing. Evidence that these are widespread phenomena generally is lacking (24, 175, 181, 206, 266).

In many respects desiccation tolerance is a primitive feature, being more prevalent in prokaryotes than in eukaryotes. This could be related to the simpler internal organization of the cells of the former, e.g. lack of cellular compartmentalization, and less complex membrane structures. Desiccation tolerance also is more common in lower plants than in higher ones, and the ability of the former to act as primary colonizers in uncompromising habitats is intimately linked to their tolerance of severe environments. The evolution of higher plants seems, to varying extents, to have been at the expense of their capacity to withstand severe water stress conditions. Adaptations to avoid or to resist drought are more common, which is not surprising, for simply the large habit of many species would make it impossible for them to reestablish a continuum of water after desiccation. Certain species of higher plants appear to have reacquired desiccation tolerance in the vegetative stages of their life cycle. Desiccation tolerance also is a property of most seeds of higher plants, although this is a feature common to propagating structures throughout the plant kingdom and hence may have "evolved" independently and perhaps, to some extent, in a different form.

Insufficient work has been done to draw any sweeping conclusions as to the nature of desiccation tolerance in plants. Both mechanical and physiological properties of cells undoubtedly play a role, although not necessarily to the same extent or in the same way in all species or structures. The challenge remains to further elucidate the nature of these properties and to determine to what extent they are common to the diverse number of species which exhibit desiccation tolerance.

Acknowledgments

I am grateful to the University of Calgary for the award of a Killam Resident Fellowship, during tenure of which much of this review was written. Support by NRC of Canada grant A6352 also is gratefully acknowledged. The writing of this review was considerably aided by several workers who sent me their published and unpublished work and who drew my attention to pertinent references; to them I extend my sincere thanks. Helpful and critical comments on the final draft were provided by Joan Krochko, Peter Halmer, and Grant Reid. Last, but by no means least, I wish to express my appreciation to Erin Smith for her industry and cheerful disposition during her typing of the manuscripts.

Literature Cited

1. Abdul-Baki, A., Baker, J. E. 1970. Changes in respiration and cyanide sensitivity of the barley floret during development and maturation. *Plant Physiol.* 45:698–702
2. Agrawal, P. K., Canvin, D. T. 1971. Respiration of developing castor bean seeds. *Can. J. Bot.* 49:263–66
3. Anderson, J. D. 1977. Adenylate metabolism of embryonic axes from deteriorated soybean seeds. *Plant Physiol.* 59:610–14
4. Anderson, J. D. 1977. Responses of adenine nucleotides in germinating soybean embryonic axes to exogenously applied adenine and adenosine. *Plant Physiol.* 60:689–92
5. App, A. A. 1969. Involvement of aminoacyl-tRNA transfer factors in polyphenylalanine synthesis by rice embryo ribosomes. *Plant Physiol.* 44: 1132–38
6. Ascaso, C., Galvan, J. 1976. The ultrastructure of symbionts of *Rhizocarpon geographicum, Parmelia conspersa* and *Umbicularia pustulata* growing under drying conditions. *Protoplasma* 87: 409–18
7. Austin, R. B., Longden, P. C., Hutchison, J. 1969. Some effects of "hardening" carrot seed. *Ann. Bot.* 33:883–95
8. Bain, J., Mercer, F. V. 1966. Subcellular organization of the cotyledons in germinating seeds and seedlings of *Pisum sativum* L. *Aust. J. Biol. Sci.* 19: 69–84
9. Barnett, L. B., Adams, R. E., Ramsey, J. S. 1974. The effect of stratification on *in vitro* protein synthesis in seeds of *Pinus lambertiana. Life Sci.* 14:653–58
10. Becquerel, P. 1942. Réviviscence et longévité de certaines algues en vie latente dans les terres déssechées des plantes des vieux herbiers. *C. R. Acad. Sci. Ser. B* 214:986–88
11. Beevers, L., Poulson, R. 1972. Protein synthesis in cotyledons of *Pisum sativum* L. I. Changes in cell-free amino acid incorporation capacity during seed development and maturation. *Plant Physiol.* 49:476–81
12. Bérard-Therriault, L., Cardinal, A. 1973. Importance de certains facteurs ecologiques sur la resistance a la desiccation des Fucales (Phaeophyceae). *Phycologia* 12:41–52
13. Berrie, A. M. M., Drennan, D. S. H. 1971. The effect of hydration-dehydration on seed germination. *New Phytol.* 70:135–42
14. Bertsch, A. 1966. Uber den CO_2-gaswechsel einiger flechten nach wasserdampfaufnahmne. *Planta* 68:157–66
15. Bertsch, A. 1966. CO_2 gaswechsel und wasserhausahlt der aerophilen grünalge *Apatococcus lobatus. Planta* 70:46–72
16. Bewley, J. D. 1972. The conservation of polyribosomes in the moss *Tortula ruralis* during total desiccation. *J. Exp. Bot.* 23:692–98
17. Bewley, J. D. 1973. Polyribosomes conserved during desiccation of the moss *Tortula ruralis* are active. *Plant Physiol.* 51:285–88
18. Bewley, J. D. 1974. Protein synthesis and polyribosome stability upon desiccation of the aquatic moss *Hygrohypnum luridum. Can. J. Bot.* 52:423–27
19. Bewley, J. D., Black, M. 1978. *Physiology and Biochemistry of Seeds in Relation to Germination. I. Development, Germination and Growth.* Berlin: Springer-Verlag. 306 pp.
20. Bewley, J. D., Dhindsa, R. S. 1977. Stability of components of the protein synthesizing complex of a plant during desiccation. In *Translation of Natural and Synthetic Polynucleotides,* ed. A. B. Legocki, pp. 386–91. Poznán: Publ. Poznán Agric. Univ.
21. Bewley, J. D., Dhindsa, R. S., Malek, L., Winner, W. E. 1975. Physiological and biochemical aspects of the resistance of lower plants (mosses) to extreme environments. *Proc. Circumpolar Conf. North. Ecol. NRC, Ottawa,* pp. 195–202
22. Bewley, J. D., Gwóźdź, E. A. 1975. Plant desiccation and protein synthesis. II. On the relationship between endogenous ATP levels and protein synthesizing capacity. *Plant Physiol.* 55:1110–14
23. Bewley, J. D., Halmer, P., Krochko, J. E., Winner, W. E. 1978. Metabolism of a drought-tolerant and a drought-sensitive moss. Respiration, ATP synthesis and carbohydrate status. In *Dry Biological Systems,* ed. J. H. Crowe, J. S. Clegg, pp. 185–203. New York: Academic.
24. Bewley, J. D., Pacey, J. 1978. Desiccation-induced ultrastructural changes in drought-sensitive and drought-tolerant plants. See Ref. 23, pp. 53–73
25. Bewley, J. D., Tucker, E. B., Gwóźdź, E. A. 1974. The effects of stress on the metabolism of *Tortula ruralis.* In *Mechanisms of Regulation of Plant Growth, R. Soc. N.Z. Bull. 12,* ed. R. L. Bieleski, A. R. Ferguson, M. M. Cresswell, pp. 395–402. Wellington: R. Soc. N.Z.
26. Bhat, S. P., Padayatty, J. D. 1974. Presence of conserved messenger RNA in

rice embryos. *Indian J. Biochem. Biophys.* 11:47–50
27. Biswas, K. P. 1952. Road slimes of Calcutta. *J. Dep. Sci. Calcutta Univ.* 7:1–8
28. Black, M., Bewley, J. D. 1979. *Physiology and Biochemistry of Seeds in Relation to Germination. 2. Viability, Dormancy and Environmental Control.* Berlin: Springer-Verlag. In press
29. Boyer, J. S. 1976. Water deficits and photosynthesis. In *Water Deficits and Plant Growth,* ed. T. T. Kozlowski, 4:153–90. New York: Academic
30. Bramlage, W. J., Leopold, A. C., Parrish, D. J. 1978. Chilling stress to soybeans during imbibition. *Plant Physiol.* 61:525–29
31. Brandle, J. R., Hinckley, T. M., Brown, G. N. 1977. The effects of dehydration-rehydration cycles on protein synthesis of black locust seedlings. *Physiol. Plant.* 40:1–5
32. Breidenbach, R. W., Castelfranco, P., Criddle, R. S. 1967. Biogenesis of mitochondria in germinating peanut cotyledons. II. Changes in cytochromes and mitochondrial DNA. *Plant Physiol.* 42:1035–41
33. Breidenbach, R. W., Castelfranco, P., Peterson, C. 1966. Biogenesis of mitochondria in germinating peanut cotyledons. *Plant Physiol.* 41:803–9
34. Brinkhuis, B. H., Tempel, N. R., Jones, R. F. 1976. Photosynthesis and respiration of exposed salt-marsh fucoids. *Marine Biol.* 34:349–60
35. Bristol, B. M. 1919. On the retention of vitality by algae from old stored soils. *New Phytol.* 18:92–107
36. Brock, T. D. 1975. Effect of water potential on a Microcoleus (Cyanophyceae) from a desert crust. *J. Phycol.* 11:316–20
37. Brock, T. D. 1975. The effect of water potential on photosynthesis in whole lichens and in their liberated algal components. *Planta* 124:13–23
38. Brooker, J. D., Tomaszewski, M., Marcus, A. 1978. Preformed messenger RNAs and early wheat embryo germination. *Plant Physiol.* 61:145–49
39. Brown, R. M., Wilson, R. 1968. Electron microscopy of the lichen *Physcia aipolia* (Ehrh.) Nyl. *J. Phycol.* 4:230–40
40. Bryant, T. R. 1972. Gas exchange in dry seeds: circadian rhythmicity in the absence of DNA replication, transcription, and translation. *Science* 178: 634–36
41. Burguillo, P. F., Nicolás, G. 1977. Appearance of an alternate pathway cyanide-resistant during germination of seeds of *Acer arietinum. Plant Physiol.* 60:524–47
42. Buttrose, M. S. 1973. Rapid water uptake and structural changes in imbibing seed tissues. *Protoplasma* 77:111–22
43. Cameron, R. E. 1962. Species of *Nostoc* Vaucher occurring in the Sonoran Desert in Arizona. *Trans. Am. Microsc. Soc.* 81:379–84
44. Cameron, R. E., Blank, G. B. 1966. Desert algae: soil crusts and diaphanous substrata as algal habitats. *NASA Jet Propul. Lab. Tech. Rep. 32–971.* 41 pp.
45. Carceller, M. S., Soriano, A. 1972. Effects of treatments given to the grain, on the growth of wheat roots under drought conditions. *Can. J. Bot.* 50:105–8
46. Carlier, A. R., Peumans, W. J. 1976. The rye embryo system as an alternative to the wheat system for protein synthesis in vitro. *Biochim. Biophys. Acta* 447:436–44
47. Chapman, V. J. 1966. The physiological ecology of some New Zealand seaweeds. *Proc. 5th Int. Seaweed Symp.,* ed. E. G. Young, J. L. McLachlan, pp. 29–54. Oxford: Pergamon
48. Chen, D., Sarid, S., Katchalski, E. 1968. The role of water stress in the inactivation of messenger RNA of germinating wheat embryos. *Proc. Natl. Acad. Sci. USA* 61:1378–83
49. Cheung, C. P., Suhadolnik, R. J. 1978. Regulation of RNA synthesis in early germination of isolated wheat (*Triticum aestivum* L.) embryo. *Nature* 271: 357–58
50. Ching, T. M. 1972. Metabolism of germinating seeds. In *Seed Biology,* ed. T. T. Kozlowski, 2:103–218. New York: Academic. 447 pp.
51. Ching, T. M. 1975. Temperature regulation of germination in crimson clover seeds. *Plant Physiol.* 56:768–71
52. Ching, T. M., Ching, K. K. 1972. Content of adenosine phosphates and adenylate energy charge in germinating Ponderosa Pine seeds. *Plant Physiol.* 50:536–40
53. Ching, T. M., Crane, J. M., Stamp, D. L. 1974. Adenylate energy pool and energy charge in maturing rape seeds. *Plant Physiol.* 54:748–51
54. Cloudsley-Thompson, J. L., Chadwick, M. J. 1964. *Life in Deserts.* Philadelphia: Dufour Ed.
55. Cocucci, S., Sturani, E. 1966. Water availability and RNA in the endosperm of ripening castor bean seeds. *G. Bot. Ital.* 73:29–30

56. Collins, D. M., Wilson, A. T. 1972. Metabolism of the axis and cotyledons of *Phaseolus vulgaris* seeds during early germination. *Phytochemistry* 11: 1931–35
57. Collins, D. M., Wilson, A. T. 1975. Embryo and endosperm metabolism of barley seeds during early germination. *J. Exp. Bot.* 26:737–40
58. Collyer, D. M., Fogg, G. E. 1955. Studies on fat accumulation by algae. *J. Exp. Bot.* 17:256–75
59. Conger, A. D., Randolph, M. L. 1968. Is age-dependent genetic damage in seeds caused by free radicals? *Radiat. Bot.* 8:193–96
60. Cooper, J. P., ed. 1975. *Photosynthesis and Productivity in Different Environments.* Cambridge: Cambridge Univ. 715 pp.
61. Crafts, A. S. 1968. Water deficits and physiological processes. In *Water Deficits and Plant Growth,* ed. T. T. Kozlowski, 2:85–133. New York: Academic
62. Crèvecoeur, M., Deltour, R., Bronchart, R. 1976. Cytological study on water stress during germination of *Zea mays. Planta* 132:31–41
63. Crittenden, P. D., Kershaw, K. A. 1978. A procedure for the simultaneous measurement of net CO_2-exchange and nitrogenase activity in lichens. *New Phytol.* 80:393–401
64. Darbyshire, B. 1974. The function of the carbohydrate units of three fungal enzymes in their resistance to dehydration. *Plant Physiol.* 54:717–21
65. Darbyshire, B., Steer, B. T. 1973. Dehydration of macromolecules. I. Effect of dehydration-rehydration on indoleacetic acid oxidase, ribonuclease, ribulosediphosphate carboxylase, and ketose-1-phosphate aldolase. *Aust. J. Biol. Sci.* 26:591–604
66. Davis, J. S. 1972. Survival records in the algae, and the survival role of certain algal pigments, fat and mucilaginous substances. *Biologist* 54:52–93
67. Dawson, M. J. 1965. Effects of seed-soaking on the growth and development of crop plants. I. Finger millet (*Eleusine coracana* Gaertn.) *Indian J. Plant Physiol.* 8:52–56
68. Delseny, M., Aspart, L., Got, A., Cooke, R., Guitton, Y. 1977. Early synthesis of polyadenylic acid, polyadenylated and ribosomal nucleic acids in germinating radish embryo axes. *Physiol. Vég.* 15:413–28
69. Delseny, M., Aspart, L., Guitton, Y. 1977. Disappearance of stored polyadenylic acid and mRNA during early germination of radish (*Raphanus sativus* L.) embryo axes. *Planta* 135:125–28
70. Deltour, R., Jacqmard, A. 1974. Relation between water stress and DNA synthesis during germination of *Zea mays* L. *Ann. Bot.* 38:529–34
71. Dhindsa, R. S., Bewley, J. D. 1976. Plant desiccation: polyribosome loss not due to ribonuclease. *Science* 191:181–82
72. Dhindsa, R. S., Bewley, J. D. 1976. Water stress and protein synthesis. IV. Responses of a drought-tolerant plant. *J. Exp. Bot.* 27:513–23
73. Dhindsa, R. S., Bewley, J. D. 1977. Water stress and protein synthesis. V. Protein synthesis, protein stability and membrane permeability in a drought-sensitive and a drought-tolerant moss. *Plant Physiol.* 59:295–300
74. Dhindsa, R. S., Bewley, J. D. 1978. Messenger RNA is conserved during drying of drought-tolerant *Tortula ruralis. Proc. Natl. Acad. Sci. USA* 75:842–46
75. Dhindsa, R. S., Cleland, R. E. 1975. Water stress and protein synthesis. II. Interaction between water stress, hydrostatic pressure, and abscisic acid on the pattern of protein synthesis in *Avena* coleoptiles. *Plant Physiol.* 55:782–85
76. Dilks, T. J. K., Proctor, M. C. F. 1974. The pattern of recovery of bryophytes after desiccation. *J. Bryol.* 8:97–115
77. Dilks, T. J. K., Proctor, M. C. F. 1976. Seasonal variation in desiccation tolerance in some British bryophytes. *J. Bryol.* 9:239–47
78. Dilks, T. J. K., Proctor, M. C. F. 1976. Effects of intermittent desiccation on bryophytes. *J. Bryol.* 9:249–64
79. Durrell, L. W. 1962. Algae of Death Valley. *J. Am. Microsc. Soc.* 81:267–73
80. Efron, D., Evenari, M., De Groot, N. 1971. Amino acid incorporation activity of lettuce seed ribosomes during germination. *Life Sci.* 10:1015–19
81. Evans, J. H. 1958. The survival of freshwater algae during dry periods. I. An investigation of the algae of five small ponds. *J. Ecol.* 46:149–76
82. Evans, J. H. 1959. The survival of fresh water algae during dry periods. II. Drying experiments. *J. Ecol.* 47:55–81
83. Evans, J. H. 1960. Further investigations of the algae of pond margins. *Hydrobiologia* 15:384–94
84. Evenari, M. 1964. Hardening treatment of seeds as a means of increasing yields under condtitions of inadequate moisture. *Nature* 204:1010–11

85. Fakunding, J. L., Hershey, J. W. B. 1973. The interaction of radioactive initiation factor IF-2 with ribosomes during initiation of protein synthesis. *J. Biol. Chem.* 248:4206–12
86. Farrar, J. F. 1973. Lichen physiology: progress and pitfalls. In *Air Pollution and Lichens,* ed. B. W. Ferry, M. S. Baddeley, D. L. Hawksworth, pp. 238–82. London: Athlone
87. Farrar, J. F. 1976. The lichen as an ecosystem: observation and experiment. In *Lichenology: Progress and Problems,* ed. D. H. Brown, D. L. Hawksworth, R. H. Bailey, pp. 385–406. London: Academic
88. Farrar, J. F. 1976. Ecological physiology of the lichen *Hypogymnia physodes.* II. Effects of wetting and drying cycles and the concept of "physiological buffering." *New Phytol.* 77:105–13
89. Farrar, J. F., Smith, D. C. 1976. Ecological physiology of the lichen *Hypogymnia physodes.* III. The importance of the rewetting phase. *New Phytol.* 77:115–25
90. Fellows, R. J., Boyer, J. S. 1978. Altered ultrastructure of cells of sunflower leaves having low water potentials. *Protoplasma* 93:381–95
91. Finean, J. B. 1969. Biophysical contributions to membrane structure. *Q. Rev. Biophys.* 2:1–23
92. Florance, E. R., Denison, W. C., Allen, T. C. 1972. Ultrastructure of dormant and germinating conidia of *Aspergillus nidulans. Mycologia* 64:115–23
93. Fork, D. C., Hiyama, T. 1973. The photochemical reactions of photosynthesis in an alga exposed to extreme conditions. *Carnegie Inst. Washington Yearb.* 72:384–88
94. Fraymouth, J. 1928. The moisture relations of terrestrial algae. III. The respiration of certain lower plants, including terrestrial algae, with special reference to the influence of drought. *Ann. Bot.* 42:75–100
95. Fridovich, I. 1975. Superoxide dismutases. *Ann. Rev. Biochem.* 44:147–59
96. Fridovich, I. 1976. Oxygen radicals, hydrogen peroxide, and oxygen toxicity. In *Free Radicals in Biology,* ed. W. A. Pryor, 1:239–77. New York: Academic
97. Gaff, D. F. 1977. Desiccation tolerant vascular plants of Southern Africa. *Oecologia* 31:95–109
98. Gaff, D. F., Churchill, D. M. 1976. *Borya nitida* Labill.—an Australian species in the Liliaceae with desiccation-tolerant leaves. *Aust. J. Bot.* 24:209–24
99. Gaff, D. F., Hallam, N. D. 1974. Resurrecting desiccated plants. See Ref. 25, pp. 389–93
100. Gaff, D. F., Latz, P. 1978. The occurrence of resurrection plants in the Australian flora. *Aust. J. Bot.* 26:485–92
101. Gaff, D. F., Zee, S.-Y., O'Brien, T. P. 1976. The fine structure of dehydrated and reviving leaves of *Borya nitida* Labill.—a desiccation-tolerant plant. *Aust. J. Bot.* 24:225–36
102. Gannutz, T. P. 1969. Effects of environmental extremes on lichens. *Bull. Soc. Bot. Fr. Mem. 1968. Colloq. Lichens 1967,* pp. 169–79
103. Gej, B. 1962. Researches on the resistance of two varieties of spring wheat to periodical water deficit. *Acta Agrobot.* 11:31–46
104. Giesen, M., Roman, R., Seal, S. N., Marcus, A. 1976. Formation of an 80S methionyl-tRNA initiation complex with soluble factors from wheat germ. *J. Biol. Chem.* 251:6075–81
105. Gillard, D. F., Walton, D. C. 1973. Germination of *Phaseolus vulgaris.* IV. Patterns of protein synthesis in excised axes. *Plant Physiol.* 51:1147–49
106. Gordon, M. E., Payne, P. I. 1976. *In vitro* translation of the long-lived messenger ribonucleic acid of dry seeds. *Planta* 130:269–73
107. Gottlieb, D. 1976. Carbohydrate metabolism and spore germination. In *The Fungal Spore. Form and Function,* ed. D. J. Weber, W. M. Hess, pp. 141–63. New York: Wiley
108. Gupta, R. K. 1976. The physiology of the desiccation resistance in bryophytes: nature of organic compounds leaked from desiccated liverwort, *Plagiochila asplenioides. Biochem. Physiol. Pflanz.* 170:389–95
109. Gupta, R. K. 1977. A study of photosynthesis and leakage of solutes in relation to the desiccation effects in bryophytes. *Can. J. Bot.* 55:1186–94
110. Gupta, R. K. 1977. An artefact in studies of the responses of respiration of bryophytes to desiccation. *Can. J. Bot.* 55:1195–1200
111. Gupta, R. K. 1977. A note on photosynthesis in relation to water content in liverworts: *Porella platyphylla* and *Scapania undulata. Aust. J. Bot.* 25:363–65
112. Gupta, R. K. 1978. The physiology of the desiccation resistance in bryophytes: effect of desiccation on water status and chlorophyll a and b in bryophytes. *Indian J. Exp. Biol.* In press
113. Gwóźdź, E. A., Bewley, J. D. 1975. Plant desiccation and protein synthesis.

An *in vitro* system from dry and hydrated mosses using endogenous and synthetic messenger RNA. *Plant Physiol.* 55:340–45
114. Gwóźdź, E. A., Bewley, J. D., Tucker, E. B. 1974. Studies on protein synthesis in *Tortula ruralis:* polyribosome reformation following desiccation. *J. Exp. Bot.* 25:599–608
115. Haas, P., Hill, T. G. 1933. Observations on the metabolism of certain seaweeds. *Ann. Bot.* 47:55–67
116. Haber, A. H., Brassington, N. 1959. Non-respiratory gas released from seeds during moistening. *Nature* 183:619–20
117. Hallam, N. D. 1972. Embryogenesis and germination in rye (*Secale cereale*). I. Fine structure of the developing embryo. *Planta* 104:157–66
118. Hallam, N. D. 1976. Anhydrous fixation of dry plant tissues using non-aqueous fixatives. *J. Microsc.* 106:337–42
119. Hallam, N. D., Gaff, D. F. 1978. Reorganization of fine structure during rehydration of desiccated leaves of *Xerophyta villosa. New Phytol.* 81:349–55
120. Hallam, N. D., Roberts, B. E., Osborne, D. J. 1972. Embryogenesis and germination in rye (*Secale cereale* L.). II. Biochemical and fine structural changes during germination. *Planta* 105:293–309
121. Hammett, J. R., Katterman, F. R. 1975. Storage and metabolism of poly(adenylic acid)-mRNA in germinating cotton seeds. *Biochemistry* 14:4375–79
122. Hanson, A. D. 1973. The effects of imbibition drying treatments on wheat seeds. *New Phytol.* 72:1063–73
123. Harman, G. E., Mattick, L. R. 1976. Association of lipid oxidation with seed ageing and death. *Nature* 260:323–24
124. Harris, G. P. 1971. The ecology of corticolous lichens. II. The relationship between physiology and the environment. *J. Ecol.* 59:441–52
125. Harris, G. P. 1976. Water content and productivity of lichens. In *Water and Plant Life. Problems and Modern Approaches,* ed. O. L. Lange, L. Kappen, E.-D. Schulze, pp. 452–68. Berlin: Springer-Verlag
126. Hecker, M., Köhler, K.-H., Wiedmann, M. 1977. Reactivation of ribonucleic acid synthesis during early germination of *Agrostemma githago* embryos. *Biochem. Physiol. Pflanz.* 171:401–8
127. Henckel, P. A. 1961. Drought resistance in plants: methods of recognition and intensification. In *Plant-Water Relationships in Arid and Semi-Arid Conditions. Proc. Madrid Symp.* 16:167–74. Paris: UNESCO
128. Henckel, P. A. 1964. Physiology of plants under drought. *Ann. Rev. Plant Physiol.* 15:363–86
129. Henckel, P. A. 1970. Role of protein synthesis in drought resistance. *Can. J. Bot.* 48:1235–41
130. Henckel, P. A., Kurkova, E. B., Pronina, N. D. 1970. Effect of dehydration on the course of photosynthesis in homeohydrous and poikilohydrous plants. *Sov. Plant Physiol.* 17:952–57
131. Henckel, P. A., Levina, V. V. 1975. Significance of cytocontraction and plasmocontraction in poikiloxerophytic plants during dehydration. *Sov. Plant Physiol.* 22:488–91
132. Henckel, P. A., Martyanova, K. L., Zubova, L. S. 1964. Production experiments on pre-sowing drought hardening of plants. *Sov. Plant Physiol.* 11:457–61
133. Henckel, P. A., Pronina, N. D. 1968. Factors underlying dehydration resistance in poikiloxerophytes. *Sov. Plant Physiol.* 15:68–74
134. Henckel, P. A., Pronina, N. D. 1969. Anabiosis with desiccation of the poikiloxerophytic flowering plant *Myrothamnus flabellifolia. Sov. Plant Physiol.* 16:745–49
135. Henckel, P. A., Satarova, N. A., Shaposhnikova, S. V. 1977. Protein synthesis in poikiloxerophytes and wheat embryos during the initial period of swelling. *Sov. Plant Physiol.* 24:737–41
136. Henriksson, E., Simu, B. 1971. Nitrogen fixation by lichens. *Oikos* 22:119–21
137. Hinshiri, H. M., Proctor, M. C. F. 1971. The effect of desiccation on subsequent assimilation and respiration of the bryophytes *Anomodon viticulosus* and *Porella platyphylla. New Phytol.* 70:527–38
138. Hitch, C. J. B., Stewart, W. D. P. 1973. Nitrogen fixation by lichens in Scotland. *New Phytol.* 72:509–24
139. Hoffman, P. 1968. Pigmentgehalt und gaswechsel von *Myrothamnus.* Blättern nach autrocknung und wiederaufsattigung. *Photosynthetica* 2:245–52
140. Holm-Hansen, O. 1963. Algae: Nitrogen fixation by Antarctic species. *Science* 139:1059–60
141. Hostetter, H. P., Hoshaw, R. W. 1970. Environmental factors affecting resistance to desiccation in the diatom *Stauroneis anceps. Am. J. Bot.* 57:512–18
142. Hsiao, T. C. 1970. Rapid changes in levels of ribosomes in *Zea mays* in re-

sponse to water stress. *Plant Physiol.* 46:281–85

143. Hsiao, T. C. 1973. Plant responses to water stress. *Ann. Rev. Plant Physiol.* 24:519–70

144. Husain, I., May, L. H., Aspinall, D. 1968. The effects of soil moisture stress on the growth of barley. IV. Responses to presowing treatment. *Aust. J. Agric. Res.* 19:213–20

145. Huss-Danell, K. 1977. Nitrogenase activity in the lichen *Stereocaulon paschale:* recovery after dry storage. *Physiol. Plant.* 41:158–61

146. Iljin, W. S. 1957. Drought resistance in plants and physiological processes. *Ann. Rev. Plant Physiol.* 8:257–74

147. Jachymczyk, W. J., Cherry, J. H. 1968. Studies on messenger RNA from peanut plants: *in vitro* polyribosome formation and protein synthesis. *Biochim. Biophys. Acta* 157:368–77

148. Jacoby, B., Oppenheimer, H. R. 1962. Pre-sowing treatment of sorghum grains and its influence on drought resistance of the resulting plants. *Phyton* 19:109–13

149. Jarvis, P. G., Jarvis, M. S. 1964. Presowing hardening of plants to drought. *Phyton* 21:113–17

150. Johnson, W. S., Gigon, A., Gulmon, S. L., Mooney, H. A. 1974. Comparative photosynthetic capacities of intertidal algae under exposed and submerged conditions. *Ecology* 55:450–53

151. Johri, M. M., Maheshwari, S. C. 1965. Studies on respiration in developing poppy seeds. *Plant Cell Physiol.* 6:61–72

152. Jones, K. 1974. Nitrogen fixation in a salt marsh. *J. Ecol.* 62:553–64

153. Jones, K. 1977. The effects of moisture on acetylene reduction by mats of blue-green algae in sub-tropical grassland. *Ann. Bot.* 41:801–6

154. Kanwisher, J. 1957. Freezing and drying in intertidal algae. *Biol. Bull.* 113:275–85

155. Kappen, L. 1973. Response to extreme environments. In *The Lichens,* ed. V. Ahmadjian, M. E. Hale, pp. 311–80. New York: Academic

156. Kappen, L., Lange, O. L., Schulze, E.-D., Evenari, M., Buschbom, U. 1975. Primary production of lower plants (lichens) in the desert and its physiological basis. See Ref. 60, pp. 133–43

157. Keller, W., Black, A. T. 1968. Preplanting treatment to hasten germination and emergence of grass seed. *J. Range Manage.* 21:213–16

158. Kershaw, K. A. 1972. The relationship between moisture content and net assimilation rate of lichen thalli and its ecological significance. *Can. J. Bot.* 50:543–55

159. Kershaw, K. A., Dzikowski, P. A. 1977. Physiological-environmental interactions in lichens. VI. Nitrogenase activity in *Peltigera polydactyla* after a period of desiccation. *New Phytol.* 79:417–21

160. Kessler, B., Frank-Tishel, J. 1962. Dehydration-induced synthesis of nucleic acids and changing of composition of ribonucleic acid: A possible protective reaction in drought-resistant plants. *Nature* 196:542–43

161. Kimura, K., Ono, T., Yanagita, T. 1965. Isolation and characterization of ribosomes from dormant and germinating conidia of *Aspergillus oryzae. J. Biochem. Tokyo* 58:569–76

162. Klein, S., Barenholz, H., Budnik, A. 1971. The initiation of growth in isolated lima bean axes. Physiological and fine structural effects of actinomycin D, cycloheximide and chloramphenicol. *Plant Cell Physiol.* 12:41–60

163. Klein, S., Pollock, B. M. 1968. Cell fine structure of developing lima bean seeds related to seed desiccation. *Am. J. Bot.* 55:658–72

164. Kluge, M. 1976. Carbon and nitrogen metabolism under water stress. See Ref. 125, pp. 243–52

165. Kollöffel, C. 1970. Oxidative and phosphorylative activity of mitochondria from pea cotyledons during maturation of the seed. *Planta* 91:321–28

166. Krochko, J. E., Bewley, J. D. 1975. Poikilohydrous plants: The extreme case of drought tolerance. *What's New in Plant Physiol. No. 8* 7:1–3

167. Krochko, J. E., Bewley, J. D., Pacey, J. 1978. The effects of rapid and very slow speeds of drying on the ultrastructure and metabolism of the desiccation-sensitive moss *Cratoneuron filicinum. J. Exp. Bot.* 29:905–17

168. Lange, O. L. 1953. Hitze- und trockenresistenz der flechten in beziehung zu ihrer verbreitung. *Flora (Jena)* 140: 39–97

169. Lange, O. L. 1969. Ecophysiological investigations in lichens of the Negev desert. I. CO_2 gas exchange of *Ramalina maciformis* (Del.) Bory under controlled conditions in the laboratory. *Flora (Jena)* 158:324–59

170. Lange, O. L. 1969. CO_2-gas exchange of mosses following water vapour uptake. *Planta* 89:90–94

171. Lange, O. L., Schulze, E.-D., Kappen, L., Buschbom, U., Evenari, M. 1975. Adaptations of desert lichens to drought and extreme temperatures. In *Environmental Physiology of Desert Organisms,* ed. N. F. Hadley, pp. 20–37. Stroudsburg, Pa: Dowden, Hutchinson & Ross
172. Lange, O. L., Schulze, E.-D., Koch, W. 1970. Experimentellokologische üntersuchungen am flechten der Negev-Wüste. II. CO_2-gaswechsel und wasserhaushalt von *Ramalina maciformis* (Del.) Bory am natürlichen standort während der sommerlichen trockenperiode. *Flora (Jena).* 159:38–62
173. Larson, D. W., Kershaw, K. A. 1976. Studies on lichen-dominated systems. XVIII. Morphological control of evaporation in lichens. *Can. J. Bot.* 54:2061–73
174. Lee, J. A., Stewart, G. R. 1971. Desiccation injury in mosses. I. Intra-specific differences in the effect of moisture stress on photosynthesis. *New Phytol.* 70:1061–68
175. Levitt, J. 1972. *Responses of Plants to Environmental Stresses.* New York: Academic. 697 pp.
176. Lott, J. N. A., Castelfranco, P. 1970. Changes in the cotyledons of *Cucurbita maxima* during germination. II. Development of mitochondrial function. *Can. J. Bot.* 48:2233–40
177. Lovett, J. S. 1976. Regulation of protein metabolism during spore germination. See Ref. 107, pp. 189–242
178. Luzzati, V., Husson, F. 1962. The structure of the liquid-crystalline phases of lipid-water systems. *J. Cell. Biol.* 12:207–19
179. Macfarlane, J. D., Maikawa, E., Kershaw, K. A., Oaks, A. 1976. Physiological-environmental interactions in lichens. I. The interaction of light/dark periods and nitrogenase activity in *Peltigera polydactyla. New Phytol.* 77:705–11
180. Maheshwari, R., Sussman, A. S. 1970. Respiratory changes during germination of urediospores of *Puccinia graminis* f. sp. *tritici. Phytopathology* 60:1357–64
181. Mahmoud, M. 1965. *Protoplasmics and drought resistance in mosses.* PhD thesis. Univ. California, Davis, Calif. 103 pp.
182. Malhotra, S. S., Solomos, T., Spencer, M. 1973. Effects of cycloheximide, D-threo-chloramphenicol, erythromycin and actinomycin D on *de novo* synthesis of cytoplasmic and mitochondrial proteins in the cotyledons of germinating pea seeds. *Planta* 114:169–84
183. Marcus, A., Feeley, J. 1964. Activation of seed protein synthesis in the imbibition phase of seed germination. *Proc. Natl. Acad. Sci. USA* 51:1075–79
184. Marcus, A., Feeley, J., Volcani, T. 1966. Protein synthesis in imbibed seeds. III. Kinetics of amino acid incorporation, ribosome activation, and polysome formation. *Plant Physiol.* 41:1167–72
185. Marcus, A., Spiegel, S., Brooker, J. D. 1975. Preformed mRNA and the programming of early embryo development. In *Control Mechanisms in Development,* ed. R. H. Meints, E. Davies pp. 1–19. New York: Plenum
186. Marcus, A., Weeks, D. P., Leis, J. P., Keller, E. B. 1970. Protein chain initiation by methionyl-tRNA in wheat embryo. *Proc. Natl. Acad. Sci. USA* 67:1681–87
187. Marinos, N. G., Fife, D. N. 1972. Ultrastructural changes in wheat embryos during a "presowing drought hardening" treatment. *Protoplasma* 74:381–96
188. Marré, E., Cocucci, S., Sturani, E. 1965. On the development of the ribosomal system in the endosperm of germinating castor bean seeds. *Plant Physiol.* 40:1162–70
189. Mathieson, A. C., Burns, R. L. 1971. Ecological studies on economic red algae. I. Photosynthesis and respiration of *Chondrus crispus* Stackhouse and *Gigartina stellata* (Stackhouse) Batters. *J. Exp. Mar. Biol. Ecol.* 7:197–206
190. Matthews, S., Rogerson, N. E. 1976. The influence of embryo condition on the leaching of solutes from pea seeds. *J. Exp. Bot.* 27:961–68
191. May, L. H., Milthorpe, E. J., Milthorpe, F. L. 1962. Pre-sowing hardening of plants to drought. An appraisal of the contributions of P. A. Henckel. *Field Crop Abstr.* 15:93–98
192. McKay, E. 1935. Photosynthesis in *Grimmia montana. Plant Physiol.* 10:803–9
193. McLean, R. J. 1967. Desiccation and heat resistance of the green alga *Spongiochloris typica. Can. J. Bot.* 45:1933–38
194. Mead, J. F. 1976. Free radical mechanisms of lipid damage and consequences for cellular membranes. See Ref. 96, pp. 51–68
195. Millerd, A., Spencer, D. 1974. Changes in RNA-synthesizing activity and template activity in nuclei from cotyledons of developing pea seeds. *Aust. J. Plant Physiol.* 1:331–41

196. Milthorpe, F. L. 1950. Changes in the drought resistance of wheat seedlings during germination. *Ann. Bot.* 14:79–89
197. Mirkes, P. E. 1974. Polysomes, ribonucleic acid, and protein synthesis during germination of *Neurospora crassa* conidia. *J. Bacteriol.* 117:196–202
198. Moreland, D. E., Hussey, G. G., Shriner, C. R., Farmer, F. S. 1974. Adenosine phosphates in germinating radish (*Raphanus sativus* L.) seeds. *Plant Physiol.* 54:560–63
199. Nawa, Y., Asahi, T. 1971. Rapid development of mitochondria in pea cotyledons during the early stage of germination. *Plant Physiol.* 48:671–74
200. Nawa, Y., Asahi, T. 1973. Biochemical studies on development of mitochondria in pea cotyledons during the early stage of germination. Effects of antibiotics on the development. *Plant Physiol.* 51:833–38
201. Nir, I., Klein, S., Poljakoff-Mayber, A. 1969. Effect of moisture stress on submicroscopic structure of maize roots. *Aust. J. Biol. Sci.* 22:17–33
202. Noailles, M.-C. 1974. Comparison de l'ultrastructure du parenchyme des tiges et feuilles d'une mousse normalement hydratée et en cours de desiccation [*Pleurozium schreberi* (Willd.) Mitt.]. *C. R. Acad. Sci. Ser. B* 278:2759–62
203. Obendorf, R. L., Marcus, A. 1974. Rapid increase in ATP during early wheat embryo germination. *Plant Physiol.* 53:779–81
204. Ogata, E. 1968. Respiration of some marine plants as affected by dehydration and rehydration. *J. Shimonoseki Univ. Fish.* 16:89–102
205. Ohmura, T., Howell, R. W. 1962. Respiration of developing and germinating soybean seeds. *Physiol. Plant.* 15: 341–50
206. Oppenheimer, H. R. 1960. Adaptation to drought: Xerophytism. See Ref. 127, pp. 105–38
207. Oppenheimer, H. R., Halevy, A. H. 1962. Anabiosis of *Ceterach officinarum* Lam et DC. *Bull. Res. Counc. Isr. D3* 11:127–47
208. Palta, J. P., Levitt, J., Stadelmann, E. J. 1978. Freezing injury in onion bulb cells. II. Post-thawing injury on recovery. *Plant Physiol.* 60:398–401
209. Pammenter, N. W., Adamson, J. H., Berjak, P. 1974. Viability of stored seed: extension by cathodic protection. *Science* 186:1123–24
210. Parker, B. C., Schanen, N., Renner, R. 1969. Viable soil algae from the herbarium of the Missouri Botanical Garden. *Ann. Mo. Bot. Gard.* 56:113–19
211. Parker, J. 1968. Drought-resistance mechanisms. See Ref. 29, 1:195–234
212. Parker, J. 1972. Protoplasmic resistance to water deficits. See Ref. 29, 3:125–76
213. Parrish, D. J., Leopold, A. C. 1977. Transient changes during soybean imbibition. *Plant Physiol.* 59:111–15
214. Paul, E. A., Myers, R. J. K., Rice, W. A. 1971. Nitrogen fixation in grassland and associated cultivated ecosystems. In *Biological Nitrogen Fixation in Natural and Agricultural Habitats,* ed. T. A. Lie, E. G. Mulder. *Plant and Soil* Spec. Ed., pp. 495–507
215. Payne, J. F., Bal, A. K. 1972. RNA polymerase activity in germinating onion seeds. *Phytochemistry* 11:3105–10
216. Payne, P. I. 1976. The long-lived messenger ribonucleic acid of flowering-plant seeds. *Biol. Rev.* 51:329–63
217. Payne, P. I. 1977. Synthesis of poly-A-rich RNA in embryos of rye during imbibition and early germination. *Phytochemistry* 16:431–34
218. Payne, P. I., Boulter, D. 1969. Free and membrane bound ribosomes of the cotyledons of *Vicia faba* (L.). *Planta* 84:263–71
219. Payne, P. I., Dobrzanska, M., Barlow, P. W., Gordon, M. E. 1978. The synthesis of RNA in imbibing seed of rape (*Brassica napus*) prior to the onset of germination: a biochemical and cytological study. *J. Exp. Bot.* 29:77–88
220. Payne, P. I., Gordon, M. E., Barlow, P. W., Parker, M. L. 1977. The subcellular location of the long-lived messenger RNA of rape seed. See Ref. 20, pp. 224–27
221. Perner, E. 1965. Elektronenmikroskopische untersuchungen an zellen von embryonen im zustand volliger somenruhe. I Mitteilung. Die zelluläre strukturordnung in der radicula lufttrockner samen von *Pisum sativum*. *Planta* 65:334–57
222. Perry, D. A., Harrison, J. G. 1970. The deleterious effects of water and low temperature on germination of pea seed. *J. Exp. Bot.* 21:504–12
223. Peumans, W. J., Carlier, A. R. 1977. Messenger ribonucleoprotein particles in dry wheat and rye embryos. In vitro translation and size distribution. *Planta* 136:195–201
224. Peveling, E. 1973. Fine structure. In *The Lichens,* ed. V. Ahmadjian, M. E. Hale, pp. 147–82. New York: Academic
225. Peveling, E. 1977. Ultrastructure of

some lichens after long dry periods. *Protoplasma* 92:129–36
226. Pradet, A., Narayanan, A., Vermeersch, J. 1968. Étude des adénosine-5'- mono, di- et tri-phosphates dans les tissus végétaux. III. Métabolisme énérgetique au cours des premiers stades de la germination des semences de laitue. *Bull. Soc. Fr. Physiol. Vég.* 14:107–14
227. Proctor, M. C. F. 1972. An experiment on intermittent desiccation with *Anomodon viticulosus* (Hedw.) Hook and Tayl. *J. Bryol.* 7:181–86
228. Pronina, N. D. 1972. Effect of dehydration on ATPase activity in poikilohydrous and homeohydrous plants. *Sov. Plant Physiol.* 19:731–32
229. Pueyo, G. 1960. Influence de l'anhydrobiose sur les lichens. Quelques résultats avec *Cladonia impexa* Harm. et *Umbilicaria pustulata* Hoffm. *Rev. Bryol. Lichénol.* 29:326–34
230. Reisener, H. J. 1976. Lipid metabolism of fungal spores during sporogenesis and germination. See Ref. 107, pp. 165–85
231. Renaut, J., Sasson, A., Pearson, H. W., Stewart, W. D. P. 1975. Nitrogen-fixing algae in Morocco. In *Nitrogen Fixation by Free-living Micro-organisms,* ed. W. D. P. Stewart, pp. 229–46. Cambridge: Cambridge Univ. IBP
232. Ried, A. 1953. Photosynthese und atmung bei xerostabilen und xerolabilen krustenflechten in der nachwerkung vorausgegangener entquellungen. *Planta* 41:436–38
233. Ried, A. 1960. Stoffwechsel und verbreitungsgrenzen von flechten. II. Wasserund assimilationshaushalt, entquellungs—und submersionsresistenz von krustenflechten benachbarter standorte. *Flora (Jena)* 149:345–85
234. Ried, A. 1960. Thallusbau und assimilationshaushalt von laub-und krustenflechten. *Biol. Zentralbl.* 79:129–51
235. Ried, A. 1960. Nachwirkungen der entquellung auf den gaswechsel von krustenflechten. *Biol. Zentralbl.* 79:657–78
236. Rogers, R. W. 1971. Distribution of the lichen *Chondropsis semiviridis* in relation to its heat and drought resistance. *New Phytol.* 70:1069–77
237. Salim, M. H., Todd, G. W. 1968. Seed soaking as a pre-sowing, drought-hardening treatment in wheat and barley seedlings. *Agron. J.* 60:179–82
238. Santarius, K. A. 1967. Assimilation of CO_2, NADP and PGA reduction and ATP synthesis in intact leaf cells in relation to water content. *Planta* 73: 228–42
239. Santarius, K. A. 1969. Der einfluβ von elektrolyten auf chloroplasten beim gefrieren und trocknen. *Planta* 89:23–46
240. Santarius, K. A. 1973. The protective effect of sugars on chloroplast membranes during temperature and water stress and its relationship to frost, desiccation and heat resistance. *Planta* 113:105–14
241. Sasaki, S., Brown, G. N. 1971. Polysome formation in *Pinus resinosa* at initiation of seed germination. *Plant Cell Physiol.* 12:749–58
242. Schneider, M. J. T., Schneider, A. S. 1972. Water in biological membranes: Adsorption isotherms and circular dichroism as a function of hydration. *J. Membr. Biol.* 9:127–40
243. Schonbeck, M. W. 1976. *A study of the environmental factors governing the vertical distribution of intertidal fucoids.* PhD thesis. Univ. Glasgow, Glasgow, Scotland. 205 pp.
244. Schonbeck, M. W., Norton, T. A. 1978. Factors controlling the upper limits of fucoid algae on the shore. *J. Exp. Mar. Biol. Ecol.* In press
245. Seibert, G., Loris, K., Zollner, J., Frenzel, B., Zahn, R. K. 1976. The conservation of poly-A-containing RNA during the dormant state of the moss *Polytrichum commune. Nucleic Acids Res.* 3:1997–2003
246. Sen, S., Osborne, D. J. 1974. Germination of rye embryos following hydration-dehydration treatments: enhancement of protein and RNA synthesis and earlier induction of DNA replication. *J. Exp. Bot.* 25:1010–19
247. Sen, S., Payne, P. I., Osborne, D. J. 1975. Early ribonucleic acid synthesis during the germination of rye (*Secale cereale*) embryos and the relationship to early protein synthesis. *Biochem. J.* 148:381–87
248. Sen Gupta, A. 1977. *Non-autotrophic CO_2 fixation by mosses.* MS thesis. Univ. Calgary, Calgary, Alberta, Canada. 92 pp.
249. Shirk, H. G. 1942. Freezable water content and the oxygen respiration in wheat and rye grain at different stages of ripening. *Am. J. Bot.* 29:105–9
250. Simon, E. W. 1974. Phospholipids and plant membrane permeability. *New Phytol.* 73:377–420
251. Simon, E. W. 1978. Membranes in dry and imbibing seeds. See Ref. 23, pp. 205–24
252. Simon, E. W., Raja Harun, R. M. 1972. Leakage during seed imbibition. *J. Exp. Bot.* 23:1076–85

253. Simon, E. W., Wiebe, H. H. 1975. Leakage during imbibition, resistance to damage at low temperatures and the water content of peas. *New Phytol.* 74:407–11
254. Siwecka, M. A., Szarkowski, J. W. 1974. Changes in distribution of ribosomes during germination of rye (*Secale cereale* L.) embryos. *Cytobios* 9:217–25
255. Slavik, B. 1965. The influence of decreasing hydration on photosynthetic rate in the thalli of the hepatic *Conocephallum conicum.* In *Water Stress in Plants, Proc. Symp. Prague 1963,* ed. B. Slavik, p. 195 Prague: Czech. Acad. Sci.
256. Smith, D. C. 1975. Symbiosis and the biology of lichenised fungi. In *Symbiosis, Symp. Soc. Exp. Biol.* 29:373–405
257. Smith, D. C., Molesworth, S. 1973. Lichen physiology. XIII. Effect of rewetting dry lichens. *New Phytol.* 72:525–33
258. Smith, D. L. 1973. Nucleic acid, protein, and starch synthesis in developing cotyledons of *Pisum arvense* L. *Ann. Bot.* 37:795–804
259. Smith, J. E., Gull, K., Anderson, J. G., Deans, S. G. 1976. Organelle changes during fungal spore germination. See Ref. 107, pp. 301–54
260. Solomos, T., Malhotra, S. S., Prasad, S., Malhotra, S. K., Spencer, M. 1972. Biochemical and structural changes in mitochondria and other cellular components of pea cotyledons during germination. *Can. J. Biochem.* 50:725–37
261. Spiegel, S., Marcus, A. 1975. Polyribosome formation in early wheat embryo germination independent of either transcription or polyadenylation. *Nature* 256:228–30
262. Spiegel, S., Obendorf, R. L., Marcus, A. 1975. Transcription of ribosomal and messenger RNAs in early wheat embryo germination. *Plant Physiol.* 56: 502–7
263. Srivastava, L. M., Paulson, R. E. 1968. The fine structure of the embryo of *Lactuca sativa.* II. Changes during germination. *Can. J. Bot.* 46:1447–53
264. Stewart, G. R., Lee, J. A. 1972. Desiccation injury in mosses. II. The effect of moisture stress on enzyme levels. *New Phytol.* 71:461–66
265. Stewart, W. D. P. 1974. Blue-green algae. In *The Biology of Nitrogen Fixation, Frontiers of Biology 33,* ed. A. Quispel, pp. 202-37. Amsterdam: North Holland
266. Stocker, O. 1960. Physiological and morphological changes in plants due to water deficiency. See Ref. 127, pp. 63–94
267. Stuart, T. S. 1968. Revival of respiration and photosynthesis in dried leaves of *Polypodium polypodioides. Planta* 83:185–206
268. Sturani, E., Cocucci, S., Marré, E. 1968. Hydration dependent polysome-monosome interconversion in the germinating castor bean endosperm. *Plant Cell Physiol.* 9:783–95
269. Swift, J. G., Buttrose, M. S. 1972. Freeze-etch studies of protein bodies in wheat scutellum. *J. Ultrastruct. Res.* 40:378–90
270. Todd, G. W. 1972. Water deficits and enzymatic activity. See Ref. 29, 3:177–216
271. Tucker, E. B., Bewley, J. D. 1974. The site of protein synthesis in the moss *Tortula ruralis* on recovery from desiccation. *Can. J. Biochem.* 52:345–48
272. Tucker, E. B., Bewley, J. D. 1976. Plant desiccation and protein synthesis. III. Stability of cytoplasmic RNA during dehydration, and its synthesis on rehydration of the moss *Tortula ruralis. Plant Physiol.* 57:564–67
273. Tucker, E. B., Costerton, J. W., Bewley, J. D. 1975. The ultrastructure of the moss *Tortula ruralis* on recovery from desiccation. *Can. J. Bot.* 53:94–101
274. Van Etten, J. L., Dunkle, L. D., Knight, R. H. 1976. Nucleic acids and fungal spore germination. See Ref. 107, pp. 243–300
275. Vieira Da Silva, J. 1976. Water stress, ultrastructure and enzymatic activity. See Ref. 125, pp. 207–24
276. Waisel, Y. 1962. Presowing treatments and their relation to growth and to drought, frost and heat resistance. *Physiol. Plant.* 15:43–46
277. Walk, R.-A., Hock, B. 1976. Mitochondrial malate dehydrogenase of watermelon cotyledons: Time course and mode of enzyme activity changes during germination. *Planta* 129:27–32
278. Walter, H. 1955. The water economy and the hydration of plants. *Ann. Rev. Plant Physiol.* 6:239–52
279. Wells, G. N., Beevers, L. 1974. Protein synthesis in the cotyledons of *Pisum sativum* L. Protein factors involved in the binding of phenylalanyl-transfer ribonucleic acid to ribosomes. *Biochem. J.* 139:61–69
280. Wilhelmsen, J. B. 1959. Chlorophylls in the lichens *Peltigera, Parmelia* and *Xanthoria. Bot. Tidsskr.* 55:30–36
281. Wilkens, J., Schreiber, U., Vidaver, W. 1978. Chlorophyll fluorescence induction: an indicator of photosynthetic ac-

tivity in marine algae undergoing desiccation. *Can. J. Bot.* 56:2787–94

282. Will, H. 1883. Uber den einfluss des einquellens und weideraustrockens auf die entwicklungfahigkeit der samen, sowie uber den Gabrauchswet "ausgenwachsener" samen als saatgut. *Landwirtsch. Vers.-Stn.* 28:51–89

283. Willis, A. J. 1964. Investigations on the physiological ecology of *Tortula ruraliformis. Trans. Br. Bryol. Soc.* 4:668–83

283a. Wilson, S. B., Bonner, W. D. Jr. 1971. Studies of electron transport in dry and imbibed peanut embryos. *Plant Physiol.* 48:340–44

284. Woodruff, D. R. 1969. Studies on presowing drought hardening of wheat. *Aust. J. Agric. Res.* 20:13–24

285. Yamamoto, N. 1977. Protein synthesis in the embryos of *Pinus thunbergii* seed. I. Influences of imbibition on seeds in the dark. *Plant Cell Physiol.* 18:287–92

286. Yoo, B. Y. 1970. Ultrastructural changes in cells of pea embryo radicles during germination. *J. Cell Biol.* 45:158–71

287. Zanefeld, J. S. 1969. Factors controlling the delimitation of littoral benthic marine algal zonation. *Am. Zool.* 9:367–91

Ann. Rev. Plant Physiol. 1979. 30:239–72

THE ROLE OF LIPID-LINKED SACCHARIDES IN THE BIOSYNTHESIS OF COMPLEX CARBOHYDRATES

❖7671

Alan D. Elbein

Department of Biochemistry, University of Texas Health Science Center, San Antonio, Texas 78284

CONTENTS

INTRODUCTION

Lipid-linked saccharides have been isolated from, or biosynthesized in, a diversity of organisms including bacteria, fungi, yeasts, protozoa, insects, birds, mammals, and higher plants (53, 82, 91, 137). There appear to be two roles for these intermediates which are probably not mutually exclusive; one is as a glycosyl donor for extracellular polysaccharides such as peptidoglycan, lipopolysaccharide, and capsular polysaccharide (89, 91), while the

0066-4294/79/0601-0239$01.00

other is as a precursor in the glycosylation of proteins (137). In either case, the rationale probably applies that the lipid-linked saccharides serve to transport the hydrophilic sugars into or through the hydrophobic or membranous environment to the location where polymerization occurs.

Plants, of course, have both cell wall (i.e. extracellular) polysaccharides and glycoproteins, and these lipid-linked saccharides have been implicated in the synthesis of both types of molecules. However, the great majority of studies on these intermediates in plants have pointed to a role in glycoprotein assembly. For that reason, this review will concentrate on glycoprotein synthesis, although other complex carbohydrates will be covered. Since much of the initial work in this field was done in animal systems, some of these early studies and the more recent findings will be presented and these data will be related to what is known (or not known) in plants. Several excellent reviews (53, 82, 137) cover much of the animal and fungal data, so in the interest of space, only selected references will be cited here. The reader is referred to other reviews for a more thorough bibliography.

This review will be presented in the following way. First of all, the structures of the oligosaccharide chains of those glycoproteins which appear to be synthesized via the lipid intermediates will be given. Then a discussion of the reactions involved in the lipid-linked pathway will be presented as well as the evidence for its role in protein glycosylation. Since there are a number of unanswered but intriguing questions about this pathway, these questions will be posed and some data relating to them will be presented. Finally, possible future trends and functional aspects will be outlined.

NATURE OF THE GLYCOPROTEINS

Glycoproteins are widely distributed in nature, and a great deal of information is available about the structures of the carbohydrate units of many animal glycoproteins. One class of glycoproteins of particular interest here has the oligosaccharide N-glycosidically linked from N-acetylglucosamine (GlcNAc) to the amide nitrogen of asparagine in the peptide (71). Many of these glycoproteins have a core structure composed of mannose and GlcNAc, and it is this core region that is synthesized by means of the lipid-linked saccharides. Figure 1 shows the types of glycopeptides that have been isolated from these glucosaminyl-asparagine glycoproteins. Both have an inner core of N,N'-diacetylchitobiose which is linked to the asparagine. Attached to the chitobiose are a variable number of mannose units. The first mannose has been shown to be linked in a β,1→4 glycosidic bond while the remaining mannose units are thought to be α-linked. In most cases, the oligosaccharide is branched at the second and third mannose units with one being attached to the β-linked mannose in an α,1→3 bond

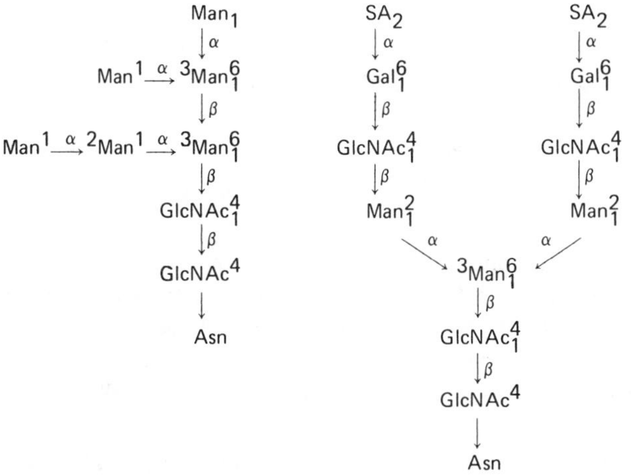

Figure 1 Structures of a "high mannose" and a "complex" glycopeptide.

and the other being joined to the β-linked mannose in an α,1→6 linkage (Figure 1). Beyond this point, the oligosaccharides may vary greatly in structure. In some cases, only mannose and GlcNAc are present and these oligosaccharides are referred to as the "high mannose" oligosaccharides, whereas in other cases additional sugars such as sialic acid, fucose, galactose, etc may occur and these are called "complex" oligosaccharides.

The obvious question which is of significance here is do plants contain these types of glycoproteins? At this time the answer is a partial yes, since the "high mannose" type has been found in plants. For example, vicillin, one of the two major storage proteins in the seeds of legumes, has been shown to be a glycoprotein containing mannose and GlcNAc. This glycoprotein probably has an oligosaccharide, $(\text{man})_x$-$(\text{GlcNAc})_x$-Asn, similar to those found in animal cells (32). Legumin, another reserve protein, also contains mannose and GlcNAc (5). Recently the detailed structure of the carbohydrate portion of soybean agglutinin was elegantly determined and shown to be a mixture of asparagine-linked oligosaccharides, one of which is shown in Figure 2 (81). The similarity between the first 5 or 6 sugars (starting at asparagine) in this structure to those shown in Figure 1 is evident, and the soybean lectin would be of the "high mannose" type. Since these types of glycoproteins do occur in plants, one would expect that the pathway of biosynthesis of the oligosaccharide would be similar to that found in animal cells. Unfortunately, at this time the plant glycoproteins synthesized via lipid intermediates have not been well characterized, so that it is not possible to point to a specific glycoprotein (such as vicillin or soybean agglutinin) formed in these reactions. Perhaps the proteins synthe-

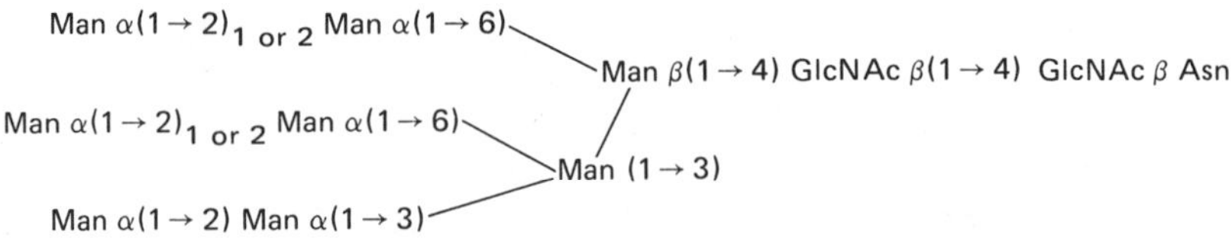

Figure 2 Structure of carbohydrate chain of soybean agglutinin.

sized in vitro must undergo other modifications before they can be recognized as known entities. In the last part of this review, some of the important functional aspects of glycoproteins will be outlined such as their possible role in host-parasite or host-symbiote interactions. The role of the carbohydrate in various types of recognition phenomena is well documented.

THE LIPID-LINKED SACCHARIDE PATHWAY

In this section, the reactions involved in the formation of the various lipid-linked saccharides will be discussed. Figure 3 presents a postulated series of reactions showing the synthesis of the individual lipid-linked saccharides that lead to the final lipid-linked oligosaccharides and ultimately to glycosylation of the protein. This scheme is a compilation of data taken from yeast, animal, and plant systems. As discussed below in more detail, each of these reactions may not occur in plants (or at least in all plants), or some that occur in plants may be absent in animals. Also, it should be pointed out that while most of the reactions shown in Figure 3 have some supportive evidence, a few of them, such as the dephosphorylation of polyprenyl-pyrophosphate, are strictly speculative. These reactions are discussed more fully in the following sections.

Monosaccharide Derivatives

The first suggestion that there might be transient types of sugar-containing lipids in eukaryotic cells came almost simultaneously in yeasts (122), animals (14, 145), and plants (62, 134). Basically these observations showed that upon incubation of GDP-[^{14}C]mannose with membrane preparations from the appropriate tissue, [^{14}C]mannose was transferred to material that was soluble in lipid solvents. In those cases that were examined in detail, the mannolipid was found to be labile to mild acid hydrolysis but stable to mild alkaline saponification, and its formation was reversed by the addition of GDP. These experiments indicated that the mannose was present in an "activated" linkage (6, 39, 138). Furthermore, the kinetics of mannose incorporation showed that the transfer of radioactivity to lipid reached a maximum very rapidly and then declined, indicating that the mannolipid was being "turned over." These properties would be in keeping with a role as an intermediate in glycosylation reactions.

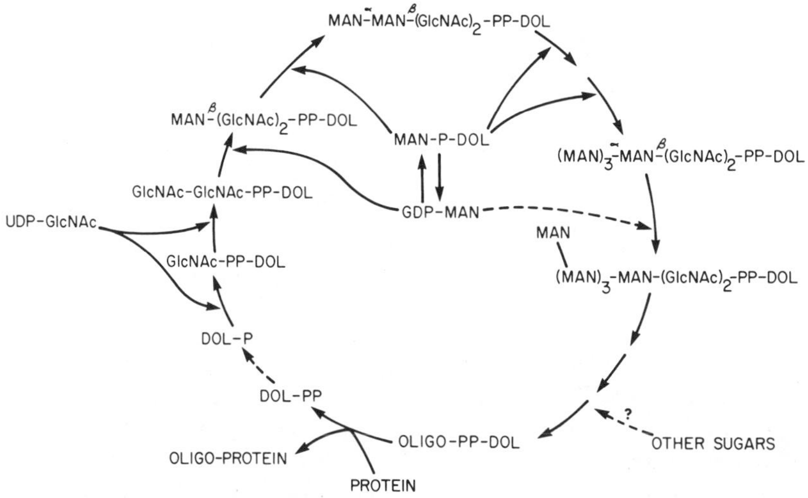

Figure 3 Reactions involved in the assembly of the lipid-linked oligosaccharides. The pathway is initiated by the transfer of GlcNAc from UDP-GlcNAc (left side) to dolichyl-phosphate (dol-P). Mannose is transferred from GDP-mannose or from mannosyl-phosphoryl-dolichol (Man-P-Dol). The lipid carrier for GlcNAc is referred to as dolichol, but this has not been demonstrated conclusively. Arrows shown with a broken line indicate reactions which have not been demonstrated.

More detailed studies in a number of animal systems suggested that these lipids were chemically related to the polyprenol types of intermediates reported earlier in bacteria (104). Thus, the animal lipids were characterized by a number of chemical and chromatographic methods as mannosyl-phosphoryl-polyprenols (15, 125, 142). However, whereas the carrier lipid in procaryotic cells was found to be the phosphomonoester of the C_{55} polyprenol undecaprenol, in eukaryotic cells, dolichyl-phosphate, a C_{80-100} polyprenol, performs this same function. In addition to having a longer hydrocarbon chain, dolichols are distinguished from the allylic alcohol undecaprenol by having an α-saturated isoprene unit (53; see Figure 4).

The isolation and purification of sufficient amounts of the mannolipid from liver allowed its definitive identification as mannosyl-phosphoryl-dolichol by mass spectrometry (6) as well as by its infrared and nuclear magnetic resonance spectra (39). Several chemical treatments have also been described to distinguish α-saturated from the unsaturated polyprenols. Thus, catalytic hydrogenation (143) or treatment with phenol (98) degrades the allylic polyprenols or their derivatives, but dolichyl-phosphoryl sugars are stable. Several investigators have therefore used these methods to establish the saturated nature of the animal lipids (45, 140). Recently, however, the reliability of catalytic hydrogenation has been questioned because this

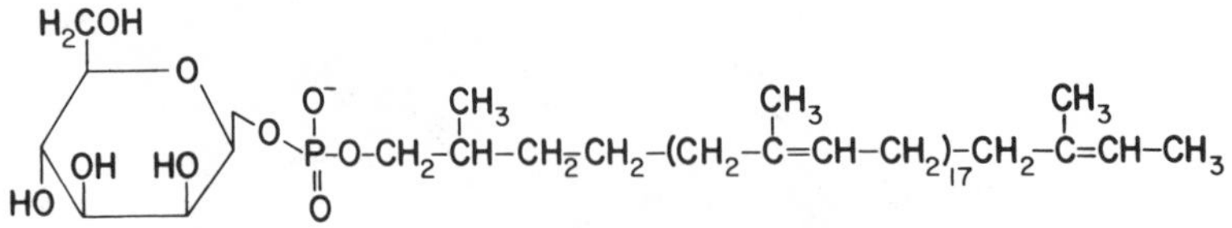

Mannosylphosphoryldolichol (dolichyl–mannosyl–phosphate)

Figure 4 Structure of mannosyl-phosphoryl-dolichol found in plant and animal tissues.

procedure was found to give variable and artifactual results (63). It is possible to label the lipid portion of these glycolipids as well as the sugar. For example, a double-labeled mannolipid was prepared from GDP-[^{14}C]-mannose and [^{3}H]dolichol in regenerating rat liver (86). In this case, rats were injected with [^{3}H]mevalonic acid, a known precursor of isoprene units, and a microsomal preparation of the [^{3}H] livers was prepared and incubated with GDP-[^{14}C]mannose. The double-labeled mannolipid had the identical properties to mannosyl-phosphoryl-dolichol. The mannosyl-phosphoryl-dolichol prepared in several animal tissues was shown to have a β-D-mannopyranosyl unit (1, 124), and the structure of the mannolipid was confirmed by chemical synthesis (141). The structure of mannosyl-phosphoryl-dolichol is shown in Figure 4.

In plant systems, the chemical properties of the mannolipid (2, 7, 33, 40, 110), as well as the fact that mannose incorporation was stimulated when various polyprenyl-phosphates were added to the incubation mixtures (13, 41, 75), strongly suggested that the lipid synthesized from GDP-[^{14}C]mannose was also a mannosyl-phosphoryl-polyprenol. The purified [^{14}C]mannosyl-phosphoryl-polyisoprenol, when incubated with the particulate enzyme from several plants in the presence of GDP, gave rise to GDP-[^{14}C]mannose, clearly demonstrating the reversibility of the reaction (41, 117). Based on a comparison of thin layer chromatographic mobilities of the mannolipid synthesized from GDP-[^{14}C]mannose and exogenous dolichyl-phosphate to the mannolipid formed from endogenous lipid, the endogenous lipid was about the same molecular size as dolichol (75). The mannolipid was identified conclusively as mannosyl-phosphoryl-dolichol by treatment with phenol, catalytic hydrogenation, and mass spectrometry (27). Recently the size of the polyprenyl-phosphoryl-mannose was also estimated by gel filtration on Sephadex G-75 in the presence of 0.5% sodium deoxycholate. Ficaprenyl-phosphoryl-mannose was well separated from the dolichol derivative, and the plant lipid eluted in the same area as dolichyl-phosphoryl-mannose (13). Thus, the evidence is quite good that the mannolipid produced in these eukaryotic systems (plant and animal) is mannosyl-phosphoryl-dolichol.

This finding is especially interesting in relation to plant systems because the major polyprenol is plants is ficaprenol, an allylic polyprenol (53). Also, ficaprenyl-phosphate does stimulate the incorporation of mannose in plants (41). It should be mentioned that the term dolichol refers to a family of α-saturated polyprenols whose chain lengths may vary from 80–110 carbons. It is possible that the variability in size could have some significance in terms of biological role. That is, dolichols of one size could function as carriers of certain sugars while those of another size could function in another way. However, there is no evidence for this speculation at this time.

Shortly after the initial reports on mannose transfer to lipid, another study demonstrated that liver microsomes catalyzed the transfer of N-acetylglucosamine 1-phosphate (GlcNAc-1-P) from UDP-GlcNAc to an endogenous lipid (88a). When the UDP-[^{14}C]GlcNAc was labeled with ^{32}P in the β-phosphate, the lipid became labeled with both ^{14}C and ^{32}P showing that both GlcNAc and phosphate were transferred. The GlcNAc-lipid, isolated from different organisms, bound more tightly to DEAE-cellulose than did the mannolipid, suggesting a pyrophosphoryl rather than a phosphoryl linkage (76, 78). This assumption of a pyrophosphoryl bridge was further strengthened by the fact that GlcNAc-lipid formation was inhibited by UMP (92). These data indicate that the lipid is a GlcNAc-pyrophosphoryl-polyprenol. Evidence for this lipid having an α-saturated isoprene unit was given by the fact that an α-saturated C_{55} polyprenyl-phosphate stimulated GlcNAc incorporation into lipid (103), as well as the finding that ozonolysis of the GlcNAc-lipid gave the same products as obtained from citronellol, a diprenol with α-saturated isoprene residues (101). N-acetylmannosamine has also been found to be associated with the lipid from pig liver, and the evidence indicated that this sugar arose by epimerization of GlcNAc while it was attached to GlcNAc-pyrophosphoryl-dolichol (92). The UDP-GlcNAc:dolichyl-phosphate GlcNAc 1-phosphate transferase was solubilized from aorta microsomes and partially purified on DEAE-cellulose (50). This enzyme catalyzed the reversible reaction and required exogenous dolichyl-phosphate for activity. However, since other polyprenyl-phosphates were not tested as GlcNAc acceptors, the specificity of the reaction is not known. This reaction has been shown to be specifically inhibited by the antibiotic tunicamycin (126), and recent studies with the soluble enzyme have shown that this inhibition is noncompetitive (52) (see section on Antibiotics).

Considerably less information is available with regard to GlcNAc incorporation in plants. Radioactivity from [^{14}C]glucosamine was shown to be incorporated into lipid by hypocotyls of *Phaseolus aureus,* but the lipids were not well characterized (106). The particulate enzyme from cotton fibers transferred GlcNAc from UDP-[^{3}H]GlcNAc into lipid-soluble material which had the following properties: (*a*) it bound more firmly to DEAE-cellulose than did the mannolipid, indicating the presence of a pyrophosphoryl linkage; (*b*) the sugar was attached to lipid in a mild-acid labile but mild-alkali stable linkage; and (*c*) the mobility of the GlcNAc-lipid on thin layer chromatography in neutral, acidic, or basic solvents was indicative of a polyprenol type of lipid (42). In the cotton system, the GlcNAc was found mainly in an N,N'-diacetylchitobiosyl-pyrophosphoryl-polyprenol, and only small amounts of GlcNAc-pyrophosphoryl-polyprenol were detected. However, in other plant systems, the major GlcNAc-lipid may be the mono-GlcNAc-pyrophosphoryl-prenol, or a mixture of mono- and di-GlcNAc-lipid (13, 44, 75). Probably the specific conditions of the incubation mixture and the state of the tissue with regard to endogenous lipid acceptors and activity of the enzymes determine whether one or two GlcNAc residues are transferred. Dolichyl-phosphate stimulated the incorporation of GlcNAc into lipids in plants (13, 75), and the GlcNAc-lipid showed the same mobility on thin layer plates as the yeast GlcNAc-pyrophosphoryl-dolichol (75). These data would imply that the carrier lipid for GlcNAc in plants is a dolichol, but this point is discussed further in a later section.

Several other monosaccharides have been found linked to polyprenols. In liver extracts, glucose from UDP-[^{14}C]glucose was transferred to dolichyl phosphate to form glucosyl-phosphoryl-dolichol (9). Glucose incorporation into lipids has also been reported in a number of plants (13, 41, 97). But a word of caution is necessary here. In many plants, most of the glucose that is transferred to lipid is actually in steryl glucoside (31), and only small amounts are transferred to polyprenols. Therefore, it is necessary to show that the biosynthetic product is charged and has the properties of a polyprenol derivative. Particulate preparations from peas have been reported to catalyze the transfer of glucose-1-phosphate from UDP-[^{14}C]glucose to endogenous acceptor to produce glucosyl-pyrophosphoryl-lipid (57). When the UDP-glucose was labeled with ^{32}P in the β-phosphate, the lipid became doubly labeled with ^{14}C and ^{32}P. These same enzyme preparations also transferred glucose alone to form glucosyl-phosphoryl-dolichol (98). Endogenous lipids were isolated from peas by solvent extraction and these lipids were purified on DEAE-cellulose (97). The purified lipids acted as glucose acceptors with enzyme preparations from either animal or plant tissues. For characterization, the glucolipid was run on thin layer plates against standards such as glucosyl-phosphoryl-ficaprenol or glucosyl-phosphoryl-dolichol, and the glucolipid was also subjected to catalytic hydrogenation and phenol treatment. All of the data showed that the acceptor lipid from peas was an α-saturated polyprenol of about the same size as dolichol (C_{80-100}) and different from the C_{55} ficaprenol. Glucose containing polyprenols are also synthesized in algae (57) and *Tetrahymena* (66). Several allylic polyprenols (i.e. undecaprenol and solanesol) were partially hydrogenated chemically and then, after chemical phosphorylation, they were compared as glucose acceptors to the unsaturated polyprenols (84). The partially hydrogenated polyprenols were better acceptors than the unsaturated derivatives.

Although early reports with plant enzymes suggested that the monosaccharide portions of UDP-xylose, UDP-arabinose, UDP-galactose, UDP-glucuronic acid, and GDP-glucose were incorporated into lipid (134), these studies have not been confirmed nor were the products identified. In oviduct tissue, however, xylose was transferred from UDP-xylose to form xylosyl-phosphoryl-polyprenol (139), but no function has been shown for this compound. Since UDP-xylose closely resembles UDP-glucose, it seems possible that the synthesis of this lipid could be caused by a lack of specificity of the glucosyl transferase which forms glucosyl-phosphoryl-dolichol. Glucuronic acid from UDP-glucuronic acid was found to be incorporated into lipid by microsomal preparations of lung, but this lipid proved to be a disaccharide with the structure glucuronyl-GlcNAc-pyrophosphoryl-dolichol (129).

Oligosaccharide-Lipids Containing Mannose

Figure 3 shows that the lipid-linked monosaccharides that are described above are precursors for the formation of the lipid-linked oligosaccharides. The lipid-linked oligosaccharides were first detected by virtue of the fact that their solubility is different than that of the lipid-linked monosaccharides (10). Thus most incubation mixtures of enzyme, metal ion (Mn^{2+} or Mg^{2+}), and sugar nucleotide (GDP-[^{14}C]mannose or UDP-[^{14}C]glucose) are treated in the following way: after incubation the reaction is terminated by the addition of chloroform and methanol in sufficient amounts to give a final mixture of chloroform:methanol:water (1:1:1). After thorough mixing, the phases are separated by centrifugation and the lower, or chloroform, phase containing the lipid-linked monosaccharides and the N,N'-diacetylchitobiosyl-pyrophosphoryl-polyprenol is removed. The lipid-linked oligosaccharides are apparently too polar to be extracted by this solvent and therefore they remain associated with the particulate material at the interface. After removal of the first chloroform extract, the particulate material is isolated by centrifugation and the lipid-linked oligosaccharides are extracted by suspending the particles in chloroform:methanol:water (10:10:3). This procedure is generally useful for separating and isolating the different classes of lipids, but the separation is not absolute. Thus, in our experience and depending on conditions in the incubation mixture, the trisaccharide-lipid and even the tetrasaccharide-lipid may be partially extracted by the chloroform:methanol:water (1:1:1) mixture, while the remainder will be removed in the chloroform : methanol : water (10 : 10 : 3). How these lipids extract probably depends on a variety of factors such as the amount of protein in the incubation mixtures, the presence of salts or other chemicals, the amount of other lipids present, and probably other as yet unknown factors.

As described in the preceding section and shown in Figure 3, the lipid-linked pathway is initiated by the formation of GlcNAc-pyrophosphoryl-polyprenol. A second GlcNAc is added from UDP-GlcNAc to form N,N'-diacetylchitobiosyl-pyrophosphoryl-polyprenol (78, 79). In some tissues, this disaccharide is the major product formed from UDP-[^{3}H]GlcNAc, whereas in other cases, the mono-GlcNAc-lipid was the major species and a second incubation with an excess of UDP-GlcNAc was necessary to make the disaccharide-lipid (19). The transfer of the second GlcNAc has been reported to be reversed by the addition of UDP to the incubation mixtures (144). The mono-GlcNAc-lipid could be separated from the di-GlcNAc-lipid by chromatography on columns of Silica Gel G-60 (103).

The next step in the pathway (Figure 3) is the addition of mannose to the N,N'-diacetylchitobiosyl-pyrophosphoryl-polyprenol to form the trisaccharide-lipid, β-man-GlcNAc-GlcNAc-pyrophosphoryl-polyprenol. When the N,N'-diacetylchitobiosyl-lipid was incubated with liver microsomes along with GDP-[^{14}C]mannose, the above-mentioned trisaccharide-lipid was formed (79). In these studies the mannosyl donor appeared to be GDP-mannose rather than mannosyl-phosphoryl-dolichol. Preincubation of hen oviduct membranes with UDP-[^{14}C]-GlcNAc and bacitracin, followed by incubation with GDP-mannose, led to the accumulation of the trisaccharide-lipid (18, 19). The enzyme that transfers the β-mannose to the di-GlcNAc-lipid was solubilized from aorta membranes with the nonionic detergent Nonidet P-40 (51). This solubilized enzyme utilized GDP-mannose rather than mannosyl-phosphoryl-dolichol as the mannosyl donor, and it showed an absolute requirement for the N,N'-diacetylchitobiosyl-pyrophosphoryl-polyprenol as the acceptor. Although the linkage of mannose to GlcNAc has not been firmly established in these lipids, it is probably a β,1→4 linkage since many glycoproteins contain this same sequence of sugars (71). The anomeric configuration of this first mannose in these lipids has been shown to be β, since mannose is released from the trisaccharide by β-mannosidase. In cotton fibers, the trisaccharide man-GlcNAc-GlcNAc was synthesized from GDP-[^{14}C]-mannose as part of a mixture of oligosaccharides, and these were released from lipid by mild acid hydrolysis. On paper chromatograms, the trisaccharide had the same mobility as the compound isolated from animal tissue (42). In extracts of *Phaseolus aureus,* the trisaccharide-lipid was synthesized by incubating N,N'-diacetylchitobiosyl-pyrophosphoryl-polyprenol with GDP-mannose (75).

The original observations which led to the discovery of the lipid-linked oligosaccharides was that glucosyl-phosphoryl-dolichol could act as a glucosyl donor in the transfer of glucose to a glycosylated endogenous acceptor (10). At first this endogenous acceptor was thought to be glucoprotein, but it was soon found to be soluble in chloroform : methanol : water (10 : 10 : 3), suggesting that it was some sort of lipid. This material bound to DEAE-cellulose that had been equilibrated in the 10 : 10 : 3 solvent, and the radioactive lipid could be eluted from the column with 25 mM ammonium formate. On the other hand, the lipid-linked monosaccharides were either washed through this column in the 10 : 10 : 3 solvent, or were eluted at very low (2.5 mM) concentrations of ammonium formate. When these lipid-linked oligosaccharides (glucosylated endogenous acceptor) were subjected to mild acid hydrolysis, the radioactivity (from UDP-[^{14}C]-glucose) was released into the aqueous phase and this radioactivity was found to be in a series of oligosaccharides (10). Upon treatment of these neutral oligosaccharides with strong alkali, two positively charged compounds were produced in a time-dependent reaction, indicating that these oligosaccharides contained amino sugars that were deacetylated by the strong alkali. This belief was strengthened by the fact that N-acetylation of the positively charged compounds restored their neutrality. Whether these oligosaccharide-lipids also contain other sugars (such as mannose, see below) is not known, but this is considered in a later section.

In liver (8, 90), mouse myeloma (59), and hen oviduct (83), a mannose containing oligosaccharide-lipid was formed when extracts of these tissues were incubated with GDP-[^{14}C]mannose. This oligosaccharide-lipid was extracted into chloroform : methanol : water (10 : 10 : 3), and like the glucose-labeled oligosaccharide-lipid, it bound to DEAE-cellulose. The oligosaccharide was obtained by mild acid hydrolysis and found to contain 7–9 glycose units of which 2 were GlcNAc, and 5–7 were mannose (59, 83). Reduction of the oligosaccharide with NaB^3H_4 and then complete acid hydrolysis gave rise to [^{3}H]glucosaminitol and [^{14}C]mannose, showing that one of the GlcNAc residues was located at the reducing terminus. On the other hand, complete acid hydrolysis first, followed by NaB^3H_4 reduction, gave both [^{3}H]mannitol (also labeled with ^{14}C) and [^{3}H]glucosaminitol in a ratio of about 5 : 2 (52). It is probably well to point out that these polyprenyl-linked saccharides are present in very low concentrations in tissue and therefore it is very difficult to obtain enough material for analysis. Therefore, structural studies have been hampered and most experiments have relied on radioactivity as a means of following structural changes.

More detailed studies showed that the intact oligosaccharide from oviduct had the structure, (α-man)$_n$-β-Man-(1→4)-β-GlcNAc-(1→4)-GlcNAc (20). The N,N'-diacetylchitobiose region of this oligosaccharide was confirmed by the fact that the oligosaccharide was cleaved by an endo-β-N-acetylglucosaminidase (123), an enzyme which hydrolyzes between two β-linked GlcNAc residues. These studies also showed that mannosyl-phosphoryl-dolichol could serve as a mannosyl donor in the transfer of mannose to the lipid-linked oligosaccharides. The role of these lipid-linked oligosaccharides in protein glycosylation is discussed below. [^{14}C]Mannose-containing oligosaccharide lipids have been isolated from a number of other mammalian tissues following an incubation with GDP-[^{14}C]mannose (64, 107, 111, 132, 133, 136).

Figure 3 shows that in the reactions from the trisaccharide-lipid to the final lipid-linked oligosaccharides, a number of mannose residues are donated by mannosyl-phosphoryl-dolichol. But since the individual reactions in this pathway have not been elucidated, the details concerning the mannosyl donors and the mannosyl acceptors are not known. It is known, however, that not all of the mannose residues come from mannosyl-phosphoryl-dolichol. In aorta, as in other tissues, the synthesis of mannosyl-phosphoryl-dolichol from GDP-mannose requires a divalent cation (Mg^{2+} or Mn^{2+}), and therefore this reaction can be blocked by the addition of a chelating agent such as EDTA. Thus, in the presence of certain concentrations of EDTA, the transfer of GDP-[^{14}C]mannose to mannosyl-phosphoryl-dolichol is inhibited more than 95%, but radioactivity is still transferred by the aorta membrane fraction to the lipid-linked oligosaccharides (16). When the radioactive oligosaccharides formed in the presence of EDTA were examined by paper chromatography, only one radioactive oligosaccharide was seen and this compound had the mobility of a heptasaccharide.

Recently it was found that the antibiotic amphomycin also inhibited the transfer of mannose from GDP-[^{14}C]mannose to mannosyl-phosphoryl-dolichol (60). In the presence of sufficient concentrations of amphomycin to essentially completely inhibit the formation of mannosyl-phosphoryl-dolichol, radioactivity was still transferred from GDP-[^{14}C]mannose to the lipid-linked oligosaccharides. Just as in the experiments with EDTA, only one oligosaccharide was labeled under these conditions and this compound migrated as a heptasaccharide (61). Therefore it seems likely that some of the mannose residues come directly from GDP-mannose. This is interesting in view of the fact that it has been assumed that mannose transfer involves an inversion of configuration. Thus the β-linked mannose (in the trisaccharide-lipid) comes from GDPα-mannose, while at least some of the α-linked mannoses (in the oligosaccharide-lipids) come from β-mannosyl-phosphoryl-dolichol. Partial characterization of some of the larger oligosaccharides (hepta to deca or dodecasaccharide) from aorta showed that they were susceptible to acetolysis indicating the presence of 1→6 branches. The radioactive products released by acetolysis were di- and trisaccharides (16).

Thus these oligosaccharides may be similar to the oligosaccharide portions of glycoproteins (Figures 1 and 2).

In plants, a series of lipid-linked oligosaccharides are formed when GDP-[^{14}C]mannose is incubated with the particulate enzyme (13, 33, 42, 75, 110). When the oligosaccharides were released by mild acid hydrolysis and examined by paper chromatography, a series of 7 or 8 radioactive peaks were observed which ranged in size from a trisaccharide to oligosaccharides having 10 to 12 glycose units (42). The smaller oligosaccharide-lipids were shown to be precursors to the larger ones, since a second incubation with unlabeled GDP-mannose chased the radioactivity from these smaller to the larger lipid-linked oligosaccharides. The oligosaccharides had GlcNAc at their reducing terminus because reduction with NaB^3H_4, followed by strong acid hydrolysis, gave rise to [^{14}C]mannose and [^3H]glucosaminitol (42). Additional evidence for GlcNAc at the reducing end was the finding that when [^3H]N,N'-diacetylchitobiosyl-pyrophosphoryl-polyprenol was isolated from incubation mixtures and then reincubated with membrane fractions along with unlabeled GDP-mannose, radioactivity was chased from the disaccharide-lipid into the lipid-linked oligosaccharides (42, 75). These studies suggested that the di-GlcNAc-lipid was acting as an acceptor of mannose residues. Other kinetic studies also showed an interaction between GDP-mannose and UDP-GlcNAc as precursors to the lipid-linked oligosaccharides (34).

In some plant and animal systems, a series of radioactive oligosaccharides is observed whereas in other cases only one oligosaccharide (attached to lipid) is seen. The reason for these variations between one tissue and another may be due to differences in the amount of endogenous lipid acceptors in these tissues. The various lipid-linked oligosaccharides present in the tissue at the time of enzyme preparation can act as acceptors of mannose in these incubations. So, depending on the relative amounts of the different oligosaccharide-lipids and on the activities of the glycosyl transferases, differences in oligosaccharide profile may occur. Most likely in the plant system as in the animal tissues, mannosyl-phosphoryl-dolichol is a precursor to some of the mannose residues, but whether this is the only mannosyl donor or whether GDP-mannose can directly donate some mannose is not known. The plant systems are probably fairly similar to the animal systems in terms of the lipid-linked oligosaccharides since the oligosaccharide-lipids isolated from pig liver are able to serve as acceptors of mannose with cell free extracts of pig aorta or of cotton fibers (43).

Oligosaccharide-Lipids Containing Glucose and Mannose

In addition to mannose and GlcNAc, glucose has also been found as a component of some of the lipid-linked oligosaccharides. The initial studies on lipid-linked oligosaccharides described above showed that glucose from either UDP-[^{14}C]glucose or from [^{14}C]-glucosyl-phosphoryl dolichol was transferred to lipid-linked oligosaccharides by membrane preparations of liver (10). Those oligosaccharides ranged in size up to one that contained about 20 glycose units, and in addition to glucose, they also contained GlcNAc. However, it was not clear whether those oligosaccharide-lipids also contained mannose. Based on more recent studies it seems likely that mannose was also present, and this was in fact suggested by these workers (8).

The first real evidence for the presence of glucose and mannose (as well as GlcNAc) in the same oligosaccharide came from studies in thyroid and several other tissues (113, 115). In these experiments, tissue slices were incubated with various radiolabeled sugars (^{14}C-mannose, ^{14}C-glucosamine) in order to completely label the oligosaccharide-lipids. The lipid-linked oligosaccharides were isolated from these tissues and purified on DEAE-cellulose. The oligosaccharide had an estimated molecular weight of about 2400 by gel filtration, and analysis showed that it was composed of about 11 mannose, 1 or 2 glucose, and 2 GlcNAc residues (115). Based on a combination of chemical (i.e. acetolysis and Smith degradation) and enzymatic (α-mannosidase) treatments, the partial structure of this oligosaccharide was proposed as shown in Figure 5. There are, of course, many questions still to be answered about this structure such as the exact location of the glucose (i.e. terminal or internal), the number and size of the branches, and so on. Tissue slices of calf kidney, pancreas, thymus, and liver also produced an oligosaccharide-lipid of similar size when incubated with [^{14}C]glucose or [^{14}C]mannose (113). In calf pancreas microsomes incubated with either GDP-[^{14}C]mannose or [^{14}C]mannosyl-phosphoryl-dolichol, a

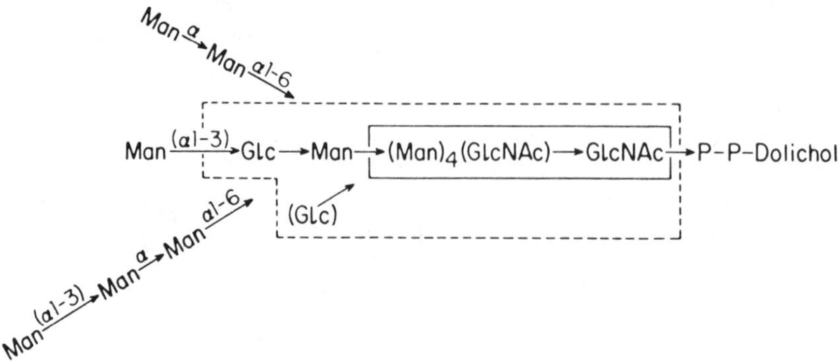

Figure 5 Proposed structure of the oligosaccharide isolated from the lipid-linked oligosaccharides of thyroid (115).

heterogeneous mixture of oligosaccharides (from lipid-linked oligosaccharides) was obtained containing GlcNAc, mannose, and glucose in ratios varying from 2:5:1 to 2:10:3. GlcNAc was at the reducing terminus of these oligosaccharides (55).

The incorporation of glucose into the oligosaccharide-lipids has been shown to cause a shift in the size of the mannose-labeled oligosaccharide-lipids (105). Thus, cell-free extracts from cultured lymphocytes incorporated mannose from GDP-[^{14}C]mannose into an oligosaccharide-lipid which appeared to contain 7–8 sugars based on its migration on Biogel P-6 columns. However, when these incubation mixtures contained unlabeled UDP-glucose in addition to GDP-[^{14}C]mannose, the radioactive oligosaccharide emerged from the P-6 column earlier in an area that indicated it was one or two sugars larger in size. This experiment suggested that glucose was being added to the same oligosaccharide-lipid as was mannose. Label from UDP-[^{3}H]glucose was also incorporated into this larger sized oligosaccharide.

Probably the most fascinating aspect of these lipid-linked oligosaccharides concerns the presence of glucose and its role. With the exception of collagen, and perhaps a few other proteins, most known glycoproteins do not appear to contain glucose. Thus these are the obvious questions: (*a*) Is the glucose-containing oligosaccharide-lipid a precursor for some specialized types of glycoproteins which contain glucose? (*b*) Is the glucose removed from these oligosaccharides during or after transfer to the protein? (*c*) Is the glucose some sort of recognition sugar which indicates the ultimate disposition of these oligosaccharides? It has been argued that glucose is a poor choice for a recognition role in animals since most mammalian tissue fluids have high concentrations of the free sugar and this would interfere with or impair recognition (46). But perhaps in the membranous environment where these oligosaccharide-lipids occur, there is no glucose to interfere. At any rate, answers to these intriguing questions are being pursued as discussed below.

Support for some kind of recognition role for the glucose in these oligosaccharide-lipids is the fact that the glucose-containing oligosaccharide-lipid is a better substrate for the transfer of oligosaccharide to protein than is the oligosaccharide-lipid which lacks glucose (114, 130). Thus in one set of experiments, the [^{14}C]-mannose-labeled oligosaccharide-lipids were synthesized from GDP-[^{14}C]mannose in cell-free extracts, and to part of the incubation UDP-glucose was added (130). The glucose-free and glucose-containing oligosaccharide-lipids were then tested for their ability to donate the oligosaccharide to protein in a cell-free system. As much as 41% of the glucose-containing oligosaccharide was transferred to protein compared to 5% of the glucose-free compound. In another set of experiments, the radioactive oligosaccharide-lipid containing 2 GlcNAc, 11 mannose, and 1 or 2 glucose residues was prepared in thyroid slices incubated with radioactive sugars (114). Approximately 50% transfer of this oligosaccharide from its lipid carrier to protein was achieved with the particulate enzyme from thyroid. Removal of 40% of the mannose from the oligosaccharide-lipid with jack bean α-mannosidase did not alter the rate of transfer. However, removal of 80% of the glucose with a thyroid glucosidase reduced the donor activity of this oligosaccharide-lipid by 80%. These experiments indicate that the glucose in the lipid-linked oligosaccharides plays some crucial role in directing or facilitating transfer to protein.

Since the glycoproteins synthesized in these reactions apparently do not contain glucose, one must ask what happens to the glucose? How and when is it removed? Apparently shortly after transfer of oligosaccharide to protein, a processing or trimming reaction(s) occurs during which a number of sugars are removed. Thus, in Chinese hamster ovary cells infected with vesicular stomatitis virus, only one glycoprotein is made and that is the viral G protein. This glycoprotein contains the complex type of oligosaccharide chain (Figure 1) with a core region of 2 GlcNAc and 3 mannose residues. The synthesis of this core oligosaccharide proceeds via lipid-linked saccharides. However, the initial oligosaccharide which is transferred to protein contains 2 GlcNAc and approximately 10 other monosaccharides of which at least 6 are mannose. Within 5 minutes after transfer of oligosaccharide to protein, processing begins and 3 monosaccharides are removed within the first 30 minutes. This is followed by the removal of 4 more monosaccharides leaving the core of $(man)_3$-GlcNAc-GlcNAc-Asn (120).

Some of the enzymes involved in processing have been identified, and they appear to be membrane bound and highly specific glycosidases. Thus, a glycosidase was solubilized from thyroid microsomes by Triton X-100. This enzyme showed maximal activity on the lipid-linked oligosaccharide, but was much less active on the glycoprotein or the glycopeptide (116). That specificity seems unusual if the function of the enzyme is to trim the protein, but perhaps the activity of the enzyme is altered upon removal from its environment, or perhaps the enzyme has two functions: one to remove glucose from the protein and another to control the level or activity of the glucose-containing oligosaccharide-lipid. An enzyme has also been found in oviduct tissue which removes glucose from the oligosaccharide chain of both endogenous protein and exogenous, glycosylated S-carboxymethylated α-lactalbumin (17). It will be interesting to determine whether the glucose is removed at the same time as oligosaccharide transfer or at later times. And perhaps the glucosidase is in close proximity, or even associated with, the enzyme which transfers the oligosaccharide to protein.

In addition to the removal of glucose from these oligosaccharides, a number of mannose residues may also be cleaved (120). So far, no one has shown a specific enzyme activity which does this. However, α-mannosidase activity, as well as other glycosidases, have been observed which are membrane-bound, and it is possible that some of these enzymes function to cleave these oligosaccharides (30). In plant systems, there is no evidence for the presence of both glucose and mannose in the same oligosaccharide-lipid. However these lipids have not been as well characterized as those of animals, so that other sugars could be present. But as shown in Figure 2, there are plant glycoproteins which contain the "high mannose" type of oligosaccharide, and for assembly of these glycoproteins, processing may not be necessary. Some of the larger mannose-containing oligosaccharide-lipids from plants may be of about the same size as the oligosaccharide chain of vicillin or soybean lectin.

One other question which is pertinent with regard to these oligosaccharide lipids is, are all of the glucose-free oligosaccharide-lipids precursors to the lipid-linked oligosaccharides which contain glucose? That is, in the earlier studies on lipid-linked saccharides in mouse myeloma (59) or oviduct (83), the lipid-linked oligosaccharide which accumulated contained only mannose and GlcNAc and was about 7 or 8 glycose units in size. Is this oligosaccharide a direct precursor to the protein or does it have to be extended by the addition of glucose before it is transferred? Some of the studies discussed below on protein glycosylation may have some bearing on this question, but probably it cannot be answered at this time. So one wonders whether there are two different kinds of oligosaccharide-lipids involved in protein glycosylation, one of which only transfers GlcNAc and mannose and another which transfers GlcNAc, mannose, and glucose. And do these different lipids function in the glycosylation of proteins with different functions?

Oligosaccharide-Lipids Containing Glucose

Early studies on the biosynthesis of cellulose in the bacterium *Acetobacter xylinum* implicated a lipid-linked glucose as an intermediate (21). But the nature of this glucolipid was not determined. More recently, extracts of this organism were shown to form lipid-pyrophosphoryl-α-glucose and lipid-pyrophosphoryl-α-cellobiose from UDP-glucose (45). In some cases, lipids containing cellotroise or higher homologs were also reported (70). Radioactivity from β-labeled ^{32}P-UDP-glucose was also transferred to these lipids (45). Particulate preparations of the algae *Prototheca zopfii* also catalyzed the incorporation of radioactivity from UDP-[^{14}C]glucose into lipids. These lipids were characterized as lipid-phosphoryl-glucose, lipid-pyrophosphoryl-glucose, and lipid-pyrophosphoryl-oligosaccharides (57). The oligosaccharides were found to be a mixture ranging in size from a disaccharide to a decasaccharide and appeared to be composed of β,1→4 linked glucoses on the basis of periodate oxidation and susceptibility to cellulase. These oligosaccharides apparently did not contain GlcNAc and gave rise to sorbitol upon reduction with $NaBH_4$, showing that glucose was at the reducing terminus. The glucolipids were proposed to be precursors of a water-soluble β,1→4 linked polymer of glucose, and in the presence of GDP-glucose this water soluble polymer became alkali (and water) insoluble. The alkali-insoluble material had the properties expected for cellulose. If these studies prove to be applicable to higher plants, they could explain the confusion that exists with regard to the glucosyl donor for cellulose. Both UDP-glucose and GDP-glucose have been reported to be precursors to cellulose, but perhaps both sugar nucleotides have a role in cellulose synthesis (89).

Transfer of Oligosaccharide to Protein

As shown in Figure 3, the ultimate step in the lipid-linked saccharide pathway is the transfer of the oligosaccharide to protein. In many of the studies which have been discussed where either UDP-[^{14}C]glucose or GDP-[^{14}C]mannose was incubated with particulate enzyme, radioactivity was transferred to material that was insoluble in lipid or aqueous solvents. This radioactivity was, however, rendered water soluble upon digestion with a proteolytic enzyme such as pronase, indicating that it was probably in glycoprotein. In several of these early studies, the transfer of oligosaccharide from lipid-linked oligosaccharide to protein was demonstrated. But the nature of the final products was not determined in these studies.

In oviduct preparations, the transfer of various oligosaccharide-lipids to protein has been studied. In the presence of Triton X-100, N,N'-diacetylchitobiose was transfered from the disaccharide-lipid to a protein of about 25,000 molecular weight. The radioactivity was released from the protein by proteolysis and the carbohydrate portion of this glycopeptide was shown to be N,N'-diacetylchitobiose. However, when the above disaccharide-lipid was incubated with the particulate enzyme in the presence of GDP-mannose, it underwent elongation, i.e. mannosylation, to the oligosaccharide-lipid before transfer occurred (119). In another study it was shown that extracts of rabbit liver catalyzed the direct transfer of GlcNAc from UDP-GlcNAc to ribonuclease A (68). In that case it was not known whether a lipid participated as an intermediate. In both of these studies it was not clear how the mannose residues were added to the protein since there was no evidence for the direct mannosylation of the protein. In addition to transfer of N,N'-diacetylchitobiose from lipid to protein, the trisaccharide β-man-

GlcNAc-GlcNAc is transferred directly from its lipid carrier to protein (18).

These results appear to conflict with others discussed above where it was shown that the glucose-containing oligosaccharide (from its lipid carrier) was transferred much more effectively than was the glucose-free oligosaccharide (130). In those experiments, 40–50% transfer of the larger oligosaccharide occured whereas removal of the glucose reduced the transfer to 5% or less. On the other hand, as much as 25% transfer to protein was reported from the trisaccharide-lipid. These studies suggest that there may be two different modes for protein glycosylation, one of which involves the glucose containing oligosaccharides and the other for the mannose-GlcNAc (or even GlcNAc alone) oligosaccharides. Perhaps these different pathways serve to glycosylate different proteins with one being a precursor for membrane proteins and another for secreted proteins.

The enzymatic transfer of the oligosaccharide from lipid-linked oligosaccharide to denatured forms of three secretory proteins—α-lactalbumin, ovalbumin, and ribonuclease A—was demonstrated using a membrane fraction from hen oviduct (96). Apparently denaturation of these proteins was necessary for them to be glycosylated. Based on a survey of 10 different proteins as acceptors of the oligosaccharide, the presence of the tripeptide asn-x-ser (thr) was necessary, but not sufficient for glycosylation (96). More recently, the tetrapeptide asn-leu-thr-lys was tested in an in vitro glycosylation system and was found to be inactive. However, after N-acetylation of this tetrapeptide, it was able to serve as an acceptor of the oligosaccharide chain (49). Thus the amino acid sequence surrounding the asparagine to be glycosylated, and probably also the conformation of the protein, is important in glycosylation. Early studies which examined the amino acid sequence around the carbohydrate region of a number of asparagine-linked glycoproteins showed that the sequence asn-x-ser (thr) occurred in this region (85).

The above studies suggest an "en bloc" transfer of the oligosaccharide chain from lipid to protein, and strongly imply that mannose and GlcNAc would be added to the protein simultaneously. But where and when this glycosylation occurs still remains an open question. Early studies in this area indicated that glucosamine was added to the elongating, nascent polypeptide chain (88). In that case, radioactive glucosamine was given to cells that were actively synthesizing proteins and the polysome fraction was isolated and examined. Generally radioactivity was found associated with polysomes as glucosamine, but the radioactivity was low so that the possibility of contamination was difficult to overcome. However, another study (12) reported that only microsomes, not polysomes, were able to incorporate GlcNAc and mannose into thyroid proteins. In both of these studies, however, evidence for the reaction either on or off the ribosome was based on the release of labeled or unlabeled nascent chains by puromycin. But puromycin has been shown to bind to membranes of the endoplasmic reticulum in rat liver microsomes and to produce nonspecific disruption of glycosyltransferase activity (127). Therefore, these studies on the location of the glycosylation events are open to reinterpretation.

A more elegant approach, isolating polysomes by immunoprecipitation, was used to show the glycosylation of ovalbumin (69). In this case, oviduct slices were incubated with [^{3}H]glucosamine or [^{3}H]mannose, and the ovalbumin nascent chains were isolated by immunoprecipitation and subsequent isolation of peptidyl tRNA by DEAE-cellulose chromatography. The data indicated that mannose and glucosamine were incorporated into nascent chains. Interestingly, the asparagine which is glycosylated is only 10 residues from the amino terminus, but the majority of the carbohydrate was attached only to those peptide chains that had been essentially completed. In fact, in some experiments in which both protein synthesis and protein glycosylation were examined in a cell-free system using a partially purified mRNA for the MOPC-46B κ glycoprotein, glycosylation of the protein occurred either very late in synthesis of the peptide or perhaps as a post-ribosomal event (128). Thus the exact time of glycosylation is not clear and in fact may differ for different proteins. Perhaps the time sequence has something to do with whether a "signal" peptide is necessary, since this extra sequence at the amino terminus has been hypothesized to interact with the inner membrane of the endoplasmic reticulum and to initiate binding of the large ribosomal subunit-nascent chain complex to the membrane (11). After the termination of protein synthesis, the completed chains containing this "signal" sequence would be discharged into the lumen of the endoplasmic reticulum where other post-translational modifications could occur. However, not all secretory proteins need necessarily have this signal sequence since ovalbumin has been reported to be synthesized without this transient, hydrophobic leader sequence (94). Perhaps this difference affects the time and location of glycosylation.

EFFECT OF ANTIBIOTICS AND OTHER DRUGS ON THE LIPID-LINKED SACCHARIDE PATHWAY

Probably one of the best ways to study a multienzyme pathway is to have inhibitors which block the sequence of reactions at different points so that one can examine the accumulation or loss of various intermediates. Several antibiotics or other inhibitors of the lipid-linked saccharide pathway have recently been described, and the use of these inhibitors has shown the role of this pathway in glycosylation of certain proteins.

The first of these antibiotics to be described was tunicamycin, a glucosamine containing compound produced by *Streptomyces lysosuperificus* (121). In calf liver rough microsomes, this antibiotic was shown to inhibit the incorporation of GlcNAc from UDP-GlcNAc into GlcNAc-pyrophosphoryl-dolichol (see Figure 3) (126). The antibiotic was shown to have a similar action in microsomal fractions of oviduct (119), brain (135), and plants (35), but it had no effect on the transfer of mannose from GDP-mannose to mannosyl-phosphoryl-dolichol. The antibiotic was also shown to have no effect on the transfer of the second GlcNAc to form N,N'-diacetylchitobiosyl-pyrophosphoryl-dolichol (77). In yeast spheroplasts, tunicamycin blocked the synthesis of the extracellular glycoproteins, invertase, and acid phosphatase, whereas synthesis of several internal proteins was unaffected (73). In oviduct slices incubated with tunicamycin, ovalbumin was synthesized at almost normal rates, but the protein was not glycosylated (119). Similar results were observed in MOPC 315 mouse plasma cells which normally secrete the glycoproteins IgA and IgE. In this case, tunicamycin inhibited the incorporation of [^{14}C]glucosamine into these proteins and also prevented their secretion (56). The antibiotic also prevented the formation of physical particles of a number of viruses presumably because it prevents the formation of viral glycoproteins (109), and it also affected the maturation of collagen (28). This antibiotic has now become a very popular and powerful tool for studying the function of certain kinds of glycoproteins in such diverse systems as conjugation, transformation, fertilization, etc. Blocking the formation of GlcNAc-pyrophosphoryl-dolichol prevents the assembly of the lipid-linked oligosaccharides and thus prevents the glycosylation of those glycoproteins that use these types of intermediates.

The mechanism of action of tunicamycin on the solubilized UDP-GlcNAc:dolichyl-phosphate GlcNAc-1-phosphate transferase was examined. The antibiotic inhibition could not be overcome by high concentrations of the substrates dolichyl-phosphate or UDP-GlcNAc, either individually or in combination, nor by increasing the cation concentration. Thus the inhibition is not of a competitive nature. Tunicamycin also inhibited the reverse reaction, that is, the formation of UDP-GlcNAc from GlcNAc-pyrophosphoryl-dolichol. Apparently the antibiotic binds to the enzyme and inactivates it (52).

Bacitracin is another antibiotic which has been used to inhibit the lipid-linked saccharides, but in this case results from various laboratories have been quite confusing. In calf pancreas microsomes, this antibiotic inhibited the formation of GlcNAc-pyrophosphoryl-dolichol but it had little effect on the synthesis of N,N'-diacetylchitobiosyl-pyrophosphoryl-dolichol or mannosyl-phosphoryl-dolichol (54). On the other hand, in membrane preparations of the yeast *Saccharomyces cerevisiae,* this antibiotic was reported to inhibit the formation of N,N'-diacetylchitobiosyl-pyrophosphoryl-dolichol but to have no effect on the formation of GlcNAc-pyrophosphoryl-dolichol (102). In aorta membrane preparations, bacitracin inhibited both the incorporation of GlcNAc from UDP-GlcNAc to dolichyl-phosphate to form GlcNAc-pyrophosphoryl-dolichol and the transfer of mannose from GDP-mannose to dolichyl-phosphate to form mannosyl-phosphoryl-dolichol (112). In these reaction mixtures, the antibiotic also inhibited the transfer of mannose, either from GDP-mannose or from mannosyl-phosphoryl-dolichol, to the lipid-linked saccharides. Similarly in plant extracts, bacitracin prevented the formation of both the GlcNAc-pyrophosphoryl-dolichol and the mannosyl-phosphoryl-dolichol (37). This antibiotic also inhibited the glycosylation of protein in vivo in several systems. For example, in carrot disc cultures, bacitracin prevented the incorporation of [^{14}C]mannose into the mannolipid and also prevented its incorporation into protein (37). Similar results were observed with yeast spheroplasts, i.e. both the sythesis of mannolipid and the mannosylation of protein were inhibited (112). In both of these tissues, bacitracin also partially blocked protein synthesis, but this reaction was less sensitive than was the protein glycosylation. In the plant system, bacitracin did not inhibit the transfer of glucose from UDP-[^{14}C]glucose to steryl glucosides, nor did it inhibit the synthesis of β,1→3 glucan (37). Thus the antibiotic may be a general inhibitor of reactions which involve polyprenols.

Amphomycin, another polypeptide antibiotic, has also been found to be an inhibitor of lipid-linked saccharides. In aorta extracts, amphomycin was found to inhibit the incorporation of mannose from GDP-[^{14}C]mannose into mannosyl-phosphoryl-dolichol and of GlcNAc from UDP-[^{3}H]-GlcNAc into GlcNAc-pyrophosphoryl-dolichol (60). As indicated in an earlier section, this antibiotic did not inhibit the direct transfer of mannose from GDP-mannose to the lipid-linked oligosaccharides, nor did it block mannose transfer from mannosyl-phosphoryl-dolichol to the oligosaccharide-lipids (61). However, the antibiotic proved to be ineffective against yeast spheroplasts, probably because it did not get into the cells, but it did inhibit glycosylation of protein and formation of mannolipid in carrot disc cultures (38).

An analog of glucose, 2-deoxyglucose, has been used in a number of studies as an inhibitor of the synthesis of complex carbohydrates. Its effects have usually been attributed either to interference with sugar transport or to prevention of synthesis by trapping uridine nucleotides as the 2-deoxyglucose intermediate. In yeast protoplasts, 2-deoxyglucose inhibited the synthesis of the glycoprotein enzymes invertase and acid phosphatase after a 20–30 min lag (72). The authors suggested that accumulation of 2-deoxyglucose 6-phosphate blocked the synthesis of cell wall polysaccharides and glycoproteins by preventing the conversion of fructose 6-phosphate to glucose 6-phosphate and mannose 6-phosphate and also by restricting the transport of fructose and maltose into the cell. This sugar analog was also used to study the synthesis and secretion of the immunoglobulin κ46 chain in mouse myeloma tumor. In single cell suspensions, 6 mM 2-deoxyglucose prevented the incorporation of glucosamine, mannose, and galactose into secreted protein while allowing the incorporation of leucine into protein to proceed at 40% the normal rate. The protein that was secreted under these conditions was shown to be the nonglycosylated form of κ46. Thus in this case, glycosylation was not necessary for secretion of the protein, although the absence of the carbohydrate chain appeared to retard the intracellular transport and the export from the cell (29).

Coumarin has been reported to inhibit cellulose synthesis without having any effect on the incorporation of glucose into pectin and hemicellulose (48). This drug inhibited the incorporation of [^{14}C]-glucose into cellulose in both 7-day-old cotton fibers and in 16-day-old cotton fibers (26). The data suggested that both the primary cell wall and the secondary cell wall synthesis share a common coumarin-sensitive step. In the alga *Prototheca zopfii,* coumarin was reported to inhibit the transfer of the oligosaccharide chain from its lipid carrier to the protein acceptor (58). This oligosaccharide-lipid is formed from UDP-glucose and then transferred to the protein acceptor which is believed to be a precursor for the cellulose chain. In the presence of GDP-glucose, the glucoprotein has been reported to become alkali-insoluble cellulose (57).

The biosynthesis of the dolichol portion of these lipid-linked oligosaccharides has been shown to share a common pathway with the synthesis of sterols. Therefore, one might anticipate that the enzyme hydroxymethylglutaryl CoA reductase would regulate the synthesis of dolichols as it does of sterols. In aorta smooth muscle cells grown in culture, 25-hydroxycholesterol inhibited the incorporation of acetate, but not of mevalonate, into cholesterol and dolichol. It also decreased the synthesis from glucose of cholesterol, dolichyl-pyrophosphoryl-oligosaccharide, and dolichol-dependent glycoproteins. The addition of mevalonate to the culture medium could restore normal synthesis of dolichol intermediates and glycoproteins. It is suggested that the reductase is a rate-controlling enzyme for dolichols as well as sterols and therefore may regulate the assembly of glycoproteins (87). These kinds of inhibitors may be useful for studies on the synthesis and function of glycoproteins.

UNRESOLVED QUESTIONS

1. *Is the same lipid carrier used for all the sugars in eukaryotic cells?* In eukaryotes, the prevailing evidence is that the lipid carriers for mannose, glucose, and GlcNAc are α-saturated polyprenols, i.e. dolichols. The mannose lipid has been identified conclusively by mass spectrometry as a dihydropolyisoprenol of at least 18 isoprene units, one of which is saturated (6). But the nature of the other carriers for glucose and GlcNAc is still circumstantial. The fact that dolichyl-phosphate stimulates the incorporation of these sugars (50, 76) and the effect of catalytic hydrogenation (13) and ozonolysis (101) make it likely that the glucose and GlcNAc polyprenols are α-saturated. But dolichols apparently vary in chain lengths from C_{80}–C_{110} or even more (53). Is it possible that different chain lengths serve different functions, such that dolichols of a certain size carry GlcNAc while others carry mannose or glucose? When the lipid portion of the yeast GlcNAc-lipid was examined by high pressure liquid chromatography, several different homologs of the polyprenols were detected, with the homolog with 15 isoprene units being most abundant (101). On the other hand, the mannose lipid of liver seems to be 18 isoprene units or larger (6).

In plants it was found that unlabeled GDP-mannose did not inhibit the incorporation of GlcNAc from UDP-[^{3}H]GlcNAc into polyprenols, nor did unlabeled UDP-GlcNAc inhibit the transfer of mannose from GDP-mannose (36). These results imply either that the carrier lipids are different and therefore do not compete, or that the glycosyl-transferases have their own pools of polyprenyl-phosphates which do not mix. In liver, UDP-glucose did not inhibit mannosyl transferase, and this was taken as evidence for two pools of dolichyl-phosphate (67). Two pools of mannosyl-phosphoryl-dolichol were also indicated by the fact that part of this compound could not be extracted in the first chloroform extract but was found in the 10:10:3 extraction (90).

However, another study suggested that Glc-, Man-, and GlcNAc-transferases all utilized the same pool of dolichyl-phosphate (132). In the plant system, the polyprenyl-phosphate isolated by mild acid hydrolysis of the GlcNAc-lipid did not stimulate mannose incorporation but it did stimulate GlcNAc transfer to lipid. Likewise, dolichyl-phosphate stimulated mannose but not GlcNAc incorporation (36). Thus the complete identification of these lipids may prove revealing and could shed light on the regulation of this pathway.

2. *What are the structures of the oligosaccharides? Do they all undergo processing (i.e. trimming)?* The structures of several of the oligosaccharides have been worked out in part as discussed in preceding sections. But more

detailed studies are necessary to determine glycosidic linkages, sugar sequences, etc. It seems likely that the structures will vary from tissue to tissue with respect to these fine details and depending on the protein to be glycosylated. In terms of processing it seems likely that the oligosaccharides containing glucose and mannose, and some of the larger mannose oligosaccharides, do undergo trimming once they are transferred. But there are apparently some "high mannose" oligosaccharides which are precursors to proteins such as soybean agglutinin (81). So far no processing reactions have been shown in plants. And the question of whether the smaller mannose oligosaccharides need to be glucosylated or whether they can donate directly to protein still remains. Such questions as how many sugars are removed or in what sequence and at what stage are all important. It seems likely that the events in processing and the possible regulation at that level will be a fascinating chapter in glycoprotein assembly.

3. *Where do these reactions take place in the cell?* It is not at all clear whether each of the reactions of Figure 3 occurs at a specific locality or whether they each can occur in many different membranes. Perhaps part of the reason for this confusion is that the lipid-linked saccharides are probably involved in the glycosylation of both secretory and membrane proteins, and these proteins are probably synthesized at different sites. In terms of secretory proteins, the polypeptide chain is synthesized on the ribosomes of the rough endoplasmic reticulum and then moves to the smooth endoplasmic reticulum and on to the Golgi apparatus. It seems clear that the distal sugars of the complex-type oligosaccharides—i.e. sialic acid, galactose, and GlcNAc—are added in the Golgi (108), but the core sugars seem to be attached either in the rough or the smooth endoplasmic reticulum (88). Recent studies suggest that glycosylation occurs either just after completion of the peptide chain or as it is nearing termination (69, 121, 131). In terms of membrane proteins, they are probably synthesized at the membrane in which they occur. i.e. at the mitochondrial or plasma membrane. Recently the membrane glycoprotein of vesicular stomatites virus was synthesized in a crude extract of wheat germ. The results were consistent with the idea that the growing membrane protein is extruded across the membrane amino terminus first and that glycosylation is restricted to the lumenal surface of the membrane (61a). If membrane proteins are synthesized at different membranous sites, this could explain the observation that lipid-linked saccharides are synthesized in various membranes including nuclear and mitochondrial (93), plasma, and so on. However, all is not that straightforward. In *Phaseolus aureus,* the enzyme which synthesizes mannosyl-phosphoryl-dolichol was mostly in the endoplasmic reticulum while the enzyme which transfers mannose to protein was in the Golgi

apparatus (74). One other confusing aspect concerning location of enzymes are the observations that intact cells can transfer mannose from GDP-[^{14}C]mannose or GlcNAc from UDP-GlcNAc to lipid-linked saccharides and glycoproteins (3, 22, 95, 118). Since sugar nucleotides do not penetrate whole cells, the enzymes involved must be at the cell surface. Do these enzymes serve a functional role in the assembly of plasma membrane glycoproteins, or are they the result of fusion of Golgi-derived vesicles with the plasma membrane (108)? And could the latter possibility account for these enzymatic activities in other membranes?

In this regard, several studies have been done on the location and synthesis of dolichol and dolichyl-phosphate. [^{3}H]Dolichol injected into rats was largely localized in the mitochondria (65). In liver, the synthesis of dolichyl-phosphate from acetate was examined and highest activity was found in mitochondrial outer membranes. But activity was also seen in nuclear membranes and microsomes, while plasma membranes had the lowest, but still detectable, activity (24). Thus dolichyl-phosphate, the acceptor of sugars, may be present in all membranes (47). But are all the homologs of dolichol synthesized in all these membranes? Dolichyl-phosphate was also synthesized by cell-free extracts of pea epicotyls, but in this case the subcellular location was not determined (25).

4. *How are these reactions regulated?* Perhaps the major unresolved question concerns the control of the lipid-linked saccharide pathway and its subsequent effect on glycosylation of proteins. Studies already discussed involving the use of inhibitors such as tunicamycin or 25-hydroxycholesterol show that if the synthesis of the lipid-linked saccharides are blocked in vivo, then certain membrane or secretory proteins are not glycosylated. But what controls the glycosylation of protein under normal circumstances and how is this linked to the control of protein synthesis? Is this pathway subject to any sort of feedback regulation? These are questions which cannot be answered now. But there is some feeling that the levels of dolichol or dolichyl-phosphate and/or the synthesis of these carriers might be one way to regulate these pathways. But again at this point there is little support for or against this hypothesis. Could the regulation somehow be linked to the different sizes of dolichols? It would seem to be beneficial to the cell to be able to control the synthesis of the mannolipid apart from synthesis of the GlcNAc-lipids or the lipid-linked oligosaccharides. If the lipid carriers are different, then it would be possible to exert such a regulation. Another possible regulatory mechanism is in terms of processing. It has been suggested that the membrane-bound glycosidases which remove glucose (and possibly other sugars) might be a control in glycosylation by regulating the rate of transfer of oligosaccharide to protein (116). These possibilities are all open for future experimentation.

5. *What is the nature of the proteins that are synthesized?* As indicated above, probably both secretory and membrane proteins are glycosylated via lipid intermediates, assuming they have an asparagine-linked core oligosaccharide. But thus far, in cell-free extracts, several endogenous proteins become radiolabeled and these all appear to be insoluble, i.e. membrane-bound proteins. Thus a distinct and known protein has not been synthesized in vitro at this point. But exogenous, denatured proteins have been glycosylated in vitro (96). As more studies accumulate more definitive information on specific proteins will certainly be forthcoming.

FUNCTIONAL ASPECTS OF GLYCOPROTEINS

Glycoproteins are important membrane constituents, both internally and at the surface of living cells. Thus they function as receptors for several biological effectors such as hormones, mitogens, viruses, and so on. They are involved in cell-cell interactions which occur during development and probably play a role in transformation. The removal of distal sugars such as sialic acid from certain serum glycoproteins causes them to be recognized by binding proteins in hepatocyte membranes and results in the removal of these asialo-glycoproteins from serum. The carbohydrate portion of cell surface proteins also has an important role in adhesion and related aspects of cell behavior (99).

In plants there are also numerous recognition phenomena which occur and these probably involve complex carbohydrates. However, the nature of the interacting molecules has not really been established. For example, plant proteins known as lectins may be involved in the specific recognition of symbiotic rhizobia by legumes since rhizobia for soybeans will not infect clover and vice versa (100). Plant exudates have been shown to cause chemotaxis of rhizobia (23). While the lectins or recognition compounds in plants have not been shown to be glycoproteins, soybean agglutinin is a glycoprotein with a “high mannose” oligosaccharide (81). Of course, the lectins are carbohydrate recognizing proteins and the carbohydrates which they recognize may well be lipopolysaccharides on the rhizobia. Recognition type phenomena in plants have also been shown in tumor induction which involves the attachment of bacteria to a site on the host plant cell wall (80) and in certain host pathogen interactions (4). These are only a few examples of these kinds of interactions in plants. Now it will be important to determine the nature of the molecules involved and to determine the mechanism of specificity. Some of these molecules undoubtedly will be synthesized via lipid-linked saccharides so that a knowledge of the pathway and its regulation should be useful for understanding these interactions.

CONCLUSIONS

It seems clear from the evidence accumulated thus far that lipid-linked saccharides play an important role in the glycosylation of certain asparagine-linked glycoproteins. But whether all glycoproteins that have a core oligosaccharide linked to asparagine require a lipid intermediate for glycosylation remains to be established. And whether there is another route for addition of sugars is an interesting question. As suggested at the beginning of this manuscript and by analogy to the microbial systems, the lipid carrier probably functions to enable the hydrophilic sugars to penetrate the hydrophobic environments. Since proteins are synthesized on membrane-bound ribosomes and there is a tight coupling between polypeptide synthesis and membrane insertion, glycosylation must occur within the membranous environment. And indeed all of the in vitro studies in this area have been done utilizing various membrane fractions. Although the general outline of reactions and the transfer of oligosaccharide to protein have been worked out, there are numerous intriguing questions still to be resolved before the real significance of this pathway, and indeed of the glycoproteins themselves, can be ascertained.

ACKNOWLEDGMENTS

The work from the author's laboratory was supported by grants from the Robert A. Welch Foundation (AQ-366), the National Science Foundation (PCM 08861, PCM 16433), and the National Institutes of Health (HL 17783). The excellent assistance of Ms. Linda Winchester in the preparation of this manuscript is gratefully acknowledged.

Literature Cited

1. Adamany, A. M., Spiro, R. G. 1975. Glycoprotein biosynthesis: Studies on the mannosyl transferases. *J. Biol. Chem.* 250:2842–54
2. Alam, S. S., Hemming, F. W. 1973. Polyprenyl phosphates and mannosyl transferases in *Phaseolus aureus. Phytochemistry* 12:1641–49
3. Arnold, D., Hommel, E., Risse, H.-J. 1976. Cell surface mannosyl transferase activity in the liver of embryonic chick. *Mol. Cell. Biochem.* 11:137–47
4. Ayers, A. R., Valent, B., Ebel, J., Albersheim, P. 1976. Host pathogen interactions. XI. Composition and structure of wall-released elicitor fractions. *Plant Physiol.* 57:766–74
5. Basha, S. M., Beevers, L. 1976. Glycoprotein metabolism in the cotyledons of *Pisum sativum* during development and germination. *Plant Physiol.* 57:93–97
6. Baynes, J. W., Hsu, A.-F., Heath, E. C. 1973. The role of mannosyl-phosphoryl-dihydropolyisoprenol in the synthesis of mammalian glycoproteins. *J. Biol. Chem.* 248:5693–5704
7. Beevers, L., Mense, R. M. 1978. Glycoprotein biosynthesis in cotyledons of *Pisum sativum. Plant Physiol.* 60:703–8
8. Behrens, N. H., Carminatti, H., Staneloni, R. J., Leloir, L. F., Cantarella, A. I. 1973. Formation of lipid-bound oligosaccharides containing mannose. Their role in glycoprotein synthesis. *Proc. Natl. Acad. Sci. USA* 70:3390–94
9. Behrens, N. H., Leloir, L. F. 1970. Dolichol monophosphate glucose: An intermediate in glucose transfer in liver.

Proc. Natl. Acad. Sci. USA 66:153–59
10. Behrens, N. H., Parodi, A. I., Leloir, L. F. 1971. Glucose transfer from dolichol monophosphate glucose: The product formed with endogenous microsomal acceptor. *Proc. Natl. Acad. Sci. USA* 68:2857–60
11. Blobel, G., Dobberstein, B. 1975. Transfer of proteins across membranes. II. Reconstitution of functional rough microsomes from heterologous components. *J. Cell Biol.* 67:852–62
12. Bouchilloux, S., Chabaud, O., Ronin, C. 1973. Cell-free peptide synthesis and carbohydrate incorporation by various thyroid particles. *Biochim. Biophys. Acta* 322:401–20
13. Brett, C. T., Leloir, L. F. 1977. Dolichol monophosphate and its sugar derivatives in plants. *Biochem. J.* 161:93–101
14. Caccam, J. F., Jackson, J. J., Eylar, E. H. 1969. The biosynthesis of mannose-containing glycoproteins: A possible lipid intermediate. *Biochem. Biophys. Res. Commun.* 35:505–11
15. Chambers, J., Elbein, A. D. 1975. Biosynthesis and characterization of lipid-linked sugars and glycoproteins in pig aorta. *J. Biol. Chem.* 250:6904–15
16. Chambers, J., Forsee, W. T., Elbein, A. D. 1977. Enzymatic transfer of mannose from mannosyl-phosphoryl-polyprenol to lipid-linked oligosaccharides by pig aorta. *J. Biol. Chem.* 252:2498–2506
17. Chen, W. W. 1978. The role of glucose-containing oligosaccharide lipids in glycosylation of protein acceptors. *Fed. Proc.* 37:2288
18. Chen, W. W., Lennarz, W. J. 1976. Participation of a trisaccharide-lipid in glycosylation of oviduct membrane glycoproteins. *J. Biol. Chem.* 251: 7802–9
19. Chen, W. W., Lennarz, W. J. 1977. Metabolism of lipid-linked N-acetylglucosamine intermediates. *J. Biol. Chem.* 252:3473–79
20. Chen, W. W., Lennarz, W. J., Tarentino, A. L., Maley, F. 1975. A lipid-linked oligosaccharide intermediate in glycoprotein synthesis in oviduct. *J. Biol. Chem.* 250:7006–13
21. Colvin, J. R. 1959. Synthesis of cellulose in ethanol extracts of *Acetobacter xylinum. Nature* 183:1135
22. Cooper, J. R., Hemming, F. W. 1977. The incorporation of [^{14}]glucosamine into dolichol-diphosphate-GlcNAc by unbroken liver cells in culture. *Eur. J. Biochem.* 78:89–94
23. Currier, W. W., Strobel, G. A. 1976. Chemotaxis of *Rhizobium* spp. to plant root exudates. *Plant Physiol.* 57:820–823
24. Daleo, G. R., Hopp, H. E., Romero, P. A., Pont-Lezica, R. 1977. Biosynthesis of dolichol phosphate by subcellular fractions from liver. *FEBS Lett.* 81:411–14
25. Daleo, G. R., Pont-Lezica, R. 1977. Synthesis of dolichol phosphate by a cell-free extract from pea. *FEBS Lett.* 74:247–50
26. Delmer, D. P. 1977. The biosynthesis of cellulose and other plant cell wall polysaccharides. *Recent Adv. Phytochem.* 11:45–77
27. Delmer, D. P., Kulow, C., Ericson, M. C. 1978. Glycoprotein synthesis in plants. II. Structure of the mannolipid intermediate. *Plant Physiol.* 61:25–29
28. Duksin, D., Bornstein, P. 1977. Changes in surface properties of normal and transformed cells caused by tunicamycin, an inhibitor of protein glycosylation. *Proc. Natl. Acad. Sci. USA* 74: 3433–37
29. Eagon, P. C., Heath, E. C. 1977. Glycoprotein biosynthesis in myeloma cells. Characterization of nonglycosylated immunoglobulin light chain secreted in presence of 2-deoxy-D-glucose. *J. Biol. Chem.* 252:2372–83
30. Elbein, A. D. 1979. *Carbohydrases in Plants. Handbook on Nutrition.* C. R. C. Press. In press
31. Elbein, A. D., Forsee, W. T. 1976. Biosynthesis of steryl glucosides and acylated steryl glucosides in plants. In *Glycolipid Methodology,* ed. L. Whitting, pp. 345–68. Am. Oil Chemists Soc.
32. Ericson, M. C., Chrispeels, M. J. 1973. Isolation and characterization of glucosamine-containing glycoproteins from cotyledons of *Phaseolus aureus. Plant Physiol.* 52:98–104
33. Ericson, M. C., Delmer, D. P. 1977. Glycoprotein synthesis in plants. I. Role of lipid intermediates. *Plant Physiol.* 59:341–47
34. Ericson, M. C., Delmer, D. P. 1978. Glycoprotein synthesis in plants. III. Interaction between UDP-N-acetylglucosamine and GDP-mannose as substrates. *Plant Physiol.* 61:819–23
35. Ericson, M. C., Gafford, J. T., Elbein, A. D. 1977. Tunicamycin inhibits GlcNAc-lipid formation in plants. *J. Biol. Chem.* 252:7431–33
36. Ericson, M. C., Gafford, J. T., Elbein, A. D. 1978. Evidence that the lipid carrier for GlcNAc is different than that

for mannose in mung beans and cotton fibers. *Plant Physiol.* 61:274–77
37. Ericson, M. C., Gafford, J. T., Elbein, A. D. 1978. Bacitracin inhibits the synthesis of lipid-linked saccharides and glycoproteins in plants. *Plant Physiol.* 62:373–76
38. Ericson, M. C., Gafford, J. T., Elbein, A. D. 1978. *In vivo* and *in vitro* inhibition of lipid-linked saccharides and glycoprotein synthesis in plants by amphomycin. *Arch. Biochem. Biophys.* 191:698–704
39. Evans, P. J., Hemming, F. W. 1973. The unambiguous characterization of dolichol phosphate mannose as a product of mannosyl transferase in pig liver endoplasmic reticulum. *FEBS Lett.* 31:335–38
40. Forsee, W. T., Elbein, A. D. 1972. Biosynthesis of acidic glycolipids in cotton fibers. *Biochem. Biophys. Res. Commun.* 49:930–39
41. Forsee, W. T., Elbein, A. D. 1973. Biosynthesis of mannosyl- and glucosyl-phosphoryl-polyprenols in cotton fibers. *J. Biol. Chem.* 248:2858–67
42. Forsee, W. T., Elbein, A. D. 1975. Glycoprotein biosynthesis in plants. Demonstration of lipid-linked oligosaccharides. *J. Biol. Chem.* 250:9283–93
43. Forsee, W. T., Elbein, A. D. 1977. Biosynthesis of lipid-linked oligosaccharides in cotton fibers. Stimulation by acceptor lipids from pig liver. *J. Biol. Chem.* 252:2444–46
44. Forsee, W. T., Valkovich, G., Elbein, A. D. 1976. Glycoprotein biosynthesis in plants. Formation of lipid-linked oligosaccharides of mannose and GlcNAc by mung bean seedlings. *Arch. Biochem. Biophys.* 174:469–79
45. Garcia, R. C., Recondo, E., Dankert, M. 1974. Polysaccharide biosynthesis in *Acetobacter xylinum.* Enzymatic synthesis of lipid diphosphate and monophosphate sugars. *Eur. J. Biochem.* 43:93–105
46. Ginsburg, V. 1964. Sugar nucleotides and the synthesis of carbohydrates. *Adv. Enzymol.* 26:35–85
47. Grange, D. K., Adair, W. L. Jr. 1977. Studies on the biosynthesis of dolichyl-phosphate. Evidence for *in vitro* formation of 2,3-dehydrodolichyl-phosphate. *Biochem. Biophys. Res. Commun.* 79: 734–40
48. Hara, M., Umetsu, N., Miyamoto, C., Tamari, K. 1973. Inhibition of the biosynthesis of plant cell wall materials, especially cellulose biosynthesis, by coumarin. *Plant Cell Physiol.* 14:11–15
49. Hart, G. W., Grant, G. A., Bradshaw, R. A. 1978. Primary structural requirement for the glycosylation of proteins. *Fed. Proc.* 37:2284
50. Heifetz, A., Elbein, A. D. 1977. Solubilization and properties of mannose and GlcNAc transferases involved in the formation of polyprenyl-sugar intermediates. *J. Biol. Chem.* 252:3057–63
51. Heifetz, A., Elbein, A. D. 1977. Biosynthesis of man-β-GlcNAc-GlcNAc-pyrophosphoryl-polyprenol by a solubilized enzyme from aorta. *Biochem. Biophys. Res. Commun.* 75:20–28
52. Heifetz, A., Keenan, R. W., Elbein, A. D. 1979. Mechanism of action of tunicamycin on the UDP-GlcNAc:dolichyl-phosphate GlcNAc-1-phosphate transferase. Submitted
53. Hemming, F. W. 1974. Lipids in glycan biosynthesis. In *Biochemistry of Lipids,* ed. T. W Goodwin, 4:39–78. London: Butterworth
54. Herscovics, A., Bugge, B., Jeanloz, R. W. 1977. Effect of bacitracin on the biosynthesis of dolichol derivatives in calf pancreas microsomes. *FEBS Lett.* 82: 215–17
55. Herscovics, A., Golovtchenko, A. M., Warren, C. D., Bugge, B., Jeanloz, R. W. 1977. Mannosyl transferase activity in calf pancreas. *J. Biol. Chem.* 252: 224–34
56. Hickman, S., Kulczycki, A. Jr., Lynch, R. G., Kornfeld, S. 1977. Studies on the mechanism of tunicamycin inhibition of IgA and IgE secretion by plasma cells. *J. Biol. Chem.* 252:4402–8
57. Hopp, H. E., Romero, P. A., Daleo, G. R., Pont-Lezica, R. 1978. Synthesis of cellulose precursors. The involvement of lipid-linked sugars. *Eur. J. Biochem.* 84:561–71
58. Hopp, H. E., Romero, P. A., Pont-Lezica, R. 1978. On the inhibition of cellulose biosynthesis by coumarin. *FEBS Lett.* 86:259–62
59. Hsu, A. F., Baynes, J. W., Heath, E. C. 1974. The role of a dolichol-oligosaccharide as an intermediate in glycoprotein biosynthesis. *Proc. Natl. Acad. Sci. USA* 71:2391–95
60. Kang, M. S., Spencer, J. S., Elbein, A. D. 1978. Amphomycin inhibits mannose and GlcNAc incorporation into lipid-linked saccharides. *Biochem. Biophys. Res. Commun.* 82:568–74
61. Kang, M. S., Spencer, J. S., Elbein, A. D. 1978. Amphomycin inhibition of mannose and GlcNAc incorporation into lipid-linked saccharides. *J. Biol. Chem.* 253:8860–66

61a. Katz, F. N., Rothman, J. E., Lingappa, V. R., Blobel, G., Ludish, H. F. 1977. Membrane assembly *in vitro:* synthesis, glycosylation, and asummetric insertion of a transmembrane protein. *Proc. Natl. Acad. Sci. USA* 74:3278–82
62. Kauss, H. 1969. A plant mannosyl-lipid acting in reversible transfer of mannose. *FEBS Lett.* 5:81–84
63. Kean, E. L. 1977. Concerning the catalytic hydrogenation of polyprenylphosphate-mannose synthesized by the retina. *J. Biol. Chem.* 252:5619–21
64. Kean, E. L. 1977. The biosynthesis of mannolipids and mannose containing complex glycans by the retina. *J. Supramol. Struct.* 7:381–85
65. Keenan, R. W., Fisher, J. B., Kruczek, M. E. 1977. The tissue and subcellular distribution of [^{3}H]dolichol in the rat. *Arch. Biochem. Biophys.* 179:634–42
66. Keenan, R. W., Kruczek, M., Fusinato, L. 1975. The role of glucolipid in the biosynthesis of glycoprotein in *Tetrahymena pyriformis. Arch. Biochem. Biophys.* 167:697–705
67. Kerr, A. K. A., Hemming, F. W. 1978. Factors affecting glucosyl and mannosyl transfer to dolichyl monophosphate by liver cell-free preparations. *Eur. J. Biochem.* 83:581–86
68. Khalkhali, Z., Marshall, R. D. 1975. Glycosylation of ribonuclease A catalyzed by rabbit liver extracts. *Biochem. J.* 146:299–307
69. Kiely, M. L., McKnight, G. S., Schimke, R. T. 1976. Studies on the attachment of carbohydrate to ovalbumin nascent chains in the oviduct. *J. Biol. Chem.* 251:5490–95
70. Kjosbakken, J., Colvin, J. R. 1973. Biosynthesis of cellulose by a particulate enzyme system from *Acetobacter xylinum.* In *Biogenesis of Plant Cell Wall Polysaccharides,* ed. F. Loweus, pp. 361–71. New York: Academic
71. Kornfeld, R., Kornfeld, S. 1976. Comparative aspects of glycoprotein structure. *Ann. Rev. Biochem.* 45:217–37
72. Kuo, S.-C., Lampen, J. O. 1972. Inhibition by 2-deoxyglucose of synthesis of glycoprotein enzymes by protoplasts of Saccharomyces. *J. Bacteriol.* 111: 419–29
73. Kuo, S.-C., Lampen, J. O. 1974. Tunicamycin—An inhibitor of yeast glycoprotein synthesis. *Biochem. Biophys. Res. Commun.* 58:287–95
74. Lehle, L., Bowles, D. J., Tanner, W. 1978. Subcellular site of mannosyl transfer to dolichyl phosphate in *Phaseolus aureus. Plant Sci. Lett.* 11:27–34
75. Lehle, L., Fartaczek, F., Tanner, W., Kauss, H. 1976. Formation of polyprenol mono- and oligosaccharides in *Phaseolus aureus. Arch. Biochem. Biophys.* 175:419–26
76. Lehle, L., Tanner, W. 1975. Formation of lipid-bound oligosaccharides in yeast. *Biochim. Biophys. Acta* 399:364–74
77. Lehle, L., Tanner, W. 1976. The specific site of tunicamycin inhibition in the formation of dolichol-bound N-acetylglucosamine derivatives. *FEBS Lett.* 71:167–70
78. Leloir, L. F., Staneloni, R. J., Carminatti, H., Behrens, N. H. 1973. The biosynthesis of an N,N'-diacetylchitobiose containing lipid by liver microsomes. *Biochem. Biophys. Res. Commun.* 52:1285–92
79. Levy, J. A., Carminatti, H., Cantarella, A. I., Behrens, N. H., Leloir, L. F., Tabora, E. 1974. Mannose transfer to lipid-linked Di-N-acetylchitobiose. *Biochem. Biophys. Res. Commun.* 60: 118–25
80. Lippincott, B. B., Whatley, M. H., Lippincott, J. A. 1977. Tumor induction by *Agrobacterium* involves attachment of the bacterium to a site on the host plant cell wall. *Plant Physiol.* 59:388–90
81. Lis, H., Sharon, N. 1978. Soybean agglutinin—A plant glycoprotein. *J. Biol. Chem.* 253:3468–76
82. Lucas, J. J., Waechter, C. J. 1976. Polyisoprenoid glycolipids involved in glycoprotein biosynthesis. *Mol. Cell. Biochem.* 11:67–78
83. Lucas, J. J., Waechter, C. J., Lennarz, W. J. 1975. The participation of lipid-linked oligosaccharide in synthesis of membrane glycoproteins. *J. Biol. Chem.* 250:1992–2002
84. Mankowski, T., Sasak, W., Chojnacki, T. 1975. Hydrogenated polyprenol phosphates. Exogenous lipids acceptors of glucose from UDP-glucose in rat liver microsomes. *Biochem. Biophys. Res. Commun.* 65:1292–97
85. Marshall, R. D. 1972. Glycoproteins. *Ann. Rev. Biochem.* 41:673–702
86. Martin, H. G., Thorne, K. J. I. 1974. The involvement of endogenous dolichol in the formation of lipid-linked precursors of glycoprotein in rat liver. *Biochem. J.* 138:281–89
87. Mills, J. T., Adamany, A. M. 1978. Impairment of dolichyl saccharide synthesis and dolichol mediated glycoprotein assembly in the aorta smooth muscle cell in culture by inhibitors of choles-

terol biosynthesis. *J. Biol. Chem.* 253: 5270–73
88. Molnar, J. 1975. A proposed pathway of plasma glycoprotein synthesis. *Mol. Cell. Biochem.* 6:3–30
88a. Molnar, J., Chao, H., Ikehara, Y. 1971. Phosphoryl-N-acetyl-glucosamine transfer to a lipid acceptor of liver microsomal preparations. *Biochim. Biophys. Acta* 239:401–10
89. Nikaido, H., Hassid, W. Z. 1971. Biosynthesis of saccharides from glycopyranosyl esters of nucleoside pyrophosphates. *Adv. Carbohydr. Chem. Biochem.* 26:352–439
90. Oliver, G. J. A., Hemming, F. W. 1975. The transfer of mannose to dolichol diphosphate oligosaccharides in pig liver endoplasmic reticulum. *Biochem. J.* 152:191–99
91. Osborn, M. J. 1969. Structure and biosynthesis of the bacterial cell wall. *Ann. Rev. Biochem.* 38:501–38
92. Palamarczyk, G., Hemming, F. W. 1975. The formation of mono-N-acetylhexosamine derivatives of dolichol diphosphate by pig liver microsomal fractions. *Biochem. J.* 148:245–51
93. Palamarczyk, G., Janczura, E. 1977. Lipid mediated glycosylation in yeast nuclear membranes. *FEBS Lett.* 77: 169–72
94. Palmiter, R. D., Gagnon, J., Walsh, K. A. 1978. Ovalbumin: A secreted protein without a transient hydrophobic leader sequence. *Proc. Natl. Acad. Sci. USA* 75:94–98
95. Patt, L. M., Grimes, W. J. 1976. Formation of mannosyl-lipids by an ectomannosyl-transferase in suspensions of BALB/c fibroblasts. *Biochim. Biophys. Acta* 444:97–107
96. Pless, D. D., Lennarz, W. J. 1977. Enzymatic conversion of proteins to glycoproteins. *Proc. Natl. Acad. Sci. USA* 74:134–38
97. Pont-Lezica, R., Brett, C. T., Martinez, P. R., Dankert, M. A. 1975. A glucose acceptor in plants with the properties of an α-saturated polyprenyl-monophosphate. *Biochem. Biophys. Res. Commun.* 66:980–87
98. Pont-Lezica, R., Romero, P. A., Dankert, M. A. 1976. Membrane bound UDP-glucose: lipid glucosyltransferases from peas. *Plant Physiol.* 58:675–80
99. Pouyssegur, J., Willingham, M., Pastan, I. 1977. Role of cell surface carbohydrates and proteins in cell behavior: Studies on the biochemical reversion of an N-acetylglucosamine-deficient fibroblast mutant. *Proc. Natl. Acad. Sci. USA* 74:243–47
100. Pueppke, S. G., Bauer, W. D., Keegstra, K., Ferguson, A. L. 1978. Role of lectins in plant microorganism interactions. *Plant Physiol.* 61:779–84
101. Reuvers, F., Boer, P., Hemming, F. W., 1978. The presence of dolichol in a lipid-diphosphate-N-acetylglucosamine from *Saccharomyces cerevisiae. Biochem. J.* 169:505–8
102. Reuvers, F., Boer, P., Steyn-Parve, E. P. 1978. The effect of bacitracin on the formation of polyprenol derivatives in yeast membrane vesicles. *Biochem. Biophys. Res. Commun.* 82:800–4
103. Reuvers, F., Habets-Williems, C., Reinking, A., Boer, P. 1977. Glycolipid intermediates involved in the transfer of N-acetylglucosamine to endogenous proteins in a yeast membrane preparation. *Biochim. Biophys. Acta* 486: 541–52
104. Richards, J. B., Hemming, F. W. 1972. The transfer of mannose from GDP-mannose to dolichyl phosphate and protein by pig liver endoplasmic reticulum. *Biochem. J.* 130:77–93
105. Robbins, P. W., Krag, S. S., Liu, T. 1977. Effects of UDP-glucose addition on the synthesis of mannosyl-lipid-linked oligosaccharides by cell-free fibroblast preparations. *J. Biol. Chem.* 252:1780–85
106. Roberts, R. M. 1975. The incorporation of glucosamine into glycolipids and glycoproteins of membrane preparations from *Phaseolus aureus* hypocotyls. *Plant Physiol.* 55:431–36
107. Ronin, C., Bouchilloux, S. 1978. Cell-free labeling in thyroid rough microsomes of lipid-linked and protein linked oligosaccharides. *Biochim. Biophys. Acta* 539:470–80
108. Roseman, S. 1970. The synthesis of complex carbohydrates by multiglycosyltransferase systems and their potential function in intercellular adhesion. *Chem. Phys. Lipids* 5:270–97
109. Schwarz, R. J., Rohrschneider, J. M., Schmidt, M. F. G. 1976. Suppression of glycoprotein formation of Semliki forest, influenza, and avian sarcoma virus by tunicamycin. *J. Virol.* 19:782–91
110. Smith, M. M., Axelos, M., Peaud-Lenoel, C. 1976. Biosynthesis of mannose and mannolipids from GDP-mannose by membrane fractions of sycamore cell cultures. *Biochimie* 58:1195–11
111. Speake, B. K., White, P. A. 1978. The formation of lipid-linked sugars as intermediates in glycoprotein synthesis in

rabbit mammary gland. *Biochem. J.* 170:273–83
112. Spencer, J. S., Kang, M. S., Elbein, A. D. 1978. Inhibition of lipid-linked saccharide synthesis by bacitracin. *Arch. Biochem. Biophys.* 190:829–37
113. Spiro, M. J., Spiro, R. G., Bhoyroo, V. D. 1976. Lipid saccharide intermediates in glycoprotein biosynthesis. *J. Biol. Chem.* 251:6420–25
114. Spiro, M. J., Spiro, R. G., Bhoyroo, V. D. 1978. Utilization of oligosaccharide-lipids in glycoprotein biosynthesis by thyroid enzyme. *Fed. Proc.* 37:2285
115. Spiro, R. G., Spiro, M. J., Bhoyroo, V. D. 1976. Lipid-saccharide intermediates in glycoprotein biosynthesis. II. Studies on the structure of an oligosaccharide-lipid from thyroid. *J. Biol. Chem.* 25:6409–19
116. Spiro, R. G., Spiro, M. J., Bhoyroo, V. D. 1978. Processing of the carbohydrate units of glycoproteins: Action of a thyroid glucosidase. *Fed. Proc.* 37:2286
117. Storm, D. L., Hassid, W. Z. 1972. The role of a mannosyl-lipid as an intermediate in the synthesis of polysaccharide in *Phaseolus aureus* seedlings. *Plant Physiol.* 50:473–76
118. Struck, D. L., Lennarz, W. J. 1976. Utilization of exogenous GDP-mannose for the synthesis of mannose containing lipids and glycoproteins by oviduct cells. *J. Biol. Chem.* 251:2511–19
119. Struck, D. L., Lennarz, W. J. 1977. Evidence for the participation of saccharide-lipids in the synthesis of the oligosaccharide chain of ovalbumin. *J. Biol. Chem.* 252:1007–13
120. Tabas, I., Schlesinger, S., Kornfeld, S. 1978. Processing of high mannose oligosaccharide to form complex type oligosaccharides on the newly synthesized polypeptides of the vesicular stomatitis virus G protein and the IgG heavy chain. *J. Biol. Chem.* 253:716–22
121. Takatsuki, A., Arima, K., Tamura, G. 1971. Tunicamycin, a new antibiotic. *J. Antibiot.* 24:215–23
122. Tanner, W. 1969. A lipid intermediate in mannan biosynthesis in yeast. *Biochem. Biophys. Res. Commun.* 35: 144–50
123. Tarentino, A. L., Maley, F. 1974. Purification and properties of an endo-β-N-acetylglucosaminidase from *Streptomyces griseus. J. Biol. Chem.* 249: 811–17
124. Tkacz, J. S., Hercovics, A. 1975. Ozonolytic cleavage of authentic and pancreatic dolichyl-mannopyranosyl phosphate. *Biochem. Biophys. Res. Commun.* 64:1009–7
125. Tkacz, J. S., Hercovics, A., Warren, C. D., Jeanloz, R. W. 1974. Mannosyltransferase activity in calf pancreas microsomes. *J. Biol. Chem.* 249:6372–81
126. Tkacz, J. S., Lampen, J. O. 1975. Tunicamycin inhibition of polyisoprenol N-acetylglucosaminyl pyrophosphate formation in calf-liver microsomes. *Biochem. Biophys. Res. Commun.* 65: 248–57
127. Treloar, M., Sturgess, J. M., Moscarello, M. A. 1974. An effect of puromycin on galactosyltransferase of Golgi-rich fractions from rat liver. *J. Biol. Chem.* 249:6628–32
128. Tucker, P., Pestka, S. 1977. *De novo* synthesis and glycosylation of the MOPC-46B mouse immunoglobulin light chain in cell-free extracts. *J. Biol. Chem.* 252:4474–86
129. Turco, S. J., Heath, E. C. 1977. Glucuronosyl-N-acetylglucosaminyl pyrophosphoryl dolichol. *J. Biol. Chem.* 252:2918–28
130. Turco, S. J., Stetson, B., Robbins, P. W. 1977. Comparative rates of transfer of lipid-linked oligosaccharides to endogenous glycoprotein acceptors *in vitro. Proc. Natl. Acad. Sci. USA* 74:4411–14
131. Vargas, V. I., Carminatti, H. 1977. Glycosylation of endogenous proteins of the rough and smooth microsomes by a lipid sugar intermediate. *Mol. Cell. Biochem.* 16:171–76
132. Vessey, D. A., Lysenko, N., Zakim, D. 1976. Evidence for multiple enzymes in the dolichol utilizing pathway of glycoprotein biosynthesis. *Biochim. Biophys. Acta* 428:138–45
133. Vijay, I. K., Fram, S. R. 1977. Role of mannosyl-phosphoryl-polyisoprenol in biosynthesis of mammary glycoproteins. *J. Supramol. Struct.* 7:251–65
134. Villamez, C. L., Clark, A. F. 1969. A particle bound intermediate in the biogenesis of plant cell wall polysaccharides. *Biochem. Biophys. Res. Commun.* 36:57–63
135. Waechter, C. J., Harford, J. B. 1977. Evidence for the enzymatic transfer of N-acetylglucosamine from UDP-N-acetylglucosamine into dolichol derivatives and glycoproteins by calf brain membranes. *Arch. Biochem. Biophys.* 181:185–98
136. Waechter, C. J., Kennedy, J. L., Harford, J. B. 1976. Lipid intermediates involved in the assembly of membrane associated glycoprotein in calf brain white

matter. *Arch. Biochem. Biophys.* 174: 726–37
137. Waechter, C. J., Lennarz, W. J. 1976. The role of polyprenol-linked sugars in glycoprotein synthesis. *Ann. Rev. Biochem.* 45:95–112
138. Waechter, C. J. Lucas, J. J., Lennarz, W. J. 1973. Membrane glycoproteins. Enzymatic synthesis of mannosyl-phosphoryl-polyisoprenol and its role as a mannosyl donor in glycoprotein synthesis. *J. Biol. Chem.* 248:7570–79
139. Waechter, C. J., Lucas, J. J., Lennarz, W. J. 1974. Evidence for xylosyl-lipids as intermediates in xylosyl transfer in hen oviduct membranes. *Biochem. Biophys. Res. Commun.* 56:343–50
140. Warren, C. D., Jeanloz, R. W. 1973. Chemical synthesis of dolichyl-α-D-mannopyranosyl phosphate and citronellyl α-D-mannopyranosyl phosphate. *Biochemistry* 12:5038–45
141. Warren, C. D., Jeanloz, R. W. 1975. Synthesis of P^1-dolichyl-P^2-α-D-mannopyranosyl-pyrophosphate. The acid and alkaline hydrolysis of polyisoprenyl-α-D-mannopyranosyl mono- and pyrophosphate diesters. *Biochemistry* 14:412–19
142. Wedgwood, J. F., Strominger, J. L., Warren, C. D. 1974. Transfer of sugars from nucleoside diphosphosugar compounds to endogenous and synthetic dolichyl phosphate in human lymphocytes. *J. Biol. Chem.* 249:6316–24
143. Wright, A. 1971. Mechanism of conversion of the salmonella O antigen by bacteriophage ϵ^{34}. *J. Bacteriol.* 105:927–36
144. Zatta, P., Zakim, D., Vessey, D. A. 1976. The lipid intermediates arising during glycoprotein biosynthesis in liver microsomes. *Biochim. Biophys. Acta* 441:103–14
145. Zatz, M., Barondes, S. H. 1969. Incorporation of mannose into mouse brain lipid. *Biochem. Biophys. Res. Commun.* 36:511–17

Ann. Rev. Plant Physiol. 1979. 30:273–88

FUSICOCCIN: A TOOL IN PLANT PHYSIOLOGY[1]

❖7672

E. Marrè

Institute of Plant Sciences, University of Milan, Milan, 20133, Italy

CONTENTS

INTRODUCTION: GENERALITIES AND SPECTRUM OF ACTION IN VIVO

Fusicoccin (FC) is the major toxin produced by *Fusicoccum amygdali* Del. and is responsible for most of the pathological symptoms induced by this fungus on peach and almond trees (20). Its structure has been worked out by Ballio (3) and by Barrow et al in 1968 (3a). Figure 1 shows the similarity of this diterpene glucoside with two other toxins: cotylenins, endowed with

[1]*Abbreviations:* ABA, abscisic acid; AIBA, α-aminoisobutyric acid; C_1/C_6 ratio of dissimilation to CO_2 of C_1 and C_6 of glucose; CBT, *Cercospora beticola* toxin; DCCD dicyclohexylcarbodiimide; DES, diethylstilbestrol; FC, fusicoccin; GA, gibberellic acid; $-\Delta H^+$, proton extrusion; ΔK^+, K^+ uptake; OP, osmotic potential; PD, transmembrane electric potential difference.

0066-4294/79/0601-0273$01.00

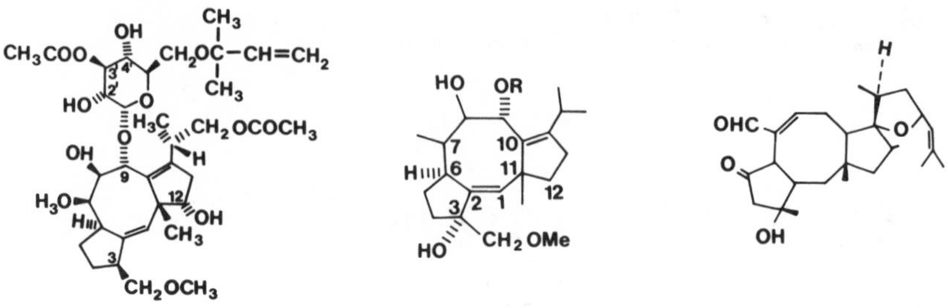

FUSICOCCIN COTYLENIN OPHIOBOLIN A

Figure 1 Structure of fusicoccin and basic structure of cotylenin and ophiobolin A.

physiological activities very similar to those of FC (3, 15, 72), and ophiobolins (75), which, on the contrary, completely lack the positive effects of FC on transport and growth which will be discussed in this article.

The use of FC as a tool in plant physiology was proposed in 1970 on the basis of its striking activity in promoting cell enlargement in the usual tests for auxin action (47). Since then, evidence has been accumulated to show that FC strongly influences a number of important physiological processes, and suggesting that its mechanism of action depends on the direct activation of a single central transport system present in all higher plants. Due to lack of space, only a limited number of the pertinent papers will be cited in this article, and the reader is referred to previous general articles (cf, for example, 8, 40, 41, 43, 44).

A list of the physiological effects of FC is presented in Table 1. Most of these effects, and in particular those on cell enlargement, acidification of the medium, potassium uptake, and stomatal opening have been observed in practically all higher plant species and materials investigated so far. The capacity of responding to FC seems generally found in all green plants from the Charophyceae to higher organization levels, while no responses have until now been observed for fungi, bacteria, and animals (43). It appears, therefore, that the action of FC (just as that of auxin and other natural plant hormones) involves some very general system, appearing, together with the capacity of responding to plant hormones, at a relatively advanced stage of plant evolution.

The present discussion on the available data on FC actions and their bearing on general physiological problems will start from the in vitro evidence—still in rapid development—that tends to localize the primary biochemical target of FC in a protein system present in the cell membrane and directly involved in the active proton extrusion.

Table 1 Effects of fusicoccin in vivo[a]

1.	Acidification of the incubation medium (H^+ extrusion) (5, 7, 8, 15, 20a, 28, 33, 42, 45, 48–50, 60, 61, 68, 69, 69a).
2.	Stimulation of the uptake of K^+ and other cations (7, 8, 10, 11, 20a, 32, 40, 41, 51, 60, 61, 69, 69a, 79).
3.	Hyperpolarization of the electric PD (2, 5, 10, 20a, 46, 54, 60, 61).
4.	Stimulation of the uptake of chloride and other anions (20a, 40, 41, 60).
5.	Stimulation of active Na^+ extrusion (44; P. Lado et al, unpublished data).
6.	Stimulation of glucose, sucrose, and aminoacid uptake (12, 16a, 39, 41, 55, 69a, 71).
7.	Stimulation of respiration and of dark CO_2 fixation into malate (28, 49, 65).
8.	Increase of pyruvate and Gl6P levels, and decrease of the C_1/C_6 ratio (41, 50, 65).
9.	Stimulation of cell enlargement in stems, coleoptiles, roots, leaves, cotyledons, and seed embryos (7a, 25, 32, 33, 42, 45, 48, 58, 68, 80).
10.	Stimulation of stomatal opening in antagonism with ABA (64, 74, 79).
11.	Promotion of seed germination in antagonism with ABA and other dormancy-inducing conditions (2, 6, 9, 11, 29, 33).

[a] Effects 1, 2, 3, 8, 9, 10, and 11 are usually stronger than for any other experimental treatment. All effects are most evident in isolated plant parts, due to the lack of active and polar transport of FC and to its very slow diffusion in tissues (M. Radice and P. Pesci, unpublished results). Effects 1, 2, and 6 have been reported also in suspension cell cultures and protoplasts (69, 69a, 71).

EFFECTS OF FC IN VITRO: MODEL OF MECHANISM OF ACTION

The major effect of FC on transport is the stimulation of electrogenic, energy-linked H^+ extrusion, associated with the uptake of K^+ (or, under suitable conditions, of other monovalent cations). In vivo and in vitro evidence suggests that this effect depends on the interaction between FC and the K^+-, Mg^{2+}-activated, DCCD-sensitive plasmalemma ATPase. In fact: (*a*) there are good reasons to believe that this enzyme is involved in K^+ transport (1, 17, 23, 35–37, 51, 78), and this process can be interpreted as dependent on, or coupled with, electrogenic H^+ extrusion (see below); (*b*) all the treatments which severely decrease the intracellular ATP level also suppress the FC-promoted H^+/K^+ exchange (43, 44, 49); (*c*) DCCD, DES, and octylguanidine, which are effective inhibitors of this ATPase in vitro, also produce rapid inhibition of the FC-stimulated H^+ extrusion and K^+ uptake (43, 51); this inhibition by DES is also true for times of treatment so short that no significant changes of ATP level or glycolysis are detectable (43, also P. Lado et al, unpublished data); (*d*) ^3H-labeled FC specifically binds to a (protein) component present in plasmalemma-enriched membrane preparations, following the gradient distribution pattern of plasmalemma markers (16, 57); (*e*) two rather specific inhibitors of FC-stimulated H^+/K^+ exchange in vivo, and of plasmalemma ATPase in vitro, DES and *Cercospora beticola* toxin (CBT) (F. Macrì and A. Vianello, unpublished data) also strongly inhibit FC binding in membrane preparations, the concentrations giving 50% inhibition on ATPase and on FC binding being practically coincident (Figure 2; 57); (*f*) FC has been shown to stimulate in vitro the cation-dependent, DCCD-sensitive ATPase activity of plasmalemma-enriched membrane preparations (4; also A. Ballio and R. Federico, unpublished). The potentially decisive result in *f* unfortunately is weakened by its still unsatisfactory reproducibility.

These findings, on the whole, suggest that FC specifically and directly activates a cell membrane system utilizing ATP energy for electrogenic H^+ extrusion. The recent finding that the FC binding component and the ATPase component present in plasmalemma preparations can be separated by solubilization with Na^+ perchlorate followed by gel filtration and electrophoresis (Figure 2) suggests a model postulating a complex of at least two subunits; the one, being responsible for the interaction with FC and with inhibitors such as DES and CBT, would be able to modulate through conformational changes the catalytic activity of the other subunit, it being endowed with the ATPase activity (57). This model, presented in Figure 3, repeats the general features of the models proposed for the ATP-driven proton pumps operating in mitochondria, chloroplasts, and bacteria (22, 26, 53, 56, 59, 73).

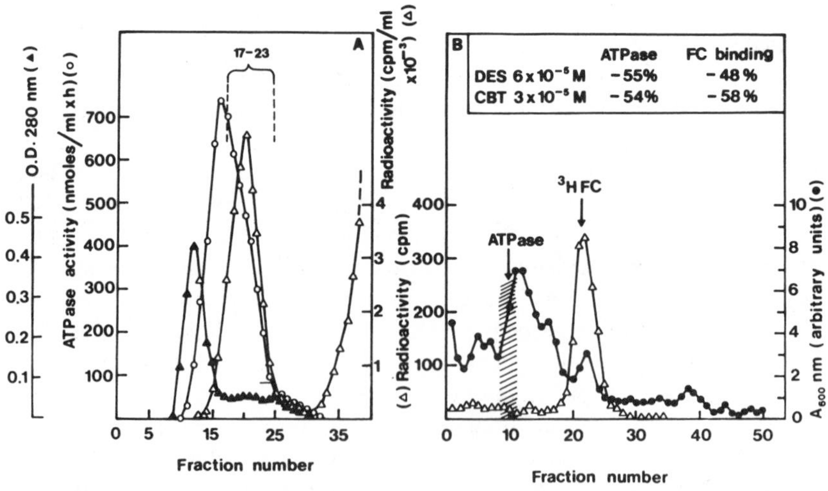

Figure 2 Resolution of the ATPase activity and of the FC-protein complex from plasmalemma-enriched membrane preparations. Solubilization by Na perchlorate followed by Sephadex G-200 fractionation (A) and by acrylamide disc electrophoresis of the pooled fractions (17–23) containing most of the radioactivity and some of the ATPase (B). The insert shows the parallelism between the inhibiting effects of DES and of *Cercospora beticola* toxin (CBT) on the DCCD-sensitive ATPase activity and on the FC binding activity of the membrane preparation (57).

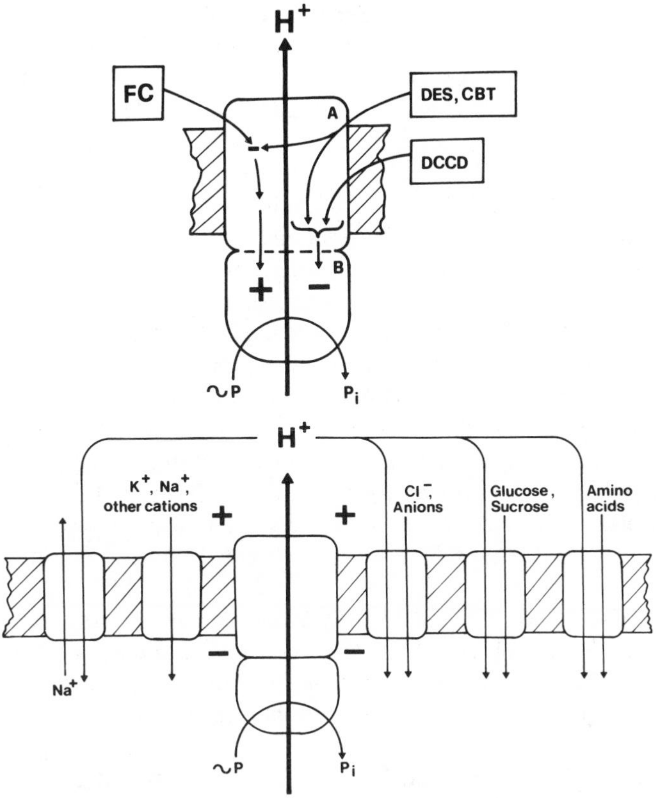

Figure 3 (*Top*) Model of the FC-activated proton pumping ATPase, with the proposed interpretation of the effects of DES and CBT on the ATPase and on the FC binding activities. (*Lower*) Model of utilization of FC-promoted H^+ extrusion for solute transport.

FC AND THE INTERDEPENDENCE OF TRANSPORT PROCESSES

The mode of integration of the various cell membrane activities represents an important general problem. If it is true that FC primarily acts on proton extrusion, the analysis of the relationship between this response and those of the other FC-stimulated transport processes might provide an insight into their interconnections.

Two major responses constantly associated with the FC-induced proton secretion are the hyperpolarization of the PD and the increase of the uptake of K^+ or Rb^+ and, although to a lesser extent, of other monovalent cations.

The rapid hyperpolarization of the PD usually ranges from 20 to 80 mV (its maximum value being observed in the absence of K^+ or other permeating cations in the medium) and is interpreted as a proof of the active, electrogenic nature of FC-induced H^+ extrusion (5, 10, 46, 54, 60, 61). This increase of the inside negative PD is also important in suggesting a reasonable interpretation of the parallel stimulation of the uptake of K^+, Rb^+, and, in general (although to a lesser extent), other monovalent cations, inasmuch as it should represent an increase of the driving force for the influx of positively charged ions by passive transport along the electrochemical gradients. In fact, the presence in the medium of K^+ (or, with lower efficiency, of Na^+ or other monovalent cations) is required for the FC-induced H^+ extrusion (40, 50, 51), and this effect of monovalent cations is correlated with the rate of their uptake and with their capacity to depolarize the PD (10, 40, 60, 61).

With pea stem or maize root segments, and with an incubation medium containing only 0.5 mM Ca^{2+}, K^+ (or Rb^+ or Na^+), and Cl^- (or $SO_4^=$ or benzensulfonate) as anions (A^-), the only relevant changes of ionic fluxes induced by FC occur at the expense of monovalent cations and of anions, and a value very close to 1 is observed for the ratio $(-\Delta H^+ + \Delta A^-)/\Delta K^+)$. The sum $-\Delta H^+$ (as titrated) $+\Delta A^-$ is conveniently interpreted as corresponding to the real total amount of extruded protons (40, 41). In fact, Cl^- or $SO_4^=$ uptake in these systems is electroneutral (10), thus suggesting anion cotransport with H^+ (or the formally equivalent antiport with OH^- or HCO_3^-) which would lead to the disappearance of a fraction of the H^+ extruded by the FC-stimulated pump. Accepting this correction, the stoichiometry of the H^+ (really extruded)/K^+ or other monovalent cation) would result very close to unity (40, 41).

A ratio close to 1 of the H^+/K^+ exchange can be interpreted by: (*a*) a chemical coupling of the two fluxes at the level of a single carrier system (54, 62), or (*b*) an electrogenic coupling, the PD energy generated by the electrogenic ~P-driven H^+ uniport being utilized for K^+ uptake through a separate channel (44, 60, 61). The second hypothesis seems supported by the findings that: (*a*) in the FC-treated systems the ratio of K^+_{out}/K^+_{in} are very close to and somewhat higher than that predicted by the Nernst equilibrium equation (10, 61); (*b*) lipophylic cations such as tributylbenzylammonium and tetraphenylphosphonium, postulated to enter nonspecifically through the lipidic component of cell membranes, can partially substitute for K^+ [as in bacteria (22)], in its promotion of the FC-induced H^+ extrusion (4a).

Besides increasing the influx of monovalent cations, FC also stimulates the uptake of various anions (Cl^-, $SO_4^=$, $HPO_3^=$, NO_3^-, benzensulfonate), amino acids (AIBA, glycine, leucine), and sugars (glucose, sucrose) (5, 12,

16a, 20a, 39, 41, 71). Here again, these effects of FC seem conveniently interpreted as consequences of the primary direct activating effect on H^+ extrusion and on the hyperpolarization of PD. In fact, evidence is accumulating that in higher plants, as in bacteria, algae, and fungi, the uptake of anions, sugars, and amino acids involve the symport with H^+ (or the antiport with OH^- or HCO_3^-), utilizing as energy source either the electrochemical proton gradient or the PD (12, 16a, 19, 22, 30, 31, 39, 55, 70, 73). The finding that FC markedly increases all of these processes gives strong experimental evidence in favor of this hypothesis.

Another important transport process consistently increased by FC is the active, K^+-stimulated Na^+ extrusion from isolated root segments (P. Lado et al, unpublished data). Here again, this effect of FC seems most conveniently interpreted as a consequence of its capacity to stimulate the proton pump, and this conclusion seems important in elucidating the mechanism of this process. In fact, two main hypotheses have been proposed for Na^+ extrusion: (*a*) a K^+/Na^+ exchange (27) or (*b*) a H^+/Na^+ exchange (66). Now the effect of FC on Na^+ extrusion is strongly synergistic with that of K^+, while it is less than additive with that of low pH of the medium. These data are consistent with the H^+/Na^+ antiport hypothesis: in this case, the synergism would be explained by the strong dependence of FC-induced H^+ extrusion on the availability of extracellular K^+, while the effect of FC in decreasing the external pH would become less and less important when the acidity of the medium is increased. In contrast, the possibility of explaining the same results with the K^+/Na^+ antiport hypothesis would implicate a number of purely speculative and apparently unlikely assumptions. The analysis of FC-induced Na^+ extrusion seems thus to provide good evidence in favor of a proton gradient-driven H^+/Na^+ antiport mechanism similar to the one demonstrated in bacteria (22). A scheme showing the proposed relationship between H^+ extrusion and other transport processes is presented in Figure 3. Another potentially interesting effect of FC, presently under investigation, is the stimulation of electro-osmotic efficiencies and transcellular water flux observed by Tazawa and Fensom in *Nitella flexilis* and *Nitella translucida* (D. F. Fensom, personal communication). This effect is probably related to that on H^+ and cation transport and might open an approach to the study of electro-osmotic coupling of water flux with ion fluxes (see 74a).

METABOLIC CHANGES ASSOCIATED WITH FC-INDUCED H^+ EXTRUSION

FC induces an early and substantial increase in the rate of respiration, in the rate of CO_2 fixation into malate, in the levels of malate, pyruvate and, to a lesser extent, of Gl6P, together with a significant decrease of the C_1/C_6 ratio (41, 65). The increase of malate synthesis from PEP and CO_2 (through oxalacetate) and that of malate level are conveniently interpreted, in agreement with the biochemical pH-stat theory (14, 67), as a response to the increased H^+ extrusion and, as a consequence, of the rise of cytoplasmic pH. This interpretation is interesting as it implicates the localization of the FC-activated H^+ pump at the plasmalemma and seems to rule out alternative possibilities [as that of mitochondria being the site of H^+ extrusion (63)]. The large increase in pyruvate in the FC-treated tissue may be tentatively interpreted as a consequence of the increase in malate, as it seems likely that NADPH, NADP, CO_2, malate, and pyruvate concentrations are maintained close to equilibrium by the high activity of the malic enzyme.

The drop of the C_1/C_6 ratio seems at least in part dependent on a decrease of Gl6P oxidation. In fact, FC not only decreases the C_1/C_6 ratio in CO_2 but it also increases the C_1/C_2 ratio in RNA ribose (from ca 0.77 to ca 1), a test which should allow a relatively reliable evaluation of the activity of the Gl6P oxidation pathway (65). As suggested in the scheme of Figure 4, the inhibition of Gl6P oxidation (possibly accounting for the slight increase of Gl6P) might be tentatively explained as due to an increase of NADPH consequent to the increase of malate.

MECHANISM OF FC ACTION ON CELL ENLARGEMENT

FC dramatically stimulates cell enlargement not only in the materials giving this response to IAA (coleoptile and stem segments) or to cytokinin (seed

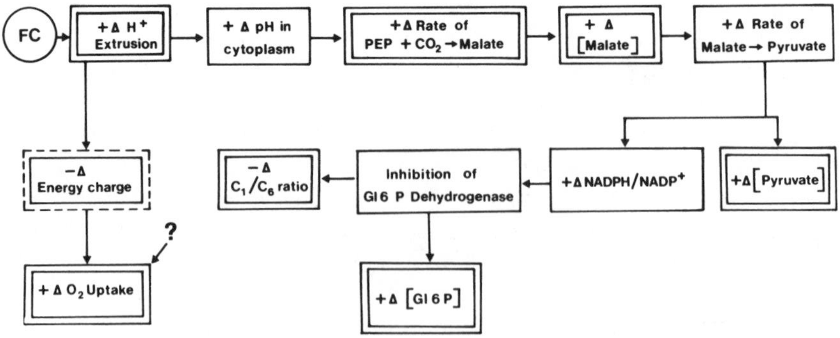

Figure 4 Tentative interpretation of the mode of action of FC on $^{Q}O_2$, dark fixation of CO_2 into malate, C_1/C_6 ratio and on the levels of pyruvate, malate, and Gl6P($+\Delta$ or $-\Delta$ mean increase or decrease). Double rules indicate the experimentally detected changes induced by FC (65).

cotyledons) but also in materials such as root and leaf segments and seed embryos (7a, 25, 32, 33, 42, 45, 48, 58, 68, 80). In root, coleoptile, and stem segments, cell enlargement is also stimulated by treatment with acid buffers (pH from 4 to 5) (7a, 32, 52, 68). The growth effects of FC and of acid buffer are not additive (32; Figure 5), thus suggesting that at least part of FC-induced cell enlargement depends on a decrease of pH in the wall space consequent to the activation of proton extrusion. This would increase the plastic extensibility of the cell wall by activating polysaccharide hydrolytic enzymes (13). This conclusion is in obvious agreement with the main thesis of the so-called "acid growth theory," according to which also the effect of auxin on cell enlargement would essentially depend on its capacity to activate H^+ extrusion and thus decrease pH in the wall space. On the other hand, it seems too simplistic, on the basis of what has been discussed in previous sections, to assume that FC or IAA activation of H^+ extrusion induces wall loosening, and thus promotes growth, only through a change of pH, as other H^+ extrusion-dependent factors, such as the concentration of various ions, colloid hydration, water activity, etc must also be affected and might variously influence the metabolism and the physical state of the wall components. This, together with the difficulty in establishing a direct relationship between the real pH in the wall and the pH measured in the medium, might explain the failure, reported by various authors, to find a

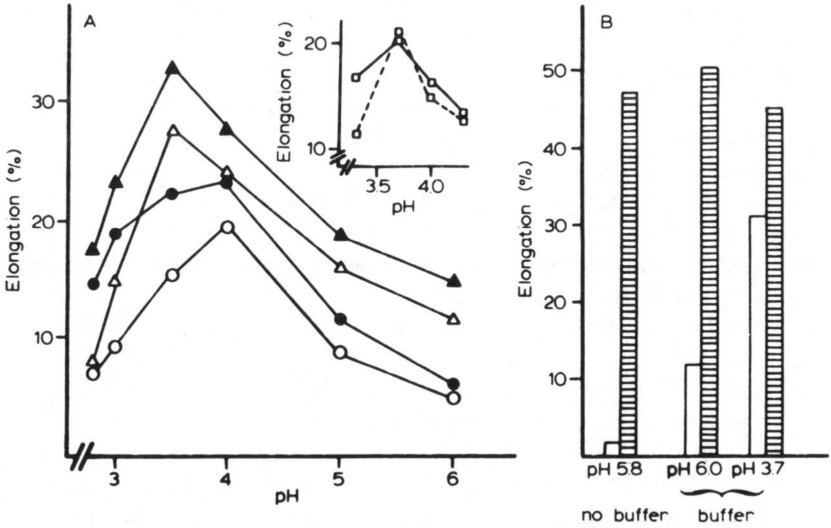

Figure 5 Elongation response of apical maize root segments to treatments: A. with acid buffer; B. with acid buffer minus (open bars) or plus FC (dashed bars), showing the lack of additivity between the effects of acid and of FC (32).

satisfactory correlation between the effect of FC and auxin on growth and that on acidification of the external medium [for a more detailed discussion of this topic, see (8, 13, 24, 42, 68, 76, 80)].

FC AND THE REGULATION OF CELL TURGOR: THE EFFECT ON STOMATAL OPENING

The stimulation of stomatal opening by light is accompanied by the increase of H^+/K^+ exchange and by malate synthesis in the guard cells; all of these responses are inhibited by ABA (38, 64). In the same system FC very efficiently substitutes for light in inducing stomatal opening and completely reverses ABA-induced inhibition (64, 74, 79). This effect of FC is interpreted as a consequence of the intracellular increase of K^+, malate (and eventually Cl^- or other anion), and thus of OP, water uptake, and cell turgor (64). Besides the intrinsic interest in the physiology of transpiration and gaseous exchanges, these results seem to provide an elegant example of the regulation of cell turgor (with its various physiological implications) by means of changes of activity of the proton pump.

FC, CELL MULTIPLICATION, AND THE BREAKING OF SEED DORMANCY

In nondormant tissues [for example, cell suspension cultures (69)], cell multiplication seems little or not at all influenced by the FC-induced changes of ion transport. Under particular physiological conditions, however, FC can very efficiently promote this type of growth. This is the case when FC induces germination of seeds either normally dormant or inhibited by treatment with ABA or by conditions such as temperature or osmotic pressure [(6, 11, 18, 33); for an extensive review see (29)].

The earliest responses of dormant seeds to FC are the increases of H^+ extrusion, K^+ uptake, PD, water uptake, and cell enlargement (2, 11, 33), together with an acceleration of the development of plasmalemma ATPase activity (9, 11), while the activation of nucleic acid synthesis and of cell multiplication seems consistently delayed in respect to seeds activated by gibberellin (18). These data can be put in relation to the known dependence of macromolecular syntheses in germinating seeds on water availability and on water uptake (see 77), and also to the recent finding that FC induces, after a lag of ca 24 hr and together with germination, a marked increase of GA in light-inhibited *Phacelia* seeds (S. Cocucci and A. M. Ranieri, unpublished data). On the other hand, a decrease of ABA in FC-treated seeds has been reported (6). The effect of FC on seed germination [and possibly also on the breaking of bud dormancy (21)] might thus be interpreted as in the scheme of Figure 6, according to which the only primary action of FC would be the activation of the proton pump. Obviously enough, this interpretation emphasizes the fundamental importance of the regulation of transport processes for the physiology of dormancy and its breaking.

FC AS A TOOL TO UNDERSTAND HORMONE ACTION

FC mimicks IAA, GA, and cytokinin, and antagonizes ABA in some of their effects; hence its interest for the study of their mode of action. This matter has been covered extensively elsewhere (8, 24, 42, 45, 48, 50, 68, 80), and only a synthetic statement of the author's opinions will be made here.

1. FC action never completely overlaps the area of influence of any particular plant hormone, and its direct effects seem limited to the promotion of ion transport. In contrast, complex differentiation phenomena involving protein synthesis are often observed among the earlier effects of the natural hormones. This is in agreement with the finding of a much stronger dependence of hormone action, as compared with that of FC, upon the integrity of protein synthesis (33a, 34, 68).

2. FC is relatively monotonous in its action and shows little or no tissue or organ specificity, while hormones display a large variety of tissue-specific effects. This suggests that the final effects of hormones on transport are mediated by long chains of intermediary events, while FC, as discussed above, seems to directly interact with its physiologically active target.

3. In the cases where the effect of hormone (IAA or ABA) on ion transport has been clearly demonstrated, its action closely repeats (IAA) or antagonizes (ABA) the main features of FC action on the various aspects of transport [H^+/K^+ exchange, CO_2 fixation into malate, PD hyperpolarization, etc (2, 11, 33, 42, 74, 79)]. This suggests that the FC-sensitive transport system may be also indirectly influenced by the relatively remote primary action of the various hormones.

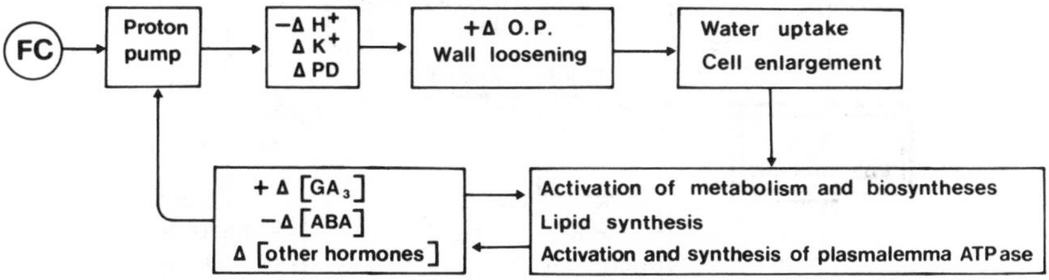

Figure 6 Interpretation of the promoting action of FC on germination of seeds either naturally dormant or inhibited by ABA.

According to these views, FC may be considered as a very powerful tool to elucidate only that part of the hormone final effects, which is concerned with ion transport, together with its physiological consequences and implications (see Figure 7).

CONCLUSIONS

The author realizes that this article represents a synthetic presentation of his personal view, rather than a detailed critical review of the existing evidence on the physiological activity of fusicoccin. This can be explained in part by shortage of space, in part by the feeling that the evidence in this field has come to a point where an effort to integrate data and theories may be more relevant than the analysis of the individual results.

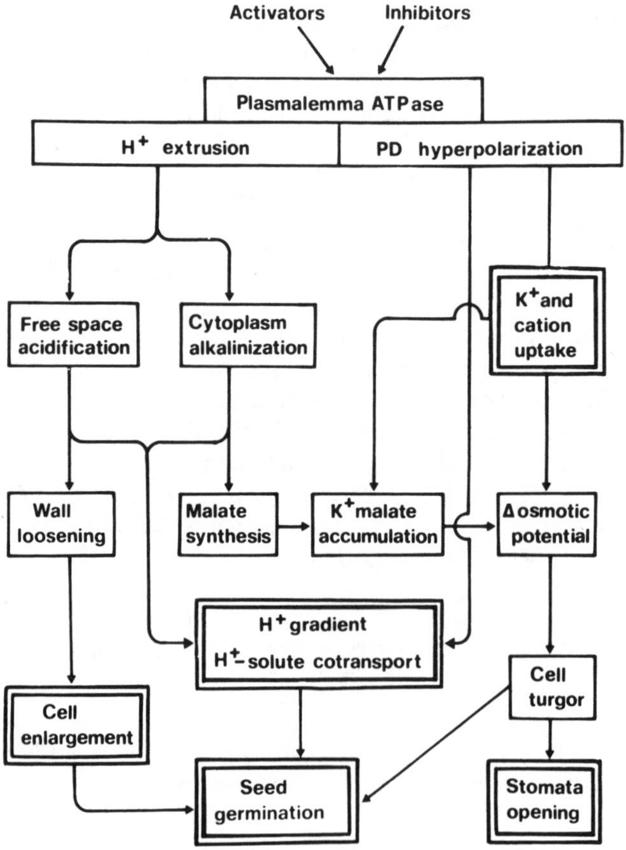

Figure 7 Scheme of the proposed relationships between FC-activated H^+ extrusion and some physiological processes. Final macroscopic responses are indicated by double rule.

This paper is centered on the hypothesis that FC specifically activates a single fundamental process, namely the conversion of ~P energy into proton electrochemical gradient energy at the plasmalemma level, and that all the other important effects of the toxin are consequences of this primary event. Hence the interest for FC as a precious tool to elucidate the cause-effect relationships in the chain of steps leading from the regulation of proton extrusion to that of a large number of physiological processes.

ACKNOWLEDGMENTS

I am grateful to many colleagues for helpful discussions and for providing access to unpublished data and information. My special thanks to Prof. Piera Lado for her friendly and unfailing collaboration and criticism.

Literature Cited

1. Balke, N. E., Hodges, T. K. 1977. Inhibition of ion absorption in oat roots: comparison of diethylstilbestrol and oligomycin. *Plant Sci. Lett.* 10:319–25
2. Ballarin Denti, A., Cocucci, M. 1978. *Effect of ABA, GA_3 and fusicoccin on the transmembrane potential during early phase of germination in radish seeds.* Presented at Fed. Eur. Soc. Plant Physiol. Meet., Edinburgh
3. Ballio, A. 1977. Fusicoccin: structure-activity relationships. In *Regulation of Cell Membrane Activities in Plants,* ed. E. Marré, O. Ciferri, pp. 217–23. Amsterdam: North-Holland. 332 pp.

3a. Barrow, K. D., Barton, D. H. R., Chain, E. B., Ohnsorge, U. F. W., Thomas, R. 1968. The constitution of fusicoccin. *Chem. Commun.* 1198–1200
4. Beffagna, N., Cocucci, S., Marrè, E. 1977. Stimulating effect of fusicoccin on K^+-activated ATPase in plasmalemma preparations from higher plant tissues. *Plant Sci. Lett.* 8:91–98

4a. Bellando, M., Trotta, A., Bonetti, A., Colombo, R., Lado, P., Marrè, E. 1979. Dissociation of H^+ extrusion from K^+ uptake by means of lipophilic cations. *Plant, Cell and Environment* 2. In press
5. Böcher, M., Novacky, A. 1978. The active component of membrane potential of *Lemna gibba* G1. Effect of fusicoccin. *Plant Physiol.* 61 Suppl. 107 (Abstr.)
6. Borkowska, B., Sing, S., Khan, A. A. 1978. Changes in ABA content of lettuce seeds by osmotic, fusicoccin and other treatments. *Plant Physiol.* 61 Suppl. 107 (Abstr.)
7. Cleland, R. E. 1976. Rapid stimulation of K^+/H^+ exchange by a plant growth hormone. *Biochem. Biophys. Res. Commun.* 69:333–38

7a. Cleland, R. E. 1976. Fusicoccin-induced growth and hydrogen ion excretion of *Avena* coleoptiles: relation to auxin responses. *Planta* 128:201–6
8. Cleland, R. E., Lomax, T. 1977. Hormonal control of H^+-excretion from oat cells. See Ref. 3, pp. 161–71
9. Cocucci, M., Cocucci, S., Ballarin Denti, A. 1978. *Effects, in vitro, of polar lipids on K^+-Mg^{++}- dependent ATPase of membrane preparation in radish seeds.* Presented at Fed. Eur. Soc. Plant Physiol. meet., Edinburgh
10. Cocucci, M., Marrè, E., Ballarin Denti, A., Scacchi, A. 1976. Characteristics of fusicoccin-induced changes of transmembrane potential and ion uptake in maize root segments. *Plant Sci. Lett.* 6:143–56
11. Cocucci, S., Cocucci, M. 1977. Effect of ABA, GA_3 and FC on the development of potassium uptake in germinating radish seeds. *Plant Sci. Lett.* 10:85–95
12. Colombo, R., De Michelis, M. I., Lado, P. 1978. 3-O-Methyl glucose uptake stimulation by auxin and by fusicoccin in plant materials and its relationships with proton extrusion. *Planta* 138: 249–56
13. Darvill, A. G., Smith, C. J., Hall, M. A. 1978. Cell wall structure and elongation growth in *Zea mays* coleoptile tissue. *New Phytol.* 80. In press
14. Davies, D. D. 1973. Control of and by pH. *Symp. Soc. Exp. Biol.* 27:513–29
15. De Michelis, M. I., Lado, P., Ballio, A. 1975. Relationship between structural

modifications of fusicoccin and their effects on cell enlargement and seed germination. *G. Bot. Ital.* 109:168–69

16. Dohrmann, U., Hertel, R., Pesci, P., Cocucci, S. M., Marrè, E., Randazzo, G., Ballio, A. 1977. Localization of "in vitro" binding of the fungal toxin fusicoccin to plasma-membrane-rich fractions from corn coleoptiles. *Plant Sci. Lett.* 9:291–99

16a. Etherton, B., Rubinstein, B. 1978. Evidence for amino acid-H^+ co-transport in oat coleoptiles. *Plant Physiol.* 61:933–37

17. Frick, H., Bauman, L. B., Nicholson, R. L., Hodges, T. K. 1977. Influence of *Helminthosporium maydis* race T toxin on potassium uptake in maize roots. II. Sensitivity of the development of the augmented uptake potential to toxin and inhibitors of protein synthesis. *Plant Physiol.* 59:103–6

18. Galli, M. G., Sparvoli, E., Caroi, M. 1975. Comparative effects of fusicoccin and gibberellic acid on the promotion of germination and DNA synthesis initiation in *Haplopappus gracilis. Plant Sci. Lett.* 5:351–57

19. Giaquinta, R. 1977. Phloem loading of sucrose. *Plant Physiol.* 59:750–55

20. Graniti, A. 1966. Azione della fusicoccina, tossina prodotta da Fusicoccum amygdali Del., sul bilancio idrico di piante recise di pomodoro. *Phytopathol. Mediterr.* 5:146

20a. Gronewald, J. W., Cheeseman, J. M., Hanson, J. B. 1979. Response of fresh and washed corn root tissue to fusicoccin. *Plant Physiol.* 63. In press

21. Guern, J., Usciati, M., Sansonetti, A. 1977. Hormonal control of apical dominance in *Cicer arietinum* seedlings: a tentative utilisation of the models of hormonal regulation of cell membrane properties. See Ref. 3, pp. 251–60

22. Harold, F. M., Papineau, D. 1972. Cation transport and electrogenesis. II Proton and sodium extrusion. *J. Membr. Biol.* 8:45–62

23. Hodges, T. K. 1976. ATPases associated with membranes of plant cells. In *Transport in Plants,* ed. U. Lüttge, M. G. Pitman, 11A:260–83. Berlin, Heidelberg, New York: Springer-Verlag. 400 pp.

24. Ilan, I., Shapira, S. 1976. On the relation between the effects of auxin on growth, pH and potassium transport. *Physiol. Plant.* 38:243–48

25. Jaccard, M., Pilet, P. E. 1978. Growth and r'.eological changes of collenchyma cells: the fusicoccin effect. *Plant Cell Physiol.* In press

26. Jagendorf, A. T. 1975. Mechanism of photophosphorylation. In *Bioenergetics of Photosynthesis,* ed. Govindjee, pp. 413–92. New York: Academic

27. Jeschke, W. D. 1977. K^+-Na^+ selectivity in roots, localization of selective fluxes and their regulation. See Ref. 3, pp. 63–78

28. Johnson, K. D., Rayle, D. L. 1976. Enhancement of CO_2 uptake in *Avena* coleoptiles by fusicoccin. *Plant Physiol.* 57:806–11

29. Khan, A. A. 1977. In *The Physiology and Biochemistry of Seed Dormancy and Germination,* ed. A. A. Khan. Amsterdam: North-Holland

30. Komor, E., Rotter, M., Tanner, W. 1977. A proton-cotransport system in a higher plant: sucrose transport in *Ricinus communis. Plant Sci. Lett.* 9:153–62

31. Komor, E., Tanner, W. 1974. Proton movement associated with hexose transport in *Chlorella vulgaris.* In *Membrane Transport in Plants,* ed. U. Zimmermann, J. Dainty, pp. 209–15. Berlin, Heidelberg, New York: Springer-Verlag. 473 pp.

32. Lado, P., De Michelis, M. I., Cerana, R., Marrè, E. 1976. Fusicoccin-induced, K^+-stimulated proton secretion and acid-induced growth of apical root segments. *Plant Sci. Lett.* 6:5–20

33. Lado, P., Rasi Caldogno, F., Colombo, R. 1975. Acidification of the medium associated with normal and fusicoccin-induced seed germination. *Physiol. Plant.* 34:359–64

33a. Lado, P., Rasi Caldogno, F., Colombo, R. 1977. Effect of cycloheximide on IAA- or FC-induced cell enlargement in pea internode segments. *Plant Sci. Lett.* 9:93–101

34. Lado, P., Rasi Caldogno, F., Pennacchioni, A., Marrè, E. 1973. Mechanism of the growth-promoting action of fusicoccin. Interaction with auxin, and effects of inhibitors of respiration and protein synthesis. *Planta* 110:311–20

35. Leonard, R. T., Hanson, J. B. 1972. Increased membrane-bound adenosine triphosphatase activity accompanying development of enhanced solute uptake in washed corn root tissue. *Plant Physiol.* 49:436–40

36. Leonard, R. T., Hodges, T. K. 1973. Characterisation of plasma membrane-associated adenosine triphosphatase activity in oat roots. *Plant Physiol.* 52:6–12

37. Lepe, B. G., Hodges, T. K. 1978. Alkylguanidine inhibition of ion absorption in oat roots. *Plant Physiol.* In press
38. MacRobbie, E. A. C., Lettau, J., Bray, M. 1978. *Ionic relations of stomatol guard cells.* Presented at Fed. Eur. Soc. Plant Physiol. meet., Edinburgh
39. Malek, T., Baker, D. A. 1978. Effect of fusicoccin on proton co-transport of sugars in the phloem loading of *Ricinus communis* L. *Plant. Sci. Lett.* 11: 233–39
40. Marrè, E. 1977. On the mechanism of auxin- and fusicoccin-stimulated proton cation exchange in plants. In *Exchanges ioniques transmembranaires chez les Végétaux,* ed. M. Thellier, A. Monnier, M. Demarty, J. Dainty, pp. 529–36. Rouen: Univ. Rouen. 607 pp.
41. Marrè, E. 1977. Effects of fusicoccin and hormones on plant cell membrane activities: observations and hypotheses. See Ref. 3, pp. 185–202
42. Marrè, E. 1977. Physiologic implications of the hormonal control of ion transport in plants. In *Plant Growth Regulation,* ed. P. E. Pilet, pp. 54–66. Berlin, Heidelberg, New York: Springer-Verlag. 305 pp.
43. Marrè, E. 1978. Membrane activities as regulating factors for plant cell functions. *Biol. Cell.* 32. In press
44. Marrè, E. 1979. Integration of solute transport in cereals. In *Recent Advances in the Biochemistry of Cereals,* ed. D. L. Laidman, R. G. Wyn Jones. London, New York: Academic. In press
45. Marrè, E., Colombo, R., Lado, P., Rasi Caldogno, F. 1974. Correlation between proton extrusion and stimulation of cell enlargement. Effects of fusicoccin and of cytokinins on leaf fragments and isolated cotyledons. *Plant Sci. Lett.* 2:139–50
46. Marrè, E., Lado, P., Ferroni, A., Ballarin Denti, A. 1974. Transmembrane potential increase induced by auxin, benzyladenine and fusicoccin. Correlation with proton extrusion and cell enlargement. *Plant Sci. Lett.* 2:257–65
47. Marrè, E., Lado, P., Rasi Caldogno, F., Colombo, R. 1971. Fusicoccin as a tool for the analysis of auxin action. *Rend. Accad. Naz. Lincei* 50:45–49
48. Marrè, E., Lado, P., Rasi Caldogno, F., Colombo, R. 1973. Correlation between cell enlargement in pea internode segments and decrease in the pH of the medium of incubation. I. Effects of fusicoccin, natural and synthetic auxins and mannitol. *Plant Sci. Lett.* 1:179–84
49. Marrè, E., Lado, P., Rasi Caldogno, F., Colombo, R. 1973. Correlation between cell enlargement in pea internode segments and decrease in the pH of the medium of incubation. II. Effects of inhibitors of respiration, oxidative phosphorylation and protein synthesis. *Plant Sci. Lett.* 1:185–92
50. Marrè, E., Lado, P., Rasi Caldogno, F., Colombo, R., Cocucci, M., De Michelis, M. I. 1975. Regulation of proton extrusion by plant hormones and cell elongation. *Physiol. Vég.* 13(4):797–811
51. Marrè, E., Lado, P., Rasi Caldogno, F., Colombo, R., DeMichelis, M. I. 1974. Evidence for the coupling of proton extrusion to K^+ uptake in pea internode segments treated with fusicoccin or auxin. *Plant Sci. Lett.* 3:365–79
52. Mentze, J., Raymond, B., Cohen, J. D., Rayle, D. L. 1977. Auxin-induced H^+ secretion in *Helianthus* and its implications. *Plant Physiol.* 60:509–12
53. Mitchell, P., Moyle, J. 1969. Translocation of some anions, cations and acids in rat liver mitochondria. *Eur. J. Biochem.* 9:149–55
54. Nelles, A. 1978. Evidence for fusicoccin-stimulated potassium ion pump in cell membranes of dwarf maize coleoptiles. *Plant Sci. Lett.* 12:349–54
55. Novacky, A., Ullrich-Eberius, C. I., Lüttge, U. 1978. Membrane potential changes during transport of hexoses in *Lemna gibba* G1. *Planta* 138:263–70
56. Papa, S. 1976. Proton translocation reactions in the respiratory chains. *Biochim. Biophys. Acta* 456:39–84
57. Pesci, P., Beffagna, N., Tognoli, L., Marrè, E. 1978. *Relationships between the FC-binding protein and the DCCD-sensitive ATPase of plasmalemma preparations from maize coleoptiles.* Presented at Fed. Eur. Soc. Plant Physiol. meet, Edinburgh
58. Pilet, P. E. 1976. Fusicoccin and auxin effects on root growth. *Plant Sci. Lett.* 7:81–84
59. Pitman, M. G. 1976. Ion uptake by plant roots. In *Transport in Plants,* ed. U. Lûttge, M. G. Pitman, 11B:95–128. Berlin, Heidelberg, New York: Springer-Verlag. 456 pp.
60. Pitman, M. G., Schaefer, N., Wildes, R. A. 1975. Relation between permeability to potassium and sodium ions and fusicoccin-stimulated hydrogen-ion efflux in barley roots. *Planta* 126:61–73
61. Pitman, M. G., Schaefer, N., Wildes, R. A. 1975. Stimulation of H^+ efflux and cation uptake by fusicoccin in barley roots. *Plant Sci. Lett.* 4:323–29

62. Poole, R. J. 1976. Transport in cells of storage tissues. See Ref. 23, pp. 229–48
63. Polevoy, V. V., Salamatova, T. S. 1977. Auxin, proton pump and cell trophics. See Ref. 3, pp. 209–16
64. Raschke, K. 1977. The stomatal turgor mechanism and its responses to CO_2 and abscisic acid: observations and a hypothesis. See Ref. 3, pp. 173–83
65. Rasi Caldogno, F., Lado, P., Colombo, R., Cerana, R., Pugliarello, M. C. 1978. *Changes in cellular metabolism associated with the activation of H^+/K^+ exchange by auxin or fusicoccin in pea internode segments.* Presented at Fed. Eur. Soc. Plant Physiol. meet., Edinburgh
66. Ratner, A., Jacoby, B. 1976. Effect of K^+, its counter anion, and pH on sodium efflux from barley root tips. *J. Exp. Bot.* 27(100):843–52
67. Raven, J. A., Smith, F. A. 1974. Significance of hydrogen ion transport in plant cells. *Can. J. Bot.* 52:1035–48
68. Rayle, D. L., Cleland, R. 1977. *Curr. Top. Dev. Biol.* 11:187–211
69. Rollo, F., Nielsen, E., Cella, R. 1977. Cell division and ion transport as tests for the discrimination between the action of 2,4-D and fusicoccin. See Ref. 3, pp. 261–66
69a. Rollo, F., Nielsen, E., Sala, F., Cella, R. 1977. Effect of fusicoccin on plant cell cultures and protoplasts. *Planta* 135:290–201
70. Rubinstein, B. 1974. Effect of pH and auxin and chloride uptake into *Avena* coleoptile cells. *Plant Physiol.* 54: 835–39
71. Rubinstein, B., Tattar, T. A. 1978. *Control of amino acid uptake into oat leaves and protoplasts.* Presented at 1978 ASPP meet., Blackburg, Virginia
72. Sassa, T., Togashi, M., Kitaguchi, T. 1975. The structures of cotylenins A, B, C, D and E. *Agric. Biol. Chem.* 39:1735–44
73. Slayman, C. L. 1974. Proton pumping and generalized energetics of transport: a review. See Ref. 31, pp. 107–19
74. Squire, G. R., Mansfield, T. A. 1972. Studies of the mechanism of action of fusicoccin, the fungal toxin that induces wilting, and its interaction with abscisic acid. *Planta* 105:71–78
74a. St. G. Ord, G. N., Cameron, I. F., Fensom, D. F. 1977. The effect of pH and ABA on the hydraulic conductivity of *Nitella* membranes. *Can. J. Bot.* 55:1–4
75. Strobel, G. A. 1974. Phytotoxins produced by plant parasites. *Ann. Rev. Plant Physiol.* 25:541–66
76. Stuart, D. A., Jones, R. L. 1978. Role of cation and anion uptake in salt-stimulated elongation of lettuce hypocotylsections.*PlantPhysiol.* 61:180–83
77. Sturani, E., Cocucci, S., Marrè, E. 1968. Hydration dependent polysome-monosome interconversion in the germinating castor bean endosperm. *Plant Cell Physiol.* 9:783–95
78. Tipton, C. L., Mondal, H. M., Benson, M. J. 1975. K^+-stimulated adenosine triphosphatase of maize roots: partial purification and inhibition by *Helmintosporium maydis* race T toxin. *Physiol. Plant Pathol.* 7:277–86
79. Turner, N. C., Graniti, A. 1969. Fusicoccin. A fungal toxin that opens stomata. *Nature* 223:1070–71
80. Yamagata, Y., Masuda, Y. 1975. Comparative studies on auxin and fusicoccin actions on plant growth. *Plant Cell Physiol.* 16:41–52

Ann. Rev. Plant Physiol. 1979. 30:289–311

INTRACELLULAR pH AND ITS REGULATION

❖7673

F. Andrew Smith

Department of Botany, University of Adelaide, Adelaide, S.A. 5001, Australia

John A. Raven

Department of Biological Sciences, University of Dundee, Dundee DD1 4HN, Scotland, U.K.

CONTENTS

INTRODUCTION

In this review we discuss mainly the factors which determine the pH of the cytoplasm and vacuole of plant cells. Differences in pH between membrane-

0066-4294/79/0601-0289$01.00

bound cytoplasmic compartments, such as mitochondria or chloroplasts, are considered only briefly. Details of mechanisms of energy-transduction involving H^+ [chemiosmotic coupling (70)] are excluded. We have already written extensively on this subject (100, 103–105, 107, 108, 121–123), but aim to avoid undue repetition. Many aspects of H^+ transport have been reviewed recently by others (22, 25, 42, 65, 76, 92, 93, 111). In addition, chapters by Davies and Marrè in the present volume contain material relevant to our theme.

The case for intracellular pH regulation can be summarized briefly as follows. The absorption and metabolism of nutrients by plants involve production or consumption of H^+ both within cells and in their bathing media. H^+ is also transferred between intracellular compartments and between the cell and its surroundings. There is experimental evidence that in spite of these processes (or, in some cases, because of them) the pH within cells is regulated within narrow limits. Of course, such regulation has long been inferred from the effects of pH changes on protein structure and hence on enzyme activity. Apart from irreversible denaturation of protein, reversible pH effects are caused by changes in ionization of any dissociable groups in an enzyme (and its substrate) that are involved in the turnover of the enzyme-substrate complex. In general, where a metabolic reaction involves H^+ as a reactant, there is pH-dependence of the apparent equilibrium constant and related parameters, such as the free energy change and (where appropriate) the redox potential. Two examples of biological importance are the hydrolysis of ATP and the oxidation of NADH or NADPH.

It must be emphasized that concentrations of H^+ cannot change without maintenance of electroneutrality. Within the cell there are changes in ionization of acids or bases, and during membrane transport of H^+ charge balance is maintained by other ion transport processes. In the present context the charge imbalance associated with membrane electric potential differences (PDs) is quantitatively insignificant (144). In the literature there is often no distinction between ionized and un-ionized forms of acids or bases, a practice which can confuse the whole issue of charge-balance and pH regulation. An example is the interchangeable use of "malic acid" and "malate" without consideration of cations such as H^+ or K^+.

The balance between biochemical (intracellular) and biophysical (membrane transport) events involved in pH regulation varies. Sometimes there are spatial constraints on external disposal of H^+ or OH^-, as in the aerial shoots or bulky underground organs of terrestrial plants. In these cases, control of pH primarily involves intracellular events, but these depend on regulated nutrient input from the external medium (106).

Finally, as well as metabolic control of pH, there is increasing interest in metabolic control by pH (15). We discuss the evidence that many developmental and related effects are a result of changes in rates of H^+ transport, in turn involving changes in cytoplasmic pH.

MEASUREMENTS OF INTRACELLULAR pH

Intracellular pH in plants has traditionally been measured with indicator dyes absorbed by the cells, or estimated from cell sap extracts (12, 39, 119, 135, 136). These values usually fall below about 5.5 and undoubtedly reflect the vacuolar pH, which is itself determined mainly by the presence of organic acids and their salts. Very low pH values are usually caused by the presence of strong (highly dissociated) acids such as oxalic acid (39, 135) and even sulfuric acid (20). Usually there is a mixture of various organic acids and their salts, and there is no simple relation between pH and titratable acidity nor (especially) between pH and "total acidity." This last quantity includes the undissociated acids and their anions. Intracellular pH responds, but not greatly, to changes in soil conditions, patterns of plant nutrition, and short-term solute transport (3, 29, 39, 48). Changes in pH are also associated with stomatal activity (69) and Crassulacean acid metabolism (CAM) (79, 95). However, although these pH measurements may be correlated with metabolism at the whole cell level, they do not indicate the state of the cytoplasm. The presence of ionizable macromolecules (giving rise to Donnan effects) and of membrane-bound organelles makes it an oversimplification to regard the cytoplasm as a uniform phase. Nevertheless, experiments with a variety of animal cells (10, 12, 143) established that it is valid to consider a "bulk" cytoplasmic pH, usually at about 7.0–7.5.

Microelectrodes sensitive to pH have been widely used for measurement of cytoplasmic pH in animals (10, 143), but until recently they have only been used for vacuolar pH studies in plants. Comparison of conventional pH measurements (using isolated sap) and measurements with microelectrodes is possible with giant algal cells such as charophytes. Similar vacuolar pH values of 5.0–5.5 have been obtained with both methods (35, 36, 56, 109, 134, 142). Roa & Pickard (114) have reported vacuolar pH from 2.0–4.0 in *Chara braunii,* using antimony-covered microelectrodes, but these measurements were made with branchlet segments rather than the axial internodal cells normally used. Bowling (8) measured the profile of vacuolar pH across the intact sunflower root, again using antimony microelectrodes. The pH rose from less than 6.0 (epidermis) to nearly 7.0 in the protoxylem. Vacuolar pH changes in the stomatal complex of *Commelina* have also been measured in this way (83); this work is discussed below.

In a pioneering study, Davis (17; also unpublished results) has used these microelectrodes to compare the pH of the chloroplast, cytoplasm, and vacuole of the hornwort *Phaeoceros.* He found that light reduced the chloroplast pH (exact phase undefined) from 6.2 to 4.6. There were only transient changes (increase then decrease) in the cytoplasm, which had a steady value of 6.6–6.8 (see Figure 1). Light decreased the vacuolar pH from about 6.6 to about 5.8. Simultaneous measurements of membrane PD were made, and light-dark responses were again obtained. Spanswick & Miller (128), using *Nitella,* have found that the antimony microelectrodes give lower cytoplasmic values (pH 6.7) than glass microelectrodes (pH 7.5), as shown in Figure 1. Similar discrepancies have been reported by animal physiologists (23), and it is clear that intracellular pH values obtained with antimony microelectrodes must be treated with caution. Spanswick & Miller (128) found that glass microelectrodes gave cytoplasmic pH values similar to those estimated from the distribution of the weak acid 5,5-dimethyloxazolidine-2,4-dione (DMO). Use of DMO to measure intracellular pH was again developed with animal cells (143). A detailed comparison of microelec-

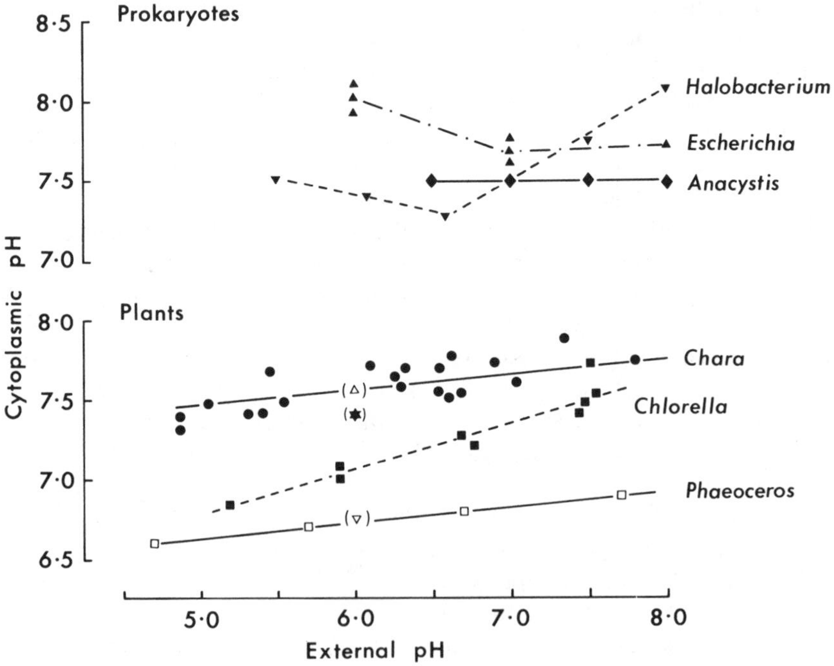

Figure 1 Cytoplasmic pH as a function of external pH. Symbols in parentheses are for *Nitella;* other organisms are identified on the figure. Data are from (2, 17, 21, 51, 80, 124, 128). Values for *Nitella* are a comparison of measurements with antimony microelectrode (▽), glass microelectrode (△) and DMO (✱). Values for *Phaeoceros* obtained with antimony microelectrodes. Other values obtained with DMO or (*Escherichia* at pH 8.0) methylamine.

trode, DMO, and also methylamine methods for measuring intracellular pH in giant barnacle muscle fibers and squid giant axons has recently been published (7). DMO has been used with bacteria (2, 80, 113), and good agreement has been obtained with the ^{31}P nuclear magnetic resonance method (cf 72, 80). The pH of isolated chloroplasts and mitochondria has also been measured with DMO (e.g. 28, 71).

DMO has been used recently to study effects of external pH on the cytoplasmic pH of *Chlorella* (51), *Chara* (109, 124, 145), and the blue-green alga (cyanobacterium) *Anacystis* (21). The results are summarized in Figure 1, along with comparable results from nonphotosynthetic bacteria (2, 80). In all cases shown, cytoplasmic pH is relatively insensitive to external pH changes, and the data provide convincing evidence for cytoplasmic pH regulation within quite narrow limits. The results with *Chara* [like those with *Nitella* (128)] were obtained after separation of vacuolar sap samples, thus allowing separate calculations for vacuolar and cytoplasmic DMO concentrations and hence pH. The DMO method rests on the assumption that only the un-ionized acid, and not the anion, permeates the plasma membrane (143). In *Chara,* at least, this assumption may not always be valid, but this factor can be taken into account when calculating pH (109). Metabolism of DMO may occur, as in cultured *Acer pseudoplatanus* cells (52), though probably not to a large extent in charophyte cells. Using DMO to estimate total intracellular pH (about 6.6) and indicator dyes for vacuolar pH (about 3.8), Kurkdjian & Guern (52) calculated that cytoplasmic pH of *Acer* cells is about 7.6.

Studies of metabolic effects on cytoplasmic pH are still in their infancy. Preliminary results with DMO suggested that light causes an increase of at most 0.2 pH units in *Chara* (145). Spanswick & Miller (128) found that azide apparently reduced cytoplasmic pH to 5.3, as measured with microelectrodes, whereas DMO gave an apparent reduction to 6.5. They suggested that the difference might reflect the maintenance of a relatively alkaline cytoplasmic compartment such as the chloroplast stroma. Cytoplasmic pH in *Chara* increases slightly with decreasing temperature over the range 25–5°C (109). It appears that the cells do not maintain constant pH, constant $[H^+]/[OH^-]$, nor constant fractional ionization of histidine in their proteins. The pH changes with temperature are closer to those needed to maintain constant fractional ionization of phosphorylated compounds or CO_2/HCO_3^-.

Finally, measurements have been made of the pH of phloem exudates, the composition of which appears to be "cytoplasmic" in nature. The pH ranges from about 7.0 to 8.5, with values in most species at about 8.0 (81, 99, 140, 150).

MECHANISMS FOR pH REGULATION

Intracellular Buffers

It is unsatisfactory to argue that preexisting buffers can effectively counter intracellular production or consumption of H^+, as the buffers are themselves produced by reactions which produce or consume H^+. In simple terms, considering a buffer as a mixture of an un-ionized organic compound and its (ionized) salt, synthesis of the compound may involve production or consumption of H^+. An example would be an amino acid starting from NH_4^+–N or NO_3^-–N (106). Furthermore, presence of the salt implies preexchange of H^+ or OH^- with inorganic counterions. There are few natural buffers operative around pH 6.0–8.0, the main examples being bicarbonate (pK_a = about 6.38), phosphate compounds (pK_a = 6.8–7.0), and some amino acids. The imidazolyl group of histidine has a pK_a of 6.0, the –SH group of cysteine a pK_a of 8.3, and an $-NH_2$ group of cystine a pK_a of 8.0. We have calculated that the cytoplasmic buffer capacity in the range pH 6.0–8.0 is at most 20 mmol H^+ $liter^{-1}$ $(pH\ unit)^{-1}$, and suggest that this is effective in countering pH changes of 0.2–0.3 units for only a few minutes (105).

In the vacuole, the buffering capacity is predetermined by events in the cytoplasm (organic acid synthesis) together with solute transport across the plasmalemma and tonoplast. As with the cytoplasm, vacuolar buffering is an effect rather than a cause of pH regulation during cell growth. During the life of a cell there may, however, be redistribution of H^+ between cytoplasm and vacuole; obvious short-term examples are stomatal guard cell metabolism and CAM.

The Biochemical pH-stat

Davies (15, 16) has proposed that cytoplasmic pH may be controlled by regulation of the number of carboxyl groups (see also 116). Davies' scheme is summarized below:

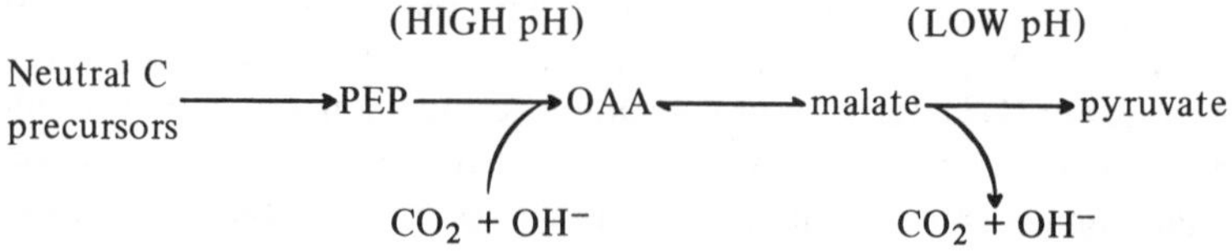

According to this scheme, excess OH^- production (or H^+ consumption) increases cytoplasmic pH sufficiently to increase the activity of phosphoenolpyruvate (PEP) carboxylase. This leads to formation and dissociation of a strong acid (malic acid). Conversely, H^+ production (OH^- consumption) lowers cytoplasmic pH, inhibiting malate formation, and activating malic enzyme, so that pyruvate, CO_2, and OH^- are formed. There is evidence that the pH responses of the enzymes in vitro are such that cytoplasmic pH could be regulated at or slightly above 7.0 (6, 15, 16, 34; see also 76). Considered in isolation from membrane transport, the biochemical pH-stat appears to be a "fine-tuning" device, coping with short-term fluctuations in cytoplasmic H^+. Over the long term (e.g. the cell's life span), H^+ production cannot be countered by decarboxylation, because the malate^{2-} (or equivalent) needed to neutralize the H^+ can only arise from previous synthesis of malic acid (or equivalent), accompanied by H^+-cation exchange. In contrast, carboxylation is of widespread importance in countering OH^- production by cells, as discussed below. In these cases, however, it is necessary to consider the role of membrane transport of ions, including H^+, as part of a combined biochemical and biophysical pH-stat.

H^+ Transport in Plants

Transport of H^+ or OH^- between the cell and its bathing medium occurs in plants, animals, and microorganisms. Figure 2 is an attempt to summarize some of the most important processes (in plants) which involve H^+ fluxes, including the carboxylation reactions discussed above. We have suggested previously that H^+ transport evolved as a mechanism for pH regulation, and have reviewed the nature of the H^+ and OH^- fluxes in plant cells. Here we summarize only the main points and review recent developments which relate to the control of fluxes of H^+ and other ions.

ACTIVE H^+ TRANSPORT Measured or extrapolated values of cytoplasmic pH and electric PDs across the plasmalemma, when combined in the Nernst equation, indicate that there are inward passive driving forces on H^+ at the plasmalemma when the external pH is below 9.0–10.0. In other words, H^+ efflux is then active (energy-dependent), while H^+ influx is passive. Fluxes of OH^- can be considered similarly, with active OH^- influx replacing active H^+ efflux. Operationally, the two are indistinguishable, and we shall generally follow the convention of considering H^+ rather than OH^-, if only because it is simpler to think of transport mechanisms in terms of protonation reactions. These considerations show that a H^+ pump drives the net H^+ efflux which occurs under many nutritional and metabolic conditions (see Figure 2).

Active H^+ transport *inward* (OH^- transport *outward*) is almost certainly involved in pH regulation in cells growing in very alkaline solutions (e.g. pH greater than 10.0). Thus the ability or otherwise to transport H^+ inward (OH^- outward) may determine the upper external pH limit tolerated by plants, just as the ability of the H^+ extrusion mechanism to cope with increasing external acidity may determine the lower pH limit.

The existence of large pH differences across the tonoplast suggests H^+ pumps at this membrane. The electric PD across the tonoplast has been measured in giant algal cells (98). The vacuole is positively charged with respect to the cytoplasm, and the PDs range from a few mV up to over 40 mV. In general, we may infer active H^+ transport from cytoplasm to vacuole.

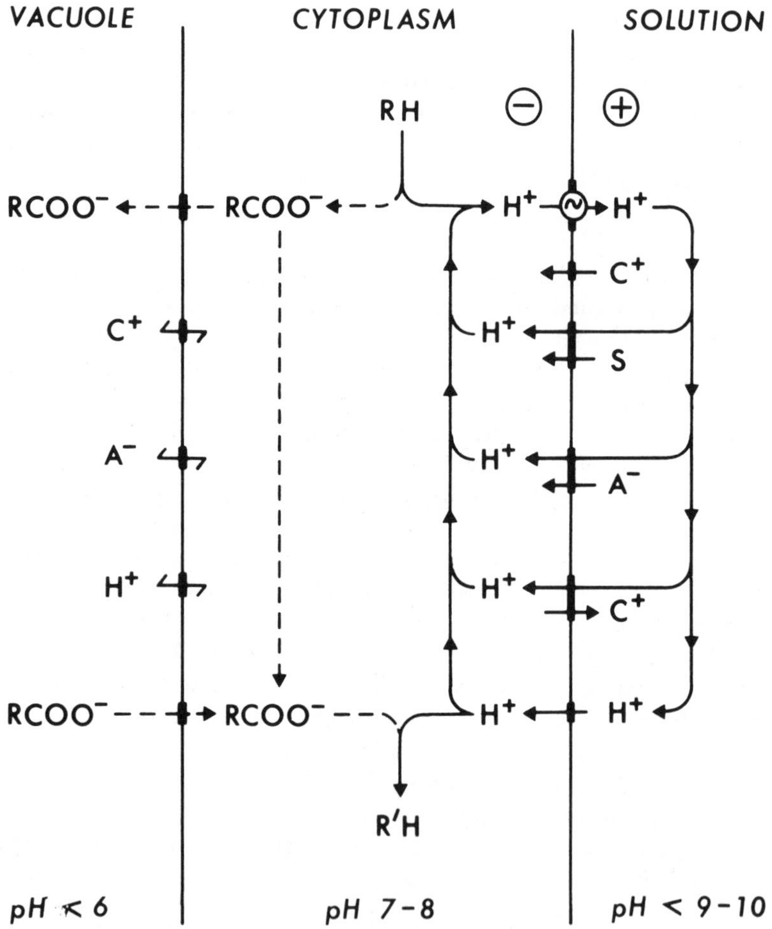

Figure 2 Composite diagram of H^+ transport between the plant cell and its bathing medium. H^+ efflux is active and H^+ influx passive when the external pH is less than 9–10 (see text). H^+ influx may be coupled to influx of uncharged solutes (S) or inorganic anions (A^-), or efflux of inorganic cations (C^+). The stoichiometry of transport is not specified. Intracellular production or consumption of H^+ from carboxylation or decarboxylation is also indicated. The nature of the transport mechanisms at the tonoplast is not specified.

H^+ is also transported across the inner (crista) membrane of the mitochondrion, and, in photosynthetic cells, the inner (envelope) membrane of the chloroplast. In both cases there is active transport of H^+ into the intermembrane phase, i.e. from the relatively alkaline mitochondrial matrix and the stroma respectively. It is an article of biochemical faith that the outer membrane of these organelles is freely permeable to small molecules and ions in vivo as in vitro (53, 68, 149). Nevertheless, the extent to which the "external" cytoplasmic pH is affected by H^+ transport in the organelles is debatable. Certainly, in many vacuolate cells, the mitochondria and chloroplasts may occupy a large fraction of the total cytoplasm. Anderson et al (1) have suggested that in carrot root cells, H^+ extruded from mitochondria might have no time to mix with H^+ in the cytoplasm before being effluxed across the plasmalemma. They consider that the local cytoplasmic pH might be low enough (less than 2.0) for the efflux across the plasmalemma to be mechanistically passive, though still dependent on respiration. Proposals (4, 64) that pH increases outside photosynthesizing cells might be a direct reflection of pH shifts across the chloroplast envelope appear unlikely (73; see also below).

H^+ TRANSPORT AND CELL ELECTROPOTENTIALS Evidence that charge-carrying (electrogenic) ion pumps contribute to membrane PDs has been reviewed elsewhere (25, 32, 93, 118, 127). The PD across the plasmalemma is now generally considered as the sum of a diffusion potential (E_D) and a potential due to the pump (E_P):

$$\text{Membrane PD} = E_D + E_P \qquad \ldots\ldots \quad 1.$$

According to the simplest models, E_D is determined by the membrane permeability to the various ions and their concentrations as in a conventional Goldman-Hodgkin-Katz voltage equation (see 1, 32, 49, 126). E_P is determined by the electrogenic flux (J) and the passive membrane conductance (g_m):

$$E_P = JF/g_m \qquad \ldots\ldots \quad 2.$$

where F is the Faraday constant. Cell membrane PDs can exceed the expected E_D value, and large and reversible decreases in PD are produced when energy metabolism is prevented (1, 32, 33). Such inhibition reduces J to zero, and thus PD = E_D (Equation 1).

The H^+ pump is most favored as the major electrogenic pump in plant cells, except for some marine algae in which the Cl^- pump appears to have this role (98). The evidence for the electrogenic H^+ pump is mostly indirect: it is usually not possible to compare membrane PD with the rate of H^+ pumping, which is not measurable due to the lack of a tracer for H^+.

Changes in membrane PD (hyperpolarization) are, however, associated with *net* H^+ efflux in some plant cells in which charge-balance is maintained by carboxylation and cation influx (66, 92, 118). Low membrane conductance (g_m) accounts for the hyperpolarization beyond E_D. If g_m were high enough, cation influx might equal H^+ efflux when the PD is not far from E_D, in effect causing an electrical short-circuit of the H^+ pump. Thus absence of hyperpolarization is not conclusive evidence against the presence of an electrogenic H^+ pump.

H^+ REENTRY There are various ways in which H^+ pumped out of the cell may be recycled, and the most important are shown in Figure 2. The simplest involve diffusion or uniport, which can for simplicity be regarded as carrier-mediated diffusion. These processes will tend to short-circuit the H^+ free-energy gradient ($\Delta\mu_{H^+}$) across the plasmalemma; membrane PD and cytoplasmic pH will tend to decrease. Decreasing membrane PD with decreasing external pH has been used as evidence for high membrane permeability to H^+ (49), implying an energetically wasteful "pump-and-leak" of H^+. An alternative scheme by Spanswick (125) involves low membrane permeability to all ions, including H^+. He suggests that the rate of H^+ pumping, and hence E_P, responds to external and internal pH changes. Metabolic effects on membrane PD, such as light-dark changes, are also explained in terms of changes in H^+ pumping.

H^+ may also move into the cell as part of a carrier complex involving other solutes. Such H^+-linked transport can be either electrogenic or electroneutral. Examples of the former include H^+ symport with sugars (51) and amino acids (74). Influx of anions can also be coupled by electrogenic symport, and efflux of cations by electrogenic antiport, provided that the stoichiometry is such that net positive charge enters the cell (124). Both the membrane PD and the transmembrane pH gradient (ΔpH) act as an energy source for electrogenic symport or antiport, and this can result in secondary active transport of solutes. As with H^+ uniport, both components of $\Delta\mu_{H^+}$ are short-circuited.

Electroneutral H^+ symport or antiport involves one-for-one stoichiometry of anions or cations with H^+, and in this case only ΔpH is the energy source for secondary active transport. Accompanied by cation influx in response to the membrane PD, the result of H^+-anion symport is accumulation of salts such as KCl or KNO_3. This mechanism has been proposed for several types of plant cells, though with emphasis on OH^--anion antiport rather than H^+-anion symport (37, 120). Effects on membrane PD will be complex because of modifications resulting from cation influx and restrictions on H^+ extrusion due to H^+ recycling. Such schemes do not depend on net H^+ efflux. Indeed, H^+ extrusion may still occur when there is net H^+ influx, as associated with NO_3^- influx and assimilation. In this case, a combination of electrogenic H^+ efflux plus H^+–NO_3^- symport (or OH^-–NO_3^- antiport) can result in hyperpolarization beyond E_D (74, 102).

The situation is further complicated by short-term metabolic effects. For example, during KNO_3 uptake by cereal roots, external pH decreases then increases (76). This suggests that malate accumulation and net H^+ efflux precede NO_3^- reduction and net H^+ influx. Because of the variety of H^+-linked transport processes, it is not surprising that characterization of the H^+ pump from its physiological manifestations (external pH changes, effects on membrane PD) is so difficult.

Control of H^+ Transport: The Biophysical pH-stat

Evidence that the energy source for H^+ transport at the plasmalemma is ATP has been obtained with *Neurospora* (118) and *Chlorella* (133). The existence of an H^+-ATPase is widely assumed, and there is some interest in trying to calculate the number of H^+ transported per ATP hydrolyzed (93). Cytoplasmic pH and membrane PD values have been used to calculate $\Delta\mu_{H^+}$ in *Chara, Phaeoceros,* and *Neurospora* (147); in each case the maximum value is about 27 kJ $(mol)^{-1}$ which is about half of the expected ΔG_{ATP}. Accordingly, a $2H^+$-ATPase working near equilibrium is energetically adequate. However, combination of measured membrane PDs with assumed cytoplasmic pH of 7.0 suggests that, when external pH is low, $\Delta\mu_{H^+}$ values of 30 to 40 kJ mol^{-1} occur in illuminated cells of *Elodea* (126) and *Vallisneria* (5). These values can best explained in terms of a 1 H^+-ATPase. In *Chlorella,* although $\Delta\mu_{H^+}$ is probably less than 25 kJ mol^{-1}, calculations of ATP consumption suggest a 1 H^+-ATPase (133). If a 1 H^+-ATPase is the general case, then control of $\Delta\mu_{H^+}$ at values less than the theoretical maximum (about 55 kJ mol^{-1}) must be exerted kinetically, i.e. by factors other than available energy. This is contrary to Spanswick's model for H^+ transport (125), and affects the restrictions on regulation of H^+ transport and hence cytoplasmic pH (147). For example, if an H^+ pump is running at thermodynamic equilibrium and with constant energy supply (e.g. ATP), then an increase in ΔpH can be achieved only at the expense of a decrease in membrane PD, and vice versa. This is what happens in *Chara* over an external pH range 5.0–7.0, resulting in almost constant cytoplasmic pH (124, 145). A change in total $\Delta\mu_{H^+}$ could be achieved only by a change in energy supply. In contrast, kinetic control allows a change in $\Delta\mu_{H^+}$ independently of changes in energy supply. This allows direct effects of metabolic "triggers," such as phytohormones, on $\Delta\mu_{H^+}$ and the rate of H^+ pumping. Kinetic control also explains lack of sensitivity of membrane PD to external pH changes, as observed in some plant cells (e.g. 5, 43). Our work with *Chara* suggests that at low temperature (5°C), H^+ transport is under kinetic control, as $\Delta\mu_{H^+}$ is reduced below $\frac{1}{2}\Delta G_{ATP}$ (109).

So far, there have been few explicit attempts to integrate the biophysical aspects of H^+ transport in plant cells with the need for regulation of cytoplasmic pH. If H^+ fluxes are to act as a biophysical pH-stat, then cytoplasmic pH must control both the primary H^+ extrusion and also the H^+ influx processes shown in Figure 2. In addition, there must be a mechanism to adjust membrane PD in relation to cytoplasmic pH. The reason is that transport of H^+ will result in a change in PD that depends on membrane capacitance and in a change in pH that depends on cytoplasmic buffer capacity (105, 147). The electric effect will be much greater and may "stall" the H^+ pump unless there is a mechanism to reduce it. The control of influx of cations (e.g. K^+ uniport) is likely to be significant here, but other ion fluxes, whether H^+-linked or otherwise, should also be taken into account where appropriate. This point has been considered in a wider context by Skulachev (117). The possibility that K^+ uniport is the "primary" flux in the pH-stat, and H^+ extrusion a secondary response, seems discounted by the rapid hyperpolarizations associated with changes in rates of K^+–H^+ exchange (66; see also below).

Bicarbonate Uptake

As an alternative to H^+ extrusion, influx of HCO_3^- might be involved in pH regulation, possibly acting directly as substrate for PEP carboxylase (38, 40, 91). HCO_3^- is certainly transported into some aquatic plants (97), in this case providing CO_2 for photosynthesis. There is an equivalent net efflux of OH^- (or influx of H^+) and net cation influx and carboxylation are not involved. In charophytes such as *Chara corallina,* HCO_3^- influx is spatially separated from OH^- efflux (H^+ influx), resulting in zones of low and high pH along the cell surface (57–59). Localized electric PDs also develop, i.e. HCO_3^- uptake is electrogenic (146). It is possible, however, that HCO_3^- uptake (in the acid zones) involves H^+-HCO_3^- symport (or OH^--HCO_3^- antiport) and that electrogenic H^+ extrusion continues, a situation analogous to that considered for net NO_3^-–OH^- exchange (102). It is thus premature to assume that HCO_3^- uptake represents a primary electrogenic process different from the orthodox H^+ pump, and that during HCO_3^- assimilation the H^+ pump is switched off.

METABOLIC IMPLICATIONS OF pH REGULATION

In this section we review briefly recent advances in the study of plant physiological processes which involve H^+ production and hence pH regulation. These processes are often considered quite separately, but have in common various aspects of the biochemical and biophysical pH-stats, as summarized in Figure 2.

Salt Accumulation and Carboxylation in Roots and Storage Tissues

It has been known for many years that these tissues can be major sites for H^+ transport and organic acid synthesis, yet until recently there has been little conclusive evidence to separate causes from effects (85). Hiatt (29–31) suggested that intracellular pH might control carboxylation via effects on enzyme activity. This was in effect an early version of the Davies pH-stat, except that the argument was based on pH measurements which must reflect vacuolar pH changes. This was pointed out by Jacoby & Laties (40), who asserted (without evidence) that cytoplasmic pH must remain constant, if not fall, during carboxylation. They proposed that cytoplasmic HCO_3^-, whether created by K^+–H^+ exchange or absorbed from the bathing solution, is the "prime mover" of organic acid synthesis. It was considered that cytoplasmic HCO_3^- would be in the range where it would limit PEP carboxylase activity, but this is questionable given the very high affinity of this enzyme for HCO_3^- (see 76). Jacoby & Laties pointed out that K^+–H^+ exchange can also be considered a "prime mover" in organic acid synthesis and maintenance of pH, and stressed the necessity for vacuolar transport of K^+ and malate for carboxylation to continue. Osmond (76) questions the validity of the Davies pH-stat, arguing that pH optima measured in vitro may be misleading. He does, however, emphasize the importance of K^+-H^+ exchange to maintain cytoplasmic pH. He also stresses the need for vacuolar transport of malate to avoid feedback inhibition of PEP carboxylase, and discusses the possibility that inorganic anions such as Cl^- might inhibit PEP carboxylase and malate dehydrogenase. This would account for accumulation of Cl^- rather than malate (or other carboxylates) by cells placed in solutions containing Cl^-. The simplest explanation in terms of the pH-stat would be that H^+-Cl^- symport lowers cytoplasmic pH below the level required for net synthesis of carboxylates.

Emphasis on electrogenic H^+ extrusion as the primary process triggering carboxylation in roots has increased following the discovery that the phytotoxin fusicoccin (FC) hyperpolarizes barley roots and greatly increases efflux of H^+ and influx of K^+ and Cl^- (86, 88, 89). FC affects a number of transport-dependent processes such as cell enlargement and stomatal opening (see below; also Marrè's chapter in this volume). The results are consistent with an alteration in kinetic control of electrogenic H^+ extrusion, resulting in both hyperpolarization and effects on malate synthesis and ion fluxes mediated by increased cytoplasmic pH.

Stomatal Turgor Mechanisms

There is now convincing evidence that decreasing osmotic potential in guard cells during stomatal opening results from K^+ influx balanced by malate (and other carboxylates), Cl^-, or a mixture of the two. Raschke (96) has argued persuasively for H^+ extrusion (or K^+–H^+ exchange) as a primary mechanism in controlling guard cell turgor, a conclusion supported by the dramatic effects of FC on H^+ efflux and K^+ influx (see 66, 96 and references therein; see also Marrè's chapter in this volume). Raschke incorporates the Davies pH-stat in his model, and attempts to explain the apparent contradiction that whereas CO_2 (strictly HCO_3^-) is required to make malate, CO_2 also causes stomatal closure. He suggests that where CO_2 is low, H^+ extrusion, K^+ influx, and vacuolar malate accumulation keep pace with formation of free malic acid in the cytoplasm. As CO_2 increases, malic acid in the cytoplasm also increases, lowering pH. In order to explain stomatal closure it is necessary to postulate that high cytoplasmic malate (or malic acid) and low pH cause leakage of K^+ and, where appropriate, Cl^-, and also loss of malate. Malate might leak from the guard cells, or it might be metabolized either by respiratory oxidation or by gluconeogenesis. The latter processes would involve net H^+ influx to the guard cells to balance K^+ efflux. The reason why low cytoplasmic pH reverses ion accumulation (as opposed to preventing further accumulation) is not specified. Raschke accounts for stomatal responses to abscisic acid (ABA) in a manner similar to that proposed for high CO_2. This reliance on the pH-stat mechanism includes recognition of the possibility of direct effects on turgor-sensing devices in cell membranes (130). Levitt (54, 55) has developed a more complex model which incorporates interactions between chloroplast (stroma) pH and cytoplasmic pH. He speculates that disappearance of malate in the dark involves the Calvin cycle, driven by NADPH synthesized during malate decarboxylation and by ATP produced by reversal of the K^+–H^+ ATPase. Bowling's "malate-switch" hypothesis (9) also involves intracellular pH changes, in this case to control movement of malate (plus K^+) between guard cells and the other cells in the stomatal complex. Bowling has used pH and K^+ measurements (82, 83) to calculate amounts of $malate^-$ and $malate^{2-}$. He suggests that pH differences could result in net movements (by diffusion or active transport) of $malate^-$, followed by reequilibration to form $malate^{2-}$ which would be "locked" in the cell. In its present form, this model does not take into account the need for maintenance of cytoplasmic pH at values quite different from the measured vacuolar pH. Travis & Mansfield (137) have shown that guard cells in epidermal strips can still synthesize malate and open even when the surrounding cells are destroyed, contrary to Bowling's model. They raise the old problem of accounting for stomatal opening in the absence of external CO_2, and on these grounds argue against the involvement of PEP carboxylase in malate synthesis. One important factor overlooked often in this argument is that the substrate for PEP carboxylase is not CO_2 but HCO_3^- (13, 14, 112). To what extent cytoplasmic HCO_3^- is removed under external "CO_2-free" conditions is not clear. It is relevant here that Splittstoesser & Beevers (129) showed that malate was still synthesized when beet disks were incubated in CO_2-free solutions at pH 7.4, and attributed this to refixation of cytoplasmic CO_2 (strictly HCO_3^-). The problem of "CO_2-free" opening is further discussed by Raschke (96).

C_4 Photosynthesis

The orthodox scheme involves spatial separation of the carboxylation and decarboxylation reactions, in mesophyll and bundle sheath cells respectively (27). The major C_4 products (malate or aspartate) then diffuse through the symplast to the bundle sheath with C_3 compounds (pyruvate or alanine) returning to the mesophyll. Membrane transport is usually discounted (78; see also 45). Overall, there is said to be an indirect CO_2 pump, increasing the activity of ribulose-1,5-bisphosphate (RuBP) carboxylase in the bundle sheath. In fact, the sequence as thus outlined is an indirect HCO_3^- pump if allowance is made for dissociation of the C_4 and C_3 compounds. As noted previously (101), H^+ must also move from mesophyll to bundle sheath (or OH^- vice versa) if large pH differences are not to develop. One possible mechanism is the triose phosphate/phosphoglycerate shuttle, usually considered as a means of transferring reductant. Electrogenic membrane transport of H^+ between mesophyll and bundle sheath is also possible in those C_4 plants without a suberized layer separating the two tissues, but quantitative aspects have not been considered.

Crassulacean Acid Metabolism (CAM)

The current interest in the biochemistry and regulation of CAM includes studies of vacuolar transport and of feedback inhibition of PEP carboxylase by malate (50, 63, 77). Carboxylation during CAM, unlike the processes discussed so far, is accompanied by reversible vacuolar disposal of H^+; it is independent of large-scale K^+–H^+ exchange across the plasmalemma. As pointed out by Lüttge & Ball (60), it is not possible to determine experimentally whether undissociated malic acid or malate (as malate$^-$ or malate^{2-}) is transported. Active transport of H^+ to the vacuole appears necessary to maintain cytoplasmic pH in the face of low vacuolar pH (e.g. 4.0) generated during CAM. Why cells in leaf slices of *Bryophyllum* can be induced to lose

H^+ and malate (malic acid?) to the bathing solution (61, 62) is not resolved; it may be a response to abnormal cytoplasmic acidification following treatments which rapidly alter water potential. PEP carboxylases isolated from some CAM plants have abnormally low pH optima: about 6.0–6.5 (50). The suggestion that this is an adaptation to low pH presumably refers to cytoplasmic pH, but is without experimental support. If cytoplasmic pH lies in the usual range of 7.0–8.0, then increases will *decrease* carboxylation and vice versa: the reverse of the situations described so far. There is a further complication in that the CAM plant *Sedum praealtum* has a "normal" PEP carboxylase, with optimum pH at about 8.5 (50). It is therefore not surprising that pH effects are not usually included among the spectrum of factors thought to control the carboxylation-decarboxylation sequence in CAM (77).

Nitrogen Assimilation

We have reviewed previously (106) the consequences (with respect to pH) of uptake and assimilation of NH_4^+ and NO_3^-. Assimilation of NH_4^+ results in H^+ production, which cannot be neutralized internally, and is excreted. In land plants, NH_4^+ assimilation takes place in the roots, and products transported to the shoot are such that no further large-scale H^+ production occurs. In contrast, NO_3^- assimilation involves production of OH^-, which may be countered in roots by net efflux (or H^+ influx). Alternatively, neutralization can result from carboxylation (19, 46, 48). Such carboxylation is the means of preventing pH imbalance during NO_3^- reduction in the shoot, though the nature of the products varies. For example, the Chenopodiaceae accumulate massive amounts of oxalate, and Na^+ is often the predominant univalent cation (75). As with other plants, the carboxylate content of CAM plants is higher when they are grown on NO_3^- than when they are grown on NH_4^+ (94). This shows that the normal cation-carboxylate balance underlies the diurnal changes in acidity (77).

Biochemical neutralization in the shoot may result in production of excess carboxylates, the recycling of which has been discussed by ourselves and others (47, 87, 106). The low transport capacity of the phloem for H^+ has also been reviewed (99).

Changes in nitrogen metabolism, including NO_3^- reduction and protein synthesis, could account for the apparent H^+ influxes associated with greening in barley and maize leaf slices (64). In intact barley leaves, greening is associated with synthesis of carboxylates (132). The implication of pH balance in other aspects of nitrogen metabolism in leaves and shoots has not been considered. Such aspects include synthesis of proline which, if derived from glutamate, appears to involve OH^- production as a result of loss of one carboxylate group.

Plant Cell Development

H^+ extrusion is a central feature of many developmental and related processes and especially those involving cell enlargement (18, 67, 84, 111, 141). A number of workers have followed Hager et al (24) in proposing models for IAA action that involve K^+–H^+ exchange plus carboxylation and result in cell wall acidification and decreased intracellular osmotic potential (111). Cytoplasmic pH changes have once more been proposed as a factor controlling carboxylation (26, 90). FC does not directly affect activity of PEP carboxylase in vitro (131). Marrè (66) has surveyed the ion fluxes which balance FC- and IAA-stimulated H^+ extrusion. He draws a distinction between direct (kinetic) effects of FC and less direct effects of IAA and other hormones including gibberellins, cytokinins, and abscisic acid. Ray (110) considers that H^+ is first pumped into the internal space of the endoplasmic reticulum and is then extruded with nascent wall materials. This poses problems of charge balance, but intracytoplasmic pH changes associated with development cannot be ruled out. Polevoy & Salamatova (90) propose that IAA acts on a redox-driven H^+ pump at the plasmalemma, and that there are effects of cytoplasmic levels of H^+ and other ions on ribosome activity and protein synthesis. Trewavas (138) has attempted to unify effects of hormones in terms of modifications in ion fluxes within the cell following activation of membrane receptors. He pays particular attention to effects of K^+ on DNA, RNA, and protein synthesis in animals and plants. The control of plant development by ionic currents, including H^+, has been studied by Jaffe (41) and reviewed in a wider context by Harold (25). Whether the resulting intracellular ionic gradients include pH differences is not known. However, there is conclusive evidence for pH increases of about 0.3 units following fertilization of sea urchin eggs (44, 115). It is suggested (115) that pH changes are a major factor controlling derepression of DNA and protein synthesis during development of these cells. Finally, it has been shown that increasing intracellular pH increases the ionic permeability of the intercellular junctions in developing amphibian embryos (139). This effect could be very relevant to the role of plasmodesmata in the development and growth of plants (11).

CONCLUSIONS

Metabolic control of pH is a fundamental physiological process, achieved by a combination of membrane transport and intracellular metabolism. Although there is now experimental evidence which helps to distinguish primary causes from secondary effects, it is too early to analyze how pH regulation in plants operates at the molecular level, e.g. by effects on specific ionizable groups of enzymes and metabolites (see 148). Metabolic control by pH (15) has been proposed in several areas of plant physiology but remains speculative. Further progress depends on refinements in techniques for measuring intracellular pH. This is important with respect to biochemical studies in vitro (e.g. control of enzymic activity), which should take into account the likely pH range in vivo. Measurements of intracellular pH and membrane PD must be combined if the mechanisms controlling H^+ transport are to be considered realistically.

In the past, the study of pH regulation has suffered somewhat from "scientific compartmentation," i.e. from lack of awareness of relevant developments in different areas of research. This is a situation which is rapidly changing, as shown by the breadth of the literature which had to be surveyed for this review.

ACKNOWLEDGMENTS

We thank N. A. Walker for many helpful discussions, especially about the biophysical implications of pH regulation, and R. F. Davis for supplying unpublished data for Figure 1. We are grateful to Carol Wilkins for producing the typescript. Research in the authors' laboratories was supported by the Australian Research Grants Committee and the Science Research Council (U.K.).

Literature Cited

1. Anderson, W. P., Robertson, R. N., Wright, B. J. 1977. Membrane potentials in carrot root cells. *Aust. J. Plant Physiol.* 4:241–52
2. Bakker, E. P., Rottenberg, H., Caplan, S. R. 1976. An estimation of the light-induced electrochemical potential difference of protons across the membrane of *Halobacterium halobium. Biochim. Biophys. Acta* 440:557–72
3. Beevers, H., Stiller, M. L., Butt, V. S. 1966. Metabolism of the organic acids. In *Plant Physiology, a Treatise,* ed. F. C. Steward. 4B: 119–262. New York: Academic. 599 pp.
4. Ben-Amotz, A., Ginzburg, B. Z. 1969. Light-induced proton uptake in whole cells of *Dunaliella parva. Biochim. Biophys. Acta* 183:144–54
5. Bentrup, F. W., Gratz, H. J., Unbehauen, H. 1973. The membrane potential of *Vallisneria* leaf cells: evidence for light-dependent proton permeability changes. In *Ion Transport in Plants,* ed. W. P. Anderson, pp. 171–82. London: Academic. 630 pp.
6. Bonugli, K. J., Davies, D. D. 1977. The regulation of potato phosphoenolpyruvate carboxylase in relation to a metabolic pH-stat. *Planta* 133:281–87
7. Boron, W. F., Roos, A. 1976. Comparison of microelectrode, DMO, and methylamine methods for measuring intracellular pH. *Am. J. Physiol.* 231:799–809
8. Bowling, D. J. F. 1974. Measurement of intracellular pH in roots using a H^+ sensitive microelectrode. In *Membrane Transport in Plants,* ed U. Zimmermann, J. Dainty, pp. 386–96. Berlin: Springer. 473 pp.
9. Bowling, D. J. F. 1976. Malate-switch hypothesis to explain the action of stomata. *Nature* 262:393–94
10. Caldwell, P. C. 1956. Intracellular pH. *Int. Rev. Cytol.* 4:229–77
11. Carr, D. J. 1976. Plasmodesmata in growth and development. In *Intercellular Communication in Plants: Studies on Plasmodesmata,* ed. B. E. S. Gunning, A. W. Robards, pp. 243–89. Berlin: Springer. 387 pp.

12. Chambers, R., Chambers, E. L. 1961. *Explorations into the Nature of the Living Cell.* Cambridge, Mass: Harvard Univ. Press. 352 pp.
13. Coombs, J., Maw, S. L., Baldry, C. W. 1975. Metabolic regulation in C_4 photosynthesis—inorganic carbon substrate for PEP carboxylase. *Plant Sci. Lett.* 4:97–102
14. Cooper, T. G., Wood, H. G. 1971. The carboxylation of phosphoenolpyruvate and pyruvate. II. The active species of "CO_2" utilized by phosphoenolpyruvate carboxylase and pyruvate carboxylase. *J. Biol. Chem.* 246:5488–90
15. Davies, D. D. 1973. Control of and by pH. *Symp. Soc. Exp. Biol.* 27:513–29
16. Davies, D. D. 1973. Metabolic control in higher plants. In *Biosynthesis and its Control in Plants,* ed. B. V. Milborrow, pp. 1–20. London: Academic. 364 pp.
17. Davis, R. F. 1974. Photoinduced changes in electrical potentials and H^+ activities of the chloroplast, cytoplasm and vacuole of *Phaeoceros laevis.* See Ref. 8, pp. 197–201
18. Dhindsa, R. S., Beasley, C. A., Ting, I. P. 1975. Osmoregulation in cotton fiber. Accumulation of potassium and malate during growth. *Plant Physiol.* 56: 394–98
19. Dijkshoorn, W. 1969. The relation of growth to the chief ionic constituents of the plant. In *Ecological Aspects of the Mineral Nutrition of Plants,* ed. I. H. Rorison, pp. 201–13. Oxford: Blackwell. 484 pp.
20. Eppley, R. W., Bovell, C. R. 1958. Sulfuric acid in *Desmarestia. Biol. Bull.* 115:101–6
21. Falkner, G., Horner, F., Werdan, K., Heldt, H. W. 1976. pH changes in the cytoplasm of the blue-green alga *Anacystis nidulans* caused by light-dependent proton flux into the thylakoid space. *Plant Physiol.* 58:717–18
22. Goldsmith, M. H. M. 1977. The polar transport of auxin. *Ann. Rev. Plant Physiol.* 28:439–78
23. Green, R., Giebisch, G. 1974. Some problems with the antimony microelectrode. *Adv. Exp. Med. Biol.* 50:43–53
24. Hager, A., Menzel, H., Krauss, A. 1971. Versuche und Hypothese zur Primärwirkung des Auxins beim Streckungswachstrum. *Planta* 100:47–75
25. Harold, F. M. 1977. Ion currents and physiological functions in microorganisms. *Ann. Rev. Microbiol.* 31:181–203
26. Haschke, H.-P., Lüttge, U. 1977. Auxin action on K^+-H^+–exchange and growth, $^{14}CO_2$-fixation and malate accumulation in *Avena* coleoptile segments. In *Regulation of Cell Membrane Activities in Plants,* ed. E. Marrè, O. Ciferri, pp. 243–48. Amsterdam: North Holland. 332 pp.
27. Hatch, M. D., Osmond, C. B. 1976. Compartmentation and transport in C_4 photosynthesis. In *Encyclopaedia of Plant Physiology, New Ser.,* ed. C. R. Stocking, U. Heber, 3:144–84. Berlin: Springer. 517 pp.
28. Heldt, H. W., Werdan, K., Milovancev, M., Geller, G. 1973. Alkalization of the chloroplast stroma caused by light-dependent proton flux into the thylakoid space. *Biochim. Biophys. Acta* 314:224–41
29. Hiatt, A. J. 1967. Relationship of cell sap pH to organic acid change during ion uptake. *Plant Physiol.* 42:294–98
30. Hiatt, A. J. 1967. Reactions *in vitro* of enzymes involved in CO_2 fixation accompanying salt uptake by barley roots. *Z. Pflanzenphysiol.* 56:233–45
31. Hiatt, A. J., Hendricks, S. B. 1967. The role of CO_2 fixation in accumulation of ions by barley roots. *Z. Pflanzenphysiol.* 56:220–32
32. Higinbotham, N. 1973. Electropotentials of plant cells. *Ann Rev. Plant Physiol.* 24:25–46
33. Higinbotham, N., Anderson, W. P. 1974. Electrogenic pumps in plant cells. *Can J. Bot.* 52:1011–21
34. Hill, B. C., Brown, A. W. 1978. Phosphoenolpyruvate carboxylase activity from *Avena* coleoptile tissue. Regulation by H^+ and malate. *Can. J. Bot.* 56:404–7
35. Hirakawa, S., Yoshimura, H. 1964. Measurements of the intracellular pH in a single cell of *Nitella flexilis* by means of micro-glass pH electrodes. *Jpn. J. Physiol.* 14:45–55
36. Hoagland, D. R., Davis, A. R. 1929. The intake and accumulation of ions by plant cells. *Protoplasma* 6:610–26
37. Hodges, T. K. 1973. ATPases associated with membranes of plant cells. *Adv. Agron.* 25:163–207
38. Hurd, R. G. 1960. An effect of pH and bicarbonate on salt accumulation by disks of storage tissue. *J. Exp. Bot.* 10:345–58
39. Hurd-Karrer, A. M. 1939. Hydrogen-ion concentration of leaf juice in relation to environment and plant species. *Am. J. Bot.* 26:834–46
40. Jacoby, B., Laties, G. G. 1971. Bicarbonate fixation and malate compartmentation in relation to salt-induced

stoichiometric synthesis of organic acid. *Plant Physiol.* 47:525–31

41. Jaffe, L. F. 1969. On the centripetal course of development, the *Fucus* egg, and self electrophoresis. *Symp. Soc. Dev. Biol.* 28:83–111
42. Jennings, D. H. 1976. Transport in fungal cells. In *Encyclopaedia of Plant Physiology, New Ser.,* ed. U. Lüttge, M. G. Pitman, 2A:189–228. Berlin: Springer. 400 pp.
43. Jeschke, W. D. 1970. Lichtabhängige Veränderungen des Membranpotentials bei Blattzellen von *Elodea densa. Z. Pflanzenphysiol.* 62:158–72
44. Johnson, J. D., Epel, D., Paul, M. 1976. Intracellular pH and activation of sea urchin eggs after fertilization. *Nature* 262:661–64
45. Karpilov, Yu. S., Bil', K. Ya. 1976. Transport of intermediate photosynthesis products through the cytoplasm of cells of assimilating tissues in C_4 plants. *Dokl. Akad. Nauk SSSR Bot. Sci.* 227:24–27
46. Kirkby, E. A. 1969. Ion uptake and ionic balance in plants in relation to the form of nitrogen nutrition. See Ref. 19, pp. 215–35
47. Kirkby, E. A. 1974. Recycling of potassium in plants considered in relation to ion uptake and organic acid accumulation. *Int. Colloq. Plant Anal. Fertilizer Problems, 7th, Hanover,* pp. 557–68
48. Kirkby, E. A., Mengel, K. 1967. Ionic balance in different tissues of the tomato plant in relation to nitrate, urea or ammonium nutrition. *Plant Physiol.* 42:6–14
49. Kitasato, H. 1968. The influence of H^+ on the membrane potential and ion fluxes of *Nitella. J. Gen. Physiol.* 52:60–87
50. Kluge, M., Osmond, C. B. 1972. Studies on phosphoenolpyruvate carboxylase and other enzymes of crassulacean acid metabolism of *Bryophyllum tubiflorum* and *Sedum praealtum. Z. Pflanzenphysiol.* 66:97–105
51. Komor, E., Tanner, W. 1974. The hexose-proton cotransport system of *Chlorella.* pH-dependent change in K_m values and translocation constants of the uptake system. *J. Gen. Physiol.* 64: 568–81
52. Kurkdjian, A., Guern, J. 1978. Intracellular pH in higher plant cells. I. Improvements in the use of the 5,5-dimethyloxazolidine-2[^{14}C],4-dione distribution technique. *Plant Sci. Lett.* 11: 337–44
53. Lemasters, J. J. 1978. Possible role of the mitochondrial outer membrane as an oncotic regulator of mitochondrial volume. *FEBS Lett.* 88:10–14
54. Levitt, J. 1974. The mechanism of stomatal movement—once more. *Protoplasma* 82:1–17
55. Levitt, J. 1976. Physiological basis of stomatal response. In *Water and Plant Life,* ed. O. L. Lange, L. Kappen, E.-D. Schulze, pp. 160–68. Berlin: Springer. 536 pp.
56. Lialin, O. O., Ktitorova, I. N. 1976. Experimental techniques of shifting intracellular acidity and effect of intracellular pH on electrogenic hydrogen pump of plant cells. *Fiziol. Rast.* 23:305–14
57. Lucas, W. J. 1975. Analysis of the diffusion symmetry developed by the alkaline and acid bands which form at the surface of *Chara corallina* cells. *J. Exp. Bot.* 26:271–86
58. Lucas, W. J. 1976. Plasmalemma transport of HCO_3^- and OH^- in *Chara corallina:* non-antiporter systems. *J. Exp. Bot.* 27:19–31
59. Lucas, W. J., Smith, F. A. 1973. The formation of alkaline and acid regions at the surface of *Chara corallina* cells. *J. Exp. Bot.* 24:1–14
60. Lüttge, U., Ball, E. 1974. Mineral ion fluxes in slices of acidified and deacidified leaves of the CAM plant *Bryophyllum daigremontianum. Z. Pflanzenphysiol.* 73:339–48
61. Lüttge, U., Ball, E. 1974. Proton and malate fluxes in cells of *Bryophyllum diagremontianum* leaf slices in relation to potential osmotic pressure of the medium. *Z. Pflanzenphysiol.* 73:326–38
62. Lüttge, U., Ball, E., Greenway, H. 1977. Effects of water and turgor potential on malate efflux from leaf slices of *Kalanchoe daigremontiana Plant Physiol.* 60:521–23
63. Lüttge, U., Kluge, M., Ball, E. 1975. Effects of osmotic gradients on vacuolar malic acid storage. A basic principle in oscillatory behaviour of crassulacean acid metabolism. *Plant Physiol.* 56: 613–16
64. Lüttge, U., Kramer, D., Ball, E. 1974. Photosynthesis and apparent proton fluxes in intact cells of greening etiolated barley and maize leaves. *Z. Pflanzenphysiol.* 71:6–21
65. MacRobbie, E. A. C. 1975. Ion transport in plant cells. *Curr. Top. Membr. Transp.* 7:1–48
66. Marrè, E. 1977. Effects of fusicoccin and hormones on plant cell membrane

activities: observations and hypotheses. See Ref. 26, pp. 185–202
67. Marrè, E., Colombo, R., Lado, P., Rasi-Caldogno, F. 1974. Correlation between proton extrusion and stimulation of cell enlargement. Effects of fusicoccin and of cytokinins on leaf fragments and isolated cotyledons. *Plant Sci. Lett.* 2:139–50
68. McCarty, R. E. 1976. Ion transport and energy conservation in chloroplasts. See Ref. 27, pp. 347–76
69. Meidner, H., Mansfield, T. A. 1968. *Physiology of Stomata.* London: McGraw-Hill. 179 pp.
70. Mitchell, P. 1966. Chemiosmotic coupling in oxidative and photosynthetic phosphorylation. *Biol. Rev.* 41:445–502
71. Moore, A. L., Rich, P. R., Bonner, W. D. Jr. 1978. Factors influencing the components of the total proton motive force in mung bean mitochondria. *J. Exp. Bot.* 29:1–12
72. Navon, G., Ogawa, S., Shulman, R. G., Yamane, T. 1977. High-resolution ^{31}P nuclear magnetic resonance studies of metabolism in aerobic *Escherichia coli* cells. *Proc. Natl. Acad. Sci. USA* 74:888–91
73. Neumann, J., Levine, R. P. 1971. Reversible pH changes in cells of *Chlamydomonas reinhardii* resulting from CO_2 fixation in the light and its evolution in the dark. *Plant Physiol.* 47:700–4
74. Novacky, A., Fischer, E., Ullrich-Eberius, C. I., Lüttge, U. 1978. Membrane potential changes during transport of glycine as a neutral amino acid and nitrate in *Lemna gibba* Gl. *FEBS Lett.* 88:264–68
75. Osmond, C. B. 1967. Acid metabolism in *Atriplex.* I. Regulation of oxalate synthesis by the apparent excess cation absorption in leaf tissue. *Aust. J. Biol. Sci.* 20:575–87
76. Osmond, C. B. 1976. Ion absorption and carbon metabolism in cells of higher plants. See Ref. 42, pp. 347–72
77. Osmond, C. B. 1978. Crassulacean acid metabolism: a curiosity in context. *Ann. Rev. Plant Physiol.* 29:379–414
78. Osmond, C. B., Smith, F. A. 1976. Symplastic transport of metabolites during C_4 photosynthesis. See Ref. 11, pp. 229–41
79. Overbeck, G. 1957. Zellphysiologische Studien an *Bryophyllum* im Zusammenhang mit dem täglichen Säurewechsel. *Protoplasma* 48:241–60
80. Padan, E., Zilberstein, D., Rottenberg, H. 1976. The proton electrochemical gradient in *Escherichia coli* cells. *Eur. J. Biochem.* 63:533–41
81. Pate, J. S. 1975. Exchange of solutes between phloem and xylem and circulation in the whole plant. In *Encyclopaedia of Plant Physiology, New Ser.,* ed. M. H. Zimmermann, J. A. Milburn, 1:451–73. Berlin: Springer. 535 pp.
82. Penny, M. G., Bowling, D. J. F. 1974. A study of potassium gradients in the epidermis of intact leaves of *Commelina communis* in relation to stomatal opening. *Planta* 119:17–25
83. Penny, M. G., Bowling, D. J. F. 1975. Direct determination of pH in the stomatal complex of *Commelina. Planta* 122:209–12
84. Pike, C. S., Richardson, A. E. 1977. Phytochrome-controlled hydrogen ion excretion by *Avena* coleoptiles. *Plant Physiol.* 59:615–17
85. Pitman, M. G. 1970. Active H^+ efflux from cells of low-salt barley roots during salt accumulation. *Plant Physiol.* 45:787–90
86. Pitman, M. G., Anderson, W. P., Schaefer, N. 1977. H^+ ion transport in plant roots. See Ref. 26, pp. 147–60
87. Pitman, M. G., Cram, W. J. 1977. Regulation of ion content in whole plants. *Symp. Soc. Exp. Biol.* 31:391–424
88. Pitman, M. G., Schaefer, N., Wildes, R. A. 1975. Relation between permeability to potassium and sodium ions and fusicoccin-stimulated hydrogen-ion efflux in barley roots. *Planta* 126:61–73
89. Pitman, M. G., Schaefer, N., Wildes, R. A. 1975. Stimulation of H^+ efflux and cation uptake by fusicoccin in barley roots. *Plant Sci. Lett.* 4:323–29
90. Polevoy, V. V., Salamatova, T. S. 1977. Auxin, proton pump and cell trophics. See Ref. 26, pp. 209–16
91. Poole, R. J. 1974. Ion transport and electrogenic pumps in storage tissue cells. *Can. J. Bot.* 52:1023–28
92. Poole, R. J. 1976. Transport in cells of storage tissues. See Ref. 42, pp. 229–48
93. Poole, R. J. 1978. Energy coupling for membrane transport. *Ann. Rev. Plant Physiol.* 29:437–60
94. Pucher, G. W., Leavenworth, C. S., Ginter, W. D., Vickery, H. B. 1947. Studies on the metabolism of crassulacean plants: the effect upon the composition of *Bryophyllum calycinum* of the form in which nitrogen is supplied. *Plant Physiol.* 22:205–7
95. Pucher, G. W., Vickery, H. B., Abrahams, M. D., Leavenworth, C. S. 1949. Studies in the metabolism of crassulacean plants: diurnal variations

of organic acids and starch in excised leaves of *Bryophyllum calycinum. Plant Physiol.* 24:610–20
96. Raschke, K. 1977. The stomatal turgor mechanism and its responses to CO_2 and abscisic acid: observations and a hypothesis. See Ref. 26, pp. 173–83
97. Raven, J. A. 1970. Exogenous inorganic carbon sources in plant photosynthesis. *Biol. Rev. Cambridge Philos. Soc.* 45:167–221
98. Raven, J. A. 1976. Transport in algal cells. See Ref. 42, pp. 129–88
99. Raven, J. A. 1977. H^+ and Ca^{2+} in phloem and symplast: relation of the relative immobility of the ions to the cytoplasmic nature of the transport paths. *New Phytol.* 79:465–80
100. Raven, J. A. 1977. The mechanisms, distribution, functions and evolution of active proton transport. *Proc. Winter Sch. Biophys. Memb. Trans., 4th, Fac. Agric., Univ. Katowice, Poland,* Part 2: 151–83
101. Raven, J. A. 1977. Ribulose bisphosphate carboxylase activity in terrestrial plants: significance of O_2 and CO_2 diffusion. *Curr. Adv. Plant Sci.* 9:579–90
102. Raven, J. A., Jayasuriya, H. D. 1977. Active nitrate influx, nitrate assimilation, and hydroxyl ion efflux by *Hydrodictyon africanum.* In *Transmembrane Ionic Exchanges in Plants,* ed. M. Thellier, A. Monnier, M. Demarty, J. Dainty, pp. 299–305. Paris: CNRS. 607 pp.
103. Raven, J. A., Smith, F. A. 1973. The regulation of intracellular pH as a fundamental biological process. See Ref. 5, pp. 271–78
104. Raven, J. A., Smith, F. A. 1974. Significance of hydrogen ion transport in plant cells. *Can. J. Bot.* 52:1035–48
105. Raven, J. A., Smith, F. A. 1976. Cytoplasmic pH regulation and electrogenic H^+ extrusion. *Curr. Adv. Plant Sci.* 8:649–60
106. Raven, J. A., Smith, F. A. 1976. Nitrogen assimilation and transport in vascular land plants in relation to intracellular pH regulation. *New Phytol.* 76: 415–31
107. Raven, J. A., Smith, F. A. 1976. The evolution of chemiosmotic energy coupling. *J. Theor. Biol.* 57:301–12
108. Raven, J. A., Smith, F. A. 1977. Characteristics, functions and regulation of active proton extrusion. See Ref. 26, pp. 25–40
109. Raven, J. A., Smith, F. A. 1978. Effect of temperature and external pH on the cytoplasmic pH of *Chara corallina. J. Exp. Bot.* 29:853–66
110. Ray, P. M. 1977. Auxin-binding sites of maize coleoptiles are localized on membranes of the endoplasmic reticulum. *Plant Physiol.* 59:594–99
111. Rayle, D. L., Cleland, R. 1977. Control of plant cell enlargement by hydrogen ions. *Curr. Top. Dev. Biol.* 11:187–214
112. Reibach, P. H., Benedict, C. R. 1977. Fractionation of stable carbon isotopes by phosphoenolpyruvate carboxylase from C_4 plants. *Plant Physiol.* 59: 564–68
113. Riebeling, V., Thauer, R. K., Jungermann, K. 1975. The internal-alkaline pH gradient, sensitive to uncoupler and ATPase inhibitor, in growing *Clostridium pasteurianum. Eur. J. Biochem.* 55:445–53
114. Roa, R. L., Pickard, W. F. 1976. The vacuolar pH of *Chara braunii. J. Exp. Bot.* 27:853–58
115. Shen, S. S., Steinhardt, R. A. 1978. Direct measurement of intracellular pH during metabolic derepression of the sea urchin egg. *Nature* 272:253–54
116. Siesjö, B. K. 1973. Metabolic control of intracellular pH. *Scand. J. Clin. Lab. Invest.* 32:97–104
117. Skulachev, V. P. 1978. Membrane-linked energy-buffering as the biological function of Na^+/K^+ gradient. *FEBS Lett.* 87:171–79
118. Slayman, C. L. 1974. Proton pumping and generalized energetics of transport: a review. See Ref. 8, pp. 107–19
119. Small, J. 1955. The pH of plant cells. *Protoplasmatologia* 2B2C:1–116
120. Smith, F. A. 1973. The internal control of nitrate uptake into barley roots with differing salt contents. *New Phytol.* 72:769–82
121. Smith, F. A., Raven, J. A. 1974. H^+ fluxes, cytoplasmic pH and the control of salt accumulation in plants. See Ref. 8, pp. 380–85
122. Smith, F. A., Raven, J. A. 1976. H^+ transport and regulation of cell pH. See Ref. 42, pp. 317–46
123. Smith, F. A., Raven, J. A. 1978. The evolution of H^+ transport and its role in photosynthetic energy transduction. In *Light Transducing Membranes: Structure, Function and Evolution,* ed. D. W. Deamer, pp. 233–51. New York: Academic. 358 pp.
124. Smith, F. A., Walker, N. A. 1976. Chloride transport in *Chara corallina* and the electrochemical potential difference for hydrogen ions. *J. Exp. Bot.* 27: 451–59

125. Spanswick, R. M. 1972. Evidence for an electrogenic ion pump in *Nitella translucens*. I. The effects of pH, K^+, Na^+, light and temperature on the membrane potential and resistance. *Biochim. Biophys. Acta* 288:73–89
126. Spanswick, R. M. 1973. Electrogenesis in photosynthetic tissues. See Ref. 5, pp. 113–28
127. Spanswick, R. M. 1974. Hydrogen ion transport in giant algal cells. *Can. J. Bot.* 52:1029–34
128. Spanswick, R. M., Miller, A. G. 1977. Measurement of the cytoplasmic pH in *Nitella translucens*. Comparison of the values obtained by microelectrode and weak acid methods. *Plant Physiol.* 59:664–66
129. Splittstoesser, W. E., Beevers, H. 1964. Acids in storage tissues. Effects of salts and ageing. *Plant Physiol.* 39:163–69
130. Steudle, E., Zimmermann, U. 1974. Turgor pressure regulation in algal cells: pressure-dependence of electrical parameters of the membrane in large pressure ranges. See Ref. 8, pp. 72–78
131. Stout, R. G., Cleland, R. E. 1978. Effects of fusicoccin on the activity of a key pH-stat enzyme, PEP-carboxylase. *Planta* 139:43–45
132. Tamàs, I. A., Yemm, E. W., Bidwell, R. G. S. 1970. The development of photosynthesis in dark-grown barley leaves upon illumination. *Can. J. Bot.* 48: 2313–17
133. Tanner, W., Komor, E., Fenzl, F., Decker, M. 1977. Sugar-proton cotransport systems. See Ref. 26, pp. 79–90
134. Taylor, C. V., Whitaker, D. M. 1928. Potentiometric determinations in the protoplasm and cell sap of *Nitella*. *Protoplasma* 3:1–6
135. Thomas, M. 1951. Vegetable acids in higher plants. *Endeavour* 10:160–65
136. Thomas, M., Ranson, S. L., Richardson, D. 1973. *Plant Physiology*, pp. 422–55. London: Longman. 1062 pp. 5th ed.
137. Travis, A. J., Mansfield, T. A. 1977. Studies of malate formation in "isolated" guard cells. *New Phytol.* 78: 541–46
138. Trewavas, A. J. 1976. Plant growth substances. in *Molecular Aspects of Gene Expression in Plants*, ed. J. A. Bryant, pp. 249–98. London: Academic. 338 pp.
139. Turin, L., Warner, A. 1977. Carbon dioxide reversibly abolishes ionic communication between cells of early amphibian embryo *Nature* 270:56–57
140. Van Die, J., Tammes, P. M. L. 1975. Phloem exudation from monocotyledonous axes. See Ref. 81, pp. 196–222
141. Van Steveninck, R. F. M. 1976. Effect of hormones and related substances on ion transport. See Ref. 42, 2B:307–42
142. Vorob'ev, L. N., Kurella, G. A., Popov, G. A. 1961. Intracellular pH of *Nitella flexilis* at rest and during excitation. *Biophysics* 6:648–56
143. Waddell, W. J., Bates, R. G. 1969. Intracellular pH. *Physiol. Rev.* 49:285–329
144. Walker, N. A. 1976. Membrane transport: Theoretical background. See Ref. 42, pp. 36–52
145. Walker, N. A., Smith, F. A. 1975. Intracellular pH in *Chara corallina* measured by DMO distribution. *Plant Sci. Lett.* 4:125–32
146. Walker, N. A., Smith, F. A. 1977. Circulating electric currents between acid and alkaline zones associated with HCO_3^- assimilation in *Chara*. *J. Exp. Bot.* 28:1190–1206
147. Walker, N. A., Smith, F. A. 1977. The H^+ ATPase of the *Chara* cell membrane: its role in determining membrane p.d. and cytoplasmic pH. See Ref. 102, pp. 255–61
148. Wilson, T. L. 1977. Theoretical analysis of the effects of two pH regulation patterns on the temperature sensitivities of biological systems in nonhomeothermic animals. *Arch. Biochem. Biophys.* 182: 409–19
149. Wiskich, J. T. 1977. Mitochondrial metabolite transport. *Ann. Rev. Plant Physiol.* 28:45–69
150. Ziegler, H. 1975. Nature of transported substances. See Ref. 81, pp. 59–100

Ann. Rev. Plant Physiol. 1979. 30:313–37

THE CONTROL OF VASCULAR DEVELOPMENT[1]

❖7674

Terry L. Shininger

Department of Biology, University of Utah, Salt Lake City, Utah 84112

CONTENTS

INTRODUCTION

Vascular development is a central theme in aspects of biology as diverse as evolution and cell biology. Our current conception of the control of development of this tissue is extremely rudimentary in comparison to the strides which have been made in other aspects of biology in recent years. Perhaps this is not surprising considering the complexity and subtleness of the

[1]Abbreviations used: AT, adenine thymidine; BAP, benzylamino purine; BUdR, bromuridine deoxyriboside; EM, electron microscope; ER, endoplasmic reticulum; FUdR, fluorouridine deoxyriboside; GA, gibberellin; NAA, naphthalene acetic acid; PAL, phenylalanine ammonia-lyase; QC, quiescent center; TdR, thymidine deoxyriboside.

0066-4294/79/0601-0313$01.00

problem. The study of the control of vascular development, however, offers an open field of investigation for the imaginative investigator interested in either basic aspects of eukaryotic developmental biology in general or in vascular plant physiology and development specifically. There have been a number of reviews of the subject from a variety of points of view: primary vascular development (29, 30); shoot apical meristem biology (140); root apical meristem biology (131); phloem regeneration (60); eukaryotic development (135); and xylem development in general (101, 102). The present review is an attempt to integrate more recent work on the developmental biology of vascularization with the earlier physiological and anatomical literature. The goal is to identify what we presently know about vascular development and to point out directions future work might take in order to expand our conceptualization of vascular development.

Vascular development involves first the formation of cells (procambium) which are committed at some point to become xylem and phloem as opposed to pith, cortex, mesophyll, and epidermis. Secondly, it involves the overt cytodifferentiation of these procambial cells into xylem, phloem, and cambium. What is the nature of these potential vascular cells which are currently called procambium during primary development or cambium during secondary development? Have they, at the time at which they become cytologically identifiable, actually progressed toward differentiation into xylem or phloem, or are they simply ordinary meristematic cells? Answers to these questions can only be obtained when the following questions are also answered: (*a*) What are the early biochemical markers of vascular development? (*b*) What regulates the spatial distribution of vascular cell formation within the stem, leaf, or root? (*c*) What mechanisms operate to prevent the formation of vascular elements in mature tissues or to stimulate their formation in mature tissues?

PROCAMBIUM: DEVELOPMENT IN THE SHOOT

The anatomical approach to vascular development in shoot apical meristems has generated some controversy concerning what should properly be called procambium. Until overt cytodifferentiation occurs, however, it is only possible to define these cells as being larger or smaller, denser or less dense, etc relative to surrounding cells. One cannot define these cells in terms of either their developmental *commitment* or their real developmental *state*. Thus, it seems presumptive at this point to state more than that one can observe some of the cytological changes in the progression of cells toward maturation as part of the vascular system. As yet, experiments have not been done which would allow one to define the stage at which the cells are committed to xylem or phloem differentiation unless they have overtly differentiated.

Esau (30) has taken the simplest view that in angiosperms the residual meristem, a residium of the shoot apical meristem, extends from the region of the apical initials into the differentiating region of the meristem where differential parenchymatization of the pith and cortex reveal its existence. Portions of the residual meristem are considered to differentiate into procambium, a meristem for the vascular system. Wardlaw (140) takes a different point of view based on his work with ferns. He observed a zone of cells which extends into the apical dome above the insertion of the youngest leaf primordium and which he considers to be procambium. In his view, vascular development is initiated prior to the development of leaf primordia. The development of the leaf primordia, in fact, causes disruption of this initially continuous procambium, resulting in the formation of parenchymatous leaf gaps in the stem. Surgical removal of the primordia resulted in elimination of gap formation (139). Thus Wardlaw believes that the cells above the leaf primordia represent procambial cells committed to vascular development unless specifically altered by influences from developing leaves. The interpretation of the histological preparations is clearly a subjective matter since one does not know from these kinds of experimental approaches whether the cells are committed to vascular development, to another fate, or not committed at all. Wardlaw's view can only be accepted when rigid criteria of procambium are developed and when those criteria are successfully applied to the cells he observed in the apical dome or when those cells are successfully isolated and induced to develop into vascular cells in vitro with relatively simple input.

Experimental approaches to the nature of procambium and the role of leaves in vascular development were taken by Helm (50) and Young (146). Both observed that in angiosperms the removal of the young leaf primordia caused the already recognizable procambium to develop into *parenchyma* and not to develop into vascular tissue nor even remain identifiable as procambium. Furthermore, Young (146) found that supplying auxin in the place of an excised leaf primordium prevented the change in the appearance of the procambium but did not stimulate its development into vascular tissue. More recently, McArthur & Steeves (75) attempted to clarify questions about the nature of procambium and the effects of leaves in *Geum chloense.* In this angiosperm the removal of the leaf primordia did not eliminate leaf gap parenchyma formation within the presumptive vascular cylinder as observed by Wardlaw (139) in ferns, although the cells did not differentiate into parenchyma as observed by Helm (50) and Young (146). McArthur & Steeves (75) concluded that the early phase of vascular development is not under specific control of the leaf since "provascular" tissue increased in longitudinal extent in the absence of the leaves. In *Geum* they found that vascular development progressed normally when auxin plus sucrose was applied in the absence of the leaf primordium but not with auxin alone. In the absence of knowledge of the effect of the auxin plus sucrose in Young's (146) work with *Lupinus,* it is not possible to make any generalizations concerning the nature of the stability of provascular or procambial tissues in the absence of developing leaves. However, from the fact that presumptive vascular tissue did not undergo parenchymatization in *Geum* while it did in *Lupinus* in the absence of leaves, it may be that there are distinct differences in the regulation of the formation and in the stability of these cells.

Procambium is considered to "differentiate" continuously acropetally (e.g. it appears to develop from differentiated regions toward undifferentiated regions in the meristem) in the shoot (29, 76, 77, 136). Ball (3) found procambium formation to be initiated near a leaf base and to then differentiate both basipetally and acropetally. Esau (30) questioned this interpretation on the grounds that it is extremely difficult to discriminate procambium without intimate knowledge of the pattern of vascular development in the given shoot. Again, identification of these cells is a relatively subjective matter. In axillary shoots procambium may develop acropetally from the main shoot to the axillary bud (29, 42), or basipetally from the bud to the shoot (44). The direction is reported to be basipetal in adventitious buds (140). To state that the procambium is propagated continuously acropetally within the meristem would be conceptually clearer, because "differentiate" has a specific meaning which has not yet been demonstrated for procambium.

If one accepts that procambium is a meristem producing the vascular tissues, then it is clear that the continued functioning of this meristem in the shoot depends upon leaf formation. The effect of the leaf is in some cases replaceable by auxin or auxin plus sucrose. It remains questionable whether or not early changes in cellular appearance represent obligate changes in the course of vascular development. Without specific characteristics attributable to procambium, e.g. a specific enzyme activity, protein complement, etc, it is virtually impossible to adequately study the development of procambium per se, but rather, one must deal with the later stages of vascular element maturation.

THE CONTROL OF PRIMARY VASCULAR DIFFERENTIATION IN THE SHOOT

Primary vascular tissue development has been pursued both descriptively and experimentally in *Coleus* by Jacobs and collaborators. They report that in *Coleus* the differentiation of the phloem precedes the differentiation of the xylem with the first formed phloem elements appearing at the base of a leaf, and then differentiation proceeds acropetally into the leaf and basipetally to connect with the mature vascular cells (reviewed in 60). A day or so later the xylem elements appear in the procambium at the same level at which the first phloem elements had earlier differentiated and the xylem too subsequently differentiates bidirectionally. Jacobs & Morrow (61) pursued the question of the role of the leaf in the induction of vascular development and found that the relative rates of auxin production by the leaf correlated well with the rates of xylem differentiation in the leaf trace. They concluded that auxin limits primary xylem differentiation in the leaf trace of *Coleus.* Wangermann (138) subsequently demonstrated that exogenous auxin could replace the leaf effect in the differentiation of the primary xylem of *Coleus.* It is of interest that the primary xylem in *Coleus* consists only of vessels, and thus this work agrees well with the results in *Phaseolus* of Jost (65), who found that xylem development stopped when leaves were excised. However, Jost (65) found that the application of auxin only restored vessel element differentiation. Leaf removal caused cessation of xylem fiber differentiation from the cambium in *Xanthium,* but vessels developed in normal numbers (106). NAA restored normal vascular development in that case (107).

Kinetin was first implicated in vascular development in intact plants as a result of the studies of Sorokin, Mathur & Thimann (122). Sorokin & Thimann (123) concluded that kinetin was responsible for the differentiation response, although the effects of the kinetin stimulation of growth were not experimentally separated from the xylogenesis response.

While it is generally believed that the leaf plays a dominant regulatory role in the overt cytodifferentiation of the procambium into vascular tissues, the various contributions of a leaf vs the rest of the shoot have not been fully revealed. Auxin is one of the principal regulatory agents supplied by the leaf. Although sucrose has been shown to improve the auxin effect, it is not likely that this is ordinarily derived from the leaf, because net exportation of sucrose from developing leaves is unlikely until after leaves are well beyond the stages under consideration in early vascular development. It seems that elucidation of the other regulatory components in primary vascular differentiation could be revealed in a rather straightforward way via meristem culture.

PROCAMBIUM: DEVELOPMENT IN THE ROOT

Vascular tissue development in the root was reviewed extensively by Torrey (131) and by Esau (30). As they have indicated, the root meristem is a less complex experimental tissue because there are no complications due to terminal appendages. Here the stages of vascular development occurred in a true linear sequence. The major questions concerning development in roots have revolved around: (*a*) the nature of the pattern-forming influences, e.g. is it from within the meristem or imposed by mature tissues; (*b*) the nuclear cytology of procambium development into xylem; and (*c*) the ultrastructural changes in the cytoplasm associated with phloem development. This section deals primarily with the pattern-forming influences. As in the shoot, procambium is distinguished early from other tissues in the root by enlargement of cells which will constitute the epidermis and cortex. This results in a central procambial cylinder rather than a ring (131). This central cylinder, whose outermost boundary is the pericycle, eventually differentiates with a variety of flexible arrangements (within and between organisms) into vascular and, in some cases, pith tissue. At a young stage the cells within the central cyclinder do not have unique cytological features at the electron microscope level which might suggest their early commitment to different fates (89).

In Torrey's view (131) the ability of the central cylinder to differentiate into vascular tissue with a variety of patterns suggests that the term procambium is appropriate. This interpretation of procambium cannot carry any connotation of a commitment to vascular development and suggests that procambium in the root, as in the shoot, ought to be considered a meristem within which vascular development may be induced, as was suggested earlier in this review.

The formation of procambium from the apical initials follows a continuous acropetal sequence (reviewed in 131). Procambium development into vascular tissue is marked by the radial enlargement of the future metaxylem vessel elements very close to the apical initials (10, 11). This produces (in cross-sectional view) a linear array of xylem (or a triarch or hexarch, etc array of xylem) with phloem developing laterally to the xylem or between the points of the triarch, hexarch, etc arrays of xylem. Development at any point in time is further advanced in the basipetally located cells producing an axial developmental sequence. Details of overt differentiation in root and shoot systems at the cytological, cytochemical, and biochemical levels are treated together in the next section.

The sequence of vascular development allows for the possibility of pattern induction either by the mature tissues or patterning inherent in the design of the apical meristem. Thoday (126) suggested that the meristem could be autonomous in pattern formation. Bünning (11) performed experiments which were perhaps overinterpreted to support that concept. Clowes (13) took a more direct experimental approach of surgical manipulation of either the apex or the mature tissues and found an effect on the pattern of vascular development (triarch/hexarch, etc) only when the apex per se was treated but not when the mature tissues were manipulated. Torrey (129, 130) and Reinhard (98) unequivocally demonstrated that the apex determines the pattern of vascular development. They cultured 0.5 mm root tips (129, 130) or 0.7 mm root tips (98) aseptically on defined medium and found that these very small root tips (though not necessarily other small pieces of the root) produced roots with vascular cells with normal organization. Torrey (130) observed that the number of xylem strands (triarch, hexarch, etc) differentiating in the procambium correlated well with the diameter of the procambial cylinder. Auxin treatment of these cultured root tips both increased the size of the procambial cylinder and converted the normal triarch xylem pattern to hexarch (130). Odhnoff (87) found that GA caused xylem to differentiate closer to the root tip and concluded that GA has a direct effect on xylogenesis.

More recently Feldman & Torrey (37) found that in *Zea mays* (cv. Kelvedon) the complexity of the vascular pattern and the size of the QC could be manipulated coordinately. When the QC is small the vascular pattern is simple, and when it is large the pattern is more complex. As a result of changing the size of the QC, there is a shift in the position of the proximal meristem from which all of the cells of the axis except the root cap are derived. Feldman (35) found that in corn the vascular pattern is evident within 63 to 110 μm of the root cap junction. He proposed that changes in the size of the QC shift the point of pattern initiation to areas of the procambium which differ in diameter, and thus more or less area is actually available for the patterning influence to operate.

CYTOLOGICAL AND BIOCHEMICAL ASPECTS OF VASCULAR DIFFERENTIATION

Cytological Aspects

Basic cytological aspects of overt vascular cell differentiation were summarized by Esau (30). The status of phloem cytology specifically at the light and EM levels was summarized by Esau (31) and Cronshaw (17). Torrey, Fosket & Hepler (135), O'Brien (86), and Roberts (102) reviewed aspects of xylem cell differentiation. The following discussion begins with an account of the cytology of overt vascular cell differentiation and then proceeds to a more detailed presentation of aspects which have received considerable attention regarding mechanisms or possible markers of differentiation.

Within the phloem the sieve elements undergo considerable elongation during differentiation from procambium (33). They develop a thickened cell wall which does not lignify. Sieve areas form at rather specific positions in the wall. Neuberger & Evert (84) consider the initiation of wall thickening to be an early diagnostic feature. Significant changes in the nucleus or nucleolus have not been reported except that nuclear degeneration may be nearly completed when wall thickening and modification are completed.

The development of callose (β-1,3-glucan) is another diagnostic feature of sieve cells. Callose develops as a pad covering the developing sieve area. At maturity the pores of the sieve area are lined with callose, but they are not plugged in functioning cells. There are also early structural changes in the plastids of sieve elements, the main change being the formation of a protein crystalloid within the plastid (83). In many species P-protein formation is diagnostic of sieve element development (28, 31). P-protein can be observed in extremely young cells and may be the first indication of the fate of the cell (32). P-protein has been identified in phloem since Hartig's (49) description of phloem "slime," but it is not present in all species and may exist in phloem cells other than the sieve elements (31, 34). Northcote & Wooding (85) observed P-protein to be bound in membrane which they interpreted as ER in origin. Sieve elements are characterized by extensive development of the ER. The cell biology and biochemistry of P-protein development could be a fertile area of research. The interested reader should consult the following accounts of phloem development: (*a*) root protophloem (33); (*b*) gymnosperm phloem (82–84); (*c*) the general review by Cronshaw (17). The basic features are similar in each plant system as well as in callus (144) and hormone-induced phloem development in excised tobacco pith (17). P-protein structure per se has been explored also (88).

Cytological aspects of xylem cell differentiation were reviewed by O'Brien (86) and by Torrey, Fosket & Hepler (135). Regardless of the origin of the cell (procambium or mature parenchyma), there are several overlapping stages of development. The problem is to ascertain which of these are critical and then to determine how they are regulated. Cell expansion is normal in development from procambium or cambium but not pronounced (if it occurs at all) in development from mature parenchyma. Thus, nuclear or nucleolar changes relevant to expansion growth in procambium may confuse studies of xylogenesis. This problem should not occur in xylogenesis in cultured parenchyma. This problem then becomes one of identifying the time and place of xylem development in order to study nuclear changes.

Secondary wall deposition and lignification are diagnostic features of xylogenesis. The control of specific patterns of wall deposition has received considerable attention (reviewed in 102, 135). Pickett-Heaps (94) observed changes in the distribution of microtubules during xylem development. He found that initially they were evenly spaced along the cell wall, but later they grouped into bands and the cell wall thickenings were deposited adjacent to these bundles of microtubules. Pickett-Heaps (94) showed that the antimicrotubular drug colchicine disrupted the microtubules and the wall pattern. Cellulose microfibril orientation was also altered by colchicine in developing xylem cells (51). How microtubules affect cell wall deposition remains an unsolved problem.

Lignin deposition is considered to be an early aspect of xylogenesis and it accompanies secondary wall deposition (145). However, as O'Brien (86) pointed out, we really do not have a stain for lignin at the EM level. Thus it is not possible to study very reliably with EM techniques how lignin deposition and wall formation are coordinated.

When secondary wall deposition and lignification near completion, portions of the primary wall may become hydrolyzed. This is followed or accompanied by protoplast autolysis. Little if anything is known about the control of these processes. At maturity the water-conducting cells (vessels, tracheids) consist of elaborately constructed cells without protoplasts but with variously modified wall regions which allow rather free liquid movement.

As noted previously, in the root the metaxylem is first detected by the radial expansion of these cells very near the apical initials (131). Large increases in nuclear DNA in these cells were reported (125). These observations generated the concept that increased ploidy was basic to xylem development (73). As will be seen, this correlation is not universal within a tissue or between organisms.

Corsi & Avanzi (15) reported an eightfold increase in the DNA of metaxylem cells of *Allium cepa* relative to pericycle cells which remained at the 2C level. They also reported changes in the DNA/histone ratios, but it is not clear that these were consistent or significant. Innocenti & Avanzi (56) interpret their observations of this system as follows: The developing metaxylem cells first undergo chromosome endo-reduplication very close to the apical initials. This is followed by amplification of the nucleolar organizer region of the DNA (rDNA coding for rRNA) which in turn is followed by extrusion of the nucleoli. Metaxylem cells showed a sixfold increase in rDNA (2). This cycle is repeated in the 2 to 5mm region from the tip (cap included in measurements) and the cells then stop DNA synthesis. During this time the metaxylem cells undergo expansion from 20 μm in length to 750 μm. Innocenti (55) reported a decrease in the histone/DNA ratio in developing metaxylem. The work cannot be evaluated adequately since the data were not presented except as ratios. Similar results were reported from work in tracheary element differentiation in the corn leaf (70, 124). As these investigators indicate, these observations need closer examination in systems more suitable for biochemical work. However, these studies suggest some experimental approaches that are not readily apparent from biochemical methods. The primary confusing aspect of these studies is the inability to separate events relevant to expansion from these relevant to overt xylogenesis. As will be seen, biochemical approaches have failed (unnecessarily) to separate replication-relevant events from tracheary element differentiation.

Biochemical Aspects

There are a few studies of vascular development at the biochemical level. Thornber & Northcote (128) studied changes in carbohydrate and lignin contents of cells developing from the cambium. The results were confusing because of the inadequate methods for such studies at that time. Jeffs & Northcote (63) found an increase in the xylose/arabinose ratio in bean callus after 1 to 2 months of hormone-stimulated xylogenesis. New methods are now available, and such analyses have been elegantly performed on primary cell walls (67) but not related to secondary wall development or vascularization.

Lignification of secondary cell walls begins very soon after secondary thickening is initiated (52, 145). The enzyme PAL catalyzes the conversion of phenylalanine to cinnamic acid and is a key reaction in the diversion of carbon from protein synthesis to lignin synthesis. PAL activity in various tissues is correlated positively with vascular development in the plant (105) and in hormone-treated callus (104). Most recently Haddon & Northcote (46) followed the time course of PAL activity during vascular nodule formation in bean callus. In this case the increased PAL activity correlated fairly well with the initiation of xylogenesis in the early stages. PAL activity declined later when it would seem that it ought to remain at least constant.

Haddon & Northcote (46, 47) also examined β-1,3-glucan synthetase activity as a marker of callose formation during phloem development. β-1,3-glucan synthetase activity followed a time course similar to PAL activity and to sieve cell differentiation. It is difficult to analyze this work since hormone-treated replicating (but nondifferentiating) controls were not examined.

Dalessandro & Northcote (21–23) initiated studies of the activities of various enzymes involved in nucleoside diphosphate sugar reactions which are expected to be involved in carbohydrate metabolism associated with vascular development. Enzyme activities were measured: (*a*) in the cambium, differentiating xylem or differentiated xylem of two angiosperms and two gymnosperms, or (*b*) in cultured Jerusalem artichoke tuber tissue in which hormone-induced vascular development occurred. Enzyme activity (cpm in product/mg protein) did not reveal consistent or large differences in the various tissues examined in *a*. The Jerusalem artichoke experiments suffered from the lack of adequate controls.

THE CONTROL OF NONPRIMARY VASCULAR DIFFERENTIATION

Differentiation from the Cambium

As the primary vascular tissues differentiate from procambium and the tissue ceases elongation, some of the procambial cells initiate replication in which the cell plate is formed in the longitudinal plane. These cells are now considered to be cambium and they will form all of the secondary vascular cells, e.g. the wood (xylem, cells differentiated interior to the cambium) and phloem (cells differentiated exterior to the cambium). In the stem the cells between the vascular (fascicular) bundles may initiate replication to form the interfascicular cambium. When this happens the cambium becomes a sheath of meristematic cells. The *initiation* of replication in the fascicular and interfascicular cambium and seasonal reactivation of the cambia are attributed to factors derived from the shoot (48, 118). Coster (16) suggested that a hormone was responsible and Snow (118) demonstrated the hormonal nature of the cambial stimulus. Snow (119) effectively replaced shoot stimulation with exogenous auxin. However, auxin is not always sufficient (96).

Siebers (112–114) studied the activation of the interfascicular cambium of *Ricinus communis*. He found that the *activation* of this cambium and the differentiation of xylem occurred in interfascicular tissue isolated and cultured on simple media without exogenous auxin. However, younger tissue required kinetin (113). He concluded that the development of this meristem was determined at a very young stage during procambium development and was not dependent upon homogenetic induction from the fascicular cambium.

Generally, studies of vascular development from the cambium have centered on explorations of the effects of combinations of hormones or photoperiod on both cambial division and differentiation in qualitative terms. Auxin stimulation of cambial division in decapitated plants may result in division and vascular differentiation (119) or cambial division without normal differentiation (65, 95, 118–121). In some cases normal vascular development may be elicited by GA (68) or may require auxin plus GA (25, 141). GA tends to be sufficient in rooted tissue but not in isolated stems.

Removal of young leaves and buds from *Xanthium* allows cambial activity to continue, once initiated, but the majority of the cells (fibers) do not differentiate. However, vessel elements, though smaller than normal in diameter, develop in normal numbers (106). In this case fiber differentiation is restored briefly by allowing one leaf to develop acropetal to this region of the stem or by adding an auxin (NAA) instead (107). The derivatives of the cambium in *Xanthium* are not programmed as a result of their cambial origin; rather, there must be hormonal input to the new cells either during cell formation or subsequent to cell formation or both. If the auxin application is delayed, the cells derived in the meantime fail to respond in terms of fiber differentiation and remain as thin-walled unlignified parenchyma (107). Similar results were observed in other systems (6, 53).

Experimentally Induced Vascular Development

WOUND-INDUCED VASCULARIZATION Vascular tissue development may occur in portions of the plant other than from procambium in root and shoot meristems. The bulk of the vascular tissues are secondary and produced by the vascular cambium. However, the literature on secondary vascular tissue formation, although extensive, reflects a great interest in the alteration of subtle characteristics of the differentiated cells, e.g. wall thickness, cell length, etc. Very little of this literature deals specifically with the problem of the control of vascular cell differentiation. The literature on the cambium was reviewed by Brown (8) and by Phillipson, Ward & Butterfield (93). While some experimental work on vascular development from the cambium will be considered in this section, the emphasis of this section will be on experimentally induced vascular development in explanted parenchymatous tissue, callus, or wounded stems.

Vascular regeneration Vascular regeneration in parenchyma of wounded stems was first reported by Crüger (18). It was observed to progress basipetally around a wound (115), and Janse (62) suggested that a hormonal stimulus was responsible. von Kaan Albest (66) demonstrated that leaves acropetal to the wound produced factors necessary for vascular regeneration. She also showed that the vascular system per se must be wounded to obtain regeneration. Xylem regeneration may involve the transformation of some cells directly into xylem without cell division (117), although one does not know what unobserved biosynthetic events may be prerequisite. Parenchyma is not converted directly to sieve elements without cell division (66).

The major break in understanding vascular regeneration came from Jacobs' work (57–59), which showed that the role of the leaf in xylem regeneration was replaceable by auxin. Subsequently, La Motte & Jacobs (71, 72) showed that phloem regeneration was similarly controlled. Surprisingly, in view of studies in other systems, sucrose did not limit in the *Coleus* system (59, 72). In *Coleus* the leaves need to be present for 48 hours after wounding to obtain a xylogenic response (100). An inductive effect was evident since the vascular strands did not complete differentiation until 72 hours after wounding (1). The dependence of phloem regeneration on roots is variable. Houck & La Motte (54) report that roots are essential but can be replaced by a cytokinin (zeatin). However, previous work had shown no role for the roots in the same clone of *Coleus* (59).

The studies on hormone involvement are, with the exception of the auxin studies, in conflict. In addition to the differences reported concerning the role of roots and cytokinin noted above, there is a clear indication that kinetin inhibits wound vessel regeneration in a different clone of *Coleus* (40). In that clone GA enhanced the auxin effect on regeneration (103). But Thompson (127) found GA to have no effect on regeneration in the Princeton clone of *Coleus*. The manner of hormone presentation may be critical because Thompson (127) found that concentrations of auxin which are effective when applied apically are not effective in regeneration when applied basally.

Ultimately, regeneration is a wound-induced redevelopment of some mature parenchyma, but it is important to establish how these newly formed cells are "plumbed" into the vascular system. Is the new vascular tissue in the wound region "plumbed" directly into the old vascular system, e.g. old cells → new cells → old cells, or is there a more extensive replacement resulting in the insertion of the regenerated cells into a new vascular strand, e.g. new cells → new (regenerated) cells → new cells? These questions were posed by Benayoun et al (5). They interpret their work to support the replacement view, e.g. new cells → new (regenerated) cells → new cells. In their work the act of wounding actually causes a reduction in the *rate* of formation of vascular cells. It is probable that a variety of patterns of regeneration may be produced (D. E. Fosket, personal communication). Jacobs (60) has reviewed the earlier work on regeneration.

VASCULARIZATION IN CULTURED TISSUE Significant progress toward understanding the control of vascular development came when Camus (12) showed that strands of xylem elements could be induced to differentiate in the parenchymatous cells of endive callus as the result of the insertion of a growing bud. Ball (4) observed a similar effect when buds were regenerated in *Sequoia sempervirons* callus. One of the roles of the inserted bud is to supply auxin (12). Wetmore & Sorokin (143) performed similar experiments with lilac callus and buds and found that auxin imitated the bud effect in forming xylem strands only when supplied from a point source. The need for sucrose was demonstrated by Wetmore & Rier (142), and it was shown in *Parthenocissus tricuspidata* callus that tracheary element production varied quantitatively with changes in sucrose concentration when auxin was held constant (99). Furthermore, the ratio of xylem to phloem varied with sucrose concentration if auxin levels were constant: low sucrose (ca 1%)

elicited relatively more xylem while high sucrose (ca 4%) elicited relatively more phloem.

Those reports of carbohydrate effects led to the investigation of effects of alternative carbohydrate supplies which are still not satisfactorily resolved. In *Phaseolus* neither glucose plus fructose nor glucose alone replaced sucrose in producing organized nodules of xylem and phloem (64). However, glucose alone caused the production of scattered xylem elements. Fructose, mannose, xylose, rhamnose, arabinose, galactose, mannitol, or α-methyl glucoside at 2% did not elicit any differentiation, while cellobiose, lactose, and raffinose (2%) all elicited some xylem differentiation but not organized nodules of xylem and phloem. The sucrose effect on nodule differentiation was partially replaced quantitatively by 2% maltose or αα-trehalose (64). Sequential treatments of sucrose then IAA, or IAA then sucrose, indicated that nodule formation occurred only when IAA preceded or was accompanied by sucrose (64).

Helianthus tuberosa tissue forms tracheids in response to NAA and BAP in the presence of sucrose or glucose or trehalose, or to a lesser extent maltose, while other carbohydrates are very much less effective (78). Soluble starch (4%) stimulated xylogenesis in *H. tuberosa* quantitatively above that produced by optimal (4–8%) sucrose (78). Phillips & Dodds (90) were unable to confirm either that observation or the trehalose stimulation. Minocha & Halperin (78) observed an inhibition of differentiation without concomitant effects on cell division when increasing amounts of glucose were presented with optimal levels of sucrose. Increased amounts of glucose or sucrose alone did not have the same effect, but data on the glucose response was not included, They postulate a competitive uptake effect which could be tested. Replotting their data (percentage of differentiation vs total cell number) indicates that the glucose effect may be a simple shift of the percentage of differentiated cells curve up on the cell number scale. Whether this means a simple delay in the initiation of xylem differentiation or a decrease in rate of formation of xylem cannot be deduced. We are not at a point where the specific effects of carbohydrate sources on differentiation can be explained satisfactorily. The situation is made even worse by the inability of investigators to see the same qualitative responses in the same tissue. However, it appears that only those carbohydrates which support significant cell division also support tracheary element formation. The disproportionate effect of carbohydrates on differentiation relative to cell division should be investigated in a time course experiment to determine how carbohydrates affect the initiation and rate of xylogenesis.

The requirement for auxin observed in vascular differentiation relative to primary, secondary, or regenerated vascular cells is also observed consistently in all forms of tissue culture, e.g. callus (12), pith (14), pea root parenchyma (134), Jerusalem artichoke tuber parenchyma (20, 78), or lettuce pith (19, 24). The specific roles of auxin have not been deciphered but may be related to aspects of cell division or gene expression or carbohydrate metabolism or all these. Fosket (39) found that IAA did not alter the time lag preceding wound vessel member formation in *Coleus* where exogenous auxin increases the numbers of vascular cells formed but is not essential for their formation. Auxin has only a small affect on DNA synthesis in *Coleus* (39) and no affect by itself in pea (116), and so it cannot be assumed to increase the numbers of differentiated cells via increasing the numbers available for differentiation. It is not clear how auxin affects RNA or protein synthesis in those systems where auxin suffices to induce vascular differentiation since "state-of-the-art" biochemical approaches have not been utilized.

Cytokinins were first shown to stimulate xylogenesis in cultured tissue in conjunction with exogenous auxin by Bergmann (7). That observation was confirmed in a qualitative manner in several systems by Torrey (132). The first quantitative analysis was achieved by Fosket & Torrey (41), who used a soybean callus which required these hormones for replication as well. Increasing concentrations of kinetin stimulated xylogenesis relatively more than replication. The soybean system used no longer exists. It is an unfortunate fact that callus cultures are notoriously capricious and develop altered hormone responses as they age. It is essential to develop and employ adequate means of tissue preservation if these studies are to achieve their true potential in developmental studies.

Torrey & Fosket (134) observed similar responses to auxin and cytokinin in pea root parenchyma. This system is known to contain a heterogeneous cell population. However, xylogenesis occurs in the hormone-stimulated replicating cortical parenchyma (134). Phillips & Torrey (91) refined the analysis by punching out the central vascular cylinder. They established that xylogenesis and replication of these cells absolutely required auxin plus cytokinin and that replication preceded xylogenesis by several days. Shininger & Torrey (111) then established that, like the soybean system, increasing concentrations of cytokinin stimulated the rate of xylem cell formation relatively more than cell replication. Furthermore, these cells showed a continuous requirement for the hormones for xylogenesis while cell divisions were induced by shorter periods of cytokinin treatment and continued after cytokinin "removal." This may not be true in the Jerusalem artichoke tissue since preliminary work suggested that xylogenesis could continue after cytokinin withdrawal (78). Cytokinin is required in lettuce pith (19), carrot tissue (80), and *Zinnia* mesophyll (69) as well as pea (91, 111, 134) and artichoke (78). The specific role of cytokinin remains to be determined in all systems.

Cell replication is observed to precede and accompany xylogenesis in nearly all culture systems. In those cases where replication is not observed it has *not* been shown that DNA synthesis did not occur. Within the plant, cell division normally accompanies primary and secondary vascular development as well as wound-induced or hormone-induced vascular development. On rare occasions tracheary element formation can also be observed in single cells in suspension cultures, but in general, replication is coincident (133). Do these observed cell divisions have any bearing on the problem of the regulation of vascular differentiation? I think they do.

In animal systems a considerable body of evidence has developed to support the concept that some forms of differentiation require cell replication (reviewed in 97). Rigorous evidence in plant systems is minimal and, like animal systems, depends upon the use of inhibitors (reviewed in 102 and 109).

In plants the first direct approach to the question was focused on the need for cell division in toto. Fosket (38) found that both FUdR and mitomycin C blocked replication and xylogenesis in *Coleus* explants. The FUdR effect was prevented by simultaneous TdR treatment. This indicates a need for some aspect of the division process. Although colchicine had similar effects in this system, it is less clear where the effect was manifest (D. E. Fosket, personal communication). Turgeon (137) concluded that DNA synthesis is not necessary for vascular differentiation, since FUdR did not prevent xylogenesis in lettuce pith. This interpretation may be premature because it was not shown rigorously that DNA synthesis was prevented. It is not adequate to demonstrate a lack of increase in cell numbers or to measure DNA microspectrophotometrically in 0.5% of the cells (when 6% will differentiate) and conclude that DNA synthesis was totally blocked. In fact, FUdR had a pronounced effect on vascular differentiation since it reduced vascular differentiation to 3% of the control level.

The evidence from gamma plantlets (45) is frequently used to support the view that vascular development does not require recent cell divisions. However, recent cell divisions have only been considered essential as a result of studies where it is known that they are occurring coincidently in time with the agents which actually induce differentiation. It is not known when induction takes place relative to overt cytodifferentiation in the gamma plantlets. We do not yet know why replication is essential although the pea system is yielding some new insight.

Endopolyploidy is frequently observed to be correlated with in vivo xylogenesis (73). Phillips & Torrey (92) and Dodds & Phillips (26) carefully eliminated this possibility as being a causal relationship in pea and Jerusalem artichoke respectively. However, the need for normal replication in the pea system remains.

Shininger (108) demonstrated a reversible FUdR inhibition of xylogenesis in the pea system, and with the use of BUdR (a thymidine analog) demonstrated a specific role of DNA synthesis per se in the differentiation process. BUdR did not block cell replication. Both FUdR and BUdR effects are prevented and reversed by simultaneous or subsequent TdR treatment. Both replication and xylogenesis were blocked within 48 hours of FUdR treatment and both recovered within 72 hours of TdR addition (108). How BUdR elicits its effect is unknown. BUdR incorporation into DNA is considered essential (74). BUdR is incorporated into the DNA of pea cells (108). In *Tetrahymena* BUdR blocks RNA synthesis in synchronized cells but only if it is available to cells during DNA synthesis and most specifically during replication of the ribosomal genes (74). Lykkesfeldt & Andersen (74) postulated that this BUdR effect results from its incorporation into AT-rich regions. Thus, the transcription of all AT-rich regions may be selectively inhibited, and this could conceivably block RNA synthesis related to xylogenesis but not that which is necessary for cell replication.

Factors other than hormones have not been extensively investigated relative to their use in studies of vascular development. Increased pressure stimulates xylem differentiation in stems (9). The mechanism has not been investigated, but Roberts (102) suggested it occurred through induced ethylene production. Doley & Leyton (27) found lowered water potential to stimulate xylogenesis and suggested that this might be the basis of the observed sucrose stimulation. Other investigators have not found osmotic agents in general to stimulate vascular development as efficiently as sucrose.

In Jerusalem artichoke tuber tissue it was observed that (*a*) reducing the nitrogen concentration of the medium, and (*b*) reducing the volume of medium are both stimulatory to xylogenesis (90). The volume effect does not seem due to a "conditioning" of the medium (90).

Light generally inhibits xylogenesis. In the Jerusalem artichoke, a brief exposure to dim white light inhibited cell replication and xylogenesis relative to a dim green light control (90). Continuous white light (relative to a brief exposure to white light) had little effect on replication but inhibited xylogenesis (90). Similar results have been observed in the pea system (T. L. Shininger, unpublished observations). In carrot culture, light can be a requirement for xylogenesis (79–81), but if so, it is replaceable by cytokinin (79). Studies of the light effect in other systems need to be conducted more conscientiously.

Temperature as a probe of vascular development has not been utilized generally. Gautheret (43) reported that in the Jerusalem artichoke callus temperatures less than 17°C were inhibitory. This was confirmed in freshly excised tubers as well (90). Naik (82) reported that a high temperature (35°C) stimulates xylogenesis in this system, but this was not confirmed by subsequent work (90). In any case, these effects have not been distinguished from effects on replication.

In a recent series of experiments with the pea system, I have found some specific effects of temperature on xylogenesis (110). Briefly, low (10°C) or high (30°C) temperatures inhibit xylogenesis relatively more than replication. Low temperature was studied extensively, and it was found to delay the initiation of xylem cell appearance by several days. However, once initiated, these cells appeared at essentially normal rates. Cell replication was not affected in this case. This is not caused by an affect on overt cytodifferentiation because transfer of cells to low temperature after the initiation of xylogenesis did not cause a drop in the rate of xylem cell appearance for at least 7 days. When cultures were initiated at low temperature and then shifted up, there was no transient increase in the rate of vascular cell appearance to suggest they had accumulated at a specific temperature-sensitive point in development. Brief low temperature treatments (24, 48 hours, etc) indicated that xylogenesis was temperature sensitive before replication of the cortical parenchyma was initiated. The results suggest that since xylogenesis can be uncoupled from the early rounds of replication, these early replications have no unique role in development. Since it is unlikely that cells committed to xylogenesis exist in the cortical parenchyma, it follows that xylogenesis in this system can be induced by defined exogenous stimuli whenever the cells are replicating at an appropriate temperature.

The pea system offers some interesting points for further analysis. At 25° both xylogenesis and nonxylem cell formation show nearly exponential kinetics. How this relates to the fate of individual replicating cells is not yet known. The possibilities are that one cell provides two daughters which differentiate or that one cell provides one cell for continued replication and one for vascular development or some complication of this scheme. Observations of sections have not yet revealed isolated single tracheary elements, but it is premature to state that they are not to be found in this system.

It is also interesting to note that in this system one can observe linearity between the log of the vascular cell number and log of the nonvascular cell number. This relationship is not disturbed by temperature changes. Nor is it altered by changes in cytokinin concentration which are known to affect the rates of xylem and nonxylem cell formation differentially. This may mean that the *rates* of these processes are coupled. If so the result of this is that changes in replication rates generate tissues with different percentages of vascular elements rather than a proportionate increase in the number of xylem and nonxylem cells. The log/log relationship is disturbed by BUdR. My reanalysis of other systems indicates that where adequate data exist the same relationship occurs (41, 90). Exactly how these responses are coupled remains to be investigated. It seems clear that this relationship indicates a basic developmental control mechanism.

CONCLUSIONS: Control of Vascularization in the Shoot and Root

The process of vascularization involves first the production of a population of cells in which vascular differentiation can be induced. The process ought to be viewed as a continuum of developmental events involving as a first identifiable step the distinguishing of the potential vascular cells from the nonvascular by differential parenchymatization. The nonparenchyma cells may now be considered to be in the pathway to vascular development but are not obligated to that fate because removal of exogenous stimuli (auxin, sucrose) allows them to divert to the parenchyma state. Procambium is a term that should be employed to describe potential vascular cells which are at this state of development. It should be possible to further break down the stages of differentiation within this concept of procambium. When these cells irreversibly change cytologically toward either the xylem or the phloem, they can be considered overtly differentiated and no longer procambial. In shoots, stimuli are derived in part from the developing leaves and in part from the mature shoot. In the root these factors may be derived in part from the terminal portions of the meristem (root cap, QC, and proximal meristem) and in part from more mature sections of the root.

In the shoot it is clear that the leaves play an important regulatory role in the formation and development of procambium as it is currently identified. However, we do not have a meaningful concept of procambium at a biochemical or ultrastructural level, and the development of concepts at these levels is important. Both auxin and sucrose are critical to procambium formation and to the formation of vascular cells in cultured tissue. We do not know the roles of either of these agents in the differentiation process.

Overt differentiation probably can be monitored at the biochemical level and certainly can be monitored at the ultrastructural level. It is conceivable that phloem differentiation could be followed immunologically in those organisms which form P-protein or biochemically with careful analyses of β-1,3-glucan synthetase activity in conjunction with cytological studies. It is not clear if significant nuclear or nucleolar changes occur in sieve element differentiation. The changes in the nuclei of differentiating tracheary elements may have been misleading, however.

Biochemical analyses of tracheary element differentiation may be more difficult until secondary cell wall composition is characterized in greater detail. Lignification probably can be used as a valid marker of xylem differentiation, especially since one really wants a marker characteristic of terminal aspects of cytodifferentiation. There exists the possibility that there is considerably more microtubular protein in differentiating xylem cells if microtubules are really associated with secondary wall deposition. Such a quantitative change might be detectable.

DNA synthesis seems critical to xylem differentiation. With biochemical markers of xylogenesis, it may be possible to determine the reason for the need for DNA synthesis.

The various roles of auxins, cytokinins, and occasionally gibberellins need meaningful exploration. We do not have an adequate understanding of these hormones even in simpler response systems. These hormones can only be understood when plant scientists generally become willing to explore hormone responses with truly rigorous and contemporary biochemical or genetic approaches. The exploitation of temperature and light may be useful and, because of the "neatness" of such experiments, the results might be extremely important.

The development of suspension cultures that will differentiate reasonably synchronously would be extremely useful. The recent work of Torrey (132) makes one optimistic comment about that approach.

The recent manipulations of the QC effect on development and attempts at physiological replacement of the QC with hormones bound to resin coated beads (36) provide impetus to explore these systems more imaginatively. Finally, the autonomous nature of the root tip makes it ideal for exploration of many of the sorts of questions which previously have been considered only in tissue culture. The chief advantage of the root tip culture is the production of vascular cells in *predictable* time and space. Thus it ought to be possible to define procambium in the root in terms of its responses to specific metabolic analogs or physical treatments in ways analogous to the early work of Torrey (132).

Literature Cited

1. Aloni, R., Jacobs, W. P. 1977. The time course of sieve tube and vessel regeneration and their relation to phloem anastomoses in mature internodes of *Coleus*. *Am. J. Bot.* 64:615–21
2. Avanzi, S., Maggini, F., Innocenti, A. M. 1973. Amplification of ribosomal cistrons during the maturation of metaxylem in the root of *Allium cepa*. *Protoplasma* 76:197–210
3. Ball, E. 1949. The shoot apex and normal plant of *Lupinus albus* L. bases for experimental morphology. *Am. J. Bot.* 36:440–54
4. Ball, E. 1950. Differentiation in a callus culture of *Sequoia sempervirens*. *Growth* 14:295–325
5. Benayoun, J., Aloni, R., Sachs, T. 1975. Regeneration around wounds and the control of vascular differentiation. *Ann. Bot.* 39:447–54
6. Benayoun, J., Sachs, T. 1976. Unusual xylem differentiation below a mature leaf of *Melia*. *Isr. J. Bot.* 25:184–94
7. Bergmann, L. 1964. Der Einfluss von Kinetin auf die Ligninbildung und Differenzierung in Gewebekulturen von *Nicotiana tabacum*. *Planta* 62:221–54
8. Brown, C. L. 1975. Secondary growth. In *Trees, Structure and Function*, ed. M. H. Zimmermann, C. L. Brown, pp. 67–123. New York: Springer-Verlag
9. Brown, C. L., Sax, K. 1962. The influence of pressure on the differentiation of secondary tissues. *Am. J. Bot.* 49: 683–91
10. Bünning, E. 1951. Über die Differenzierungs-vorgänge in der Crucifernwurzel. *Planta* 39:126–53
11. Bünning, E. 1952. Weitere Untersuchungen über die Differenzierungsvorgänge in Wurzeln. *Z. Bot.* 40:385–406

12. Camus, G. 1949. Recherches sur le rôle des bourgeons dans les phénomènes de morphógènese (Extrait). *Rev. Cytol. Biol. Veg.* 11:1–199
13. Clowes, F. A. L. 1953. The cytogenerative centre in roots with broad columellas. *New Phytol.* 52:48–57
14. Clutter, M. E. 1960. Hormonal induction of vascular tissues in tobacco pith in vitro. *Science* 132:548–49
15. Corsi, G., Avanzi, S. 1970. Cytochemical analyses on cellular differentiation in the root tip of *Allium cepa. Carylogia* 23:381–94
16. Coster, Ch. 1927. Zur Anatomie und Physiologie der Zuwachszonen—und Jahresringbildung in den Tropen. *Ann. Jard. Bot. Buitenzorg* 37:49–160
17. Cronshaw, J. 1974. Phloem differentiation and development. In *Dynamic Aspects of Plant Ultrastructure,* ed. A. W. Robards, pp. 391–413. London-New York: McGraw Hill
18. Crüger, H. 1855. Zur Entwickelungsgeschichte der Zellenwand. *Bot. Ztg.* 13: 601–29
19. Dalessandro, G. 1973. Hormonal control of xylogenesis in pith parenchyma explants of *Lactuca. Ann. Bot.* 37: 375–82
20. Dalessandro, G. 1973. Interaction of auxin, cytokinin, and gibberellin on cell division and xylem differentiation in cultured explants of Jerusalem artichoke. *Plant Cell Physiol.* 14:1167–76
21. Dalessandro, G., Northcote, D. H. 1977. Changes in enzymic activities of nucleoside diphosphate sugar interconversions during differentiation of cambium to xylem in sycamore and poplar. *Biochem. J.* 162:267–79
22. Dalessandro, G., Northcote, D. H. 1977. Changes in enzymic activities of nucleoside diphosphate sugar interconversions during differentiation of cambium to xylem in pine and fir. *Biochem. J.* 162:281–88
23. Dalessandro, G., Northcote, D. H. 1977. Changes in enzymic activities of UDP-D-glucuronate decarboxylase and UDP-D-xylose 4-epimerase during cell division and xylem differentiation in cultured explants of Jerusalem artichoke. *Phytochemistry* 16:853–59
24. Dalessandro, G., Roberts, L. W. 1971. Induction of xylogenesis in pith parenchyma explants of *Lactuca. Am. J. Bot.* 58:378–85
25. Digby, J., Wareing, P. F. 1966. The effect of growth hormones on cell division and expansion in liquid suspension cultures of *Acer pseudoplatanus. J. Exp. Bot.* 17:718–25
26. Dodds, J. H., Phillips, R. 1977. DNA and histone content of immature tracheary elements from cultured artichoke explants. *Planta* 135:213–16
27. Doley, D., Leyton, L. 1970. Effects of growth regulating substances and water potential on the development of wound callus in *Fraxinus. New Phytol.* 69:87–102
28. Esau, K. 1947. A study of some sieve-tube inclusions. *Am. J. Bot.* 34:224–33
29. Esau, K. 1954. Primary vascular differentiation in plants. *Biol. Rev. Cambridge Philos. Soc.* 29:46–86
30. Esau, K. 1965. *Vascular Differentiation in Plants.* New York: Holt, Rinehart & Winston. 160 pp.
31. Esau, K. 1969. The phloem. In *Handbuch der Pflanzenanatomie,* Vol. 2. Berlin: Gebrüder Borntraeger. 505 pp.
32. Esau, K., Gill, R. H. 1970. Observations on spiny vesicles and P-protein in *Nicotiana tabacum. Protoplasma* 69: 373–88
33. Esau, K., Gill, R. H. 1972. Nucleus and endoplasmic reticulum in differentiating root protophloem of *Nicotiana tabacum. Ultrastruct. Res.* 41:160–75
34. Evert, R. F., Eschrich, W., Eichhorn, S. E. 1973. P-protein distribution in mature sieve elements of *Cucurbita maxima. Planta* 109:193–210
35. Feldman, L. J. 1977. The generation and elaboration of primary vascular tissue patterns in roots of *Zea. Bot. Gaz.* 138:393–401
36. Feldman, L. J. 1978. Cytokinin biosynthesis in corn roots. *Plant Physiol. Suppl.* 61:11
37. Feldman, L. J., Torrey, J. G. 1975. The quiescent center and primary vascular tissue pattern in cultured roots of *Zea mays. Can. J. Bot.* 53:2796–2803
38. Fosket, D. E. 1968. Cell division and the differentiation of wound-vessel members in cultured stem segments of *Coleus. Proc. Natl. Acad. Sci. USA* 59:1089–96
39. Fosket, D. E. 1970. The time course of xylem differentiation and its relation to DNA synthesis in cultured *Coleus* stem segments. *Plant Physiol.* 46:64–68
40. Fosket, D. E., Roberts, L. W. 1964. Induction of wound vessel differentiation in isolated *Coleus* stem segments in vitro. *Am. J. Bot.* 51:19–25
41. Fosket, D. E., Torrey, J. G. 1969. Hormonal control of cell proliferation and xylem differentiation in cultured tissues

of *Glycine max* var Biloxi. *Plant Physiol.* 44:871–80
42. Garrison, R. 1949. Origin and development of axillary buds, *Syringa vulgaris* L. *Am. J. Bot.* 36:205–13
43. Gautheret, R. J. 1961. Action de la lumiére et de la température sur la néoformation de racines par des tissues de Topinambour cultives in vitro. *C. R. Acad. Sci.* 250:2791–96
44. Gregory, F. G., Veale, J. A. 1957. A re-assessment of the problem of apical dominance. *Soc. Exp. Biol. Symp.* 11: 1–20
45. Haber, A. H. 1968. Ionizing radiations as research tools. *Ann. Rev. Plant Physiol.* 19:463–89
46. Haddon, L. E., Northcote, D. H. 1975. Quantitative measurement of the course of bean callus differentiation. *J. Cell Sci.* 17:11–26
47. Haddon, L., Northcote, D. H. 1976. The influence of gibberellic acid and abscisic acid on cell and tissue differentiation of bean callus. *J. Cell Sci.* 20: 47–55
48. Hartig, T. 1853. Uber die Entwicklung des Jahringes der Holzpflanzen. *Bot. Ztg.* 11:553–60
49. Hartig, T. 1854. Über die Querscheidewände zwischen den einzelnen Gleidern der Siebröhren in *Cucurbita pepo*. *Bot. Ztg.* 12:51–54
50. Helm, J. 1932. Über die Beeinflussung der Sprossgewebe-Differenzierung durch Entfernen der jungen Blattanlagen. *Planta* 16:607–21
51. Hepler, P. K., Fosket, D. E. 1971. The role of microtubules in vessel member differentiation in *Coleus*. *Protoplasma* 72:213–36
52. Hepler, P. K., Fosket, D. E., Newcomb, E. H. 1970. Lignification during secondary wall formation in *Coleus:* An electron microscopic study. *Am. J. Bot.* 57:85–96
53. Hess, T., Sachs, T. 1972. The influence of a mature leaf on xylem differentiation. *New Phytol.* 71:903–14
54. Houck, D. F., La Motte, C. E. 1977. Primary phloem regeneration without concomitant xylem regeneration: Its hormone control in *Coleus*. *Am. J. Bot.* 64:799–809
55. Innocenti, A. M. 1975. Cyclic changes of histones/DNA ratio in differentiating nuclei of metaxylem cell line in *Allium cepa* root tip. *Caryologia* 28: 225–28
56. Innocenti, A. M., Avanzi, S. 1971. Some cytological aspects of the differentiation of metaxylem in the root of *Allium cepa*. *Caryologia* 24:283–92
57. Jacobs, W. P. 1952. The role of auxin in the differentiation of xylem around a wound. *Am. J. Bot.* 39:301–9
58. Jacobs, W. P. 1954. Acropetal auxin transport and xylem regeneration—a quantitative study. *Am. Nat.* 88:327–37
59. Jacobs, W. P. 1959. What substance normally controls a given biological process? 1. Formulation of some rules. *Dev. Biol.* 1:527–33
60. Jacobs, W. P. 1970. Regeneration and differentiation of sieve tube elements. *Int. Rev. Cytol.* 28:239–73
61. Jacobs, W. P., Morrow, I. 1957. A quantitative study of xylem development in the vegetative shoot apex of *Coleus*. *Am. J. Bot.* 44:823–42
62. Janse, J. M. 1921. La polarité des cellules cambiennes. *Ann. Jard. Bot. Buitenzorg* 31:167–80
63. Jeffs, R. A., Northcote, D. H. 1966. Experimental induction of vascular tissue in an undifferentiated plant callus. *Biochem. J.* 101:146–52
64. Jeffs, R. A., Northcote, D. H. 1967. The influence of indol-3yl acetic acid and sugar on the pattern of induced differentiation in plant tissue culture. *J. Cell Sci.* 2:77–88
65. Jost, L. 1939. Zur Physiologie der Geffässbildung. *Z. Bot.* 35:114–50
66. Kaan Albest, A. von. 1934. Anatomische und physiologische Untersuchungen über die Entstehung von Siebröhrenverbindungen. *Z. Bot.* 27:1–94
67. Keegstra, K., Talmadge, K. W., Bauer, W. D., Albersheim, P. 1973. The structure of plant cell walls. *Plant Physiol.* 31:188–96
68. Kiermayer, O. 1959. Gesteigerte Xylementwicklung bei *Solanum nigrum* durch Einfluss von Gibberellinsäure. *Ber. Dtsch. Bot. Ges.* 72:343–48
69. Kohlenbach, H. W., Schmidt, B. 1975. Cytodifferenzierung in Form einer direkten Umwandlung isolieiter Mesophyllzellen zu Tracheiden. *Z. Pflanzenphysiol.* 75:369–74
70. Lai, V., Srivastava, L. M. 1976. Nuclear changes during differentiation of xylem vessel elements. *Cytobiologia* 12:220–43
71. La Motte, C. E., Jacobs, W. P. 1962. Quantitative estimation of phloem regeneration in *Coleus* internodes. *Stain Technol.* 37:63–73
72. La Motte, C. E., Jacobs, W. P. 1963. The role of auxin in phloem regeneration in *Coleus* internodes. *Dev. Biol.* 8:80–98

73. List, A. Jr. 1963. Some observations on DNA content and cell and nuclear volume growth in the developing xylem cells of certain higher plants. *Am. J. Bot.* 50:320–29
74. Lykkesfeldt, A. E., Andersen, H. A. 1975. Inhibition of rRNA synthesis following incorporation of 5-bromodeoxyuridine into DNA of *Tetrahymena pyriformis. J. Cell Sci.* 17:495–502
75. McArthur, I. C. S., Steeves, T. A. 1972. An experimental study of vascular differentiation in *Geum-chloense. Bot. Gaz.* 133:276–87
76. McGahan, M. W. 1955. Vascular differentiation in the vegetative shoot of *Xanthium chinense. Am. J. Bot.* 42: 132–40
77. Millington, W. F., Gunkel, J. E. 1950. Structure and development of the vegetative shoot of *Liriodendron tulipifera. Am. J. Bot.* 37:326–35
78. Minocha, S., Halperin, W. 1974. Hormones and metabolites which control tracheid differentiation with or without concomitant effects on growth in cultured tuber tissue of *Helianthus tuberosa* L. *Planta* 116:319–31
79. Mizuno, K., Komamine, A. 1978. Isolation and identification of substances inducing formation of tracheary elements in cultured carrot-root slices. *Planta* 138:59–62
80. Mizuno, K., Komamine, A., Shimokoriyama, M. 1971. Vessel element formation in cultured carrot-root phloem. *Plant Cell Physiol.* 12:823–30
81. Mizuno, K., Komamine, A., Shimokoriyama, M. 1973. Isolation of substances inducing vessel element formation in cultured carrot root slices. In *Plant Growth Substances,* pp. 111–18. Proc. 8th Int. Conf. Plant Growth Substances. Tokyo: Hirokawa
82. Naik, G. G. 1965. *Studies on the effects of temperature on the growth of plant tissue cultures.* MSc thesis. Univ. Edinburgh, Scotland (cited in Ref. 90)
83. Neuberger, D. S., Evert, R. F. 1974. Structure and development of sieve-element protoplast in the hypocotyl of *Pinus resinosa. Am. J. Bot.* 61:360–74
84. Neuberger, D. S., Evert, R. F. 1976. Structure and development of sieve cells in the primary phloem of *Pinus resinosa. Protoplasma* 87:27–37
85. Northcote, D. H., Wooding, F. B. P. 1966. Development of sieve tubes in *Acer pseudoplatanus. Proc. R. Soc. Ser. B* 163:524–37
86. O'Brien, T. P. 1974. Primary vascular tissues. See Ref. 17, pp. 414–40
87. Odnoff, C. 1963. The effect of gibberellin and phenylboric acid on xylem differentiation and epidermal cell elongation in bean roots. *Physiol. Plant.* 16:474–83
88. Parthasarathy, M. V., Mühlethaler, K. 1969. Ultrastructure of protein tubules in differentiating sieve elements. *Cytobiologie* 1:17–36
89. Phillips, H., Torrey, J. G. 1974. The ultrastructure of the quiescent center in the apex of cultured roots of *Convolulus arvensis* L. *Am. J. Bot.* 61:871–78
90. Phillips, R., Dodds, J. H. 1977. Rapid differentiation of tracheary elements in cultured explants of Jerusalem artichoke. *Planta* 135:207–12
91. Phillips, R., Torrey, J. G. 1973. DNA synthesis, cell division and specific cytodifferentiation in cultured pea root cortical explants. *Dev. Biol.* 31:336–47
92. Phillips, R., Torrey, J. G. 1974. DNA levels in differentiating tracheary elements. *Dev. Biol.* 39:322–25
93. Phillipson, W. R., Ward, J. M., Butterfield, B. G. 1971. *The Vascular Cambium: Its Development and Activity,* London: Chapman & Hall. 182 pp.
94. Pickett-Heaps, J. D. 1967. The effects of colchicine on the ultrastructure of dividing plant cells, xylem wall differentiation and distribution of cytoplasmic microtubules. *Dev. Biol.* 15:206–36
95. Rehm, S. 1936. Zur Entwicklungsphysiologie der Gefässe und des trachealen Systems. *Planta* 26:255–74
96. Reinders-Gouwentak, C. 1965. Physiology of the cambium and other secondary meristems of the shoot. *Handb. Pflanzenphysiol.* 15:1077–1105
97. Reinert, J., Holtzer, H., eds. 1975. *Cell Cycle and Cell Differentiation.* New York: Springer. 331 pp.
98. Reinhard, E. 1954. Beobachtungen an in vitro kultivierten beweben aus dem Vegetationskegel der *Pisum* Wurzel. *Z. Bot.* 42:353–76
99. Rier, J. P., Beslow, D. T. 1967. Sucrose concentration and the differentiation of xylem in callus. *Bot. Gaz.* 128:73–77
100. Roberts, L. W. 1960. Experiments on xylem regeneration in stem wound responses in *Coleus. Bot. Gaz.* 121:201–8
101. Roberts, L. W. 1969. The initiation of xylem differentiation. *Bot. Rev.* 35: 201–50
102. Roberts, L. W. 1976. *Cytodifferentiation in Plants.* Cambridge Univ. Press. 160 pp.
103. Roberts, L. W., Fosket, D. E. 1966. Interaction of gibberellic acid and indoleacetic acid in the differentiation of

wound vessel members. *New Phytol.* 65:5–8
104. Rubery, P. H., Fosket, D. E. 1969. Changes in phenylalanine ammonia-lyase activity during xylem differentiation in *Coleus* and soybean. *Planta* 87:54–62
105. Rubery, P. H., Northcote, D. H. 1968. Site of phenylalanine ammonia-lyase activity and synthesis of lignin during xylem differentiation. *Nature* 219: 1230–34
106. Shininger, T. L. 1970. The production and differentiation of secondary xylem in *Xanthium pennsylvanicum. Am. J. Bot.* 57:769–81
107. Shininger, T. L. 1971. The regulation of cambial division and secondary xylem differentiation in *Xanthium* by auxins and gibberellin. *Plant Physiol.* 47: 417–22
108. Shininger, T. L. 1975. Is DNA synthesis required for the induction of differentiation in quiescent root cortical parenchyma? *Dev. Biol.* 45:137–50
109. Shininger, T. L. 1978. Hormone regulation of development in plant cells. *In Vitro* 14:31–50
110. Shininger, T. L. Quantitative analysis of temperature effects on xylem and non-xylem cell formation in cytokinin-stimulated root tissue. *Proc. Natl. Acad. Sci. USA.* In press
111. Shininger, T. L., Torrey, J. G. 1974. The roles of cytokinins in the induction of cell division and cytodifferentiation in pea root cortical tissue *in vitro.* In *Mechanisms of Regulation of Plant Growth,* ed. R. L. Bieleski, A. R. Ferguson, M. M. Cresswell, pp. 721–28. Wellington: Bull. 12 R. Soc. N.Z.
112. Siebers, A. M. 1971. Initiation of radial polarity in the interfascicular cambium of *Ricinus communis* L. *Acta Bot. Neerl.* 20:211–20
113. Siebers, A. M. 1971. Differentiation of isolated interfascicular tissue of *Ricinus communis. Acta Bot. Neerl.* 20:343–55
114. Siebers, A. M. 1972. Vascular bundle differentiation and cambial development in cultured tissue blocks excised from the embryo of *Ricinus communis* L. *Acta Bot. Neerl.* 21:327–42
115. Simon, S. 1908. Experimentelle Untersuchungen über die Entstehung von Gefässverbindungen. *Ber. Dtsch. Bot. Ges.* 26:364–96
116. Simpson, S., Torrey, J. G. 1977. Hormonal control of deoxyribonucleic acid synthesis and protein synthesis in pea root cortical explants. *Plant Physiol.* 59:4–9
117. Sinnott, E. W., Bloch, R. 1945. The cytoplasmic basis of intercellular patterns in vascular differentiation. *Am. J. Bot.* 32:151–56
118. Snow, R. 1933. The nature of the cambial stimulus. *New Phytol.* 34:288–96
119. Snow, R. 1935. Activation of cambial growth by pure hormones. *New Phytol.* 34:347–60
120. Söding, H. 1936. Uber den Einfluss von Wuchstoff auf das Dickenwachstum der Bäume. *Ber. Dtsch. Bot. Ges.* 54:291–304
121. Söding, H. 1940. Weitere Untersuchungen über die Wuchestoffregulation der Kambiumtätigkeit. *Z. Bot.* 36:8–141
122. Sorokin, H. P., Mathur, S. N., Thimann, K. V. 1962. The effects of auxins and kinetin on xylem differentiation in the pea epicotyl. *Am. J. Bot.* 49:444–54
123. Sorokin, H. P., Thimann, K. V. 1964. The histological basis for inhibition of axillary buds in *Pisum sativum* and the effects of auxins and kinetin on xylem development. *Protoplasma* 59:326–50
124. Srivastava, L. M., Singh, A. P. 1972. Certain aspects of xylem differentiation in corn. *Can. J. Bot.* 50:1795–1804
125. Swift, H. 1950. The constancy of deoxyribose nucleic acid in plant nuclei. *Proc. Natl. Acad. Sci. USA* 36:643–54
126. Thoday, D. 1939. The interpretation of plant structure. *Rep. Br. Assoc. Adv. Sci.* 1:84–104. Cited in Ref. 131
127. Thompson, N. P. 1965. *The influence of auxin on regeneration of xylem and sieve tubes around a stem wound.* PhD thesis. Princeton Univ., Princeton, NJ
128. Thornber, J. P., Northcote, D. H. 1961. Changes in the chemical composition of a cambial cell during its differentiation into xylem and phloem tissue in trees. *Biochem. J.* 81:449–55
129. Torrey, J. G. 1954. The role of vitamins and micronutrient elements in the nutrition of the apical meristem of pea roots. *Plant Physiol.* 29:279–87
130. Torrey, J. G. 1955. On the determination of vascular patterns during tissue differentiation in excised pea roots. *Am. J. Bot.* 42:183–98
131. Torrey, J. G. 1965. Physiological bases of organization and development in the root. In *Encyclopedia of Plant Physiology* 15(1):1256–1327
132. Torrey, J. G. 1968. Hormonal control of cytodifferentiation in agar and cell suspension cultures. In *Biochemistry and Physiology of Plant Growth Substances,* ed. F. Wightman, G. Setterfield, pp. 843–55, Ottawa: Runge

133. Torrey, J. G. 1975. Tracheary element formation from single isolated cells in culture. *Physiol. Plant* 35:158–65
134. Torrey, J. G., Fosket, D. E. 1970. Cell division in relation to cytodifferentiation in cultured pea root segments. *Am. J. Bot.* 57:1072–80
135. Torrey, J. G., Fosket, D. E., Hepler, P. K. 1971. Xylem formation: A paradigm of cytodifferentiation in higher plants. *Am. Sci.* 59:338–52
136. Tucker, S. 1959. Ontogeny of the inflorescence and flower in *Drimys winteri* var. dulensis. *Univ. Calif. Publ. Bot.* 30:257–336
137. Turgeon, R. 1975. Differentiation of wound vessel members without DNA synthesis, mitosis or cell division. *Nature* 257:800–8
138. Wangermann, E. 1967. The effect of the leaf on differentiation of primary xylem in the internode of *Coleus blumei* benth. *New Phytol.* 66:747–54
139. Wardlaw, C. W. 1947. Experimental investigations of the shoot apex of *Dryopteris aristata. Philos. Trans. R. Soc. Ser. B* 232:343–84
140. Wardlaw, C. W. 1965. The organization of the shoot apex. In *Encyclopedia of Plant Physiology* 15(1):966–1076
141. Wareing, P. F. 1958. Interaction between indoleacetic acid and gibberellic acid in cambial activity. *Nature* 181: 1744–45
142. Wetmore, R. H., Rier, J. P. 1963. Experimental induction of vascular tissues in callus of angiosperms. *Am. J. Bot.* 50:418–30
143. Wetmore, R. H., Sorokin, H. P. 1955. On the differentiation of xylem. *Arn. Arb. J.* 36:305–17
144. Wooding, F. B. P. 1969. P-protein and microtubular systems in *Nicotiana* callus phloem. *Planta* 85:284–98
145. Wooding, F. B. P., Northcote, D. H. 1964. The development of the secondary wall of the xylem in *Acer pseudoplatanus. J. Cell Sci.* 23:327–36
146. Young, B. S. 1954. The effects of leaf primordia on differentiation in the stem. *New Phytol.* 53:445–60

Ann. Rev. Plant Physiol. 1979. 30:339–67

EXPLANATORY MODELS IN CROP PHYSIOLOGY

❖7675

R. S. Loomis

Department of Agronomy and Range Science, University of California, Davis, California 95616

R. Rabbinge

Department of Theoretical Production Ecology, The Agricultural University, Wageningen, The Netherlands

E. Ng

Department of Agronomy and Range Science, University of California, Davis, California 95616

CONTENTS

0066-4294/79/0601-0339$01.00

"We can claim to understand the plant when we can express it all in a mathematical model."

Folke Skoog, over coffee, 1955

INTRODUCTION

An inherent feature of biological science is the conceptualization of complex systems into organization levels—from lower levels such as molecules, organelles, and cells to higher levels such as communities, populations, and ecosystems. Each of those hierarchic levels possesses a characteristic behavior resulting from integration of sublevel processes under influences from the external environment. Classical plant physiology explores the mechanistic basis for that behavior by reductionist techniques, by seeking to isolate each sublevel process from the influences of higher levels and from competing elements at the same level. Quantitative integration of those mechanisms into an explanation of system behavior, however, remains a task for integrative physiology. In that task, the interactions within and between levels become the foci of research. Particularly concerned with that problem are crop physiologists, whose task is to explain the behavior of vegetation in a variable environment.

Some types of physiological information are readily extrapolated from lower to higher levels; others are not. An understanding of certain qualitative phenomena, such as photoperiodism or the phase changes of lipids with temperature, may be used more or less directly in interpreting and predicting organism behavior. In other cases, the extrapolations may be frustrated by the very complexity of the interactions and their quantitative and temporal natures. As an example, the adequacy of a nutrient uptake system depends on variations in the activity of ion carriers in root membranes, ion availability in the soil, root surface area and distribution, degree of suberization, cortical and xylem transport resistances, ion assimilation capacity, and sink demand. The system is further complicated by its dynamic character —each of those factors is subject to diurnal and seasonal change.

Mathematical modeling is used increasingly as a method for effecting such integrations. That approach has been favored by the absence of other effective methodologies, by the emerging formalism of systems analysis, and by computers. The efforts in animal science are rather advanced. Major areas of work include explanatory models for thermoregulation, blood circulation, morphogenetic control, neurological functioning, and even artificial intelligence. It has even been possible to deal with the biochemical kinetics of ruminant digestion as a basis for organismal growth (5, 98). The basis for such work in many instances has been feedback theory developed in the 1940s (6, 72, 100), enzyme kinetics (42a, 49, 105), and compartmentalization concepts (2, 49).

The plant sciences have lagged well behind except in the physical aspects of the plant environment and community physiology. We find expanding interest in mathematical formulations of specific biological processes, with major attention given to such subjects as leaf growth and phyllotaxy (35, 69, 124, 128), carrier kinetics (20), photosynthesis (15, 17, 48, 119), and catenary diffusion sequences (84). The biomathematical analysis of physiological problems has been given extended treatments by Nobel (84), Riggs (99), and Thornley (117). In contrast, the integrative systems approach has been limited largely to the higher organization levels. That seems to reflect the quantitative concerns of systems ecologists and others, particularly of agronomists, for the behavior of vegetation. The grassland (56, 88) and tundra (12) biome studies, as examples, are impressive for their scope but are short in physiological detail. Workers in those areas have drawn more on the concepts of systems analysis and environmental physics as the starting point for their work. The systems level is also the arena for our own work on physiological models, emphasized in this essay. Our models focus on the organismal and systems levels, but the approach outlined is also applicable to integration at lower levels.

We distinguish two broad categories of crop models: same-level descriptive models and multilevel explanatory models. A wide range of descriptive models exist. Multivariate regression models, for example, are used widely for the important task of yield prediction in variable climates (10, 81, 82, 93, 96, 116). Such models may be static, i.e. involving no concept of time. Variables in that case are integrated seasonal totals of yield, rainfall, and temperature. Sophistication is improved by introducing some concept of time based, for example, on the calculation of developmental rate as a function of temperature during the season (101, 108) and by sharpening the environmental parameters, e.g. use of a soil moisture balance rather than rainfall as an input variable (3, 10).

The explanatory approach emphasized here is considerably more sophisticated. It employs dynamic models of the system hierarchy in an effort to provide prediction and explanation of integrated behavior from more detailed knowledge of the underlying physiological and morphological processes (26, 28). All such knowledge becomes descriptive at the ultimate level of reduction. While crop models do not go that far, they do become descriptive where knowledge is lacking or simplification is required. However, with a hierarchic structure, description at lower levels becomes explanatory of higher levels. A dynamic structure also aids in explanation, and the capability for continuous printout of many variables contrasts to experimental work generally providing observations only for discrete times.

In dynamic models, a system is described by a set of *state variables* (such as the weights of various organs) that are updated at each iteration of the model by *rate variables* (such as the flow of carbon in photosynthesis and respiration) defining changes in the state variables (Figure 1, left). The rate variables are considered to be constant during the iteration interval (DELT) so that the change in state is DELT X rate. DELT must be small enough that the assumed constancy is reasonably accurate. Models with too large an iteration interval give wrong results and may develop oscillations because of repeated overshoot and undershoot. The calculation of rate variables depends upon information from external *forcing variables* (such as air temperature) and internal *auxiliary variables* (such as the meristematic status of an organ) drawn from the current state of the system (Figure 1, right). A state variable and its determining processes represent a minimum hierarchy that becomes explanatory when the rate processes are calculated by rules based on the biological, chemical, or physical mechanism involved. A feature of most models is the use of balance equations at each level to ensure that conservation of mass and energy is observed. The method is not limited to integration from lower to higher levels, although that is the approach used with crop models. Recognizing that controls operate both upward and downward in biological hierarchies, one might employ simplified-ecosystem or organismal-level models to provide the controlling environment for a detailed organ or tissue model.

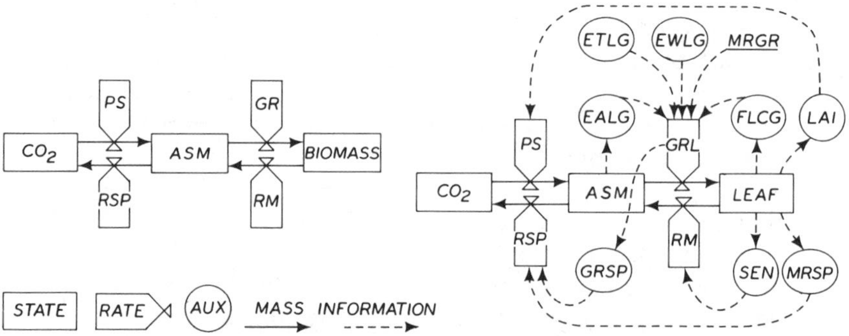

Figure 1 (*left*) A relational diagram for the transfer of carbon from atmospheric CO_2 to new assimilates (ASM) by photosynthesis (PS) and then by growth (GR) to new BIOMASS. Also shown are the reverse transfers by remobilization (RM) and respiration (RSP). (*right*) The hypothesis is extended with auxiliary variables controlling the growth rate of a leaf (GRL). The effect of assimilate supply on leaf growth rate (EALG) is shown to depend upon the ASM level; effectors for temperature (ETLG) and water status (EWLG) would be calculated in sub-routines. The meristematic fraction of the leaf capable of growth (FLCG) interacts with a maximum relative growth rate parameter (MRGR) to determine potential leaf growth. Other attributes of the leaf feedback to other processes: leaf area index (LAI) to photosynthesis; senescence state (SEN) to remobilization; and leaf weight and development state to maintenance respiration (MRSP). GRSP represents a calculation of the respiration associated with the growth achieved during each iteration of the model.

Although explanatory hierarchic modeling is still in its infancy and has not been subject to extensive development by systems analysts (70), the method holds great potential for plant physiology. Starting in the mid-1960s (11, 22, 114), it has become an active area of research by crop physiologists. For example, Milthorpe & Moorby's *Crop Physiology* (73) derives from their efforts in dynamic modeling, and the Trebon (107) and Long Ashton (62) conference volumes show a heavy modeling content. Much of the current activity is reported at workshops and in limited distribution publications. *Annals of Botany, Agricultural Meteorology, Journal of Theoretical Biology, Crop Science,* and the Dutch *Simulation Monographs* are among the major publications for botanical models. Rather than attempting an exhaustive review, the following essay draws largely from our own work. Our plan is: first, to present a limited background on the state-variable approach to systems analysis; second, to outline some of the special problems and attributes found in crop models at community, organismal, and cellular levels; and then to close with a survey of applications.

THE MODELING PROCESS

Model building should begin with a clear formulation of objectives concerning the use of the completed model. Biological systems are so complex that their models always represent a simplification or abstraction of the real system. That contrasts with some cases in engineering where the realized machine may be only an approximation of the perfection visualized in the model or plan. The objectives provide a basis for decisions about necessary simplifications. A second task, involving identification of the variables and processes that define the system, is aided by relational diagrams for the main variables (such as Figure 1). That task is coupled with the formulation of mathematical expressions for rate variables (i.e. differential equations). The choice and structure of those equations constitute a set of implicit assumptions about the system which should be carefully defined. The same is true in the choice of parameters for the equations. Taken together, the model with its parameterized equations represents a collective hypothesis about the real system. That leads directly to the construction of a computer program to execute the model and then to the critical step of validation (122).

Validation is distinguished from verification, which means testing to see that the computer program in fact operates on input data in the intended way. In addition to an a priori analysis of the model's structure (81a), validation generally involves comparison of model predictions with results from independent experiments relating to both processes (e.g. photosynthesis rate) and system states (e.g. biomass levels) (see Figure 2, p. 347). Some modelers use such comparisons as a basis for calibrating or "tuning" their models. That usually involves empirical adjustments of parameters to bring model performance into correspondence with standard behavior. Calibration can create a model useful for mimicking reality but is a dangerous practice for explanation. Departures from realism in model behavior usually represent either errors or incompleteness in the implicit assumptions on basic processes which should be given direct attention. de Wit (25) distinguished among real systems as repeatable, recurrent, or unique in terms of validation. Repeatable systems, such as fields of corn or manufactured cars, can easily be done again in independent validation experiments. Peat formation and forest successions, however, although recurrent in time, are too slow to repeat, and validation must be made on submodels or through comparison with a series of real systems in different stages of development. Examples of unique systems are the Mississippi River and biospheric cyling of carbon dioxide. In those cases, experimental perturbations of the real system may be hazardous, impractical, or socially unacceptable. Although validation may be possible from historic knowledge of past great events, it is clear that very strict criteria are needed in such constructions, particularly when they are used in forming public policy.

Validation may be extended through behavioral analyses (response of the model system to some pertubation, e.g. leaf pruning or climate change) and sensitivity analyses (response of the model system to systematic variations in model structure or of one parameter or input variable; Figure 2). They reveal the degree of truthfulness or realism with which the model handles the intended problem. They also tell us about the importance of various components in achieving that truthfulness and thus provide an objective basis for the simplification of complex models.

Those stages in model building are not mutually exclusive, and iteration and feedback among the stages is considerable (4). The process is actually little different from that used in experimental research, with the model hypotheses accepted or rejected through validation tests. Many subjective decisions are involved, and the quality of the model depends greatly on the skill and knowledge of the modeler. Almost invariably, deficiencies are found in the information base that define needs for additional experiments. That in itself has been one of the most rewarding features of modeling.

The choice of an appropriate time interval for iterations of a model is closely linked to objectives and to the levels of the system being modeled. In principle, the iteration interval must be only 0.1 to 0.2 times as long as the time required for a system to recover from a small perturbation. In practice, a sensitivity analysis with shorter intervals is usually necessary to determine the effect on accuracy. Higher levels and/or larger systems usually respond more slowly, and crop growth can be modeled with daily or weekly advances. But an interval of 1 to 2 hours must be used if sensitivity to diurnal events is required, while stomatal closure and some cellular processes require intervals of minutes or even seconds for accurate simulation.

Those facts cause coupling difficulties of the "stiff-system" sort when very fast subsystems are used to explain the behavior of the slower whole. If computer time is freely available, that problem is overcome by operating with an iteration interval appropriate to the fastest subsystem. Costs can be reduced sharply with special integration routines which allow the use of longer time intervals (43a), but those have yet to be used widely by crop modelers. Crop modelers sometimes use empirical submodels with slower time constants on the basis that less error results from that than from other features of their models. Alternatively, one can avoid the problem by limiting the hierarchic structure to two or three levels and thus restricting the range of time constants within the memory and computational capacities of their computers (and budgets). Either approach tends to compromise our objective of developing explanatory detail. At any level of detail, coupling problems are reduced by the introduction of negative feedback control since slight overshoot in one part of the system in one iteration tends to be corrected by a slight undershoot in the next.

Modeling can be facilitated with special simulation languages designed for use with state-variable models (9, 14, 39, 94). Such languages include integration routines, Gaussian generators, timing and array devices, automatic input/output formating, function generators for interpolation of tabular data, and a selection of more specialized intrinsic functions that can be accessed easily. Such languages are more expensive of computer time but save effort in programming. More important, the simplified programs can serve as a means for communication between modeler and experimenter. Thus far, plant modelers generally have not selected a universal simulation language. That plus the fact that few crop models are directed to the same objective means that very few standard program modules are shared by different modelers.

HIERARCHIC LEVELS IN CROP GROWTH MODELS

The principal focus of the output of crop growth models is community behavior. Such models simulate the production of new photosynthates, the partitioning of that material to growth, respiration and storage, and the related morphogenesis. The greatest attention is given to the state variables that define the environment and the age, weight, and morphology of the main elements of the biomass. That may be done with perhaps 50 to 100 state variables (21, 22, 87, 118). Computer models with 100 state variables are quite large and carry the danger that the model will be insensitive to incorrect opinions about structure and parameters. But a detailed hierarchic structure may require a very large number of variables, and some crop models concerned with integrative physiology have employed much larger numbers of variables [BACROS (11, 27, 29); SUBGOL (37, 38, 55); POTATO: Ng, unpublished].

The level of detail is determined by the aim of the modeling effort and the relative importance of various subprocesses to system behavior. Important processes should be developed with more detail. However, large models such as BACROS and SUBGOL also tend to reflect the present state of knowledge, providing detail on what is known and retreating to simple mimicking efforts on subjects such as morphogenesis where less is known. Models of limited size may be sufficient for many objectives. For that purpose, the highly detailed models can be simplified, following sensitivity analyses. It is also possible in that way to formulate simple algorithms of detailed submodels for use in more comprehensive models (37, 121).

The following sections outline some of the characteristics of model structure found with the various hierarchic levels of crop-growth models. Such models represent limited ecosystems consisting of the producer community and relevant abiotic components.

Community/Ecosystem Level

The milieu for crop growth is determined by environmental processes that function largely at the ecosystem level. The key elements are radiation interception and exchange, evaporation and transpiration, aerodynamic transport, and microclimate profiles as well as water and nutrient supplies. Each of those is some function of the area occupied by the vegetation and can be modeled as a vertical distribution. Environmental physicists have developed highly detailed explanatory models for most of those topics. For example, detailed models are available on infiltration and movement of water in soil (115, 123), including in some cases the influence of an expanding root system (53, 60, 63). Radiation interception by foliage also can be approached with rather sophisticated light distribution models (19, 24, 33, 44, 64, 76, 77, 102), and microweather within the vegetation can be simulated by coupling such models into net radiation budgets (latent and sensible heat exchanges and radiation balance) and eddy transport models (44, 83, 109, 126).

A useful approach for both aerial and soil environments is to subdivide the systems into horizontal layers, considering balances for each property within each layer, and using transport equations to calculate vertical fluxes between layers (Figure 2 shows the results of such calculations for air temperature and humidity). Transport equations (flux = gradient X conductivity) are used also for nutrient and water fluxes into roots and for water and carbon dioxide exchange by the leaves. Several interesting issues develop here. The conductivity term can be defined explicitly for a small system—in the case of roots, a single cell, or a small root segment—but it takes on a more general, empirical context when applied to a whole root system. But subdivision into smaller parts or layers can introduce a stiff-system problem. Radiation, for example, is absorbed at the surface soil layer and within leaves by very thin strata with a low capacity for heat storage. Those strata change temperature very rapidly in contrast to mixed air and the rest of the soil, which as large systems have much greater heat capacities and change temperature more slowly with time. Goudriaan (44) modeled that with a "bypassing" method in which the fast system is iterated to steady state and then abandoned (assumed to remain in steady state) until a new iteration is made of the higher level.

Organismal/organ-level elements, such as the size and characteristics of the foliage and root systems, enter directly into some of the physical processes. Most crop models have dealt only with random or homeogeneous distributions of roots and leaves within each layer, although other distributions may be important in nature. Those other arrangements, e.g. with plants in rows, can become quite complex (16, 42, 44). Microclimate models frequently include biological processes such as stomatal behavior, which

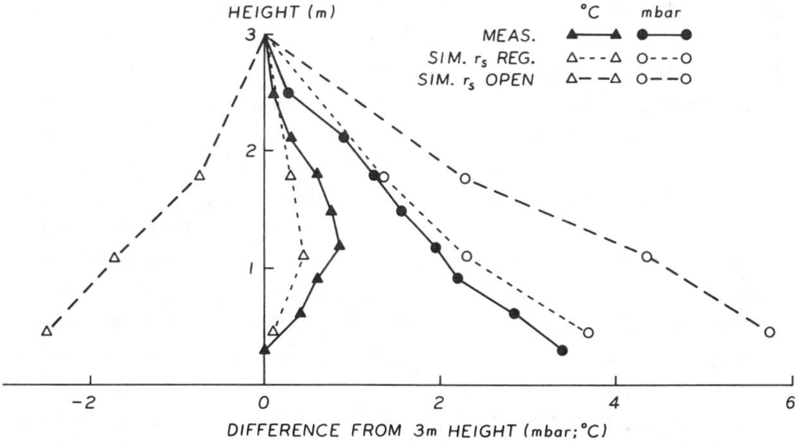

Figure 2 Validation of the BACROS, flux-dependent, microclimate simulator for air temperature and humidity through comparisons with profiles measured within a corn crop at Wageningen, the Netherlands. The profiles indicate the difference in temperature and humidity within the canopy as compared to that observed above the canopy at 3 m. Simulations with and without stomatal regulation represent a sensitivity analysis for the necessity of a submodel on stomatal control [adapted from (29)].

regulates water loss, and "suberization," which modifies root permeability to water and nutrients. The importance of stomatal control is illustrated clearly in Figure 2. Detailed models of stomatal action also have been developed (97, 110), and many ecosystem models include stomatal dependence upon the current levels of CO_2, water, and radiation.

Photosynthesis also is best treated as an ecosystem process because of its close dependence upon foliage display, radiation interception, and eddy transport. The more advanced models include all of those features. Shawcroft et al (109), for example, started with Duncan's (33) light distribution model for a layered leaf canopy. Ross (102) and Goudriaan (44) also presented highly detailed light models. Coupling light distribution with a leaf photosynthesis model achieves a simulation of canopy production. The simpler approaches employ a measured light-response curve for leaf photosynthesis; for more detail, a leaf model (15, 17, 48, 119) incorporating attributes such as quantum efficiency and a dependence of the saturation rate on temperature and CO_2 internal can be used. CO_2 internal can be simulated with an eddy transport model (to give CO_2 external) and a stomatal model (44).

Ecosystem models of that type offer a high degree of realism and accuracy in simulations of the main processes in crop productivity—photosynthesis, transpiration, and respiration of crop canopies. But it is not yet practical to employ all details available on ecosystem behavior in crop growth models which explore the partitioning of photosynthate during growth and development. The result would be a vast model very taxing to both computers and researchers. Submodels for the ecosystem parts can be simplified in various ways. BACROS retains considerable explanatory detail in the environmental and photosynthesis modules while using only rudimentary plant growth sections. In contrast, SUBGOL and POTATO employ simplified environmental modules while expanding on plant growth and development. In both cases, air and soil temperatures are taken as daylength-dependent, sinusoidal (or other) variations of reported screen temperatures, and evapotranspiration is estimated from a modified Penman equation (75) taking into account dew-point temperature and daily wind run with a submodel to describe (rather than simulate) stomatal conductance. BACROS simulates microweather within the canopy (Figure 2) whereas SUBGOL does not. BACROS approaches photosynthesis with a description of canopy architecture and a simple radiation penetration model coupled with transport equations for estimating the movement of CO_2 into the leaf based on stomatal conductance and the CO_2 gradient between leaf and air. In the Davis models, photosynthesis was reduced to a tabular presentation by running the Duncan photosynthesis simulator for 100 combinations of leaf-area distributions and solar altitude for clear and overcast skies, using a standard light-response function for leaf photosynthesis (37). The clear and overcast tables are interpolated at each hour according to solar angle and the ratio of potential and current daily total radiation (thus adjusting for cloudiness). Those hourly rates are then reduced by temperature and water-deficit functions, but CO_2 variations are ignored. Both approaches provide fast and reasonably accurate simulations of the photosynthate supply available for plant growth.

Organ/Organismal Level

The simplified vegetation models provide a framework within which the whole-plant level can be developed. An important aspect of that coupling is that a simulation of competitive effects due to varying plant density is achieved. At the plant level we can focus on detailed morphological descriptions of roots, stems, and leaves and their growth and ontogeny. Such models serve as means to explore partitioning and developmental processes and as a basis for integrative explanations of vegetation-level processes. Gutierrez et al (46) and Wang et al (127) incorrectly characterized crop growth models of that type as "single-plant" models (and also their limited-ecosystem model as a "population" model). Rather clearly, vegetation processes are simulated in such models at the ecosystem level, and those processes provide the photosynthate supply, water and nutrient status, and external environment which serve as forcing and auxiliary variables for the plant level. The multilevel model thus becomes reductionist as well as integrative.

Options exist to divide the plant into functional morphological classes (leaves, stems, and roots) and model each class en masse, or to model each individual leaf, internode, tuber, or fibrous root. Even when successive organs are considered separately, that is usually done for a "standard plant" so that the organs of all plants are identical. The en masse method may be used when ecosystem behavior is the principal interest, but the individual organ approach is usually required when integrative physiology is the aim. Some models take an intermediate approach by simulating the bulk behavior of all leaves or roots within specific "age" classes. That facilitates distinction of physiological capabilities (e.g. growth or senesence) according to developmental state and is a common approach for root systems. The age classes can be retained in programming devices known as "pushdown tables," advanced by an aging or developmental-rate submodel.

Modeling the initiation, growth, and development of individual organs in an explanatory way is not always easy or in some cases even possible. Little information exists about the mechanisms controlling the morphogenesis of individual organs. In many cases, the modeling becomes descriptive—for example, by using a temperature-dependent plastochron to control the initiation of successive leaves. Difficult questions also arise with the morphogenetic rules for integration and coordination of organs into whole plants. The rules center on partitioning (the distribution of new assimilates to growth centers) and physiological age. One approach to partitioning is to set the model "genotype" into descriptive allocation patterns (74, 96, 120) which may be drawn from real plants. But fixed allocations are likely to fail when the simulations are placed in a new environment. A more explanatory approach requires simulating morphological and physiological plasticity in response to density and other features of the environment, using variables that introduce the properties of apical dominance, photosynthate and water supplies, microclimate position, and age.

Photosynthate supply is made a central factor for organismal integration by following the nutritional-control approach of Brouwer and de Wit (11, 27) in which organ growth is dependent upon the concentration of available assimilates. The assimilate pool visualized in Figure 1 includes all readily available carbon fractions. It is fed by rates of photosynthesis and remobilization of materials from senescing organs, and is depleted by rates of growth, respiration, and storage. The growth of one organ then indirectly affects all others by altering that common pool. A division of the general pool into compartments for each growth center according to transport resistance (117) or axial position (52) attempts an explanatory basis compatible with morphological concepts but opens the issue of how to model vascular transport as a variable function of growth and development. Thornley (117) and Goeschl et al (131) have developed very nice models of phloem transport based on the number and dimension of cells but without a good solution to the developmental aspects. We have settled on a empirical "priority" concept in which each class of organs is given a different response function for substrate dependence. Those functions incorporate qualitative properties of transport resistance or position, as deduced from shading and pruning experiments with whole plants (37, 38), and the quantitative nature of substrate dependence of growth when unconstrained by transport (54). That approach gives more explanation to the simulation than descriptive allometry.

Similar pools can be established for water status and nutrient supplies. With water, which affects growth through variations in turgor, we have used a bulked water-status parameter such as plant relative water content (RWC), which is simulated from a balance between water uptake and transpiration.

The interaction of two or more pools, plus other variables relating to age and environment, establishes a dynamic pattern of partitioning that may include a balancing of root and shoot functions (11, 37). The balancing can be done with a model analogy (Figure 3) to real plants in which shoot

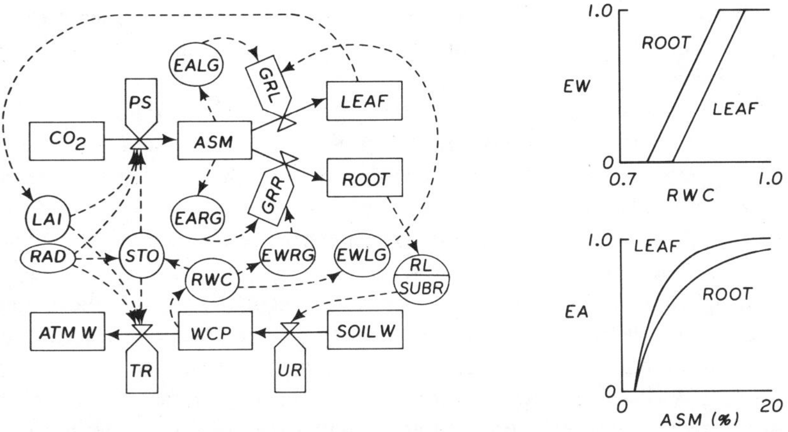

Figure 3 A relational diagram for a hypothesis about the functional balance of root and leaf growth based on carbon and water. The transfer of water from the soil through the plant to the air determines the current water content of the plant (WCP) and its relative water content (RWC). RWC in turn regulates root and leaf growth rates, according to the response functions (EW) shown to the right, and stomates. Also on the right are the response functions for the effects of assimilate supply (EA) on growth rates. Stomatal status (STO) is influenced by current radiation (RAD) and RWC and in turn regulates transpiration (TR) and photosynthesis (PS). The feedback from ROOT weight to water uptake rate (UR) involves root length (RL) and its suberization (SUBR) with age.
Structure similar to this is used in BACROS, SUBGOL, and POTATO.

growth is more affected than root growth by water or nutrient status (root supply functions) and root growth is more affected than shoot growth by assimilate status (a shoot supply function). Such *functional balances* represent hierarchic (across level) feedback loops, and their inclusion provides realistic organismal integration and greatly increases the power of the model. In addition, the door is then open to validation against data from root pruning and defoliation experiments (27, 38).

The state variables of interest in simulating the growth of an organ are its weight at present (W_t) and its rate of growth (GR). A common formulation is:

$$GR = MRGR \cdot F(AGE) \cdot W_t \cdot MIN(EA, EW, ET)$$

where MRGR is the maximum unrestrained relative growth rate ($g\ g^{-1}t^{-1}$), and F(AGE) is an "age"-dependent fraction of organ weight still capable of growth. MIN(EA, EW, ET) indicates the use of Liebig's law of the minimum to choose among the response functions for the most limiting of the effects of assimilate supply (EA), water status (EW), or temperature (ET) in that iteration. The new weight of the organ at the future time can be simply $W_t + GR \times DELT$. Figure 1, right, diagrams that scheme while Figure 4 shows the operation of such effectors during the growing season for sugar beet.

The cleanest conception of a response function is as the relative effect of one factor on the rate of a process with all other factors maintained near optimal levels. Thus, we would have the relative effect of temperature on growth rate, with assimilate, nutrients, and water nonlimiting. Such idealizations are not easy to achieve in experiments but can sometimes be approached with isolated systems (54, 79).

Liebig's law seems to hold in many cases, particularly when short time steps (1 hr) are employed. But in other cases, the effects of several factors may be additive or multiplicative and a multivariate approach [e.g. F(EA, EW, ET)] is required for limiting factor interactions. That is always necessary when growth rates are calculated for long time intervals (1 day). With either method, the modeling becomes descriptive at this level. The response functions and their interactions result from complex cell-level processes that cannot be modeled in detail largely because we do not understand the systems. For example, the basis of the response function for the temperature dependence of growth rate is unknown, and the manner and

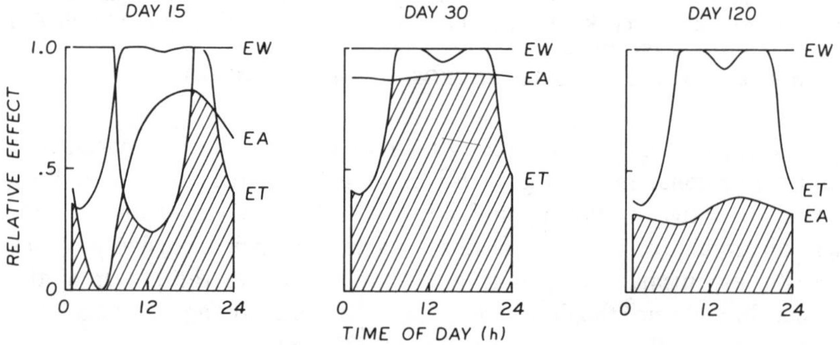

Figure 4 Operation of the Liebig's law analogy in SUBGOL for the effects of water status (EW), assimilate supply (EA), and temperature (ET) on sugar-beet leaf growth rate. The effect factors are derived from response functions such as those shown in Figure 3; a value of 1.0 indicates that the factor is not limiting to growth rate. The diurnal course of the effectors is shown for 15, 30, and 120 days after emergence on May 16, at Davis, California (38 N) with 7 plants m^{-2}; LAI, m^2m^{-2}/total biomass, g m^{-2} were 0.14/33, 1.18/180, and 4.82/2520 respectively at 15, 30, 120 days.

The shaded area shows the course of the most limiting factors. Inadequate root length leads to a water deficit (low EW) on day 15 but is not a factor thereafter in this well watered crop. On day 30, sink capacity is limited by the small size of the storage root and EA is near saturation for leaf growth; low night temperature (ET) is the main limiting factor. But at day 120, the system is source-limited (low EA; the storage root is very large and has a high capacity for growth) and temperature (ET), surprisingly, is not directly limiting to leaf growth rate.

mechanisms of how that may interact with substrate supply and other factors are also unknown.

Similar problems arise with the developmental concepts embodied in F(AGE). We can visualize F(AGE) as dependent upon the fractions of the organ that remain capable of further weight additions through division, expansion, or differentiation. With the sugar beet model, we approached that with descriptive cell division/differentiation generators to gain realistic simulations of the size of successive leaves (see Figure 5 later) (68). With wheat, Morgan (78) employed a simulation of apex size and primordium generation for that purpose. While the meristematic fraction of a particular organ depends to a considerable extent on past growth, other developmental events depend more on age or inductive conditions. Most crop growth models include aging routines in which chronological time is converted to "physiological time," or experience, with a temperature-response function such as has been found for plastochron events (37, 51), or as a "heat sum" (46). Suberization of roots, senescence, and "maturity" can be mimicked in that way while phasic development may require an additional dependence upon photoperiod (61, 117). Processes such as germination can be modeled with "dispersed delay" routines to generate a distribution of developmental states (28, 57, 95). But all of those approaches are only descriptive of developmental rate.

Leaf initiation rate is a key developmental control for organismal integration because that sets the rate of production of new leaves and lateral potential (67). Modeling of lateral branch initiation and growth has not been well developed. Frijters has expressed branching (41) and inflorescence (40) rule information in analytical equations, but the few crop models which deal with branching (78, 129; Ng unpublished with POTATO) have taken a simpler approach. A potential branching rule is set which is then limited at each iteration by assimilate status and physiological age (i.e. old axillary buds, long suppressed by lack of assimilate, lose their potential for growth). Similar aging/stress routines can be invoked for shedding of plant parts such as the flowers and bolls of cotton (32, 129).

The number, dry weight, and physiological age of various organs thus simulated represent a basic morphological description of the plant. It is also important to know something about the size and disposition of those organs. With leaves, for example, disposition in a foliage canopy influences mutual shading, affecting both production rate and leaf senescence. Crop models generally have depended on descriptive translation techniques using morphological response functions for converting a simulated increment of dry weight into an increment of size. With leaves, the key expression is area and can be translated from the weight of a leaf using an area/weight ratio expressed as a function of temperature and radiation environments, age, and assimilate status by means of the same "effector" approach we showed for the calculation of growth rates. Fortunately, crop ecology provides considerable information on how area/weight ratios vary with internal and external conditions (but not on why). Information is much less satisfactory on the variations in physiological capability during development.

Cell/Tissue Level

The effort given to modeling tissue and lower levels of organization has been much less than we saw at the higher levels. Crop modelers, with their principal focus on community behavior, find that cellular submodels not only tax their competence but also lead to unmanageably large models. More seriously, we have a very poor understanding of how organ behavior is determined by cellular processes. As a result, tissue-level information in organismal models frequently consists of descriptive functions.

Considering the great amount of information which exists on metabolic pathways, the kinetic properties of enzymes, and biomathematics of component elements, the modeling of tissues does not appear difficult. Some progress has been made with integration of differential equations for uncompartmentalized biochemical components of such systems (43, 71). But placing that into a physiological model of a tissue is another matter (2, 49, 85). Morphological and developmental description is just as essential at the cellular level as at the organismal level. We need to work toward an ability to simulate the changes in metabolic ability and compartmentation which occur during development, and an explanatory approach will require a simulation of the controls over cell differentiation (130).

Critical in our current work with a simulator of nitrate metabolism (85) are the size of the cytosol and vacuolar compartments and the membrane transport capacity between them. No explanatory basis exists now for simulating those entities over time, and they must be dealt with descriptively. Indeed, the word vacuole has become rare in indices of plant physiology texts. We also must use descriptive generators to translate experimental data on the kinetics of biochemical processes in vitro (generally per unit tissue fresh weight or per unit protein) into cellular-level physiological process. Despite those problems, the insights gained from the nitrate model are quite intriguing: while the biochemical model explains the dynamics of certain intermediates, organismal properties (supplies of nitrate and carbon, sinks for amino acids) rather than enzyme kinetics provide the principal means of regulation. Except where branched pathways occur, it seems that metabolic systems can be simplified to single operators, or a few sequentially linked operators, each performing a transfer function according to certain rules.

Models dealing with the stoichiometry of biochemical processes have been more successful than kinetic models. The best example is the elegant respiration model of Penning de Vries and coworkers (91). In that, a simple set of assumptions regarding synthesis via least-cost pathways, degree of respiratory coupling, and "tool" maintenance allows the model to calculate substrate use, O_2 requirement, and CO_2 production for the respiration associated with the biosynthesis of specific end products. The model has been subjected to validation tests (89, 92) with reasonable success, and unknown elements such as tool maintenance (cost of enzyme and mRNA turnover) were subjected to sensitivity analyses (91).

The Penning de Vries model has been used to calculate biological efficiency in the formation of complex organs (7, 111). And it has been simplified for use in crop growth models (29, 55), where it adds a great deal of explanation to the simulations, and season-long consequences of biomass composition can be evaluated. Its success in those instances results from treatment of the new growth en masse, without attention to cellular detail except to specify the biochemical composition of existing and newly formed biomass.

That approach holds that the respiration costs of biosyntheses and growth are independent of temperature and that temperature operates only through an influence on the rates of biosynthesis. The respiratory costs of cell maintenance must be approached more empirically. In Penning de Vries' (90) analysis, the explanatory basis of maintenance respiration lies principally in coupling to lipid and nonenzymic protein turnover and maintence of ion concentrations. Those processes cannot be assessed in detail, but under normal conditions they can be estimated (90) to require 15 to 25 mg glucose g^{-1} (dry weight) day^{-1} for leaf tissue. That amount falls within the range of observed values. The maintenance respiration load is a critical factor to productivity—variations in its rate between 1 and 3% per day lead to large difference in predictions of organ growth (104) and seasonal productivity (55).

SPECIAL ISSUES IN PHYSIOLOGICAL MODELING

We have noted how the concept of physiological age and the relations between structure and function introduce difficulties for modeling crop growth. In those cases, the modeling efforts provide a new viewpoint for experimental studies on uncommon topics in plant physiology. The following sections comment on two other biological issues of similar promise and on the ways in which modelers have coped with them.

Stochastic Versus Deterministic Simulation

Real biological systems display a great deal of variation at all levels of organization. One part of that variation is the result of the plasticity that plants show with variations in environment; another arises from the genotypic variability of plant populations. There are good arguments for stochastic treatment of environment and biological response, but the models we have described are mostly deterministic in that all plants are of a single genotype and are exposed to a single starting time and a single environment. The deterministic approach provides a prediction of mean behavior, following the law of large numbers. On a small scale (one cell, one plant), however, there can be large departures from that mean. Methods for introducing probabilistic elements into initial conditions and rate variables are readily available in computer languages. Their use can quickly become a meaningless exercise, however. A community model of stochastic plants requires three-dimensional treatment of space with lateral duplication of the organismal model—a prohibitively large problem. More fundamentally, the "noise" generated by realistic stochastic treatment of 100 variables over 1000 iterations can exceed by many orders of magnitude the variation found in real systems. The explanation for that difference is that real systems are strongly constrained by feedback, functional balances, and other homeostatic mechanisms. Stochastic variations in individual processes are strongly damped or eliminated in the integrated system. Thus, explanatory models must also include feedback mechanisms if they are to achieve realism. The only alternative is to provide arbitrary limits to the course of the simulation, and that degenerates to description and leads to a loss of predictive value. One way of studying variation is to introduce distributive (57) or stochastic generators into only selected processes. A large number of simulations would then generate a "genetic" or "environmental" population sample. We have done little of that because of high cost and questions in interpretation.

Those questions bear closely on the problems associated with the important issue of simulations of mixed vegetation (8, 12, 56, 88). That problem has yet to be studied seriously with detailed plant growth models, but it is easier in some respects since specific spatial arrays can be established and a limited number of genetically different organismal models can be linked laterally. Complementary models of that sort must give attention to the vertical distribution of leaves and roots so that central issues of interference among species for radiation, nutrients, and water are simulated properly.

Adaptation

Some crop growth models automatically predict the larger aspects of climate-induced physiological and morphological change. Since the simulations are dynamic, the current state of the system (number, size, and age of organs) represents a condition with adaptation to the environmental history used in the simulation. Also, as noted above, morphological translations can be made subject to the current environment (e.g. sun vs shade for leaves) for each increment of growth.

The deterministic aspects of genetic adaptation are handled easily by a change in model structure to a new "genotype." But physiological adaptation involving alterations in physiological capability per unit tissue (in contrast to the weight of tissue capable of a function) must be handled carefully. Some physiological adjustments occur rapidly. Those can be accommodated by broadening the physiological functions so that optimum performance occurs over a broader range of conditions than might be found with short-term observations [e.g. the photosynthesis-temperature relation in BACROS (29)]. Phenomena such as hardening, with their attendant slow changes in anatomy and physiological capability of both existing and new tissues, are more difficult. If a reasonable data base can be found (which unfortunately is generally not the case), the process can be described in a manner analogous to that of physiological age, using an integrator of stress experience. That indicator can then be used in modifying physiological processes and morphological translations to produce a hardened state. Crop modelers have yet to give serious attention to physiological adaptation. Most are still focused on developing realistic simulations of "normal" plants, well watered and well supplied with nutrients, but eventually we must also come to an ability to simulate acclimatory processes (112).

APPLICATIONS

Crop physiologists have long sought some means for applying physiological information to quantitative interpretations of plant growth in agricultural systems. State-variable models with hierarchic structure deal directly with the translation to the field of mechanisms elucidated in the laboratory. Progress toward interpretation of field behavior has been slow, however, largely because of the nature and infancy of the method. The rather special kinds of physiological and morphological information required as input come largely from specialized-organ and organismal-level research in which the information base is weak. Thus, the modeling efforts couple poorly with the current mainstreams of cellular-level research. In our own programs, we find that 50 to 80% of our effort goes into experiments to fill such information gaps. Other problems arise from the interdisciplinary and subjective nature of the work. Good biology is essential, but biologists generally are not very skilled in systems analysis and the best systems analysts may be poor biologists.

Fall-out benefits, however, such as Penning de Vries' respiration studies (90, 91), the erect-leaf hypothesis (31, 33) and the Buringh-van Heemst (13) analysis of world food production have been significant, and the crop models themselves have been highly useful in certain applications. Some of the major areas merit brief review.

Productivity and Bioclimatology

Many crop models have had yield prediction as a principal objective. In some cases that objective has been attained consistently and well, in others, accuracy is poor. The estimates of gross photosynthesis provided by ecosystem-level models, when corrected for respiration, provide good predictions of primary productivity (1, 24). Economic yields can be derived from that using generalized partitioning factors, and our best current estimates of global food production under various agricultural strategies have been obtained in that way (13).

The addition of an organismal level adds additional environmental dependence and accuracy to the prediction. But most dynamic models with organismal level submodels are aimed more at optimal conditions than at usual commercial conditions. As a result, the multivariate regression models reviewed earlier are still the principal means for yield prediction. With proper tuning, such models accommodate better to average field conditions since the historic data include the effects of variations in plant stand, disease and pests, and nutrient and water supply which may be the principal determinants of yield. Such regression models perform best in predicting the mean performance of a population of fields, whereas the dynamic models may work best with the individual field. Among dynamic models, the Gutierrez cotton model, when parameterized for normal production practice, performs well in prediction (46). BACROS and SUBGOL, which give emphasis to the achievement of realism through hierarchic structure rather than to accuracy, have done surprisingly well in prediction for optimum conditions. That success probably derives from accuracy in simulating photosynthetic productivity, which under optimal conditions varies chiefly with radiation (1).

The orientation of existing crop growth models toward optimal conditions limits their use in crop management research. Exceptions occur with varietal-choice and timing aspects of management. With forage crops, for example, management decisions center on timing and intensity of clipping or grazing. The SIMED alfalfa simulator (52) handles the recovery from such defoliation quite well. The consequences of various management strategies can be given in graphical displays useful for extension education and research.

Dynamic models, either complex or simplified, will be particularly useful in climatological assessments. Preliminary results including the prediction that thermoperiodism can result from the integration of growth processes under diurnal regimes are promising (66). Such models can be used in regional climate analyses (66, 113, 121) and as a basis for upgrading the multivariate methods.

Integrative Physiology and Ideotype Evaluation

The integrative physiological aspects of crop models have been directly relevant to some of the major issues in crop physiology. The greatest advantage comes from quantitative integration over time of simple physiological and morphological traits in source-sink relations. Such models constitute organized bodies of knowledge about whole-plant physiology. The simulations provide a way to describe and explain the consequences of increasing or decreasing photosynthate supply and the number of meristematic centers or their capability for growth. Similarly, one can explore the effects of specific weather sequences.

In the thermoperiodism case (66), large diurnal fluctuations in air temperature placed the growing leaves at temperatures unfavorable for growth for many hours each day. Soil temperature fluctuated less and the roots remained at temperatures favorable for growth throughout the day. Larger sugar beet storage roots were obtained in such simulations.

Since the opinions in the models can be viewed as "genetic traits," the models can serve in a similar way for formulating and evaluating genetic combinations (ideotypes) more suitable in crop production than existing strains (30, 65). At present, plant breeders have little basis other than trial and error for combining quantitative physiological and morphological traits into new phenotypes. An integrative tool is needed because yield improvements through plant breeding almost invariably have come through changes in partitioning rather than through improvements in photosynthetic capability (23, 36). In one example, Duncan et al (34) have shown with a simulation model of peanut that dramatic yield increases in cultivars of that species came solely from changes in flowering time and other aspects of partitioning. In other cases, the simulations allow clear identification of features such as crop duration and slow development of leaf area as principal limits to seasonal yield. It seems likely that future improvements in most crops will also come through changes in partitioning rather than in photosynthesis. That certainly is the case with cotton and sugar beet. But many of the simple partitioning traits such as lodging resistance in small grains have been well explored, and progress will rest more on combinations of quantitative traits. The possible combinations of traits can be very large, but by simulation, certain optimal hypotheses can be identified as breeding objectives.

It is surprising how many model predictions for ideotype concepts are counter-intuitive—a low maintenance respiration requirement may translate into a greater respiration loss over a season [because more biomass is accumulated early (55)]—or outside of conventional wisdom. One hypothesis about sugar beet (from shading experiments) was that storage beet growth was accomplished from surplus assimilate not used by leaves. That opinion yielded a decent sugar beet simulator, and by changing leaf growth potentials (genetic switch), sugar beet, chard, and mangel phenotypes were generated (67, 68). The simulated result was similar (Figure 5), however, when the control was placed on root growth, i.e. chard leaves became large because their roots had a low capacity for growth. That conflict led to experiments with reciprocal grafts between chard and beet that confirmed the concomitant operation of both hypotheses (68). As another example, the models generally predict that source, not sinks, is limiting to production rate in closed stands. That shows clearly in Figure 4 where growth rate at 120 days is surprisingly independent of temperature. Under those conditions, a high capacity for growth as characterized in hybrid vigor is quickly negated by areal restraints (radiation, water and CO_2 flux). High densities of small plants are shown to do as well as low densities of large, heterotic plants.

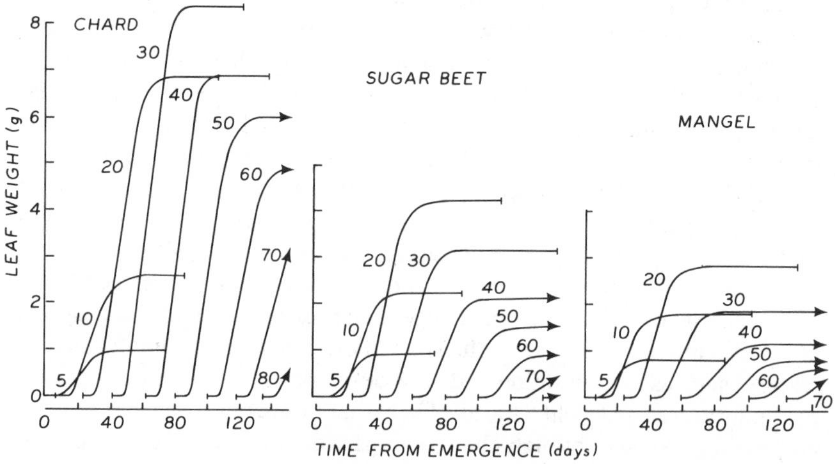

Figure 5 Simulations with SUBGOL of the weights of successively numbered genetically identical sugar-beet leaves. The storage-root submodel was modified to mimic the small roots of chard and the large roots of mangel-wurzeln. Leaves grown in competition with normal sugar beet storage roots are shown in the center, in competition with chard roots to the left, and with mangel-wurzeln roots to the right. Emergence on May 16 with 7 plants m^{-2} at Davis, California (38 N).

The principal features of the juvenile-adult leaf sequence in these varieties is duplicated: small, slow growing leaves early followed by large leaves and then smaller leaves, reflecting the effectors displayed in Figure 4. However, the later leaves are smaller than real leaves, indicating some defect in the model in simulating the plasticity to intra- and interplant competition [adapted from (68)].

Plant-Herbivore Relations

There is an increasing use of crop growth models across trophic levels in which the crop model serves as a dynamic description of the substrate and environment for grazing animals. The emphasis in pest management studies is reflected in Ruesink's review (103). Most entomological models emphasize the description of pest population dynamics, with stochastic submodels to simulate the infection and spread of the insects. The amount and specific sites for insect feeding or disease damage become important. Combined models of that sort have proved to be reliable predictors of pest or disease development and expected injury to the crop, and they are useful in the study of control strategies. Combined models have been developed for many situations with those for cotton (46, 47, 58, 59, 127), alfalfa (45), and apple (95) serving as examples. In some cases, those have led to simplified econometric models for decisions about spraying or praying. There have been similar efforts toward the simulation of plant diseases (125). The combined models also offer a means, little used as yet, for examining biological efficiency (energy and nutrient transfers) in host-parasite couplings.

Similar activity is found with researchers concerned in analyses of the grazing of vegetation by large animals. Much of that work centers on range and pasture management (18, 88, 106) and ruminant nutrition (98), but the issue is met also in studies of natural grazing of tundra (12) and grassland (56, 88). Here again, somewhat simplified vegetation models characterize the supply and nutritional status of feed and the physical environment of the animals. Such models frequently must deal with mixed vegetation, such as the grass-legume combination, and with variations in animal preferences for the various forages.

AN ASSESSMENT: SENSE OR NONSENSE

Several things are now clear about the future of systems analysis in plant physiology. The unbridled enthusiasm that many of us displayed during our early euphoria with the method must now be tempered. A great deal of hard work remains, and "grand" models are not about to substitute for real plants and real experiments. Still, in many ways the modeling is ahead of the information base, and it is likely to remain there as computer capacities increase and costs decline. That is particularly true at the whole-plant level which has not been emphasized in plant physiology research. We can expect the modeling efforts to continue as sources of innovative questions (and sometimes of answers) about those gaps in our knowledge of plant life.

Passioura (86) raised a storm among modelers with a pungent and thoughtful essay on "Sense and Nonsense in Crop Simulation." One of his suggestions was that a little clear thinking about systems problems would contribute more to the advance of our science than complex models. Passioura was not alone among the cautionaries. Crop modeling has shared the criticism directed at other areas of modern ecology as lacking in depth and unifying concepts and subject to excessive jargon. But just as conceptual models such as "carriers" and "genes" and simple analytical models such as Fick's Law and the Michaelis-Menten expression are now integral features of plant physiology, so are hierarchic simulation models. Their *raison d'être* is that the problem is there. No other means exists as powerful for the integrative physiology of plants as adaptive control systems. Quantitative assessments of the importance of various physiological and morphological traits, extrapolations from laboratory to field, conduct of otherwise impossible experiments, and the exploration of integrative controls are all within their domain (4, 80). It was once hoped that phytotrons would fill that role. But phytotrons have not been used effectively for that, and it now seems that modelers will be the principal consumers of phytotron results (37, 50) and that models will be the integrative tool.

We feel also that there is considerable promise for the use of systems analysis for integration at lower levels. The early efforts are promising but nothing in plant physiology yet approaches the detail and sophistication of the models of cellular processes found in animal research. One limitation is the evident lack of a physiological systems view in plant biochemistry for processes other than photosynthesis; modern plant biochemistry texts reflect this in their focus on natural product classes rather than plants. Cell physiologists are well equipped to fill the serious information gap between cell and whole-plant physiology. Progress with tissue-level models would provide considerable help for crop modelers, who have generally proceeded from the top down.

The fact that hierarchic models have been limited more by our knowledge and conceptualizations of the system than by computing facilities, software, and system theory is a natural reflection of the need for at least a few more years of effort in plant physiology research. But the more we learn through reductionist research, the greater the need and opportunity for integrative research. Our conviction is that systems methods will become more and more central to plant physiology.

ACKNOWLEDGMENTS

We thank C. T. de Wit, W. G. Duncan, and P. C. Miller for inspiration but would deny them responsibility for inadequacies in the product. R. Rabbinge received support from the Dutch Organization for the Advancement of Pure Sciences, and E. Ng received support from a cooperative agreement of the University of California with the USDA and the University of Idaho (12–14–5001–287) during this work. E. Ng's present address is: Mars Ltd., Slough, U.K.

Literature cited

1. Alberda, T., Sibma, L. 1968. Dry matter production and light interception of crop surfaces. III. Actual herbage production in different years compared with potential values. *J. Br. Grassl. Soc.* 23:206–15
2. Atkins, G. L. 1969. *Multicompartment Models for Biological Systems.* London: Methuen. 153 pp.
3. Baier, W., Robertson, G. W. 1968. The performance of soil moisture estimates as compared with the direct use of climatological data for estimating crop yields. *Agric. Meteorol.* 5:17–31
4. Baker, C. H., Curry, R. B. 1976. Structure of agricultural simulators: a philosophical view. *Agric. Syst.* 1:201–18
5. Baldwin, R. L., Koong, K. J., Ulyatt, M. J. 1977. A dynamic model of ruminant digestion for evaluation of factors affecting nutritive value. *Agric. Syst.* 2:255–88
6. Baylis, L. E. 1966. *Living Control Systems.* San Francisco: Freeman. 189 pp.
7. Bhatia, C. R., Rabson, R. 1976. Bioenergetic considerations in cereal breeding. *Science* 194:1418–21
8. Botkin, D. B., Janak, J. F., Wallis, J. R. 1972. Some ecological consequences of a computer model of forest growth. *J. Ecol.* 60:849–72
9. Brennan, R. D., de Wit, C. T., Williams, W. A., Quattrain, V. E. 1970. The utility of a digital simulation language for ecological modeling. *Oecologia (Berlin)* 4:113–32
10. Bridge, D. W. 1976. A simulation model approach for relating effective climate to winter wheat yields on the Great Plains. *Agric. Meteorol.* 17:185–94
11. Brouwer, R., de Wit, C. T. 1969. A simulation model of plant growth with special attention to root growth and its consequences. In *Root Growth,* ed. W. J. Whittington, pp. 224–44. London: Butterworth. 450 pp.
12. Brown, J., ed. 1979. *An Arctic Ecosystem: The Coastal Tundra of Northern Alaska.* Stroudsburg, PA: Dowden, Hutchinson & Ross. In press
13. Buringh, P., van Heemst, H. D. J. 1977. *An Estimation of World Food Production Based on Labour-Oriented Agriculture.* Wageningen: Cent. World Food Market Res. 46 pp.
14. Buxton, J. N. 1968. *Simulation Programming Languages.* Amsterdam: North Holland. 464 pp.
15. Charles-Edwards, D. A., Ludwig, J. L. 1974. A model for leaf photosynthesis by C_3 plant species. *Ann. Bot.* 38: 921–30
16. Charles-Edwards, D. A., Thorpe, M. R. 1976. Interception of diffuse and direct-beam radiation by a hedgerow apple orchard. *Ann. Bot.* 40:603–13
17. Chartier, P., Prioul, J. L. 1976. The effects of light, carbon dioxide and oxygen on the net photosynthetic rate of the leaf: a mechanistic model. *Photosynthetica* 10:20–24
18. Christian, K. R., Freer, M., Donnelly, J. R., Davidson, J. L., Armstrong, J. S. 1978. *Simulation of Grazing Systems.* Wageningen: Pudoc. 115 pp.
19. Cowan, I. R. 1968. The interception and absorption of radiation in plant stands. *J. Appl. Ecol.* 5:367–79
20. Cram, W. J. 1976. Negative feedback regulation of transport in cells. The maintenance of turgor, volume and nutrient supply. In *Encyclopedia of Plant Physiology,* new ser., ed. U. Luttge, M. G. Pitman, 2A:284–316. Berlin: Springer-Verlag. 394 pp.
21. Curry, R. B., Baker, C. H., Streeter, J. G. 1975. SOYMOD I. A dynamic simulator of soybean growth and development. *Trans. ASAE* 18:963–74
22. Curry, R. B., Chen, L. H. 1971. Dynamic simulation of plant growth—Part II. Incorporation of daily weather and partitioning of net photosynthate. *Trans. ASAE* 14:1170–74
23. de Vries, C. A., Ferwerda, J. D., Flach, M. 1967. Choice of food crops in relation to actual and potential production in the tropics. *Neth. J. Agric. Sci.* 15:241–48
24. de Wit, C. T. 1965. Photosynthesis of leaf canopies. *Agric. Res. Rep. 663,* Wageningen. 57 pp.
25. de Wit, C. T. 1978. Simulatie van le-

vende systemen. *Landbouwk. Tijdschr.* 90:237–40
26. de Wit, C. T., Arnold, G. W. 1976. Some speculation on simulation. In *Critical Evaluation of Systems Analysis in Ecosystems Research and Management,* ed. G. W. Arnold, C. T. de Wit, pp. 3–9. Wageningen: Pudoc. 108 pp.
27. de Wit, C. T., Brouwer, R., Penning de Vries, F. W. T. 1970. The simulation of photosynthetic systems. See Ref. 107, pp. 47–70
28. de Wit, C. T., Goudriaan, J. 1974. *Simulation of Ecological Processes.* Wageningen: Pudoc. 159 pp.
29. de Wit, C. T., et al. 1978. *Simulation of Assimilation, Respiration and Transpiration of Crops.* Wageningen: Pudoc. 141 pp.
30. Donald, C. M. 1968. The breeding of crop ideotypes. *Euphytica* 17:385–403
31. Duncan, W. G. 1971. Leaf angles, leaf area, and canopy photosynthesis. *Crop. Sci.* 11:482–85
32. Duncan, W. G. 1972. SIMCOT: A simulator of cotton growth and yield. In *Proc. Workshop on Tree Growth Dynamics and Modeling,* ed. C. Murphy et al, pp. 115–18. Durham, NC: Duke Univ.
33. Duncan, W. G., Loomis, R. S., Williams, W. A., Hanau, R. 1967. A model for simulating photosynthesis in plant communities. *Hilgardia* 4:181–205
34. Duncan, W. G., McCloud, D. W., McGraw, R. L., Boote, K. J. 1978. Physiological aspects of peanut yield improvement. *Crop. Sci.* 18:1015–20
35. Erickson, R. O. 1976. Modeling of plant growth. *Ann. Rev. Plant Physiol.* 27:407–34
36. Evans, L. T. 1975. The physiological basis of crop yield. In *Crop Physiology,* ed. L. T. Evans, pp. 327–55. Cambridge, UK: Cambridge Univ. Press. 374 pp.
37. Fick, G. W., Loomis, R. S., Williams, W. A. 1975. Sugar beet. See Ref. 36, pp. 259–95
38. Fick, G. W., Williams, W. A., Loomis, R. S. 1973. Computer simulation of dry matter distribution during sugar beet growth. *Crop. Sci.* 13:413–17
39. Forrester, J. W. 1961. *Industrial Dynamics.* Cambridge, Mass: MIT Press. 464 pp.
40. Frijters, D. 1978. Principles of simulation of inflorescence development. *Ann. Bot.* 42:549–60
41. Frijters, D. 1978. Mechanisms of developmental integration of *Aster novae-angliae* L. and *Hieracium murorum* L. *Ann. Bot.* 42:561–75
42. Fukai, S., Loomis, R. S. 1976. Leaf display and light environments in row-planted cotton communities. *Agric. Meteorol.* 17:353–79
42a. Garfinkel, D., Garfinkel, L., Moore, W. T. 1977. Computer simulation as a means of physiological integration of biochemical systems. In *Mathematical Models in Biological Discovery,* ed. D. L. Solomon, C. Walter, pp. 147–73. *Lecture Notes in Biomathematics,* Vol. 13. New York: Springer-Verlag. 240 pp.
43. Garfinkel, D., Williamson, J. R., Olson, M. S. 1969. Simulation of the Krebs cycle. *Simulation* 11:43–48
43a. Gear, C. W. 1971. *Numerical Initial Value Problems in Ordinary Differential Equations.* New York: Prentice-Hall
44. Goudriaan, J. 1977. *Crop Micrometeorology: A Simulation Study.* Wageningen: Pudoc. 250 pp.
45. Gutierrez, A. P., Christensen, J. B., Merritt, C. M., Loew, W. B., Summers, C. G., Cothran, W. R. 1976. Alfalfa and Egyptian alfalfa weevil (Coleoptera, Curailionidae). *Can. Entomol.* 108: 635–48
46. Gutierrez, A. P., Falcon, L. A., Loew, W., Leipzig, P. A., van den Bosch, R. 1975. An analysis of cotton production in California: A model of Acala cotton and the effects of defoliators on its yields. *Environ. Entomol.* 4:125–36
47. Gutierrez, A. P., Leigh, T. F., Wang, Y., Cave, R. D. 1977. An analysis of cotton production in California: *Lygus hesperus* injury—an evaluation. *Can. Entomol.* 109:1375–86
48. Hall, A. E., Bjorkman, O. 1975. Model of leaf photosynthesis and respiration. In *Perspectives of Biophysical Ecology,* ed. D. M. Gates, R. B. Schmerl, pp. 55–72. New York: Springer-Verlag. 609 pp.
49. Heinmats, F. 1970. *Quantitative Cellular Biology.* New York: Dekker. 327 pp.
50. Hesketh, J. D., Associates. 1975. The role of phytotrons in constructing plant growth models. In *Phytotronics in Agricultural and Horticultural Research,* Phytotronics III, ed. P. Chouard, N. de Bilderling, pp. 117–29. Paris: Gauthier-Villars. 410 pp.
51. Hesketh, J. D., Baker, D. N., Duncan, W. G. 1972. Simulation of growth and yield in cotton: II. Environmental control of morphogenesis. *Crop. Sci.* 12:436–39
52. Holt, D. A., Bula, R. J., Miles, G. E., Schreiber, M. M., Peart, R. M. 1975.

Environmental physiology, modeling and simulation of alfalfa growth: I. Conceptual development of SIMED. *Purdue Agric. Exp. Sta. Res. Bull.* 907. 26 pp.
53. Huck, M. G. 1977. Root distribution and water uptake patterns. In *The Belowground Ecosystem,* ed. J. K. Marshall, pp. 215–26. Fort Collins, Colo: Colorado State Univ., Range Sci. Dep. Sci. Ser. 26. 351 pp.
54. Hunt, W. F., Loomis, R. S. 1976. Carbohydrate-limited growth kinetics of tobacco (*Nicotiana rustica* L.) callus. *Plant Physiol.* 57:802–5
55. Hunt, W. F., Loomis, R. S. 1979. Respiration modelling and hypothesis testing with a dynamic model of sugar beet growth. *Ann. Bot.* In press
56. Innis, G. S., ed. 1978. *Grassland Simulation Model.* New York: Springer-Verlag. 298 pp.
57. Janssen, J. G. M. 1974. Simulation of germination of winter annuals in relation to microclimate and microdistribution. *Oecologia (Berlin)* 14:197–228
58. Jones, J. W. 1975. *A simulation model of boll weevil population dynamics as influenced by the cotton crop status.* PhD thesis. North Carolina State Univ., Raleigh, NC. 254 pp.
59. Jones, J. W., Thompson, A. C., McKinnion, J. M. 1975. Developing a computer model with various control methods for eradication of boll weevils. *Proc. Beltwide Cotton Prod. Res. Conf., Dallas,* p. 118
60. Lambert, J. R., Penning de Vries, F. W. T. 1973. Dynamics of water in the soil plant atmosphere system: A model named TROIKA. In *Physical Aspects of Soil, Water and Salts in Ecosystems,* ed. A. Hadas, D. Swartzendruber, P. E. Rijtema, M. Fuchs, B. Yaron. Berlin: Springer-Verlag. 460 pp.
61. Landsberg, J. J. 1977. Effects of weather on plant development. See Ref. 62, pp. 289–307
62. Landsberg, J. J., Cutting, C. V. 1977. *Environmental Effects on Crop Physiology.* New York: Academic. 388 pp.
63. Landsberg, J. J., Fowkes, N. D. 1978. Water movement through plant roots. *Ann. Bot.* 42:493–508
64. Lemeur, R., Blad, B. L. 1974. A critical review of light interception models for estimating the short wave radiation of plant communities. *Agric. Meterol.* 14:255–86
65. Loomis, R. S. 1978. Ideotype concepts for sugar beet improvement. *J. Am. Soc. Sugar Beet Technol.* In press
66. Loomis, R. S., Ng, E. 1978. Influences of climate on photosynthetic productivity of sugar beet. In *Photosynthesis 77,* Proc. 4th Int. Congr. Photosynth., ed. D. O. Hall, J. Coombs, T. W. Goodwin, pp. 259–68. London: Biochem. Soc. 827 pp.
67. Loomis, R. S., Ng, E., Hunt, W. F. 1976. Dynamics of development in crop production systems. In *CO_2 Metabolism and the Productivity of Plants,* ed. R. H. Burris, C. C. Black, pp. 269–86. Baltimore: Univ. Park Press. 431 pp.
68. Loomis, R. S., Rapoport, H. 1977. Productivity of root crops. In *Proc. 4th Symp. Int. Soc. Trop. Root Crops,* ed. J. Cock, R. MacIntyre, M. Graham, pp. 70–84. Ottawa: Int. Dev. Res. Cent. 277 pp.
69. Maksymowych, R. 1973. *Analysis of Leaf Development.* Cambridge, UK: Cambridge Univ. Press. 109 pp.
70. Mesarovic, M. D., Macko, D., Takahara, Y. 1970. *Theory of Hierarchical, Multilevel Systems.* New York: Academic. 294 pp.
71. Milstein, J. 1975. *Estimation of the dynamical parameters of the Calvin photosynthesis cycle, optimization and ill-conditioned inverse problems.* PhD thesis. Univ. California, Berkeley, Calif. 241 pp.
72. Milsum, J. H. 1966. *Biological Control Systems Analysis.* New York: McGraw Hill. 466 pp.
73. Milthorpe, F. L., Moorby, J. 1974. *Crop Physiology.* Cambridge, UK: Cambridge Univ. Press. 202 pp.
74. Monsi, M., Murata, Y. 1970. Development of photosynthetic systems as influenced by the distribution of matter. See Ref. 107, pp. 115–29
75. Monteith, J. L. 1964. Evaporation and the environment. *Symp. Soc. Exp. Biol.* 19:205–34
76. Monteith, J. L. 1965. Light distribution and photosynthesis in field crops. *Ann. Bot.* 29:17–37
77. Monteith, J. L. 1973. *Principles of Environmental Physics.* New York: Elsevier. 241 pp.
78. Morgan, J. M. 1976. *A simulation model of the growth of the wheat plant.* PhD thesis. Macquarie Univ., North Ryde, N.S.W. 192 pp.
79. Morgan, P. H., Mercer, L. P., Flodin, N. W. 1975. General model for nutritional responses of higher organisms. *Proc. Natl. Acad. Sci. USA* 72:4327–31
80. Morley, F. H. W. 1974. Avoiding nonsense in simulation. *J. Aust. Inst. Agric. Sci.* 40:43–44

81. Murata, Y. 1975. Estimation and simulation of rice yield from climatic factors. *Agric. Meteorol.* 15:117–31
81a. Naylor, T. H., Finger, J. M. 1967. Verification of computer simulation models. *Manage. Sci.* 14:B92–B101
82. Nelson, W. L., Dale, R. F. 1978. A methodology for testing the accuracy of yield predictions from weather-yield regression models for corn. *Agron. J.* 70:734–40
83. Ng, E., Miller, P. C. 1977. Validation of a model of the effects of tundra vegetation on soil temperatures. *Arct. Alp. Res.* 9:89–104
84. Nobel, P. S. 1974. *An Introduction to Biophysical Plant Physiology.* San Francisco: Freeman. 488 pp.
85. Novoa, R. 1979. *A preliminary dynamic model of nitrogen metabolism in higher plants.* PhD thesis. Univ. California, Davis, Calif.
86. Passioura, J. B. 1973. Sense and nonsense in crop simulation. *J. Aust. Inst. Agric. Sci.* 39:181–83
87. Patefield, W. M., Austin, R. B. 1971. A model for the simulation of the growth of *Beta vulgaris* L. *Ann. Bot.* 35: 1227–50
88. Pendleton, D. F., Menke, J. W., Williams, W. A., Woodmansee, R. G. 1979. Annual grassland ecosystem model. *Hilgardia.* In press
89. Penning de Vries, F. W. T. 1975. Use of assimilates in higher plants. In *Photosynthesis and Productivity in Different Environments,* Int. Biol. Prog. 3, ed. J. P. Cooper, pp. 459–80. Cambridge, UK: Cambridge Univ. Press. 715 pp.
90. Penning de Vries, F. W. T. 1975. The cost of maintenance processes in plant cells. *Ann. Bot.* 39:77–92
91. Penning de Vries, F. W. T., Brunsting, A. H. M., van Laar, H. H. 1974. Products, requirements and efficiency of biosynthesis: a quantitative approach. *J. Theor. Biol.* 45:339–77
92. Penning de Vries, F. W. T., van Laar, H. H. 1977. Substrate utilization in germinating seeds. See Ref. 62, pp. 217–28
93. Pitter, R. L. 1977. The effect of weather and technology on wheat yields in Oregon. *Agric. Meteorol.* 18:115–31
94. Pritsker, A. A. B. 1974. *The GASP IV Simulation Language.* New York: Wiley. 451 pp.
95. Rabbinge, R. 1976. *Biological Control of Fruit-Tree Red Spider Mite.* Wageningen: Pudoc. 234 pp.
96. Raeuber, A., Engel, K. H. 1966. Untersuchungen uber de Verlauf der Massenzunahme bei Kartoffeln (*Sol. tuberosum* L.) in Abhangigkeit von Umwelt- und Erbguteinflussen. *Abh. Meteorol. Dienstes Dtsch. Demokr. Repub.* No. 76
97. Raschke, K. 1975. Stomatal action. *Ann. Rev. Plant. Physiol.* 26:309–40
98. Rice, R. W., Morris, J. G., Maeda, B. T., Baldwin, R. L. 1974. Simulation of animal functions in models of production systems: ruminants on the range. *Fed. Proc.* 33:188–95
99. Riggs, D. S. 1963. *The Mathematical Approach to Physiological Problems.* Cambridge, Mass: MIT Press. 445 pp.
100. Riggs, D. S. 1970. *Control Theory and Physiological Feedback Mechanisms.* Baltimore: Williams & Wilkins. 599 pp.
101. Robertson, G. W. 1973. Plant response to climate factors. In *Plant Response to Climatic Factors,* Proc. Uppsala Symp., ed. R. O. Slatyer, pp. 327–43. Paris: Unesco. 574 pp.
102. Ross, J. K. 1975. *Radiacionnyj Rezim i Archtektonika Rastitel nogo Pokrova.* Leningrad: Gidrometeoizdat. 342 pp.
103. Ruesink, W. G. 1976. Status of the systems approach to pest management. *Ann. Rev. Entomol.* 21:27–44
104. Ryle, G. J. A., Brockington, N. R., Powell, C. E., Cross, B. 1973. The measurement and prediction of organ growth in a uniculm barley. *Ann. Bot.* 37:233–246
105. Segal, I. H. 1975. *Enzyme Kinetics.* New York: Wiley-Interscience. 957 pp.
106. Seligman, N. G. 1976. A critical appraisal of some grassland models. See Ref. 26, pp. 60–97
107. Setlik, I., ed. 1970. *Prediction and Measurement of Photosynthetic Productivity.* Proc. IBP/PP Technol. Meet., Trebon. Wageningen: Pudoc. 632 pp.
108. Shaw, L. M. 1964. The effect of weather on agricultural output: a look at methodology. *J. Farm Econ.* 46:218–30
109. Shawcroft, R. W., Lemon, E. R., Allen, L. H. Jr., Stewart, D. W., Jensen, S. E. 1974. The soil-plant-atmosphere model and some of its predictions. *Agric. Meteorol.* 14:287–307
110. Shiraz, G. A., Stone, J. F., Bacon, C. M. 1976. Oscillatory transpiration in a cotton plant. II. A model. *J. Exp. Bot.* 27:619–33
111. Sinclair, T. R., de Wit, C. T. 1975. Photosynthate and nitrogen requirements for seed production by various crops. *Science* 189:565–67
112. Sitaraman, V., Rao, N. J. 1977. Hierarchical modeling of acclimatory processes. *J. Theor. Biol.* 67:25–47
113. Splinter, W. E. 1974. Modeling plant

growth for yield prediction. *Agric. Meteorol.* 14:243–63
114. Stapleton, H. N., Meyers, R. P. 1971. Modeling subsystems for cotton—the cotton plant simulation. *Trans. ASAE* 14:950–53
115. Stroosnijder, L., van Keulen, H., Vachaud, G. 1972. Water movement in layered soils. 2. Experimental confirmation of a simulation model. *Neth. J. Agric. Sci.* 20:67–72
116. Thompson, L. M. 1969. Weather and technology in the production of corn in the U.S. Corn Belt. *Agron. J.* 61:453–56
117. Thornley, J. H. M. 1976. *Mathematical Models in Plant Physiology.* New York: Academic. 318 pp.
118. Thornley, J. H. M., Hurd, R. G. 1974. An analysis of the growth of young tomato plants in water culture at different light integrals and CO_2 concentrations. II. A mathematical model. *Ann. Bot.* 38:389–400
119. Tooming, H. 1967. Mathematical model of plant photosynthesis. *Photosynthetica* 1:233–40
120. Vanderlip, R. L., Arkin, G. F. 1977. Simulating accumulation and distribution of dry matter in grain sorghum. *Agron. J.* 69:917–23
121. van Keulen, H. 1975. *Simulation of Water Use and Herbage Growth in Arid Regions.* Wageningen: Pudoc. 176 pp.
122. van Keulen, H. 1976. Evaluation of models. See Ref. 26, pp. 22–29
123. van Keulen, H., van Beck, G. G. E. M. 1971. Water movement in layered soils: a simulation model. *Neth. J. Agric. Sci.* 19:138–53
124. Veen, A. H., Lindenmayer, A. 1977. Diffusion mechanisms for phyllotaxis. Theoretical physicochemical and computer study. *Plant Physiol.* 60:127–39
125. Waggoner, P. E., Horsfall, J. G., Lukens, R. J. 1972. EPIMAY. A simulator of southern corn leaf blight. *Conn. Agric. Exp. Sta. Bull.* 729. 84 pp.
126. Waggoner, P. E., Reifsnyder, W. E. 1968. Simulation of the temperature, humidity and evaporation profiles in a leaf canopy. *J. Appl. Meteorol.* 7:400–9
127. Wang, Y., Gutierrez, A. P., Oster, G., Daxl, R. 1977. A population model for plant growth and development: coupling cotton-herbivore interaction. *Can. Entomol.* 109:1359–74
128. Williams, R. F. 1975. *The Shoot Apex and Leaf Growth.* Cambridge, UK: Cambridge Univ. Press. 256 pp.
129. Wilson, J. L. 1975. *Growth simulation: an application of integrative theory to cotton crop physiology.* PhD thesis. Univ. California, Davis, Calif. 332 pp.
130. Wolpert, L. 1969. Positional information and the spatial pattern of cellular differentiation. *J. Theor. Biol.* 25:1–47

Added in proof:

131. Goeschl, J. D., Magnuson, C. E., DeMichele, D. W., Sharpe, P. J. H. 1976. Concentration-dependent unloading as a necessary assumption for a closed form mathematical model of osmotically driven pressure flow in phloem. *Plant Physiol.* 58:556–62

Ann. Rev. Plant Physiol. 1979. 30:369–404

BIOSYNTHESIS OF TERPENOIDS

❖7676

T. W. Goodwin

Department of Biochemistry, University of Liverpool,
Liverpool L69 3BX, England

CONTENTS

INTRODUCTION

My brief was originally a wide one, "Biosynthesis of Terpenoids," but I intend to restrict myself to two main groups of terpenoids—sterols and carotenoids. To add further restrictions, only those aspects which appear

0066-4294/79/0601-0369$01.00

ready for detailed treatment will be considered, and in all the discussion emphasis will be laid on the photosynthetic organisms and on mechanisms of biosynthesis.

It is well known that the C_6 compound mevalonic acid (MVA) is the first specific percursor of terpenoids (66); perhaps slightly less well known is that it has three prochiral centers (C-2, C-4, and C-5) (Formula 1). At these centers the *pro-R* and *pro-S* hydrogens[1] are treated individually during the conversion of MVA into terpenoids. The outstanding work of the Cornforths (46, 47) in synthesizing the six species of MVA stereospecifically labeled with either deuterium or tritium at C-2, C-4, and C-5 allowed Popják & Cornforth (125) by 1966 to solve, with one exception, every stereochemical problem posed in the biosynthetic pathway from MVA to cholesterol in animals. As will soon become apparent, the outstanding problem has now been solved. MVA is converted by a well-established pathway into isopentenyl pyrophosphate (IPP), and then IPP is isomerized to dimethyl allyl pyrophosphate (DMAPP) (see 24). The mechanism outlined in Figure 1 indicates the chirality (2*R*) of the hydrogen removed from C-2 of IPP during the isomerization.[2] This was the first stereochemical question answered by Popják and Cornforth, but proof that the incoming proton attacks the *re,re* face[3] of the double bond of IPP (Figure 1) was only solved in 1972 (42).

Farnesyl pyrophosphate was synthesized by a pig liver preparation in the presence of [2(*R*),2-$^3H_1$] MVA and D_2O. After hydrolysis the free farnesol was ozonized to yield acetone, which on appropriate manipulation was converted without change of configuration into chiral acetate.[4] The chirality of this acetate was *R*. As [2(*R*),2-$^3H_1$] MVA yields *cis*-IPP (125), *R*-acetate

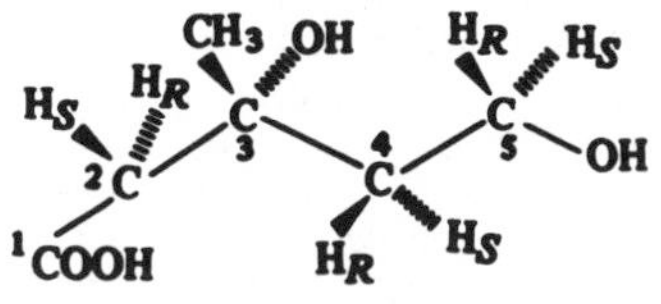

Formula 1 Mevalonic acid

[1]See (66) for a simple discussion of the *R, S* nomenclature which defines the absolute configuration at a chiral center.

[2]In the *RS* convention the 2*R* hydrogen of IPP is the 4*S* hydrogen of MVA—an example of the pitfalls which await the unwary in using the *RS* convention.

[3]The *RS* convention for defining the two faces of a double bond (66). The opposite face is the *si, si* face.

[4]Acetate with two hydrogens of the methyl group replaced stereospecifically by one deuterium and one tritium (see 66).

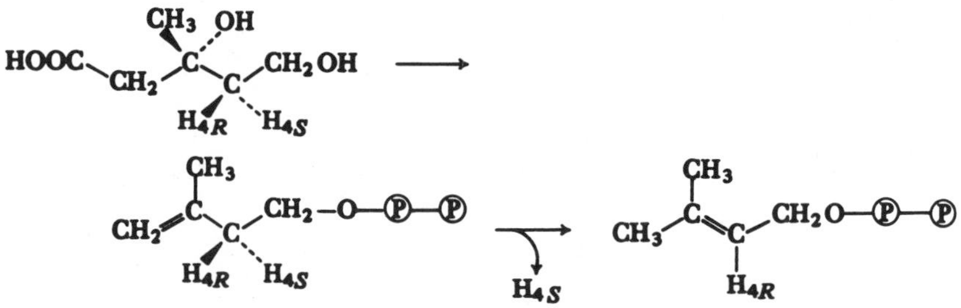

Figure 1 The chirality of the hydrogen removed from C-4 of MVA (C-2 of IPP) during the isomerization of IPP to DMAPP.

can be generated only by the addition of a deuteron from D_2O to the *re, re* face of the double bond of IPP (Figure 2).

Chain elongation, started by addition of IPP to DMAPP, is formally similar to the IPP isomerase but in which DMAPP replaces a proton. However, the reactions differ stereochemically in that the allylic group is added at the *si, si* face of IPP. Thus, in this case, the mechanism cannot be concerted as in the case of the isomerase because the incoming and outgoing groups are on the same side of the double bond; that is, a *cis* elimination would occur. To avoid this mechanistic difficulty, it has been proposed that an "X" group (active site of enzyme) attacks C-3 of IPP by *trans* addition across the double bond; this is followed by a *trans* elimination to form the new double bond in DMAPP (Figure 3). The hydrogen eliminated is the 2-*pro-R* hydrogen of IPP (125). At each further elongation the 2-*pro-R* hydrogen of the incoming IPP is also eliminated.

In the case of sterol biosynthesis, the first C_{30} hydrocarbon precursor squalene is formed from two molecules of farnesyl pyrophosphate (C_{15}). The first C_{30} compound produced by squalene synthetase is presqualene pyrophosphate, which is converted into squalene by the enzyme in the presence of NADPH. The stereochemistry of this complex reaction has been fully investigated (125). The incoming hydrogen from NADPH takes up the *pro-R* position at C-15 of squalene. The mechanism of the formation of squalene is still not completely settled, but the most likely sequence of reactions is indicated in Figure 4.

The participation of presqualene pyrophosphate in squalene synthesis in plants has been demonstrated in a cell-free system from bramble tissue cultures; in the absence of NADPH the pyrophosphate accumulates in place of squalene (75). Its direct conversion into squalene has been achieved in cell-free systems from peas (13) and etiolated french beans (71). In the bean preparation the enzyme activity is localized in the endoplasmic reticulum. C-2 of MVA retains its individuality at the end of the squalene mole-

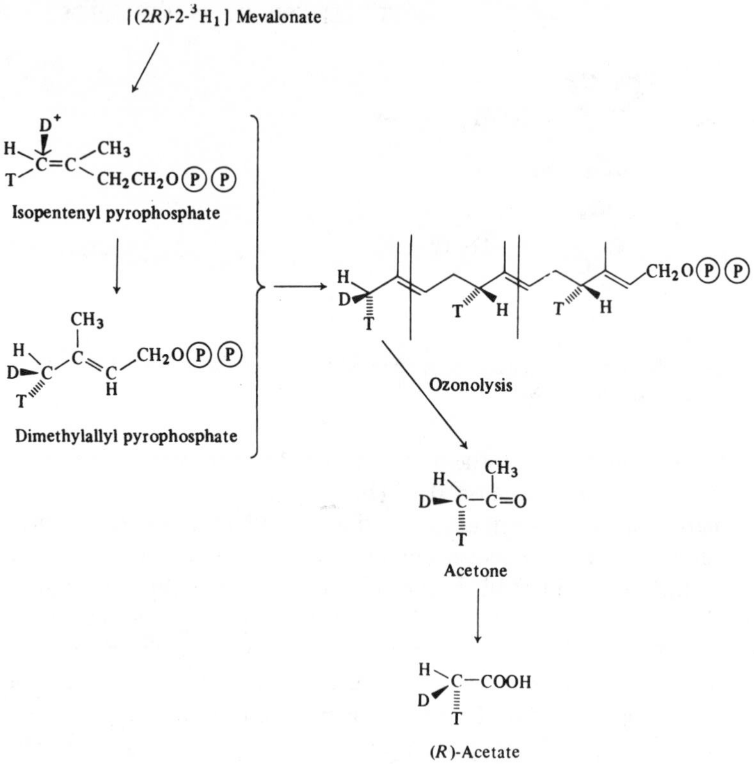

Figure 2 Demonstration that hydrogen is added to the *re-re* face of the double bond of isopentenyl pyrophosphate in the formation of dimethylallyl pyrophosphate (42).

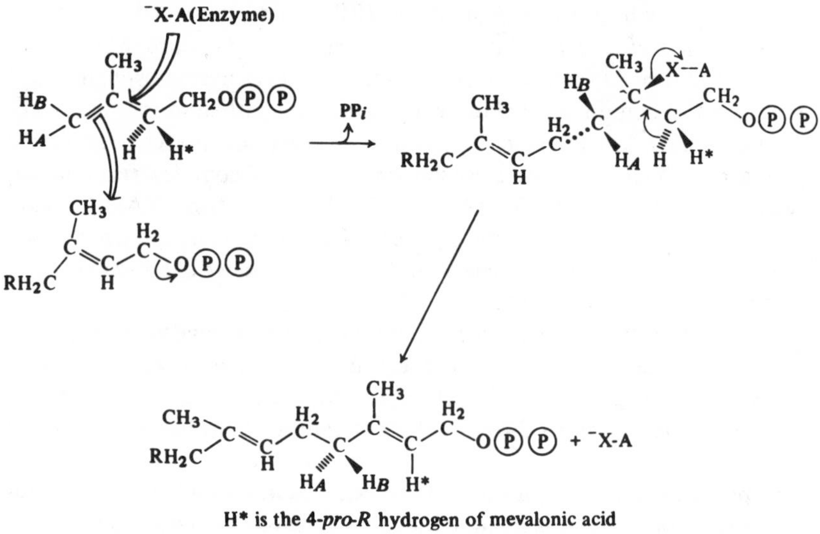

Figure 3 Probable mechanism for condensation of C_5 units to form polyisoprenoids (125).

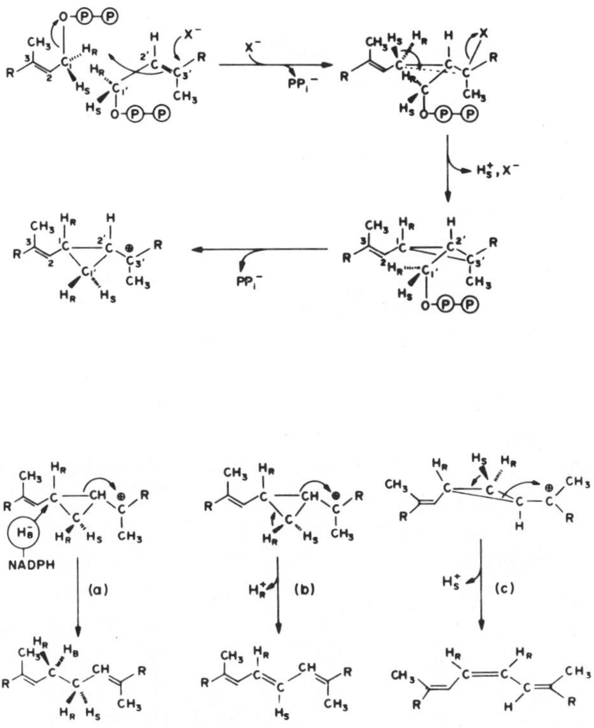

Figure 4 Probable mechanism for the formation of (*a*)squalene from presqualene pyrophosphate. (R =), (*b*) *trans*-phytoene and (*c*) 15-*cis*-phytoene from prephytoene pyrophosphate (R =) (70).

cule synthesized by the pea enzyme in that it is confined to the *cis* terminal methyl groups (153).

The first C_{40} hydrocarbon precursor of the carotenoids is phytoene which, in contrast to squalene, has an additional double bond at the center of the molecule. It is produced from the C_{40} prephytoene pyrophosphate (3, 8), itself formed by the condensation of two geranylgeranyl pyrophosphate (C_{20}) molecules (70); NADPH is not involved in the reaction (40, 106, 140), but an additional problem arises because all *trans* phytoene and 15-*cis* phytoene are produced by different organisms. In higher plants (32) and *Phycomyces blakesleeanus* (70) the *cis*-isomer is produced, during which time the *pro-S* hydrogen is lost from C-1 of each participating geranylgeranyl pyrophosphate molecule; on the other hand, in a *Mycobacterium* species the *trans* isomer is the major species formed and one *pro-R* and one *pro-S* hydrogen is lost from the participating pyrophosphates (70). The mechanism suggested in Figure 4 accommodates these experimental observations.

The conversion of squalene into triterpenes and sterols and of phytoene into carotenoids will now be considered separately.

STEROLS

Cyclization of Squalene

Squalene is first converted into squalene 2,3-oxide by a mono-oxygenase. The labeled oxide was first noted in plants after feeding [1-^{14}C] acetate to tissue cultures of *Nicotiana tabacum* (16) and to the latex of *Euphorbia cyparissias* (123). It also accumulates in the presence of the inhibitors SKF-7989 (73, 132), iminosqualene (44), and AMO 1618 and CCC (53).

Cyclization of squalene 2,3-oxide occurs in animals and fungi with the formation of lanosterol (1-A)[5] which is then converted into sterols. In the case of higher plants and algae, cycloartenol (2-A) and not lanosterol is formed as the first cyclic triterpenoid precursor of sterols (13, 54, 60, 130). The general view that the cycloartenol pathway is strictly confined to photosynthetic organisms requires slight modification because it exists in *Astasia longa,* a white nonphotosynthetic naturally occurring mutant of *Euglena gracilis* (133), in the parasitic plants *Cuscuta europaea, C. epithymum,* and *Orobanche lutea,* which are capable of synthesizing their own sterols (134), and probably also in the fungus *Saprolegnia ferax* (33a). The last observation, which is based on the ability of the organism to open the cyclopropane ring of cycloartenol rather than on the detection of cycloartenol itself as a metabolite, is unexpected but complements the view that the Oomycetes, of which *Saprolegnia* is a member, can be considered colorless algae. This view is also supported by the observation that *Saprolegnia* also synthesizes C_{29} sterols (33a), a rare occurrence in fungi but very characteristic of algae.

The proposed mechanism of cyclization of squalene epoxide (Figure 5), which involves the chair-boat-chair-boat folding of the molecule (130), demands the participation of a nucleophilic intermediate X^- (enzyme). Otherwise the final migration of the hydrogen of the C-19 methyl group would be *cis* to the C-9 hydrogen transfer, which is contrary to the biogenetic isoprene rule (137).The neutralization of the charge generated at C-9 during the backward rearrangement of the C-20 cation would allow removal of X^- with formation of the cyclopropane ring in a *trans* manner.

[5]In order to reduce the number of formulae in the text, sterols are designated with a number and a letter according to the nature of the steroid nucleus and the C-17 side chain, respectively (60). The nuclei are summarized in Scheme 1 and the side chains in Scheme 2. For example, lanosterol is (1-A) and cycloartenol is (2-A). The general numbering of the steroid molecule is given in Formula 2.

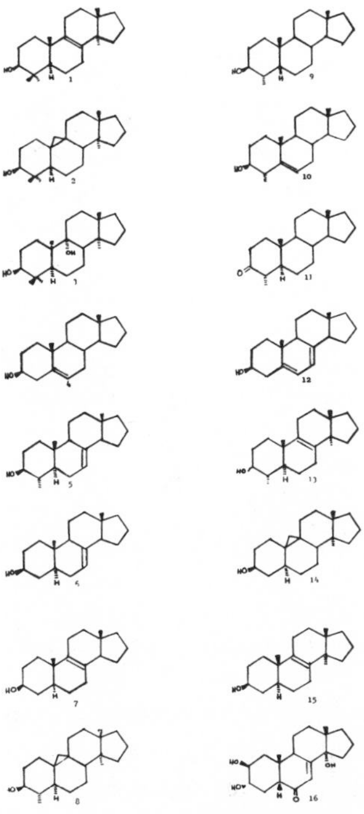

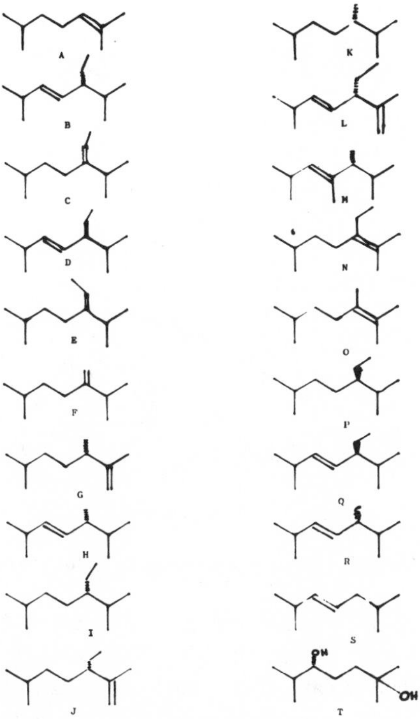

Scheme 1 (*left*) The various sterol nuclei discussed in the text.
Scheme 2 (*right*) The various sterol side-chains discussed in the text.

This mechanism also requires the uptake of a proton at C-19, and this has been demonstrated by isolating and examining by mass spectrometry the cycloartenol formed by cotyledons and embryos of pea seedlings germinating in D_2O (37).

The possibility exists that the initial C-20 carbonium ion could be neutralized by water at an anionic enzyme center to yield a protosterol (3-A) which could then reorient on the enzyme so that OH^- is removed to yield cycloartenol (45, 79). Indeed, (3-A) is effectively incorporated into poriferasterol

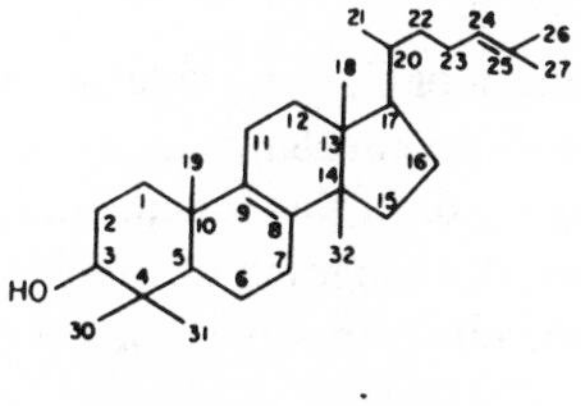

Formula 2 Lanosterol

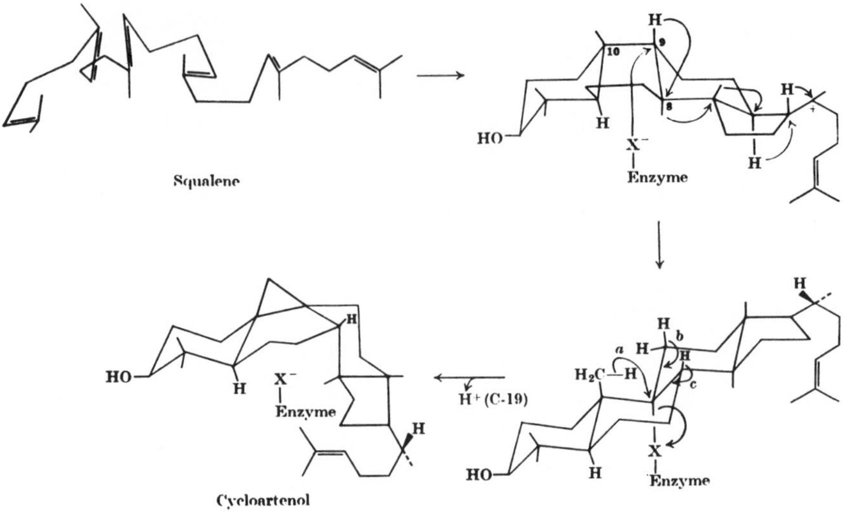

Figure 5 Proposed mechanism of cyclization of squalene epoxide to form cycloartenol (130).

(4-B) by *Ochromonas malhamensis* but insignificantly incorporated into higher plants, the alga *Trebouxia* or yeast (120a). The one well-authenticated case of the existence of lanosterol in plants is in *Euphorbia helioscopia* latex (123), but the possibility of two cyclases—one producing lanosterol and one cycloartenol—being involved is reduced by the observation that cycloartenol is isomerized to lanosterol in *E. lathyris* (124).

Alkylation Reactions

The reactions unique to the pathway from cycloartenol to plant sterols are opening of the cyclopropane ring and alkylation at C-24; the latter results in the presence of supernumerary C_1 or C_2 side chains with either α- or β-chirality.[6] As the enzyme which opens the ring (isomerase) acts only on alkylated intermediates, the alkylation reactions will be discussed first. However, it should be remembered that occasionally the isomerase will act on nonalkylated compounds, for example, the enzyme in *E. lathyris* just discussed which isomerizes cycloartenol itself.

MECHANISM The addition of C_1 or C_2 side chains involves either a single or double methylation, and the mechanisms involved have been studied in detail and outlined in Figure 6. In this scheme only the reactions involving the side chain are considered; changes in the nucleus of the molecules which precede or follow the side-chain modifications will be considered later.

[6] α- and β-chirality at C-24 are indicated  and , respectively. See (113) for a full discussion.

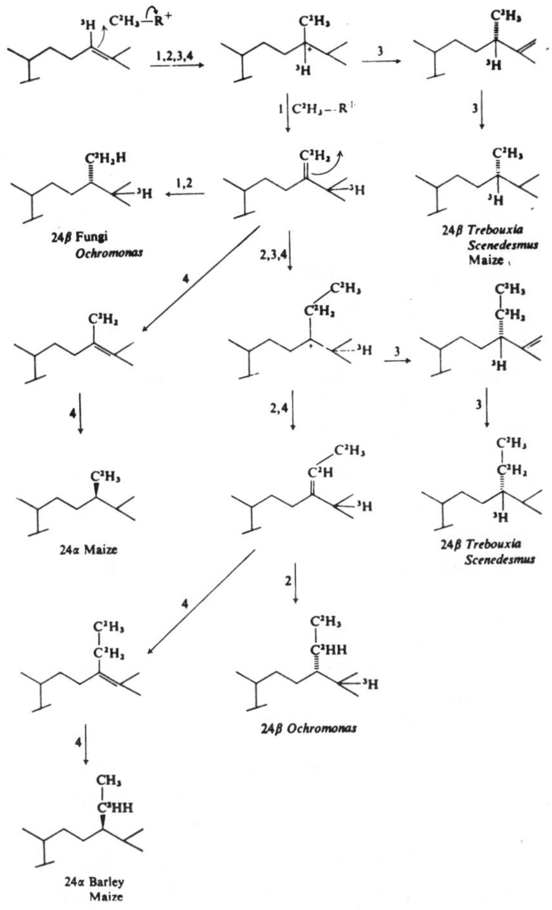

Figure 6 Mechanisms involved in alkylation of the sterol side-chain (67).

The single and double methylations involving one and two molecules, respectively, of *S*-adenosylmethionine were first demonstrated in higher plants in *Pisum sativum* (39), *Salvia officinalis* (118), and *Menyanthes trifoliata* (7), and in the algae *Laminaria saccharina* (160), *Fucus spiralis* (62), and *Ochromonas* spp. (97, 150). The mechanism proposed by Nes and his colleagues (39) [pathway 1 for C_{28} sterols and pathway 2 for C_{29} sterols (Figure 6)] was shown to hold in the case of *Ochromonas* spp. on the following grounds: (*a*) methylene cycloartanol is converted into poriferasterol (50a). (*b*) Poriferasterol (4-B) synthesized in the presence of $[CD_3]$-methionine exhibited molecular ions with m/e +1, +2, and +4; no ions m/e +3 or +5 were encountered as they would have been if methylene and ethylidene intermediates had not been involved (97, 150). (*c*) Three ethylidene derivatives, isofucosterol ($\Delta^{24(28)}Z$)[7] (4-C), ethylidene lophenol (5-C), and Δ^7-avenasterol (6-D) were effectively converted into poriferasterol (4-B) (89, 90, 99). Fucosterol[7] ($\Delta^{24(28)}E$) (4-E), the geometrical isomer of isofucosterol, is also converted into poriferasterol but much less effectively than the *Z* isomer (99). This suggests a lack of specificity in the reductase involved, but this cannot be the full explanation (see later). Furthermore, isofucosterol is widely distributed in trace amounts, which suggests that it is a true biosynthetic intermediate. On the other hand, fucosterol is not widely distributed, but when it is found it is the major sterol component, as in the Phaeophyceae (64), which suggests that it is an end product rather than an intermediate. (*d*) When synthesized in the presence of [4*R*, 4-$^3H_1$]-MVA, cycloartenol retains a tritium atom at C-24. In the formation of poriferasterol (4-B) and brassicasterol (4-H), this tritium moves to C-25 (149). The same migration has been observed in clionasterol (4-I), synthesized by the Xanthophyte *Monodus subterraneus* (112), and in sterols synthesized by *Euglena* spp. (64). It appears that this mechanism is of very limited distribution and that the major route for the synthesis of 24β sterols involves pathway 3 (Figure 6) for C_{28} sterols and pathway 4 for C_{29} sterols. Experiments with *Chlorella vulgaris* showed that five deuteriums were incorporated from [CD_3]methionine into the ethyl group of chondrillasterol (6-B) and its 22-dihydroderivative (157). The same observations were made on poriferasterol (4-B) formation in *Trebouxia* sp. (166), *Scenedesmus obliquus* (166), and *C. ellipsoidea* (1, 158). Furthermore, in a similar experiment the 24-methyl sterol brassicasterol (4-H) contained three deuteriums in its methyl group (166). A full investigation with *Trebouxia* and *Scenedesmus* revealed the pathway indicated in Figure 6 (166). Evidence additional to that obtained with [CD_3]methionine, which also ruled out the intermediate formation of C-24 methylene and C-24 ethylidene derivatives in the formation of C_{28} and C_{29} sterols in these algae, included: (*a*) the radioisotopic detection of both 24-methylene cycloartanol (2-F) and its isomer cyclolaudenol (2-G) (steps 1 and 2) after feeding [$^{14}CH_3$]methionine; (*b*) [^{14}C]-31-norcyclolaudenol (8-G) is converted only into the C_{28} sterol brassicasterol (4-H), and not into C_{29} sterols; on the other hand, tritiated cycloeucalenol (8-F) is converted only into the C_{29} sterol poriferasterol (4-B) and not into C_{28} sterols. Furthermore, clerosterol (4-J) and codisterol (4-G), both of which are β sterols with a double bond at C-25, are naturally occurring in algae (135), and the former is converted into clionasterol (4-I) and poriferasterol (4-B) (95). Also, 24-methylene cholesterol (4-F) is converted into clionasterol (4-I) and poriferasterol (4-B) in *Chlorella ellipsoidea*

[7]The designations *Z* (isofucosterol, 4-C) and *E* (fucosterol, 4-E) define the geometrical configuration around the C-24(28) double bond in the two molecules (see 66).

and not significantly into brassicasterol (4-H) or ergost-5-en-3β-ol (4-K) (159). Support for this pathway comes indirectly from inhibitor studies with *C. sorokiniana;* both AY-9944 (41) and triarimol cause the accumulation of C_{28} Δ^{25} sterols, which suggest that these compounds are intermediates in C_{28} sterol synthesis. The absence of $\Delta^{24(28)}$ methylene sterols from inhibited cultures suggests that they are not normal intermediates and provides an explanation why this organism does not synthesize C_{29} sterols.

It has not yet been possible to show in green algae that the hydrogen originally present at C-24 in cycloartenol remains in that position in the alkylated derivatives; this is because the impermeability of the algae to MVA rules out experiments with [4*R*, 4-$^3H_1$]MVA. However, the higher plant *Clerodendrum campbellii* synthesizes the 24β ethyl sterol 22-dehydroclerosterol (4-L) which has a double bond at C-25, and experiments with [4*R*, 4-$^3H_1$]MVA show that a tritium is retained at C-24 in this sterol (20, 21). This, taken with the observation noted above that clerosterol can be converted into poriferasterol, is good supporting evidence that the mechanism is as indicated. This pathway may be common to many algae which characteristically synthesize 24β alkyl sterols. However, 24α alkyl sterols are present in some diatoms (e.g. *Phaeodactylum tricornutum* (*Nitzschia closterium* f. *minutissima*) (135) and some dinoflagellates (144, 164). In the case of the latter group, *Crypthecodinium cohnii* synthesizes dinosterol (9-M), dehydrodinosterol (10-M), and dinosterone (11-M) (164). So far it has been established from experiments with [CD_3]methionine that the C-24 methyl is introduced via a C-24 methylene intermediate, as are the 24α methyls in higher plant sterols (see later), while C-33 is introduced as an intact methyl group (165). Further details are being investigated.

With only a few exceptions, e.g. leaves of *Clerodendrum* spp. (20) and *Kalanchoë daigremontiana* (116) and seeds of *Cucurbita pepo* (117,151,152), higher plants appear to produce 24α ethyl sterols exclusively. [*C. pepo* is particularly interesting in this respect because the 24β isomers are confined to the seeds; the pericarp of the fruit and the leaves contain the expected 24α isomers (117)]. On the other hand, recent surveys have shown that 24α methyl sterols, although almost always abundant in higher plants, are generally accompanied by small amounts of 24β isomers [see (117) for a detailed review]. The presence of both 24α and 24β methyl sterols in higher plants has raised the question of whether or not they are synthesized by the same pathway. The elucidation of the pathway of 24α sterol synthesis is still not completely clear, but the putative pathway for 24α methyl sterols and 24α ethyl sterols are indicated as pathway 4 in Figure 6. Experiments with [2-^{14}C, 4*R*, 4-$^3H_1$]MVA in *Nicotiana tabacum* and *Dioscorea tokoro* (156), *Larix decidua* (129), *Spinacea oleracea* (5), *Medicago sativa* (5), and *Cyathula capitata* (19) showed that no tritium was retained at C-24 or C-25 in the alkylated sterols, although in barley embryos isofucosterol was formed and still retained a tritium at C-25 (98). Furthermore, work with barley embryos growing in the presence of [CD_3] methionine produced methyl and ethyl sterols containing only two and four deuterium atoms, respectively (98), which requires the formation of methylene and ethylidene intermediates. Finally, 24-methylene lophenol (5-F) and 24-ethylidene lophenol (5-C) were isolated from barley embryos (98) and the incorporation of tritiated 24-ethylidene lophenol was demonstrated (94). Earlier isofucosterol had been reduced to sitosterol in *Pinus pinea* (114), and it was later isolated as an intermediate in barley embryos after feeding [2-^{14}C]MVA (98). Recently the stereochemistry of hydrogen migration from C-24 to C-25 during isofucosterol biosynthesis has been elucidated in *P. pinea* seedlings (119). The migration, which takes place at the 24-methylene level, is such that the *cis*-pro-E-methyl group of the nonalkylated precursor takes up the pro-*R* position thus:

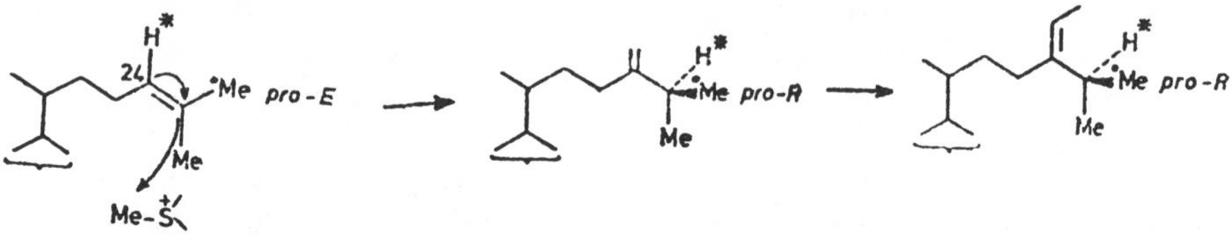

Proof that a $\Delta^{24(25)}$ intermediate is involved in this pathway includes (*a*) the possible occurrence of a $\Delta^{24(25)}$ ethyl sterol (6-N) in sunflowers (77), pumpkin seeds (117), and Solanaceae seeds (81), the presence of a $\Delta^{24(25)}$ methyl sterol (4-O) in *Withania somnifera* (102) and Solanaceae seeds (81), and the detection of two 4α-$\Delta^{24(25)}$ dimethyl sterols also in Solanaceae seeds (80, 81); (*b*) the label from [24-^{3}H] lanosterol, which is incorporated into plant sterols although not an in vivo precursor, is not incorporated into 24α methyl sterols but is incorporated into 24β methyl sterols in seedlings of *Pinus pinea* (111), indicating the loss of the C-24 hydrogen in the α-pathway but not in the β-pathway. However, feeding [26-^{14}C] stigmasta-5,24-dien-3β-ol (4-N) to barley seedlings resulted in a very low incorporation into sitosterol (4-P), and attempts to trap the labeled diene after feeding [^{14}C]MVA were unsuccessful (94). Experiments with labeled (4-N) of much greater specific activity than that used in the original experiments in cell-free systems are needed before the role of $\Delta^{24(25)}$ sterols in 24α alkyl sterol biosynthesis in higher plants is finally settled.

If 24β methyl sterols are formed in higher plants in the same way as in green algae, then one would expect a ^{14}C:^{3}H ratio varying between 5:3 and 5:2 in the isolated "mixed" 24-methyl sterols when [2-^{14}C, 4*R*, 4-^{3}H$_1$] was the starting substrate, according to the relative amounts of 24α- and 24β-epimers present (the ratio 5:3 is expected when tritium is retained at

C-24 or C-25 and 5 : 2 is expected when it is not). Recent experiments with maize (see 67) have borne this idea out. The observed $^{14}C:{}^{3}H$ ratio of the mixed 24-methyl sterol was 5 : 2.82, which fitted in well with that calculated from the known amounts of the α- and β-isomers present as determined by NMR spectroscopy.

Again, if the biosynthetic pathway for synthesis of the β-methyl sterols in higher plants is similar to that in green algae, then the early precursor cyclolaudenol (2-G) should exist alongside 24-methylene cycloartanol (2-F). Recent labeling experiments have demonstrated the presence of both compounds in maize seedlings (see 67; L. J. Goad and M. Zakelj, unpublished observations). It had already been shown that in the fern *Polypodium vulgare* cyclolaudenol is formed by the mechanism indicated for 24β methyl sterols in algae. An exception to the mechanism indicated for 24α alkyl sterol synthesis apparently occurs in the slime mould *Physarum polycephalum,* which incorporates five deuteriums from [CD_3]methionine into sitosterol (4-P) and stigmasterol (4-Q) (96).

ENZYMOLOGY S-Adenosyl-L-methionine-cycloartenol methyl transferase activity has been demonstrated in cell-free extracts from peas (136), *Scenedesmus obliquus* (166), *Trebouxia* sp. (166), bramble (*Rubus fructicosus*) suspension cultures (57), and maize coleoptiles (72). In all cases cycloartenol (2-A) was the best substrate, and in the algal preparations the products of the reaction were 24-methylene cycloartanol (2-F) and cyclolaudenol (2-G), but it is not known whether or not the activity is due to one or two enzymes (166). The enzyme preparations are particulate, and in the case of maize coleoptiles activity is specifically associated with the endoplasmic reticulum-enriched membranes (71, 72).

The second methylation step was for some time difficult to demonstrate in vitro, but recently it has been achieved in bramble cell cultures. The enzyme is microsomal and is specific for 24-methylene lophenol (5-F); that is, the cyclopropane ring must first be cleaved. The product is 24-ethylidene lophenol (5-C) (57). A similar enzyme, for which lanosterol was the substrate, was noted in the uredospores of *Uromyces phaseoli* (101); this is one of the few fungi which synthesize C_{29} sterols (100).

A $\Delta^{24(28)}$ reductase, which is NADPH-dependent and which reduces ergosta-5,7,22,24(28)-tetraenol to ergosterol is present in the microsomal fractions from yeast (82).

Opening of the Cyclopropane Ring

The enzyme which cleaves the cyclopropane ring and thus isomerizes cycloeucalenol (8-F) to obtusifoliol (13-F) has been demonstrated in cell-free preparations from tissue culture of bramble (74) and in microsomes of *Zea mays* embryos; the preparation from maize is specific for cycloeucalenol (8-F) (127, 128). Cycloartenol (2-A) and 24-methylene cycloartanol (2-F) do not act as substrates, and this explains the absence of lanosterol and 24-methylene lanosterol from the great majority of most plants studied up to now. If the isomerization is carried out in D_2O with cycloeucalenol as substrate, the incorporation of one deuterium into C-19 of the product obtusifoliol occurs (127, 128). This observation confirms the earlier proposed mechanism for the opening of the ring in which the carbonium ion generated at C-9 following proton attack is stabilized by a suitable electrophilic group in the enzyme. A *trans* antiperiplanar elimination of the 8β hydrogen then results in the formation of obtusifoliol (13-F) (Figure 7).

The Position up to Now

The first steps in the conversion of cycloartenol into sterols can now be summarized with some confidence (Figure 8), although variations may well be found to exist. Reactions involved in this sequence which have not yet been dealt with are (*a*) the removal of a methyl group from C-4; (*b*) the isomerization of a Δ^8 to a Δ^7 intermediate; and (*c*) the removal of the C-14 methyl group. In plants the first C-4 methyl group appears generally to be

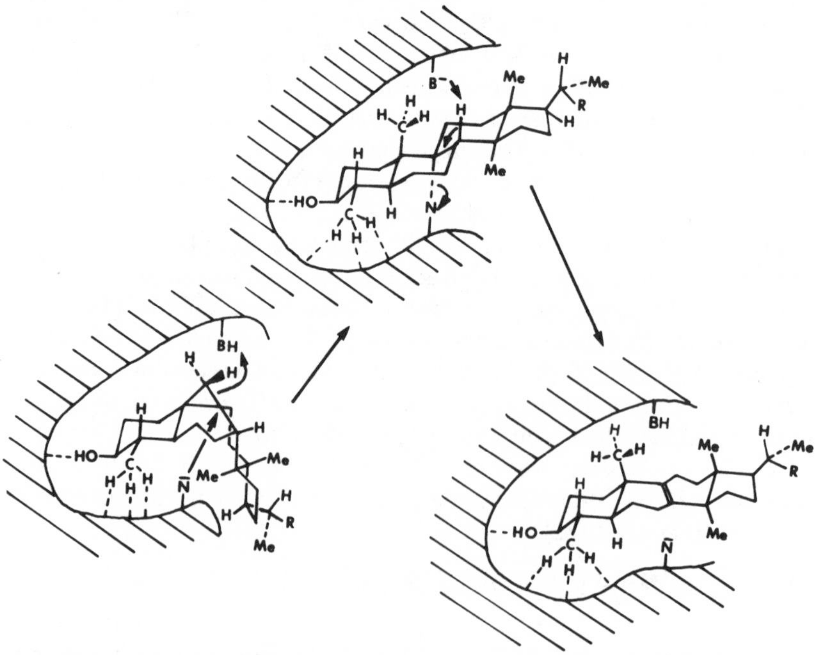

Figure 7 Mechanism of opening of cyclopropane ring of cycloeucalenol (127).

removed before the C-14 methyl group (65), whereas in animals the situation is reversed in the biosynthesis of cholesterol (142, 143). Studies on the fern *Polypodium vulgare* showed that the 4α methyl group of cycloartenol arose specifically from C-2 of mevalonic acid and that it is specifically removed on conversion of cycloartenol into 31-norcycloartenol (8-A). In the process the original 4β methyl group takes up the 4α position (58). It was also demonstrated in this and other investigations that the 3α hydrogen is exchanged during the demethylation, indicating that a 3-keto intermediate is involved (58, 92, 131). Thus the overall demethylation reaction of the first C-4 methyl group is similar to that observed in animals. The loss of a 4α methyl group has also been demonstrated in the formation of cycloeucalenol (8-F) from 24-methylene cycloartenol (2-F) and in the conversion of cyclolaudenol (2-G) into 31-norcyclolaudenone (11-G) by the banana (91, 92). Little is known about the point at which a Δ^8 precursor isomerizes to a Δ^7 product.

The loss of the 14α methyl group appears to take place at the obtusifoliol stage, but there have been no reports of a detailed study up to now. The mechanism involved is still not clear, but in *Calendula,* (38), peas (38), yeast (35), and rats (see 59) a hydrogen originally at C-15 is lost, indicating a Δ^{14} intermediate. In all the systems examined the 15α hydrogen is lost while the 15β hydrogen is retained and epimerized. Compounds with such a Δ^{14} double bond have been isolated as minor components of yeast sterols

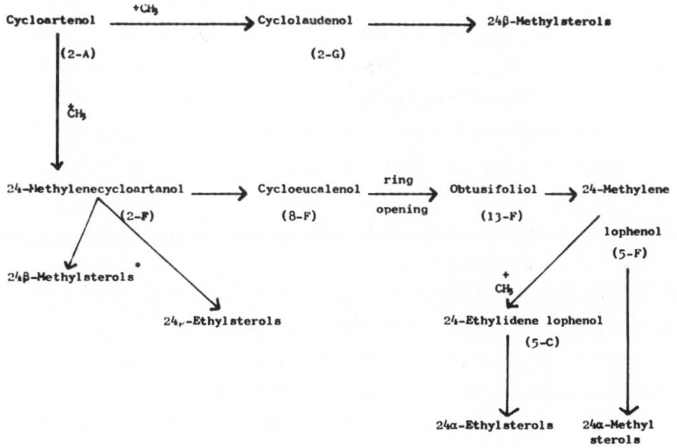

*Restricted to Ochromonas and a few other algae.

Figure 8 Probable first steps in the conversion of cycloartenol into sterols.

(10) and as major components of the sterols of *Methylococcus capsulatus,* the only bacterium known which unequivocally synthesizes sterols de novo (18, 23). Possible intermediates are also $\Delta^{8,14}$ dienes, and such compounds are observed in organisms growing in the presence of certain inhibitors, although the difficulties in interpreting such studies are considerable.

Later Stages

In converting methylene lophenol and ethylidene lophenol into C_{28} and C_{29} sterols, the following major changes have to occur in addition to the reduction of the ethylidene and methylene groups already discussed: (*a*) removal of the remaining α methyl group at C-4; (*b*) conversion of a Δ^7 derivative into a Δ^5 derivative; and (*c*) (frequently) the insertion of a double bond in the side chain at C-22. Although a considerable amount is known about the way in which these transformations occur, the number of possible variations in the sequence involved is awe inspiring, and one can only conclude from the reported isolation from plants of sterols with almost every possible structure that most of the variations do occur somewhere in the plant kingdom. There is clearly a multiplex of interdigitating pathways, but one must try to define the normal pathway to the major sterols such as poriferasterol (4-B) and stigmasterol (4-Q).

REMOVAL OF THE SECOND 4α METHYL GROUP With regard to the removal of the remaining 4α methyl group, the conversion of [2,2,4-$^3H_3$] obtusifoliol (13-F) into poriferasterol (4-B) in *Ochromonas* resulted in the tritium in the axial (4β) position of obtusifoliol being inverted into the equatorial (4α) position in poriferasterol (87, 88). A possible mechanism for the reaction is indicated in Figure 9.

In spite of the impressive evidence previously quoted, that the 14α methyl group is removed after the first 4α methyl group and before the second 4α methyl group, evidence exists in some microorganisms that the 14α methyl is the first methyl removed, as is the case in mammals. For example, *Chlorella emersonii* grown in the presence of triparanol accumulates 14α methyl sterols such as (14-F) and (15-F) (122), but as just suggested, the stress exerted on the organism by the inhibitor might deviate the metabolic pathway from its normal route. On the other hand, the bacterium *Methylococcus capsulatus* accumulates 4,4-dimethyl sterols and 4α methyl sterols, which strongly suggests that the dominating biosynthetic pathway in this organism involves removal of the 14α methyl group before the two 4-methyl groups (18, 23).

CONVERSION OF Δ^8 STEROLS INTO Δ^5 STEROLS The pathway well established in animals is $\Delta^8 \rightarrow \Delta^7 \rightarrow \Delta^{5,7} \rightarrow \Delta^5$ (see 59). It is usually

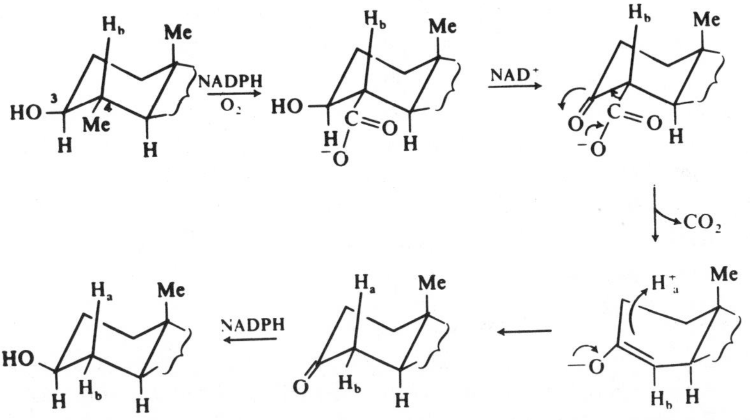

Figure 9 Suggested mechanism for the demethylation of 4α-methyl sterols (88).

assumed that such a progression occurs in plants, although it has not been directly demonstrated. However, a $\Delta^8 \rightarrow \Delta^7$ step involves loss of hydrogen from C-7, and it has been found that it is the 7β hydrogen which is lost in the formation of sterols in *Ochromonas malhamensis* (147), *Camellia sinensis* (see 63), *Larix decidua* (61), *Clerodendrum campbellii* (see 63), *Calendula officinalis* (146), and in β-ecdysone (16–T) in *Taxus baccata* and *Polypodium vulgare* (43). This stereospecificity is the same as that found in rats (see 59). However, in yeast (2, 36), *Aspergillus niger* (see 17) and *Fucus spiralis* (see 63), it is the 7α hydrogen which is lost. Furthermore, in cultures of bramble cells the fungicide AY-9944 specifically inhibits the $\Delta^8 \rightarrow \Delta^7$ isomerase, and a series of Δ^8 sterols accumulate in place of the usual Δ^5 series (138). In contrast to those clear-cut observations, the effect of AY-9944 on algae is complex. In *Chlorella emersonii* there was inhibition of removal of the C–14 methyl group, of the C–28 transmethylation, and of desaturation at C–22, while in *C. ellipsoidea* the inhibition was mainly on the Δ^{14} reductase and to a lesser extent on the $\Delta^8 \rightarrow \Delta^7$ isomerase (52, 122).

Insertion of the Δ^5 double bond involves the stereospecific removal of the 6α hydrogen along with the 5α hydrogen in *O. malhamensis* (148), *Larix decidua* (61), yeast (17), *Aspergillus fumigatus* (17), *Blakeslea trispora* (see 63), *Calendula officinalis* (146), and *Clerodendrum campbellii* (see 63). The same stereochemical removal is found in the rat, but the mechanism involved is still unknown (see 59).

Nothing is known about the reduction of the C–7 double bond in higher plants, but by analogy with the mammalian process it should involve a *trans* addition of 7α and 8β hydrogen derived from NADPH (B-face) and water (proton) respectively (see 60). In many fungi the reductase must be absent or present only in traces because the major biosynthetic end products are $\Delta^{5,7}$ dienes such as ergosterol (12–H).

The stereochemistry of the elimination of hydrogen from C–22 and C–23 in the formation of Δ^{22} double bonds has been examined in detail using the appropriate stereospecifically labeled species of MVA. Interesting variations are observed, but in all cases the hydrogens involved are in the *cis* configuration (see 63). Once again it must be said that the mechanism involved has not yet been clarified.

SEQUENCE OF TRANSFORMATIONS OF 24-METHYLENE AND 24-ETHYLIDENE LOPHENOL INTO C_{28} and C_{29} STEROLS As indicated earlier, the number of possible branches involved in these transformations is great. Evidence for the putative pathway in various algae and higher plants is still not fully compelling and is still based mainly on feeding experiment in intact systems, on inhibitor experiments, and on the isolation of likely intermediates; little enzymic work has yet been published.

In the case of *Ochromonas* the pathway in Figure 10 is based on the observed incorporation of stigmasta-7*Z* 24,28-dien-3β-ol (6-C), stigmasta-5,7,*Z*24(28)trien-3-β-ol (12-C) and isofucosterol (4-C) into poriferasterol (4-B) (89, 91, 96, 99). However, the point in the sequence at which the desaturation of the side chain at C–22 takes place is still not clear and may be different in different systems as exemplified by comparing the situation in the red alga *Porphyridium cruentum,* in which 22-dehydrocholesterol (4-F) represents 63% of the total sterols (12, 14), with that in cultured bramble cells in which there is no indication of the existence of Δ^{22} sterols (138).

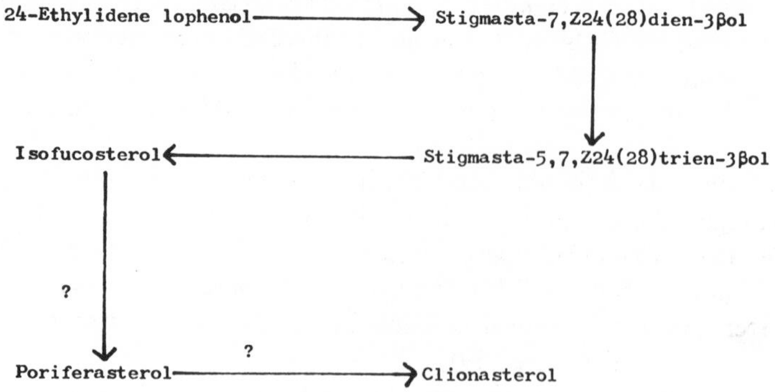

Figure 10 Final steps in the formation of sterols in *Ochromonas.*

In the case of the C_{28} sterols the only firm observation is that, contrary to earlier views, the appearance of the C–22 double bond occurs at a relatively early stage in the biosynthetic pathway. In *O. malhamensis* 31-norcyclolaudenol (8-G) is converted into 22-dihydrobrassicasterol (4-K), but this compound is not desaturated at C–22 to form brassicasterol (4-H) (64). Furthermore, 5α-ergosta-8,14-dien-3β-ol and 4α-methyl-5α-ergosta-8-en-3β-ol are both converted into dihydrobrassicasterol in *Chlorella* (51). The final step in brassicasterol synthesis in these organisms is therefore probably not the reduction of the Δ^{22} double bond. In ergosterol biosynthesis in yeast the reaction also occurs early in the sequence at the stage from 5α-ergosta-7,24-dien-3β-ol to 5α-ergosta-7,22,24-trien-3β-ol (9–11). The existence of 4α,24 ξ-dimethyl-5α-cholesta-8,22-dien-3β-ol[8] (13-R) in *Porphyridium* spp. also indicates an early desaturation at C–22 (12, 14).

Evidence for early desaturation in the formation of C_{29} sterols is less substantial than for the C_{28} sterols. It is mainly circumstantial and based on the observation that in *Ochromonas malhamensis* isofucosterol (4-C) (*Z*-isomer) is a better precursor of poriferasterol (4-B) than is fucosterol (4-E) (*E*-isomer) (99). The proximity of the C–29 methyl group of the *E*-isomer to C–22 and C–23 would reasonably be expected to hinder the enzymic desaturation at C–22,23 while the C–29 methyl group of the *Z* isomer would have no such effect (89). This view is supported by the observation that 5α-stigmasta-7-*Z*-24(28)dien-3β-ol (6-C) is as good a precursor as isofucosterol in *O. malhamensis* (89) and that in *Chlorella ellipsoidea* fucosterol is reduced to clionasterol (4-I) but is not converted into poriferasterol (4-B) (159). The latter result further suggests that the C–22 desaturase is much more specific in *Chlorella* than in *Ochromonas.* This idea is also borne out by experiments with the protozoan *Tetrahymena pyriformis* which cannot itself synthesize sterols but which can modify sterols added to the growth medium. Thus it desaturates isofucosterol (*Z*-isomer) at C–22 but fails to desaturate fucosterol (*E*-isomer) at this position (115).

Sterol Synthesis During Seed Germination

Recent experiments (110) have modified the conclusion from earlier experiments (34, 114) on the sterol biosynthetic capacity of germinating pea seedlings. Although in the early stages, up to 9 hours, squalene and β-amyrin synthesis preponderate in the axis tissue sterol is also synthesized during this period. On the other hand sterol is not synthesized by the cotyledons in which the triterpene β-amyrin is predominantly formed. It is

[8] ξ indicates that the chirality at C–24 has not been determined.

possible that as whole seedlings were used in the earlier experiments the uptake of exogeneous MVA into the cotyledons and thus into β-amyrin, resulted in undetectable amounts finding their way into the axis tissue and thus the sterol fraction there would be unlabelled. It is reasonable to assume that sterols are synthesized in the developing axis for membrane formation but physiological reasons for the rapid synthesis of β-amyrin are still not clear.

In intact pine seedlings germinating in [2–^{14}C]-MVA, the pattern of incorporation into the embryo and endosperm (dissected out after the end of the incubation period) is similar in both tissues. Incorporation during the first 24 hours is confined to squalene but thereafter there is a rapid incorporation into sterols while the relative amount of label in squalene falls precipitately. It would appear that the enzymes for the steps subsequent to squalene are either derepressed or actively synthesized during the first 3 days' germination. UDPG: sterol glucosyltransferase has been studied in etiolated pea seedlings (56).

CAROTENOIDS

The problem of sterol biosynthesis in yeasts and fungi was not considered in detail in the previous section in order to keep within the limitations of space. However, in considering the later stages of carotenogenesis, most of the work has been carried out on these organisms and on bacteria, and thus the emphasis in this section will be somewhat different.

Desaturation of Phytoene to Lycopene

The general pathway from phytoene to lycopene involves a stepwise desaturation (Figure 11) in which a previously isolated double bond is brought into conjugation.[9] This idea first arose some 30 years ago following studies on tomato mutants (126), but only recently have the details become clearer. For example, the stereochemistry of desaturation is the same for all double bonds (108, 161, 162; Figure 12). As indicated earlier, 15 *cis*-phytoene (*E,Z,E*-phytoene) is the main isomer in tomatoes, carrots, and fungi, and also accumulates in these tissues in the presence of inhibitors which prevent its desaturation. However, all *trans* phytoene is generally present as a minor component of a number of nonphotosynthetic bacteria (see 24, 68). The reason for the formation of *cis*-phytoene is not clear, but it can be either enzymically isomerized to all-*trans*-phytoene, which is then converted into lycopene, or desaturated to *cis*-phytofluene, the main component of tomato

[9]The general numbering of the carotenoid molecule is illustrated by γ-carotene (Formula 3) (one β- and one ψ- end group).

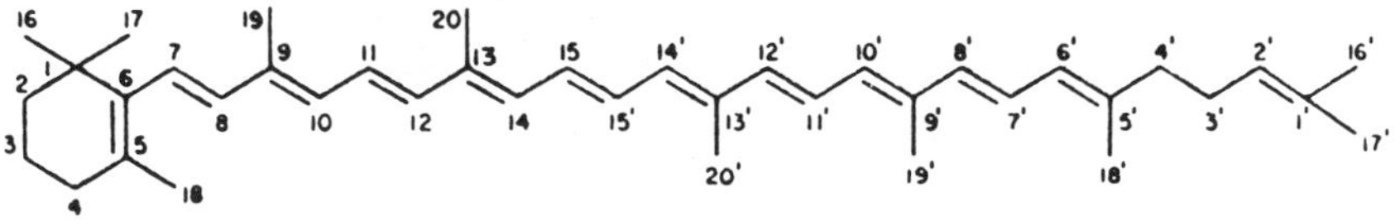

Formula 3 γ-Carotene

"phytofluene." This in turn is isomerized to *trans*-phytofluene before being desaturated to form eventually lycopene. The overall position is summarized in Figure 13.

The nature of the enzymes concerned in the desaturation is not known, but genetic studies on mutants of *Phycomyces blakesleeanus* suggest that the complete desaturation sequence is controlled by the product of one gene (*car R*) with four copies of its product acting in a complex to convert phytoene into lycopene (4, 50a). A leaky *car B* mutant produces a large amount of phytoene and progressively smaller amounts of phytofluene, ζ-carotene, neurosporene, and lycopene; this suggests that the complex contains four copies of the same enzyme (4, 55). However, later studies suggest that such a complex is only functional during exponential growth of the fungus (78).

Mutants of organisms that normally synthesize cyclic carotenoids are obtained which accumulate either lycopene, ζ-carotene, or phytoene (see 68); as far as is known, none has been isolated in which phytofluene or neurosporene is the major polyene synthesized. This and other information (see under Cyclization) suggest that the enzyme complex controlled by

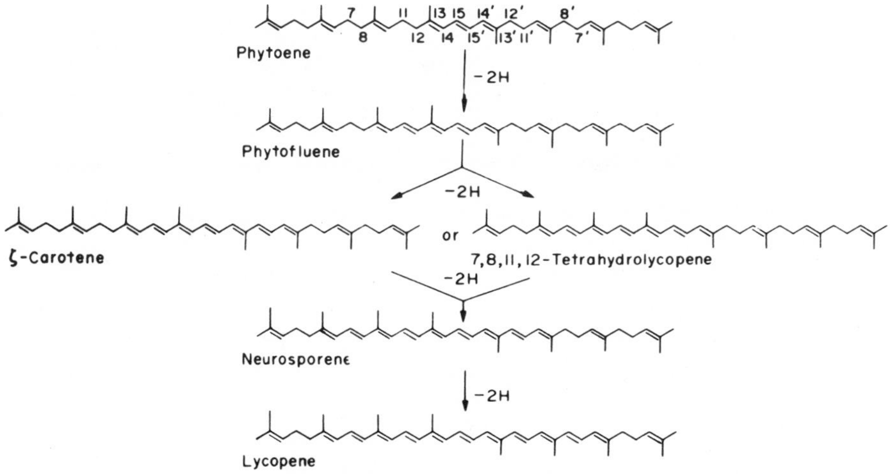

Figure 11 Stepwise desaturation of phytoene to lycopene.

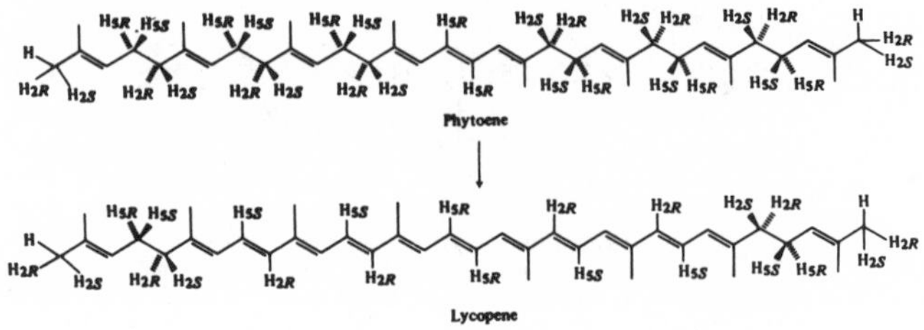

Figure 12 The stereochemistry of the desaturation of phytoene to lycopene.

one gene carries out two desaturations, i.e. phytoene → [phytofluene] → ζ-carotene and ζ-carotene → [neurosporene] → lycopene. Possibly under normal conditions the intermediates in square brackets never leave the enzyme complex. Clearly some photosynthetic bacteria which accumulate neurosporene and its derivatives (see 68) have a different mechanism for desaturation which involves 7,8,11,12-tetrahydrolycopene (Figure 11) and not ζ-carotene (48), the usual intermediate in other organisms.

Further Metabolism of Lycopene

Lycopene undergoes a number of reactions which involve its terminal double bonds. These reactions are (*a*) cyclization, (*b*) hydration, (*c*) hy-

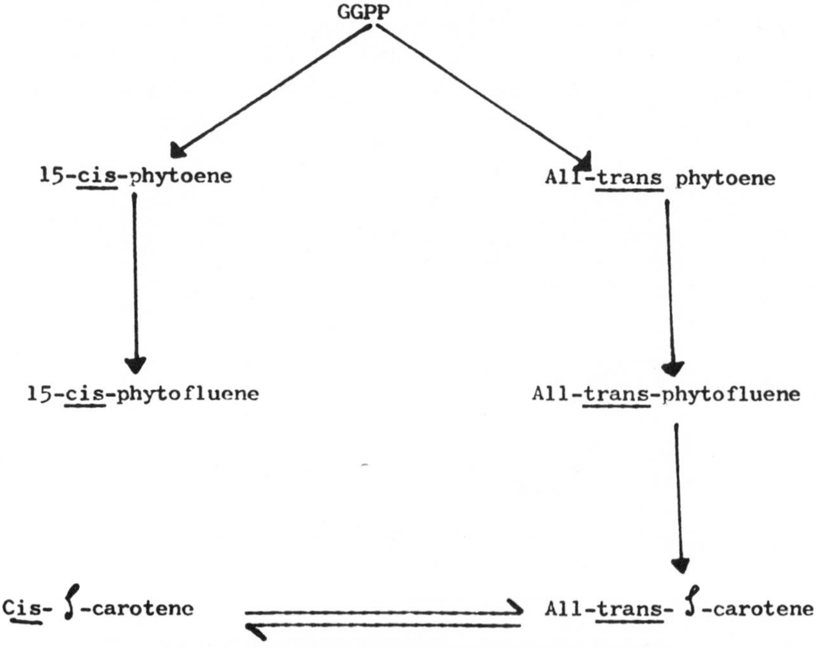

Figure 13 Some transformations of *cis*- and *trans*-phytoene.

drogenation, and (*d*) addition of a further C_5 isoprenoid unit which can occur with or without simultaneous cyclization.

CYCLIZATION Direct evidence for the cyclization of lycopene comes from its conversion into cyclic carotenes in soluble preparations from spinach chloroplasts (76), tomato plastids (121), *Phycomyces blakesleeanus* (50), and a *Flavobacterium* sp. (31). Furthermore, in the presence of nitrogenous bases (120), in particular nicotine and the herbicide CPTA (see 68 and references therein), cyclization is inhibited and lycopene accumulates. On removal of the inhibitor, the lycopene is converted into cyclic carotenoids (see 68). The mechanism involved in the inhibition is not known, but it is apparently the cyclizing reaction which is blocked rather than synthesis of the cyclizing enzymes (27, 28). In *Myxococcus fulvus* the conversion of the lycopene accumulated in the presence of inhibitors into cyclic pigments can only take place in the absence of a source of earlier substrate (84–86). This suggests an assembly line with access at a late point (lycopene) only when the supply of earlier substrates is exhausted.

In the presence of high concentrations of nicotine, cyclization at both ends of the molecule is inhibited, while at lower concentrations only one end of the molecule is affected. This has led to the view that the "substrate" for the cyclizing enzyme(s) is a "half molecule" of lycopene and that the enzyme has two active centers with the center which carries out the second cyclization more susceptible to nicotine than that carrying out the first cyclization (25). The relative sensitivity of various reactions of the terminal double bonds of lycopene to nicotine has recently been reviewed (48a).

The generally accepted view of the cyclization mechanism (Figure 14) involves first a proton attack with the formation of a transitory "carbonium

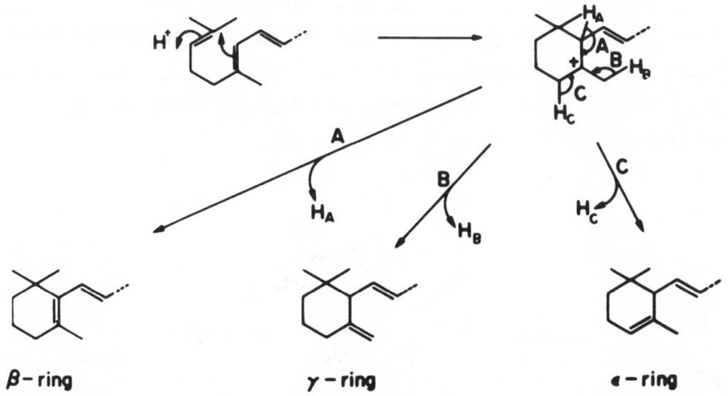

Figure 14 Mechanism of cyclization of lycopene to yield β, ε, and γ-rings.

ion" which is neutralized by ejection of either H_A, H_B, or H_C to yield β, ϵ-, and γ-rings (Figure 14). The reality of the proton attack has only recently been directly demonstrated. The mutant PG1 of *Scenedesmus obliquus* has been isolated which when grown in darkness accumulates ζ-carotene. When the dark-grown cells are illuminated, the ζ-carotene is converted into lycopene which is cyclized to produce the usual group of chloroplast carotenoids [α-carotene (one β- and one γ-end groups) and β-carotene (two β-end groups), zeaxanthin (3,3,'-dihydroxy-β-carotene), and neoxanthin] (29). In fact, illumination brings about the development of normal functional chloroplasts; other ζ-carotene-accumulating mutants are, on the other hand, killed on illumination in the presence of oxygen (28).

If dark-grown cells of mutant PG1 are illuminated when suspended in D_2O instead of H_2O, the β-carotene with a molecular weight 536 (normal) +2 is formed, and the two deuteriums which have been incorporated during cyclization of lycopene are located in the molecule at C–2 and C–2'. The same results have been obtained with *Rhodomicrobium vannielii* (see 68). The stereochemistry of proton attack has also been demonstrated by similar experiments with a *Flavobacterium* sp. The incoming proton was found to attack the *re, re* face of the C–1,2 double bond (Figure 15). This information was obtained by growing the organism in the presence of nicotine, which causes the accumulation of lycopene, washing out the inhibitor and resuspending the cells aerobically in D_2O when the lycopene cyclizes by deuteron attack to form β-carotene which is itself then hydroxylated to zeaxanthin. The chirality at C–2 which results from the insertion of deuterium at this position in the formation of zeaxanthin was shown by NMR studies (270 KHz) to be *S* as indicated in Figure 15 (26).

Although lycopene 1,2-epoxide, corresponding to squalene epoxide, occurs naturally in tomatoes (15), there is no evidence that it is involved in cyclization reactions. However, it is possible that it might be involved in the formation of the caroten-2-ols found in *Trentepohlia iolithus* (83) and blue-green algae (see 68). It is significant that in the two organisms mentioned the chirality at C–2 is opposite, so that if the mechanism suggested is functioning, the enzymes involved must be stereospecific for the two epimers of the epoxide or for the incoming proton.

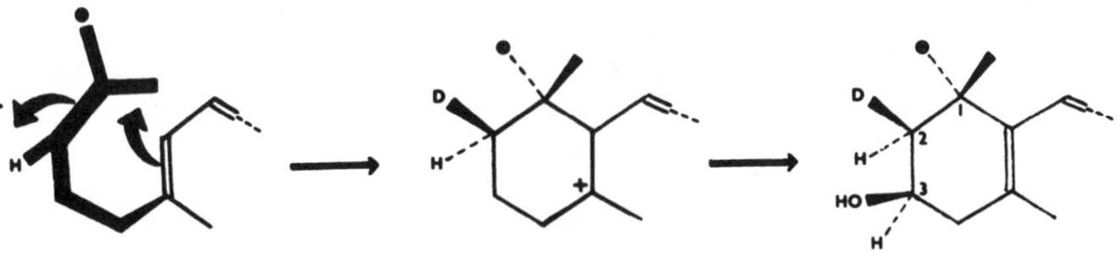

Figure 15 The full stereochemistry of the formation of the β-ring in zeaxanthin (26).

Experiments with stereospecifically labeled species of mevalonic acid have demonstrated that the β- and ϵ-rings of cyclic carotenes are formed from a common intermediate, indicated formally as a carbonium ion and not by isomerization after ring formation (see Figure 14). Experiments with [2-^{14}C,(4*R*)-4-^{3}H$_1$]MVA in a variety of tissues demonstrated that a tritium is lost from C–6 in the formation of a β-ring but is retained in the formation of an ϵ-ring; if the ϵ-ring were formed from the β-ring, then there would be no tritium at C–6 in the former (69, 163; Figure 17); similar experiments with [2-^{14}C,2-^{3}H$_2$]MVA demonstrated that the β-ring was not formed by isomerization of the ϵ-ring. Direct formation of the β-ring results in retention of the two labeled hydrogens at C–4 while isomerization of a preformed ϵ-ring would result in only one being present (Figure 16). Again experiments with a variety of tissues demonstrated the retention of both tritiums (163).

In the formation of the ϵ-ring of α-carotene in tomatoes, experiments with geranylgeranyl pyrophosphate stereospecifically labeled from either (2*R*-2^{3}H$_1$) or (2*S*-2,^{3}H$_1$)MVA revealed that it is the pro-*S* hydrogen originating from C–2 of MVA which is lost from C–4' of α-carotene (see 24). The same stereochemistry was demonstrated in the formation of decaprenoxanthin by a cell-free system from *Flavobacterium dehydrogenans* (55a).

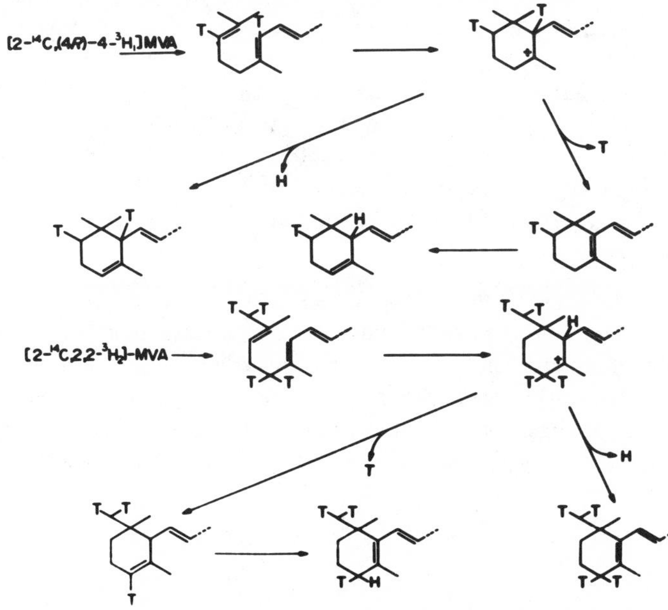

Figure 16 Distribution of tritium from [2-^{14}C, (4*R*)-4-^{3}H$_1$]MVA and [2-^{14}C, 2,2-^{3}H$_2$]MVA in β- and ϵ-rings of carotenoids.

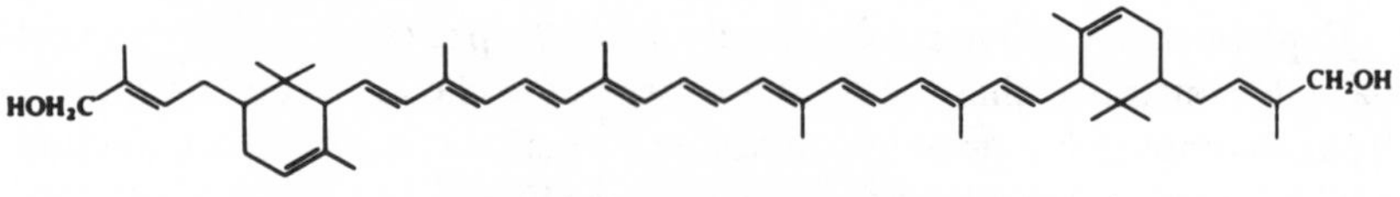

Formula 4 Decaprenoxanthin

Cyclic end-groups which have an additional branched C–5 unit at C–2 are found in some C_{45} and C_{50} bacterial carotenoids, e.g. decaprenoxanthin (Formula 4). It has been suggested that the C_5 units are added to the conventional C_{40} precursor lycopene; this is substantiated by the observation that in the presence of nicotine, lycopene and other precursors accumulate in *Halobacterium salinarum*. On removal of the inhibitor these disappear and the final product is formed (93). In the cyclization, the C_5 unit presumably takes the place of H^+ as the electrophilic attacking species.

A stereochemical problem which has been solved recently is the orientation of the methyl groups at C–1. Early experiments on the biosynthesis of torularhodin (16'-carboxy-3', 4'-dehydro-γ-carotene) (155) showed that the gem dimethyl groups of the acylic end of the molecule retained their individuality and that the one methyl originating from C–2 of MVA was specifically converted into the carboxyl group of torularhodin. Recently it has been shown by ^{13}C NMR studies that the configuration in the β-ring of zeaxanthin synthesis by *Flavobacterium* is as indicated in Figure 15. The methyl group labeled • arises from C–2 of mevalonic acid. This direct demonstration of the orientation of the methyl groups is opposite to that inferred from work on trisporic acid synthesis in *Blakeslea trispora* (6, 33). When [2-^{14}C]MVA is incorporated into trisporic acid, only the methyl group at C–1 is labeled in Formula 5. As the configuration at C–1 of trisporic acid is known and as β-carotene is said to be converted into trisporic acid (6, 33), it follows that C–2 of MVA must have become the 1-*proR* methyl group of β-carotene and not the 1-*pro-S* group as indicated in Figure 15.

CYCLIZATION OF PRECURSORS OTHER THAN LYCOPENE. The existence in nature of compounds such as α- and β-zeacarotene (7'8'-dihydro-γ-carotene) in maize and in tomato mutants (delta) (see 68) indicates that neurosporene can cyclize at the end of the molecule carrying a double bond at C–7,8. The suggestion that unsaturation at C–7 is mandatory for cyclization is supported by the observation that in diphenylamine-inhibited cul-

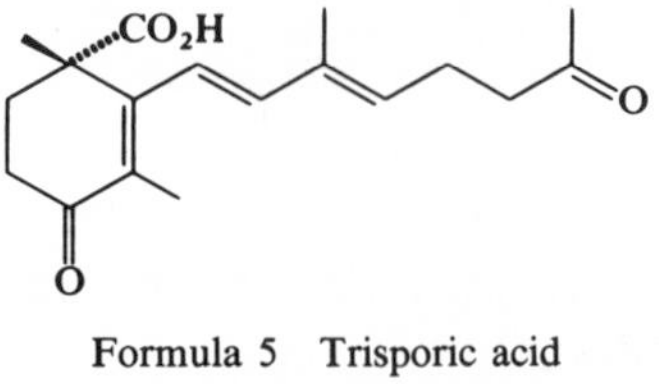

Formula 5 Trisporic acid

tures of *P. blakesleeanus,* a cyclic isomer of 7,8,11,12-tetrahydrolycopene exists, while a cyclic isomer of ζ-carotene could not be detected (49).

HYDRATION AND HYDROGENATION It has been shown mainly by using the inhibitors nicotine and CPTA that the formation of rhodopin (1-hydroxy-1,2-dihydrolycopene) in photosynthetic bacteria involve reactions at the C–1,2-double bond of lycopene and not of earlier precursors. When *Rhodomicrobium vannielii* is cultured in the presence of either inhibitor, the major carotenoid in normal cultures, rhodopin, is replaced by lycopene. On removal of the inhibitors, lycopene is converted into rhodopin, that is, hydration occurs across the C–1,2 double bond (24, 107). Furthermore, labeled lycopene is converted into rhodopin by a cell-free system from *Rm. vannielii* (145), although in this instance hydration only occurs after insertion of the 7,8 double bond as is the case in the cyclization reactions. The existence of hydroxyspheroidenone (Formula 6) in *Rhodopseudomonas* spp. indicates that this is not mandatory in all cases and under the stress of diphenylamine inhibition many hydrated compounds which do not carry such a double bond are formed (103, 105). There is also no evidence for desaturation at C–3,4 before hydration at C–1,2. A similar situation exists with the hydrogenation reaction at C–1,2 in *Rhodopseudomonas viridis* (68, 105), although in this case saturation occurs before desaturation at C–7 because the main pigment is 1,2-dehydroneurosporene.

Direct Insertion of Oxygen

The insertion of oxygen at C–3 in cyclic carotenoids has been studied intensively, particularly in the zeaxanthin-producing *Flavobacterium* spp. Hydroxylation follows cyclization, but the differential effect of low and high concentrations of nicotine in cyclization suggests that more than one pathway may be involved. As indicated earlier, at high concentration cyclization at both ends of the molecule is inhibited, while at lower concentrations inhibition occurs only at one end of the molecule and rubixanthin (3-hydroxy-γ-carotene) accumulates (109). This suggests alternative pathways (24; Figure 17). It has not been decided yet whether all reactions function under normal growth conditions, and if so, which one is quantitatively the most important.

The stereochemistry of hydroxylation at C–3 and C–3' has been determined for zeaxanthin formation in *Flavobacterium* by the use of [2-^{14}C,5-(R)-5-$^{3}H_1$]mevalonic acid. The results indicate a direct insertion of the hydroxyl group (30, Figure 15). The requirement for O_2 and the stereo-

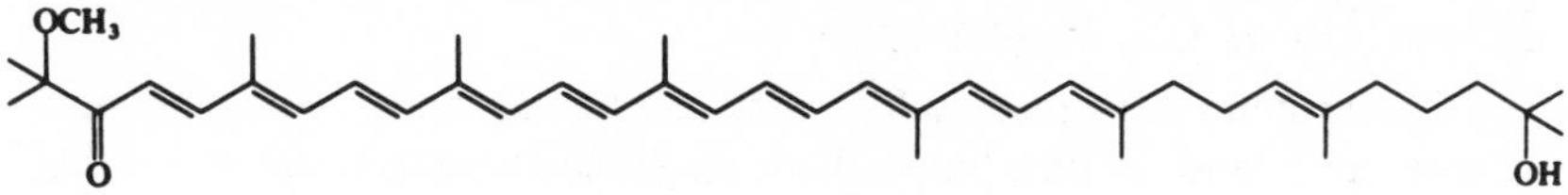

Formula 6 Hydroxyspheroidenone

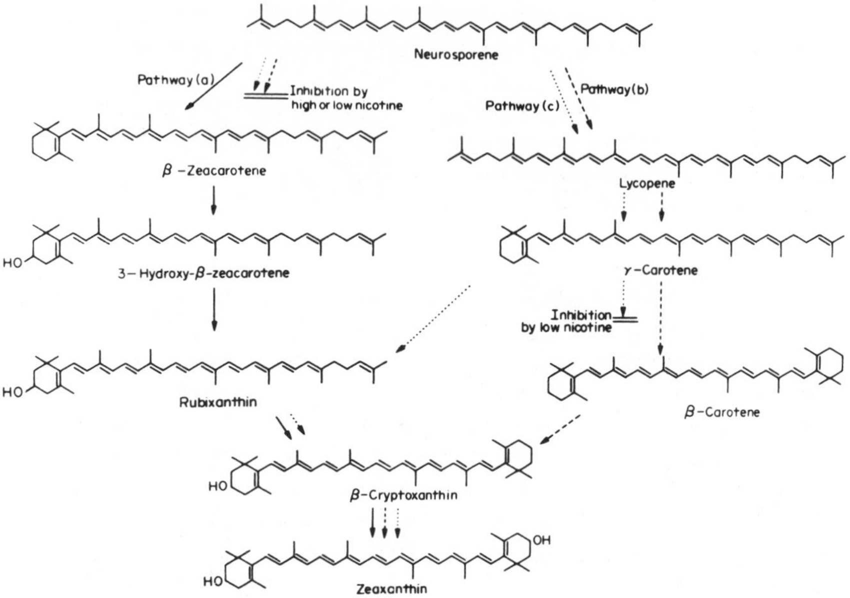

Figure 17 Alternative pathways of xanthophyll synthesis in a *Flavobacterium* spp. (24).

chemistry of the reaction suggest that the enzyme involved is a monooxygenase. However, formation of β-cryptoxanthin (3-hydroxy-β-carotene) has been demonstrated unexpectedly but unequivocally in the strict anaerobic *Rhodomicrobium vannielii* (30). Hydroxy derivatives of phytoene are present in cultures of *Rsp. rubrum* cultured in the presence of diphenylamine (104).

Formation of Acyclic C_{45} and C_{50} Carotenoids

The formation of cyclic C_{45} and C_{50} carotenoids has already been discussed, but acyclic C_{45} and C_{50} pigments are also present in some microorganisms, e.g. bacterioruberin in *Halobacterium salinarum* (22). In the presence of nicotine, lycopene accumulates, which suggests that the C_5 units are added to a conventional C_{40} unit lycopene (93) after desaturation to 3,4,3',4'-tetradehydrolycopene and that the insertion of these double bonds effectively prevents cyclization. The proposed pathway is given in Figure 4, in which the C_5 unit replaces H^+ as the species attacking the terminal double bond of lycopene.

Formation of C_{30} Carotenoids

The C_{30} carotenoids formed by *Streptococcus* spp. probably have symmetrical structures, and the final biosynthetic stages can be considered as starting with 4,4'-diapophytoene and proceeding through the unknown compounds to 4,4'-diapolycopene-4-al and the glucoside of 4-hydroxy-4,4'-diaponeurosporene (50). The sequence is based on isolation from *Streptococcus* of the constituent compounds. It should be noted that the 4,4'-diapophytoene formed is the *cis* isomer while the remaining members of the series are all-*trans* isomers (50).

In the case of *Staphylococcus aureus* the biosynthetic pathways cannot be considered in detail until the structure of the major pigment is settled. A decision still has to be made between an esterified glucoside of either the symmetrical 4,4'-diaponeurosporen-4-oic acid ester (50) or an unsymmetrical analog (105a).

CONCLUSION

This review demonstrates that during the past decade sophisticated and subtle biochemical methods have been deployed in attacking the problem of sterol and carotenoid biosynthesis in plants and microorganisms and this is indeed true for other terpenoids as well. Many of the biosynthetic sequences have been revealed and the mechanism of the reactions together with the attendant stereochemistry demonstrated. Progress in the study of the enzymology of the processes has been substantial but in this area there is still all to play for and glittering prizes are available to the first researcher who, for example, can demonstrate at the molecular level the difference in the active sites of the two different squalene 1,2-oxide cyclases—one producing lanosterol and one cycloartenol—or can explain in similar terms how one end of a lycopene molecule is cyclized to a β-ring while the other isomerizes to an ϵ-ring. Some limited progress has been made in studying the control of sterol and carotenoid synthesis in plants, but here again, biochemists are poised for the attack rather than fully committed to the battle. It is in these fields, together with a genetic attack which is gaining momentum particularly in carotenoid biosynthesis, that the next reviewer of terpenoid biosynthesis should be able to report substantial and exciting progress.

Literature Cited

1. Adler, J. H., Patterson, G. W. 1974. The metabolism of potential sterol precursors in *Chlorella ellipsoidea*. *Plant Physiol.* 53:5–13
2. Akhtar, M., Rahimtula, A. D., Wilton, D. C. 1970. The stereochemistry of hydrogen elimination from C–7 in cholesterol and ergosterol biosynthesis. *Biochem. J.* 117:539–42
3. Altmann, L. J., Ash, L., Kowerski, R. C., Epstein, W. W., Larsen, B. R., Rilling, H. C., Muscio, F., Gregonis, D. E. 1972. Prephytoene pyrophosphate, a new intermediate in the biosynthesis of carotenoids. *J. Am. Chem. Soc.* 94: 3257–59
4. Aragon, C. M. G., Murillo, F. J., de la Guardia, M. D., Cerdá-Olmedo, E. 1976. An enzyme complex for the dehydrogenation of phytoene in *Phycomyces*. *Eur. J. Biochem.* 63:71–75
5. Armarego, W. L. F., Goad, L. J., Goodwin, T. W. 1973. Biosynthesis of α-spinasterol from (2-^{14}C (4*R*)-4-$^{3}H_1$) mevalonic acid by *Spinacea oleracea* and *Medicago sativa*. *Phytochemistry* 12:2181–87
6. Austin, D. J., Bu'Lock, J. D., Drake, D. 1970. Biosynthesis of trisporic acids from β-carotene via retinal and trisporal. *Experientia* 26:348–49
7. Bader, S., Guglielmetti, L., Arigoni, D. 1964. The origin of the ethyl side chain in spinasterol. *Proc. Chem. Soc.* 16
8. Barnes, F. J., Qureshi, A. A., Semmler, E. J., Porter, J. W. 1973. Prelycopersene pyrophosphate and lycopersene. *J. Biol. Chem.* 248:2768–73
9. Barton, D. H. R., Corrie, J. E. T., Marshall, P. J., Widdowson, D. A. 1973. Unified scheme for the biosynthesis of ergosterol in *Saccharomyces cerevesiae*. *Bioorg. Chem.* 3:363–73
10. Barton, D. H. R., Corrie, J. E. T., Widdowson, D. A., Bard, M., Woods, R. A. 1974. Biosynthetic implications of the sterol content of ergosterol-deficient mutants of yeast. *J. Chem. Soc. Chem. Commun.* 30–31
11. Barton, D. H. R., Harrison, D. M., Moss, G. P., Widdowson, D. A. 1970. Investigations on the biosynthesis of steroids and terpenoids. Part II. Role of 24-methylene derivatives in the biosynthesis of steroids and terpenoids. *J. Chem. Soc. C.*, 775–85
12. Beastall, G. H., Rees, H. H., Goodwin, T. W. 1971. Sterols in *Porphyridium cruentum*. *Tetrahedron Lett.* 52: 4935–38
13. Beastall, G. H., Rees, H. H., Goodwin, T. W. 1971. Properties of a 2,3-oxidosqualene-cycloartenol cyclase from *Ochromonas malhamensis*. *FEBS Lett.* 18:175–78
14. Beastall, G. H., Tyndall, A. M. Rees, H. H., Goodwin, T. W. 1974. Sterols in *Porphyridium* species. *Eur. J. Biochem.* 41:301–9
15. Ben-Aziz, A., Britton, G., Goodwin, T. W. 1973. Carotene epoxides of *Lycopersicon esculentum*. *Phytochemistry* 12: 2759–64
16. Benveniste, P., Massy-Westropp, R. A. 1967. Mise en evidence de l'epoxide-2,3 de squalène dans les tissues de tabac. *Tetrahedron Lett.* 37:3553–56
17. Bimpson, T., Goad, L. J., Goodwin, T. W. 1969. The stereochemistry of hydrogen elimination at C–6, C–22 and C–23 during ergosterol biosynthesis by *Aspergillus niger* Fres. *Chem. Commun.* 297–98
18. Bird, C. W., Lynch, J. M., Pirt, F. J., Reid, W. W., Brooks, C. J. W., Middleditch, B. S. 1971. Steroids and squalene in *Methylococcus capsulatus* grown on methane. *Nature* 230:473
19. Boid, R., Rees, H. H., Goodwin, T. W. 1975. Studies in insect moulting hormone biosynthesis. Biosynthesis of cyasterone in the plant *Cyathula capitata*. *Biochem. Physiol. Pflanz.* 168:27–40
20. Bolger, L. M., Rees, H. H., Ghisalberti, E. L., Goad, L. J., Goodwin, T. W. 1970. Biosynthesis of 24-ethylcholesta-5,22,25-trien-3β-ol, a new sterol from *Clerodendrum campbellii*. *Biochem. J.* 118:197–200
21. Bolger, L. M., Rees, H. H., Ghisalberti, E. L., Goad, L. J., Goodwin, T. W. 1970. Isolation of two new sterols from *Clerodendrum campbelli*. *Tetrahedron Lett.* 35:3043–46
22. Borch, G., Norgärd, S., Liaaen-Jensen, S. 1972. Circular dichroism and relative configuration of C_{50}-carotenoids. *Acta Chem. Scand.* 26:402–3
23. Bouvier, P., Rohmer, M., Benveniste, P., Ourisson, G. 1976. $\Delta^{8(14)}$-Steroids in the bacterium *Methylococcus capsulatus*. *Biochem. J.* 159:267–71
24. Britton, G. 1976. Biosynthesis of carotenoids. In *Chemistry and Biochemistry of Plant Pigments*, ed. T. W., Goodwin, 1:262–327. London: Academic. 2nd ed.
25. Britton, G. 1976. Later stages of carotenoid biosynthesis. *Pure Appl. Chem.* 47:223–36

26. Britton, G., Lockley, W. J. S., Mundy, A. P., Patel, N. J., Goodwin, T. W., Englert, G. 1979. Stereochemistry of cyclization in carotenoid biosynthesis: use of ^{13}C-labelling to elucidate the stereochemical behaviour of the C–1 methyl substituents during zeaxanthin biosynthesis in a *Flavobacterium*. *J. Chem. Soc. Chem. Commun.* In press
27. Britton, G., Lockley, W. J. S., Powls, R., Goodwin, T. W., Heyes, L. M. 1977. Carotenoid transformations during chloroplast development in *Scenedesmus obliquus* PG1 demonstrated by deuterium labelling. *Nature* 268:81
28. Britton, G., Powls, R. 1977. Phytoene, phytofluene and ζ-carotene isomers from a *Scenedesmus obliquus* mutant. *Phytochemistry* 16:1253–55
29. Britton, G., Powls, R., Schultze, R. M. 1977. The effect of illumination on the pigment composition of the ζ-carotenic mutant PG1 *Scenedesmus obliquus*. *Arch. Microbiol.* 113:281–84
30. Britton, G., Singh, R. K., Goodwin, T. W., Ben-Aziz, A. 1975. The carotenoids of *Rhodomicrobium vannielii* (Rhodospirillaceae) and the effect of diphenylamine on the carotenoid composition. *Phytochemistry* 14:2427–33
31. Brown, D. J., Britton, G., Goodwin, T. W. 1975. Carotenoid biosynthesis by a cell-free preparation from a *Flavobacterium* species. *Biochem. Soc. Trans.* 3:741–42
32. Buggy, M. J., Britton, G., Goodwin, T. W. 1969. Stereochemistry of phytoene biosynthesis by isolated chloroplasts. *Biochem. J.* 114:641–43
33. Bu'Lock, J. D., Austin, D. J., Snatzke, G., Hruban, I. 1970. Absolute configuration of trisporic acid and the stereochemistry of cyclization in β-carotene biosynthesis. *Chem. Commun.* 255–56
33a. Bu'Lock, J. D., Osagie, A. U. 1976. Sterol biosynthesis via cycloartenol in *Saprolegnia*. *Phytochemistry* 15:1249–51
34. Capstack, E., Baisted, D. J., Newschwander, W. W., Blondin, G., Rosin, N. L., Nes, W. R. 1962. The biosynthesis of squalene in germinating seeds of *Pisum sativum*. *Biochemistry* 1:1178–83
35. Caspi, E., Moreau, J., Ramm, P. J. 1974. Sterol biosynthesis from (3*RS*,-2*R*)-[2-^{14}C,2-^{3}H] mevalonic acid in a yeast homogenate. Stereochemistry of the C-15 tritium atom. *J. Am. Chem. Soc.* 96:8107–8
36. Caspi, E., Ramm, P. J. 1969. Stereochemical differences in the biosynthesis of C_{27} Δ^7-steroidal intermediates. *Tetrahedron Lett.* 3:181–85
37. Caspi, E., Sliwowski, J. 1975. On the role of cycloartenol on the formation of phytosterols. Biosynthesis of [19-^{2}H]sitosterol in deuterium oxide germinated peas. *J. Am. Chem. Soc.* 97:5032–34
38. Caspi, E., Sliwowski, J., Robichaud, C. S. 1975. Biosynthesis of sitosterol from (2*R*)-and(2S)-[2-^{3}H] mevalonic acid in the pea. The incorporation of a 15α-tritium atom derived from (3*RS*,2*R*)-[2-^{14}C,2^{3}H] mevalonic acid. *J. Am. Chem. Soc.* 97:3820–22
39. Castle, M., Blondin, G. A., Nes, W. R. 1963. Evidence for the origin of the ethyl group of β-sitosterol. *J. Am. Chem. Soc.* 85:3306–7
40. Charlton, J. M., Treharne, K. J., Goodwin, T. W. 1967. Incorporation of [2-^{14}C] mevalonic acid into phytoene in isolated chloroplasts. *Biochem. J.* 105: 205–12
41. Chin, P. L., Patterson, G. W., Dutky, S. R. 1976. Effect of AY–9944 on sterol biosynthesis in *Chlorella sorokiniana*. *Phytochemistry* 15:1907–10
42. Clifford, K., Cornforth, J. W., Mallaby, R., Phillips, G. J. 1971. The stereochemistry of isopentenyl pyrophosphate isomerase. *Chem. Commun.* 1599–1600
43. Cook, I. F., Lloyd-Jones, J. G., Rees, H. H., Goodwin, T. W. 1973. The stereochemistry of hydrogen elimination from C–7 during biosynthesis of ecdysones in insects and plants. *Biochem. J.* 136: 135–45
44. Corey, E. J., Ortiz de Montellano, P. R. 1967. Enzymic synthesis of β-amyrin from 2,3-oxidosqualene. *J. Am. Chem. Soc.* 89:3362–63
45. Cornforth, J. W. 1968. Olefin alkylation in biosynthesis. *Angew Chem.* 7:903–11
46. Cornforth, J. W., Cornforth, R. 1970. Chemistry of mevalonic acid. *Biochem. Soc. Symp.* 29:5–15
47. Cornforth, J. W. Ross, F. P. 1970. Synthesis of 5*S*-5[$^{3}H_1$] mevalonic acid. *Chem. Commun.* 1395–96
48. Davies, B. H. 1970. Alternative pathways of spirilloxanthin biosynthesis in *Rhodospirillum rubrum*. *Biochem. J.* 116:101
48a. Davies, B. H. 1979. *Pure Appl. Chem.* In press
49. Davies, B. H., Rees, A. F. 1973. 7,8,11,12-tetrahydro-ζ-carotene: a novel carotenoid from *Phycomyces blakesleeanus*. *Phytochemistry* 12:2745–50
50. Davies, B. H., Taylor, R. F. 1976. Carotenoid biosynthesis—the early steps. *Pure Appl. Chem.* 47:211–21
50a. de la Guardia, M. D., Aragon, C. M. G., Murillo, F. J., Cerdá-Olmedo, E.

1971. A carotenogenic enzyme aggregate in *Phycomyces:* Evidence from quantitative complementation. *Proc. Natl. Acad. Sci. USA* 68:2012–15

51. Dickson, L. G., Patterson, G. W. 1972. Inhibition of sterol biosynthesis in *Chlorella ellipsoidea* by AY–9944 [trans-1,4-bis(2-chlorobenzylaminomethyl cyclohexane dihydrochloride) *Lipids* 7:635–43
52. Dickson, L. G., Patterson, G. W., Cohen, C. F., Dutky, S. R. 1972. Two novel sterols from inhibited *Chlorella ellipsoidea. Phytochemistry* 11:3473–77
53. Douglas, T. J., Paleg, L. G. 1974. Plant growth retardants as inhibitors of sterol biosynthesis in tobacco seedlings. *Plant Physiol.* 54:238–45
54. Eppenberger, U., Hirth, L., Ourisson, G. 1968. Anaerobische Cyclisierung von squalene-2,3-epoxide zu cycloartenol in Gewebecultur von *Nicotiana tabacum* L. *Eur. J. Biochem.* 8:180–83
55. Esclava, A. P., Cerdá-Olmedo, E. 1974. Genetic control of phytoene dehydrogenation in *Phycomyces. Plant Sci. Lett.* 2:9–14

55a. Fahey, D., Milborrow, B. V. 1978. The stereochemistry of biosynthesis of decaprenoxanthin in a cell-free system. *Phytochemistry* 17:2077–82

56. Fang, T. Y., Baisted, D. J. 1976. UDPG: sterol glucosyltransferase in etiolated pea seedlings. *Phytochemistry* 15:273–78
57. Fonteneau, P., Hartmann-Bouillon, M. A., Benveniste, P. 1977. A 24-methylene lophenol C–28 methyl transferase from suspension cultures of bramble cells. *Plant Sci. Lett.* 10:147–55
58. Ghisalberti, E. L., de Souza, N. J., Rees, H. H., Goad, L. J., Goodwin, T. W. 1969. Biological removal of the 4α-methyl group during the conversion of cycloartenol into 31-norcycloartenol in *Polypodium vulgare* Linn. *Chem. Commun.* 1403–8
59. Goad, L. J. 1970. Sterol biosynthesis. *Biochem. Soc. Symp.* 29:45–77
60. Goad, L. J. 1977. In *Lipids and Lipid Polymers in Higher Plants,* ed. M. Tevini, H. K. Lichtenthaler, pp. 116–68. Berlin: Springer
61. Goad, L. J., Gibbons, G. F., Bolger, L. M., Rees, H. H., Goodwin, T. W. 1969. Incorporation of [2-^{14}C,(5-*R*)-5-^{3}H$_1$] mevalonic acid into cholesterol by a rat liver homogenate and into β-sitosterol and fucosterol in *Larix decidua* leaves. *Biochem. J.* 114:885–92
62. Goad, L. J., Goodwin, T. W. 1969. Studies on phytosterol biosynthesis: observation on the biosynthesis of fucosterol in the marine brown alga *Fucus spiralis. Eur. J. Biochem.* 7:502–8
63. Goad, L. J., Goodwin, T. W. 1972. The biosynthesis of plant sterols. *Prog. Phytochem.* 3:113–98
64. Goad, L. J., Lenton, J. R., Knapp, F. F., Goodwin, T. W. 1974. Phytosterol side-chain biosynthesis. *Lipids* 9: 582–95
65. Goad, L. J., Williams, B. L., Goodwin, T. W. 1967. The presence of 4α,14α-dimethyl-$\Delta^{8,24(28)}$-ergostadien-3β-ol in grapefruit peel and its cooccurrence with cycloeucalenol in higher plant tissues. *Eur. J. Biochem.* 3:232–36
66. Goodwin, T. W. 1974. Prochirality in biochemistry. *Essays Biochem.* 9: 104–60
67. Goodwin, T. W. 1977. Sterol alkylation in higher plants and micro-organisms. *Biochem. Soc. Trans.* 5:1252–55
68. Goodwin, T. W. 1979. *Comparative Biochemistry of Carotenoids,* Vol. 1. London: Chapman & Hall. 2nd ed.
69. Goodwin, T. W., Williams, R. J. H. 1965. A mechanism for the cyclization of an acyclic precursor to form β-carotene. *Biochem. J.* 94:5C–7C
70. Gregonis, D. E., Rilling, H. C. 1974. The stereochemistry of *trans*-phytoene synthesis. Some observations on lycopersene as a carotene precursor and a mechanism for the synthesis of *cis*-and *trans*-phytoene. *Biochemistry* 13:1538–42

70a. Hall, J., Smith, A. R. H., Goad, L. J., Goodwin, T. W. 1969. The conversion of lanosterol, cycloartenol and 24-methylene cycloartenol into poriferasterol by *Ochromonas malhamensis. Biochem. J.* 112:129–30

71. Hartmann, M. A., Ferne, M., Gigot, C., Brandt, R., Benveniste, P. 1973. Isolement, characterization et composition en sterols de fractions subcellularies de feuilles etoilées de Haricot. *Physiol. Veg.* 11:209–30
72. Hartmann, M. A., Fonteneau, P., Benveniste, P. 1977. Subcellular localization of sterol synthesizing enzymes in maize coleoptiles. *Plant Sci. Lett.* 10:147–55
73. Heintz, R., Benveniste, P. 1970. Cyclization de l'epoxyde-2,3 de squalène par des microsomes extraits de tissues de tabac cultives *in vitro. Phytochemistry* 9:1499–1503
74. Heintz, R., Benveniste, P. 1974. Plant sterol metabolism. Enzymatic cleavage of the 9β,19β cyclopropane ring of cy-

clopropylsterols in bramble tissue cultures. *J. Biol. Chem.* 249:4267–74

75. Heintz, R., Benveniste, P., Robinson, W. H., Coates, R. M. 1972. Demonstration and identification of a biosynthetic intermediate between farnesyl PP and squalene in a higher plant. *Biochem. Biophys. Res. Commun.* 49:1547–53
76. Hill, H. M., Calderwood, S. K., Rogers, L. J. 1971. Conversion of lycopene to β-carotene by plastids isolated from higher plants. *Phytochemistry* 10: 2051–58
77. Homberg, E. E., Schiller, H. P. R. 1973. Neue sterine in *Helianthus annuus. Phytochemistry* 12:1767–73
78. Hsu, W-J., Ailion, D. C., Delbrück, M. 1974. Carotenogenesis in *Phycomyces. Phytochemistry* 13:1463–68
79. Immer, H., Huber, K. 1971. Synthese und biologische Ausvertung von 3β,-20(R)-dihydroxy-protost-24en. *Helv. Chim. Acta* 54:1346–60
80. Itoh, T., Ishii, T., Tamura, T., Matsumoto, T. 1978. Four new and other 4α-methyl sterols in the seeds of Solanaceae. *Phytochemistry* 17:971–77
81. Itoh, T., Tamura, T., Matsumoto, T. 1977. 4-Desmethylsterols in the seeds of Solanceae. *Steroids* 30:425–33
82. Jarman, T. R., Gunatilaka, A. A. I., Widdowson, D. A. 1975 Biosynthesis of steroids and terpenoids XI. 24-Methylene sterol 24(28)-reductase of *Saccharomyces cerevesiae. Bioorg. Chem.* 4:202–11
83. Kjösen, H., Arpin, N., Liaaen-Jensen, S. 1972. The carotenoids of *Trentepohlia iolithus.* Isolation of β,β-carotene-20-ol, β-ε-carotene-2-ol and β,β-carotene-2-2-diol. *Acta Chem. Scand.* 26:3053–67
84. Kleinig, H. 1974. Inhibition of carotenoid synthesis in *Myxococcus fulvus* (Myxobacteriales). *Arch. Mikrobiol.* 97:217–226
85. Kleinig, H. 1975. On the utilization *in vivo* of lycopene and phytoene as precursors for the formation of carotenoid glucoside ester and on the regulation of carotenoid biosynthesis in *Myxococcus fulvus. Eur. J. Biochem.* 57:301–8
86. Kleinig, H., Reichenbach, H. 1973. Biosynthesis of carotenoid glucoside esters in *Myxococcus fulvus* (Myxobacteriales): inhibition by nicotine and carotenoid turnover. *Biochim. Biophys. Acta* 306:249–56
87. Knapp, F. F., Goad, L. J., Goodwin, T. W. 1973. Inversion of the 4β-hydrogen during the conversion of the sterol obtusifoliol into poriferasterol by *Ochromonas malhamensis. Chem. Commun.* 149–50
88. Knapp, F. F., Goad, L. J., Goodwin, T. W. 1977. Stereochemistry of C-4 demethylation during conversion of obtusifoliol into poriferasterol by *Ochromonas malhamensis. Phytochemistry* 16:1677–81
89. Knapp, F. F., Goad, L. J., Goodwin, T. W. 1977. The conversion of 24-ethylidene sterols into poriferasterol by the alga *Ochromonas malhamensis. Phytochemistry* 16:1683–88
90. Knapp, F. F., Grieg, J. B., Goad, L. J., Goodwin, T. W. 1971. The conversion of 24-ethylidene sterols into poriferasterol by *Ochromonas malhamensis. Chem. Commun.* 707–9
91. Knapp, F. F., Nicholas, H. J. 1970. Phytosterol biosynthesis in banana peel. Initial removal of the 4β-methyl group of 24-methylene cycloartanol during its conversion into cyclolaudenol in *Musa sapientum. Chem. Commun.* 399–400
92. Knapp, F. F., Nicholas, H. J. 1971. The biosynthesis of 31-norcyclolaudenone in *Musa sapientum. Phytochemistry* 10:97–102
93. Kushwaha, S. C., Kates, M. 1976. Effect of nicotine on biosynthesis of C_{50} carotenoids in *Halobacterium cutirubrum. Can. J. Biochem.* 54:824–29
94. Largeau, C., Goad, L. J., Goodwin, T. W. 1977. Biosynthesis of sitosterol in barley seedlings. *Phytochemistry* 16: 1925–30
95. Largeau, C., Goad, L. J., Goodwin, T. W. 1977. Conversion of a 24β-ethyl-25-methylene intermediate into poriferasterol by *Trebouxia* species. *Phytochemistry* 16:1931–33
96. Lecompte, M. F., Lenfant, M. 1978. Biosynthèse des ramifications en C-24 des phytosterols de *Physarum polycephalum* et *Ochromonas danica. Phytochemistry* 17:1123–26
97. Lederer, E. 1969. Some problems concerning biological C-alkylation reactions and phytosterol biosynthesis. *Q. Rev. London* 23:453–81
98. Lenton, J. R., Goad, L. J., Goodwin, T. W. 1975. Sitosterol biosynthesis in *Hordeum vulgare. Phytochemistry* 14:1523–28
99. Lenton, J. R., Hall, J. L., Smith, A. R. H., Ghisalberti, E. L., Rees, H. H., Goad, L. J., Goodwin, T. W. 1971. The utilization of potential phytosterol precursors by *Ochromonas malhamensis. Arch. Biochem. Biophys.* 143:664–74
100. Liu, H. K., Knoche, H. W. 1974. Origin

of sterols in uredospores of *Uromyces phaseoli. Phytochemistry* 13:1795–99

101. Liu, H. K., Knoche, H. W. 1976. A sterol methyltransferase from bean rust uredospores, *Uromyces phaseoli. Phytochemistry* 15:683

102. Lockley, W. J. S., Roberts, D. P., Rees, H. H., Goodwin, T. W. 1974. 24-Methylcholesta-5,24(25)-dien-3β-ol: a new sterol from *Withania somnifera. Tetrahedron Lett.* 43:3773–76

103. Malhotra, H. C., Britton, G., Goodwin, T. W. 1970. The mono- and dimethoxy-carotenoids of diphenylamine-inhibited cultures of *Rhodospirillum rubrum. Phytochemistry* 9:2369–75

104. Malhotra, H. C., Britton, G., Goodwin, T. W. 1970. The occurrence of hydroxy derivatives of phytoene in diphenylamine-inhibited cultures of *Rhodospirillum rubrum. FEBS Lett.* 6:334–36

105. Malhotra, H. C., Britton, G., Goodwin, T. W. 1970. Occurrence of 1,2-dihydrocarotenoids in *Rhodopseudomonas viridis. Chem. Commun.* 127–28

105a. Marshall, J. H., Wilmoth, G. J. 1978. Biosynthesis of triterpenoid carotenoids in *Staphylococcus aureus. Abstr. 5th Int. Symp. Carotenoids, Wisconsin,* p. 36

106. Maudinas, B., Bucholz, M. L., Porter, J. W. 1975. The partial purification and properties of a phytoene synthetase complex isolated from tomato fruit plastids. *Abstr. 4th Int. Symp. Carotenoids, Berne,* p. 41

107. McDermott, J. C. B., Ben-Aziz, A., Singh, R. K., Britton, G., Goodwin, T. W. 1973. Recent studies of carotenoid biosynthesis in bacteria. *Pure Appl. Chem.* 35:29–45

108. McDermott, J. C. B., Britton, G., Goodwin, T. W. 1973. Carotenoid biosynthesis in a *Flavobacterium* sp: Stereochemistry of hydrogen elimination in the desaturation of phytoene to lycopene, rubixanthin and zeaxanthin. *Biochem. J.* 134:1115–17

109. McDermott, J. C. B., Brown, D. J., Britton, G., Goodwin, T. W. 1974. Alternative pathways of zeaxanthin biosynthesis in a *Flavobacterium* species. *Biochem. J.* 144:231–43

110. McKean, M. L., Nes, W. R. 1977. Delayed conversion of squalene to sterols during development of *Pinus pinea* seedlings. *Lipids* 12:382–85

111. McKean, M. L., Nes, W. R. 1977. Evidence for separate intermediates in the biosynthesis of 24α- and 24β-sterols in tracheophytes. *Phytochemistry* 16: 683–88

112. Mercer, E. I., Harries, W. B. 1975. The mechanism of alkylation at C-24 during clionasterol biosynthesis in *Monodus subteranneus. Phytochemistry* 14:439–43

113. Nes, W. R. 1977. The biochemistry of plant sterols. *Adv. Lipid Res.* 15:233–324

114. Nes, W. R., Baisted, D., Capstack, E., Newschwander, W. W., Russell, P. T., 1967. In *Biochemistry of Chloroplasts,* ed. T. W. Goodwin, 2:273. New York: Academic

115. Nes, W. R., Malya, P. A. G., Mallory, F. B., Ferguson, K. A., Landrey, J. R., Conner, R. L. 1971. Conformation analysis of the enzyme-substrate complex in the dehydrogenation of sterols in *Tetrahymena pyriformis. J. Biol. Chem.* 246:561–68

116. Nes, W. R., Krevitz, K., Behzadan, S. 1976. Configuration at C-24 of 24-methyl and 24-ethylsterols in Tracheophytes. *Lipids* 11:118–26

117. Nes, W. R., Krevitz, K., Joseph, J., Nes, W. D., Harris, B., Gibbons, G. F., Patterson, G. W. 1977. The phylogenetic distribution of sterols in Tracheophytes. *Lipids* 12:511–27

118. Nicholas, H. J., Moriarty, S. 1963. Biosynthesis of β-sitosterol. *Fed. Proc.* 22:529

119. Nicotra, F., Ronchetti, F., Russo, G., Lugaro, G., Casellato, M. 1977. Stereochemistry of hydrogen migration from C-24 to 2-25 during isofucosterol biosynthesis in *Pinus pinea. Chem. Commun.* 889–90

120. Ninet, L., Renaut, J., Tissier, R. 1969. Activation of the biosynthesis of carotenoids by *Blakeslea trispora. Biotechnol. Bioeng.* 11:1195–1210

120a. Palmer, M. A., Goad, L. J., Goodwin, T. W., Cosey, D. B., Boar, R. B. 1978. The conversion of 5α-lanost-24-ene-3β 9α-diol and parkeol into poriferasterol by the alga *Ochromonas malhamensis. Phytochemistry* 17:1577–80

121. Papastephanou, C., Barnes, F. J., Briedis, A. V., Porter, J. W. 1973. Enzymatic synthesis of carotene by cell-free preparation of fruit of several genetic selections of tomatoes. *Arch. Biochem. Biophys.* 157:415–25

122. Patterson, G. W., Doyle, P. J., Dickson, L. G., Chan, J. T. 1974. Effect of triparanol and AY-9944 upon sterol biosynthesis in *Chlorella. Lipids* 9: 567–74

123. Ponsinet, G., Ourisson, G. 1967. Biosynthesis *in vitro* des triterpenes dans

le latex d'*Euphorbia. Phytochemistry* 6:1235–43
124. Ponsinet, G., Ourisson, G. 1968. Aspects particulars de la biosynthetic des triterpenes dan le latex d'Euphorbia. *Phytochemistry* 6:1235–43
125. Popjak, G., Cornforth, J. W. 1966. Substrate stereochemistry in squalene biosynthesis. *Biochem. J.* 101:553–68
126. Porter, J. W., Lincoln, R. E. 1950. Lycopersion selections containing a high content of carotenes and colorless polyenes. The mechanism of carotene biosynthesis. *Arch. Biochem.* 27:390–404
127. Rahier, A., Benveniste, P., Cattel, L. 1976. *In vitro* incorporation of ^2H at the C-10 methyl group of obtusifoliol during the enzymatic cleavage of the 9β,19β-cyclopropane ring of cycloeucalenol. *Chem. Commun.* 287–88
128. Rahier, A., Cattel, L., Benveniste, P. 1977. Mechanism of the enzymatic cleavage of the 9β,19β-cyclopropane ring of cycloeucalenol. *Phytochemistry* 16:1187–92
129. Randall, P. J., Rees, H. H., Goodwin, T. W. 1972. Mechanism of alkylation during sitosterol biosynthesis in *Larix decidua. Chem. Commun.* 1295–96
130. Rees, H. H., Goad, L. J., Goodwin, T. W. 1968. Studies in phytosterol biosynthesis. Mechanism of biosynthesis of cycloartenol. *Biochem. J.* 107:417–26
131. Rees, H. H., Mercer, E. I., Goodwin, T. W. 1966. The stereospecific biosynthesis of plant sterols and α- and β-amyrin. *Biochem. J.* 99:726–34
132. Reid, W. W. 1968. Accumulation of squalene-2,3-oxide during inhibition of phytosterol biosynthesis in *Nicotiana tabacum. Phytochemistry* 7:451–52
133. Rohmer, M., Brandt, R. D. 1973. Les sterols et les precurseurs chez *Asteria longa* Pringsheim. *Eur. J. Biochem.* 36:446–54
134. Rohmer, M., Ourisson, G., Benveniste, P., Bimpson, T. 1975. Sterol biosynthesis in heterotrophic plant parasites. *Phytochemistry* 14:727–30
135. Rubinstein, I., Goad, L. J. 1974. Occurrence of (24S)-24-methyl-cholesta,5,22E-dien-3β-ol in the diatom *Phaeodactylum tricornutum. Phytochemistry* 13:485–87
136. Russell, P. T., van Aller, R. T., Nes, W. R. 1967. The mechanism of introduction of alkyl groups at C-24 of sterols. The necessity for the Δ^{24}-bond. *J. Biol. Chem.* 242:5802–6
137. Ruzicka, L. 1959. History of the isoprene rule. *Proc. Chem. Soc.* 341–60
138. Schmitt, P., Benveniste, P. 1979. Effect of AY-9944 on sterol biosynthesis in suspension cultures of bramble cells. *Phytochemistry.* In press
139. Schroepfer, G. J., Lutsky, J. B. N., Martin, J. A., Huntoon, S., Fourcans, B., Lee, W. H., Vermilion, J. 1972. *Proc. Roy. Soc.* 180B:125
140. Shah, D. V., Feldbrügge, D. H., Houser, A. R., Porter, J. W. 1968. The enzymatic synthesis of phytoene. *Arch. Biochem. Biophys.* 127:124–31
141. Sharma, R. K. 1970. Biosynthesis of plant sterols: stereochemistry of hydrogen elimination at C-7 in α-spinasterol. *Chem. Commun.* 543
142. Sharpless, K. B., Snyder, T. E., Spencer, T. A., Maheshwari, K. K., Guhn, G., Clayton, R. B. 1968. Biological demethylation of 4,4-dimethyl sterols. Initial removal of the 4α-methyl group. *J. Am. Chem. Soc.* 90:6874–75
143. Sharpless, K. B., Snyder, T. E., Spencer, T. A., Maheshwari, K. K., Nelson, J. A., Clayton, R. B. 1969. Biological demethylation of 4,4-dimethyl sterols. Evidence for the enzymic epimerization of the 4β-methyl group prior to its oxidative removal. *J. Am. Chem. Soc.* 91:3394–96
144. Shimizu, Y., Alam, M., Kobayashi, A. 1975. Dinosterol the major sterol with a unique side chain in the toxic dinoflagellate *Gonyaulax tamarensis. J. Am. Chem. Soc.* 98:1059–60
145. Singh, R. K., Ben-Aziz, A., Britton, G., Goodwin, T. W. 1977. Carotenoid biosynthesis in *Rhodomicrobium vannielii.* Experiments with nicotine and 2-(4-chlorophenylthio)triethylammonium chloride (CPTA). *Biochim. Biophys. Acta* 488:475–83
146. Sliwowski, J., Kasprzyk, Z. 1974. Stereospecificity of sterol biosynthesis in *Calendula officinalis* flowers. *Phytochemistry* 13:1451–57
147. Smith, A. R. H., Goad, L. J., Goodwin, T. W. 1968. The stereochemistry of hydrogen elimiation at C(7) and C(22) in phytosterol biosynthesis by *Ochromonas malhamensis. Chem. Commun.* 926–27
148. Smith, A. R. H., Goad, L. J., Goodwin, T. W. 1968. The stereochemistry of hydrogen elimination at C-6 and C-23 in phytosterol biosynthesis by *Ochromonas malhamensis. Chem. Commun.* 1259–60
149. Smith, A. R. H., Goad, L. J., Goodwin, T. W. 1972. Incorporation of stereospecifically labelled mevalonic acid into

poriferasterol by *Ochromonas malhamensis. Phytochemistry* 11:2775–81
150. Smith, A. R. H., Goad, L. J., Goodwin, T. W., Lederer, E. 1966. Phytosterol biosynthesis: evidence for a 24-ethylidene intermediate during sterol formation in *Ochromonas malhamensis. Biochem. J.* 104:56C–58C
151. Sucrow, W., Reimerdes, A. 1968. Δ^7-Sterine an *Cucurbitaceen. Z. Naturforsch.* 23b:42–45
152. Sucrow, W., Slopianka, M., Kircher, H. W. 1976. The occurrence of C_{29} sterols with different configurations at C-24 in *Cucurbita pepo* as shown by 270 NMR. *Phytochemistry* 15:1533–35
153. Suga, T., Shishibor, T. 1975. The stereospecificity of biosynthesis of squalene and β-amyrin in *Pisum sativum. Phytochemistry* 14:2411–17
154. Sunder, R., Ayengar, K. N. N., Rangasami, S. 1976. Structures of four new triterpenes from the rhizomes of *Polypodium juglandifolium. J. Chem. Soc. Perkin Trans.* 1:117–21
155. Tefft, R. E., Goodwin, T. W., Simpson, K. L. 1970. Aspects of the stereochemistry of torularhodin biosynthesis. *Biochem. J.* 177:921–27
156. Tomita, Y., Uomori, A. 1970. Mechanism of the biosynthesis of the ethyl side chain at C-24 of stigmasterol in tissue cultures of *Nicotiana tabacum* and *Dioscorea tokoro. Chem. Commun.* 1416–17
157. Tomita, Y., Uomori, A., Minato, H. 1970. Biosynthesis of the methyl and ethyl groups at C-24 of phytosterols in *Chlorella vulgaris. Phytochemistry* 9:555–60
158. Tomita, Y., Uomori, A., Sakurai, E. 1971. Biosynthesis of the ethyl group at C-24 of poriferasterol and Δ^5-ergosterol in *Chlorella ellipsoidea. Phytochemistry* 10:573–77
159. Tsai, L. B., Patterson, G. W., Cohen, C. F., Klein, P. D. 1974. Metabolism of 2,4-^{3}H-14α-methyl-5α-ergost-8-enol and 2,4-^{3}H-5α-ergosta-8,14-dienol in *Chlorella ellipsoidea. Lipids* 9:1014–17
160. Villanueva, V., Barbier, M., Lederer, E. 1964. Sur la biosynthèsis de la chaine laterale ethylidene du fucosterol par double methylation par la méthionine. *Bull. Soc. Chim. Fr.* 1423–24
161. Williams, R. J. H., Britton, G., Charlton, J. M., Goodwin, T. W. 1967. The stereospecific biosynthesis of phytoene and polyunsaturated polyenes. *Biochem. J.* 104:767–77
162. Williams, R. J. H., Britton, G., Goodwin, T. W. 1966. Stereospecificity in carotene biosynthesis. *Biochem. J.* 101:7P
163. Williams, R. J. H., Britton, G., Goodwin, T. W. 1967. The biosynthesis of cyclic carotenes. *Biochem. J.* 105:99–105
164. Withers, N. W., Tuttle, R. C., Goad, L. J., Goodwin, T. W. 1979. Dinosterol side chain biosynthesis in a marine dinoflagellate, *Crypthecodinium cohnii. Phytochemistry* 18:71–73
165. Withers, N. W., Tuttle, R. C., Holz, G. C., Beach, D. H., Goad, L. J., Goodwin, T. W. 1978. Dehydrodinosterol, dinosterone and related sterols of a non-photosynthetic dinoflagellate, *Crypthecodinium cohnii. Phytochemistry* 17: 1987–89
166. Wojciechowski, Z. A., Goad, L. J., Goodwin, T. W. 1973. S-Adenosyl-L-methionine-cycloartenol methyltransferase activity in cell-free systems from *Trebouxia* sp. and *Scenedesmus obliquus. Biochem. J.* 136:405–12

Ann. Rev. Plant Physiol. 1979. 30:405–23

DNA PLANT VIRUSES ❖7677

Robert J. Shepherd

Department of Plant Pathology, University of California,
Davis, California 95616

CONTENTS

Interest in DNA plant viruses has been stimulated by recent developments in recombinant DNA technology and the potential this has for improving crop species. The endeavor will require DNA molecules with powers of autonomous replication in plant cells, to which foreign DNAs can be attached, and the simple genomes of the DNA plant viruses may have some advantages not offered by other types of genetic vectors. Consequently, one can expect that the properties and biology of these viruses will be thoroughly explored in the next few years, and feasibility studies will be made to evaluate their potential as genetic vehicles.

Of the several hundred known plant viruses, which fall into 20 odd taxonomic groups, only two (or perhaps three) distinctive types have DNA as their genomic material. Why so few plant viruses have DNA chromosomes is not known. All other host-allied types of viruses, such as those of

0066-4294/79/0601-0405$01.00

higher animals, arthropods, and bacteria, are abundantly represented by both RNA and DNA genomes.

Each of the two types of DNA plant viruses comprises a taxonomic group. The first taxon to be described was given the name *caulimovirus* (43). The best known member of this group, the cauliflower mosaic virus, was the first plant virus found to contain DNA (104). About a half-dozen similar viruses are known, all of which contain double-stranded DNA.

The second group of DNA viruses is exemplified by bean golden mosaic with single-stranded DNA (37, 38). The name *geminivirus* has been proposed for this group (42).

A third type of virus reported to contain DNA is the potato leafroll virus (94), which has not been well investigated. Unfortunately, the virus is present in plants in almost vanishing amounts and quantities for adequate characterization have not been obtained.

The intention of this article is to review the current state of knowledge of the DNA plant viruses and to provide some information for recombinant DNA experiments. Although more information is needed before the usefulness of any of the viruses can be predicted, it will be obvious that the biological features of some will limit their potential as gene-cloning vehicles.

CAULIMOVIRUSES

These viruses are common and widely distributed throughout the temperate regions of the world. Three serologically related viruses, cauliflower mosaic, dahlia mosaic, and carnation etched ring viruses (13–15, 46), together with figwort mosaic virus, have been found to contain DNA (32–34, 36, 63, 104). They have several expedient features not exhibited by some of the other caulimoviruses. For example, they are transmissible mechanically to easily propagated herbaceous plants, their free DNA genomes are infectious (101), and they have no tissue restrictions.

Biological Properties

The caulimoviruses have moderately restricted host ranges. In nature most of the viruses are found infecting a few species within a single plant family and, experimentally, little if any overlap of host ranges exists among the various viruses. Cauliflower mosaic virus (CaMV), for example, infects only species of the Cruciferae in nature. The caulimoviruses and their host ranges are listed in Table 1.

The caulimoviruses induce mosaic-mottle types of diseases that are not much different in appearance than diseases caused by several other types of plant viruses. Illustrations of infected plants, desirable propagation species, and other properties are given in the *Descriptions of Plant Viruses*

Table 1 The caulimoviruses and their host ranges[a]

Virus[b]	Susceptible species and family	References
Carnation etched ring virus	*Dianthus caryophyllus, D. barbatus, Saponaria* sp., and *Silene* sp. (in Caryophyllaceae)	33, 46, 63, 66, 89
Cauliflower mosaic virus	Many species of the Cruciferae: Some strains infect *Nicotiana clevlandii* and *Datura stramonium* of the Solanaceae.	12, 44, 70, 98, 99
Dahlia mosaic virus	*Dahlia variabilis, Verbesina encelioides, Zinnia elegans,* and other species of the Compositae, some species of Amaranthaceae, Chenopodiaceae, and Solanaceae	11, 13–15
Figwort mosaic virus	Various sp. of Scrophulariaceae and Chenopodiaceae	J. E. Duffus, personal communication
Mirabilis mosaic virus	*Mirabilis* sp. (Nyctaginaceae)	17
Strawberry vein banding virus	*Fragaria* sp. (Rosaceae)	30, 56

[a]All of the viruses have aphid vectors in nature and can be mechanically transmitted.
[b]Cassava vein mosaic (57, 58) and petunia vein-clearing viruses (67) may be members of this group.

published by the British Association of Applied Biologists and Commonwealth Mycological Society (14, 66, 98). These should be consulted for a more complete account of the individual viruses. Diagnostic features of these viruses are their morphology, transmissibility, and characteristic inclusion bodies (99, 100).

Caulimoviruses are spread by insects and by infected vegetatively propagated planting material. A single type of sucking insect, aphids (Aphididae, Hemiptera, Insecta), are the vectors in nature. These insects are common in greenhouses where unintentional transmissions may occur frequently. This problem can be prevented by use of nonaphid-transmitted strains which have been isolated by the author.

When transmitted by insect vectors, the viruses are picked up on the mouthparts (stylet bundle) of feeding aphids and carried to healthy plants during later feeding activity. After acquisition, transmissibility is retained by some aphid species for several hours and occasionally for 1–2 days. Vector transmission of caulimoviruses has been reviewed (99).

The caulimoviruses specify the production of one or more gene products in infected plants which function in the insect transmission process, as shown by the following observations. Nonaphid transmissible (defective) isolates of CaMV have been found which become transmissible if aphids are allowed to feed first on plants infected with a transmissible strain, or if plants are mixedly infected with aphid-transmissible and the nonaphid-transmissible isolate of the virus (71, 72). This shows that aphid-transmissible strains produce something in infected plants that cause defective strains to become transmissible. Furthermore, when ordinary purified CaMV is fed to aphids through membranes, the virus is not transmitted even when strains are used that are readily transmitted from plants (83). However, when aphids have been allowed to probe first on infected plants, they do acquire virus by feeding through membranes. Thus the virus specifies some vector transmission factor that is produced in infected plants independently of the virus particles. It is not known how the transmission factor functions, whether in acquisition or inoculation (71, 72).

Physical and Chemical Properties of Caulimoviruses

Cauliflower mosaic virus is easier to propagate and occurs in larger amounts in infected tissue than other caulimoviruses. Although very stable during manipulation, the virus is difficult to liberate from inclusion bodies and has a marked tendency to aggregate during purification. Lack of release from inclusion bodies also complicates the assay of virus in crude leaf extracts (3). These limitations have been largely overcome by use of a detergent and urea to release virus from inclusion bodies (36, 51).

The spherical particles of CaMV are 50 nm in diameter. They appear highly hydrated (51) and may have a large hydrated center, as suggested by an extreme tendency to collapse when dried in air (84, 85). The particles are penetrated by potassium phosphotungstate, but none of the electron-dense stains reveal any readily discernible external structure. Additional properties of the virions of CaMV are given in Table 2.

STRUCTURAL PROTEINS The virions of CaMV are extremely stable structures which resist treatments that readily degrade other plant viruses. When treated with 1% sodium dodecyl sulfate (SDS) at 50–70°C, for example, the virus swells to produce more slowly sedimenting forms without loss of DNA. These revert to their original, more slowly sedimenting form when the SDS is removed (52). Boiling in 1% SDS or 6M guanidine ·HCl are effective in dissociating virions.

Several studies on the structural proteins of CaMV have revealed two major coat proteins plus small amounts of other polypeptides (16, 49, 54, 111). The two coat proteins differ almost twofold in molecular weight and about fivefold in their molar ratios in the CaMV virion. The smaller protein has been variously reported to be 32,000 to 42,000 in molecular weight (16, 54, 111), but it may suffer proteolysis during virus isolation. The larger coat protein is about 64,000 daltons (49). Together the two polypeptides comprise more than 90% of the protein associated with the virus particle. From

Table 2 Properties of cauliflower mosaic virus and its double-stranded DNA

Virus		Reference
Sedimentation coefficient	202.2 ± 1.1	51, 52
Diffusion coefficient	0.75×10^{-7} cm^2/s	51
Morphology	50.3 ± 1.4 nm sphere	35, 51, 84, 91
Phosphorus content	1.63%	51
DNA content	16%	51, 101, 104
Partial specific volume	0.704 ± 0.007 g/ml	51
Coat proteins	Two, molecular weights 64,000 and 37,000	16, 49, 54, 110
Amino acid content	18% lysine, 5% arginine	16
Molecular weight	22.8×10^6 daltons	51
Viral DNA		
Molecular weight	4.5–5 $\times 10^6$ daltons	75, 92, 102
Nucleotide composition	GC = 43%	101
Melting temperature	87.2°	101
Buoyant density	1.702 g/ml	101
Contour length	2.31 μm	92, 102
RNA content	Less than 1%	50

the molecular weight equivalent of the virion, 22.8×10^6 daltons, calculated from the diffusion and sedimentation coefficients and partial specific volume (51) given in Table 2, one can estimate that each virion has somewhat over 400 copies of the smaller protein and 55–60 copies of the larger protein. This and other evidence (110) suggests a T = 7 structure for the exterior capsid made up of the smaller protein, and a T = 1 core composed of the larger polypeptide (49).

The minor proteins in CaMV may not be virus-specified polypeptides. Similar small DNA viruses of animals such as polyoma and SV40 contain host-derived histones (31).

NUCLEIC ACID The DNA of the caulimoviruses is double-stranded, as shown by a cooperative-type melting curve, nonreactivity with formaldehyde, and buoyant density and nucleotide ratio typical of double-stranded DNA (101).

No unusual bases have been found in CaMV DNA in nucleotide analyses. After hydrolysis with 90% formic acid to preserve 5-hydroxymethyl-cytosine (115), neither it or 5-methyl cytosine, common in chloroplast DNA (109), were found (101). Nearest neighbor frequency analysis of CaMV DNA showed an array of sequences similar to that of host-plant DNA (92).

CaMV DNA in solution exhibits two sedimenting forms (17.1 and 19.OS) plus a small amount of more rapidly sedimenting material that is more heterogeneous than the rest of the material (50). Electrophoresis in 2.5% polyacrylamide gels also resolves two major components in addition to some aggregated material near the origin. After limited DNase digestion, the slower major electrophoretic component is transformed into the faster. Infectivity is associated only with the slower component (50). The two sedimenting and electrophoretic species correspond to the linear and circular molecules observed in electron micrographs (19, 64, 92, 102, 112). Although considerable tangling of circular molecules occurs, suggesting some supercoiling, the presence of only one isopycnic species in ethidium bromide-containing cesium chloride gradients indicates no supercoiling of CaMV DNA (101). The density of this component is consistent with a relaxed molecule.

Freshly prepared CaMV DNA consists of 80–90% circular molecules which appears to be the native infectious form of the DNA (50, 112). Linear DNA is probably a breakage product of the circular form.

The DNA of CaMV shows an interesting sort of electrophoretic heterogeneity that has not been accounted for by any particular structural feature. In an extensive investigation of this matter, Volovitch, Drugeon & Yot (112) have found that none of the electrophoretic components differ significantly in contour length, in restriction endonuclease pattern, or in single-stranded fragments released by alkaline, heat, or dimethylsulfoxide denaturation. The denaturation products, related to the locations of specific single-stranded gaps in the genome, are identical for the different types of native DNA molecules resolved by gel electrophoresis. They suggested that electrophoretic heterogeneity was probably due to local conformational differences in the different types of molecules (112).

The location of single-stranded gaps in the genome of CaMV was also investigated by Volovitch et al (112, 113). Electrophoresis in alkaline gels revealed three linear fragments and no circular molecules. Thus there are three discontinuities in the molecule and neither DNA strand is covalently closed. The well-defined sizes of the single-stranded fragments showed that the gaps are not randomly located. S_1 nuclease treatment yielded three double-stranded fragments. Mapping of the cleavage sites for a particular strain of CaMV shows that the S_1 sites are identical with the location of the interruptions in the complementary strand. The termini at the single-stranded gaps have free 3' hydroxyl groups (112).

Hull & Howell (48) have observed pairs of linear DNA molecules of about 5.0 and 4.0 daltons which are believed to arise by strand breakage of circular molecules at specific sites near one another. One of these sites appeared to coincide with a single-stranded gap in the molecule that is also cleaved by S_1 nuclease. Although somewhat less certain, the other site probably does not coincide with the second S_1 site. In fact, some strains which lack the second S_1 site show both types of linear molecule.

An interesting feature of CaMV DNA is the presence of a small amount of covalently linked RNA (50). Denaturation in alkali for short periods gives a pattern of sedimenting forms similar to formaldehyde denaturation, but treatment for periods up to 18 hr, which would destroy RNA, gives more slowly sedimenting forms. Hence, the alkali treatment seems to be cleaving the nucleic acid strands. Neither polyoma nor Col E1 plasmid DNA is affected after the same alkali treatment, showing that nicking of phosphodiester bonds of DNA is not occurring during the prolonged exposure to alkali (50). RNase does not remove RNA after H^3-uridine labeling of CaMV DNA, which confirms its covalent nature.

The function of the RNA in CaMV DNA is unknown. It may be misincorporation of nucleotides or RNA primers left in situ after DNA synthesis, although the latter generally are removed enzymatically after DNA synthesis (18). Short RNA segments covalently attached to viral and plasmid DNAs are common (45, 88, 105, 114) and may be present in eukaryotic nuclear DNA as well (29, 41).

Unusual bases in CaMV DNA could account for its progressive scission in alkali. The replacement of some thymine residues with α-putrescinylthymine confers alkali sensitivity on phage ϕW-14 DNA (68). A similar abnormal base in CaMV DNA could account for its alkali sensitivity, although no similar discrepancy between buoyant density and melting temperature, as with ϕW-14 DNA (62), is shown.

Restriction endonuclease mapping of CaMV DNA A physical map of the CaMV genome based on cleavage by restriction endonucleases has been determined (48, 75, 113). Although overlapping fragments have been used for major features of the map, the examination of selected fragments cloned in *E. coli* (75) has helped to sort out problems caused by sequence heterogeneity in the DNA.

Uncommon restriction endonuclease cleavage sites are present in the DNA of native strains of CaMV. Cleavage of DNA by *Sal* I, for example, gives less than a 5% conversion of circular DNA molecules to linears, showing that a minor fraction of the DNA has a *Sal* I sensitive site (75). Four such variable restriction sites, widely spaced on the viral genome, were examined in molecularly cloned fragments of the DNA. When the restriction sites flanking these variable cleavage sites were examined, in both native and also in plasmid cloned fragments without the variable sites, the flanking sites were spaced at the same distance indicating that no extensive deletions, insertions, or inversions of DNA had occurred in these segments. Base substitutions (misincorporation) or modifications, inversions, or very small deletions may account for the sequence variations in CaMV DNA (75). Similar heterogenity has been observed in plant mitochondrial DNA (86).

The restriction mapping of the genomes of several strains of CaMV has revealed differences that suggest a particular part of the genome has a higher frequency of sequence variation than the rest of the molecule. A small deletion may occur in this region in a nonaphid-transmissible strain (48).

Cloning of CaMV DNA in the Escherichia coli system An unspecified strain of CaMV was cloned by Szeto et al (108) in *E. coli* after ligation to a plasmid. When amplified viral DNA was excised with the same restriction endonuclease used for insertion in the plasmid, it showed no infectivity for plants, even when ligated into circular molecules. These investigators suggested that a small but essential fragment of the genome may have been lost after restriction endonuclease digestion. The loss of the specifically located interruptions in the CaMV genome may be an alternative suggestion to account for the lack of infectivity of viral DNA cloned in the *E. coli* system.

The entire virus genome and various genome fragments have been cloned in *E. coli* minicells after plasmid insertion by Meagher et al (75, 76) in order to examine expression of the viral DNA in bacterial cells. When a 2.6×10^6 dalton fragment of CaMV obtained by *Eco* RI cleavage was ligated to Col E1 plasmid, the recombinant directed the production of a new 37,000 dalton protein. The latter comigrated during gel electrophoresis with native 37,000 dalton coat protein from CaMV virions. However, examination of the cyanogen bromide cleavage products and immunological analysis showed that the bacterial and native virion proteins were not the same. Proteins of 43,000 and 40,000 induced by another CaMV DNA recombinant were similarly unrelated to native virus proteins (76).

The cloning experiments with CaMV DNA prove that this is a useful method for amplification of viral DNA but not for its expression. Although the results indicate that new proteins are specified by CaMV DNA in the *E. coli* system, with an impressive amount of the new proteins being produced, these proteins are not the same as those produced in infected plants. Thus the bacterial system is not useful for identifying viral genes by their primary translation products (76).

Cytopathological Observations and Replication of the Caulimoviruses

The caulimoviruses cause conspicuous inclusion bodies in infected cells which appear as highly refractive, ovoid, or irregularly lobed masses in the cytoplasm when viewed with the light microscope. Inclusion bodies occur throughout the epidermal, palisade, and spongy parenchyma and to a lesser extent in young tracheary and phloem companion cells (35, 87). Though varying with different strains of the virus, they seem to increase progressively in size during the course of infection.

Early investigators with the light microscopy suggested that inclusions sometimes occur in the nucleus. Now they are known to be strictly cytoplasmic as found by numerous investigations with the electron microscope. These bodies are unique structures unlike those induced by any other type of plant viruses. The bulk of these structures consists of an electron-dense, granular matrix in which virions are embedded (35, 82). Vacuole-like areas that may contain many virions occur throughout the matrix. In some areas without vacuoles a very sparse network of fine fibrils can be seen under very high resolution (60).

Virus that can be visualized within the cell appears to be almost wholly in association with the inclusion (35, 82). A very small amount occurs free in the cytoplasm near the inclusions and infrequently in plasmodesmata (20, 35, 59). When virions are observed in plasmodesmata, these structures become enlarged to about twice their normal diameter (20, 59, 65), as if this were an effect associated with intercellular movement of the viruses. Wall thickenings have also been associated with infections of several of the caulimoviruses (2, 17, 20, 59).

Carnation etched ring virus (CERV) may not be as restricted in its distribution in infected cells (Table 1) as is CaMV in cruciferous plants. Virions are found individually in cytoplasm and perhaps nuclei of *Dianthus barbatus* (65) and *Saponaria vaccaria* (66, 89). Inclusion bodies of CERV in the cytoplasm have less granular electron dense material and proportionately more virions (32, 65, 89, 90).

The close association of virions with inclusion bodies suggests that these are sites for virus synthesis, or at least for virus assembly. Other cytological observations indicate that viral DNA replication occurs in the inclusion body.

When tritiated thymidine is administered to mature leaves of *Brassica* infected with CaMV, and tissue sections subjected to autoradiographic analysis, the active sites for thymidine uptake are centered over the inclusion bodies indicating these are loci for DNA replication (28, 53). These observations, plus the sequential development of the components of viral inclusions, implicate these bodies as sites for DNA replication and virion assembly.

The granular matrix material of caulimovirus inclusion bodies appears in the cytoplasm prior to virion formation. The incipient inclusions, several of which may occur in a single cell, appear as minute patches of electron dense material surrounded by numerous ribosomes as if these are sites of intensive protein synthesis. As the electron dense patches reach a certain size, virions appear in the center embedded in the matrix of the developing inclusion (60, 65, 74).

Inclusion bodies of CaMV have been isolated and found to consist mainly of a single protein of about 55,000 daltons (103). This polypeptide is believed to represent the matrix of the inclusion body. With some strains of the virus, inclusion bodies lose virus particles during the course of isolation so that purified inclusions are almost exclusively matrix protein. Isolated inclusion bodies also contain about 0.5% by weight of unencapsidated viral DNA.

The inclusion bodies of different strains of CaMV have distinctive properties suggesting they may be virus-specified and that the matrix protein is a virus-coded product. Chemical and immunological differences in the matrix proteins of different strains of CaMV should be sought to determine if it is virus- or host-specified.

The successful infection of protoplasts achieved recently with CaMV may provide a useful system for obtaining synchronized infections (47). This will facilitate greatly the study of early events during virus replication. The virus is unusually slow in its development, however, being detected only after 4 days of infection.

A CaMV transcription product that may be messenger RNA has been detected in virus-infected protoplasts (47). Although the material hybridized specifically to CaMV DNA, its identity has not been well established, and nothing was done to rid the protoplast extracts of unencapsidated viral DNA. The transcript was estimated to represent about 70% of one strand of the genome. The material may be polyadenylated as indicated by an affinity for poly(U)-Sepharose.

GEMINIVIRUSES

The name for this group was proposed by Harrison et al (42) because of the frequent occurrence of paired particles in electron micrographs. In addition to this peculiarity, the viruses have smaller and less complex particles with single-stranded DNA genomes that may be multipartite in nature. The viruses reproduce mainly in nuclei of phloem parenchyma cells and probably have a different reproductive strategy than the caulimoviruses with their double-stranded genomes and cytoplasmic replication sites.

A notable feature of the geminiviruses is their association with the phloem rather than being generally distributed throughout somatic tissues as most plant viruses. Some of the viruses may be wholly restricted to the phloem (4, 79), as they cannot be mechanically transmitted to plants in spite of the very stable nature of the virus in plant extracts. In these examples only the insect vector can inoculate plants with any degree of efficiency. Such lack of mechanical transmissibility is a serious handicap in the experimental manipulation of these viruses. However, the bean golden mosaic virus can be mechanically transmitted (77).

Biological Properties

The geminiviruses are fairly heterogeneous biologically with members of the group diverging in host range, insect vector, and other properties (Table 3). They induce a variety of diseases which have some features in common. Chlorosis, for example, is a prominent feature of the diseases induced by some of these viruses. Such generalized yellowing may be from the accumulation of carbohydrates in leaves because of phloem necrosis. Other symptoms are frequently associated with the veins, which is not surprising perhaps since the viruses seem to be mainly phloem parasites (4, 26, 27, 79).

Geminiviruses have either leafhopper or whitefly vectors. Maize streak and beet curly top viruses have leafhopper vectors; bean golden mosaic and euphorbia mosaic viruses have whitefly vectors.

After acquisition the viruses are retained for long periods by the insect vectors, frequently for the life of the insect (4, 7). Either acquisition or inoculation can occur with feeding periods of less than 1 hr, although longer feeding times favor efficiency of transmission. A brief latent period of perhaps 4–8 hr is required for the virus to become circulative throughout the

Table 3 Geminiviruses and their host ranges[a]

Virus	Host Range[b]	Vector	Reference
Bean golden mosaic virus	*Phaseolus vulgaris, P. lunatus* (Leguminosae)	Whitefly (*Bemisia tabaci*)	6, 77
Beet curly top virus	44 plant families, 300 species, including many cultivated species	Leafhopper (*Circulifer tenellus*)	4
Cassava latent virus	*Manihot utilissima* in Euphorbiaceae, and various Solanaceae	Whitefly (*Bemisia tabaci*)	8, 69
Euphorbia mosaic virus	*Euphorbia prunifolia* (Euphorbiaceae) and some Solanaceae and Leguminosae sp.	Whitefly (*Bemisia tabaci*)	6
Maize streak virus	*Zea mays, Saccharum officinarum*, and many other Gramineae	Leafhoppers (*Cicadulina mbila*)	106, 107
Tobacco leaf curl virus	*Nicotiana tabacum, N. glutinosa, Datura stramonium, Lycopersicum esculentum*, and other Solanaceae	Whitefly (*Bemisia tabaci*)	80

[a] Some of the other whitefly-borne viruses mentioned by Bird & Maramorosch (6) probably belong in this group.

[b] The host ranges of some geminiviruses may be far more extensive than indicated here. This group of viruses has only recently been defined, and the host ranges of some of the viruses have not been adequately explored.

body of the vector before transmission occurs (5, 106). The virus is absorbed through the gut wall into the hemolymph of the insect and is secreted by the salivary glands (4, 7). The persistence of virus within the body of the vector is related to the duration of acquisition feeding, an indication that the viruses do not multiply in the insect vector.

Physical and Chemical Properties of Geminiviruses

The isometric particles of geminiviruses have two unusual features when viewed with the electron microscope. The particles are very small, only 16–18 nm in diameter, and frequently appear in pairs. The latter is a unique characteristic.

Paired particles have been reported for beet curly top (78), bean golden mosaic (40), maize streak (10), cassava latent (42), and tobacco leaf curl viruses (80). The paired particles appear five-sided with the contiguous side appearing longer than the others. For example, the single particles are 16–18 nm in diameter and paired particles are about 15–20 X 25–30 nm in size when negatively stained. The center-to-center distance of intracellular packed arrays of bean golden mosaic virus is 18 nm (55).

Substantial numbers of single particles accompany the paired particles in purified preparations. With maize streak virus the ratio of paired to single particles varies from 1 : 3 to 1 : 8 depending on the method of purification (7). Goodman et al (40) suggested pairing of virions might occur from fixation with aldehydes which were required to prevent dissociation by heavy metal salts used for negative staining before electron microscopy. In some cases when viewed in situ in fixed sections of infected tissue, the particles of geminiviruses are not obviously paired (10, 27, 55). However, in spinach infected with beet curly top virus the particles in nuclei are occasionally found in extensive ribbon-like arrays that seem to be monolayers of paired particles (26). The tobacco leaf curl virus in nuclei of its solanaceous hosts occurs in symmetrical paired arrays of virions which form large rigid rod-like structures (80).

Pairs of particles of maize streak and cassava latent viruses sediment at about 76S, whereas single particles sediment at about 54S (7, 42), but the two types of particles have not been separated and separately tested for infectivity. The virions of bean golden mosaic virus sediment at about 69S (38) and beet curly top virus at 56S (22).

Little is known about the structural proteins of the geminiviruses except that some of the viruses have one protein and others have two of almost the same molecular weight. Maize streak, cassava latent, and bean golden mosaic viruses each have a single capsid protein of about 28,000, 34,000, and 31,000 daltons, respectively (9, 55). Beet curly top has two proteins of 32,800 and 36,400 daltons (22).

Nucleic Acid

Goodman (37–39) has identified the nucleic acid of bean golden mosaic virus as single-stranded DNA. The nucleic acid gives a positive diphenylamine test, is sensitive to bovine pancreatic DNase I and S_1 nuclease, but is unaffected by RNase A or 0.3 N NaOH. The nucleic acid melts gradually with a hyperchromicity of about 15% over a temperature range of 20–70°C without exhibiting a cooperative-type transition. It reacts with formaldehyde to give a 20% hyperchromic effect with a slight shift in the maximum absorption to longer wavelengths. The nucleic acids of maize streak and cassava latent viruses show similar properties upon treatment with alkali and nucleases and thus must be single-stranded DNAs also (42). The nucleic acid of beet curly top virus has not been as well investigated, but the purified virus gives a positive diphenylamine test for DNA (73).

The molecular weight of the DNA genome of the geminiviruses is surprisingly small. That for bean golden mosaic virus, for example, has been estimated by sedimentation velocity experiments at 6.5–7.8 $\times 10^5$ daltons and by equilibrium cesium chloride gradient centrifugation at 6.6–9.5 $\times 10^5$ daltons (38, 39). The molecular weight of maize streak and cassava latent viruses has been estimated from their contour lengths in the electron microscope at 0.7×10^6 and 0.80×10^6, respectively (42). These low values for the geminiviruses suggest the viruses may have multipartite genomes in which two or more small essential pieces of the genome are encapsidated in separate particles. The latter is a common feature of small RNA viruses of plants. Alternatively, the geminiviruses may be new examples of single-stranded DNA viruses with overlapping genes like those of phage ϕX174 or G4. Portions of these single-stranded DNA genomes are transcribed in more than one reading frame so that the same stretch of DNA codes for more than one protein (1, 93, 97). Some small double-stranded DNA viruses also have overlapping genes (21). Perhaps the genomes of the geminiviruses are similar examples of compressed genetic information.

An observation that suggests the geminiviruses have a multipartite genome is the presence of two components during gel electrophoresis of the DNA (42, 55). However, electron microscopic examination of maize streak and cassava latent virus DNAs reveal circular and linear molecules of the same contour length, suggesting they may be identical except for a break in one molecule (42). Until the two DNA molecules are separated and their infectivities evaluated alone and in combination, it will not be known if the molecules are separate genetic entities.

Cytopathological Effects of Geminiviruses

Geminiviruses cause anatomical and cytological disturbances of phloem and phloem-associated cells, a reflection of their tissue distribution in their plant hosts. The anatomical changes associated with beet curly top virus, for example, involve cell hypertrophy, hyperplasia, and necrosis throughout the phloem tissues of the plant including the root system. Hyperplastic growth of the phloem is particularly conspicuous with the tissue developing an abnormal number of cells differentiated as sieve elements (23–25). Although less studied, other geminiviruses also cause phloem proliferation and necrosis.

Degenerative changes at the cellular level accompany the more conspicuous anatomical effects of the viruses. Some of these effects coincide with the appearance of virus particles in the phloem. For example, in plants infected with beet curly top virus, phloem parenchyma cells become prominently vacuolated and frequently disintegrate as virus particles appear in the nucleus. The nuclear chromatin in such cells is progressively depleted until the virus particles nearly fill the nucleus (26, 27). The virus never appears in sieve elements or companion cells even in hyperplastic phloem. Neither curly top virus (26) nor bean golden mosaic virus (55) has been found in mesophyll, xylem, or epidermal tissues.

With bean golden mosaic virus, particles appear as aggregated masses or hexagonal arrays in the nuclei of both phloem parenchyma and young sieve elements. These cells show hypertrophy of nucleoli with segregation of the granular and fibrillar regions. The latter, which become arranged into ring-like areas, are composed of electron-dense fibrils that seem to be finer and more compact than those of fibrillar regions of normal nuclei. As infection progresses, these fibrillar rings increase in size and number and are accompanied by the appearance of virus particles in nearby areas of the nucleoplasm (55). In mature sieve elements in which nuclei degenerate, virus particles are found throughout the cytoplasm, probably as a result of release from the nucleus (55). Crystalline arrays of maize streak virus (10) and tobacco leaf curl (80) occur in nuclei of their infected hosts.

POTATO LEAFROLL VIRUS

This virus was reported by Sarkar (94) to contain DNA. Although the virus is remarkably similar biologically to several viruses with single-stranded RNA, it is set apart by its small double-stranded DNA genome. For this reason it is taxonomically distinct from any other plant virus.

Potato leafroll virus causes yellowing and leafrolling symptoms and is associated mainly with the phloem of infected plants (81). The virus is transmitted by aphids in a persistent manner. It is not transmissible by mechanical methods. It is difficult to isolate in amounts sufficient for chemical characterization (61).

The virion of potato leafroll is a small quasi-isometric particle about 23 nm in diameter with 40% double-stranded DNA. The DNA has a T_m =

87.4 in SSC and a density of 1.698 g/ml in cesium chloride (95). Its contour length has been reported to be 0.65 μm, corresponding to a molecular weight of 1.3×10^6. The virion has a single protein of about 15,000 daltons (96).

FINAL COMMENTS

From the foregoing it is obvious that the two main groups of DNA viruses are quite different, biologically and biochemically. Of these, the caulimoviruses are the most attractive candidates for recombinant DNA experiments. They have double-stranded genomes that can be directly cleaved and restructured with restriction endonucleases for physical and functional mapping and for recombinant DNA experiments. These viruses replicate independently in the cytoplasm without any requirement for integration in the host chromosome. Perhaps there would be less likelihood of repression of new genes or interference with host functions as with some plasmids that become integrated in the chromosome. Moreover, these viruses provide a more simple genetic system than most plasmids, and it will be easier to identify and map their essential functions.

Unfortunately, a knowledgeable approach to recombinant DNA experiments is not possible now with any of the DNA viruses. Information is lacking on essential virus functions or where these are located on the genome, so that insertions can be made without destroying the ability of the DNA to replicate. Although the vector acquisition gene of CaMV may represent a nonessential region where functional insertions could be made, its location has not been definitely established.

The replication origin(s) and promotor site(s) are other features of the viral genomes which should be identified. Insertion of foreign DNA at any nonessential site in a viral genome may be adequate for cloning of chimeric DNA, but the insertion must occur downstream from a functional promotor sequence in order to be expressed. In addition, complete transcriptional maps of the viral genomes should be developed using messenger RNA from infected plants.

Certain features of the geminiviruses make them less attractive for recombinant DNA experiments. Their association with the phloem, for instance, suggesting they may not move well in other plant tissues, might be an obstacle to their use as vectors of recombinant DNA. Lack of mechanical transmission of some of the viruses would be another disadvantage, but their wider host ranges might be a great asset in the future. The single-stranded genome of these viruses probably is not a serious disadvantage since the replicative form can be used for restriction endonuclease and transcriptional mapping.

Literature Cited

1. Barrell, B. G., Air, G. M., Hutchison, C. A. III. 1976. Overlapping genes in bacteriophage ϕX174. *Nature* 264: 34–41
2. Bassi, M., Favali, M. A., Conti, G. G. 1974. Cell wall protrusions induced by cauliflower mosaic virus in Chinese cabbage leaves: a cytochemical and autoradiographic study. *Virology* 60: 353–58
3. Beier, H., Shepherd, R. J. 1978. Serologically specific electron microscopy in the quantitative measurement of two isometric viruses. *Phytopathology* 68: 533–38
4. Bennett, C. W. 1971. The curly top disease of sugar beet and other plants. *Am. Phytopathol. Soc. Monogr.* 7. 81 pp.
5. Bennett, C. W., Wallace, H. E. 1938. Relation of the curly top virus to the vector, *Eutettix tenellus. J. Agric. Res.* 56:31–52
6. Bird, J., Maramorosch, K. 1978. Viruses and virus diseases associated with whiteflies. *Adv. Virus Res.* 22:55–110
7. Bock, K. R. 1974. Maize streak virus. *Commonw. Mycol. Inst. Descr. Plant Viruses* 133. 4 pp.
8. Bock, K. R., Guthrie, E. J. 1976. In *African Cassava Mosaic,* ed. B. L. Nestel, pp. 11–16. Ottawa: Int. Dev. Res. Center
9. Bock, K. R., Guthrie, E. J., Meredith, G., Barker, H. 1977. RNA and protein components of maize streak and cassava latent viruses. *Ann. Appl. Biol.* 85:305–8
10. Bock, K. R., Guthrie, E. J., Woods, R. D. 1974. Purification of maize streak virus and its relationship to viruses associated with streak diseases of sugarcane and *Panicum maximum. Ann. Appl. Biol.* 77:289–96
11. Brierley, P., Smith, F. F. 1950. Some vectors, host, and properties of dahlia mosaic virus. *Plant Dis. Reptr.* 34:363
12. Broadbent, L. 1957. *Investigations of Virus Diseases of Brassica Crops.* Cambridge Univ. Press. 94 pp.
13. Brunt, A. A. 1966. Partial purification, morphology and serology of dahlia mosaic virus. *Virology* 28:778–79
14. Brunt, A. A. 1971. Dahlia mosaic virus. *Commonw. Mycol. Inst. Descr. Plant Viruses,* 51. 4 pp.
15. Brunt, A. A. 1971. Some hosts and properties of dahlia mosaic virus. *Ann. Appl. Biol.* 67:357–68
16. Brunt, A. A., Barton, R. J., Tremaine, J. H., Stace-Smith, R. 1975. The composition of cauliflower mosaic virus protein. *J. Gen. Virol.* 27:101–6
17. Brunt, A. A., Kitajima, E. W. 1973. Intracellular location and some properties of *Mirabilis* mosaic, a new member of the cauliflower mosaic group of viruses. *Phytopathol. Z.* 76:265–75
18. Brutlag, P., Schekman, R., Kornberg, A. 1971. A possible role for RNA polymerase in the initiation of M13 DNA synthesis. *Proc. Natl. Acad. Sci. USA* 68:2826–29
19. Civerolo, E. L., Lawson, R. H. 1978. Topological forms of cauliflower mosaic virus nucleic acid. *Phytopathology* 68: 101–9
20. Conti, G. G., Vegetti, G., Bassi, M., Favali, M. A. 1972. Some ultrastructural and cytochemical observations on Chinese cabbage leaves infected with cauliflower mosaic virus. *Virology* 47: 694–700
21. Contreras, R., Rogiers, R., Van de Voorde, A., Fiers, W. 1977. Overlapping of the VP_2-VP_3 gene and the VP_1 gene in the SV40 genome. *Cell* 12: 529–38
22. Egbert, L. N., Egbert, L. D., Mumford, D. L. 1976. Physical characteristics of sugarbeet curly top virus. *Ann. Meet. Am. Soc. Microbiol.* (Abstr.) 258 pp.
23. Esau, K. 1930. Studies of the breeding of sugar beets for resistance to curly top. *Hilgardia* 4:415–40
24. Esau, K. 1933. Pathological changes in the anatomy of leaves of the sugar beet, *Beta vulgaris* L., affected by curly top. *Phytopathology* 23:679–712
25. Esau, K. 1934. Cell degeneration in relation to sieve-tube differentiation in curly-top beets. *Phytopathology* 24: 303–5
26. Esau, K. 1977. Virus-like particles in nuclei of phloem cells in spinach leaves infected with the curly top virus. *J. Ultrastruct. Res.* 61:78–88
27. Esau, K., Hoefert, L. L. 1973. Particles and associated inclusions in sugarbeet infected with the curly top virus. *Virology* 56:454–64
28. Favali, M. A., Bassi, M., Conti, G. G. 1973. A quantitative autoradiographic study of intracellular sites for replication of cauliflower mosaic virus. *Virology* 53:115–19
29. Filippidis, E., Meneghini, R. 1977. Evidence for alkali-sensitive linkers in DNA of African green monkey kidney cells. *Nature* 269:445–47
30. Frazier, N. W. 1955. Strawberry vein

banding virus. *Phytopathology* 45: 307–12
31. Frearson, P. M., Crawford, L. V. 1972. Polyoma virus basic proteins. *J. Gen. Virol.* 14:141–55
32. Fujisawa, I., Rubio-Huertos, M., Matsui, C. 1971. Incorporation of thymidine-^{3}H into carnation etched ring virus. *Phytopathology* 61:681–84
33. Fujisawa, I., Rubio-Huertos, M., Matsui, C. 1972. Deoxyribonuclease digestion of the nucleic acid from carnation etched ring virus. *Phytopathology* 62: 810–11
34. Fujisawa, I., Rubio-Huertos, M., Matsui, C. 1974. Deoxyribonucleic acid in dahlia mosaic virus. *Phytopathology* 64:287–90
35. Fujisawa, I., Rubio-Huertos, M., Matsui, C., Yamaguchi, A. 1967. Intracellular appearance of cauliflower mosaic virus particles. *Phytopathology* 57: 1130–32
36. Goméc, B. 1973. *Properties and biochemical characterization of dahlia mosaic virus.* PhD thesis. Univ. California, Davis, Calif. 44 pp.
37. Goodman, R. M. 1977. Infectious DNA from a whitefly-transmitted virus of *Phaseolus vulgaris. Nature* 266:54–55
38. Goodman, R. M. 1977. Single-stranded DNA genome in a whitefly-transmitted plant virus. *Virology* 83:171–79
39. Goodman, R. M. 1977. Single-stranded DNA genome of a whitefly-transmitted virus of bean. *Proc. Am. Phytopathol. Soc.* 4:142
40. Goodman, R. M., Bird, J., Thongmeearkom, P. 1977. An unusual virus-like particle associated with golden yellow mosaic of beans. *Phytopathology* 67:37–42
41. Grossman, L. I., Watson, R., Vinograd, J. 1973. The presence of ribonucleotides in mature closed-circular mitochondrial DNA. *Proc. Natl. Acad. Sci. USA* 70:3339–43
42. Harrison, B. D., Barker, H., Bock, K. R., Guthrie, E. J., Meredith, G. 1977. Plant viruses with circular single-stranded DNA. *Nature* 270:760–63
43. Harrison, B. D., Finch, J. T., Gibbs, A. J., Hollings, M., Shepherd, R. J., Valenta, V., Wetter, C. 1971. Sixteen groups of plant viruses. *Virology* 45: 356–63
44. Hills, G. J., Campbell, R. N. 1968. Morphology of broccoli necrotic yellows virus. *J. Ultrastruct. Res.* 24:134–44
45. Hirsch, I., Vonka, V. 1974. Ribonucleotides linked to DNA of herpes simplex virus. *J. Virol.* 13:1162–68
46. Hollings, M., Stone, O. M. 1969. Carnation viruses. *Rep. Glasshouse Crops Res. Inst.* 1968:102
47. Howell, S. H., Hull, R. 1978. Replication of cauliflower mosaic virus and transcription of its genome in turnip leaf protoplasts. *Virology* 86:468–81
48. Hull, R., Howell, S. H. 1978. Structure of the cauliflower mosaic virus genome. II. Variation in DNA structure and sequence between isolates. *Virology* 86: 482–93
49. Hull, R., Shepherd, R. J. 1975. The coat proteins of cauliflower mosaic virus. *Virology* 70:217–20
50. Hull, R., Shepherd, R. J. 1977. The structure of cauliflower mosaic virus genome. *Virology* 79:216–30
51. Hull, R., Shepherd, R. J., Harvey, J. D. 1976. Cauliflower mosaic virus: an improved purification procedure and some properties of the virus particles. *J. Gen. Virol.* 31:93–100
52. Itoh, T., Matsui, C., Hirai, T. 1969. Conformational changes in cauliflower mosaic virus. *Virology* 39:367–72
53. Kamei, T., Rubio-Huertos, M., Matsui, C. 1969. Thymidine-^{3}H uptake by X-bodies associated with cauliflower mosaic virus infection. *Virology* 37: 506–8
54. Kelley, D. C., Cooper, V., Walkey, D. G. A. 1974. Cauliflower mosaic virus structural proteins. *Microbios* 10: 239–45
55. Kim, K. S., Shock, T. L., Goodman, R. M. 1978. Infection of *Phaseolus vulgaris* by bean golden mosaic virus: ultrastructural aspects. *Virology* 89:22–33
56. Kitajima, E. W., Betti, J. A., Costa, A. S. 1973. Strawberry vein-banding virus, a member of the cauliflower mosaic virus group. *J. Gen. Virol.* 20:117–19
57. Kitajima, E. W., Costa, A. S. 1966. Particulas esferaidois associados ao virus do mosaico das nervuras da mandioca. *Bragantia* 25:211–21
58. Kitajima, E. W., Costa, A. S. 1973. Morphology of virus and mycoplasma that infect cassava in the American continent and the ultrastructure of diseased tissues. *Int. Congr. Plant Pathol., 2nd.* Abstr. 0927
59. Kitajima, E. W., Lauritis, J. A. 1969. Plant virions in plasmodesmata. *Virology* 37:681–85
60. Kitajima, E. W., Lauritis, J. A., Swift, H. 1969. Fine structure of zinnia leaf tissue infected with dahlia mosaic virus. *Virology* 39:240–49
61. Kojima, M., Shikata, E., Sugawara, M., Murayama, D. 1969. Purification and

electron microscopy of potato leafroll virus. *Virology* 39:162–74

62. Kropinski, A. M. B., Bose, R. J., Warren, R. A. J. 1973. 5-(4-aminobutylaminomethyl) uracil, an unusual pyrimidine from the deoxyribonucleic acid of bacteriophage ϕW-14. *Biochemistry* 12:151–57
63. Lawson, R. H., Civerolo, E. L. 1976. Purification of carnation etched ring virus and comparative properties of CERV and cauliflower mosaic virus. *Acta Hortic.* 59:49–59
64. Lawson, R. H., Civerolo, E. L. 1978. Carnation etched ring virus: purification, stability of inclusions, and properties of the nucleic acid. *Phytopathology* 68:181–88
65. Lawson, R. H., Hearon, S. S. 1973. Ultrastructure of carnation etched ring virus-infected *Saponaria vaccaria* and *Dianthus caryophyllus. J. Ultrastruct. Res.* 48:201–15
66. Lawson, R. H., Hearon, S. S., Civerolo, E. L. 1977. Carnation etched ring. *Commonw. Mycol. Inst. Descr. Plant Viruses* 182. 4 pp.
67. Lesemann, D., Casper, R. 1973. Electron microscopy of petunia vein-clearing virus, an isometric plant virus associated with specific inclusions in petunia cells. *Phytopathology* 63:1118–24
68. Lewis, H. A., Miller, R. C. Jr., Stone, J. C., Warren, R. A. J. 1975. Alkali lability of bacteriophage ϕW-14 DNA. *J. Virol.* 16:1375–79
69. Lister, R. M. 1959. Mechanical transmission of cassava brown streak virus. *Nature* 183:1588–89
70. Lung, M. C. Y., Pirone, T. P. 1972. *Datura stramonium,* a local lesion host for certain isolates of cauliflower mosaic virus. *Phytopathology* 62:1473–74
71. Lung, M. C. Y., Pirone, T. P. 1973. Studies on the reason for differential transmissibility of cauliflower mosaic virus isolates by aphids. *Phytopathology* 63:910–14
72. Lung, M. C. Y., Pirone, T. P. 1974. Acquisition factor required for aphid transmission of purified cauliflower mosaic virus. *Virology* 60:260–64
73. Magyarosy, A. C., Schurmann, P., Buchanan, B. B., Finlay, A. 1977. Purification and some properties of curly top virus. *Proc. Am. Phytopathol. Soc.* 4:161
74. Martelli, G. P., Costellano, M. A. 1971. Light and electron microscopy of the intracellular inclusions of cauliflower mosiac virus. *J. Gen. Virol.* 13:133–40
75. Meagher, R. B., Shepherd, R. J., Boyer, H. W. 1977. The structure of cauliflower mosaic virus: I. A restriction endonuclease map of cauliflower mosaic virus DNA. *Virology* 80:362–75
76. Meagher, R. B., Tait, R. C., Betlach, M., Boyer, H. W. 1977. Protein expression in *Escherichia coli* minicells by recombinant plasmids. *Cell* 10:521–36
77. Meiners, J. P., Lawson, R. H., Smith, F. F. Diaz, A. J. 1975. Mechanical transmission of whitefly (*Bemisia tabaci*)-borne disease agents of beans in El Salvador. ed. J. Bird, K. Maramorosch, pp. 61–69. In *Tropical Diseases of Legumes,* New York: Academic, 171 pp.
78. Mumford, D. L. 1974. Purification of curly top virus. *Phytopathology* 64: 136–39
79. Mumford, D. L., Thornley, W. R. 1977. Intracellular location of curly top virus antigen as revealed by fluorescent antibody staining. *Proc. Am. Phytopathol. Soc.* 4:144
80. Osaki, T., Inouye, T. 1978. Resemblance in morphology and intranuclear appearance of viruses isolated from yellow dwarf diseased tomato and leaf curl diseased tobacco. *Ann. Phytopathol. Soc. Jpn.* 44:167–78
81. Peters, D. 1970. Potato leafroll virus. *Commonw. Mycol. Inst. Descr. Plant Viruses* 36. 4 pp.
82. Petzold, H. 1968. Elektronenmikroskopische Untersuchungen an von Dahlienmosaik-virus infizierten Pflanzen. *Phytopathol. Z.* 63:201–18
83. Pirone, T. P., Megahed, E-S. 1966. Aphid transmissibility of some purified viruses and viral RNAs. *Virology* 30: 631–37
84. Pirone, T. P., Pound, G. S., Shepherd, R. J. 1960. Purification and properties of cauliflower mosaic virus. *Nature* 186:656–57
85. Pirone, T. P., Pound, G. S., Shepherd, R. J. 1961. Properties and serology of purified cauliflower mosaic virus. *Phytopathology* 51:541–46
86. Quetier, F., Vedel, F. 1977. Heterogeneous population of mitochondrial DNA molecules in higher plants. *Nature* 268:365–68
87. Robb, S. M. 1964. Location, structure, and cytochemical staining reactions of the inclusion bodies found in *Dahlia variabilis* infected with dahlia mosaic virus. *Virology* 23:141–44
88. Rosenkranz, H. S. 1973. RNA in coliphage T5. *Nature* 242:327–29
89. Rubio-huertos, M., Castro, S., Fujisawa, I., Matsui, C. 1972. Electron microscopy of the formation of carnation etched ring virus intracellular in-

clusion bodies. *J. Gen. Virol.* 15:257–60

90. Rubio-Huertos, M., Castro, S., Morena, R., Lopez, D. 1968. Ultrastructura de celulas de *Dianthus caryophyllus* infectadas por dos virus al mismo tiempa. *Microbiol. Esp.* 21:1–9
91. Rubio-Huertos, M., Matsui, C., Yamaguchi, A., Kamei, T. 1968. Electron microscopy of X-body formation in cells of cabbage infected with *Brassica* virus 3. *Phytopathology* 58:548–49
92. Russell, G. J., Follett, E. A. C., Subak-Sharpe, J. H., Harrison, B. D. 1971. The double-stranded DNA of cauliflower mosaic virus. *J. Gen. Virol.* 11:129–38
93. Sanger, F., Air, G. M., Barrell, B. G., Brown, N. L., Coulson, A. R., Fiddes, J. C., Hutchison, C. A. III, Slocombe, P. M., Smith, M. 1977. Nucleotide sequence of bacteriophage ϕX174 DNA. *Nature* 265:687–95
94. Sarkar, S. 1973. DNA-like properties of the nucleic acid of potato leafroll virus. *Naturwissenschaften* 60:480–81
95. Sarkar, S. 1976. Potato leafroll virus contains a double-stranded DNA. *Virology* 70:265–73
96. Sarkar, S. 1978. Characterization of four isolates of the potato leafroll virus. *Third Int. Congr. Plant Pathol., Munich, Germany* (Abstr.), p. 54
97. Shaw, D. C., Walker, J. E., Northrop, F. D., Barrell, B. G., Godson, G. N., Fiddes, J. C. 1978. Gene K, a new overlapping gene in bacteriophage G4. *Nature* 272:510–15
98. Shepherd, R. J. 1970. Cauliflower mosaic virus. *Commonw. Mycol. Inst. Descr. Plant Viruses* 24. 4 pp.
99. Shepherd, R. J. 1976. DNA viruses of higher plants. *Adv. Virus Res.* 20:205–339
100. Shepherd, R. J. 1977. Cauliflower mosaic virus (DNA virus of higher plants). In *The Atlas of Insect and Plant Viruses,* ed. K. Maramorosch, pp. 159–66. New York: Academic. 478 pp.
101. Shepherd, R. J., Bruening, G. E., Wakeman, R. J. 1970. Double-stranded DNA from cauliflower mosaic virus. *Virology* 41:339–47
102. Shepherd, R. J., Wakeman, R. J. 1971. Observation on the size and morphology of cauliflower mosaic virus deoxyribonucleic acid. *Phytopathology* 61:188–93
103. Shepherd, R. J., Wakeman, R. J. 1977. Isolation and properties of the inclusion bodies of cauliflower mosaic virus. *Proc. Am. Phytopathol. Soc.* 4:145
104. Shepherd, R. J., Wakeman, R. J., Romanko, R. R. 1968. DNA in cauliflower mosaic virus. *Virology* 36:150–52
105. Speyer, J. F., Chao, J., Chao, L. 1972. Ribonucleotides covalently linked to deoxyribonucleic acid in T4 bacteriophage. *J. Virol.* 10:902–9
106. Storey, H. H. 1928. Transmission studies of maize streak disease. *Ann. Appl. Biol.* 15:1–25
107. Storey, H. H., Thomson, G. M. 1961. Streak disease. In *Sugar-cane Diseases of the World,* ed. J. P. Martin, E. F. Abbott, C. G. Hughes, 1:461–72. Amsterdam: Elsevier
108. Szeto, W. W., Hamer, D. H., Carlson, P. S., Thomas, C. A., Jr. 1977. Cloning of cauliflower mosaic virus (CLMV) in *Escherichia coli. Science* 196:210–12
109. Tewari, K. K., Wildman, S. G. 1966. Chloroplast DNA From tobacco leaves. *Science* 153:1264–71
110. Tezuka, N., Taniguchi, T. 1972. Stepwise degradation of cauliflower mosaic virus by pronase. *Virology* 47:142–46
111. Tezuka, N., Taniguchi, T. 1972. Structural protein of cauliflower mosaic virus. *Virology* 48:297–99
112. Volovitch, M., Drugeon, G., Yot, P. 1978. Studies on the single-stranded discontinuities of the cauliflower mosaic virus genome. *Nucleic Acid Res.* 5: 2913–25
113. Volovitch, M., Dumas, J. P., Drugeon, G., Yot, P. 1977. Single-stranded interruptions in cauliflower mosaic virus DNA. In *Acides Nucleiques et Synthese dez Proteines chez les Vegetraux,* ed. L. Bogorad, J. H. Weil, pp. 635–41. Colloq. No. 261 Centre National de la Recherche Sientifique, Paris
114. Williams, P. H., Boyer, H. W., Helinski, D. R. 1973. Size and base composition of RNA in supercoiled plasmid DNA. *Proc. Natl. Acad. Sci USA* 70:3744–48
115. Wyatt, G. R., Cohen, S. S. 1953. The bases of the nucleic acids of some bacterial and animal viruses: the occurrence of 5-hydroxy-methylcytosine. *Biochem. J.* 55:774–82

Ann. Rev. Plant Physiol. 1979. 30:425–84

PLANT CELL FRACTIONATION ❖7678

Peter H. Quail[1]

Carnegie Institution of Washington, Stanford, California 94305

CONTENTS

[1]Present address: Department of Botany, University of Wisconsin, Madison, Wisconsin 53706.

0066-4294/79/0601-0425$01.00

INTRODUCTION

Some 15 years ago De Duve provided a conceptual framework for cell fractionation studies by bringing into sharp focus the contrasting goals and relative merits of preparative and analytical fractionation procedures (64, 65). Preparative cell fractionation aims at the isolation of a structurally or morphologically identifiable entity for subsequent analysis, the emphasis being on purity at the expense of yield. In the absence of prior biochemical analysis, the criteria for assessing purity are strictly morphological. Analytical fractionation requires a meticulous, quantitative accounting of the distribution of each biochemical entity of interest among all individual fractions derived from the initial homogenate. This latter approach establishes the distribution of each biochemical entity unencumbered by any predilections regarding the subcellular composition of the derived fractions. With preparative fractionation the question posed is: "What biochemical activities or properties does a given structural entity possess?" With analytical fractionation the strategy is reversed and the question posed is: "On which (if any) of the array of identifiable (or even yet to be identified) subcellular components does a particular biochemical entity reside?" Clearly the confidence with which the first question can be answered rests on the quantitative reliability of the criteria of purity used. The confidence with which the second question can be answered rests on the reliability with which each structural entity can be qualitatively identified and quantitatively measured in the separate fractions. At the heart of both questions is the concept of markers (64, 65, 224).

For the purposes of this review, plant cell fractionation studies are divided into two broad categories: (*a*) those concerned with the *delineation* of markers for specific cell organelles and membranes; and (*b*) those concerned with the *use* of these (established or accepted) markers as the basis for assigning a subcellular location to a particular activity or molecule of previously unknown subcellular distribution. The bulk of the article focuses on studies in the former category. Because of the constant accumulation of new information regarding the distribution of cellular constituents, the distinction between the two classes of study is somewhat arbitrary and in several instances is based more on historical precedent than fundamental differences in experimental approach. Both preparative and analytical procedures have been used in both cases. Nonetheless, there is a catalog of "accepted" markers that are widely employed as the basis for identifying specific plant subcellular components. An attempt is made here to evaluate critically the primary evidence on which the designated subcellular localization of each of these markers has been established. Space limitations have precluded more than a brief survey of studies in the second category.

Except for comparative purposes, only investigations with higher plants are considered. Recent reviews of several of the individual membranes or organelles are indicated in the relevant sections. The general principles, conventional methodology, and specific practical problems peculiar to the preparation, fractionation, and analysis of subcellular fractions from plant tissue have been expounded by many authors (91, 129, 166, 169, 187, 256, 320). Specific methodological innovations worthy of emphasis or reemphasis include the successful use of: protoplasts as starting material (79, 237, 239, 242, 329); thiol and other reagents (91, 187), protease inhibitors (56), and lipase inhibitors (91, 216, 253); vertical rotors (98); a variety of alternate gradient media such as Ficoll (120, 292), Metrizamide (124, 136, 173), Renografin (201, 268), and silica sols (135, 256, 257, 289); column chromatography of particulate fractions (24, 146, 186); phase partition (165); and free flow electrophoresis (109).

THE MARKER CONCEPT

Principles and Concepts

The marker concept is central to all cell fractionation studies (65, 224). It seems useful as a framework for this review to highlight briefly some of the main considerations to emerge from a recent comprehensive workshop and debate on "Markers for Membranous Cell Components" (224). The documented outcome of this debate represents an attempt to achieve a consensus among workers in the field regarding acceptable terminology and principles in marker use with the ultimate goal of establishing a uniform "code of practice." It is recommended reading for researchers in this area.

TERMINOLOGY The term *component* designates a structural or morphological entity (chloroplast, mitochondrion). The term *constituent* designates a biochemical entity (enzyme, phospholipid) (2, 64).

TYPES OF MARKERS Two basic types of markers are utilized—morphological and biochemical. In either case the diagnostic feature may be either endogenous (e.g. an enzyme activity, ribosome-studded vesicles) or extrinsically imposed (a surface label, e.g. radioactivity or ferritin). Clearly the simultaneous use of both biochemical and morphological markers greatly enhances the confidence with which a given subcellular component can be monitored. Correct deployment of quantitative morphometry is a powerful complement to, although not a substitute for, biochemical markers (11, 22) and its routine use wherever possible is strongly recommended (224). Two central points regarding the assessment and experimental use of morphometric procedures can be profitably emphasized: (*a*) The amount of material represented by even 10 to 100 electron micrographs is of the order of 10^3 less than that used for a biochemical assay of a subcellular fraction (G. Siebert, personal communication); and (*b*) care is needed in the preparation and processing of samples for electron microscopy to ensure that unbiased, statistically valid electron micrographs of the fractions are obtained. The preparation of thin pellicles from particle suspensions by filtration on Millipore filters, followed by vertical sectioning and photographing of the entire depth of the pellicle, is the method of choice (11, 22). This approach largely obviates the problems of statistical analysis inherent in the marked multidirectional stratification of particles that occurs in pellets prepared by centrifugation. If pellets are prepared by centrifugation, they should also be thin, sectioned vertically, and the entire depth of the pellet analyzed (169). It should be stressed, however, that satisfactory resolution of these sampling problems does not preclude the subjectivity inherent in the scoring process itself. Despite the use of procedures ensuring random selection of the membrane profiles to be scored on a micrograph, the actual classification of each scored profile still involves subjective visual identification.

USES OF MARKERS Any given marker may be used in either of two different modes (224). A given activity or feature is referred to as a *positive* marker when the objective is to locate those fractions *enriched* in the subcellular component for which the activity is a marker. A given activity is called a *negative* marker when the objective is to establish the *absence* or *low level* of a particular subcellular component in fractions enriched for other organelles or membranes, i.e. where the objective is to assess contamination by unwanted components.

SCOPE OF INFORMATION Appropriately chosen markers, in addition to simply establishing the presence or absence of a given organelle, can provide an index of: (*a*) the intactness and/or functional integrity of the organelle as a whole (e.g. ribulose bisphosphate carboxylase is a marker for intact plastids); (*b*) the presence or absence of specific suborganellar components in multicompartment organelles (e.g. galactolipid-synthesizing enzymes are markers for the plastid envelope); (*c*) the presence or absence of specific submembrane domains (patches or regions) in multidomain membranes (e.g. RNA content of rough versus smooth ER). Indeed it is imperative to be aware of the suborganelle location of any given marker when interpreting distribution data. Cytochrome *c* oxidase is a marker of the inner mitochondrial membrane but is not alone a marker of intact mitochondria.

LIMITATIONS TO THE USE OF MARKERS Absolute markers (those completely restricted to and uniformly distributed throughout the entire population of a given subcellular component) may be the exception rather than the rule. For morphological markers this problem is manifested as the disintegration of the initial recognizable structural form accompanied by assimilation of the derivative fragments into the pool of morphologically indistinguishable smooth-surfaced vesicles. For biochemical markers two principally different types of deviation from the ideal (opposite in direction) are encountered: (*a*) One type results from heterogeneity in the plane of the membrane giving rise to domains within a single membrane species or, in the special case of Golgi, across the maturing individual cisternae of the dictyosome. A constituent that is confined to a given domain will no longer be a valid representative marker of the parent membrane upon nonuniform fragmentation. Such a "domain marker" might be considered to "under-represent" the parent membrane in contrast to a "universal marker" that is homogeneously distributed throughout that entire membrane. (*b*) The second type of deviation occurs when the same constituent is present in more than one morphologically distinct membrane type or region ("over-representation"). It has been argued that this situation should be particularly prevalent in the endomembrane system as the inevitable consequence of membrane flow and differentiation (222, 224). Apparent multimodal distributions of this nature should be examined closely for isozymes, however. There are numerous examples of differential subcellular compartmentation of distinct isozymes (231, 286, 295, 296, 331).

PRACTICAL MARKER USE In practice, constituents that are located *predominantly* in a single component or domain are satisfactory for use as positive markers but not as negative markers. In general, the use of multiple markers of a single component coupled with multiple separation procedures (e.g. both rate-zonal and isopycnic gradient (s-ρ) centrifugation; density perturbation) is highly desirable (256).

Reporting Data

The most prominent consensus recommendation to emerge from the "Markers Debate" (224) is that relating to the reporting of data. It is recommended, in keeping with the concepts and principles of quantitative cell fractionation (64, 65), that the provision of complete balance sheets be mandatory for publication of subcellular fractionation studies. The data presented should include (or at least permit calculation of) both the *total activity* and the *specific activity* of the constituent or marker under study, both for the original *total homogenate* and for *all fractions* subsequently derived from that homogenate including the final supernatant.

When practical, the procedure of choice is that used routinely by Beevers and coworkers (191, 208, 237), namely, direct fractionation of the entire crude homogenate on a single linear sucrose gradient followed by assay and an accounting for each activity of interest in every gradient fraction including the pellet. This procedure has the additional advantage that it avoids the demonstrably destructive effects of the pelleting and resuspension operations associated with prior differential centrifugations.

MARKERS FOR SPECIFIC SUBCELLULAR COMPONENTS

Table 1 lists some of the markers currently used in fractionation studies of plant subcellular components. The list is not exhaustive and the rigor with which the diagnostic value of each marker has been tested varies considerably.

It is worth emphasizing that the strategies, problems, and constraints involved in the preparation of satisfactory fractions of the large and/or spatially discrete organelles (nuclei, plastids, mitochondria, microbodies, Golgi, vacuoles, storage organelles) clearly will differ from those for the isolation of the initially sheet-like components of the endomembrane system (endoplasmic reticulum, plasma membrane, tonoplast). In the former case the preservation of the original structural and, ideally, functional integrity is a primary concern. In the latter case fragmentation and vesicularization are expected and exploited as the inevitable consequence of cell disruption. It is likewise axiomatic that the pool of smooth-surfaced, morphologically indistinguishable vesicles derives not only from the sheets of the endomembranes but also to a variable and often significant extent from other organelles disrupted during homogenization and fractionation (22). This problem is central to cell fractionation studies.

Nuclei

An impressive array of tissue disruption methods from "pea poppers" (272) to protoplast production (200, 242, 350) have been tried with varying success in attempts to isolate whole plant nuclei (for review see 322). All procedures exploit the large size of the organelle for morphological identification and fractionation based on selective sedimentation at low *g* forces. Various stains provide a cytochemical marker at the light microscope level (184, 200, 273). The distinctive double membrane and nuclear pore complexes are unequivocally diagnostic at the EM level, not only for the whole organelle but also for membrane fragments (86, 87, 253). While DNA might at first sight appear to be a definitive biochemical marker, its use requires qualification. First, plastid and mitochondrial DNA need to be accounted for, especially if DNA is to be used as a negative marker (242). While plastid

Table 1 Densities and commonly used markers of plant subcellular components and their derivative parts

Subcellular fraction	Suborganelle fraction	Density[a] (g cm^{-3})	Markers[b] Morphological or cytochemical	Biochemical	Ref.
Nuclei	Intact	1.32	Large size Double membrane + nuclear pore complexes	DNA	86, 87, 102, 242, 253
	Envelope	1.21–1.24	Double membrane + nuclear pore complexes	Attached DNA	86, 87, 253
	Nucleoplasm	—	Methyl green pyronin Y	DNA	184, 200, 242
	Nucleoli	—	Methyl green pyronin Y	RNA polymerase I	184, 200
Mitochondria	Intact	1.18–1.20	LM[c]-Phase EM[d]-Double membrane cristae	Cytochrome *c* oxidase Fumarase Succinate dehydrogenase Succinate: cytochrome *c* reductase	10, 19, 51, 73, 119, 168, 248, 305
	Outer envelope	1.10	EM-folded transparent membrane bag negative stain	Antimycin A-insensitive NADH-cytochrome *c* reductase	77, 217, 305
	Mitoplast (inner membrane + matrix)	1.19	EM-swollen single bounding membrane + internal cristae vesicles	Cytochrome *c* oxidase Succinate dehydrogenase Fumarase Succinate: cytochrome *c* reductase	10, 119, 305
	Inner membrane	1.14	—	Cytochrome *c* oxidase Succinate dehydrogenase Succinate: cytochrome *c* reductase	10, 119, 305
	Stroma	—	—	Fumarase	305
Chloroplasts	Intact	1.21–1.24	LM-Phase contrast dull appearance -CHL[e] fluorescence EM-double envelope thylakoids	Chlorophyll RuBP[f]-carboxylase (C_3)[g] NADP-triose phosphate dehydrogenase Pyruvate Pi dikinase (C_4)[h]	102, 160 169, 239 291
	Envelopes (inner + outer)	1.12	—	Galactosyl transferase	72, 75
	Thylakoids	1.16–1.18	LM-Phase contrast shiny appearance -CHL fluorescence EM-granal stacks	Chlorophyll	169, 207, 277, 291
	Stroma	—	—	RuBP-carboxylase (C_3) Pyruvate P_i dikinase (C_4) NADP-triose phosphate dehydrogenase	160, 239, 291
Etioplasts	Intact	1.26	LM-PCHL[i] fluorescence EM-Double membrane + PLB[j]	RuBP carboxylase Protochlorophyll(ide) Carotenoids	100, 102 137, 152 264, 267
	Envelopes (inner + outer)	1.12	—	Galactosyl transferase	72, 75
	Prolamellar bodies	1.12–1.14	EM-PLB crystalline structure	Protochlorophyll(ide) Carotenoids	100, 264 267, 293, 338

Table 1 *(Continued)*

Subcellular fraction	Suborganelle fraction	Density[a] (g cm^{-3})	Markers[b] Morphological or cytochemical	Markers[b] Biochemical	Ref.
Etioplasts (cont.)	Stroma	—	—	RuBP carboxylase Isozyme 1 pentose phosphate cycle enzymes (?)[k]	295, 296
Proplastids	Intact	1.23	EM-Double membrane rudimentary inner membranes	RuBP-carboxylase Fatty acid synthetase Nitrite reductase Acetolactate synthetase	34, 207 237, 328
	Envelopes	—	—	Galactosyl transferase (?)	72, 75
	Internal membranes	—	—	Protochlorophyll (?)	34
	Stroma	—	—	RuBP carboxylase Fatty acid synthetase	237, 328
Glyoxysomes	Intact	1.25	DAB[l] stain	Catalase Isocitrate lyase Malate synthetase	15, 16, 130, 236
	Membrane	1.21	—	Monoglyceride-specific alkaline lipase (castor bean) Malate synthetase	130, 232
	Lumen	—	DAB stain	Catalase Isocitrate lyase	15, 16, 130, 236
Peroxisomes	Intact	1.25	DAB stain	Catalase Hydroxypyruvate reductase	130, 236 239, 320
	Membrane	—	—	—	—
	Lumen	—	DAB stain	Catalase Hydroxypyruvate reductase Glycolate oxidase	130, 236 239, 320
Endoplasmic reticulum (ER)	Membrane + ribosomes (RER only)	1.15–1.18	EM-rough surfaced membranes	16S and 26S RNA	2, 68, 99
	Membrane	1.11–1.12	—	Phospholipid synthesizing enzymes Antimycin A-insensitive NAD(P)H-cyt *c* reductase	191, 211 215, 234
Golgi	Intact (dictyosome)	1.12–1.15	Dictyosome stack IDPase	Latent IDPase[m] Glucan synthetase I	60, 226, 269, 270
	Secretory vesicles	1.14	IDPase	Latent IDPase Glucan synthetase I	269, 270
Plasma membrane	—	1.13–1.18	Periodic acid phosphotungstate chromate stain	K^+-ATPase[n] Glucan synthetase II	120, 128, 176, 233, 268, 325
Vacuole	Intact	<1.03 to >1.18	Large size ± pigments Neutral red stain	RNase[o] Phosphodiesterase Pigments	24, 25 174, 237, 330
	Tonoplast	1.10 (?)	—	—	24, 25
	Lumen	—	RNase immunocytochemically	RNase Phosphodiesterase Pigments	13, 24, 25, 174, 237

Table 1 *(Continued)*

Subcellular fraction	Suborganelle fraction	Density[a] (g cm^{-3})	Markers[b]: Morphological or cytochemical	Markers[b]: Biochemical	Ref.
Protein bodies	Intact	1.26–1.32 (sucrose) 1.36–1.39 (hexane)	Dense matrix ± globoid and crystalloid inclusions Single membrane	Lectins Storage proteins	80, 279, 323, 343, 346
	Membrane	1.15	—	—	346
		1.22	—	—	206
	Matrix	—	PAS[p] stain	Lectins	323, 346
	Crystalloids	1.30	Size and shape	Storage globulins	323
	Globoids	>1.46	Size and shape	Phytin	5, 323
Lipid bodies	Intact	0.96	Sudan black B Osmiophillic	Acid lipase (castor bean) Storage triglycerides	232, 246 345
	Membrane	1.12 (wax body)	Half-unit membrane	Acid lipase (castor bean)	232, 344, 345
	Matrix	—	Sudan black B Osmiophillic	Neutral lipids	134, 345

[a]Density at isopycnic equilibrium in sucrose.
[b]Diagnostic value of markers listed is highly variable. Consult relevant section of text for discussion of merits and limitations of each.
[c]LM = light microscope.
[d]EM = electronmicroscope.
[e]CHL = chlorophyll(ide).
[f]RuBP = ribulose bisphosphate.
[g](C_3) = for plants with C_3 photosynthetic pathway.
[h](C_4) = for plants with C_4 photosynthetic pathway.
[i]PCHL = protochlorophyll(ide).
[j]PLB = prolamellar body.
[k](?) = likely, but not rigorously established.
[l]DAB = diaminobenzidine.
[m]IDPase = inosinediphosphatase.
[n]K^+-ATPase = K^+-stimulated *increment* in adenosinetriphosphatase activity *above* the Mg^{2+}-dependent rate.
[o]RNase = ribonuclease.
[p]PAS = periodic acid Schiff carbohydrate stain.

DNA usually accounts for only about 5% of the total, it may be as high as 20% in some cases (W. F. Thompson, personal communication.) Second, nuclear DNA occurs both apparently free in the nucleoplasm and firmly bound to the nuclear envelope even after fragmentation (86, 87, 253). Third, up to 80% of the total DNA may be recovered in fractions other than nuclear pellets, especially when tissue is disrupted with power-driven homogenizers, presumably as a result of fragmentation of the nuclei and chromatin (139). This fragmentation, coupled with the propensity of chromatin to interact nonspecifically with other cellular constituents (M. G. Murray, personal communication), raises the possibility that the nucleic acid will redistribute artifactually to other particulate components. Fourth, less fragmented chromatin freed from disrupted nuclei will be pelleted itself under the appropriate differential centrifugation conditions (184). Thus the simple presence of DNA in a pellet or membrane fraction is not, in the absence of other corroborative data, evidence of the presence of intact nuclei or nuclear membranes. These points are of particular importance to analytical fractionation studies. The high intrinsic density of intact nuclei [in excess of 1.32 g cm^{-3} (117, 161)] should permit their ready separation from other organelles under standard isopycnic sucrose gradient centrifugation conditions (117). The reported densities of separated nuclear membrane fragments (1.21 to 1.235 g cm^{-3}) (253, 308), however, overlap the range in which intact plastids are normally located (17, 129, 169, 320).

Preparative studies purporting to isolate and characterize plant nuclei vary considerably as regards the rigor with which yield, purity, and structural or functional integrity are assessed. Two distinct classes of protocol are apparent—those that use detergent treatments (42, 59, 200, 309) and those that don't (85, 161, 253, 313). Mascarenhas et al (200) have surveyed some of the detailed preparative methodology with particular emphasis on the former. Detergent treatment of crude, nuclei-containing pellets apparently very effectively solubilizes contaminating cytoplasmic membranes but, at the same time, also removes at least the outer nuclear membrane (59, 200, 242, 273). The treatment has been asserted to remove the outer membrane selectively and leave the inner one intact but this point is controversial (86, 87). In any case, the residual structures, while being suitable for some purposes (42, 47, 273), are clearly of restricted value for complete characterization studies (86).

An alternative approach that uses octanol and gum arabic in the preparative media (85, 161, 313) results in much improved structural preservation with no apparent sacrifice in yield (108, 253). A major disadvantage of this latter procedure in its original form (161, 313), however, is the prescribed 14 to 20 hr preincubation period for the tissue in the octanol medium before homogenization. More recent studies have shortened or eliminated this preincubation in attempts to minimize autolysis (253). A 30 sec pretreatment of tissue in cold ether has also been used and reported to produce high yields of nuclei in the absence of gum arabic or any other pretreatment (108). It is not clear from the data, however, whether the structural integrity of the double envelope is retained after this treatment.

Reported yields of nuclei based on either cell number or DNA content of the parent tissue range from a few percent to as high as 90% depending on the procedure used. Adequate quantitative assessment of purity of the fractions, however, is almost nonexistent in the literature. It is frequently not reported at all (200, 273, 308) or is reported only as a qualitative visual assessment with the light or electron microscope (59, 117, 313, 350). A few authors report numbers of light-microscope-visible contaminating particles (108, 161). Compositional analyses of such fractions are therefore of debatable value.

The most thorough quantitative preparative study available is that of Philipp et al (253), who used combined morphometric and biochemical procedures. These authors obtained preparations of intact nuclei containing 81% nuclear membrane, the remainder being contaminants. Isolated membranes were also prepared from this fraction and their gross chemical composition and several enzyme activities determined. The data indicate a close biochemical similarity between the nuclear envelope and the ER. This result is consistent with the in situ observed direct continuity of the outer nuclear envelope with the ER cisternae and the presence of ribosomes on the outer envelope (86, 87, 102, 241). No enzymatic activity or other biochemical parameter with the potential to serve as a diagnostic marker for the nuclear envelope has been encountered (253, 308). The only reliable marker presently available appears to be the morphologically identifiable nuclear pore complexes (86, 87).

Isolation and characterization of nucleoli has been reported by workers using methyl green pyronin Y to monitor the preparations (184). The authors conclude from their data that RNA polymerase I is localized in the nucleoli whereas RNA polymerase II is in the nucleoplasm. These enzymes are clearly potential nuclear markers, although their routine use for determining distribution of subnuclear components might be precluded by the lengthy separation and assay protocol required.

Mitochondria

Both general (26, 110, 168) and more restricted (248, 304) reviews on plant mitochondria are available. A variety of preparative procedures based on the specific needs of investigations in the broad areas of bioenergetics, biogenesis, and carbon metabolism are to be found in the literature (73, 76, 84, 149, 166, 167, 169, 305). Much of the work on electron transport and oxidative phosphorylation has been performed with crude differential or step gradient centrifugation fractions with little or no rigorous quantitation of potential membranous contaminants (62, 166). Emphasis in these studies is placed on the retention of maximal functional integrity as determined by respiratory control and ADP : O ratios, with some authors seeing little need for anything more than differential centrifugation (248). Such undefined fractions are clearly unsuited to localization studies.

The difficulty of preparing a mitochondrial fraction free of contamination from the thylakoids of broken chloroplasts (151, 169, 207, 208, 256, 291, 293) has resulted in the use of etiolated or storage tissue for most studies. This difficulty with green tissue arises because at isopycnic equilibrium the mitochondria (1.18 g cm^{-3}) are sandwiched between and/or overlap with intact (1.21 g cm^{-3}) and ruptured (1.16 g cm^{-3}) chloroplasts (151, 277, 291,

293). Attempts to overcome the problem include combined differential and isopycnic centrifugation (149, 277), rate-zonal centrifugation (207, 208), silica sol gradients (135), and phase partition (165). The most effective separations thus far reported have been obtained with leaf protoplasts (239).

There is ample evidence that plant mitochondria resemble mammalian mitochondria in most fundamental respects both structurally and functionally (168, 248). There are also differences, however, notably the capacity of plant mitochondria to oxidize exogenous NADH and, in many tissues, the possession of a cyanide-insensitive respiratory pathway, both features lacking in mammalian mitochondria (168, 248, 304). Despite these differences, the familiar mitochondrial ultrastructure provides an obvious morphological marker, while the well-documented mitochondrial location of enzymes of the tricarboxylic acid (TCA) cycle and components of the oxidative electron transport chain provide biochemical markers for matrix and inner membrane respectively.

In practice, cytochrome *c* oxidase, succinate: cytochrome *c* reductase and succinic dehydrogenase are the most convenient and commonly used inner membrane markers (10, 19, 168, 239, 248, 305) while fumarase serves as a suitable matrix marker (15, 51, 215, 305). Other TCA cycle enzymes such as citrate synthetase and malate dehydrogenase are less suitable matrix markers because they also occur in other compartments, notably the microbodies (16, 51, 239, 277; see also 149). The mitochondria do contain, however, a distinct isozyme of malate dehydrogenase that may be useful in some circumstances (331). Cytochrome *c* oxidase, as normally assayed, signals the presence of the inner membrane but gives no information on mitochondrial integrity or recovery. A pH 9 ATPase that coincides with cytochrome *c* oxidase on gradients is likewise considered to represent the inner membrane (119, 120, 177). Fumarase is released to the soluble phase upon disruption of the inner mitochondrial compartment so that the recovery of this enzyme in mitochondrial fractions provides an index of the yield of intact inner compartments.

None of these activities, however, indicate the degree of retention of the outer membrane. Likewise, the conventional demonstration that the isolated organelles have a high degree of respiratory control and high ADP : O ratios (166–168, 248), while indicating retention of the integrity of the matrix and electron transport chain, does not alone establish that the outer membrane is present. Several direct tests of mitochondrial integrity have been developed, the most straightforward of which are based on the impermeability of the intact outer membrane to cytochrome *c* (26, 73, 77, 166, 168, 248). Completely intact mitochondria will not reduce exogenous cytochrome *c* in the presence of succinate and KCN. Rupture of the outer membrane gives the exogenous cytochrome *c* access to the electron transport chain of the inner membrane resulting in succinate-driven, antimycin-sensitive reduction of the added cytochrome (26, 73). This assay can be quantitated to provide an estimate of the proportion of intact mitochondria in a fraction by measurement before and after deliberate disruption. Coupling this estimate to determinations of fumarase or cytochrome *c* oxidase recovery in a fraction permits the overall yield of completely intact mitochondria to be calculated.

Numerous reports from Beevers and coworkers (16, 51, 190, 215, 237, 327) have demonstrated that, with care, virtually 100% of the cytochrome oxidase and fumarase can be recovered in a fraction essentially uncontaminated by plastids, microbodies, or ER. No attempt was made in these studies to assess quantitatively the level of outer membrane retained. Douce et al (73, 76, 77) have reported the preparation of mitochondrial fractions that are >95% intact, as determined by a variety of tests, contain relatively low levels of NADPH-cytochrome *c* reductase, lack detectable ascorbate-reducible *b*-type cytochrome (163), and appear enriched for mitochondria in unquantitated electron micrographs. These date indicate low ER contamination, but the presence of other cellular membranes is not ruled out.

Outer membrane fractions have been prepared from plant mitochondrial fractions following controlled osmotic lysis (28, 67, 76, 77, 216, 217, 305). The outer membrane is estimated to comprise 10–15% of the total mitochondrial membrane complement (216). Sparace & Moore (305) have demonstrated that subfractionation of lysed castor bean mitochondria on sucrose gradients resolves three fractions. The densest fraction at 1.19 g cm^{-3} contains 90% of the remaining particulate fumarase, about half the succinate:cytochrome *c* reductase, half the antimycin A-sensitive NADH-cytochrome *c* reductase, and appears as single-membrane-bounded vesicles in electron micrographs. This fraction represents the mitoplasts (inner membrane plus matrix). A second, intermediate fraction at 1.14 g cm^{-3} has slightly lower levels of the two reductases but little fumarase and contains primarily small vesicles. These data and the presence of only low levels of antimycin A-*in*sensitive NADH-cytochrome *c* reductase suggest that the fraction is enriched in separated inner membrane vesicles with little or no enclosed matrix. The data also verify the inner membrane location of the antimycin-sensitive NADH-cytochrome *c* reductase (76, 217). The third fraction at 1.10 g cm^{-3} is devoid of detectable fumarase, succinate:cytochrome *c* reductase, and antimycin-sensitive NADH-cytochrome *c* reductase. The membranes present have the characteristic "empty-sac" appearance of outer mitochondrial membranes in negative stain (76, 217) and have an associated NADH-cytochrome *c* reductase that is totally insensitive to antimycin A.

Several studies have now provided definitive evidence that this latter activity is associated with the outer mitochondrial membrane (67, 76, 77, 217, 305) and is not present as a result of contaminating ER, the major site of antimycin-insensitive NADH-cytochrome *c* reductase activity in the cell (191, 215). The evidence is that several components that have a verified ER location are demonstrably absent from the starting mitochondrial fraction and/or the isolated outer membrane fraction itself. These components include NADPH-cytochrome *c* reductase (76, 77, 217), ascorbate-reducible *b* cytochrome (76, 77, 163), oleoyl-coenzyme A desaturase (67), and enzymes involved in the synthesis of CDP-diglyceride, phosphatidylglycerol, and phosphatidylcholine via methylation (76, 212, 213, 305). The last three enzymes occur in both the ER and mitochondria but in the latter case are found to be exclusively localized on the inner membrane (76, 305).

Both NADH- and NADPH-cytochrome *c* reductases have been reported present in nuclear envelope preparations (253). This fraction is unlikely to account for the activity attributed to the outer mitochondrial membrane because of the absence of NADPH-dependent activity in the latter fraction, and because of the widely disparate densities of the two fractions (1.10 and 1.21–1.24 g cm^{-3} for mitochondrial and nuclear membranes respectively). Note also that the use of rotenone by Philipp et al (253) does not, in contrast to antimycin, preclude participation of the inner membrane in the measured NADH-dependent activity (248). A report of the presence of NADPH-cytochrome *c* reductase activity in outer membrane fractions (31) would appear to indicate ER contamination.

The outer membrane fraction also contains a cytochrome b_{555} (76, 217), but its utility as a marker would appear limited given its close spectral resemblance to cytochrome b_5 of the ER. Monamine oxidase and kynurenine hydroxylase, markers of mammalian mitochondrial outer membrane, are both reported to be undetectable in plant preparations (217). Various compositional analyses (67, 71, 197, 216) have provided no unique markers for the outer membrane. Cardiolipin, a marker of mammalian mitochondrial inner membrane, is likewise at high levels in the plant organelle (71), but its suborganelle distribution does not appear to have been documented. Galactolipid, considered to be exclusive to the plastids (75), is reportedly undetectable in carefully prepared mitochondrial fractions (216), indicating a lack of plastid contamination.

It would appear, therefore, that there is presently no absolute marker for the plant outer mitochondrial membrane. Both the antimycin A-insensitive NADH-cytochrome *c* reductase and cytochrome b_{555} detected in the preparations have counterparts on the ER where they are present in substantially greater absolute amounts (76, 77, 191, 215, 234). These quantitative considerations and the nearness in the densities of outer membrane (1.10 g cm^{-3}) and ER (1.11–1.12 g cm^{-3}) indicate that positive identification of separated outer membrane in an analytical fractionation scheme is unlikely.

Digitonin treatment of crude, 12,000 X *g* pellets (63) or mitochondria-rich sucrose gradient fractions (61) releases antimycin-insensitive NADH-cytochrome *c* reductase to a 30,000 X *g* supernatant from which it can be pelleted at 144,000 X *g* for 5 hr. This pelleted material is asserted to represent the outer membrane, but its mitochondrial (as opposed to ER) origin, its membranous character, and its value as a fraction representative of the parent membrane are yet to be documented.

Plastids

The complexity of the plastid is reflected in the catalog of preparative procedures in the literature. Procedures differing in the balance achieved between yield, purity, and overall structural and functional integrity have evolved in response to the need for differing optimal conditions for a variety of experimental purposes.

CHLOROPLASTS Jensen (143) has reviewed this organelle. For many studies on the mechanism of photosynthesis, simple preparative differential centrifugation protocols yielding enriched but by no means pure plastid preparations suffice. For the study of specific aspects of electron transport and photophosphorylation, naked thylakoids, resulting from organelle rupture, intentional or otherwise, are not only sufficient but necessary (169). Compilations of some of the various procedures for the preparation of chloroplast fractions are available (33, 169, 239, 276, 282). Hall (103) has proposed a nomenclature for describing chloroplasts possessing varying degrees of structural and functional integrity. Both morphological and biochemical criteria are used to assess this integrity (169). This nomenclature, however, does not consider the level of contamination by other membranes or organelles.

Conventional mechanical disruption of green tissue followed by isopycnic sucrose gradient centrifugation successfully separates broken and intact chloroplasts, but a considerable degree of cross-contamination with microbodies and mitochondria results (151, 291). Rate-zonal sedimentation with (277) or without (208) subsequent isopycnic banding substantially reduces this contamination. By far the most impressive separations thus far obtained with green tissue, however, have resulted from the use of protoplasts as starting material (239). High yields of structurally intact and contamination-free chloroplasts (as judged by 80 to 90% retention of the stromal enzymes RuBP carboxylase and NADP-triose phosphate dehydrogenase and virtually complete separation from mitochondrial and peroxisomal marker enzymes) are obtained by isopycnic sucrose gradient centrifugation of gently ruptured protoplasts (239).

Morphological and photochemical integrity are preserved during sucrose gradient centrifugations such as these (147, 169, 171, 277, 330), but plastids prepared in this way are functionally inactive as determined by CO_2 fixation capacity (169, 239). Fully functional chloroplasts with CO_2 fixation rates equivalent to those in the intact cell are obtainable by simple differential centrifugation (79, 332), but these preparations are clearly contaminated by other organelles. A compromise between these two extremes has been sought with the use of rapid rate zonal sucrose gradients (147, 171), silica sol gradient centrifugation (219, 257), and by phase partition (165). Plastids obtained by these procedures are morphologically intact with moderate to high CO_2 fixation rates. In the latter two cases, the degree of contamination by other organelles is yet to be quantitated. Thus it appears that no single preparative procedure so far reported provides chloroplasts that are pure and also retain full structural and functional integrity. Full functional competence is not, however, a prerequisite for successful fractionation studies concerned with delineating useful organelle and suborganelle markers. Sucrose gradient centrifugation, coupled with determinations of structural integrity, is adequate for this purpose.

The evidence for the specificity of the morphological, stromal, and thylakoid markers listed in Table 1 is of long standing (75, 169) and requires no elaboration here. Isolation and characterization of the envelope is comparatively more recent (for review see 75). Carefully prepared chloroplast envelope fractions are devoid of detectable chlorophyll, NADPH-cytochrome *c* reductase, CDP-choline:diglyceride transferase, and *b*-type cytochromes, indicating the absence of thylakoid, ER, and mitochondrial contamination. Efforts to delineate biochemical or morphological parameters to permit the separation of inner and outer envelope membranes have thus far failed. Although the envelope preparations exhibit an unusual gross chemical composition (e.g. high violoxanthin levels, no phosphatidyl ethanolamine), none of these features are useful as positive markers for analytical fractionation studies (140).

There is now substantial evidence that a galactosyl transferase activity (UDP-galactose:diacylglycerol transferase) is predominantly if not exclusively located in the chloroplast envelope fraction and appears, therefore, to be an acceptable marker for that membrane (75). It should be noted, however, that although the *specific* activity of this enzyme is up to eightyfold higher in the envelope than in the thylakoid fraction, the bulk of the *total* activity (60–80%) is in fact found in the much more abundant thylakoid fraction (72, 147). While this latter activity may represent envelope contamination of the thylakoids, genuine thylakoid-localized activity is yet to be definitively eliminated. This possibility poses a potential quantitative problem for analytical fractionation studies because the absolute amount of thylakoid membrane protein is more than fiftyfold that of the envelope (147). A similar argument applies to oleoyl-CoA-synthetase activity present at eightyfold higher specific activity in the envelope than in the thylakoid fraction (281). The variability observed in galactosyl transferase activity between different envelope preparations (259) should be noted.

A N,N'-dicyclohexyl-carbodiimide (DCCD)-insensitive Mg^{2+}-ATPase has also been proposed as a marker for the chloroplast envelope (72, 74). This proposal requires qualification. First, a complete balance sheet for the distribution of this activity among the original subfractions of the total homogenate has not been published. The possibility that the activity has a multimodal distribution or derives from minor contamination from a non-plastid membrane (other than ER or mitochondria) therefore remains. Such an activity has been reported for a putative tonoplast fraction (184). Second, this ATPase is not observed in envelope preparations from all species (259). The galactosyl transferase would, therefore, appear to be the most reliable marker for envelope membranes presently available.

ETIOPLASTS Schiff (288) has reviewed etioplasts and proplastids. Both rate-zonal centrifugation (100, 137, 171, 195) and a Sephadex G-50 column procedure (339) have been used to prepare etioplast-rich fractions. Quantitation of yield and structural integrity is variable between reports and adequate quantitation of the purity of preparations is rare. Parameters for functional integrity are ill defined, but in vitro light-induced ultrastructural changes in the prolamellar body reminiscent of those observed early after the onset of irradiation in vivo have been reported (159, 339). Preparative density gradient procedures have been documented that remove 99% of the bacteria (137, 171) and mitochondria [succinate dehydrogenase (171)] initially present in a crude plastid pellet while preserving plastid structure. No quantitative tests for other potential membrane contaminants were reported, however. Other authors report that the original morphological integrity of the etioplasts is lost upon sucrose gradient centrifugation, possibly because of osmotic effects (339).

Passage of a crude plastid pellet through Sephadex G-50 allows preservation of the etioplast morphology, but does not, contrary to the unquantitated claims of the original report (339), separate intact plastids from mitochondria. About 15% of the original homogenate cytochrome *c* oxidase is present in the fraction both before and after passage through the column (262). Moreover, recent studies show that the elution profiles of all marker enzymes tested (including an ER marker) are identical, because all particulate material above the exclusion limit of the gel (30,000 daltons) comigrates in the void volume (126, 199, 262). Studies purporting to assign activities or transport functions to the etioplasts in these fractions (337) should therefore be viewed with caution.

Despite the firm evidence that the prolamellar body of etioplasts is the exclusive location of protochlorophyll(ide) in the cell (21, 152, 162), this pigment appears to have been used infrequently as a marker (137). Carotenoids have been used as markers for etioplasts (144, 264, 267), but the justification for this use appears to have been derived more from extrapolation from established chloroplast data (20, 75, 115) than from direct quantitative analysis. The recently measured complete coincidence of protochlorophyll and carotenoid profiles on sucrose gradients (D. Marmé, personal communication) strengthens this assumption (see also 52).

Schnarrenberger and coworkers have attempted to establish stromal markers for etioplasts (293, 296). Particulate forms of triosephosphate isomerase, NADPH-glyoxylate reductase (293), and specific isozymes of several oxidative pentose phosphate cycle enzymes (295, 296) were demonstrated to be chloroplast-localized in greened tissue. Particulate forms of these activities were also found in etiolated tissue of identical age, and the conclusion was drawn that the activities were etioplast-located (295, 296). While the *exclusive* localization of these various activities in the chloroplast of the greened system has not been rigorously established, the data do strongly indicate that the fraction of the activities that is particulate is located in the stroma of intact chloroplasts. The extrapolation of this argument to the etioplasts of the nonirradiated system likewise seems reasonable but not unequivocal because neither an independent marker for etioplasts, such as protochlorophyll(ide), nor a morphological parameter was used for correlation. A cross check of this nature might have helped eliminate the still feasible possibility that the high density (1.26 g cm^{-3}) of putative etioplast activities and their strong coincidence with the microbody markers (295, 296) result from nonspecific adsorption of soluble activity to the microbodies. On the other hand, etioplasts have been reported previously to have a density of 1.25–1.26 g cm^{-3}, higher than mature chloroplasts (1.21 g cm^{-3}) (21, 258).

Reports of the subfractionation of etioplasts into envelope and prolamellar body fractions (48, 50) should be viewed in the context of the demonstrated contamination of the parent plastid fraction (126, 199, 262). Nevertheless, the presence of a Ca^{2+}-ATPase in, and visually identifiable coupling factor particles on, sedimented prolamellar body-rich fractions (336) offers a potentially powerful tandem marker for this membrane system. The level of contamination of putative envelope fractions by vesicles derived from fragmented prolamellar bodies appears to be very low if the Ca^{2+}-ATPase activity is accepted as a prolamellar body marker (48). These detached prolamellar body vesicles are indistinguishable morphologically from envelope vesicles when stained conventionally (336). Protochlorophyll could also be used to estimate prolamellar body contamination of the envelope fraction but this has not been documented. It is unclear how reports of a putative etioplast envelope, nonlatent Mg^{2+}-ATPase (48, 50) are related to the DCCD-insensitive enzyme of the chloroplast envelope reported by Douce (72) because the uncoupler was apparently not used in assaying the etioplast activity (48, 50).

Galactosyl transferase is a logical, but as yet apparently untested, candidate for an etioplast envelope marker (75), tentatively assigned a density of 1.12 g cm^{-3} here (6, 48). The isopycnic density of naked prolamellar bodies does not appear to have been precisely documented. However, a density of <1.14 g cm^{-3} is suggested by extrapolation from the observation that stripped thylakoids have this density 6 hr from the onset of greening and that this value increased to the normal 1.16 to 1.17 g cm^{-3} with further greening (293). In agreement with this suggestion, prolamellar bodies from intentionally lysed etioplasts band isopycnically at the 0.9 M/1.2 M interface of a step sucrose gradient, indicating a density between 1.12 and 1.16 g cm^{-3} (6, 338).

PROPLASTIDS This term is used here for plastids derived from tissue that is not normally photosynthetic. Proplastid-rich fractions of varying degrees of purity have been prepared from roots (207, 208, 319) and from castor bean endosperm during development (297, 300) and during germination (237, 247, 328) with rate-zonal and/or isopycnic sucrose gradient centrifugations.

Nishimura & Beevers (237), in an analytical fractionation study, have demonstrated that essentially 100% of the RuBP carboxylase in germinating castor bean endosperm is located in a fraction of density 1.22 g cm^{-3} clearly separated from mitochondrial and glyoxysome markers. Based on the observed density, previous electron micrographs of this fraction (34), and the well-established location of the carboxylase in green tissue, there is little doubt that the fraction represents highly purified proplastids. This fraction is also the exclusive location of assayable fatty acid synthetase in the same tissue even when conventional preparative procedures that cause demonstrable plastid rupture and loss of stromal RuBP carboxylase are used (328). These last data are more indicative of enzyme inactivation upon plastid rupture than a membrane-bound activity, however, because any fragmentation or loss of stromal content would be expected to lead to a decrease in density (75) and this decrease is not observed.

The behavior of acetyl-CoA carboxylase parallels that of RuBP carboxylase in both the developing (300) and germinating (328) castor bean endosperm, suggesting that the former enzyme is also a reasonable candidate for a proplastid stromal marker. Coincidence with triosephosphate isomerase on sucrose gradients provides evidence that both nitrite reductase and acetolactate synthetase are stromal enzymes of plastids from several sources (207). No marker of the proplastid envelope has been reported, but by analogy with chloroplasts, galactosyl transferase is a potential candidate (75). The red fluorescence earlier observed in the castor bean proplastid band (34) suggests the possible use of protochlorophyll as a marker for the few internal membranes apparently present.

Microbodies

Evidence justifying the use of the markers listed in Table 1 is included by Beevers in his review of microbodies (16). The following points, however, are worthy of emphasis in the context of fractionation studies. The lumenal enzymes isocitrate lyase (glyoxysomal), NADH-hydroxypyruvate reductase (peroxisomal), and catalase (common) are reliable positive markers of their respective intact organelles when coupled with suitable separation procedures. The coincident pelleting or isopycnic banding of catalase and cytochrome *c* oxidase observed in some systems (61, 120, 349) has been suggested to indicate mitochondrially localized catalase or result from nonspecific adsorption of the enzyme to mitochondria. However, Miflin & Beevers (208), using rate-zonal centrifugation, have obtained clear separation of these two activities, indicating that the catalase was located in microbodies inadequately separated from the mitochondria by the previous protocols. Malate synthetase is a peripheral (302) protein and is a reliable marker of the glyoxysomal membrane [density 1.21 g cm^{-3} when isolated (130, 131)] provided conditions (such as high ionic strength) known to dislodge such proteins are avoided.

It should be emphasized that choice of tissue and stage of development are crucial parameters in the reliability of the above markers. The phenomenal successes achieved by Beevers and coworkers with castor bean endosperm have rarely been duplicated by other investigators using different systems. Malate synthetase and catalase appear to be absolute markers for glyoxysomes from castor bean at later stages of germination. Early in germination, however, significant levels of catalase are soluble and a significant amount of malate synthetase is (transiently) located in the ER (96, 189). Moreover, while alkaline lipase is an integral protein primarily located in the glyoxysomal membrane of castor bean endosperm (130) and peanut (60%) (131), it is predominantly located in the ER of other tissues from *Ricinus* (130) and of other species (316) and in the spherosomes of rape seed (315, 334).

The preservation and separation of peroxisomes from the other organelles of green tissue has been less successful in the past than for glyoxysomes from etiolated tissue (16, 291). More recently, however, Nishimura et al (239) have achieved excellent separation of peroxisomes (30% of which remain intact) from chloroplasts and mitochondria with direct sucrose gradient centrifugation of extracts from disrupted protoplasts. Note that high phosphate levels may contribute to poor separations by causing aggregation of peroxisomes with intact chloroplasts on gradients (290). Alkaline lipase is not detected in the membrane of peroxisomes (16).

Endoplasmic Reticulum (ER)

Reviews of plant ER are available (44, 88). The characteristic appearance of ribosome-coated membrane vesicles (2, 22, 68, 253) provides an obvious morphological marker for rough ER. In addition, the dramatic decrease in density of such vesicles induced by deliberate "stripping" of the ribosomes from the membrane surface (2, 186, 191, 260, 267, 299) provides an invaluable diagnostic tool for identifying ER membrane-associated constituents. Establishment of biochemical markers for plant ER has been based both on (*a*) direct analogy with mammalian tissues, and (*b*) studies concerned with the question of membrane biogenesis.

Antimycin A-insensitive NADH- and NADPH-cytochrome *c* reductases are well-documented activities of mammalian ER (66). These activities are predominantly particulate in plant homogenates when assayed under the appropriate conditions (see below) (119, 191). Evidence that the activities are localized primarily on the ER of plants derives from their density-shift behavior on sucrose gradients in the presence and absence of Mg^{2+}. The inclusion of millimolar levels of Mg^{2+} in extraction and gradient media [although exogenous Mg^{2+} is not necessary in all tissues (234, 260)] results in NAD(P)H-cytochrome *c* reductase assuming a broad distribution over the density range 1.15–1.18 g cm^{-3} in most systems (191, 211, 234, 260, 267, 299). The fractions in this region are rich in morphologically identifiable rough ER and contain RNA of ribosomal origin (99, 186, 191, 234, 267). Omission of Mg^{2+} and/or inclusion of EDTA causes a large decrease in the isopycnic density of the reductases to 1.10–1.12 g cm^{-3} accompanied by a redistribution of much of the previously membrane-bound RNA to a sedimenting, non-membrane–bound fraction presumed to consist of dissociated ribosomes (68, 99, 150, 191, 234, 260, 269). The reductase-containing membrane fractions then consist of smooth-surfaced vesicles (191, 234, 269).

This "Mg^{2+}-shift" phenomenon when used with care provides strong evidence for ER-localized constituents. The level of Mg^{2+} is critical, however (191, 267). Too high a level leads to nonspecific aggregation of membranes, the reversal of which upon Mg^{2+} withdrawal might generate density shifts unrelated to the dissociation of ribosomes from rough ER. This effect is exacerbated by pelleting prior to gradient centrifugation. For this reason the demonstration of the absence of a density shift of non-ER components on the same or similar gradients is a mandatory control (191, 211, 234, 267, 269, 299) not reported in all studies (55, 172).

The NADPH-cytochrome *c* reductase activity would appear to be exclusively confined to the ER (191). The high soluble activity observed in several tissues is heat insensitive and presumed to be nonenzymatic (J. M. Lord, personal communication; P. H. Quail, unpublished). The NADPH-dependent activity would thus appear to be a more specific marker for the ER than the NADH-dependent activity (which has a multiple location) except that the former is present at often less than 1/10 the activity of the latter (28, 119, 191, 267). The significant NADH-dependent activity associated with the mitochondrial fraction (28, 76, 77, 119, 191, 212, 215, 217, 234) is largely (70–95%) inhibited by antimycin A, whereas that associated with the ER is insensitive to the drug (28, 191, 215, 234). As discussed above, the antimycin A-insensitive activity remaining in the mitochondrial fraction is associated with the outer membrane of that organelle (305). In absolute terms, antimycin A-insensitive NADH-dependent activity appears to be predominantly (~90%) localized in the ER (191, 234). Nagahashi & Beevers (234) have examined this question in some detail.

Philipp et al (253) report the presence of both NADH- and NADPH-cytochrome *c* reductases (rotenone insensitive) in preparations of nuclei (81% pure by morphometry) at enzymatic specific activities comparable to those in ER fractions (83% pure by morphometry) from the same homogenate. The data do not permit estimation of the quantitative contribution that such nuclear material could make to the total reductase activity but independent morphometric estimates (D. J. Morré, personal communication) indicate that ER membranes from the same tissue are about eight times more abundant than nuclear membranes. Moreover, both nuclei and nuclear envelopes have considerably higher densities (1.32 g cm^{-3} and 1.21–1.24 g cm^{-3} respectively) than ER or rough ER, virtually precluding overlap. It would seem, therefore, that both NADPH- and antimycin A-insensitive NADH-cytochrome *c* reductases can serve under controlled conditions as reasonably unambiguous positive ER markers.

Cytochrome b_5 (191, 253) and P_{450} (191), both components of mammalian ER (66), have been reported present in plant ER fractions, although another report failed to detect P_{450} in an ER-rich preparation (253; see 88, 168 for discussion of distribution and confusing nomenclature of plant *b*-type cytochromes). The isopycnic distributions of P_{450} (19) and of an NADH-reducible cytochrome postulated (but not documented) to be b_5 (19, 144) have been reported to coincide with that of NAD(P)H-cytochrome *c* reductase on sucrose gradients. There is evidence that the cytochrome P_{450} is an integral component of the multienzyme complex, *trans*-cinnamic acid 4-hydroxylase, a mixed function oxidase, whose activity profile on sucrose gradients matches that of the P_{450} itself, as well as those of the other coincident activities mentioned above (19, 19a, 284, 347). This enzyme has the potential to be a useful positive ER marker in plant systems, but definitive verification of a predominant ER localization from sedimentation velocity and/or Mg^{2+}-shift experiments is needed in each new system because of detectable activity observed in other fractions in the castor bean system (347).

Based both on morphological observations that the ER appears to function as a growth center for other membranes (222, 227) and on precedent from mammalian systems (324), it has been anticipated that plant ER would play a major role in membrane phospholipid synthesis. Direct evidence that this is indeed the case for plants derives from the combined use of in vivo pulse-labeling with phospholipid precursors (especially ^{14}C-choline) and in vitro enzymatic determinations in isolated fractions (145, 150, 191, 211, 220, 229). Using this approach, Beevers and coworkers have provided strong evidence that several of the enzymes responsible for the terminal steps in the synthesis of the major membrane phospholipids are exclusively localized in the ER of castor bean endosperm (150, 191, 215; see 16 for detailed discussion). In particular, the terminal enzyme in phosphatidylcholine synthesis, CDP-choline: diglyceride transferase has been shown to have a distribution profile on sucrose gradients precisely coincident with that of NADPH-cytochrome *c* reductase and to respond in concert with the reductase in Mg^{2+}-density shift experiments (191). Combined with the accompanying microscopic evidence, there is little doubt that both these activities are essentially absolute markers of the ER in this system.

There is evidence, however, that the phospholipid-synthesizing enzymes in other plant systems are not exclusively ER-localized. Montague & Ray (211) have observed that about 25% of the total CDP-choline: diglyceride transferase from peas coincides with Golgi IDPase activity and does not change in density in Mg^{2+}-shift experiments when the remainder of the transferase and all of the NADH-cytochrome *c* reductase do so in parallel. The second of the apparent bifunctional activities of this transferase (188), CDP-enthanolamine: diglyceride transferase, exhibits a similar dual distribution (211). Although the total activity of this enzyme is lower in the Golgi than in the ER, the specific activities are similar in both membrane fractions. It seems clear, therefore, that the terminal phospholipid-synthesizing enzymes cannot be used as unequivocal ER markers in all systems. The higher total activities in the ER indicate, however, that these enzymes can be used with reasonable utility as positive ER markers, especially when combined with the cytochrome *c* reductases under Mg^{2+}-shift conditions. The absence of significant amounts of Golgi in the castor bean endosperm may account for the distribution pattern observed in that tissue (16).

Nagahashi & Beevers (234), using Mg^{2+}-shift and coincidence with antimycin-insensitive NADH-cytochrome *c* reductase as criteria, have recently provided evidence that enzymes capable of transferring *N*-acetylglucosamine from UDP-*N*-acetylglucosamine to both lipid and protein acceptors are principally localized in the ER of developing pea cotyledons. Some activity is, however, also detected in an IDPase-containing fraction (235). It is suggested that these glycosyl transferases are functional in glycoprotein synthesis via a lipid-linked sugar intermediate. Similar, but less well controlled evidence, indicates an ER-localized enzyme capable of mannosyl transfer to dolichyl phosphate, a polyprenyl phosphate also considered to be an intermediate in glycoprotein synthesis (172). The above data are in contrast to other reports that have provided evidence for the presence of glycoprotein glycosyl transferases in Golgi fractions while being absent from the ER (92, 254).

Ray (267) has provided conclusive evidence that the membrane-associated sites that specifically and reversibly bind naphthaleneacetic acid (NAA) in homogenates of corn coleoptiles are predominantly localized on the ER. This activity might also serve as a useful positive ER marker, although its use as a negative marker is restricted by its apparent presence at low levels in non-ER fractions (10, 267). A preparative study comparing ER and nuclear membranes has drawn attention to several similarities in gross composition, lipid patterns, and content of certain enzymes (253). As indicated above, however, contamination of ER fractions by nuclear membranes bearing similar marker activities is likely to be quantitatively inconsequential.

Golgi

The unique ultrastructure of the dictyosome elements of the Golgi apparatus, both in negative stain and thin section, provides an unmistakable morphological marker for isolation of this organelle (35, 210, 220, 225, 226, 228). The structure is extremely fragile, however, being readily disrupted by such operations as mortar and pestle grinding and pelleting and resuspension (221, 225, 270). The pretreatment of tissue with cellulase (35) and the inclusion of glutaraldehyde (220, 221, 227, 269), bovine serum albumin (268), and/or coconut milk and dextran (220, 225) in the extraction medium have all been reported to preserve structural integrity. Glutaraldehyde inactivates associated enzyme activities, however, so that a compromise between structural and functional integrity is necessary (269). In the absence of stabilizing agents and mild extraction conditions, the dictyosomes tend to become unstacked into single cisternae and the peripheral secretory vesicles detach and become part of the morphologically unidentifiable pool of smooth membranes (225, 269). Available marker evidence suggests that the vesicles and single cisternae assume the same isopycnic density as the parent dictyosomes (269, 270). A range of peak densities falling mainly between 1.12 and 1.15 g cm^{-3} have been recorded (90, 214, 250, 269, 270, 299). Maximum separation from other identifiable organelles has been achieved with coupled rate-zonal and isopycnic equilibrium centrifugations (269, 270). The presence of unidentified smooth vesicles, however, has thus far precluded apparent purities much in excess of 50% dictyosomes as determined morphometrically (89, 223).

The development of suitable biochemical markers derives (*a*) from the empirical cytochemical observation that an inosinediphosphatase (IDPase) activity is localized in the Golgi (60, 226, 270), and (*b*) from the postulated dual secretory and membrane differentiation functions of the organelle (43, 222, 225, 228).

The cytochemical localization of IDPase in situ indicates that activity is associated with the Golgi but not exclusively so (60, 88, 89, 93). Fractionation studies indicate likewise [see (89) for critical evaluation]. Where recorded, substantial amounts of activity (up to 75–80%) are present in nonparticulate form (90, 122, 127, 214, 226, 299). The particulate portion of the IDPase is in most cases associated with a relatively discretely banding fraction that has a distribution different from mitochondrial and ER markers (90, 127, 144, 172, 214, 270, 299). A reported correlation between the distribution of IDPase and NAD(P)H-cytochrome *c* reductase (28) is unconvincing, and the data presented do not exclude the possible secretory vesicle localization of the IDPase measured. The absence of detectable IDPase in barley aleurone, a tissue with very low Golgi content (186), is consistent with a Golgi localization but not with an ER localization of this activity. Nevertheless, the existence of high levels of soluble activity precludes the conclusion that total IDPase is uniquely Golgi localized. It has been suggested that this soluble activity might represent nonspecific phosphatases (127) and there is some evidence consistent with this idea.

The initial fractionation study implicating IDPase as a Golgi enzyme reported that the activity was "latent," developing to a maximum activity over several days at 4°C or activatable by detergent treatment (270). Morré et al (226) have more recently shown that despite the high level of soluble IDPase the latent *increment* in activity (the activity after 4 days storage minus the activity at extraction) is totally particulate, being about equally distributed between 10,000 and 20,000 X *g* pellets. The incremental activity was confined to fractions in the density range 1.10–1.13 g cm^{-3} upon step gradient centrifugation. These fractions contained morphologically recognizable dictyosomes, but no quantitative morphometric-enzymatic distribution correlation was reported. In vitro cytochemically detectable IDPase was associated in these fractions both with identifiable dictyosomes and with vesicles. The latter were interpreted to be Golgi derived, implying an exclusive Golgi location for the IDPase, but the data presented do not alone unequivocally support this contention. Other evidence discussed below is consistent with the notion that the particulate, "latent increment" in IDPase activity can function as an exclusive Golgi marker. A direct quantitative correlation between the distribution of this activity and morphologically identifiable organelles has been reported (111). It should be noted, however, that in one report no latency of the IDPase present was detected (234).

The apparent function of the Golgi in the synthesis, transport, and secretion of cell wall precursors and other export macromolecules, initially suggested by histochemical and autoradiographic studies (for review see 43), predicts that polysaccharide synthesizing activities will be associated with the organelle and could therefore act as markers. Further evidence indicates that the Golgi are the primary locus of cell wall *matrix* polysaccharide (pectins, hemicelluloses) and glycoprotein (extensin) synthesis in vivo, whereas cellulose synthesis appears to occur principally at the cell surface and is presumably a plasma membrane (PM)-localized function (43, 92, 269, 275, 325). This apparent functional specialization suggests that Golgi and PM membranes should be distinguishable on the basis of the products of the polysaccharide synthesizing activities they are expected to bear.

In practice, several glycosyltransferase activities have been detected in Golgi-rich fractions (29, 43, 92, 172, 254, 269, 270, 325) as well as in putative PM fractions (325) and ER (172, 234). Most attention has been focused on the β-glucan synthetases present (268). Two β-glucan synthetase activities differing in their responsiveness to Mg^{2+}, UDP-glucose concentration, products formed and subcellular distribution have been reported (267–269, 325). The activity expressed at low (μM) UDPG/high (10 mM)Mg^{2+} concentrations (glucan synthetase I) forms primarily, if not entirely, β-1,4 linkages and appears to be Golgi associated (269). A second activity expressed at high (mM)UDPG/low (none) Mg^{2+} concentrations (glucan synthetase II) forms predominantly β-1,3 linkages, and is thought to be localized primarily on the PM (111, 127, 128, 325). It is held that these dual activities represent separate enzymes with low (glucan synthetase I) and high (glucan synthetase II) K_ms for UDPG, and that the latter enzyme is activated by high UDPG concentrations (271). Ray (268) has recently investigated these activities in some detail.

Two main types of evidence indicate that glucan synthetase I is localized principally in the Golgi. First, there are several reports that fractions enriched in glucan synthetase I are also qualitatively enriched for morphologically identifiable dictyosomes (28, 35, 225, 269, 270). Morphometrically determined quantitative correlations with the distribution of the enzyme appear to have been rarely documented, however (111, 325). Second, the distribution of glucan synthetase I coincides precisely, on both rate-zonal and isopycnic gradient centrifugation, with the membrane associated, radioactively labeled, matrix polysaccharides that are formed during a prior, brief (8 min) pulse in vivo with ^{14}C-glucose (269). There is comprehensive in vivo and in vitro autoradiographic and cell fractionation evidence that these in vivo labeled polysaccharide products are predominantly localized in the dictyosomes and Golgi-derived secretory vesicles under these conditions, and as such serve as markers for these subcellular components (43, 249, 250, 269, 275). It should be stressed, however, that despite this coincidence in distribution, the products of the glucan synthetase I assayed in vitro indicate that this enzyme is not primarily responsible for the major incorporation products detected in the Golgi following labeling in vivo.

There is evidence derived from in vivo pulse-chase labeling and fractionation experiments that the dictyosome elements can be separated from the secretory vesicles by the former's more rapid sedimentation on rate-zonal centrifugation (269). Both components band ultimately at the same isopycnic density (1.14 g cm^{-3}), however. A higher ratio of glucan synthetase I to labeled product is observed in the dictyosome than in the vesicle fraction, and this has been interpreted to indicate that the enzyme is confined to the cisternal region of the Golgi and is not transported to the PM via the secretory vesicle membrane (269). This conclusion, however, is not unequivocally supported by the data. It is unclear whether there are equal amounts of membrane in the two fractions or whether the different ratios simply reflect the lower surface to lumenal volume ratio expected of the vesicles relative to the flattened cisternae of the dictyosomes. Glucan synthetase I cannot, therefore, necessarily be dismissed as a marker for the secretory vesicles as well as for dictyosomes. Earlier studies interpreted to indicate that both short-term, in vivo pulse-label products (30) and glucan synthetase activity (299) are ER-associated did not preclude the possibility that the activities measured were in fact Golgi localized. There is now convincing evidence that neither the in vivo products (269) nor glucan synthetase I (114, 144, 267) are ER localized.

The function of this enzyme in vivo is a matter of debate (269–271, 275, 299, 325). It seems clear that glucan synthetase I is not responsible for cellulose synthesis in the dictyosome in vivo despite its capacity to form almost entirely β-1,4 glucans in the in vitro assay (269), since cellulose synthesis apparently occurs at the cell surface (43). It has been suggested that the enzyme might either participate in xyloglucan synthesis in the Golgi (269, 325) or represent a proenzyme form of the genuine in vivo cellulose synthetase en route to the cell surface where it becomes fully activated upon incorporation into the PM (35, 43, 325).

Regardless of its function, however, glucan synthetase I would appear to provide a reasonable positive Golgi marker. Moreover, the reported high correlation in the distributions of glucan synthetase I and latent IDPase (211, 270) provides mutual reinforcement for the notion that these activities both reside principally in the Golgi. Glucan synthetase I should nevertheless be used with care, given the ill-defined boundary between its activity and that of the putative PM-localized activity detected at high UDPG concentrations (267, 268, 325).

Data analogous to those obtained for polysaccharide synthesis indicate that the Golgi are involved in the synthesis and secretion of the cell wall gylcoprotein extensin (43). Gardiner & Chrispeels (92) have shown that in vivo labeled, hydroxyproline-rich, glycoprotein precursors are localized in a membrane fraction rich in morphologically identifiable dictyosomes. The enzyme UDP-arabinose arabinosyl transferase, shown to be involved in the glycosylation of the hydroxyproline-containing protein, is likewise almost exclusively localized in this fraction with a distribution profile that coincides on both rate-zonal and isopycnic gradients with both the in vivo labeled glycoprotein products and with IDPase. These data are interpreted to indicate that extensin is glycosylated in the Golgi in transit to the cell wall and that the enzyme responsible can serve as a marker for this organelle.

The second of the postulated dual functions of the Golgi apparatus—namely, a locus of dynamic, steady-state membrane differentiation where ER-like membranes are progressively transformed into PM-like membranes by insertion or modification of constituents during migration across the dictyosome stack (29, 35, 222, 227, 228)—predicts that the enzymes responsible for such membrane modifications should be localized in this organelle. Relatively few studies have directly addressed this question to date. Glucosylation of sterols and polyprenolphosphate has recently been reported to occur in a particulate fraction on sucrose gradients with a peak coincident with IDPase activity (29, 183). The data presented are insufficient alone, however, to conclude that this glucosylating activity is confined to the Golgi, and other locations in different systems have been reported (114, 325).

A major aspect of the proposed dual membrane modification and secretory functions of the Golgi is the implied physical transfer of membrane constituents from ER to Golgi to PM. Inherent in this membrane flow concept (89, 222, 227) is the prediction that the molecular constituents of the Golgi membrane will overlap significantly with those of both presumptive parent (ER) and derived (PM) membranes. Morré and colleagues have assembled considerable evidence consistent with this view (89, 222, 227). At the same time, however, it is clear that the molecular differences that underlie the obvious structural differences between these components have the potential to provide unique biochemical markers.

Plasma Membrane (PM)

Leonard & Hodges (179) have reviewed this topic. Three conceptually independent approaches to establishing markers for the plant plasma membrane (PM) are discernable in the literature: 1. Those based empirically on the use of the cytochemical phosphotungstate-chromate (PTA-CrO_3) stain, demonstrably selective for the PM in situ under defined conditions (278). 2. Those based on established or anticipated cell surface functions deduced from physiological experiments and/or analogy with animal systems, including (*a*) ion transport (127), (*b*) cell wall synthesizing activities (325), and (*c*) hormone binding or transport (124). 3. Those based on attempts to impose a nonpenetrant, surface-specific label on the outer surface of the PM of intact cells to facilitate subsequent identification in homogenates (90, 263). Various combinations of these approaches have been used in several studies.

PTA-CrO_3 STAIN Roland and coworkers (278) first reported that PTA-CrO_3 would selectively stain the PM of plants in situ. Many studies have extrapolated this observation to the identification of presumptive PM-derived vesicles in fractionated tissue homogenates (111, 128, 158, 176, 299, 311, 325, 341). This extrapolation has been controversial. The reliability of the stain for unequivocally identifying PM vesicles in isolated membrane fractions without the benefit of the cell architecture has been challenged on two levels: First, the specificity per se has been questioned on the basis that cellular components other than PM have been found to become stained in situ. These components include other membranous structures such as the tonoplast (106, 314, 317), chloroplasts (314), ER (317) and prolamellar bodies (265), and nonmembranous elements such as lipid droplets and ribonucleoprotein material from ribosomes (265) that have a recognized propensity to become redistributed and adsorbed to membranous components upon homogenization. Any of these components have the potential to generate PTA-CrO_3-positive profiles in cell-free homogenates and thus be falsely scored as PM. Second, it has been argued that there is no a priori reason to assume that the basis for the specificity toward the PM displayed in situ under critically exacting staining conditions will be preserved unaltered in homogenates (119, 144, 263, 265).

Recently, Nagahashi et al (233) have competently addressed both of these issues. Concerning selectivity per se in situ, these authors strongly emphasize the need for strict adherence to the experimental protocol in conducting the staining procedure (182) and identify critical operations in the procedure that have the potential to generate nonselective staining of tonoplast and other membranes. Because of this potential for technically generated lack of selectivity, the authors further indicate that the routine inclusion of sections from fixed, intact tissue for processing through the staining procedure together with the isolated particulate fractions is an obvious minimum and mandatory control. While clearly not guaranteeing selective staining of the PM, this control ensures that one recognized cause of nonselective staining can be identified and eliminated. Nagahashi et al (233) argue further that the morphology and density of those elements (such as prolamellar body fragments, ribonucleoprotein material, and lipid bodies) that do appear to stain when other non-PM membranes do not (265) are sufficiently different to preclude their contributing significantly to the PTA-CrO_3-positive profiles in the fractions designated as PM-rich. False positives resulting from surface adsorption of nonmembranous materials do, however, still clearly pose a potential problem.

Nagahashi et al (233) acknowledge that the fundamental assumption that the PM retains its specific staining properties following homogenization and fractionation has not been verified unequivocally. However, they argue that this assumption appears to be reasonable as PTA-CrO_3-positive vesicles are observed in the same isolated fractions, adjacent to morphologically identifiable organelles such as mitochondria, Golgi, nuclei, and plastids that stain neither before nor after fractionation. While this internal control does not constitute direct evidence *for* the retention of staining properties by the PM itself, it is clearly evidence *against* any general, homogenization-induced collapse of staining specificity. It indicates additionally that fragmentation of the major organelles is unlikely to contribute to the pool of PTA-CrO_3–positive profiles. The stain's selectivity in situ for the PM relative to the majority of other cellular membranes has now been documented for a variety of different plant tissues (278; see 233 for references). Taylor & Hall, on the other hand (314), reported that silicotungstate-CrO_3 (an analog of PTA-CrO_3) fails completely to stain the PM in freshly isolated protoplasts, suggesting that the stain may not be useful in this case. The PM of aged protoplasts regain the capacity to stain, perhaps explaining the apparent contradiction between this (314) and an earlier (205) report.

Thus the specificity of the PTA-CrO_3 stain for the PM is clearly not absolute, even when optimized (182, 233, 265). It would appear, however, that provided the specified controls for identifying and minimizing recognized technical sources of nonselective staining are coupled with the rational use of other morphological and fractionation criteria, the stain can provide tentative identification of PM vesicles in isolated subcellular fractions. An additional factor favoring this view is the emerging evidence that the tonoplast, perhaps the major, potential PTA-CrO_3–positive contaminant, has a buoyant density of ~1.10 g cm^{-3} (see below) whereas the bulk of the detectable PTA-CrO_3–positive vesicles appear to be distributed over the range 1.13 to 1.18 g cm^{-3} (111, 179, 265).

Based on acceptance of the qualitative selectivity of the stain, morphometric procedures are used to quantitate the proportion of PTA-CrO_3–positive vesicles in each subcellular fraction. The need for adequate sampling and scoring procedures has been emphasized (11, 22, 169, 325). It should also be stressed that the primary data generated by morphometric analysis of isolated fractions is equivalent to a *specific* activity (score of PTA-CrO_3–positive profiles per total score of membrane profiles). Absolute distribution between the various isolated fractions can be estimated by multiplying the proportion of PTA-CrO_3–positive vesicles in each fraction by the total number of milligram protein or phospholipid in that fraction. A complete balance sheet of this nature has only recently been provided (223). Most authors report only the "specific content" of presumptive PM vesicles in each separate fraction (111, 158, 182, 233, 265, 325, 341). On this basis, maximal purities of 75–80% presumptive PM in isolated fractions have been reported (128, 179, 233). Attainment of higher purities has generally been considered difficult because of unavoidable cross-contamination from the pool of other unidentifiable smooth-surfaced vesicles of similar size and density (179). Compositional analysis of these (128) or yet less well-defined fractions (113, 170, 301) are clearly of questionable value.

K^+-ATPase Based on the rationale that energy-dependent ion uptake in plant roots is likely to be driven by a PM-localized ATPase, Hodges and coworkers have sought to determine whether such an activity can be identified and localized in homogenates by cell fractionation procedures (127, 128, 178). Tissue homogenates contain a substantial number of ATP-hydrolyzing activities that are distinguishable on the basis of their subcellular distribution and response to varying pH, cation concentration, and inhibitors (177). A particulate ATPase that is activated by Mg^{2+}, further stimulated by K^+, and has a pH optimum of 6.5 was initially resolved in a microsomal fraction from oat roots and found to be localized in a membrane fraction having a mean isopycnic buoyant density in the range 1.14–1.20 g cm^{-3} (128, 177). The enzyme is not inhibited by oligomycin, in contrast to mitochondrial activities (8, 180), but is sensitive to DCCD, diethylstilbestrol, and octylguanidine (8, 95, 180).

Three main lines of evidence have been offered in support of the argument that this activity is localized on the plasma membrane and is responsible for cation transport in vivo. First, the kinetic behavior of the K^+-stimulated increment of the fractionated ATPase is very similar to that for K^+ influx into intact roots when the two are compared in Eadie-Hofstee plots (178, 180). Second, cytochemical staining indicates the preferential association of ion-stimulated ATPase with the PM in situ (342). Third, the distribution of K^+-stimulated ATPase activity is correlated with that of PTA-CrO_3–positive vesicles in fractions prepared by differential or step-gradient centrifugation (128, 182, 233). Evaluation of these and other related data requires an awareness of the distinction between the following measured and derived activities: (*a*) basal ATPase (measured without added Mg^{2+} or K^+), (*b*) Mg^{2+}-ATPase (exogenous Mg^{2+} only), (*c*) ($Mg^{2+} + K^+$)-ATPase (exogenous Mg^{2+} and K^+), and (*d*) K^+-ATPase (the K^+-stimulated *increment* in activity derived by subtracting the Mg^{2+}-ATPase activity from the ($Mg^{2+} + K^+$)-ATPase activity.)

Although suggestive, the kinetic correlation of the K^+-ATPase activity in vitro with its presumptive in vivo function (178, 180) does not unequivocally establish that the enzyme is *uniquely* localized on the PM. The association of cytochemically detectable ATPase with the PM in situ has been established for some time (18, 104–106). However, apart from the technical difficulties associated with the procedure (83, 104), general ATPase activity is simultaneously observed in several cellular membranes, and few attempts have been made to determine whether the characteristics of the in situ PM-localized activity match those unique to the K^+-ATPase in isolated fractions. Those data that are available are inconsistent with one another. Winter-Sluiter et al (342), using barley roots, have provided what appears to be the only attempt to obtain cytochemical evidence of a K^+-stimulated increment in ATPase activity in situ preferentially localized in the PM. Even had these data been quantitated, the difficulty remains that the ATPase of barley root membrane fractions, in contrast to oats and corn, is hardly stimulated by K^+(233). Moreover, while oligomycin does not inhibit the K^+-ATPase of isolated corn membrane fractions (180), this inhibitor completely suppresses cytochemically determined, in situ PM-associated ATPase in this species (196).

Correlations in the distribution of PTA-CrO_3–positive vesicles and K^+-ATPase activity have been reported and interpreted as strong, direct evidence that this enzyme is PM-localized (128, 182). This conclusion is clearly dependent on the reliability of the PTA-CrO_3 stain for identifying PM vesicles. A further qualification is that these quantitative correlations are based on data from relatively crude differential and step gradient fractionations. The resolving power of these procedures is poor, possibly accounting for the observation that both K^+-ATPase and (Mg^{2+} + K^+)-ATPase in the same fractionation experiment correlate individually with PTA-CrO_3–positive material while not always correlating well with one another distributionally (182). Moreover, the correlations observed between PTA-CrO_3–positive vesicles and K^+-ATPase do not extrapolate through zero, suggesting that the ATPase is not uniquely associated with such vesicles (182).

Information obtained from further fractionation studies has likewise not provided a consensus view that K^+-ATPase can serve as a universal and unequivocal marker of the PM. Apart from the complication posed by the presence of a large number of ATP-hydrolyzing activities in homogenates, including nonspecific phosphatases, the distribution of K^+-stimulated activity per se has been the subject of some controversy. A degree of confusion has been generated by the fact that at least three different activities—the originally described K^+-ATPase (3, 55, 119, 120, 128, 178), a (K^+ + Mg^{2+})-ATPase (10, 181, 182, 233), and even Mg^{2+}-ATPase (27, 90, 154)—have all been interpreted by different authors to represent the PM. While in some cases the distribution of K^+-ATPase and (Mg^{2+} + K^+)-ATPase activities between fractions appears to be reasonably well correlated (128, 234), in others it is not (182), raising the question as to which if either is PM localized. In addition, the degree of K^+ stimulation above the Mg^{2+}-activated level is not constant between species, ranging from a mere 8% for barley (233), 30–50% for corn and peas (119, 181, 234), to two- to threefold for oats (178).

The association of putatively PM-localized ATPase activities [K^+-ATPase and (Mg^{2+} + K^+)-ATPase] with multiple membrane fractions differing in sedimentation and density properties have been recorded by several authors (55, 119, 175, 177, 182, 234). In particular, the K^+-ATPase that pellets with the initial mitochondrial fraction ($\leqslant 13{,}000 \times g$) from corn has a substantially higher density upon subsequent sucrose gradient centrifugation than the membrane-associated K^+-ATPase in the postmitochondrial supernatant (119, 120). Much of the published literature purporting to demonstrate that this enzyme is PM-localized has utilized only the subpopulation in the postmitochondrial supernatant [$<$ 50% of the total activity (119, 120, 182, 234)] for further fractionation and kinetic analysis (8, 127, 128, 178, 180). While these results might reflect the expected heterogeneous fragmentation of the PM during homogenization and the data obtained may be representative of the whole membrane, the possibility that the activities do indeed have either a multiple membrane location or are rather domain markers of the PM cannot be excluded.

It also seems likely that the assay normally utilized measures only the ATPase on inside-out or possibly leaky PM vesicles because exogenously supplied ATP would not be expected to penetrate the intact membrane. The marked enhancement of activity in the presence of Triton X-100 (57, 95) suggests that this is indeed the case. The possibility that the subcellular distributions of putative PM ATPases thus far determined are those of an unrepresentative subpopulation of vesicles is yet to be precluded. Taken together, all of the above considerations would suggest that ATPase activity has not been established as an unequivocal marker of the PM and should be used with caution in conjunction with other markers.

GLUCAN SYNTHETASE II Substantial evidence from in situ autoradiographic and other studies that cellulose synthesis occurs directly at the cell surface (3, 35, 43, 269) has led to the expectation that isolated PM should contain a demonstrable cellulose synthetase activity. Despite much effort this expectation is yet to be realized. The glucan synthetases thus far detected in particulate fractions are clearly not responsible for cellulose synthesis in vivo (3, 116, 267, 269, 271, 275, 325). As discussed above, the apparently Golgi-localized glucan synthetase I may function in xyloglucan synthesis (269), whereas the glucan synthetase II activity to be discussed here appears to form predominantly β-1,3 linkages, suggesting that its primary function is callose synthesis (3, 268, 271, 325). The presence of such an activity at the cell surface is not entirely unexpected, as callose synthesis continues in sieve tube cells after most of the internal organelles have been degraded (35).

Glucan synthetase II activity is expressed under high (mM) UDPG/low (none) Mg^{2+} conditions (111, 267, 271, 325). Evidence suggesting that this activity is primarily localized on the PM derives mainly from the correlated distributions of the activity and PTA-CrO_3–positive vesicles observed upon differential and sucrose gradient centrifugation (111, 325). Consistent with these data are reports that UDPG supplied to tissue segments leads to extracellular β-1,3 glucan formation (3, 43, 116, 271). Anderson & Ray (3) present convincing indirect evidence that the in vivo glucan synthesis in their system occurs at the external surface of intact cells rather than on internal membranes or the inner surface of the PM of damaged cells at the excision surface, although Raymond et al (271) favor the opposite interpretation in similar experiments.

The question of whether the enzyme responsible for the observed in vivo callose formation is identical with the glucan synthetase II activity assayed in vitro awaits direct experimental verification. Without this independent evidence, the conclusion that glucan synthetase II is preferentially PM-localized relies heavily on the validity of the PTA-CrO_3 strain. Moreover, the ambiguities posed by the inherent overlap in the activities of glucan synthetases I and II dictate that neither can serve as an absolute marker for their respective membranes (267, 268). This limitation is clearly more serious for use of the enzymes as negative markers than as positive ones. A further consideration is that, as for the K^+-ATPase, the sidedness of the membrane vesicles may determine the apparent distribution of the activity detected. Only those vesicles with the enzyme oriented to the outside (outside-out?) (3) would be expected to be measured in the assay as it is performed.

The few attempts that have been made to correlate directly the distribution of glucan synthetase II with K^+-ATPase or $(Mg^{2+} + K^+)$-ATPase have met with varied success. Correlations ranging from reasonable (27, 127) to poor (10, 28, 55, 123) have been reported. The most thorough investigation is that of Hendriks (120, 121). He reports reasonably good distributional coincidence on sucrose gradients of the two activities previously sedimented at 10,000 X *g* but significant differences for a 10,000–153,000 X *g* pellet. The latter fraction is analogous to that used in many studies of putative PM-localized activities (127, 128, 180, 325). Considerations of sidedness might suggest that the two assays detect differently oriented vesicles—the glucan synthetase II measuring inside-in and the K^+-ATPase measuring inside-out vesicles. Regardless of the cause, however, the general inconsistencies apparent in the PTA-CrO_3-K^+-ATPase-glucan synthetase II triangle emphasize the need for caution in the use of these markers.

NAPHTHYLPHTHALAMIC ACID (NPA) BINDING Reasoning that a probable target site for the physiologically observed inhibition of polar auxin transport by NPA might be the PM, Hertel and coworkers sought and found a particulate fraction that specifically binds the inhibitor (125, 176, 318). Fractionation on step gradients provided a correlation between the binding activity and the content of PTA-CrO_3–positive vesicles in the fractions, a result interpreted as support for the initial proposal. Based on this evidence, several studies have employed NPA-binding as a marker for the PM (69, 70, 113, 114, 125, 136, 267). Independently derived, direct evidence that the binding activity is primarily PM-localized is lacking. The validity of NPA-binding as a PM marker is strictly dependent, therefore, on the validity of the PTA-CrO_3 stain. Reports of correlated distributions of NPA-binding and glucan synthetase II (113, 114, 144, 267) would be expected since both are themselves primary correlates of PTA-CrO_3–positive vesicles (176, 325). Nevertheless, there is evidence that these two activities are not completely coincident (124, 317; D. Marmé, personal communication), suggesting that at least not all molecules of both reside on a common particle. Poor correlations between the distributions of NPA-binding and K^+-ATPase have also been reported (54, 317).

There is no documented study in which all four of the above most commonly used nominal PM markers—PTA-CrO_3, K^+-ATPase, glucan synthetase II, and NPA-binding—have been quantitatively compared in a single fractionation. The most generous interpretation of the discrepancies that do exist is that the PTA-CrO_3 is a "universal" marker capable of detecting all PM vesicles equally, whereas one or more of the other three are domain markers leading to internally noncoincident distributions upon heterogeneous PM fragmentation. This effect would become increasingly more pronounced with decreasing vesicle size. If this suggestion were true, all three biochemical markers should fall within the broader PTA-CrO_3 distribution "envelope." It would then presumably have to be argued that the original separate correlations observed between each individual activity and PTA-CrO_3–positive vesicles lacked the resolving power needed to distinguish these internal differences. Meanwhile, in the absence of direct documentation to this effect, the reliability of the morphometric-biochemical correlations obtained by the procedures used must remain open to question. The widespread practice of preparing crude step gradients and expressing the activities in each fraction as a specific activity in the absence of information permitting calculation of absolute distributions (41, 111, 113, 114, 127, 128, 226, 240) is particularly undesirable. This practice greatly increases the probability that *enrichment* will be confused with *distribution*. Examples of this confusion are not infrequent in the literature.

SURFACE LABELING There are now several reports of attempts to label selectively the outer surface of the PM. Both plant tissue segments (55, 107, 118, 263, 348) and protoplasts (27, 53, 90, 251, 314) have been used. Because all these labels have the potential to react indiscriminantly with any protein (or glycoprotein) to which they have access, their selectivity for the cell surface depends, in principle, on their inability (physically or kinetically) to penetrate the intact PM.

In practice there are two further considerations. First, it can be calculated that the protein exposed on the external face of the plant PM constitutes no more than 2–3% of the total cell protein (65; D. J. Morré, personal communication). Thus, penetration and labeling of the cytoplasmic contents of only a very small number of ruptured or leaky cells in the population can reduce or eliminate any measurable preferential labeling of the PM. The potential exists for both direct contamination of PM fractions by other labeled cytoplasmic organelles or nonspecific adsorption of highly labeled cytosolic proteins to membrane fractions. Intense internal labeling of this kind has been observed by autoradiography to occur in small numbers of individual cells in populations of mammalian cells (148) and cell wall-less *Chlamydomonas* (P. H. Quail, T. Hendricks and A. Browning, unpublished). In the absence of autoradiographic data, preferential labeling of a discrete membrane fraction relative to soluble proteins and identifiable cytoplasmic organelles is generally interpreted as evidence *against* significant, nonselective labeling of the intracellular components of damaged or leaky cells and *for* the labeling of the PM (40, 148). Several studies purporting to demonstrate successful selective PM labeling have cited biochemical evidence of this nature (27, 90, 118, 348). As detailed below, however, such evidence is not always unequivocal.

The second practical consideration may apply uniquely to surface labeling studies with excised tissue segments. It has been observed autoradiographically that a nonplasma membrane, extracellular component (possibly phloem exudate protein), becomes heavily labeled upon lactoperoxidase-catalyzed iodination of excised tissue segments (263). Upon cell fractionation, this component behaves as a preferentially labeled, discrete, membranous fraction having a density of 1.15 g cm^{-3} and a distribution profile distinct from mitochondrial and ER markers. No significant labeling of the PM could be detected. The membrane-like behavior of the iodinated material upon fractionation is attributed to the posthomogenization association of this material preferentially with a particular membrane fraction.

Both these potential artifacts indicate the need to verify directly, preferably with autoradiographic or cytochemical evidence, that putatively selective surface labels do in fact preferentially label the PM in the system under investigation. Failure to provide such evidence has thus far led to two demonstrably premature claims of having identified PM in tissue homogenates by surface labeling (118, 348). Lack of preferential surface labeling due to reagent penetration has been reported for fluorescamine (55) pyridoxal phosphate-NaB^3H_4 (107; P. H. Quail, unpublished) with tissue segments, and for acetic anhydride (D. H. Cribbs and D. A. Stelzig, personal communication) and certain other reagents (90) with protoplasts. A report claiming to demonstrate selective surface labeling with pyridoxal phosphate-NaB^3H_4 provides no direct verification of the asserted lack of penetration by the probe (311). The preferential labeling with ^{35}S-sulfanilic acid (90) and ^{14}C-concanavalin A (27) of a discrete protoplast-derived membrane fraction has been reported. In each case the fraction was separable from other organelles such as mitochondria and ER and coincident with other putative PM markers. No autoradiographic evidence of surface selectivity was provided in either study.

In a combined histochemical and fractionation study, Taylor & Hall (314) recently reported that $LaCl_3$ binds apparently irreversibly to the external surface of intact protoplasts with no detectable penetration and binding to internal organelles. Fractionation of prelabeled protoplasts yielded lanthanum stained membrane fragments in isolated fractions in the absence of any detectable staining of identifiable cytoplasmic organelles. This is evidence against indiscriminate redistribution of label during homogenization. While subject to the same basic limitations regarding surface selectivity as other labels, $LaCl_3$ appears to offer the potential of a morphologically identifiable probe that can be monitored relatively easily in both the labeled intact cell and the fractions derived directly from the same material. Ferritin- or hemocyanin-labeled lectins offer similar apparent potential (27, 164, 340) but do not appear to have been successfully exploited for fractionation studies thus far.

In an innovative approach to surface labeling, Anderson & Ray (3) have recently exploited an intrinsic, presumptively PM-localized, callose synthesizing activity to incorporate glucose, exogenously supplied to pea tissue segments as UDPG, into nascent callose polymers that are thought to remain attached to the PM upon subsequent homogenization. The authors offer controls and arguments that appear to eliminate the possible participation in the enzymatic process of the internal membranes of severed cells at the excision surface. The labeled membrane fraction has a broad density distribution of 1.16–1.20 g cm^{-3}, coincides with a K^+-ATPase activity, and has a distribution distinct from mitochondrial and ER markers. Unfortunately no direct comparison of the distribution of the membrane-bound, in vivo labeled reaction product with that of the in vitro assayable glucan synthetase II activity has yet been published. The successful use of polylysine-coated beads for erythrocyte, HeLa cell, and *Dictyostelium* PM isolation (49, 138, 153) and of high lectin concentration for the isolation of PM ghosts from *Neurospora* and yeast protoplasts (32, 78, 287) suggests that these techniques have significant potential for the isolation of PM from plant protoplasts.

A variety of experiments suggests the presence of a toxin-binding protein on the external surface of the PM of susceptible sugar cane cells (310, 311). Antisera against previously purified toxin-binding protein (*a*) provides apparently complete protection when administered to tissue segments against subsequently administered host-specific toxin, and (*b*) causes agglutination of isolated protoplasts. In addition, exogenously presented purified binding protein confers toxin susceptibility upon isolated protoplasts from resistant sugar cane or tobacco. Fractionation studies show that both toxin-binding (311, 317) and antigenic (311) activities are high in membrane fractions in the density range 1.10–1.17 g cm^{-3}. Only one of these studies reports the use of quantitative markers (317), and the data are insufficient to conclusively establish any correlations with the putative PM-localized activities tested. Although none of the data thus far reported unequivocally establish that the toxin-binding protein is *exclusively* PM-localized, this possibility is clearly worthy of further investigation. Elicitors from the cell wall of the pathogen *Phytophthora* have also been reported to cause protoplast agglutination at moderate but not low or high concentrations, indicating that these too have binding sites at the outer surface of the PM (252).

From the above discussion it seems clear that there is no single, rigorously documented, absolute marker for the plant PM. On the other hand, the bulk of the accumulated information gleaned from the use of different approaches is, broadly speaking, consistent. It would seem, therefore, that the PTA-CrO_3 stain, when used cautiously (233) in combination with K^+-ATPase (120, 128), glucan synthetase II (120, 267, 268, 325), or NPA binding (176, 267), or ideally all three, can provide a positive marker package of some utility for tentative PM identification in membrane fractions. Surface labeling with UDPG, coupled with adequate reaction product analysis (3), should provide a useful new tool in the quest for a definitive PM marker.

A further potentially powerful aid to the identification of PM-localized activities is the "digitonin shift." This is the increase in density observed for the high cholesterol-containing mammalian PM as a result of controlled binding of digitonin to the sterol (2, 65). If the plant PM is likewise high in sterols as claimed (113, 114, 128), a similar digitonin-induced density shift might be observed. A poorly documented attempt to characterize the PM starting with a cell wall preparation has been reported (9). This approach has intriguing potential, however. As much as half of the total PM surface area may be involved in lining the plasmodesmata of meristematic cells (274).

Tonoplast (Central Vacuole)

Matile has recently reviewed this topic, clearly documenting the dual function of vacuoles as storage and lytic organelles (203, 204). Boller & Kende (24) have further suggested a defense function against invading pathogens. The discussion here is restricted primarily to the large, central vacuole of mature cells. Although specialized storage organelles such as protein bodies are, strictly speaking, a class of vacuole (203), they are considered separately below.

The gentle tissue disruption procedures mandatory for the isolation of intact central vacuoles in substantial numbers have been achieved using automated, low-shear tissue slicing (173, 174) or protoplast preparation followed by controlled lysis brought about osmotically (24, 25, 39, 185, 283, 330), mechanically (238), or by poly-base treatment (38). The enormous size of the organelle provides an obvious morphological marker and aids separation by very low centrifugal forces [Buser & Matile (38) used unit gravity sedimentation]. Both naturally pigmented vacuoles (173, 330) and vacuole-accumulated, exogenously supplied neutral red (38, 238, 283) have also been employed to simplify identification and provide a readily measurable criterion of integrity.

Yields based on the starting tissue are rarely given but $< 0.5\%$ for both the mechanical procedure (173) and for a protoplast preparation (333) have been reported. Based on the starting number of protoplasts, vacuole yields of between 10–15% (24, 39, 330) and 42% (283) have been documented. The contents of the vacuolar lumen vary considerably with the precise metabolic function served by the organelle in different tissues (203, 204). This variability is reflected in the intrinsic densities reported for intact vacuoles isolated from different sources, ranging from < 1.03 g cm^{-3} (192, 333) to > 1.18 g cm^{-3} (238) in sucrose. This diversity in function and origin likewise tempers the search for biochemical markers common to all vacuoles.

The anticipated lytic function of the organelle has prompted most investigators to examine the hydrolase activities in their vacuolar preparations. Several reports of the presence of such activities in intact vacuoles have appeared, but comparative assessment of the data is complicated for several reasons. First, some authors provide quantitative estimates of the purity of their preparations using negative markers for other cellular organelles (24, 25, 76, 238), others do not (38, 39, 185, 192, 330). Second, many of the activities have been tested by only one laboratory, precluding cross-comparison. Third, the cell-wall degrading enzymes used for protoplast preparation have been shown to contain high levels of many of the putative vacuolar hydrolases, to the extent that some activities (especially proteinase) read substantially higher in the protoplast preparation than in the initial tissue (24, 25). Fourth, apparently conflicting reports regarding vacuolar localization of some specific enzymes (24, 25, 39, 174, 238) may simply reflect real differences in the functional roles of the vacuoles in the different source tissues.

Using vacuoles virtually uncontaminated by other cellular constituents, Nishimura & Beevers (238) have convincingly demonstrated the predominant, if not exclusive vacuolar localization of several hydrolases and of sucrose in castor bean endosperm. Boller & Kende (24) have provided similar evidence for cultured cells and flower petals, although for the most part with different hydrolases. These two studies agree that intracellular RNase, proteinase, and phosphodiesterase are primarily located in the vacuolar lumen (phosphodiesterase being also found in the cell wall). Immunocytochemical localization of RNase supports this view (13). An earlier report of RNase in a vacuolar preparation did not quantitatively document with negative markers the level of contamination potentially present from other organelles (39). There is strong complimentary autoradiographic (285) and fractionation (283) evidence for the exclusive location of the cyanogenic glucoside dhurrin in the vacuole of *Sorghum*. It also seems likely that proteinase inhibitor I (333) and malate (38) are confined to the vacuole of the tissues tested.

There is a lack of consensus as to whether the following activities are primarily vacuolar: acid phosphatase (24, 39, 238, 333); carboxypeptidase

(39, 238, 333); acid protease (24, 39, 238); ATPase (24, 174, 185); hexokinase (24, 94). A report that NADH-cytochrome *c* reductase might be vacuole-associated (173) should be examined with caution. Although this activity is a significant proportion of that in the initial 2000 X *g* pellet used for further vacuole purification, it may be calculated that the absolute activity in the final vacuole band is only 0.5% of the total particulate activity. This level is little different from that of cytochrome *c* oxidase (0.4% of the total particulate), which is considered by the authors to represent insignificant mitochondrial contamination. Other investigators have found insignificant levels of NAD(P)H-cytochrome *c* reductase in their vacuole preparations (24, 238).

All of the above discussion has referred to activities determined in intact vacuole preparations. With the exceptions to be discussed below, all of the activities demonstrated to be genuinely vacuolar have been found to be soluble upon vacuole lysis and membrane sedimentation. These activities are thus considered to be lumenal, and thereby potentially useful for assessments of vacuole yield and integrity. It is clear, however, that the inconsistency with which many of these activities have been found to be exclusively vacuolar dictates that caution be exercised in extrapolating the findings of any one study to a new system.

There are two reports of a tonoplast-located activity of potential marker value. The first is an ATPase (39). As no quantitation of negative markers was reported in this study, the possibility that this activity is attributable to a contaminant cannot be excluded. A similar DCCD-insensitive ATPase has been reported to be present in plastid envelopes (74, 75). Moreover, the higher activities observed by Butcher et al (39) in the vacuolar fraction than in the parent protoplasts suggest that the measurements are potentially unreliable. An initial report that phosphodiesterase might be mainly tonoplast associated (23) has not been verified by further investigation (24, 25).

There is at present, therefore, no unequivocal report of a unique tonoplast marker. Despite this lack, the major membrane band obtained following lysis of vacuole preparations is consistently observed in different systems by several investigators to assume an isopycnic density of about 1.10 g cm^{-3} on sucrose gradients (24, 25, 174; G. Wagner, personal communication). It seems probable that this band is predominantly tonoplast.

Specialized Storage Organelles

PROTEIN BODIES These organelles contain the reserve protein of seeds (often 50–70% of the total cellular protein) in a dense matrix enclosed by a single limiting membrane (see 5, 193, 209 for reviews). There are three morphologically identifiable types based on the kind of inclusion observed in the amorphous protein matrix (279): Type 1, no inclusions; Type 2, globoid inclusion only; Type 3, globoid and crystalloid inclusions. The mature protein body of dry seeds is by definition a transient organelle. It represents both (*a*) the end product of a phase of storage protein deposition in vacuoles (7) or distending ER cisternae (156) during development (for review see 209); and (*b*) the future locus of mobilization of these reserves during germination (for review see 5) via proteolysis within the lumen (14, 43). As germination proceeds, the matrix proteins disappear and the protein bodies enlarge and coalesce, eventually forming the tonoplast-bounded central vacuole (5, 112). Protein body fractions have been prepared from developing (37, 234), from mature, dry (112, 343, 346), and from germinating (45, 80, 112, 294) seeds. As a result of their high protein content, these organelles assume a relatively high buoyant density (1.26–1.30 g cm^{-3}) on sucrose gradients (234, 323).

Protein bodies, particularly those from oil seeds, are often unstable when conventional preparative procedures are used (141, 142, 323, 346). Several less conventional methods have evolved in attempts to maintain the integrity of the organelle during isolation. These include the use of nonaqueous media such as organic solvents (1, 321, 346) and glycerol (112, 141, 142, 323, 343) and the use of low pH (12, 321). The potential for denaturation and/or redistribution of cellular components through coprecipitation at pH 4.5 (12) or from acetone or hexane extraction (1, 321) is clearly of concern for localization studies.

Mature protein bodies isolated from developing or ungerminated seeds retain, to a greater or lesser extent, their amorphous, electron-dense matrix with (323) or without (37, 80) inclusions and are readily identified microscopically. Morphological identification later in germination may be less certain but does not appear to have been specifically documented. The degree of complete structural preservation achieved by the various preparative procedures is variable but often poor as judged by the proportion of unruptured limiting membranes (1, 80, 326).

Assessment of the purity of protein body preparations appears in the majority of cases to have been confined to qualitative, microscopic examination (80, 245, 321, 323, 326, 346) and/or measurements of soluble cystolic enzyme activities (112, 343). Direct, quantitative demonstrations of the absence of significant levels of contaminating cytoplasmic organelles or membranes are rare (142, 230, 294). Sheer quantitative considerations clearly render such a demonstration superfluous with respect to the question of storage protein localization in dry seeds at least. Since both the morphologically identifiable protein bodies (which in situ often occupy a large fraction of the total cellular volume and possess the bulk of the cytochemically detectable protein) and the major storage proteins (which often constitute 50–70% of the total extractable protein) are both almost exclusively localized in the same particulate fraction following extraction (43, 80, 112, 321, 323, 326, 343), there is little doubt that the storage proteins reside principally in the matrix of the protein bodies. Direct immunocytochemical evidence to this effect has been provided for vicilin and legumin in beans (97). Fractionation studies with hemp (307) and castor bean (323) provide similar direct evidence.

The potential presence of contaminating elements is, however, important to studies on the localization of nonstorage protein activities in protein bodies. The postulated autolytic capacity of the mature organelle (202), based on in situ ultrastructural observations that hydrolysis of the matrix appears to proceed within the intact, limiting membrane during germination (5), has led to several reports of the presence of proteolytic and other hydrolytic activities in protein body-rich fractions (1, 12, 45, 112, 141, 230, 244, 245, 343). Chrispeels and coworkers, however, in an elegant series of publications (12, 14, 45, 46, 112), have conclusively demonstrated that, in mung bean at least, the low level of acid protease activity detectable in the isolated mature protein body fraction is incapable of degrading the endogenous legumin and vicilin storage protein present. Mobilization of these proteins must await de novo synthesis of a vicilin endopeptidase in as yet unidentified cytoplasmic loci (shown immunofluorometrically), followed by presumptive transport into the protein bodies (14). At that time antigenic activity is detected in both the cytoplasmic foci and protein bodies consistent with the dual distribution of endopeptidase activity observed in fractionation studies (45). Thus mature protein bodies from ungerminated mung bean are not autolytic, whereas those from germinating seeds appear to have acquired this potential.

This inability of the proteases associated with the mature protein body fraction to hydrolyze the storage protein (45) might be consistent with a nonprotein body locale of these activities. The high density of the protein bodies might normally be considered adequate to preclude contamination from other organelles (230), but the unorthodox use of low pH and nonaqueous media in many studies leaves open the possibility that the presence of some of the reported activities (1, 244, 245) results from cross-aggregation and cosedimentation of other organelles with the protein bodies. Membranous material that may or may not have been derived from ruptured protein bodies is evident in some published micrographs (80, 326) and has been quantitated in one study (141). There is, on the other hand, convincing evidence from a combined histochemical and fractionation study that acid phosphatase is primarily localized in the protein bodies from cotton seed (343). Histochemically detectable acid phosphatase has also been observed on the surface of protein bodies in wheat seed extracts (141).

Subfractionation of oil seed protein bodies has established that phytin is principally located in the globoid inclusions (5) and the major storage globulins are located in the crystalloid inclusions (307, 323, 346). In addition, the matrix proteins in castor bean have been shown to constitute the water-soluble albumins of the protein bodies and to consist primarily of two proteins, the lectins ricin and phytohemaglutinin (323, 346). These lectins are therefore potential biochemical markers for the presence and integrity of protein bodies in this tissue. The castor bean studies (206, 323, 346) demonstrate, in addition, the potential fate of protein bodies from that source upon homogenization in aqueous media. Transfer of the nonaqueously isolated fraction to buffer produces four fractions on subsequent sucrose gradients: matrix lectins in the supernatant, the presumptive limiting membrane at 1.15 g cm^{-3} [hexane isolations (346)] or 1.22 g cm^{-3} [glycerol isolations (206)], the crystalloid at 1.30 g cm^{-3}, and the globoid at $>$ 1.46 g cm^{-3}. Clearly, detection of storage protein at a high density does not constitute evidence of the presence of intact protein bodies. Differences between the relative distributions of acid hydrolase activities observed for wheat aleurone fractionated in glycerol and in aqueous media have been attributed to protein body fragmentation in the aqueous media, leading to redistribution to other cellular fractions of these putatively protein body-bound activities (141).

Thus identifiable storage proteins provide an obvious marker for protein bodies with the proviso that organelle integrity be independently assessed and with the limitations that detection is nonenzymatic and the individual storage proteins are tissue-specific (209). In addition, storage protein, phytin, and lectins are markers of crystalloid, globoid, and matrix subcompartments respectively, where these have been documented (307, 323, 346). No marker for the limiting membrane appears to have been reported. The specific localization of certain hydrolases in mature protein bodies appears to be distinctly possible but, with the possible exception of acid phosphatase in cotton (343), rigorous evidence to this effect is yet to be presented.

LIPID BODIES These organelles, also called spherosomes or oleosomes (for review see 4; 303, 345), are the principle site of lipid storage in oil seeds (134, 218, 246, 306) but also occur in low amounts in other tissues (157, 345). They consist of a half-unit membrane (334, 344; see 157 for contrary view) surrounding the matrix triglycerides [oil seeds (344, 345)] or waxes [jojoba (218, 280)]. The spherical shape and the lipophilic (141, 280) and osmiophilic (345) staining capacities provide ready criteria for morphological identification that are retained unaltered in vitro. Isolation has generally involved centrifugation of conventional buffer homogenates and physical collection of the "fat pad" that forms on the surface of the aqueous phase

(134, 218, 232, 246, 344, 345). The density of the intact organelle has been estimated to be 0.92 g cm^{-3} (134). Spherosomes from wheat aleurone with unusual apparent densities of 1.06 and 1.16–1.18 g cm^{-3} have been reported (141). The yield of lipid bodies, where reported, is generally high, representing in excess of 90% of the original seed lipid (134, 246, 306). Purity of fractions has in most cases been assessed qualitatively from micrographs (134, 218, 246, 344), but morphometric (141), histochemical (345), and biochemical (306) procedures have also been employed.

Like protein bodies, the unique density, distinctive morphology, and coupled histochemical and biochemical lipid determinations leave no doubt that the storage lipid is essentially exclusively localized in the matrix of the spherosomes and is in effect a biochemical marker. The definitive localization of other, particularly enzymatic, activities in the lipid body requires quantitative deployment of negative markers. Structural integrity appears to be fully retained on isolation (134, 218, 246, 344, 345).

The proposed autolytic capacity of the organelle (141) has led several authors to seek evidence for the presence of lipolytic and other hydrolytic enzymes in isolated spherosome fractions (131, 132, 134, 232, 246, 345). No general pattern has emerged from these studies. The exclusive localization of an enzymatic activity in the lipid body fraction has only been rigorously documented for acid lipase in castor bean endosperm (232, 246). This activity is present in the organelles of the ungerminated seed and declines rapidly over the first 2 days from imbibition. An active alkaline lipase has been detected in the wax bodies of jojoba seedlings (132, 218). This activity is not present, however, in the ungerminated seed, but rather is acquired by the organelle during germination. The coincidence of about half the lipase activity with the ER in the germinating seed has been interpreted to represent the ghosts of ruptured or empty spherosomes (218). An alternative possibility is that the lipase is synthesized on the ER and transferred to the wax body. This alternative is reminiscent of the acquisition of autoproteolytic activity by the protein bodies of mung bean during germination (14).

There is evidence that the acid lipase of castor bean spherosomes is active at the surface membrane (243, 246), an observation consistent with the requirement for water for hydrolysis of the glyceride substrate. The alkaline lipase of the jojoba wax bodies has also been shown to be membrane localized (132, 218). This membrane has an isopycnic density of 1.12 g cm^{-3} on sucrose after extraction of the wax from the organelles with diethyl ether. Lipid bodies from dry rape seed have no lipase activity (316). During germination there is an increase in alkaline lipase activity, 90% of which is found to be associated with a membrane fraction of density 1.085 g cm^{-3}. Unquantitated morphological studies have been interpreted to suggest that the lipase is localized in "membrane appendages" that originate from the spherosomes as two half-unit membranes separated by a thick intermediate layer of storage lipid. In several other oil seeds tested, no lipase was found to be associated with the spherosomes either in the dry seed or during germination (131, 134). Likewise acid lipase from dry wheat aleurone was found not to reside primarily in the lipid bodies (141). Acid phosphatase has been shown to be absent from the spherosomes of several species (141, 345). It appears, therefore, that the presence of the acid lipase in ungerminated castor bean spherosomes may be the exception rather than the rule, and it is not yet clear whether even in this case the activity confers an autolytic capacity on the organelle (131, 232). Thus there is at present no general marker for the lipid body membrane.

Cytosol

Selection of an unequivocal cytosolic marker requires cognizance of the dual distribution of several of the glycolytic and pentose phosphate pathway enzymes (88, 231). Although the differentially compartmentalized activities often represent distinct isozymes (231, 286, 295, 296), the need for additional separations reduces their utility. Alcohol dehydrogenase is one readily assayable activity that apparently has not been reported to occur in particulate form in plants (286, 286a; J. G. Scandalios, personal communication).

ASSIGNMENT OF LOCATION

The markers discussed above have been used in numerous studies seeking to ascertain the subcellular localization of cellular constituents of previously undetermined distribution. Examples of such constituents include: a variety of enzymes (58, 82, 158, 298, 335); lectins (28); lipids (113, 114); a Ca^{2+} transport function (101); the hormones, abscisic acid (194), and certain gibberellins (50, 81, 266); binding sites for the hormones (for review see 155), auxin (10, 56, 57, 69, 125, 133, 136, 154, 240, 267, 341) and cytokinin (312); fusicoccin binding sites (70); phytochrome (for reviews see, 198, 199, 255, 261); and blue-light induced absorbance changes in *b*-type cytochromes (for review see 36).

Studies in this category cannot be examined in detail here, but certain generalizations are appropriate. Strict adherence to the principles of analytical cell fractionation provides information concerning the distribution of the constituent in question that is independent of preconceptions regarding the subcellular composition of the separated fractions (64, 65). The confidence attached to the definitive step of assigning the constituent to a host-particle(s) then is subject to the limitations imposed: (*a*) by the reliability of the chosen markers, whether morphological or biochemical, for identifying and quantitating the various subcellular components in the separated fractions; and (*b*) by the resolving power of the fractionation procedure used for separating these components.

The proportion of studies that fail to provide complete balance sheets of constituent, positive, and negative markers is substantial. The practice of performing "partial" analytical fractionations on restricted subfractions preselected by differential centrifugation without analysis of the discarded fractions is not uncommon. The conclusions that may be validly drawn in such cases are clearly restricted, simply because the data are incomplete. The uncertain diagnostic value of several of the common markers suggests that more caution than is sometimes encountered should be exercised in the interpretation of locational assignment studies. In particular, assertions that constituents are PM-localized should be evaluated in the context of the discrepancies that exist between the various putative PM markers. The low resolving power of differential centrifugation is generally recognized. Apparently less well appreciated is that the use of discontinuous gradients substantially reduces the resolving power of density gradient centrifugation while simultaneously creating the illusion of clear-cut separation (65). Correlations obtained with isopycnic step gradients should be assessed with the real number of separated fractions in mind. Also worthy of reemphasis is the point that, despite frequent conclusions to the contrary, the *specific* activities of constituents and markers in such fractions do not represent the *distribution* of the molecules in question.

CONCLUDING REMARKS

The primary purpose of this review has been twofold. First, to focus attention on the need for rigorous and complete analytical quantitation in the execution of cell fractionation experiments whether the goal is to define unique markers of specific subcellular components or to use such markers to assign a location to a constituent of previously unknown distribution (64, 65, 224). Second, to provide an assessment of the merits and limitations of the common markers in current use. Some organelles and membranes such as the nuclear envelope and tonoplast still lack an unequivocal biochemical marker, while the internal inconsistencies among the array of putative markers attributed to the plasma membrane are yet to be resolved. Still other components, such as the outer mitochondrial membrane, lack a verified *unique* marker. Despite these deficiencies, there has been a clear and systematic accumulation of increasingly refined information on plant subcellular components in recent years. This information, judiciously employed, provides a sound framework for further cell fractionation studies.

ACKNOWLEDGMENTS

I am grateful to Drs. W. R. Briggs, H. Beevers, D. J. Morré, W. F. Thompson, and M. G. Murray for helpful comments on the manuscript and to the many colleagues who provided reprints, preprints, and unpublished information relevant to this review.

Literature Cited

1. Adams, C. A., Novellie, L. 1975. Acid hydrolases and autolytic properties of protein bodies and spherosomes isolated from ungerminated seeds of *Sorghum bicolor* (Linn.) Moench. *Plant Physiol.* 55:7–11
2. Amar-Costesec, A., Wibo, M., Thinés-Sempoux, D., Beaufay, H., Berthet, J. 1974. Analytical study of microsomes and isolated subcellular membranes from rat liver. *J. Cell Biol.* 62:717–45
3. Anderson, R. L., Ray, P. M. 1978. Labeling of the plasma membrane of pea cells by a surface-localised glucan synthetase. *Plant Physiol.* 61:723–30
4. Appleqvist, L. A. 1975. Biochemical and structural aspects of storage and membrane lipids in developing oil seeds. In *Recent Advances in the Chemistry and Biochemistry of Plant Lipids,* ed. T. Galliard, E. I. Mercer, pp. 89–103. New York: Academic
5. Ashton, F. M. 1976. Mobilization of storage proteins of seeds. *Ann. Rev. Plant Physiol.* 27:95–117
6. Bahl, J., Francke, B., Monéger, R. 1976. Lipid composition of envelopes, prolamellar bodies and other plastid membranes in etiolated, green and greening wheat leaves. *Planta* 129:193–201
7. Bailey, C. J., Cobb, A., Boulter, D. 1970. A cotyledon slice system for the electron autoradiographic study of the synthesis and intracellular transport of the seed storage protein of *Vicia faba. Planta* 95:103–18
8. Balke, N. E., Hodges, T. K. 1977. Inhibition of ion absorption in oat roots: comparison diethylstilbestrol and oligomycin. *Plant Sci. Lett.* 10:319–25
9. Bartholomew, L., Mace, K. D. 1972. Isolation and identification of phospholipids from root tip cell plasmalemma of *Phaseolus limensis. Cytobios* 5:241–47
10. Batt, S., Venis, M. A. 1976. Separation and localization of two classes of auxin binding sites in corn coleoptile membranes. *Planta* 130:15–21
11. Baudhuin, P. 1974. Morphometry of subcellular fractions. *Methods Enzymol.* 22:3–20
12. Baumgartner, B., Chrispeels, M. J. 1976. Partial characterization of a protease inhibitor which inhibits the major endopeptidase present in the cotyledons of mung beans. *Plant Physiol.* 58:1–6
13. Baumgartner, B., Matile, P. 1976. Immunocytochemical localization of acid ribonuclease in morning glory flower tissue. *Biochem. Physiol. Pflanz.* 170: 279–85
14. Baumgartner, B., Tokuyasu, K. T., Chrispeels, M. J. 1978. Localization of vicilin peptohydrolase in the cotyledons of mung bean seedlings by immunofluorescence microscopy. *J. Cell Biol.* 79:10–17
15. Beevers, H. 1975. Organelles from castor bean seedlings: Biochemical roles in gluconeogenesis and phospholipid biosynthesis. See Ref. 4, pp. 287–99
16. Beevers, H. 1979. Microbodies in higher plants. *Ann. Rev. Plant Physiol.* 30:159–93
17. Beevers, H., Breidenbach, R. W. 1974. Glyoxysomes. *Methods Enzymol.* 31: 565–71
18. Bentwood, B. J., Cronshaw, J. 1978. Cytochemical localization of adenosine triphosphatase in the phloem of *Pisum sativum* and its relation to the function of transfer cells. *Planta* 140:111–20
19. Benveniste, I., Salaün, J. P., Durst, F. 1978. Phytochrome-mediated regulation of a monooxygenase hydroxylating cinnamic acid in etiolated pea seedlings. *Phytochemistry* 17:359–63

19a. Billett, E. E., Smith, H. 1978. Cinnamic acid 4-hydroxylase from gherkin tissues. *Phytochemistry* 17:1511–16

20. Boardman, N. K., Anderson, J. M. 1967. Fractionation of the photochemical systems of photosynthesis. *Biochim. Biophys. Acta* 143:187–203
21. Boardman, N. K., Wildman, S. G. 1962. Identification of proplastids by fluorescence microscopy and their isolation and purification. *Biochim. Biophys. Acta* 59:222–24

22. Bolender, R. P., Paumgartner, D., Losa, G., Muellener, D., Weibel, E. R. 1978. Integrated stereological and biochemical studies on hepatocytic membranes. I. Membrane recoveries in subcellular fractions. *J. Cell Biol.* 77: 565–83
23. Boller, T., Kende, H. 1978. Vacuolar enzymes from cultured tobacco cells. *Plant Physiol.* 61:97 (Abstr.)
24. Boller, T., Kende, H. 1979. Hydrolytic enzymes in the central vacuole of plant cells. *Plant Physiol.* In press
25. Boller, T., Kende, H. 1979. The tonoplast of tobacco cells: Density in sucrose gradients and search for a marker enzyme. *Plant Physiol.* In press
26. Bonner, W. D. 1973. Mitochondria and plant respiration. *Phytochemistry* 3:221–61
27. Boss, W. F., Ruesink, A. W. 1979. Isolation and characterization of concanavalin A labeled plasma membranes of carrot protoplasts. *Plant Physiol.* In press
28. Bowles, D. J., Kauss, H. 1976. Characterization, enzymatic and lectin properties of isolated membranes from *Phaseolus aureus. Biochim. Biophys. Acta* 443:360–74
29. Bowles, D. J., Lehle, L., Kauss, H. 1977. Glucosylation of sterols and polyprenolphosphate in the Golgi apparatus of *Phaseolus aureus. Planta* 134:177–81
30. Bowles, D. J., Northcote, D. H. 1974. The amounts and rates of export of polysaccharides found within the membrane system of maize root cells. *Biochem. J.* 142:139–44
31. Bowles, D. J., Schnarrenberger, C., Kauss, H. 1976. Lectins as membrane components of mitochondria from *Ricinus communis. Biochem. J.* 160:375–82
32. Bowman, B. J., Slayman, C. W. 1977. Characterization of plasma membrane adenosine triphosphatase of *Neurospora crassa. J. Biol. Chem.* 252:3357–63
33. Bradbeer, J. W. 1977. Chloroplasts structure and development. In *The Molecular Biology of Plant Cells,* ed. H. Smith, pp. 64–104. Oxford: Blackwell
34. Breidenbach, R. W., Kahn, A., Beevers, H. 1968. Characterization of glyoxysomes from castor bean endosperm. *Plant Physiol.* 43:705–13
35. Brett, C. T., Northcote, D. H. 1975. The formation of oligoglucans linked to lipid during synthesis of β-glucan by characterized membrane fractions isolated from peas. *Biochem. J.* 148: 107–17
36. Britz, S. J., Briggs, W. R. 1979. News on a blue light-absorbing photoreceptor system. In *Plant Growth and Light Perception,* ed. B. Deutch. In press
37. Burr, B., Burr, F. A. 1976. Zein synthesis in maize endosperm by polyribosomes attached to protein bodies. *Proc. Natl. Acad. Sci. USA* 73:515–19
38. Buser, C., Matile, P. 1977. Malic acid in vacuoles isolated from *Bryophyllum* leaf cells. *Z. Pflanzenphysiol.* 82:462–66
39. Butcher, H. C., Wagner, G. J., Siegelman, H. W. 1977. Localization of acid hydrolases in protoplasts. Examination of the proposed lysosomal function of the mature vacuole. *Plant Physiol.* 59:1098–1103
40. Carraway, K. L. 1971. Covalent labeling of membranes. *Biochim. Biophys. Acta* 415:379–410
41. Cassagne, C., Lessire, R., Carde, J. P. 1976. Plasmalemma enriched fraction from leek (*Allium porrum* L.) epidermal cells. *Plant Sci. Lett.* 7:127–35
42. Chen, Y. M., Lin, C. Y., Chang, H., Guilfoyle, T. J., Key, J. L. 1975. Isolation and properties of nuclei from control and auxin-treated soybean hypocotyl. *Plant Physiol.* 56:78–82
43. Chrispeels, M. J. 1976. Biosynthesis, intracellular transport, and secretion of extracellular macromolecules. *Ann. Rev. Plant Physiol.* 27:19–38
44. Chrispeels, M. J. 1979. Endoplasmic reticulum. In *The Biochemistry of Plants,* ed. P. K. Stumpf, E. E. Con, Vol. 1. New York: Academic. In press
45. Chrispeels, M. J., Baumgartner, B., Harris, N. 1976. Regulation of reserve protein metabolism in the cotyledons of mung bean seedlings. *Proc. Natl. Acad. Sci. USA* 73:3168–72
46. Chrispeels, M. J., Boulter, D. 1975. Control of storage protein metabolism in the cotyledons of germinating mung beans: Role of endopeptidase. *Plant Physiol.* 55:1031–37
47. Clay, W. F., Katterman, F. R. H., Bartels, P. G. 1975. Chromatin and DNA synthesis associated with nuclear membrane in germinating cotton. *Proc. Natl. Acad. Sci. USA* 72:3134–38
48. Cobb, A. H., Wellburn, A. R. 1974. Changes in plastid envelope polypeptides during chloroplast development. *Planta* 121:273–82
49. Cohen, C. M., Kalish, D. I., Jacobson, B. S., Branton, D. 1977. Membrane isolation on polylysine-coated beads. Plasma membrane from HeLa cells. *J. Cell Biol.* 75:119–34

50. Cooke, R. J., Kendrick, R. E. 1976. Phytochrome controlled gibberellin metabolism in etioplast envelopes. *Planta* 131:303–7
51. Cooper, T. G., Beevers, H. 1969. Mitochondria and glyoxysomes from castor bean endosperm. Enzyme constituents and catalytic capacity. *J. Biol. Chem.* 244:3507–13
52. Costes, C., Burghoffer, C., Carrayol, E. 1976. Occurrence of carotenoids in non-plastidial materials from potato tuber cells. *Plant Sci. Lett.* 6:253–59
53. Cribbs, D. H., Stelzig, D. A. 1978. Labeling, isolation and characterization of plasma membrane proteins from potato leaf protoplasts. *Plant Physiol.* 61:113 (Abstr.)
54. Cross, J. W., Briggs, W. R. 1976. An evaluation of markers for plasma membranes in membrane fractions from *Zea mays. Carnegie Inst. Washington Yearb.* 75:379–83
55. Cross, J. W., Briggs, W. R. 1977. Labeling of membranes from erythrocytes and corn with fluorescamine. *Biochim. Biophys. Acta* 471:67–77
56. Cross, J. W., Briggs, W. R. 1978. Properties of a solubilized microsomal auxin-binding protein from coleoptiles and primary leaves of *Zea mays. Plant Physiol.* 62:152–57
57. Cross, J. W., Briggs, W. R., Dohrmann, U. C., Ray, P. M. 1978. Auxin receptors of maize coleoptile membranes do not have ATPase activity. *Plant Physiol.* 61:581–84
58. Czichi, U., Kindl, H. 1977. Phenylalanine ammonia lyase and cinnamic acid hydroxylases as assembled consecutive enzymes on microsomal membranes of cucumber cotyledons: Cooperation and subcellular distribution. *Planta* 134: 133–43
59. D'Alessio, G., Trim, A. R. 1968. A method for the isolation of nuclei from leaves. *J. Exp. Bot.* 19:831–39
60. Dauwalder, M., Whaley, W. G., Kephart, J. E. 1969. Phosphatases and differentiation of the Golgi apparatus. *J. Cell Sci.* 4:455–97
61. Day, D. A., Arron, G. P., Laties, G. G. 1978. Enzyme distribution in potato mitochondria. *J. Exp. Bot.* In press
62. Day, D. A., Hanson, J. B. 1977. On methods for the isolation of mitochondria from etiolated corn shoots. *Plant Sci. Lett.* 11:99–104
63. Day, D. A., Wiskich, J. T. 1975. Isolation and properties of the outer membrane of plant mitochondria. *Arch. Biochem. Biophys.* 171:117–23
64. De Duve, C. 1964. Principles of tissue fractionation. *J. Theor. Biol.* 6:33–59
65. De Duve, C. 1971. Tissue fractionation. Past, present and future. *J. Cell Biol.* 50:20D–55D
66. DePierre, J. W., Ernster, L. 1977. Enzyme topology of intracellular membranes. *Ann. Rev. Biochem.* 46:201–62
67. Dizengremel, P., Kader, J. C., Mazliak, P., Lance, C. 1978. Electron transport and fatty acid synthesis in microsomes and outer mitochondrial membranes of plant tissues. *Plant Sci. Lett.* 11:151–57
68. Dobberstein, B., Volkmann, D., Klämbt, D. 1974. The attachment of polyribosomes to membranes of the hypocotyl of *Phaseolus vulgaris. Biochim. Biophys. Acta* 374:187–96
69. Dohrmann, U., Hertel, R., Kowalik, H. 1978. Properties of auxin binding sites in different subcellular fractions from maize coleoptiles. *Planta* 140:97–106
70. Dohrmann, U., Hertel, R., Pesci, P., Coccucci, S. M., Marrè, E., Randazzo, G., Ballio, A. 1977. Localization of "in vitro" binding of the fungal toxin fusicoccin to plasma-membrane-rich fractions from corn coleoptiles. *Plant Sci. Lett.* 9:291–99
71. Donaldson, R. P., Beevers, H. 1977. Lipid composition of organelles from germinating castor bean endosperm. *Plant Physiol.* 59:259–63
72. Douce, R. 1974. Site of biosynthesis of galactolipid in spinach chloroplasts. *Science* 183:852–53
73. Douce, R., Christensen, E. L., Bonner, W. D. 1972. Preparation of intact plant mitochondria. *Biochim. Biophys. Acta* 275:148–60
74. Douce, R., Holz, R. B., Benson, A. A. 1973. Isolation and properties of the envelope of spinach chloroplasts. *J. Biol. Chem.* 248:7215–22
75. Douce, R., Joyard, J. 1979. Structure and function of the plastid envelope. *Adv. Bot. Res.* In press
76. Douce, R., Mannella, C. A., Bonner, W. D. 1972. Site of the biosynthesis of CDP-diglyceride in plant mitochondria. *Biochem. Biophys. Res. Commun.* 49:1504–9
77. Douce, R., Mannella, C. A., Bonner, W. D. 1973. The external NADH dehydrogenases of intact plant mitochondria. *Biochim. Biophys. Acta* 292:105–16
78. Durán, A., Bowers, B., Cabib, E. 1975. Chitin synthetase zymogen is attached to the yeast plasma membranes. *Proc. Natl. Acad. Sci. USA* 72:3952–55

79. Edwards, G. E., Robinson, S. P., Tyler, N. J. C., Walker, D. A. 1978. Photosynthesis by isolated protoplasts, protoplast extracts, and chloroplasts of wheat. Influence of orthophosphate, pyrophosphate, and adenylates. *Plant Physiol.* 62:313–19
80. Ericson, M. C., Chrispeels, M. J. 1973. Isolation and characterization of glucosamine-containing storage glycoproteins from the cotyledons of *Phaseolus aureus. Plant Physiol.* 52:98–104
81. Evans, A., Smith, H. 1976. Localization of phytochrome in etioplasts and its regulation in vitro of gibberellin levels. *Proc. Natl. Acad. Sci. USA* 73:138–42
82. Feierabend, J., Brassel, D. 1977. Subcellular localization of shikimate dehydrogenase in higher plants. *Z. Pflanzenphysiol.* 82:334–36
83. Firth, J. A. 1978. Cytochemical approaches to the localization of specific adenosine triphosphatases. *Histochem. J.* 10:253–69
84. Forde, B. G., Oliver, R. J. C., Leaver, C. J. 1978. Variation in mitochondrial translation products associated with male-sterile cytoplasms in maize. *Proc. Natl. Acad. Sci. USA* 75:3841–45
85. Franke, W. W. 1966. Isolated nuclear membranes. *J. Cell Biol.* 31:619–23
86. Franke, W. W. 1974. Structure, biochemistry, and functions of the nuclear envelope. *Int. Rev. Cytol. Suppl.* 4:72–236
87. Franke, W. W. 1974. Nuclear envelopes. *Philos. Trans. R. Soc. London Ser. B* 268:67–93
88. Franke, W. W., Jarasch, E. D., Herth, W., Scheer, U., Zerban, H. 1974. General and molecular cytology. *Prog. Bot.* 36:1–21
89. Franke, W. W., Morrè, D. J., Herth, W., Zerban, H. 1976. General and molecular cytology. See Ref. 88, 38:1–16
90. Galbraith, D. W., Northcote, D. H. 1977. The isolation of plasma membrane from protoplasts of soybean suspension cultures. *J. Cell Sci.* 24:295–310
91. Galliard, T. 1974. Techniques for overcoming problems of lipolytic enzymes and lipoxygenases in the preparation of plant organelles. *Methods Enzymol.* 31:520–28
92. Gardiner, M., Chrispeels, M. J. 1975. Involvement of the Golgi apparatus in the synthesis and secretion of hydroxyproline-rich cell wall glycoproteins. *Plant Physiol.* 55:536–41
93. Goff, C. W. 1973. Localization of nucleoside diphosphatase in the onion root tip. *Protoplasma* 78:397–416
94. Goldschmidt, E. E., Branton, D. 1977. Vacuoles as sugar storage and transformation organelles in the beet root. *Plant Physiol.* 59:104 (Abstr.)
95. Gomez-Lepe, B., Hodges, T. K. 1978. Alkylguanidine inhibition of ion absorption in oat roots. *Plant Physiol.* 61:865–70
96. Gonzalez, E., Beevers, H. 1976. Role of endoplasmic reticulum in glyoxysome formation in castor bean endosperm. *Plant Physiol.* 57:406–9
97. Graham, T. A., Gunning, B. E. S. 1970. Localization of legumin and vicilin in bean cotyledons using fluorescent antibodies. *Nature* 228:81–82
98. Gregor, H. D. 1977. A new method for the rapid separation of cell organelles. *Anal. Biochem.* 82:255–57
99. Gressel, J., Quail, P. H. 1976. Particle-bound phytochrome: Differential pigment release by surfactants, ribonuclease and phospholipase C. *Plant Cell Physiol.* 17:771–76
100. Griffiths, W. T. 1975. Characterization of the terminal stages of chlorophyll(ide) synthesis in etioplast membrane preparations. *Biochem. J.* 152:623–35
101. Gross, J., Marmé, D. 1978. ATP-dependent Ca^{2+} uptake into plant membrane vesicles. *Proc. Natl. Acad. Sci. USA* 75:1232–36
102. Gunning, B. E. S., Steer, M. W. 1975. *Ultrastructure and the Biology of Plant Cells.* London: Arnold
103. Hall, D. O. 1972. Nomenclature for isolated chloroplasts. *Nature New Biol.* 235:125–26
104. Hall, J. L., Al-Azzawi, M. J., Fielding, J. L. 1977. Microscopic cytochemistry in enzyme localization and development. In *Regulation of Enzyme Synthesis and Activity in Higher Plants,* ed. H. Smith, pp. 329–65. London: Academic
105. Hall, J. L., Davie, A. M. 1975. Fine structure and localization of adenosine triphosphatase in the halophyte *Suaeda maritima. Protoplasma* 83:209–16
106. Hall, J. L., Flowers, T. J. 1976. Properties of membranes from the halophyte *Suaeda maritima.* I. Cytochemical staining of membranes in relation to the validity of membrane markers. *J. Exp. Bot.* 27:658–71
107. Hall, J. L., Roberts, R. M. 1975. Biochemical characteristics of membrane fractions isolated from maize (*Zea mays* L.) roots. *Ann. Bot.* 39:983–93

108. Hamilton, R. H., Künsch, U., Temperli, A. 1972. Simple and rapid procedures for isolation of tobacco leaf nuclei. *Anal. Biochem.* 49:48–57
109. Hannig, K., Heidrich, H. G. 1974. The use of continuous preparative free-flow electrophoresis for dissociating cell fractions and isolation of membranous components. *Methods Enzymol.* 31: 746–60
110. Hanson, J. B., Day, D. A. 1979. Mitochondria. See Ref. 44
111. Hardin, J. W., Cherry, J. H., Morré, D. J., Lembi, C. A. 1972. Enhancement of RNA polymerase by a factor released by auxin from plasma membrane. *Proc. Natl. Acad. Sci. USA* 69:3146–50
112. Harris, N., Chrispeels, M. J. 1975. Histochemical and biochemical observations on storage protein metabolism and protein body autolysis in cotyledons of germinating mung beans. *Plant Physiol.* 56:292–99
113. Hartmann, M. A., Normand, G., Benveniste, P. 1975. Sterol composition of plasma membrane enriched fractions from maize coleoptiles. *Plant Sci. Lett.* 5:287–92
114. Hartmann-Bouillon, M. A., Benveniste, P. 1978. Sterol biosynthetic capability of purified membrane fractions from maize coleoptiles. *Phytochemistry* 17: 1037–42
115. Hashimoto, H., Murakami, S. 1975. Dual character of lipid composition of the envelope membrane of spinach chloroplasts. *Plant Cell Physiol.* 16:895–902
116. Heiniger, U., Delmer, D. P. 1977. UDP-glucose: glucan synthetase in developing cotton fibers. II. Structure of the reaction product. *Plant Physiol.* 59:719–23
117. Hendriks, A. W. 1972. Purification of plant nuclei using colloidal silica. *FEBS Lett.* 24:101–5
118. Hendriks, T. 1976. Iodination of maize coleoptiles: A possible method for identifying plant plasma membranes. *Plant Sci. Lett.* 7:347–57
119. Hendriks, T. 1977. Multiple location of K-ATPase in maize coleoptiles. *Plant Sci. Lett.* 9:351–63
120. Hendriks, T. 1978. The distribution of glucan synthetase in maize coleoptiles: A comparison with K-ATPase. *Plant Sci. Lett.* 11:261–67
121. Hendriks, T. 1978. Isolation of plasma membrane from a maize coleoptile homogenate: Possible use of a mitochondrial fraction. *Z. Pflanzenphysiol.* 89:461–66
122. Hendriks, T. 1979. Cell fractionation of iodinated maize coleoptiles: Relative distributions of iodide and various organelle markers. *Plant Sci. Lett.* In press
123. Hendrix, D. L., Kennedy, R. M. 1977. Adenosine triphosphatase from soybean callus and root cells. *Plant Physiol.* 59:264–67
124. Hertel, R. 1979. Auxin receptors in plant membranes: Subcellular fractionation and specific binding assays. In *Plant Organelles. Methodological Surveys in Biochemistry,* ed. E. Reid, 9. In press
125. Hertel, R., Thomson, K. St., Russo, V. E. A. 1972. *In vitro* auxin binding to particulate cell fractions from corn coleoptiles. *Planta* 107:325–40
126. Hilton, J. R., Smith, H. 1978. Substantiating evidence for the phytochrome mediation of GA-like substances in etioplast fractions. *Proc. Ann. Eur. Symp. Photomorphogen.,* p. 46 (Abstr.)
127. Hodges, T. K., Leonard, R. T. 1974. Purification of a plasma membrane-bound adenosine triphosphatase from plant roots. *Methods Enzymol.* 32:392–406
128. Hodges, T. K., Leonard, R. T., Bracker, C. E., Keenan, T. W. 1972. Purification of an ion-stimulated adenosine triphosphatase from plant roots: association with plasma membranes. *Proc. Natl. Acad. Sci. USA* 69:3307–11
129. Honda, S. I. 1974. Fractionation of green tissue. *Methods Enzymol.* 31: 544–53
130. Huang, A. H. C., Beevers, H. 1973. Localization of enzymes within microbodies. *J. Cell Biol.* 58:379–89
131. Huang, A. H. C., Moreau, R. A. 1978. Lipases in the storage tissues of peanut and other oil seeds during germination. *Planta* 141:111–16
132. Huang, A. H. C., Moreau, R. A., Liu, K. D. F. 1978. Development and properties of a wax ester hydrolase in the cotyledons of jojoba seedlings. *Plant Physiol.* 61:339–41
133. Ihl, M. 1976. Indole-acetic acid binding proteins in soybean cotyledon. *Planta* 131:223–28
134. Jacks, T. J., Yatsu, L. Y., Altschul, A. M. 1967. Isolation and characterisation of peanut spherosomes. *Plant Physiol.* 42:585–97
135. Jackson, C., Dench, J. E., Halliwell, B., Hall, D. O., Moore, A. L. 1979. Separation of chloroplasts from mitochondria utilising silica sol gradient centrifugation. See Ref. 124

136. Jacobs, M., Hertel, R. 1978. Auxin binding to subcellular fractions from *Cucurbita* hypocotyls: In vitro evidence for an auxin transport carrier. *Planta* 142:1–10
137. Jacobson, A. B. 1968. A procedure for isolation of proplastids from etiolated maize leaves. *J. Cell Biol.* 38:238–44
138. Jacobson, B. S., Branton, D. 1977. Plasma membrane: Rapid isolation and exposure of the cytoplasmic surface by use of positively charged beads. *Science* 195:302–4
139. Jaworski, A., Key, J. L. 1974. Distribution of ribosomal deoxyribonucleic acid in subcellular fractions of higher plants. *Plant Physiol.* 53:366–69
140. Jeffrey, S. W., Douce, R., Benson, A. A. 1974. Carotenoid transformations in the chloroplast envelope. *Proc. Natl. Acad. Sci. USA* 71:807–10
141. Jelsema, C. L., Morré, D. J., Ruddat, M., Turner, C. 1977. Isolation and characterization of the lipid reserve bodies, spherosomes, from aleurone layers of wheat. *Bot. Gaz.* 138:138–49
142. Jelsema, C. L., Ruddat, M., Morré, D. J., Williamson, F. A. 1977. Specific binding of gibberellin A_1 to aleurone grain fractions from wheat endosperm. *Plant Cell Physiol.* 18:1009–19
143. Jensen, R. G. 1979. Chloroplasts. See Ref. 44
144. Jesaitis, A. J., Heners, P. R., Hertel, R., Briggs, W. R. 1977. Characterization of a membrane fraction containing a *b*-type cytochrome. *Plant Physiol.* 59: 941–47
145. Johnson, K. D., Kende, H. 1971. Hormonal control of lecithin synthesis in barley aleurone cells: regulation of the CDP-choline pathway by gibberellin. *Proc. Natl. Acad. Sci. USA* 68:2674–77
146. Jose, A. M. 1977. Gel filtration of particle-bound phytochrome. *Planta* 134: 287–93
147. Joyard, J., Douce, R. 1976. Préparation et activités enzymatiques de l'enveloppe des chloroplastes d'Épinard. *Physiol. Veg.* 14:31–48
148. Juliano, R. L., Behar-Bannlier, M. 1975. An evaluation of techniques for labeling the surface proteins of cultured mammalian cells. *Biochim. Biophys. Acta* 375:249–67
149. Kagawa, T., Hatch, M. D. 1975. Mitochondria as a site of C_4 acid decarboxylation in C_4-pathway photosynthesis. *Arch. Biochem. Biophys.* 167:687–96
150. Kagawa, T., Lord, J. M., Beevers, H. 1973. The origin and turnover of organelle membranes in castor bean endosperm. *Plant Physiol.* 51:61–65
151. Kagawa, T., Lord, J. M., Beevers, H. 1975. Lecithin synthesis during microbody biogenesis in watermelon cotyledons. *Arch. Biochem. Biophys.* 167: 45–53
152. Kahn, A., 1968. Developmental physiology of bean leaf plastids. II. Negative contrast electron microscopy of tubular membranes in prolamellar bodies. *Plant Physiol.* 43:1769–80
153. Kalish, D. I., Cohen, C. M., Jacobson, B. S., Branton, D. 1978. Membrane isolation on polylysine-coated glass beads. Asymmetry of bound membrane. *Biochim. Biophys. Acta* 506:97–110
154. Kasamo, K., Yamaki, T., 1976. In vitro binding of IAA to plasma membrane-rich fractions containing Mg^{++}-activated ATPase from mung bean hypocotyls. *Plant Cell Physiol.* 17:149–64
155. Kende, H., Gardner, G. 1976. Hormone binding in plants. *Ann. Rev. Plant Physiol.* 27:267–90
156. Khoo, V., Wolf, M. J. 1970. Origin and development of protein granules in maize endosperm. *Am. J. Bot.* 57:1042–50
157. Kleinig, H., Steinki, C., Kopp, C., Zaar, K. 1978. Oleosomes (spherosomes) from *Daucus carota* suspension culture cells. *Planta* 140:233–37
158. Koehler, D. E., Leonard, R. T., Van der Woude, W. J., Linkins, A. E., Lewis, L. N. 1976. Association of latent cellulase activity with plasma membranes from kidney bean abscission zones. *Plant Physiol.* 58:324–30
159. Kohn, S., Klein, S. 1976. Light-induced structural changes during incubation of isolated maize etioplasts. *Planta* 132: 169–75
160. Ku, S. B., Edwards, G. E. 1975. Photosynthesis in mesophyll protoplasts and bundle sheath cells in various types of C_4 plants. IV. Enzymes of respiratory metabolism and energy utilizing enzymes of photosynthetic pathways. *Z. Pflanzenphysiol.* 77:16–32
161. Kuehl, L. 1964. Isolation of plant nuclei. *Z. Natuforsch.* 19b:525–32
162. Lafléche, D., Bové, J. M., Duranton, J. 1972. Localization and translocation of the protochlorophyllide holochrome during the greening of etioplasts in *Zea mays* L. *J. Ultrastruct. Res.* 40:205–14
163. Lambowitz, A. M., Bonner, W. D. 1974. The *b*-cytochromes of plant mitochondria. A spectrophotometric and potentiometric study. *J. Biol. Chem.* 249:2428–40

164. Larkin, P. J. 1978. Plant protoplast agglutination by lectins. *Plant Physiol.* 61:626–29
165. Larsson, C., Andersson, B. 1979. Two-phase methods for chloroplasts, chloroplast elements and mitochondria. See Ref. 124
166. Laties, G. G. 1974. Isolation of mitochondria from plant material. *Methods Enzymol.* 31:589–600
167. Leaver, C. J., Harmey, M. A. 1973. Plant mitochondrial nucleic acids. *Biochem. Soc. Symp.* 38:175–93
168. Lee, D. C. 1977. Plant mitochondria. See Ref. 33, pp. 105–35
169. Leech, R. M. 1977. Subcellular fractionation techniques in enzyme distribution studies. See Ref. 104, pp. 289–327
170. Lees, G. L., Thomson, J. E. 1975. The effects of senescence on the protein complement of plasma membranes from cotyledons. *New Phytol.* 75: 525–32
171. Leese, B. M., Leech, R. M., Thomson, W. W. 1971. Isolation of plastids from different regions of developing maize leaves. *Int. Congr. Photosynth., 2nd,* pp. 1485–94
172. Lehle, L., Bowles, D. J., Tanner, W. 1978. Subcellular site of mannosyl transfer to dolichyl phosphate in *Phaseolus aureus. Plant Sci. Lett.* 11:27–34
173. Leigh, R. A., Branton, D. 1976. Isolation of vacuoles from root storage tissue of *Beta vulgaris* L. *Plant Physiol.* 58:656–62
174. Leigh, R. A., Branton, D., Marty, F. 1979. Methods for the isolation of intact plant vacuoles and fragments of tonoplast. See Ref. 124
175. Leigh, R. A., Williamson, F. A., Wyn Jones, R. G. 1975. Presence of two different membrane-bound, KCl-stimulated adenosine triphosphatase activities in maize roots. *Plant Physiol.* 55:678–85
176. Lembi, C. A., Morré, D. J., Thomson, K. St., Hertel, R. 1972. N-l-napthylphthalamic-acid-binding activity of a plasma membrane-rich fraction from maize coleoptiles. *Planta* 99:37–45
177. Leonard, R. T., Hansen, D., Hodges, T. K. 1973. Membrane-bound adenosine triphosphatase activities of oat roots. *Plant Physiol.* 51:749–54
178. Leonard, R. T., Hodges, T. K. 1973. Characterization of plasma membrane-associated adenosine triphosphatase activity of oat roots. *Plant Physiol.* 52:6–12
179. Leonard, R. T., Hodges, T. K. 1979. The plasma membrane. See Ref. 44
180. Leonard, R. T., Hotchkiss, C. W. 1976. Cation-stimulated adenosine triphosphatase activity and cation transport in corn roots. *Plant Physiol.* 58:331–55
181. Leonard, R. T., Hotchkiss, C. W. 1978. Plasma membrane-associated adenosine triphosphatase activity of isolated cortex and stele from corn roots. *Plant Physiol.* 61:175–79
182. Leonard, R. T., Van der Woude, W. J. 1976. Isolation of plasma membranes from corn roots by sucrose density gradient centrifugation. An anomalous effect of ficoll. *Plant Physiol.* 57:105–14
183. Lercher, M., Wojciechowski, Z. A. 1976. Localization of plant UDP-glucose: sterol glucosyl-transferase in the Golgi membrane. *Plant Sci. Lett.* 7:337–40
184. Lin, C. Y., Guilfoyle, T. J., Chen, Y. M., Key, J. L. 1975. Isolation of nucleoli and localization of ribonucleic acid polymerase I from soybean hypocotyl. *Plant Physiol.* 56:850–52
185. Lin, W., Wagner, G. J., Siegelman, H. W., Hind, G. 1977. Membrane-bound ATPase of intact vacuoles and tonoplasts isolated from mature plant tissue. *Biochim. Biophys. Acta* 465:110–17
186. Locy, R., Kende, H. 1979. The mode of secretion of a-amylase in barley aleurone layers. *Planta* 143:89–99
187. Loomis, W. D. 1974. Overcoming problems of phenolics and quinones in the isolation of plant enzymes and organelles. *Methods Enzymol.* 31:528–44
188. Lord, J. M. 1975. Evidence that phosphatidylcholine and phosphatidylethanolamine are synthesized by a single enzyme present in the endoplasmic reticulum of castor-bean endosperm. *Biochem. J.* 151:451–53
189. Lord, J. M. 1979. Developmental relationship between the endoplasmic reticulum and glyoxysomes. See Ref. 124
190. Lord, J. M., Kagawa, T., Beevers, H. 1972. Intracellular distribution of enzymes of the cytidine diphosphate choline pathway in castor bean endosperm. *Proc. Natl. Acad. Sci. USA* 9:2429–32
191. Lord, J. M., Kagawa, T., Moore, T. S., Beevers, H. 1973. Endoplasmic reticulum as the site of lecithin formation in castor bean endosperm. *J. Cell Biol.* 57:659–67
192. Lörz, H., Harms, C. T., Potrykus, I. 1976. Isolation of 'vacuoplast' from protoplasts of higher plants. *Biochem. Physiol. Pflanz.* 169:617–20

193. Lott, J. 1979. Protein bodies. See Ref. 44
194. Loveys, B. R. 1977. The intracellular location of abscisic acid in stressed and non-stressed leaf tissue. *Physiol. Plant.* 40:6–10
195. Lütz, C. 1975. Biochemische und cytologische Untersuchungen zur Chloroplastenentwicklung. I. Die chemische Charakterisierung der Prolamellarkörper aus Etioplasten von *Avena sativa* L. *Z. Pflanzenphysiol.* 75:346–59
196. Malone, C. P., Burke, J. J., Hanson, J. B. 1977. Histochemical evidence for the occurrence of oligomycin-sensitive plasmalemma ATPase in corn roots. *Plant Physiol.* 60:916–22
197. Mannella, C. A., Bonner, W. D. 1975. Biochemical characteristics of the outer membranes of plant mitochondria. *Biochim. Biophys. Acta* 413:213–25
198. Marmé, D. 1977. Phytochrome: Membranes as possible sites of primary action. *Ann. Rev. Plant Physiol.* 28: 173–98
199. Marmé, D. 1979. Intracellular localization of phytochrome: A review. See Ref. 36
200. Mascarenhas, J. P., Berman-Kurtz, M., Kulikowski, R. R. 1974. Isolation of plant nuclei. *Methods Enzymol.* 31: 558–65
201. Matile, P. 1966. Enzyme der Vakuole aus Wurzelzellen von Maiskeimlingen. Ein Beitrag zur funktionellen Bedeutung der Vakuole bei der intrazellulären Verdauung. *Z. Naturforsch.* 21b:871–78
202. Matile, P. 1974. Lysosomes. In *Dynamic Aspects of Plant Ultrastructure,* ed. A. W. Robards, pp. 178–218. London: McGraw Hill
203. Matile, P. 1976. Vacuoles. In *Plant Biochemistry,* ed. J. Bonner, J. E. Varner, pp. 189–224. New York: Academic
204. Matile, P. 1978. Biochemistry and function of vacuoles. *Ann. Rev. Plant Physiol.* 29:193–213
205. Mayo, M. A., Cocking, E. C. 1969. Detection of pinocytic activity using selective staining with phosphotungstic acid. *Protoplasma* 68:231–36
206. Mettler, I. J., Beevers, H. 1978. Isolation and characterization of the surrounding membrane of protein bodies from castor beans. *Plant Physiol.* 61:115 (Abstr.)
207. Miflin, B. J. 1974. The location of nitrite reductase and other enzymes related to amino acid biosynthesis in the plastids of root and leaves. *Plant Physiol.* 54:550–55
208. Miflin, B. J., Beevers, H. 1974. Isolation of intact plastids from a range of plant tissues. *Plant Physiol.* 53:870–74
209. Millerd, A. 1975. Biochemistry of legume seed proteins. *Ann. Rev. Plant Physiol.* 26:53–72
210. Mollenhauer, H. H., Morré, D. J., Van der Woude, W. J. 1975. Endoplasmic reticulum-Golgi apparatus associations in maize root tips. *Mikroskopie* 31: 257–72
211. Montague, M. J., Ray, P. M. 1977. Phospholipid-synthesizing enzymes associated with Golgi dictyosomes from pea tissue. *Plant Physiol.* 59:225–30
212. Moore, T. S. 1974. Phosphatidylglycerol synthesis in castor bean endosperm. Kinetics, requirements, and intracellular localization. *Plant Physiol.* 54:164–68
213. Moore, T. S. 1976. Phosphatidylcholine synthesis in castor bean endosperm. *Plant Physiol.* 57:383–86
214. Moore, T. S., Beevers, H. 1974. Isolation and characterization of organelles from soybean suspension cultures. *Plant Physiol.* 53:261–65
215. Moore, T. S., Lord, J. M., Kagawa, T., Beevers, H. 1973. Enzymes of phospholipid metabolism in the endoplasmic reticulum of castor bean endosperm. *Plant Physiol.* 52:50–53
216. Moreau, F., Dupont, J., Lance, C. 1974. Phospholipid and fatty acid composition of outer and inner membranes of plant mitochondria. *Biochim. Biophys. Acta* 345:294–304
217. Moreau, F., Lance, C. 1972. Isolement et propriétés des membranes externes et internes de mitochondries végétales. *Biochimie* 54:1335–48
218. Moreau, R. A., Huang, A. H. C. 1977. Gluconeogenesis from storage wax in the cotyledons of jojoba seedlings. *Plant Physiol.* 60:329–33
219. Morgenthaler, J. J., Price, C. A., Robinson, J. M., Gibbs, M. 1974. Photosynthetic activity of spinach chloroplasts after isopycnic centrifugation in gradients of silica. *Plant Physiol.* 54:532–34
220. Morré, D. J. 1970. *In vivo* incorporation of radioactive metabolites by Golgi apparatus and other cell fractions of onion stem. *Plant Physiol.* 45:791–99
221. Morré, D. J. 1971. Isolation of Golgi apparatus. *Methods Enzymol.* 22: 130–48
222. Morré, D. J. 1975. Membrane biogenesis. *Ann. Rev. Plant Physiol.* 26:441–81
223. Morré, D. J. 1979. Isolation of plant Golgi apparatus. See Ref. 124

224. Morré, D. J., Bridges, J. W., Cline, G. B., Coleman, R., Glaumann, H., Headon, D. R., Reid, E., Siebert, G., Widnell, C. C. 1979. Markers for membranous cell components. *Nature*. In preparation
225. Morré, D. J., Lembi, C. A., Van der Woude, W. J. 1974. Isolation of Golgi apparatus and related cell components. In *Praktikum der Zytologie*, ed. G. Jacobi, pp. 146–71. Stuttgart: Thieme Verlag
226. Morré, D. J., Lembi, C. A., Van der Woude, W. J. 1977. A latent inosine-5'-diphosphatase associated with Golgi apparatus-rich fractions from onion stem. *Cytobiologie* 16:72–81
227. Morré, D. J., Mollenhauer, H. H. 1974. The endomembrane concept: A functional integration of endoplasmic reticulum and Golgi apparatus. See Ref. 202, pp. 84–137
228. Morré, D. J., Mollenhauer, H. H., Bracker, C. E. 1970. Origin and continuity of Golgi apparatus. In *Results and Problems in Cell Differentiation. Origin and Continuity of Cell Organelles*, ed. J. Reinert, H. Ursprung, 2:82–126. Berlin: Springer-Verlag
229. Morré, D. J., Nyquist, S., Rivera, E. 1970. Lecithin biosynthetic enzymes of onion stem and the distribution of phosphorylcholine-cytidyl transferase among cell fractions. *Plant Physiol.* 45:800–4
230. Morris, G. F. I., Thurman, D. A., Boulter, D. 1970. The extraction and chemical composition of aleurone grains (protein bodies) isolated from seeds of *Vicia faba*. *Phytochemistry* 9:1707–14
231. Mülbach, H., Schnarrenberger, C. 1978. Properties and intracellular distribution of two phosphoglucomutases from spinach leaves. *Planta* 141:65–70
232. Muto, S., Beevers, H. 1974. Lipase activities in castor bean endosperm during germination. *Plant Physiol.* 54:23–28
233. Nagahashi, G., Leonard, R. T., Thomson, W. W. 1978. Purification of plasma membranes from roots of barley. Specificity of the phosphotungstic acid-chromic acid stain. *Plant Physiol.* 61:993–99
234. Nagahashi, J., Beevers, L. 1978. Subcellular localization of glycosyl transferases involved in glycoprotein biosynthesis in the cotyledons of *Pisum sativum* L. *Plant Physiol.* 61:451–59
235. Nagahashi, J., Mense, R. M., Beevers, L. 1978. Membrane-associated glycosyl transferases in cotyledons of *Pisum sativum*. Differential effects of Mg^{2+} and Mn^{2+}. *Plant Physiol.* 62:766–72
236. Newcomb, E. H., Becker, W. M. 1974. The diversity of plant organelles with special reference to the electron cytochemical localization of catalase in plant microbodies. *Methods Enzymol.* 31:489–500
237. Nishimura, M., Beevers, H. 1978. Isolation of intact plastids from protoplasts from castor bean endosperm. *Plant Physiol.* 62:40–43
238. Nishimura, M., Beevers, H. 1978. Hydrolases in vacuoles from castor bean endosperm. *Plant Physiol.* 62:44–48
239. Nishimura, M., Graham, D., Akazawa, T. 1976. Isolation of intact chloroplasts and other cell organelles from spinach leaf protoplasts. *Plant Physiol.* 58: 309–14
240. Normand, G., Hartmann, M. A., Schuber, F., Benveniste, P. 1975. Caractérisation de membranes de coléoptiles de mais fixant l'auxine et l'acide N-naphtyl phtalamique. *Physiol. Veg.* 13:743–61
241. Northcote, D. H. 1974. Complex envelope system. Membrane systems of plant cells. *Philos. Trans. R. Soc. London Ser. B.* 268:119–28
242. Ohyama, K., Pelcher, L. E., Horn, D. 1977. A rapid, simple method for nuclei isolation from plant protoplasts. *Plant Physiol.* 60:179–81
243. Ory, R. L. 1969. Acid lipase of the castor bean. *Lipids* 4:177–85
244. Ory, R. L. 1972. Enzyme activities associated with protein bodies of seeds. In *Symposium: Seed Proteins*, ed. G. E. Inglett, pp. 86–98. Westport, Conn: Avi
245. Ory, R. L., Henningsen, K. W. 1969. Enzymes associated with protein bodies isolated from ungerminated barley seeds. *Plant Physiol.* 44:1488–98
246. Ory, R. L., Yatsu, L. Y., Kircher, H. W. 1968. Association of lipase activity with the spherosomes of *Ricinus communis*. *Arch. Biochem. Biophys.* 264:255–64
247. Osmond, C. B., Akazawa, T., Beevers, H. 1975. Localization and properties of ribulose diphosphate carboxylase from castor bean endosperm. *Plant Physiol.* 55:226–30
248. Palmer, J. M. 1976. The organization and regulation of electron transport in plant mitochondria. *Ann. Rev. Plant Physiol.* 27:133–57
249. Paull, R. E., Jones, R. L. 1975. Studies on the secretion of maize root cap slime. *Planta* 127:97–110
250. Paull, R. E., Jones, R. L., 1976. Studies

on the secretion of maize root cap slime. *Plant Physiol.* 57:249–56
251. Perlin, D. S., Spanswick, R. M. 1978. Isolation of plasma membranes from corn leaf protoplasts. *Plant Physiol.* 61:113 (Abstr.)
252. Peters, B. M., Cribbs, D. H., Stelzig, D. A. 1978. Agglutination of plant protoplasts by fungal cell wall glucans. *Science* 201:364–365
253. Philipp, E. I., Franke, W. W., Keenan, T. W., Stadler, J., Jarasch, E. D. 1976. Characterization of nuclear membranes and endoplasmic reticulum isolated from plant tissue. *J. Cell Biol.* 68:11–29
254. Powell, J. T., Brew, K. 1974. Glycosyltransferases in Golgi membranes of onion stem. *Biochem. J.* 142:203–9
255. Pratt, L. H. 1978. Molecular properties of phytochrome. *Photochem. Photobiol.* 27:81–105
256. Price, C. A. 1974. Plant cell fractionation. *Methods Enzymol.* 31:501–19
257. Price, C. A., Bartolf, M., Ortiz, W., Reardon, E. M. 1979. Isolation of chloroplasts in silica sol gradients. See Ref. 124
258. Price, C. A., Hirvonen, A. P. 1967. Sedimentation rates of plastids in an analytical zonal rotor. *Biochim. Biophys. Acta* 148:531–38
259. Priestly, D. A., Woolhouse, H. W. 1978. The chloroplast envelope of *Phaseolus vulgaris* L. II. Enzymic characteristics. *J. Exp. Bot.* In press
260. Quail, P. H. 1975. Particle-bound phytochrome: Association with a ribonucleoprotein fraction from *Cucurbita pepo* L. *Planta* 123:223–34
261. Quail, P. H. 1975. Interaction of phytochrome with other cellular components. *Photochem. Photobiol.* 22:299–301
262. Quail, P. H. 1977. How 'pure' are G·50 plastids? *Proc. Ann. Eur. Symp. Photomorphogen.*, p. 78
263. Quail, P. H., Browning, A. 1977. Failure of lactoperoxidase to iodinate specifically the plasma membrane of *Cucurbita* tissue segments. *Plant Physiol.* 59:759–66
264. Quail, P. H., Gallagher, E. A., Wellburn, A. R. 1976. Membrane-associated phytochrome: Non-coincidence with plastid membrane marker profiles on sucrose gradients. *Photochem. Photobiol.* 24:495–98
265. Quail, P. H., Hughes, J. E. 1977. Phytochrome and phosphotungstate-chromate-positive vesicles from *Cucurbita pepo* L. *Planta* 133:169–77
266. Railton, I. D. 1977. 16,17-Dihydro 16,17-dihydroxy gibberellin A_9: A metabolite of (^{3}H)-gibberellin A_9 in chloroplast sonicates from *Pisum sativum* var. <Alaska>. *Z. Pflanzenphysiol.* 81:323–29
267. Ray, P. M. 1977. Auxin-binding sites of maize coleoptiles are localized on membranes of the endoplasmic reticulum. *Plant Physiol.* 59:594–99
268. Ray, P. M. 1979. Glucan synthetases associated with Golgi and plasma membranes. See Ref. 124
269. Ray, P. M., Eisinger, W. R., Robinson, D. G. 1976. Organelles involved in cell wall polysaccharide formation and transport in pea cells. *Ber. Dtsch. Bot. Ges.* 89:121–46
270. Ray, P. M., Shininger, T. L., Ray, M. M. 1969. Isolation of β-glucan synthetase particles from plant cells and identification with Golgi membranes. *Proc. Natl. Acad. Sci. USA* 64:605–12
271. Raymond, Y., Fincher, G. B., Maclachlan, G. A. 1978. Tissue slice and particulate β-glucan synthetase activities from *Pisum* epicotyls. *Plant Physiol.* 61:938–42
272. Rho, J. H., Chipcase, M. I. 1962. Incorporation of tritiated cytidine into ribonucleic acid by isolated pea nuclei. *J. Cell Biol.* 14:183–93
273. Rizzo, P. J., Pederson, K., Cherry, J. H. 1978. Transcription and RNA polymerase stimulatory activity in nuclei isolated from soybean. *Plant Sci. Lett.* 12:133–43
274. Robards, A. W. 1975. Plasmodesmata in higher plants. In *Intercellular Communication in Plants,* ed. B. E. S. Gunning, A. W. Robards, pp. 15–57. Berlin: Springer-Verlag
275. Robinson, D. G., Eisinger, W. R., Ray, P. M. 1976. Dynamics of the Golgi system in wall matrix polysaccharide synthesis and secretion by pea cells. *Ber. Dtsch. Bot. Ges.* 89:147–61
276. Robinson, S. P., Walker, D. A. 1979. Established methods for large organelles. See Ref. 124
277. Rocha, V., Ting, I. P. 1970. Preparation of cellular plant organelles from spinach leaves. *Arch. Biochem. Biophys.* 140:398–407
278. Roland, J. C., Lembi, C. A., Morré, D. J. 1972. Phosphotungstic acid-chromic acid as a selective electron-dense stain for plasma membranes of plant cells. *Stain Technol.* 47:195–200
279. Rost, T. L. 1972. The ultrastructure and physiology of protein bodies and lipids from hydrated dormant and nondormant embryos of *Setaria lutescens* (Gramineae). *Am. J. Bot.* 59:607–16

280. Rost, T. L., Patterson, K. E. 1978. Structural and histochemical characterization of the cotyledon storage organelles of jojoba (*Simmondsia chinensis*). *Protoplasma* 95:1–10
281. Roughan, P. G., Slack, C. R. 1977. Long-chain acyl-coenzyme A synthetase activity of spinach chloroplasts is concentrated in the envelope. *Biochem. J.* 162:457–59
282. San Pietro, A. 1971. *Methods in Enzymology,* Vol 23. New York: Academic
283. Saunders, J. A., Conn, E. E. 1978. Presence of the cyanogenic glucoside dhurrin in isolated vacuoles from *Sorghum. Plant Physiol.* 61:154–57
284. Saunders, J. A., Conn, E. E., Lin, C. H., Shimada, M. 1977. Localization of cinnamic acid 4-monooxygenase and the membrane-bound enzyme system for dhurrin biosynthesis in *Sorghum* seedlings. *Plant Physiol.* 60:629–34
285. Saunders, J. A., Conn, E. E., Lin, C. H., Stocking, C. R. 1977. Subcellular localization of the cyanogenic glucoside of sorghum by autoradiography. *Plant Physiol.* 59:647–52
286. Scandalios, J. G. 1974. Isozymes in development and differentiation. *Ann. Rev. Plant Physiol.* 25:225–58
286a. Scandalios, J. G. 1977. Isozymes: Genetic and biochemical regulation of alcohol dehydrogenase. See Ref. 104, pp. 129–53
287. Scarborough, G. A. 1975. Isolation and characterization of *Neurospora crassa* plasma membranes. *J. Biol. Chem.* 250:1106–11
288. Schiff, J. A. 1979. Proplastids and etioplasts. See Ref. 44
289. Schmitt, J. M., Herrmann, R. G. 1977. Fractionation of cell organelles in silica sol gradients. *Methods Cell Biol.* 15:177–200
290. Schnarrenberger, C., Burkhard, C. 1977. In-vitro interaction between chloroplasts and peroxisomes as controlled by inorganic phosphate. *Planta* 134: 109–14
291. Schnarrenberger, C., Fock, H. 1976. Interactions among organelles involved in photorespiration. In *Encyclopedia of Plant Physiology, New Series,* ed. C. R. Stocking, U. Heber, 3:185–234. Berlin: Springer-Verlag
292. Schnarrenberger, C., Herbert, M. 1979. Isolation of cell organelles from CAM plants by the use of Ficoll. See Ref. 124
293. Schnarrenberger, C., Oeser, A., Tolbert, N. E. 1972. Isolation of plastids from sunflower cotyledons during germination. *Plant Physiol.* 50:55–59
294. Schnarrenberger, C., Oeser, A., Tolbert, N. E. 1972. Isolation of protein bodies on sucrose gradients. *Planta* 104:185–94
295. Schnarrenberger, C., Oeser, A., Tolbert, N. E. 1973. Two isozymes each of glucose-6-phosphate dehydrogenase and 6-phosphogluconate dehydrogenase in spinach leaves. *Arch. Biochem. Biophys.* 154:438–48
296. Schnarrenberger, C., Tetour, M., Herbert, M. 1975. Development and intracellular distribution of enzymes of the oxidative pentose phosphate cycle in radish cotyledons. *Plant Physiol.* 56:836–40
297. Scott Burden, T., Canvin, D. T. 1975. The effect of Mg^{2+} on density of proplastids from the developing castor bean and the use of acetyl-CoA carboxylase activity for their cytochemical identification. *Can. J. Bot.* 53:1371–76
298. Shargool, P. D., Steeves, T., Weaver, M., Russell, M. 1978. The localization within plant cells of enzymes involved in arginine biosynthesis. *Can. J. Biochem.* 56:273–79
299. Shore, G., Maclachlan, G. A. 1975. The site of cellulose synthesis. Hormone treatment alters the intracellular location of alkali-insoluble β-1,4-glucan (cellulose) synthetase activities. *J. Cell Biol.* 64:557–71
300. Simcox, P. D., Reid, E. E., Canvin, D. T., Dennis, D. T. 1977. Enzymes of the glycolytic and pentose phosphate pathways in proplastids from the developing endosperm of *Ricinus communis L. Plant Physiol.* 59:1128–32
301. Sinensky, M., Strobel, G. 1976. Chemical composition of a cellular fraction enriched in plasma membranes from sugarcane. *Plant Sci. Lett.* 6:209–14
302. Singer, S. J. 1974. The molecular organisation of membranes. *Ann. Rev. Biochem.* 43:805–33
303. Smith, C. G. 1974. The ultrastructural development of spherosomes and oil bodies in the developing embryo of *Crambe abyssinica. Planta* 119:125–42
304. Solomos, T. 1977. Cyanide-resistant respiration in higher plants. *Ann. Rev. Plant Physiol.* 28:279–97
305. Sparace, S. A., Moore, T. S. 1979. Phospholipid metabolism in plant mitochondria: Submitochondrial sites of synthesis. *Plant Physiol.* In press
306. Spichiger, J. U. 1969. Isolation und Charakterisierung von Sphärosomen

und Glyoxisomen aus Tabakendosperm. *Planta* 89:56–75
307. St. Angelo, A. J., Yatsu, L. Y., Altschul, A. M. 1968. Isolation of edestin from aleurone grains of *Cannabis sativa*. *Arch. Biochem. Biophys.* 124:199–205
308. Stavy (Rodeh), R., Ben-Shaul, Y., Galun, E. 1973. Nuclear envelope isolation in peas. *Biochim. Biophys. Acta* 323:167–77
309. Stout, T. T., Katovich Hurley, C. 1977 Isolation of nuclei and preparation of chromatin from plant tissues. *Methods Cell Biol.* 16:87–96
310. Strobel, G. A., Hapner, K. D. 1975. Transfer of toxin susceptibility to plant protoplasts via the helminthosporoside binding protein in sugarcane. *Biochem. Biophys. Res. Commun.* 63:1151–56
311. Strobel, G. A., Hess, W. M. 1974. Evidence for the presence of the toxin-binding protein on the plasma membrane of sugar cane cells. *Proc. Natl. Acad. Sci. USA* 71:1413–17
312. Sussman, M. R., Kende, H. 1978. *In vitro* cytokinin binding to a particulate fraction of tobacco cells. *Planta* 140:251–59
313. Tautvydas, K. J. 1971. Mass isolation of pea nuclei. *Plant Physiol.* 47:499–503
314. Taylor, A. R. D., Hall, J. L. 1978. An ultrastructural comparison of lanthanum and silicotungstic acid/chromic acid (STAC) plasma membrane stains of isolated protoplasts. *Plant Sci. Lett.* In press
315. Theimer, R. R., Rosnitschek, I. 1978. Development and intracellular localization of lipase activity in rapeseed *Brassica napus* L. cotyledons. *Planta* 139: 249–56
316. Theimer, R. R., Rosnitschek, I., Wanner, G. 1979. Characterization of a distinct lipolytic membrane fraction from oil seed cotyledons. See Ref. 124
317. Thom, M., Laetsch, W. M., Maretzki, A. 1975. Isolation of membranes from sugarcane cell suspensions: Evidence for a plasma membrane enriched fraction. *Plant Sci. Lett.* 5:245–53
318. Thomson, K. St., Hertel, R., Müller, S., Tavares, J. E. 1973. 1-N-Naphthylphthalamic acid and 2,3,5-triiodobenzoic acid. *In-vitro* binding to particulate cell fractions and action on auxin transport in corn coleoptiles. *Planta* 109:337–52
319. Thomson, W. W., Foster, P., Leech, R. M. 1972. The isolation of proplastids from roots of *Vicia faba*. *Plant Physiol.* 49:270–72
320. Tolbert, N. E. 1974. Isolation of subcellular organelles of metabolism on isopycnic sucrose gradients. *Methods Enzymol.* 31:734–46
321. Tombs, M. P. 1967. Protein bodies of the soybean. *Plant Physiol.* 42:797–813
322. Trewavas, A., Jordan, E. G., Timis, J. 1979. Nucleus. See Ref. 44
323. Tully, R. E., Beevers, H. 1976. Protein bodies of castor bean endosperm. Isolation, fractionation, and the characterization of protein components. *Plant Physiol.* 58:710–16
324. Van den Bosch, H. 1974. Phosphoglyceride metabolism. *Ann. Rev. Biochem.* 43:243–77
325. Van der Woude, W. J., Lembi, C. A., Morré, D. J., Kindinger, J. I., Ordin, L. 1974. β-glucan synthetases of plasma membrane and Golgi apparatus from onion stem. *Plant Physiol.* 54:333–40
326. Varner, J. E., Schidlovsky, G. 1963. Intracellular distribution of protein in pea cotyledons. *Plant Physiol.* 38:139–44
327. Vick, B., Beevers, H. 1977. Phosphatidic acid synthesis in castor bean endosperm. *Plant Physiol.* 59:459–63
328. Vick, B., Beevers, H. 1978. Fatty acid synthesis in endosperm of young castor bean seedlings. *Plant Physiol.* 62: 173–78
329. Wagner, G. J., Butcher, H. C., Siegelman, H. W. 1978. The plant protoplast: A useful tool for plant research and student instruction. *Bioscience* 28:95–101
330. Wagner, G. J., Siegelman, H. W. 1975. Large-scale isolation of intact vacuoles and isolation of chloroplasts from protoplasts of mature plant tissues. *Science* 190:1298–99
331. Walk, R. A., Michaeli, S., Hock, B. 1977. Glyoxysomal and mitochondrial malate dehydrogenase of watermelon (*Citrullus vulgaris*) cotyledons. I. Molecular properties of the purified isoenzymes. *Planta* 136:211–20
332. Walker, D. A., Herold, A. 1977. Can the chloroplast support photosynthesis unaided? *Photosynthetic Organelles. Special Issue Plant Cell Physiol.* 3:295–310
333. Walker-Simmons, M., Ryan, C. A. 1977. Immunological identification of proteinase inhibitors I and II in isolated tomato leaf vacuoles. *Plant Physiol.* 60:61–63
334. Wanner, G., Theimer, R. R. 1978. Membranous appendices of spherosomes (oleosomes). Possible role in fat utilization in germinating oil seeds. *Planta* 140:163–69
335. Wardale, D. A., Lambert, E. A., Galliard, T. 1978. Localization of fatty acid hydroperoxide cleavage activity in

membranes of cucumber fruit. *Phytochemistry* 17:205–12
336. Wellburn, A. R. 1977. Distribution of chloroplast coupling factor (CF_1) particles on plastid membranes during development. *Planta* 135:191–97
337. Wellburn, A. R., Hampp, R. 1976. Movement of labelled metabolites from mitochondria to plastids during development. *Planta* 131:17–20
338. Wellburn, A. R., Quail, P. H., Gunning, B. E. S. 1977. Examination of ribosome-like particles in isolated prolamellar bodies. *Planta* 134:45–52
339. Wellburn, A. R., Wellburn, F. A. M. 1971. A new method for the isolation of etioplasts with intact envelopes. *J. Exp. Bot.* 22:972–79
340. Williamson, F. A., Fowke, L. C., Constabel, F. C., Gamborg, O. L. 1976. Labelling of concanavalin A sites on the plasma membrane of soybean protoplasts. *Protoplasma* 89:305–16
341. Williamson, F. A., Morré, D. J., Hess, K. 1977. Auxin binding activities of subcellular fractions from soybean hypocotyls. *Cytobiologie* 16:63–71
342. Winter-Sluiter, E., Läuchli, A., Kramer, D. 1977. Cytochemical localization of K^+-stimulated adenosine triphosphatase activity in xylem parenchyma cells of barley roots. *Plant Physiol.* 60:923–27
343. Yatsu, L. Y., Jacks, T. J. 1968. Association of lysosomal activity with aleurone grains in plant seeds. *Arch. Biochem. Biophys.* 124:466–71
344. Yatsu, L. Y., Jacks, T. J. 1972. Spherosome membranes. Half unit-membranes. *Plant Physiol.* 49:937–43
345. Yatsu, L. Y., Jacks, T. J., Hensarling, T. P. 1971. Isolation of spherosomes (oleosomes) from onion, cabbage, and cottonseed tissues. *Plant Physiol.* 48: 675–82
346. Youle, R. J., Huang, A. H. C. 1976. Protein bodies from the endosperm of castor bean. Subfractionation, protein components, lectins, and changes during germination. *Plant Physiol.* 58: 703–9
347. Young, O., Beevers, H. 1976. Mixed function oxidases from germinating castor bean endosperm. *Phytochemistry* 15:379–85
348. Yu, R., Carter, J., Osawa, T. 1976. Separation of particulate phytochrome and plasma membranes of maize coleoptiles by density gradient centrifugation. *J. Exp. Bot.* 27:294–302
349. Zschoche, W. C., Ting, I. P. 1973. Malate dehydrogenase of *Pisum sativum*. Tissue distribution and properties of particulate forms. *Plant Physiol.* 51: 1076–81
350. Zuily-Fodil, Y., Passaquet, C., Esnault, R. 1978. High yield isolation of nuclei from plant protoplasts. *Physiol. Plant.* 43:201–4

Ann. Rev. Plant Physiol. 1979. 30:485–531

THE CELL BIOLOGY OF PLANT-ANIMAL SYMBIOSIS

❖7679

R. K. Trench

Department of Biological Sciences and The Marine Science Institute, University of California, Santa Barbara, California 93106

CONTENTS

INTRODUCTION

The intimate association between living plant cells and animal cells or tissues is a phenomenon of widespread occurrence. Although plant-animal symbioses have been under study for over a century, progress has been remarkably slow, probably because these phenomena have often been relegated to the realm of "biological curiosities." By comparison, other areas of symbiosis such as parasitism have made many advances because of medical applications. The fact that mutualistic symbioses do not result in

0066-4294/79/0601-0485$01.00

harmful effects, and seldom involve human partners, has been the most important factor in the relatively low priority given to the investigation of such problems. The mutualistic association of algae and animals is a system of cells and tissues composed of two genetically distinct cell types which proliferate harmoniously, resulting in a stable unit (28). Such situations are in contrast to many parasitic, oncogenic, or other pathologic conditions where interactions between genetically or developmentally divergent cells often result in disequilibrium.

By analyzing mutualistic systems, new insight into many basic parameters of intercellular interactions might be revealed. The underlying theme of this review will be to illustrate the possible use of plant-animal symbioses as model systems through which basic problems of intercellular relationships may be experimentally analyzed. In this article, I shall restrict myself to a review of recent developments in plant-animal symbiosis, placing the emphasis on intracellular associations between unicellular algae and invertebrates.

Reduced to its basic elements, intracellular symbiosis represents the coexistence of at least two genomes of divergent evolutionary origins occupying the same cytoplasmic environment. Foreign genomes (as infectious agents) within a cell may take the form of DNA-containing particles, such as lamda, kappa, mu and pi particles in *Paramecium auralia* (85, 102), or as double-stranded RNA particles in fungi (50), organelles such as plant chloroplasts in animal cells (73, 125), or complete organisms such as malaria symbiotes in erythrocytes (19, 120) or algae in invertebrates (64, 65, 113, 114). All these associations have several basic features in common.

First, one cell, by convention the host cell, serves as an environment within which the foreign genome expresses itself, either throughout its entire life history or for a portion thereof. The symbiont genome, therefore, interacts with the host's cytoplasmic environment, and constraints placed on the symbiont could originate from the host's genome.

Second, since the symbiont represents a foreign entity, its ability to exist within the host's cells must be predicated by some mechanism whereby the symbiont avoids being recognized as "nonself" by the host. If avoidance of recognition is not utilized, then avoidance of host responses aimed at the elimination of recognized foreign entities must be brought into play.

Finally, following the establishment of an association, the two partners of the consortium must mutually adapt their separate functions resulting in the increased fitness of the consortium.

All extant symbiotic associations represent the result to date of a series of events which brought two originally genetically independent entities together to produce a more or less integrated unit (126). In extreme cases a completely integrated association may result, as depicted in mutually obligate symbioses. In many such instances, the line separating a "semiautonomous endosymbiont" from a "semiautonomous organelle" might become vanishingly thin (92, 131). When investigating current associations, it must be recognized that the origins of the association are shrouded in the evolutionary past, and the future direction the association may take with respect to the degree of the integration expressed by the two components is unpredictable (22, 126). The aspects of plant-animal symbiosis that I shall discuss in this review will include the inception of associations between plant and animal cells or tissues and their subsequent integration, i.e. the physiological and biochemical adaptations which result in improving the "biological fitness" of the symbiotic unit or consortium.

As is the case in the study of the physiology and biochemistry of any biological group, the greatest advance is usually made after conditions for the controlled laboratory maintenance of the organisms involved have been developed. In this context, *Hydra viridis* and *Paramecium bursaria* were, and still are, important contributors to the analysis of plant-animal symbioses (41, 56). More recently, several marine systems have also been maintained under laboratory conditions. For example, *Convoluta roscoffensis* was brought into culture by Provasoli (86), *Amphiscolops langerhansi* by Taylor (111), and now *Aiptasia tagetes* and the scyphistomae of *Cassiopeia xamachana* are being maintained in our laboratories. In addition to the culture of the invertebrate hosts, the culture of the algal symbionts also represented a major advance. Although the algal symbionts of *H. viridis* have yet to be brought into culture, the symbionts of *P. bursaria* have been in culture for some time (41). *Platymonas convolutae* was brought into culture by Parke (80). The pioneering work of McLaughlin & Zahl (59, 60) was the prime force in bringing the dinoflagellate endosymbionts of marine invertebrates into culture (113, 115).

The plants and animals involved in symbiotic associations include the full range of taxa: the cyanobacteria, diatoms, green algae, and dinoflagellates on the one hand and invertebrate hosts ranging from the Protozoa to the Urochordata on the other. Examples of the major groups are presented in Table 1. More comprehensive lists may be found elsewhere (60, 64, 115). The general morphology of some of the symbiotic algae in situ in their respective hosts is illustrated in Figures 1–9.

INCEPTION

Since any extant intracellular symbiosis was initiated sometime in the evolutionary past, there is very little experimental work that can be conducted which would illuminate the mechanisms involved in the initiation of intracellular plant-animal associations. There are very few examples of plant-animal associations which have progressed from initial inception to an obligate association within the lifespan of any investigator. One possible exception is the association between the foraminiferan *Metarotaliella simplex* and chloroplasts of diatoms (E. Lanners, personal communication). Apparently cultures of *M. simplex* were maintained in the laboratory by feeding them diatoms over a number of years. Subsequently, seemingly intact diatom chloroplasts were observed in the cytoplasm of these forams. Unfortunately, to my knowledge, very little is known about functional aspects of this association, so it is unclear whether a "symbiosis" between the diatom chloroplasts and the foraminiferans has been established (51). However, some recent studies (13) suggest very strongly that functional diatom chloroplasts may be retained by foraminiferans under natural conditions. Such a situation may well parallel that observed in *Mesodinium rubrum* (5).

There are several instances wherein symbiotic associations between algae and animal hosts have to be reestablished with each succeeding generation of the host. Although with respect to the individual host or the algae infecting it the association is new for each new generation, one may assume that the potential to form such an association is probably based on a long evolutionary history of an existing association. Therefore, the inception of

Table 1 Algae-invertebrate associations[a]

Algal group	Identity	Host group	Host identity	Habitat	References
Cyanobacteria (blue-green algae)	*Cyanocyta korschikoffiana*	Protozoa (Fagellata)	*Cyanophora paradoxa*	Freshwater	29, 131
	Aphanocapsa	Porifera	*Ircinia*	Marine	90
	Prochloron	Urochordata	Several didamnid ascidians	Marine	53, 54
Bacillariophyceae	*Licmophora*	Platyhelminthes	*Convoluta convoluta*	Marine	1
Cryptophyceae		Protozoa (Sarcodina)	*Archaias*	Marine	49
Dinophyceae	*Amphidinium klebsii*	Platyhelminthes	*Amphiscolops langerhansi*	Marine	111
	Symbiodinium (= *Gymondinium*) *microadriaticum*	Various, from Protozoa to Mollusca	Some Radiolaria, many Cnidaria, Tridacnid clams	Marine	115
Chlorophyceae	*Platymonas convolutae*	Platyhelminthes	*Convoluta roscoffensis*	Marine	86
	Chlorella	Protozoa	*Paramecium*	Freshwater	69
		Porifera	*Spongilla*		
		Cnidaria	*Hydra*		
(Chloroplast symbioses)					
Rhodophyceae	*Griffithsia flosculosa*	Mollusca (Sacoglossa)		Marine	116
Bacillariophyceae		Protozoa (Sarcodina)		Marine	13
Chlorophyceae	*Codium fragile*	Mollusca (Sacoglossa)	*Elysia viridis*	Marine	125

[a] This list is not intended to be comprehensive, but should indicate examples within the range of associations.

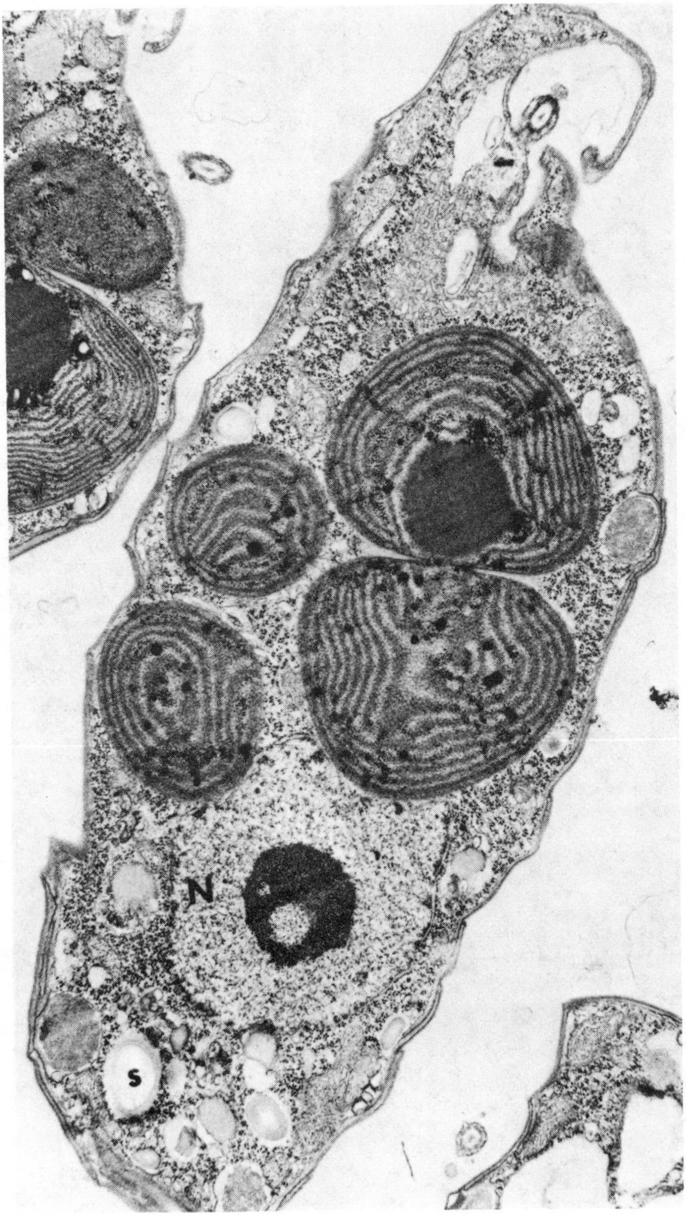

Figure 1 Transmission electron micrograph of *C. paradoxa.* N, nucleus with prominent nucleolus; S, starch. One symbiotic cyanobacterium is shown undergoing binary fission. Magnification approximately 15,000X.

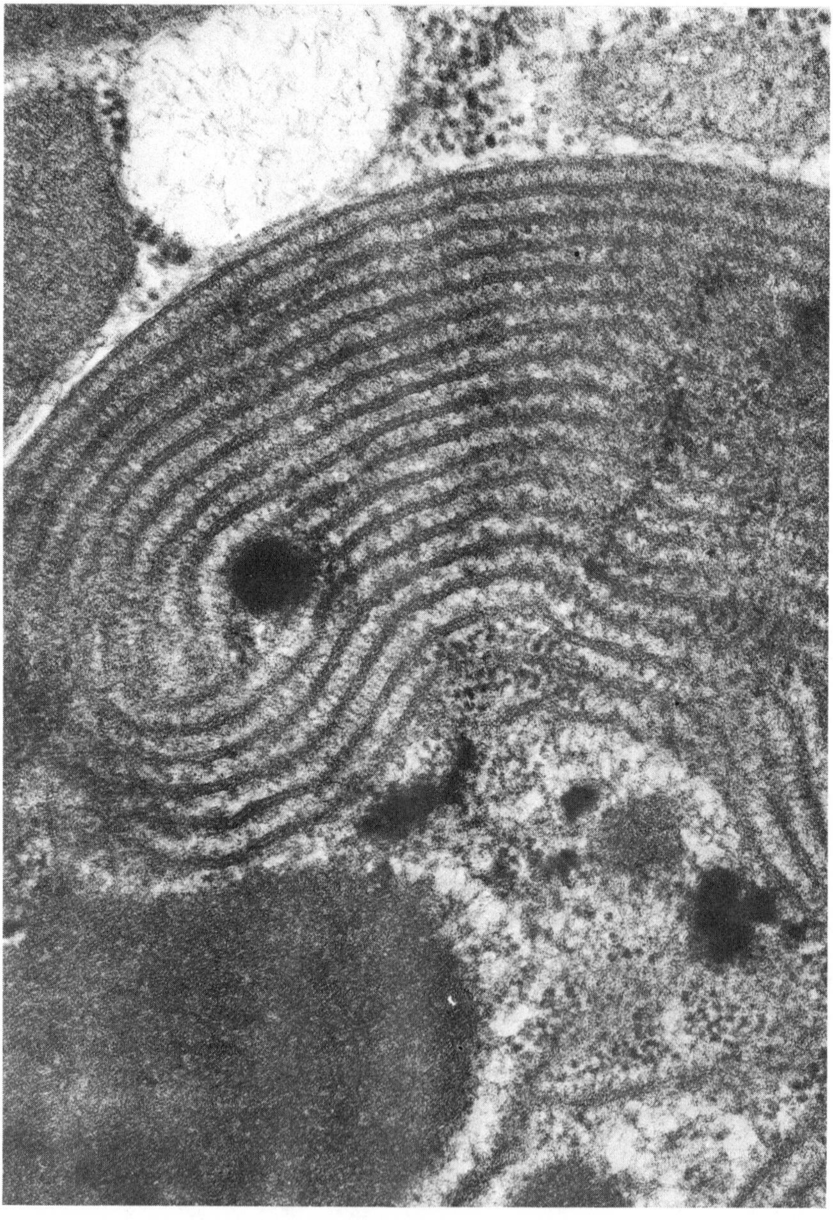

Figure 2 Transmission electron micrograph of the symbiotic cyanobacterium *C. korschikoffiana* showing details of the thylakoid structure and the attached phycobilisomes. Magnification approximately 170,000X.

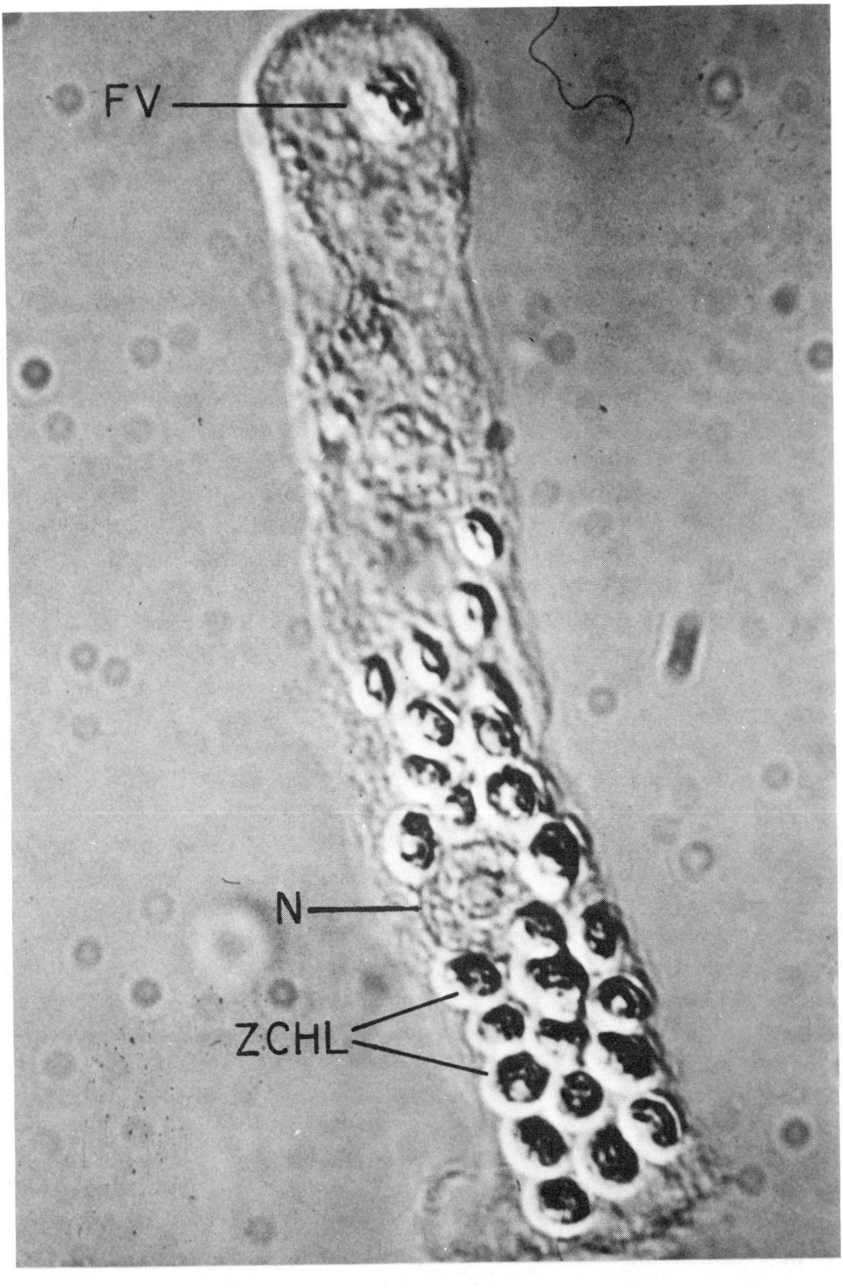

Figure 3 A photomicrograph taken with Nomarski optics of an isolated endoderm cell from *H. viridis* showing the intracellular location of the symbiotic *Chlorella.* ZCHL, "zoochlorellae"; N, nucleus, FV, food vacuole. Photograph taken by R. L. Pardy. Magnification approximately 900X.

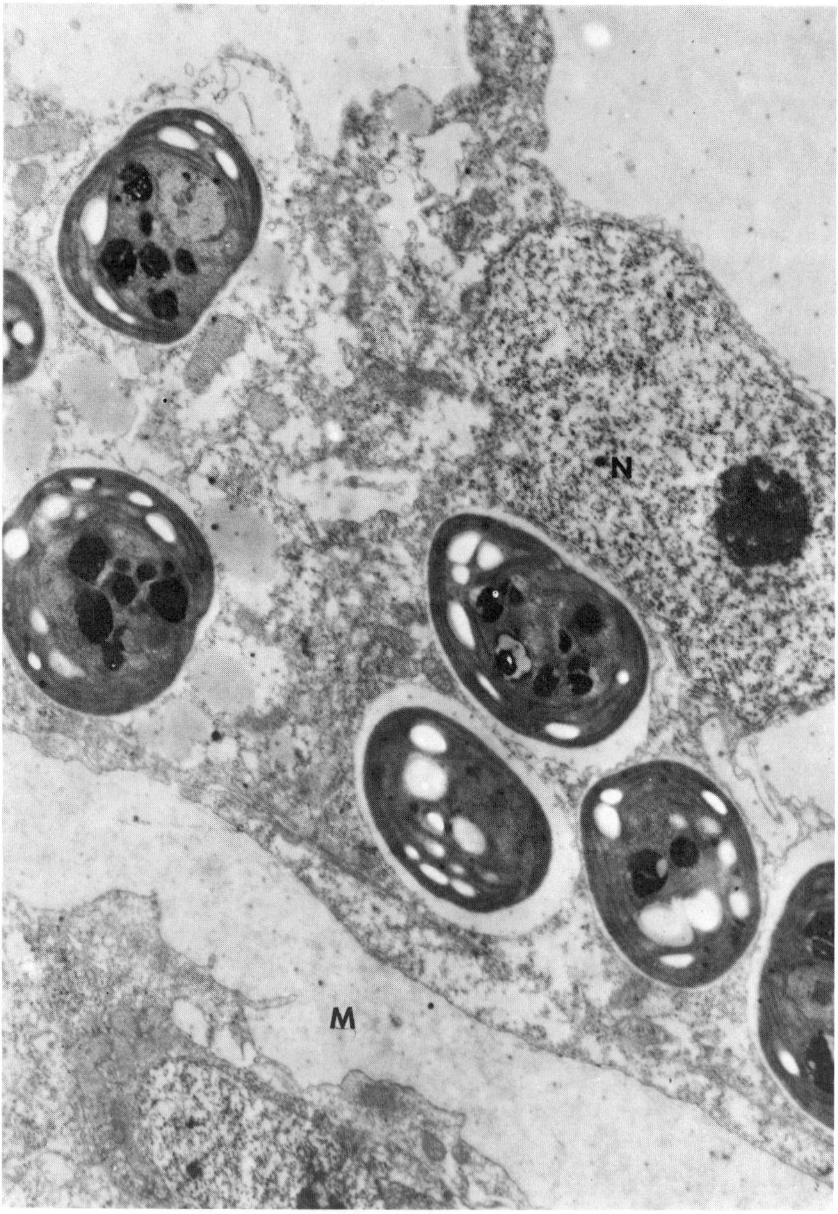

Figure 4 Transmission electron micrograph of an endoderm cell of *H. viridis* showing the symbiotic *Chlorella* in individual vacuoles at the base of the cell. N, nucleus of the endoderm cell; M, mesoglea. Photograph taken by T. Hohman. Magnification approximately 10,000X.

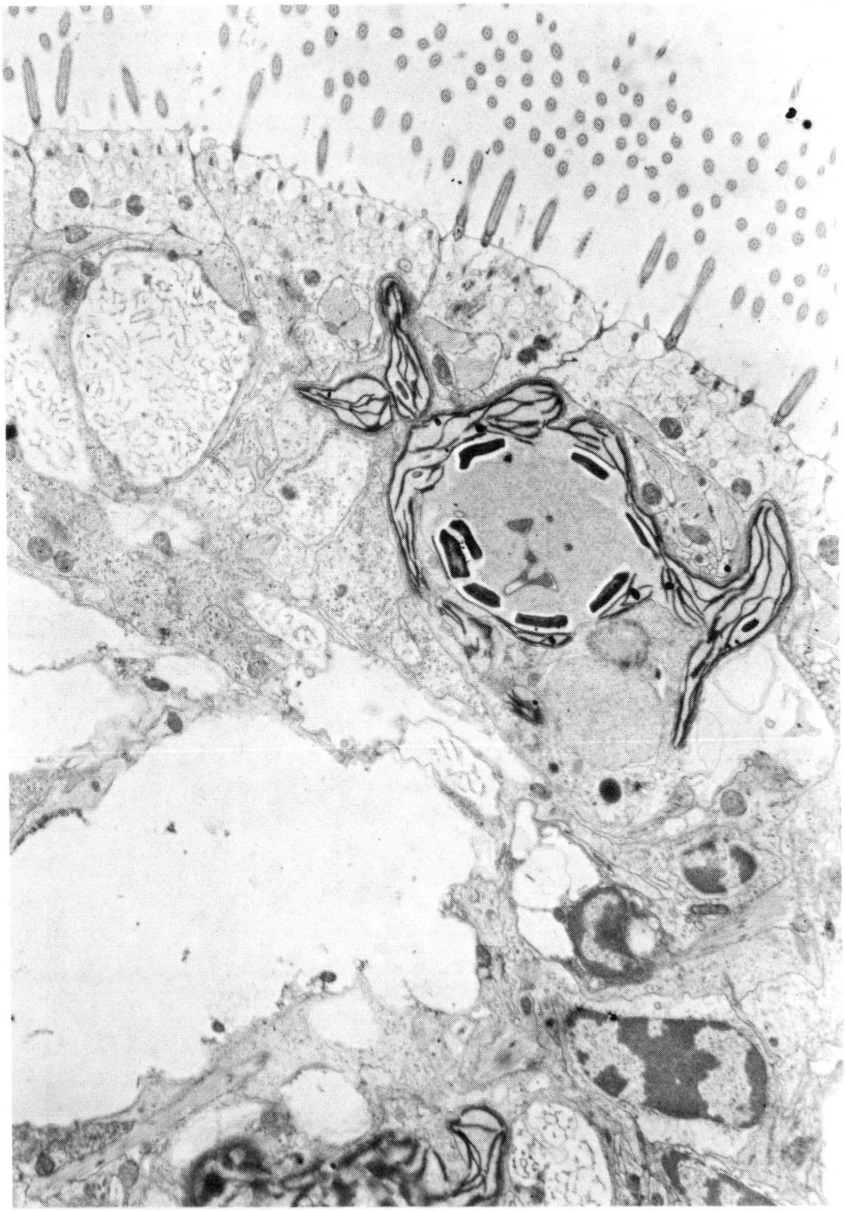

Figure 5 Transmission electron micrograph of *C. roscoffensis* showing a *Platymonas* located between the parenchyma cells of the host. Photograph courtesy of L. Provasoli. Magnification approximately 7,000X.

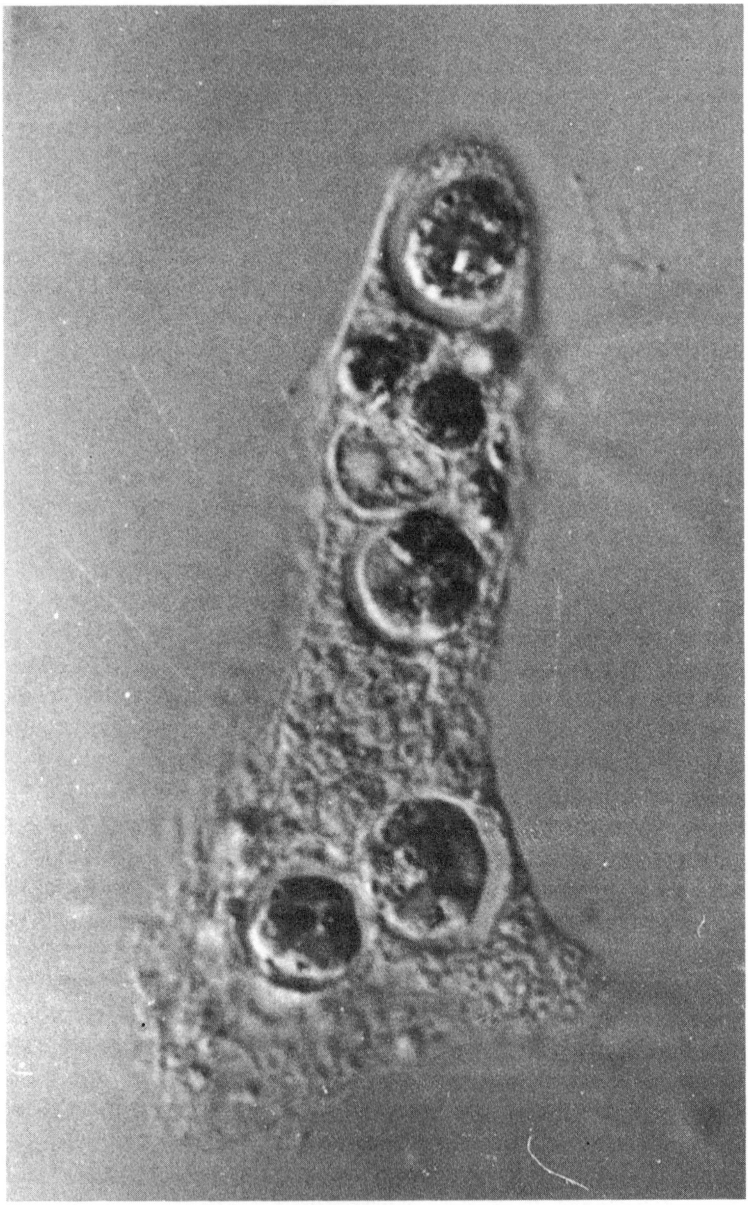

Figure 6 Light micrograph of an isolated endoderm cell from the marine hydroid *Myrtionema amboinense* showing the intracellular location of the symbiotic dinoflagellate *S.* (=*G.*) *microadriaticum.* Note that the algae are in individual vacuoles. Photograph taken with Nomarski optics. Preparation by W. Fitt. Magnification approximately 1,000X.

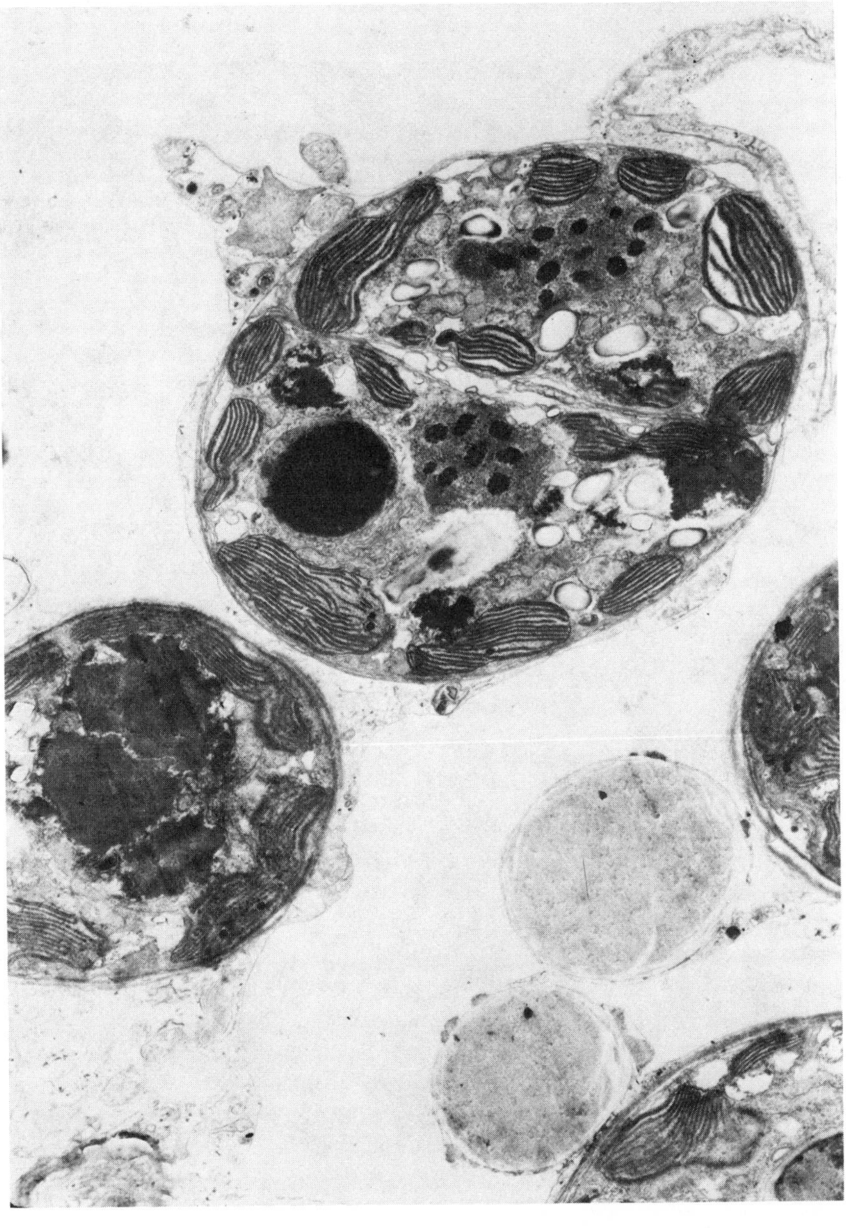

Figure 7 Transmission electron micrograph showing *S.* (=*G.*) *microadriaticum* in an endoderm cell of *M. amboinense.* A recently divided alga is illustrated. Magnification approximately 10,000X.

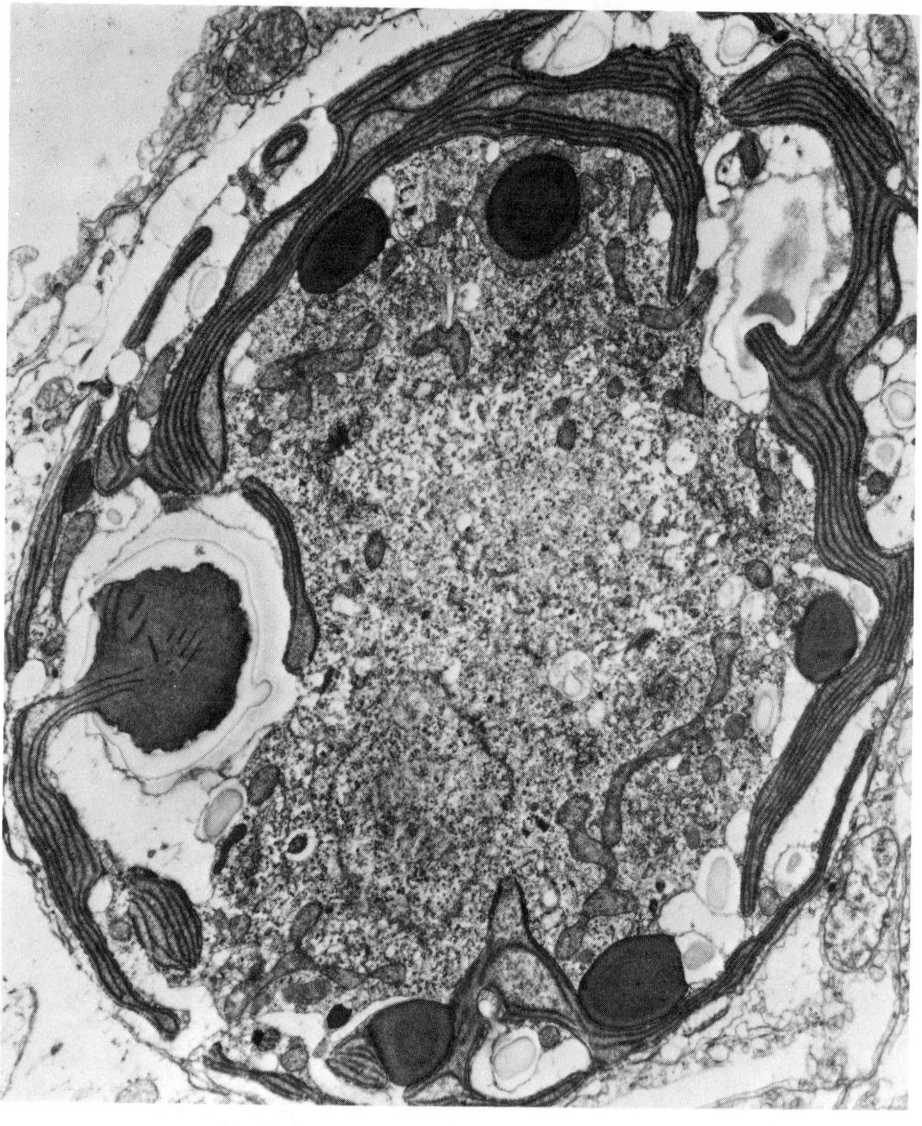

Figure 8 Transmission electron micrograph of the symbiotic dinoflagellate *Endodinium* (= *Amphidinium?*) *chattoni* in the "sail" *Velella velella.* Magnification approximately 12,000X.

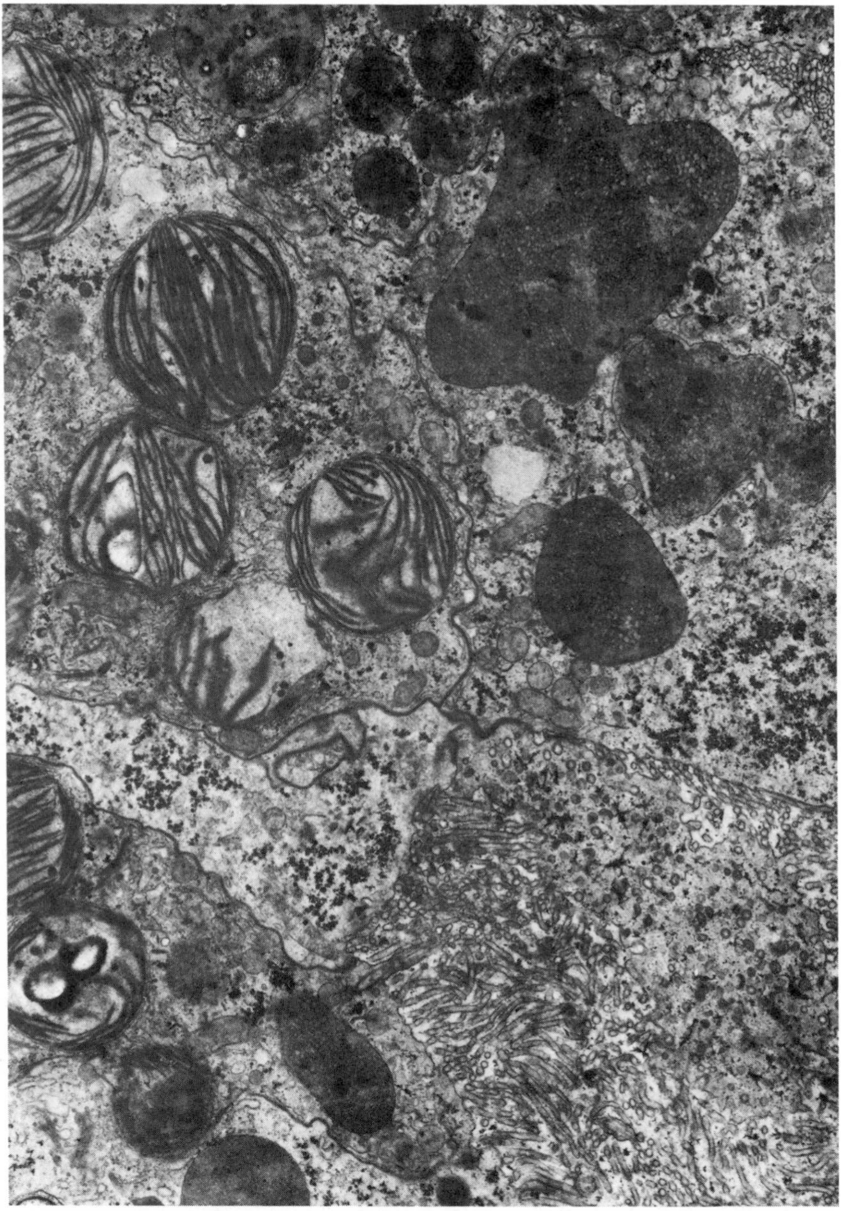

Figure 9 Transmission electron micrograph of digestive cells of the slug *Elysia viridis* showing the intracellular location of the symbiotic chloroplasts derived from the seaweed *Codium fragile.* Photograph from Trench et al (128), by permission of the Royal Society. Magnification approximately 10,000X.

intracellular symbioses in the systems I describe below is not homologous with what may have occurred during the initial establishment of an association. Nonetheless, it is possible to use such associations where new generations establish symbioses de novo to obtain information on the mechanisms involved in the process of symbiont acquisition by hosts.

In several examples of plant-animal symbioses, the algae are passed on from one generation to another via maternal inheritance. For example, in many symbiotic corals, marine hydroids, and zoanthids, as well as in *Hydra viridis,* the algae become incorporated into the egg shortly after fertilization (9, 26). The larvae then leave the parent organism, carrying with them a population of algae derived directly from the parent. Hence inheritance is cytoplasmic, maternal, and therefore non-Mendelian (126).

By contrast, there are many examples of symbiotic parents producing offspring which in the larval stage are devoid of algal symbionts. Such is the case in the flatworm *Convoluta roscoffensis* (86) and *Amphiscolops langerhansi* (111), the rhizostome medusae *Cassiopeia* and *Mastigias* (2, 57, 109), and tridacnid clams (47). In all these cases the postlarval stage of the new generation must acquire algal symbionts from the ambient environment. The mechanisms through which infection is achieved are very poorly understood. The question, therefore, becomes: How do potential algal symbionts locate hosts? As stated, this question implies that the active component in the location process is the algal symbiont. Such is not necessarily the case. For example, in the symbiosis between sea slugs (*Elysia viridis, Tridachia crispata*) and chloroplasts of siphonaceous algae, the juvenile slugs acquire plastids by feeding on the appropriate plant (73, 125). In this case, the plant is stationary, the slug mobile.

One of the underlying features of intracellular plant-animal symbioses is that at the cellular level, acquisition of symbionts is almost inseparable from phagotrophy (142). In all intracellular symbiotic associations examined to date, the algae are phagocytosed superficially just as any food particle might be (61, 68, 91). Even in the intercellular symbiosis in *C. roscoffensis,* it is believed (76) that the algae pass through an intracellular stage before finally taking up residence in the interstitial spaces between the host's parenchyma cells. In some of the recently described associations between algae and foraminiferans (13, 51), the algae are phagocytosed and all components but the chloroplasts are presumed to be digested. Thus, establishment of intracellular symbioses could be viewed, at least in part, as failure of the host's intracellular digestive function.

Location (Mechanisms Whereby Symbionts Locate Hosts)

In *C. roscoffensis* the eggs are deposited in egg cases in the sandy beaches where the adult worms occur. The algal symbiont *Platymonas convolutae* also occurs in the same environment. Several investigators of this association believe that the free-swimming *P. convolutae* are attracted to the egg cases of the young *C. roscoffensis* through chemotaxis (37). However, the substance in the egg cases responsible for eliciting chemotaxis by the algae has not been identified. In fact, chemotaxis (as opposed to a random chance contact event) has never been unambiguously demonstrated.

The adults of the symbiotic rhizostome medusae *Cassiopeia* and *Mastigias* produce aposymbiotic (free of algae) planula larvae. These larvae settle and metamorphose into the scyphistoma stage of the life cycle. If they become infected with an appropriate strain of the dinoflagellate *Symbiodinium* (=*Gymnodinum*) *microadriaticum*, the polyps then strobilate producing juvenile jellyfish (W. Fitt and R. Trench, unpublished). In this case, understanding the mechanism of infection in the natural environment is more difficult than is the case with *C. roscoffensis.* The algal symbionts have never been recognized free living in the habitat where the scyphistomae occur. However, under laboratory conditions it has been possible to demonstrate infection of scyphistomae with algae isolated from adult hosts (2, 57, 109). W. Fitt and R. Trench (unpublished) observed that the cultured gymnodinioid swarmers entered the polyps via the mouth, but no taxis could be demonstrated. In fact, the process appears to be one of random chance, but further studies are currently underway to resolve this problem. A similar phenomenon was reported by Kinzie (46) in the infection of postlarval stages of the gorgonian *Pseudopterogorgia bipinnata* with motile cells of *S.* (=*G.*) *microadriaticum.*

The situation is even more unclear in the case of *Tridacna.* These clams are generally regarded as specialized herbivores (141). The veligers produced by the adults are presumably planktonic before they settle to metamorphose into juveniles which are aposymbiotic. It is thought that the algae are acquired from the ambient environment. Assuming this to be the case, the only mechanism through which they could enter the clam is through the feeding apparatus and the gut. As the algae eventually come to reside in the haemal sinuses of the hypertrophied siphonal tissues, there is presumably some route whereby they avoid digestion and obtain access to the blood spaces (88, 134). Although the details of this process are totally unknown, repeated observation of algae passing from the siphonal tissue to the digestive tract from whence they are voided, often intact, suggests that a mechanism for the reverse process probably also exists.

Recognition, Endocytosis, and Sequestration

In many instances of plant-animal symbioses, the algae are acquired anew by each new host generation, and the motile stage in the alga's life history probably represents a dispersal phase as well as the infectious stage. Thus it is very likely that under natural circumstances, recognition of the motile algae by potential hosts may be the most important process in determining the continued success of a given symbiotic association.

Intercellular recognition phenomena are known to play a very important role in the establishment of specific associations between a variety of organisms, for example legumes and root nodule-forming bacteria. Through the use of serological tests (for surface antigenic determinants) and lectins (which demonstrate specific affinities for sugar moieties associated with surface membrane proteins), the molecular basis for recognition and specificity in several associations has been analyzed with a great deal of success (20, 81, 89, 97). The state of the art with respect to algae-invertebrate symbioses has not yet progressed to the same level of refinement, but these powerful molecular biological tools are now being brought to bear on problems of recognition and specificity in plant-animal endosymbioses (84). I shall, therefore, review the current status of recognition phenomena in algae-invertebrate symbioses, treating selected examples separately. Some aspects of recognition merge insensibily with "specificity." The latter topic is discussed in a later section.

CONVOLUTA ROSCOFFENSIS AND *PLATYMONAS CONVOLUTAE* The symbiosis between the marine flatworm *C. roscoffensis* and the green flagellate *P. convolutae* represents an intercellular association, since the algal symbionts reside between the host cells (73, 76).

In their study on the resynthesis of the symbiosis between *C. roscoffensis* and *P. convolutae,* Provasoli, Yamasu & Manton (86) illustrated the subtleties in aspects of the "recognition" and "specificity" of algal-invertebrate associations. As stated previously, the eggs of *C. roscoffensis* are free of algae when they are deposited in egg cases on the beach. Taking advantage of this, the investigators were able to obtain aposymbiotic juveniles which they used as a test system through which to assay the capabilities of different prasinophycean algae to "infect" the worms and establish a stable association with them. The results they obtained indicated that there was no obvious specificity in the process of "infection" since worms provided with different species of algae which were not "normal" symbionts became infected by those algae "as readily" as with their normal algae, *P. convolutae.* However, the unnatural algae proliferated less rapidly in the worms than did *P. convolutae,* and the worms infected with *P. convolutae* grew more rapidly than those infected with other algal species. The most significant observation was that when unnatural algae had infected and become established in the worms, they could be quantitatively displaced by subsequent infection by the normal symbiont.

PARAMECIUM BURSARIA AND *CHLORELLA* Although the term "zoochlorella" has been used to refer to those green algae found as endosymbionts in freshwater invertebrates, the term has no real taxonomic meaning. It has not been determined whether the "zoochlorella" in *P. bursaria, H. viridis,* and *Spongilla* are the same species, different species, or subspecies, varieties, or strains. Hence a major underlying problem in the analysis of recognition and specificity in associations involving "zoochlorellae" pivots on the genetic characterization of the algae.

In their studies of the reinfection of *P. bursaria* with cultured algae, Karakashian & Karakashian (43) found that *Chlorella* isolated from other hosts (e.g. *Hydra, Spongilla*), some strains of free-living *Chlorella vulgaris,* and a strain of *Scenedesmus* could all "infect" aposymbiotic *P. bursaria.* In the artificial associations so produced, those formed with free-living algae were found to be less stable than those formed with symbiotic algae. The free-living algae were able to sustain growth of the host less well than the symbiotic forms. Bomford (6) also infected *P. bursaria* with free-living *Scenedesmus* and demonstrated that subsequent infection with native *Chlorella* from *P. bursaria* resulted in the elimination of the *Scenedesmus* cells. These observations, suggesting a lack of recognition and discrimination, are in disagreement with the conclusion reached by Weis & Berry (138), who proposed that each stock of *P. bursaria* is host to a single host-specific species of *Chlorella.*

HYDRA VIRIDIS AND *CHLORELLA* Probably the most significant studies on intercellular recognition phenomena in plant-animal symbiosis have been those reported by Muscatine's laboratory at UCLA (68, 73, 83, 84) using the *Hydra viridis-Chlorella* system.

The digestive endoderm cells of *H. viridis* readily phagocytose *Chlorella* cells when the latter are injected into the coelenteron of the hydra. The endoderm cells usually respond to the presence of the algae by accumulating them in one or more large vacuoles which remain at the distal end of the cells. A variety of "strains" of *Chlorella* may thus be phagocytosed. With the exception of *Chlorella* isolated from *Hydra* and those from *P. bursaria* (NC64A), the algae usually disappear from the endoderm cells within 24 hr after phagocytosis, presumably through exocytosis.

By contrast, algae from *H. viridis* and those from *P. bursaria* are endocytosed into single vacuoles, and they are rapidly transposed from the distal (bordering the coelenteron) to the proximal (basal) end of the cell.

The events described above define the "recognition" system in *Hydra viridis.* Muscatine et al (68) distinguished five phases in the process of the reinfection of *Hydra* by symbiotic *Chlorella:* (*a*) the contact phase, (*b*) the engulfment phase, (*c*) a recognition phase, (*d*) an intracellular migration phase (sequestration), and (*e*) repopulation (increase in the number of algae).

When symbiotic *Chlorella* are injected into the coelenteron of *Hydra,* they make contact with the endocytotic microvilli of the endoderm cells. The points of contact between animal cell microvilli and algae are often marked by electron dense regions in the animal cell membrane. These electron dense regions are interpreted as diffuse glycocalices (17, 69) similar to those seen during phagocytosis in amoebae (11) and in macrophages (99, 108). Scanning electron microscopy of the endocytotic event shows that microvilli of the *Hydra* endoderm cell initially extend and form a meshwork over the surface of the algae before phagocytosis. This is in contrast to what occurs during the phagocytic uptake of heat-killed *Chlorella* or latex beads, wherein a veil of cytoplasm forming a tubular extension of the endodermal cell membrane is produced, and this envelopes the particle (P. McNeil, personal communication). The endocytotic microvilli of *Hydra* endoderm cells which are involved in endocytosis of the algae are devoid of coated vesicles, in marked contrast to the microvilli which phagocytose food particles (17). When food particles or killed *Chlorella* cells are phagocytosed, the vacuoles thus formed are large, may contain several algae or food particles, and tend to remain at the distal end of the cell, while other vacuoles, each containing a single symbiotic *Chlorella,* are transported proximally. However, P. McNeil (personal communication) has found evidence for the movement of food vacuoles from distal to proximal ends of *Hydra* cells. Nonetheless, the observations cited above suggest very strongly that at the cellular level the endocytosis of symbiotic algae is a unique event and may not be homologous with phagocytosis of food particles. Whether recognition of the algae by the endoderm cells of *Hydra* occurs at the point of intercellular contact or after the endocytotic event is unclear. That several different algae or even heat-killed symbiotic *Chlorella* are phagocytosed might suggest that the initial uptake process is nondiscriminatory. However, since the initial and ultimate fate of the foreign or killed algae and of food particles appear distinct from those of symbiotic *Chlorella,* there is a strong suggestion that recognition is initiated during intercellular contact. The process of segregation and sequestration of the algae may then be a continuation of the initial recognition process.

Microtubules have been implicated in the process of the transfer of the symbiotic algae, enclosed within individual endosomes, from the distal to the proximal end of the endoderm cells. The process is inhibited by a variety of drugs known to disrupt microtubule assembly, e.g. colchicine and vinblastine (C. Cook, personal communication) and β-peltatin and podophylotoxin (16, 18). That these drugs specifically disrupt microtubule function in *Hydra* has not been ascertained.

Taking advantage of the discriminatory aspects of the algal uptake process in *Hydra,* Pool (83, 84) tested the hypothesis that the *Hydra* endoderm cells were able to discern differences on the surface of the algae. Using the technique of microcomplement fixation, a series of elegant experiments was conducted which demonstrated that antisera produced against intact algae isolated from *H. viridis* were specific to those algae and reacted with the surface of the algal cells, thereby establishing that the algae possessed surface antigenic determinants.

Since microcomplement fixation can illustrate both qualitative and quantitative differences in the distribution of antigenic determinants on cell surfaces (52), Pool (83, 84) then compared algae from *Hydra* (Florida strain) with a variety of other "strains" of symbiotic *Chlorella* freshly isolated from their respective hosts (e.g. English strain and Carolina strain). It was observed (Figure 10) that all the *Hydra* algae, with the single exception of the English strain, contained qualitatively similar surface antigenic determinants. However, on introducing algae from the English *Hydra* into aposymbiotic individuals of the Florida strain and allowing the algae to proliferate in such a heterologous association for from 3–21 months, subsequent comparison of the surface antigenicity of algae from the heterologous and the homologous associations demonstrated that the algae from the English *Hydra* had acquired antigenic determinants qualitatively similar to those of the normal symbionts from Florida *Hydra* (Figure 11).

When similar analyses were conducted comparing the algae from the Florida *Hydra* with NC64A grown in culture, the latter demonstrated

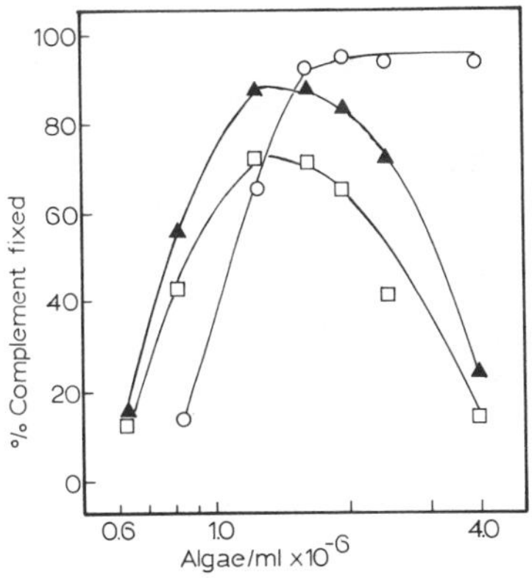

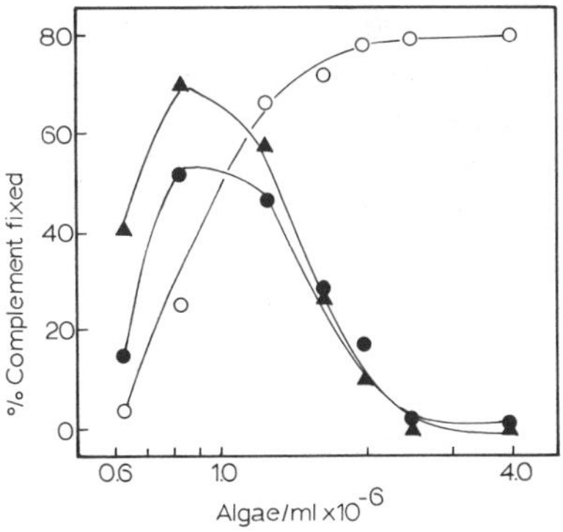

Figure 10 (*left*) Microcomplement fixation assay using antiserum prepared against *Chlorella* isolated from *H. viridis* (Florida strain) to assay "Florida" algae (closed triangles), "English strain" algae (open circles), and "Carolina strain" algae (open squares). All algae represent homologous associations. Data by permission from R. R. Pool (83).

Figure 11 (*right*) Microcomplement fixation assay using antiserum prepared against *Chlorella* isolated from *H. viridis* (Florida strain) to assay "Florida" algae (closed triangles), "English strain" algae grown in "English *Hydra*" (open circles) and "English strain" algae grown in "Florida strain" *Hydra* (closed circles). Data by permission from R. R. Pool (83).

neither qualitative nor quantitative similarity to the former. When NC64A was introduced into *Hydra,* allowed to proliferate, and then compared with the normal *Hydra* symbionts, the NC64A were observed to have acquired some antigenic characteristics similar to those of the normal symbionts (Figure 12).

The data on surface antigenicity could be interpreted as demonstrating that the *Chlorella* used in these studies are probably strains or variants of the same species which possess adaptive capabilities allowing each different "strain" to acquire at least a critical minimum number of surface antigenic determinants in common with "normal" *Hydra* algae when in the *Hydra* cell environment. This conclusion would imply that surface characteristics play an important, if not central, role in the process of recognition of potential algal symbionts by *Hydra* endoderm cells.

Pool (83, 84) tested the hypothesis that a relationship exists between the presence of antigenic determinants on the surface of *Chlorella* and the rate of uptake and sequestration by *Hydra* endoderm cells, by comparing rates of uptake of antigenically characterized *Hydra* algae and NC64A. His results indicate quite strongly that the rate of uptake and sequestration of *Hydra* algae and of NC64A, previously grown in *Hydra,* was reduced by greater than 60% when such algae were first treated with antiserum prepared against "normal" *Hydra* algae. This result implies that masking of the antigenic sites on the surface of the algae markedly reduced the capacity for intercellular recognition. That some algae were nonetheless phagocytosed and sequestered is significant. It is possible that the antibodies complexed to some algal surfaces were digested away within the coelenteron before intercellular contact was made. A possible alternative is that the surface antigenic sites on a certain proportion of the algae were not masked by the antibodies, thereby leaving some recognition sites available. The possibility of the existence of genetic heterogeneity in the algal population in *Hydra* cannot be ruled out.

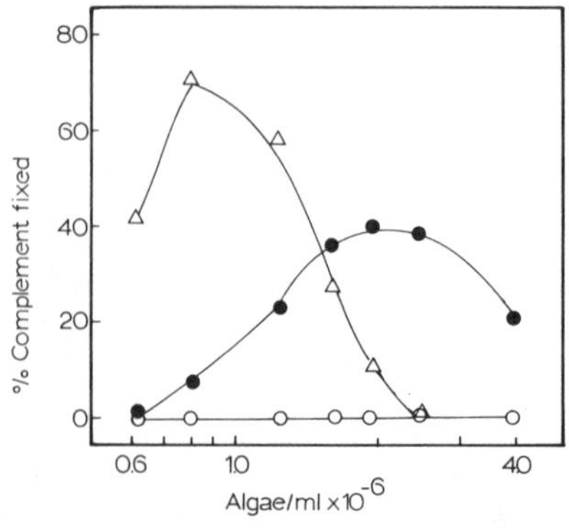

Figure 12 Microcomplement fixation assay using antiserum prepared against *Chlorella* isolated from *H. viridis* (Florida strain) to assay "Florida" algae (closed triangles), NC64A grown in "Florida strain" *H. viridis* (closed circles) and NC64A grown in culture (open circles). Data by permission from R. R. Pool (83).

MARINE INVERTEBRATES AND DINOFLAGELLATES Two groups of dinoflagellates are known to occur as endosymbionts with marine invertebrates. Amphidinioid dinoflagellates, e.g. *Amphidinium klebsii,* are found as intercellular symbionts in the acoel flatworm *Amphiscolops langerhansi* (111) and as intracellular symbionts, e.g. *A. chattoni,* in *Velella velella* (112, 115; cf 36). By far the most frequently occurring dinoflagellate endosymbionts are the "gymnodinioid" dinoflagellates. These algae, commonly referred to as "zooxanthellae," occur among marine invertebrates ranging phylogenetically from the Protozoa to the Mollusca (115). They also possess an extensive geographic range, and although found principally in the tropics in association with coral reefs, they also occur in temperate regions, e.g. the west and northeast coasts of the United States and the United Kingdom. Until recently (93) it has been a widely held view that all the gymnodinioid dinoflagellates associated with benthic marine invertebrates were one species, *Gymnodinium microadriaticum* (112–115). A thorough discussion of the taxonomy of zooxanthellae is outside the scope of this review. A summary of the evidence supporting the view that the "gymnodinioid" dinoflagellates are not identical will be presented, however, because it is relevant to a discussion of recognition and discrimination of the algae by potential hosts. A more extensive treatment may be found in Schoenberg & Trench (94–96).

Zooxanthellae isolated from a variety of marine invertebrates were brought into axenic culture in a defined medium and maintained under identical conditions of temperature and illumination for up to 2 years. Aqueous extracts of the algae were prepared, and the extracts were analyzed for isoenzymes by starch gel electrophoresis and for general soluble proteins by polyacrylamide gel disc electrophoresis. The resulting isoenzyme and protein electrophoretic patterns were used as phenotypic characters in estimating similarity among the isolates. Algae isolated from 17 different hosts were resolved into 12 electrophoretic strains (see Figure 13). Some of these strains resembled each other more closely than others, and the strains could be separated into three groups. The biochemical differences observed were corroborated by morphological and ultrastructural differences (95). Experimental analyses showed that the biochemical and morphological characteristics were intrinsic and stable in that they did not vary when the algae were artificially placed into heterologous associations. Similarily, recent studies (S. Chang, unpublished) of the peridinin-chlorophyll *a*-protein (PCP) complexes of different strains of *S.* (=*G.*) *microadriaticum* have shown that the proteins produce distinct isofocusing patterns (Figure 14) which may be correlated with differences in the amino acid composition of the proteins, suggesting differences in the genetic constitution of the different strains. The possible use of PCP as a genetic marker is under investigation.

Since intrinsic differences were evident in different isolates of *S.* (=*G.*) *microadriaticum,* Schoenberg & Trench (96) tested the possibility that such differences might be detected by a given potential host organism. Clones of the sea anemone *Aiptasia tagetes* were used as the experimental host, and the anemones were offered various strains of algae including those isolated from *A. tagetes.* This latter assay served as control. The rate of infection and subsequent repopulation of the host was monitored by counting the number of algae present in excised tentacles.

The results (Table 2) demonstrated that algae from *A. tagetes* infected aposymbiotic anemones most rapidly. They also proliferated and established maximum densities within 10 days. Those strains of algae found to be biochemically closely related to the algae from *A. tagetes* also infected rapidly but proliferated more slowly. The more dissimilar strains were much less infective and achieved population densities up to three orders of magnitude lower than those from *A. tagetes.* Several strains demonstrated no infectivity, and infectivity appeared to have little direct correlation with the electrophoretic group into which the algae fell or their gross morphology (see Table 3), suggesting that some other property of the algae, probably molecular surface characteristics, may play an important role in recognition in dinoflagellate associations as they appear to in the *Hydra-Chlorella* system. The molecular basis of recognition in marine inverte-

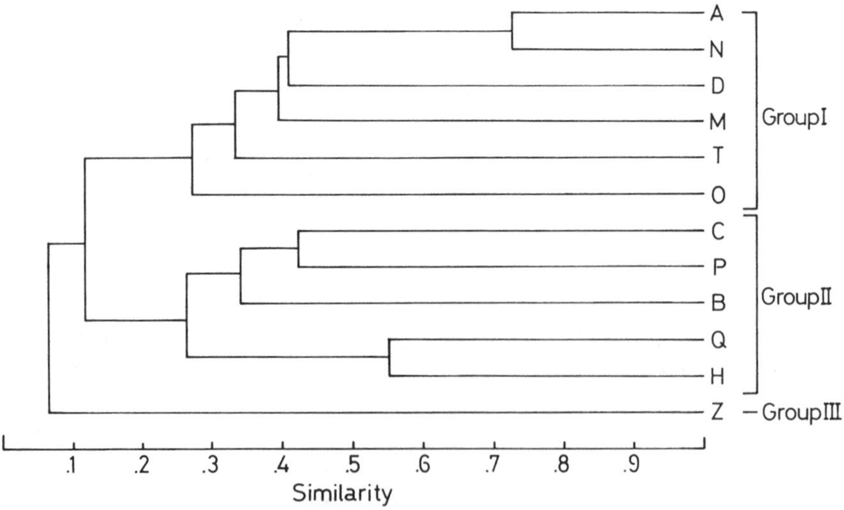

Figure 13 A dendrogram of similarity coefficients based on isoenzyme patterns obtained by electrophoresis of extracts of cloned *Symbiodinium* (=*Gymnodinium*) *microadriaticum* isolated from different invertebrate hosts. Data from Schoenberg & Trench (93).

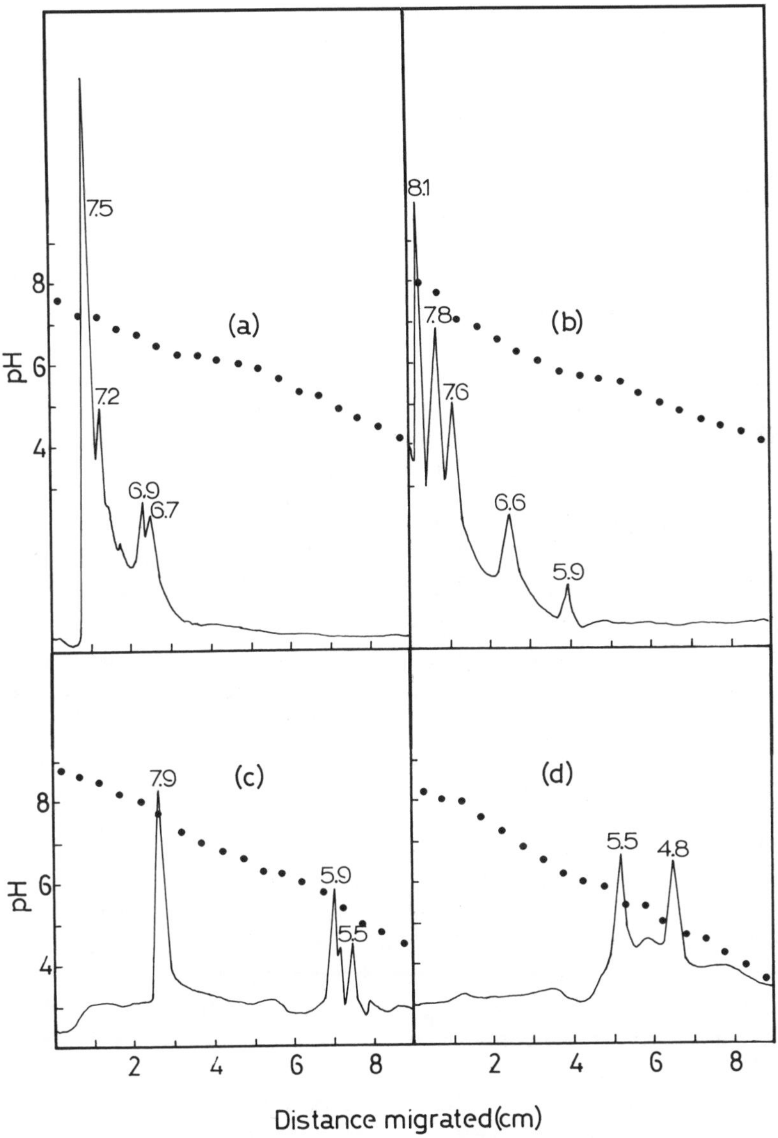

Figure 14 Isoelectric focusing patterns of peridinin-chlorophyll *a*-proteins (PCP) extracted from *Symbiodinium* (=*Gymnodinium*) *microadriaticum* isolated from (a) *Anthopleura elegantissima,* (b) *Zoanthus sociatus,* (c) *Aiptasia tagetes,* and (d) *Oculina diffusa.* The numbers associated with the peaks are the isoelectric points of those proteins. Data collected by S. Chang.

Table 2 Infectivity and repopulation of aposymbiotic *Aiptasia tagetes* by different "strains" of *S.* (=*G*) *microadriaticum* isolated from different hosts and grown axenically under identical conditions[a]

Algal "strain"	Original host	Mean no. of algae/mm tentacle, 0 + 5 days (± range)	N	Mean no. of algae/mm tentacle, 0 + 50 days (± range)	N
A	*Aiptasia tagetes*	3×10^2 (± 100)	12	2×10^4 (± 1600)	12
N	*Mussa angulosa*	1×10^1 (± 5)	5	7×10^3 (± 400)	5
B	*Bartholomea annulata*	1×10^1 (± 2)	5	2.5×10^3 (± 370)	5
H	*Heteractis lucida*	7×10^1 (± 31)	5	2.3×10^2 (± 80)	5
O	*Oculina diffusa*	1.5×10^2 (± 10)	5	2×10^2 (± 90)	5
Z	*Zoanthus sociatus*	0	5	0	5

[a] Data from Schoenberg (92a). See also Schoenberg & Trench (96).

brate-gymnodinioid dinoflagellate associations is currently being investigated in our laboratory through the combined use of serology and lectin-binding techniques.

Algal Persistence and Avoidance of Host Destruction

It is apparent that the early events in the initiation of an intracellular symbiosis, that is, endocytosis and intracellular transport of algal symbionts (sequestration), result in the incorporation of a "foreign" entity (the algae) into the cytoplasmic environment of the host cell. In order to persist in the host cell, the potential endosymbiont must either be recognized by the host as "self" or somehow be able to counteract or prohibit the host cell from destroying or eliminating it.

It seems clear from the capability of several hosts to discriminate between different algae that some algae are recognized as compatible while others are not. However, there is no evidence to date, with the possible exception of the system involving (Florida) *Hydra* and algae from (English) *Hydra*, which suggests that the algae acquire unique characteristics by virtue of residing in the cell of a given host. Masking recognition sites in order to avoid host immune responses is known to exist in some parasitic symbioses (117).

As a general rule, intracellular symbiotic algae appear to resist host destruction by avoiding digestion, either by counteracting the fusion of lysosomes with vacuoles containing the algae, a situation analogous to that seen in the interaction between macrophages and *Mycobacterium* (120), or possibly by possessing cell walls which are resistant to hydrolytic attack by the host (73).

Avoiding digestion by counteracting lysosomal fusion with endosomal vacuoles can be inferred from two unrelated systems. First, Karakashian &

Table 3 Comparison of biochemical and morphological characteristics of "strains" of *S.* (= *G*) *microadriaticum* with their ability to infect and repopulate *A. tagetes*[a]

Electrophoretic group	Size group	Strain	Infectivity and repopulation (with respect to *A. tagetes*)
1	small	A	high
	(<10μm × 10μm)	N	intermediate
		M, O, T	low
		D	none
2	large	B, C, H, Q	low
	(>10μm × 10μm)	P	none
3	large	Z	none
	(>10μm × 10μm)		

[a] Data from Schoenberg (92a) and Schoenberg & Trench (93).

Karakashian (40) showed that in *P. bursaria,* the symbiotic *Chlorella* are not digested under normal circumstances, but heat-killed *Chlorella* are. However, when heat-killed and living *Chlorella* are simultaneously provided to paramecia, both may be taken into the same endosome, but digestion of neither ensues. Eventually, the live *Chlorella* cells are sequestered into separate vacuoles from the heat-killed ones. At this point, the heat-killed cells are subject to lysosomal degradation (39), while the live algae remain undigested.

A second example comes from the symbiosis between algal chloroplasts and sea slugs. Trench (125) and Muscatine et al (73) reported that newly endocytosed chloroplasts in the digestive cells of *Tridachia crispata* and *Elysia viridis* remain relatively briefly in the vacuoles. The vacuolar membranes soon disappear, and the plastids then come to lie freely in the animal cell cytoplasm. As long as they retain their morphological and functional integrity, they are retained by the slug's cells. However, when they become defunct, because of their lack of total biochemical autonomy (130), vacuoles interpreted as autophagic vacuoles are formed around the moribund plastids, and lysosomes containing acid phosphatase fuse with such vacuoles and appear to effect hydrolytic degradation of the plastids. This sequence of events is not unlike that found in vaccina virus, *Toxoplasma* or *Mycobacterium,* after endocytosis by macrophages (38, 73).

Although digestion of algae by animal hosts has often been proposed as the mechanism whereby animals derive nutritional benefit from their algae, unambiguous evidence for the intracellular digestion of symbiotic algae is still lacking in most systems. Muscatine et al (68) postulated that *Chlorella* in *Hydra viridis* were resistant to lysosomal degradation by virtue of the presence of the polycarotenoid sporopollenin in the cell wall.

Fankboner (24) interpreted the results of his electron microscopic histochemical study as demonstrating lysosomal degradation of zooxanthellae in blood amoebocytes in *Tridacna maxima,* but this interpretation has not been corroborated by other investigators. Ricard & Salvat (88) reported defaecation of apparently intact zooxanthellae by *T. maxima* while Trench et al (134) have found morphologically intact, photosynthetically active zooxanthellae in the intestine and faeces of *T. maxima.* These defaecated algae were viable and could be brought into culture.

Recently, Steele & Goreau (107) found a substance in the ruff, oral disc, and tentacles of the anemone *Phyllactis flosculifera*, which when incubated in vitro with algae isolated from *P. flosculifera* resulted in a reduction in the numbers of algae. The active ingredient from the anemone was $(NH_4)_2SO_4$-precipitable and nondialyzable and was assumed to be protein. No direct cause and effect relation between the presumed degradative protein and the occurrence of pycnotic algae in the animal's digestive cells was established.

Despite all of the above, several investigators have observed the presence of pycnotic algae in the digestive cells of their hosts, particularly in marine invertebrate-zooxanthellae associations. In observing what appeared to be intracellular degradation of algae, but finding no concrete evidence for intracellular digestion of algae in *Zoanthus sociatus,* Trench (124) raised the possibility that symbiotic algae may undergo senescence and autolytic degradation while in the host cell. The apparent sequential degradation of organelles within the algae while the limiting membranes remain intact tends to support the notion of "self degradation" by the algae. These two possible alternatives, autolytic degradation versus host digestion of algal endosymbionts, remain to be resolved.

INTEGRATION

Physical Location—Tissue Specificity

In most associations between marine invertebrates and dinoflagellates, the algae are located within the hosts' cells. Although this view is generally held by many investigators, there is no consensus. Kawaguti (44), based on interpretation of electron micrographs, concluded that the algae in some coral species were intercellular. On the other hand, Yonge (141), Fankboner (24), and Goreau, Goreau & Yonge (27) concluded that the algae in *Tridacna* were in blood amoebocytes, but this conclusion has not been corroborated by recent ultrastructural studies (3, 134). Instead it would appear that the algae in the hypertrophied siphonal tissue of *Tridacna* lie freely in the blood sinuses. One clear example of an intercellular association is that found between *Amphidinium klebsii* and *Amphiscolops langerhansi* (112).

In the cases involving green algae, again there is a dichotomy. The *Chlorella* in *Paramecium, Spongilla* and *Hydra* are all intracellular. *P. convolutae* in *C. roscoffensis* and *Licmophora* (a diatom) in *C. convoluta* are intercellular (1, 86). Similarly, in associations involving cyanobacteria (blue-green algae), there are examples of intracellular and intercellular associations. The cyanobacteria in *Cyanophora paradoxa* and *Glaucocystis nostochinearum,* for example, are intracellular (29, 30, 82, 131) as are the cells of *Aphanocapsa* in the sponge *Ircinia,* while the cyanobacteria, the *Prochlaron* of Lewin (53, 54), associated with ascidians appear to be intercellular.

There is, therefore, no general pattern that can be perceived. Different associations establish distinct morphological relationships. Thus, among most coelenterates associated with either green algae or dinoflagellates the algae are usually located within the endoderm cells. However, there are exceptions to this rule, because the dinoflagellates in many zoanthids, particularly those belonging to the genus *Palythoa,* are found in the endoderm, the mesoglea, and the ectoderm as well (121). A similar situation may exist in *Cassiopeia xamachana.*

Morphological Modifications in Symbiotic Algae

Most of the reports on changes in the morphology of algae following establishment of symbiosis are incidental and anecdotal. Nonetheless, an overview of the available information strongly indicates that in this respect also there is no generality. Different algae respond differently to existence in symbiosis. Some algae show no obvious morphological modification between the free-living and symbiotic state, while others demonstrate marked changes (73).

Symbiotic *Chlorella* appear to demonstrate no morphological changes nor changes in the life cycle as a result of an intracellular association with invertebrates (42, 76). Similarly, *Amphidinium klebsii* retain all their free-living morphological characteristics while residing in *Amphiscolops langerhansi* (111).

By contrast, there are several examples of marked morphological differences between some algae in the free-living state and in symbiosis. Although the free-living equivalent of the cyanobacteria in *Glaucocystis nostochinearum* and *Cyanophora paradoxa* is unknown, comparison of the ultrastructure of symbiotic cyanelles with free-living cyanobacteria (23, 29, 30, 92, 131) illustrates marked reduction in the peptidoglycan cell wall of the symbiotic cyanobacteria. Similarly, ultrastructural comparisons of *S.* (= *G.*) *microadriaticum* in culture and in symbiosis have revealed marked differences. Schoenberg & Trench (95) found that the algae in situ (in their respective hosts) demonstrated a markedly reduced theca or amphiesma.

When the same algae were brought into culture, they developed a "robust" pellicle and a morphologically complex amphiesma. In addition, several stages in the life history of the algae observed in culture were undetectable in the host. The stage, following establishment of an association, at which the morphological changes occur and the cellular events regulating such changes are completely unknown.

Probably the most dramatic example of structural modifications occurring as a result of symbiosis is found in *C. roscoffensis.* Free-living *P. convolutae* possess a characteristic cell wall, flagella, and eye spot. These structures are all lost when the alga takes up its final position between the host's cells, and they are readily produced anew when the algae are brought into culture (86).

Metabolic Interactions

INTERMEDIARY METABOLITES The movement of metabolites between partners in plant-animal symbioses is directly analogous to the movement of metabolites between cytoplasmic compartments of a cell. Metabolic interactions between plants and animals in intracellular symbioses lend themselves readily to experimental analysis of the exchange of metabolites between the components of the association, since ^{14}C as $^{14}CO_2$ or $H^{14}CO_3^-$ can be easily administered and incorporation through photosynthesis readily occurs, allowing the subsequent transport of substances from algae to animal host to be ascertained. Hence data on the amounts of the different kinds of metabolites moving between symbionts could be obtained.

Similarly, the movement of metabolites from the animals' tissues to the algae can be followed by the introduction of labeled food substrate to the animal hosts followed by subsequent assay of the algae. Thus food substrates labeled with ^{35}S, ^{3}H, or ^{15}N have been used in "black transport" studies (58, 126).

The bilateral movement of metabolites between photosynthetic endosymbionts and their animal hosts represents a system of intercellular (or interorganellar) transport of selected molecules across the interfaces of symbionts and host cells. Under most circumstances, the boundary between symbionts and hosts can be represented by the limiting membrane of the symbiont and either the limiting membrane (in intercellular associations) or the vacuolar (endosome) membrane (in most intracellular symbioses) of the host cell (73). In symbioses with plant chloroplasts, contact is direct because the plastids often lie freely in the animal cell cytoplasm (125).

Since aspects of the transport of organic carbon from photosynthetic endosymbionts to animal hosts have been reviewed extensively (64, 65, 100, 101, 113–115, 125), mostly within the context of the nutrition of the heterotrophic partner in the association, in this paper I shall place the emphasis on a comparison of the intermediary metabolism of the algae or chloroplasts in symbiosis and in isolation.

Many studies have been conducted on photosynthetic $^{14}CO_2$ fixation by the symbiotic dinoflagellate *S.* (=*G.*) *microadriaticum* (62, 67, 72, 122–124). However, much remains to be learned about the biochemistry of carbon fixation in these algae. Whether the algae demonstrate a C_3 or a C_4 pathway or a combination of these remains unresolved. C_3 metabolism in symbionts from the clam *Tridacna maxima* has been inferred from $\delta^{13}C$ values (4) which are within the same range (–23 to –33) as those found in terrestrial C_3 plants (99, 136, 140). The same algae also showed about 20% photosynthetic inhibition in oxygen concentrations of 60% saturation (21). In addition, these algae and those from a variety of other marine hosts synthesize and excrete appreciable quantities of glycolic acid (62, 122).

In opposition to the data above suggesting C_3 metabolism, the following pieces of evidence are consistent with C_4 metabolism. In preliminary studies of short-term photosynthesis (10 to 60 sec) under ambient O_2 tensions, the algae from *Anthopleura elegantissima* do not appear to incorporate ^{14}C into 3-PGA as the first product (R. Trench, unpublished; L. Muscatine, personal communication). In no instance reported to date is there any evidence of the rapid incorporation of phosynthetic ^{14}C into serine and glycine. In fact, von Holt (137) concluded from his studies that the algae from *Zoanthus flos-marinus* (=*sociatus*) were incapable of synthesizing glycine and were dependent on the supply of this amino acid by the host. Activities of NADH malate dehydrogenase (MDH) in algae from several clams was found to be reasonably high, and phosphoenolpyruvate (PEP) carboxylase activity was also readily demonstrable (119).

Symbiotic *Chlorella* probably fixes CO_2 via the C_3 pathway as do the symbiotic chloroplasts from siphonaceous algae in the cells of marine slugs (127, 128) and *Platymonas* in *C. roscoffensis.*

Regardless of the pathway of carbon fixation used by the different algae in symbiosis, they all uniformly release various quantities of photosynthetically fixed carbon to the tissues of their respective hosts (Table 4). The evidence indicates quite clearly that different algae release different proportions of their fixed carbon as different compounds.

In one of the few studies of photosynthetic products released by symbiotic cyanobacteria, Trench et al (131) found that the isolated cyanelles from *Cyanophora paradoxa* released about 20% of the carbon fixed, predominantly as glucose. However, it was pointed out that since the association is photoautotrophic, this contribution could not possibly represent the total quantitative or qualitative contribution by the cyanobacteria.

Table 4 Small molecular weight metabolites moving from symbiotic algae (or chloroplasts) to their hosts

Algae	Host organism	% transported[b]	Major metabolites transported[a]				References
			Glycerol	Glucose	Alanine	Maltose	
Cyanocyta korschikoffiana	*Cyanophora paradoxa*	ca 20		70			131
Chlorella sp.	*Paramecium bursaria*	5–86		0.5–1.0		95–99	69
	Spongilla lacustris	3–5		99			10, 69
	Hydra viridis	6–85 (pH dependent)		12		84	
Platymonas convolutae	*Convoluta roscoffensis*	8–50	(a variety of amino acids, particularly glutamine)				8, 37, 66
Symbiodinium (= *Gymnodinium*) *microadriaticum*	*A. elegantissima*	49–58	43	0.5	1.2		67, 72 122, 124
	A. pulchella	35	24	1.3	5.4		
	M. cookii	40	43	21	5.6		
	Z. sociatus	36–41	75	16	1.2		
	Z. pacificus	42	95	1	1.1		
	C. frondosa	23	41	15	7.3		
	M. papua	20	21	2	4.3		
	F. scutaria	25	61	5	4.5		
	P. damicornis	38–46	90	6	4.0		
	T. crocea	36–47	80	4	9.0		
	A. agaricites	56	85	4	10		
Chloroplasts of *Codium fragile*	*Elysia viridis*	36–50		75	25(?)		34, 128

[a] ^{14}C in each metabolite as % of total transported ^{14}C.
[b] ^{14}C transported as % of total ^{14}C fixed in photosynthesis.

The products released by symbiotic green algae are either glucose or maltose, with the exception of *Platymonas convolutae,* wherein it is believed (8) that the algae release amino acids. It should be borne in mind constantly that very often the algae isolated from their hosts may not necessarily perform metabolically as they do in the host, and this in turn may well be a function of the isolation methods as well as the incubation media in which the assay is conducted in vitro.

In the many studies conducted on the release of photosynthetic products by *S.* (=*G.*) *microadriaticum,* it was found that glycerol was uniformly the major compound released. Following the studies of Trench (122, 123) and Muscatine et al (72), it became apparent that the algae in isolation from the host rapidly modified their metabolic activities and either released less of the compounds they synthesized or released different compounds.

In studies of the chloroplasts of *Codium fragile* symbiotic with *Elysia viridis,* Trench et al (127, 128) found evidence which led them to conclude that the plastids from *Codium* in artificial media released very small quantities of their photosynthetic products, but that the same plastids in slug tissues released an estimated 36–50% of the carbon fixed, either as glucose or a precursor thereof. Many of these findings have been corroborated by later studies (34, 35).

Lewis & Smith (55) in their studies on corals showed the influence of NH_4^+ on enhancing movement of alanine from algae to animals. Recently, using ^{15}N, Marrian (58) has found direct evidence for the uptake of NO_3^- by *S.* (=*G.*) *microadriaticum* in vivo and its subsequent transfer to the host tissue.

Thus far I have placed the emphasis on movement of metabolites from the algae (or chloroplasts) to animals. The movement of organic metabolites in the opposite direction has not yet been studied to the same extent, but some evidence does exist that the phenomenon is real.

In his studies with *Aiptasia pulchella* and *Hydra viridis,* Cook (14, 15) found evidence for movement of organic material from animals to algae. Incorporation of ^{35}S, provided as organic sulfur to the animal, could be traced to the algae. However, his data did not make it clear whether the sulfur was acquired from the animals as organic or inorganic sulfur.

In unpublished studies conducted in my laboratory, we found that when organic-^{35}S was fed to the anemone *Anthoplerua elegantissima,* up to 40% became incorporated into the algae in both ethanol soluble and insoluble pools over a period of 6 days. The evidence obtained by chromatographic analysis of the extracted algae and animal tissues after physical separation suggested that cystine, methionine, and cystathione were probably being shunted back and forth between the host and the algae. It has been know for some time (60, 115) that *S.* (=*G.*) *microadriaticum* in culture can use a variety of organic N sources as well as organic P.

Finally, Trench & Gooday (129) found evidence, using electron microscopic autoradiography, for the incorporation of ^3H-leucine supplied exogenously to the slug *E. viridis* by the chloroplasts in the animals' cells. A general scheme depicting the bilateral movement of metabolites between symbionts is presented in Figure 15.

MACROMOLECULAR SYNTHESIS AND REGULATION OF GENE EXPRESSION A major feature of plant-animal intracellular mutualistic symbioses is that the two partners in the association demonstrate harmonious proliferation. Seldom, if ever, do the algae divide out of control resulting in overpopulation of the host, and seldom do the hosts outgrow the algae. The mechanism whereby such regulation is achieved at the cellular level is not known, yet it is probably at the macromolecular level that control is exerted. Very few studies have been conducted on macromolecular biosynthesis and the regulation of gene expression in symbiotic algae or their hosts, and these associations could represent excellent model systems for the analysis of how one genome regulates expression in the other.

In his studies of *S.* (=*G.*) *microadriaticum* in *Zoanthus flos-marinus* (= *sociatus*), von Holt (137) observed the release of nucleoside polyphosphates by the algae in vitro and interpreted his findings as evidence for the possible

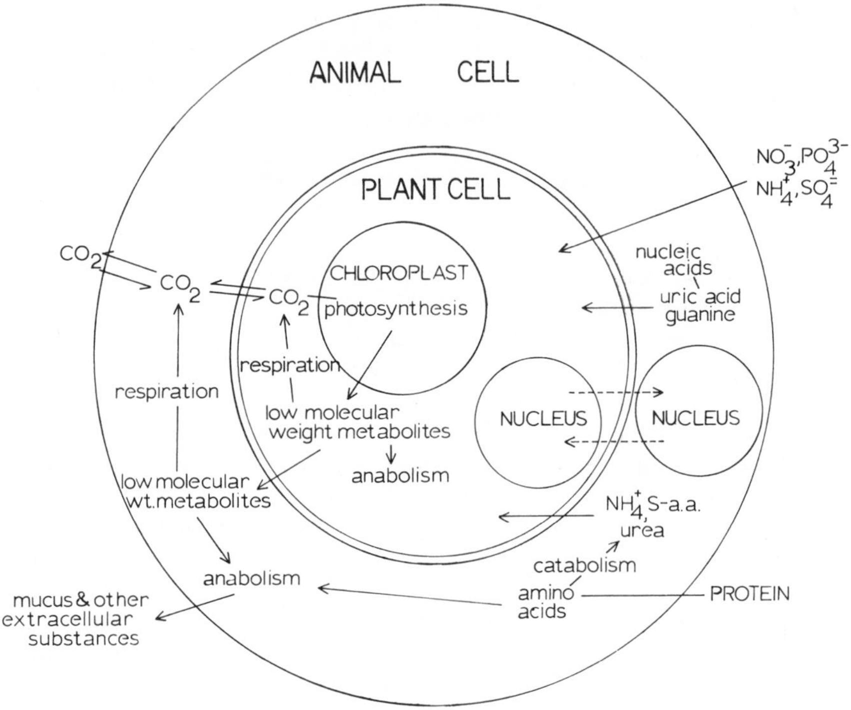

Figure 15 Schematic representation of the interchange of metabolites between endosymbiotic algae and the cell of their host animal. The alga is depicted as being within a host cell vacuole. Pathways which are supported by experimental evidence are depicted with bold arrows while uncertain pathways are indicated by broken lines.

transfer of genetic information from symbiont to host. However, these experiments have not to my knowledge been repeated, and so the conclusions drawn need corroboration.

Another aspect of possible interaction between host and symbiont genome comes from studies on the excretion of photosynthetic products by *S.* (=*G.*) *microadriaticum.* Several workers (62, 72, 123) have shown that these symbiotic algae liberate larger quantities of their photosynthetic products in vitro when they are incubated with ^{14}C in the presence of a homogenate of the tissues of the host. Although not exhaustively studied, and certainly not chemically characterized, the active component in the animal tissue homogenate was found to be heat labile and oxygen sensitive and is thought to be a protein-like substance.

In his studies with the algae from *A. elegantissima,* Trench (123) found that the stimulatory "factor" was present only in anemones which possessed symbiotic algae. The homogenates from aposymbiotic anemones did not stimulate release of photosynthate by the algae until they had been artificially infected with algae and had acquired a "normal" population density of the symbionts. These observations were interpreted as implicating algal "induction" of the synthesis of a protein-like substance by the host. The fact that this substance then enhances translocation of photosynthate from the algae might imply a regulatory "feedback" mechanism in operation, since the increased flow of metabolites from the algae could potentially reduce their gowth and proliferation.

Yet another example comes from observations on the importance of *S.* (=*G.*) *microadriaticum* in the morphogenesis of rhizostome medusae. The life history of the animal hosts involves an alteration between a sexually reproductive symbiotic stage and an asexually reproducing aposymbiotic polyp stage. Development of the ephyrae by the polyp stage is essential to the completion of the life cycle, but this morphogenetic event is arrested unless the polyps become infected by an appropriate strain of *S.* (=*G.*) *microadriaticum* (126). Although iodine and thyroxin stimulate strobilation in polyps of scyphozoans (75, 103), these substances have no effect on the scyphistomae of symbiotic rhizostomes. The stimulus produced by the algae is unknown, and it is unlikely that the active component would be small molecular weight metabolites released by the algae.

The final example comes from some recent studies on *Cyanophora paradoxa* (98, 131–133). The organism known as *C. paradoxa* is actually an intracellular consortium between an apochlorotic cryptomonad and a cyanobacterium called *Cyanocyta korschikoffiana.* The evidence available indicates that the association is mutually obligate, and because of the apparently small genome size of the cyanobacteria (31), it has been suggested that they are actually chloroplasts (104).

Studies have indicated that all the photosynthetically active pigments in *C. paradoxa* are located within the endosymbiotic cyanobacteria (132). The pigments include chlorophyll *a,* β-carotene, and zeaxanthin as well as the two bilichromoproteins, allo- and c-phycocyanin. Analysis of the pigments indicated that allo-phycocyanin is a homodimer of about 28×10^3 daltons, each subunit being about 13×10^3 daltons. By contrast, c-phycocyanin is a heterodimer of about 31×10^3 daltons, the subunits being about 13×10^3 and 16×10^3 daltons, respectively (Figure 16).

Analysis of rRNA from *C. paradoxa* revealed the presence of four major molecular species: 25s and 18s species characteristic of the eukaryotic host, and 23s and 16s species characteristic of the cyanobacteria (98). Studies of the incorporation of ^{33}P into rRNA indicated essentially equimolar synthesis of the different molecular species. However, in the presence of rifampicin ($40 \mu g \cdot ml^{-1}$), incorporation of ^{33}P into the prokaryotic rRNA species was virtually completely inhibited. D(–) threo chloramphenicol (300 μg·ml^{-1}) also inhibited ^{33}P incorporation into the prokaryotic rRNA species. By contrast, cycloheximide (3 μg·ml^{-1}) only affected the host rRNA (133).

However, neither rifampicin nor chloramphenicol inhibited the incorporation of photosynthetically fixed ^{14}C into chlorophyll as much as cycloheximide did (Figure 17). Again, rifampicin was found to completely inhibit the incorporation of ^{14}C into both biliproteins, but cycloheximide inhibited the incorporation of ^{14}C into the small subunit of c-phycocyanin specifically (Figure 18). Taken together, these data suggest that the host cytoplasm plays an important role in the macromolecular biosynthetic events occurring within the symbiont. In fact, these observations are in many ways analogous to those reported in plant cell-chloroplast systems wherein the small subunit of ribulose-diphosphate carboxylase is synthesized on cytoplasmic ribosomes (7, 45).

In contrast with the above, evidence from studies on symbiotic slug-chloroplast systems demonstrate minimal interaction between slug genome and plastid genome. Attempts at illustrating macromolecular biosynthesis by symbiotic chloroplasts proved unrewarding (130), and even lipid synthesis by the chloroplasts in slug cells is absent, as compared with the same chloroplasts in the plant cell (Figure 19). These observations have provided further support to the concept that chloroplasts are semiautonomous organelles (125).

Regulation of Algal Numbers

The mechanism whereby the number of algal cells in a given animal host cell is regulated is very poorly understood. Based on observations of several

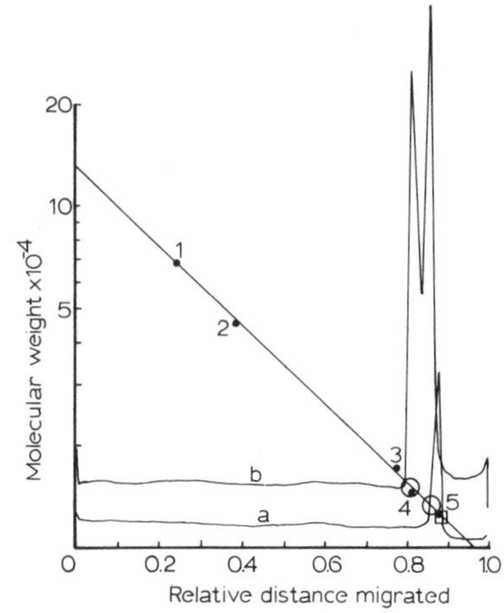

Figure 16 Densitometer scans of SDS-polyacrylamide gel electrophoretograms of (a) allophycocyanin (AP) and (b) c-phycocyanin (CP) from *C. paradoxa.* The scans are superimposed on a semilogarithmic plot of the molecular weight of subunits of AP and CP against the relative migration distance of the polypeptides in acrylamide gels. Solid points represent molecular weight markers; 1, BSA; 2, ovalbumin; 3, myoglobin; 4, lysozyme; 5, cytochrome *c.* Open circles represent the heavy and light subunits of CP. Open squares represent the subunits of AP. Data from Trench & Ronzio (132).

different kinds of associations, it is possible to state that the numbers of algal cells found in a host cell vary depending on the association and on the specific location of the host cell. For example, in *Hydra* there are about 20 algae/endoderm cells in the central growth region and about 8 algae/cells in other regions of the host (77). W. Fitt and R. Trench (unpublished) have found that in the marine hydroid *M. amboinense,* the number of *S.* (=*G.*) *microadriaticum* per host endoderm cell varied from 1 to 56, the greatest densities being found in the tentacles with intermediate to low densities in the cells of the hydranth and stalk, respectively (126). A well-populated

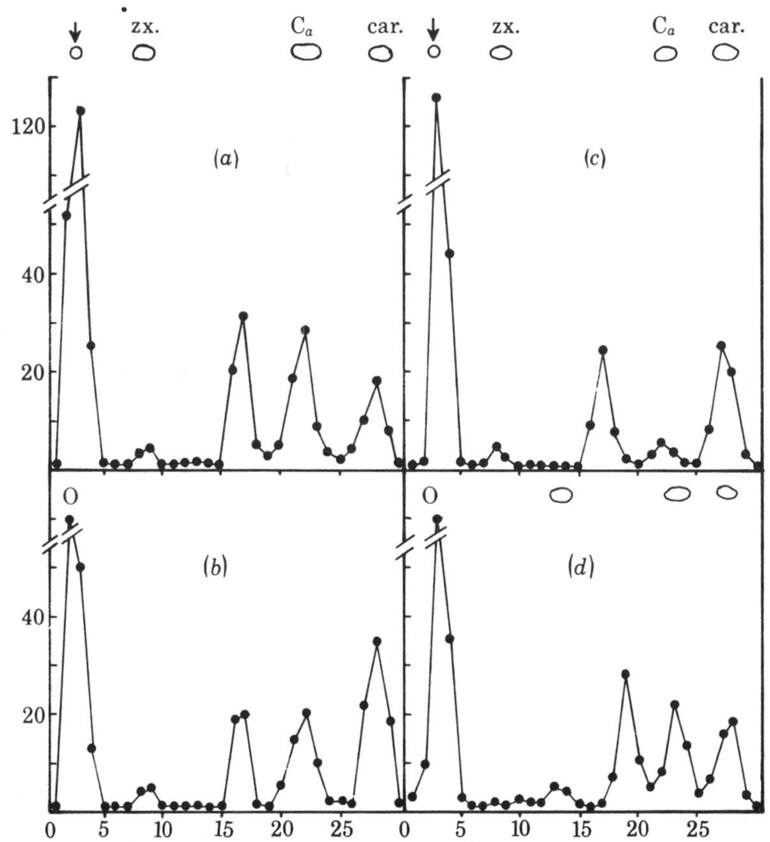

Figure 17 The distribution of photosynthetically fixed ^{14}C in ether soluble pigments from *C. paradoxa* in the presence of various metabolic inhibitors. (a) control untreated cells; (b) cells treated with 300 $\mu g/ml$ chloramphenicol; (c) cells treated with 3 $\mu g/ml$ cycloheximide; (d) cells treated with 40 $\mu g/ml$ rifampicin. Data from Trench and Siebens (133), by permission of the Royal Society.

P. bursaria may contain well over 300 *Chlorella* (39) while *C. paradoxa* in stationary phase may contain 8 cyanelles (131).

It is quite apparent that the volume of a given host cell is finite. Given that the algal endosymbionts grow and divide, if such proliferation should proceed unchecked, there must come a time when the volume of the algae will exceed the carrying capacity of the host cell. Under such circumstances, there are three possibilities whereby the population densities of algae in animal cells could be "normalized."

First, the host cells and the algae could divide in a manner approaching synchrony. This does not mean that each event of cytokinesis by the algae must be immediately followed by host cell division; but if a given host cell with maximum number of algae were to divide and the algae were distributed equally between the progeny, subsequent growth of the animal cell, with concomitant retardation of division by the algae, would tend to lower the population density of the host cells. Such a situation may exist in *Hydra* (78) but the details of the process are unclear. The situation in *Hydra* is rendered somewhat more complex since the symbiotic *Chlorella* divide by the production of tetraspores.

A simpler system which lends itself more readily to quantitative analysis is found in *C. paradoxa* (Figure 20). When cells from stationary phase were transferred to fresh media, the lag phase of growth was found to be a period of readjustment of cyanelle numbers (131). The proportion of *C. paradoxa*

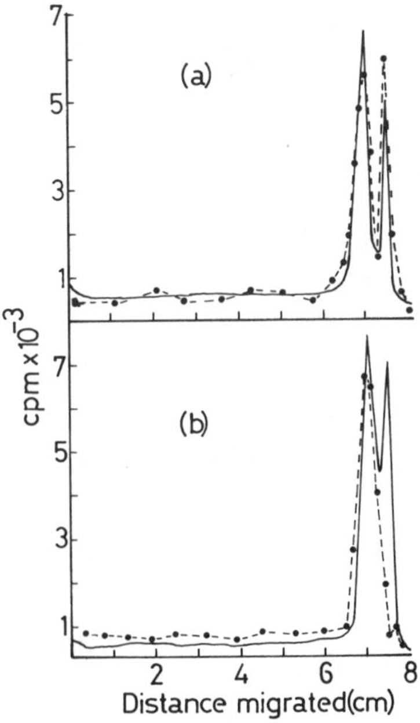

Figure 18 Scans of SDS electrophoretograms of CP from *C. paradoxa* showing the distribution of ^{14}C (broken lines) in the subunit polypeptides in (a) untreated controls and (b) cycloheximide (3 μg/ml) treated cells.

containing two or more cyanelles decreased. As the host cells moved into log phase growth, the cyanelles appeared to become equipartitioned such that about half the cells contained two cyanelles and the other half contained one. Hence, division of the host and cyanobacteria was not in strict synchrony. In fact, the data show that log phase growth of the cyanelles follows log phase growth of the host cells by about 10 days and begins in

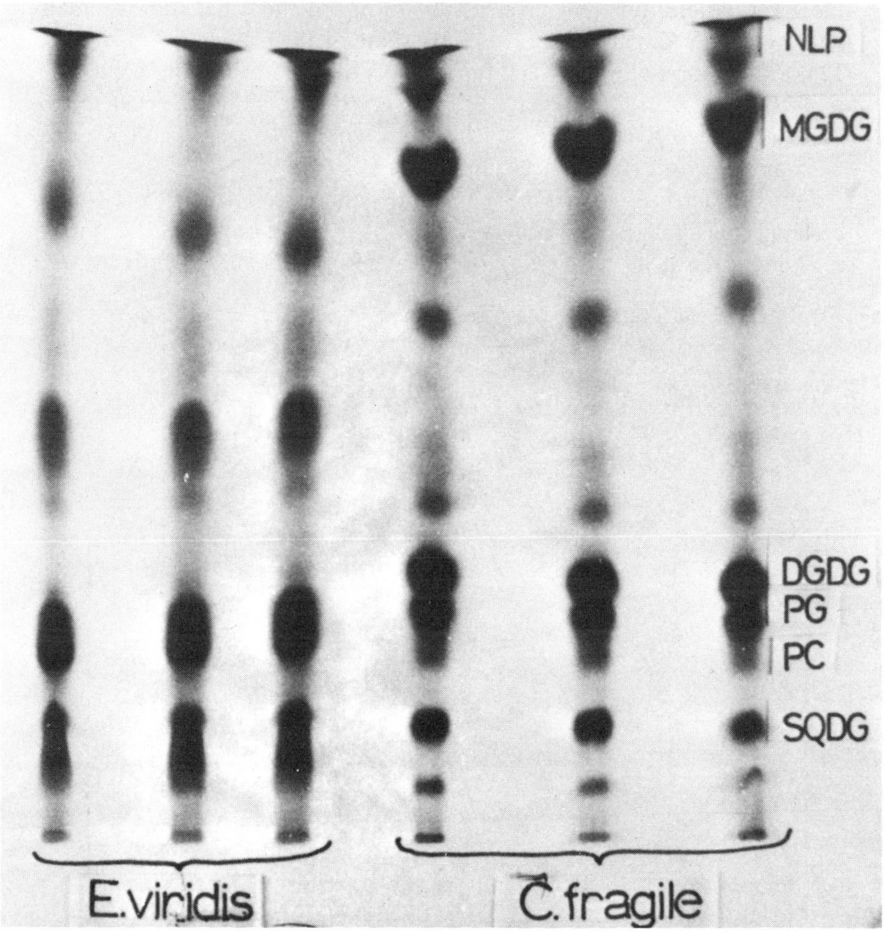

Figure 19 Autoradiograph of a thin layer chromatogram of ^{14}C-lipids extracted from *C. fragile* and *E. viridis* after 10 hr photosynthesis in $NaH^{14}CO_3$. NLP, neutral lipids and pigment; MGDG, monogalactosyl diglyceride; DGDG, digalactosyl diglyceride; PG, phosphatidyl glycerol; PC, phosphatidyl choline; SQDG, sulphoquinovosyl diglyceride. Note that the substance in the extract from *E. viridis* with relative mobility similar to SQDG separates from it on two-dimensional chromatograms. Data from Trench et al (128), from whom details may be obtained.

earnest when the host cells have attained their asymptote. At this point, the number of host cells containing two or more cyanelles increases dramatically.

The second possibility is that a host cell with its maximum complement of algae exocytoses the excess algal cells which may subsequently be phagocytosed by other host cells. Good evidence for this process is lacking. However, several different investigators have observed that some animals with symbiotic dinoflagellates "expell" some of their algae (87, 105, 106, 124), often morphologically and functionally intact. In addition, nonmotile *S.* (=*G.*) *microadriaticum* have been observed free within the coelenteric cavity of the soft coral *Xenia,* the stony coral *Plerogyra,* and the sea anemone *Aiptasia* (R. Trench, unpublished).

Finally, there is the possibility that the animals selectively "weed out' certain algae by digestion. This process was suggested as part of the "farming" of zooxanthellae in *Tridacna* (141) and in *Anemonia* (110). However,

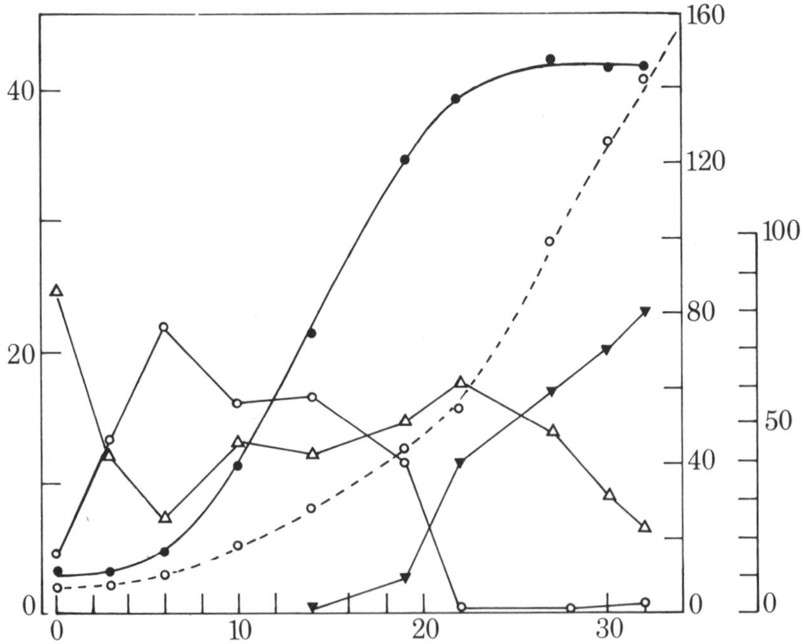

Figure 20 Growth characteristics of *C. paradoxa* and *C. korschikoffiana* in CY II medium. ●——●, *C. paradoxa* cells (left-hand scale); o- - -o, *C. korschikoffiana* cells (inner right-hand scale, also x10^{-8} cells/liter); o——o, % *C. paradoxa* cells containing one cyanelle; △——△, % *C. paradoxa* cells containing two cyanelles; ▼——▼, % *C. paradoxa* cells containing four cyanelles. Scale on outer right. Data from Trench et al (131), by permission from the Royal Society.

as stated before, good evidence for digestion of zooxanthellae by animal hosts is lacking (124).

An alternative mechanism of regulation of algal numbers by hosts could be host control of the availability of potential growth-promoting nutrients to the algae. Since nitrogenous end products of protein catabolism produced by the animal hosts may be utilized by the algae, regulation of nitrogen, sulfur, and phosphate metabolism by the animal could influence the availability of these substances to the algae and thereby regulate their growth (83, 143).

SPECIFICITY

The problem of specificity in intracellular associations between plant cells and animals is intimately linked to the systematic and genetic identity of the two interacting associants. One of the major drawbacks in the analysis of plant-animal associations is the rather primitive nature of our understanding of the genetic and, therefore, the specific identity of the algae involved in many symbioses.

Students of algal symbioses are often impressed by the large number of examples of plant cells living in animal cells, but often overlook the details of the randomness of both the geographic and phylogenetic distribution of such associations. For example, in the case of symbiotic *Chlorella* found in *Hydra, Spongilla,* and *Paramecium,* it is not known whether these algae are strains, variants, or different species. It has been suggested, based on observed morphologic plasticity demonstrated by symbiotic *Chlorella* when in different environments, that they might be genetically closely related (42, 79). The observation of alterations in surface antigenicity as a function of host cell environment (83) tends to support this view even further. The point that is often overlooked, however, is that there are many freshwater ciliates, sponges, and hydroids which do not form symbioses with *Chlorella* or any other alga for that matter, and attempts to form artificial associations between nonsymbiotic animals and symbiotic algae have not been generally successful.

Similarly, in the associations between "gymnodinioid" dinoflagellates and marine invertebrates, our inadequate knowledge of the genetics of the algae have led to conflicting views on the phylogenetic and geographic distribution of these algae in their respective hosts.

Several years of observations of the ultrastructure of symbiotic dinoflagellates in situ and in culture has led to one view that as many as 80 invertebrate hosts, representing 3 phyla and distributed through the Caribbean, Indo-Pacific, and Mediterranean, are associated with a single algal species, *Gymnodinium microadriaticum* (Freudenthal) (115). These algae are recognized as being distinct from the amphidinioid dinoflagellates which occur as symbionts in the marine flatworm *Amphiscolops* (111) and *Velella* (112).

There are several theoretical reasons for questioning the view that the symbiotic gymnodinioid dinoflagellates represent a single species population. First, if these associations originated prior to the Triassic (230M ybp), one could well ask whether they were initiated from a single algal species population. Even if they were, have the algae remained evolutionarily conservative in light of the marked speciation and divergence of the host animals?

Second, it has been proposed that the reason why one algal species can occupy the cytoplasmic environment of so many different invertebrate phyla is that the hosts possess poorly developed immune systems (113). However, it is now becoming increasingly apparent that invertebrates do possess the capability to distinguish between self and nonself (12, 25, 32, 33, 48, 118, 135), even though the mechanisms employed are often unclear. In addition, that different symbiotic invertebrates can distinguish between different algae is also becoming more apparent (84, 93, 96, 126).

Finally, there are many invertebrates on coral reefs which are not symbiotic. Similarly, of all the anemones on the coast of Britain, only *Anemonia* is symbiotic, and only *Anthopleura* is symbiotic on the west coast of the United States.

The specificity of algae-invertebrate symbioses goes well beyond a situation involving "one alga-one host." The concepts which appear most consistent with the available evidence are those proposed by Weiss (139) and Dubos & Kessler (22), which stress the relative nature of the phenomenon and attempt to place emphasis equally on the two interacting partners, such that specificity is viewed as the "overall pattern of adjustment between the two components and is an expression of complementariness of all their dynamically interacting attributes."

In *A. langerhansi* it has been shown (111) that the larvae will acquire a few selected amphidinioid and gymnodinioid dinoflagellates (as well as some green algae) other than that with which it usually associates, *A. klebsii,* but only the latter alga permits completion of the worm's life history. This observation is not unlike that observed by Provasoli et al (86) in *C. roscoffensis.* Similarly, it has been shown that although different strains of *S.* (=*G.*) *microadriaticum* may "infect" *A. tagetes* (93, 96, 126), the proliferation of the heterologous algae in the host is markedly reduced compared to the normal algae. It remains to be determined what, if any, effect the difference in algal proliferation has on the metabolism or growth of the host. A clearer situation exists in the case of the rhizostome *C. xamachana,* where in the experiments conducted to date, only one strain of *S.* (=*G.*) *microadriaticum* appears suitably compatible to permit rapid morphogenesis of the host (126) and completion of its life history.

Within the context of specificity, it is also significant that there are few reported examples of animal hosts which are simultaneously infected by more than one algal type. The exception to this is *A. xanthogramica* on the northwest coast of the United States which may harbor a dinoflagellate related to *S.* (=*G.*) *microadriaticum* as well as a green alga (63, 74).

In their studies with different strains of *S.* (=*G.*) *microadriaticum,* Schoenberg & Trench (94) found that the isoenzyme patterns obtained from crude isolates from a given host remained the same after the isolates were cloned, suggesting that the hosts possessed a population of algae which, within the limits of detection, were equivalent to a clone. Mechanisms whereby specificity between algae and host could be maintained include the inheritance of algae by offspring (a closed system) (94). In open systems where offspring must be infected de novo, a system of recognition and of discrimination at the cellular level may well be the most important factor in maintaining specificity in plant-animal associations.

CONCLUSIONS

In this article I have tried to illustrate that (*a*) there are many similarities between aspects of the cell biology of intracellular symbioses and other associations such as classical parasitism (120) and (*b*) that many plant-animal associations readily lend themselves to experimental analysis of phenomena of importance to cell biology, e.g. cell-cell recognition and specificity (89) and metabolite transport.

Despite the many advances made over the past decade, there are numerous problems which remain to be resolved. Among the important ones are the genetics of the infective process. Since certain organisms appear to readily form associations and closely related organisms do not, it is probable that susceptibility to infection may be a genetic trait. Thus, malfunction of cellular recognition or defense could render a host susceptible to invasion (126).

Although we have now approached the level whereby the basis of the recognition process has partially been defined, we have only touched the tip of the iceberg. Recognition at the limiting membrane of the host cell is obviously of importance, but there are obvious examples wherein intracellular recognition is equally important. For example, how do coelenterate endoderm cells recognize pycnotic algae which become exocytosed, and how do the digestive cells of slugs recognize the difference between intact functional chloroplasts and defunct ones (125, 126)?

The problem of regulation of symbiont numbers is also basic, and our knowledge of this phenomenon is extremely primitive. Such regulation is probably closely linked to interactions between host and symbiont genomes, but we know virtually nothing about these phenomena.

Finally, the aspect of symbiosis that is fundamental to both interacting organisms is the selective advantage of the association. Although it has been postulated that autotroph-heterotroph symbioses permit exploitation of nutrient depleted environments (55), documentation of selective advantage to individual hosts in terms of growth or reproductive potential is lacking in most instances with the exception of *Hydra* (70, 71), *Convoluta* (86), and *Amphiscolops* (111). The apparent indispensability of the algae to the morphogenesis of *Cassiopeia* and *Mastigias* may also be an example, but it is unclear from an ecological point of view whether sexuality is of greater selective advantage than asexual reproduction.

It is apparent that there are many facets to the study of plant-animal symbiosis, and it is encouraging to see that interest is still being generated among young biologists, particularly those with interests in cell and molecular biology.

Acknowledgments

Some of the materials drawn upon in the preparation of this article include unpublished observations. I am grateful to all my colleagues and students, past and present, who freely permitted inclusion of their unpublished work.

Much of the data from my laboratory were obtained because of support from the following sources: NIH-5RO1-GM-20177 and NIH-5RO1-GM-24014, NSF-DES74-14672, and GB-34211. Assistance during the preparation of the manuscript was provided by the Marine Science Institute and by the Biomedical Sciences Support Fund (NIH-RR-07099) from the Department of Biological Sciences, University of California, Santa Barbara.

Literature Cited

1. Apelt, G. 1969. Die Symbiose zwischen dem acoelen Turbellar *Convoluta convoluta* und Diatomeen de Galtung *Licmophora*. *Mar. Biol.* 3:165–87
2. Bigelow, R. P. 1900. The anatomy and development of *Cassiopeia xamachana*. *Mem. Bost. Soc. Nat. Hist.* 5:191–236
3. Bishop, D. G., Bain, J. M., Downton, W. J. S. 1976. Ultrastructure and lipid composition of zooxanthellae from *Tridacna maxima*. *Aust. J. Plant Physiol.* 3:33–40
4. Black, C. C. Jr., Bender, M. M. 1976. $\delta^{13}C$ values in marine organisms from the Great Barrier Reef. *Aust. J. Plant Physiol.* 3:25–32
5. Blackbourn, D. J., Taylor, F. J. R., Blackbourn, J. 1973. Foreign organelle retention by ciliates. *J. Protozool.* 20:451–60
6. Bomford, R. 1965. Infection of alga-free *Paramecium bursaria* with strains of *Chlorella, Scenedesmus* and a yeast. *J. Protozool.* 12:221–24
7. Boulter, D., Ellis, R. J., Yarwood, A. 1972. Biochemistry of protein synthesis in plants. *Biol. Rev.* 47:113–75
8. Boyle, J. E., Smith, D. C. 1975. Biochemical interactions between the symbionts of *Convoluta roscoffensis*. *Proc. R. Soc. London Ser. B* 189:121–35
9. Brien, P., Reniers-Decoen, M. 1950. Etude d'*Hydra viridis* (Linneus), la blastogénèse, la spermatogénèse, l'ovogénèse. *Ann. Soc. R. Zool. Belg.* 81:33–110
10. Cernichiari, E., Muscatine, L., Smith, D. C. 1969. Maltose excretion by the symbiotic algae of *Hydra viridis*. *Proc. R. Soc. London Ser. B* 173:557–76

11. Chapman-Andresen, C. 1977. Endocytosis in fresh water amebas. *Physiol. Rev.* 57:371–85
12. Cheng, T. C. 1975. Functional morphology and biochemistry of molluscan phagocytes. *Ann. NY Acad. Sci.* 266: 343–79
13. Christensen, E. 1979. Algal chloroplasts in the protoplasm of three species of benthic foraminifera: taxonomic identity, viability and persistence. *Mar. Biol.* In press
14. Cook, C. B. 1971. Transfer of ^{35}S-labelled material from food ingested by *Aiptasia* sp. to its endosymbiotic zooxanthellae. In *Experimental Coelenterate Biology,* ed. H. Lenhoff, L. Muscatine, L. Davis, pp. 218–24. Honolulu: Univ. Hawaii Press
15. Cook, C. B. 1972. Benefit to symbiotic zoochlorellae from feeding by green hydra. *Biol. Bull.* 142:236–42
16. Cook, C. B. 1979. The infection of invertebrates by algae. *Ann. Colloq. 5th, Ohio State Univ., Columbus.* In press
17. Cook, C. B., D'Elia, C., Muscatine, L. 1978. Endocytic mechanisms of the digestive cells of *Hydra viridis.* I. Morphological aspects. *Cytobios.* In press
18. Cooper, C. G., Margulis, L. 1978. Delay in migration of symbiotic algae in *Hydra viridis* by inhibitors of microtubule protein polymerization. *Cytobios* 19:7–19
19. Cox, F. E. G. 1975. Factors affecting infections of mammals with intraerythrocytic protozoa. *Symp. Soc. Exp. Biol.* 29:429–51
20. Dazzo, F. B., Hubbell, D. H. 1975. Cross-reactive antigens and lectin as determinants of symbiotic specificity in the rhizobium-clover association. *Appl. Microbiol.* 30:1017–33
21. Downton, W. J. S., Bishop, D. G., Larkum, A. W. D., Osmond, C. B. 1976. Oxygen inhibition of photosynthetic oxygen evolution in marine plants. *Aust. J. Plant Physiol.* 3:73–79
22. Dubos, R., Kessler, A. W. 1963. Integrative and disintegrative factors in symbiotic associations. *Symp. Soc. Gen. Microbiol.* 13:1–11
23. Echlin, P. 1967. The biology of *Glaucocystis nostochinearum.* I. The morphology and fine structure. *Br. Phycol. Bull.* 3:225–39
24. Fankboner, P. V. 1971. Intracellular digestion of symbiotic zooxanthellae by host amoebocytes in giant clams (Bivalvia, Tridacnidae), with a note on the nutritional role of the hypertrophied siphonal epidermis. *Biol. Bull.* 141: 222–34
25. Francis, L. 1973. Clone specific segregation in the sea anemone *Anthopleura elegantissima. Biol. Bull.* 144:64–72
26. Fraser, E. A. 1931. Observations on the life history and development of the hydroid *Myrionema amboinense. Sci. Rep. Great Barrier Reef Exped.* 3:135–44
27. Goreau, T. F., Goreau, N. I., Yonge, C. M. 1973. On the utilization of photosynthetic products from zooxanthellae and of dissolved amino acids in *Tridacna maxima* f. *elongata* (Roeding). *J. Zool. London* 169:417–54
28. Gregory, F. G. 1952. A discussion on symbiosis involving micro-organisms (pp. 202–3). Under the leadership of H. G. Thornton. *Proc. R. Soc. London Ser. B* 139:170–207
29. Hall, W. T., Claus, G. 1963. Ultrastructural studies on the blue-green algal symbiont in *Cyanophora paradoxa* Korschikoff. *J. Cell Biol.* 19:551–63
30. Hall, W. T., Claus, G. 1966. Fine structure of the blue-green algal symbiont in *Glaucocystis nostochinearum* Itz. *J. Phycol.* 3:37–51
31. Herdman, M., Stanier, R. Y. 1977. The cyanelle: chloroplast or endosymbiotic prokaryote? *FEMS Lett.* 1:7–12
32. Hildemann, W. H. 1974. Some new concepts in immunological phylogeny. *Nature* 250:116–20
33. Hildemann, W. H., Raison, R. L., Hull, C. J., Akaka, L., Okumoto, J., Cheung, G. 1977. Tissue transplantation immunity in corals. *Proc. Int. Coral Reef Symp.* 3:537–43
34. Hinde, R. 1978. The metabolism of photosynthetically fixed carbon by isolated chloroplasts from *Codium fragile* (Chlorophyta: Siphonales) and by *Elysia viridis* (Mollusca: Sacoglossa). *Biol. J. Linn. Soc.* 10:329–42
35. Hinde, R., Smith, D. C. 1975. The role of photosynthesis in the nutrition of the mollusc *Elysia viridis. Biol. J. Linn. Soc.* 7:161–71
36. Hollande, A., Carré, D. 1974. Les xanthelles des radiolaires sphaerocollides, des acanthaires et de *Velella velella:* infrastructure-cytochimie-taxonomie. *Protistologica* 10:573–601
37. Holligan, P. M., Gooday, G. W. 1975. Symbiosis in *Convoluta roscoffensis. Symp. Soc. Exp. Biol.* 29:205–27
38. Jones, T. C. 1974. Macrophages and intracellular parasitism. *J. Reticuloendothelial Soc.* 15:439–50
39. Karakashian, M. W. 1975. Symbiosis in

Paramecium bursaria. Symp. Soc. Exp. Biol. 29:229–65
40. Karakashian, M. W., Karakashian, S. J. 1973. Intracellular digestion and symbiosis in *Paramecium bursaria. Exp. Cell Res.* 81:111–19
41. Karakashian, S. J. 1963. Growth of *Paramecium bursaria* as influenced by the presence of algal symbionts. *Physiol. Zool.* 36:52–68
42. Karakashian, S. J. 1970. Morphological plasticity and evolution of algal symbionts. *Ann. NY Acad. Sci.* 175:474–87
43. Karakashian, S. J., Karakashian, M. W. 1965. Evolution and symbiosis in the genus *Chlorella* and related algae. *Evolution* 19:368–77
44. Kawaguti, S. 1964. Zooxanthellae in the coral are intercellular symbionts. *Proc. Jpn. Acad. Sci.* 40:545–48
45. Kawashima, N. 1970. Non-synchronous incorporation of ^{14}C into amino acids of the two subunits of fraction I protein. *Biochem. Biophys. Res. Commun.* 38:119–24
46. Kinzie, R. A. III. 1974. Experimental infection of aposymbiotic gorgonian polyps with zooxanthellae. *J. Exp. Mar. Biol. Ecol.* 15:335–45
47. LaBarbera, M. 1975. Larval and post-larval development of the giant clams *Tridacna maxima* and *Tridacna squamosa* (Bivalvia, Tridacnidae). *Malacologia* 15:69–79
48. Lang, J. C. 1973. Interspecific aggression by scleractinian corals. II. Why the race is not only to the swift. *Bull. Mar. Sci.* 23:260–77
49. Lee, J. J., Zucker, W. 1969. Algal flagellate symbiosis in formanifer *Archaias. J. Protozool.* 16:71–80
50. Lemke, P. A. 1976. Viruses of eucaryotic microorganisms. *Ann. Rev. Microbiol.* 30:105–45
51. Leutenegger, S. 1977. Ultrastructure de foraminiferes perfores et imperfores ainsi que de leur symbiotes. *Cah. Micropol.* 3:1–31
52. Levine, L. 1967. Micro-complement fixation. In *Handbook of Experimental Immunology,* ed. D. M. Weis, pp. 707–19
53. Lewin, R. A. 1976. Prochlorophyta as a proposed new division of algae. *Nature* 261:697–98
54. Lewin, R. A. 1977. Prochloron, type genus of the Prochlorophyta. *Phycologia* 16:217
55. Lewis, D. H., Smith, D. C. 1971. The autotrophic nutrition of symbiotic marine coelenterates with special reference to hermatypic corals. I. Movement of photosynthetic products between the symbionts. *Proc. R. Soc. London Ser. B* 178:111–29
56. Loomis, W. F., Lenhoff, H. M. 1956. Growth and sexual differentiation of hydra in mass culture. *J. Exp. Zool.* 132:555–74
57. Ludwig, L. D. 1969. Die zooxanthellen bei *Cassiopeia andromeda* Eschscholz 1929 (Polyp-Stadum) und ihre Bedentung fur die Strobilation. *Zool. Jahrb. Abt. Anat.* 86:238–77
58. Marrian, R. E. 1977. *Assimilation and fate of nitrate nitrogen in the reef coral Pocillopora.* Presented at West. Soc. Nat., Univ. California, Santa Cruz (Abstr.)
59. McLaughlin, J. J. A., Zahl, P. A. 1959. Axenic zooxanthellae from various invertebrate hosts. *Ann. NY Acad. Sci.* 77:55–72
60. McLaughlin, J. J. A., Zahl, P. A. 1966. Endozoic algae. In *Symbiosis,* ed. S. M. Henry, 1:257–97. New York: Academic
61. McLean, N. 1976. Phagocytosis of chloroplasts in *Placida dendritica* (Gastropoda: Sacoglossa). *J. Exp. Zool.* 197:321–30
62. Muscatine, L. 1967. Glycerol excretion by symbiotic algae from corals and *Tridacna* and its control by the host. *Science* 156:516–19
63. Muscatine, L. 1971. Experiments on green algae coexistent with zooxanthellae in sea anemones. *Pac. Sci.* 25:13–21
64. Muscatine, L. 1973. Nutrition of corals. In *Biology and Geology of Coral Reefs,* ed. O. A. Jones, R. Endean, 2(1):77–115. New York, London: Academic
65. Muscatine, L. 1974. Endosymbiosis of Cnidarians and algae. In *Coelenterate Biology,* ed. L. Muscatine, H. M. Lenhoff, pp. 359–95. New York, San Francisco, London: Academic
66. Muscatine, L., Boyle, J. E., Smith, D. C. 1974. Symbiosis of the acoel flatworm *Convoluta roscoffensis* with the algae *Platymonas convolutae. Proc. R. Soc. London Ser. B* 187:221–24
67. Muscatine, L., Cernichiari, E. 1969. Assimilation of photosynthetic products of zooxanthellae by a reef coral. *Biol. Bull.* 137:506–23
68. Muscatine, L., Cook, C. B., Pardy, R. L., Pool, R. R. 1975. Uptake, recognition and maintenance of symbiotic *Chlorella* by *Hydra viridis. Symp. Soc. Exp. Biol.* 29:175–203
69. Muscatine, L., Karakashian, S. J., Karakashian, M. W. 1967. Soluble extracellular products of algae with a cili-

ate, a sponge and a mutant hydra. *Comp. Biochem. Physiol.* 20:1–12
70. Muscatine, L., Lenhoff, H. 1963. Symbiosis: On the role of algae symbiotic with hydra. *Science* 142:956–58
71. Muscatine, L., Lenhoff, H. 1965. Symbiosis of hydra and algae. II. Effects of limited food and starvation on growth of symbiotic and aposymbiotic hydra. *Biol. Bull.* 129:316–28
72. Muscatine, L., Pool, R. R., Cernichiari, E. 1972. Some factors influencing selective release of soluble organic material by zooxanthellae from reef corals. *Mar. Biol.* 13:298–308
73. Muscatine, L., Pool, R. R., Trench, R. K. 1975. Symbiosis of algae and invertebrates: Aspects of the symbiont surface and the host-symbiont interface. *Trans. Am. Microsc. Soc.* 94:450–69
74. O'Brien, T. 1978. An ultrastructural study of zoochlorellae in a marine coelenterate. *Trans. Am. Microsc. Soc.* 97:320–29
75. Olman, J. E., Webb, K. 1974. Metabolism of ^{131}I in relation to strobiliation in *Aurelia aurita* L. (Scyphozoa). *J. Exp. Mar. Biol. Ecol.* 16:113–22
76. Oschman, J. L. 1966. Development of the symbiosis of *Convoluta roscoffensis* Graff and *Platymonas* sp. *J. Phycol.* 2:105–11
77. Pardy, R. L. 1974. Some factors affecting the growth and distribution of the algal endosymbionts of *Hydra viridis. Biol. Bull.* 147:105–18
78. Pardy, R. L. 1974. Regulation of the endosymbiotic algae in hydra by digestive cells and tissue growth. *Am. Zool.* 14:583–88
79. Pardy, R. L. 1976. The morphology of green *Hydra* endosymbionts as influenced by strain and host environment. *J. Cell Sci.* 20:665–69
80. Parke, M., Manton, I. 1967. The specific identity of the algal symbiont of *Convoluta roscoffensis. J. Mar. Biol. Assoc. UK* 47:445–64
81. Pfeiffer, S. E., Herschman, H. R., Lightbody, J. E., Sata, G., Levine, L. 1971. Modification of cell surface antigenicity as a function of culture conditions. *J. Cell. Physiol.* 78:145–52
82. Pickett-Heaps, J. 1972. Cell division in *Cyanophora paradoxa. New Phytol.* 71:561–67
83. Pool, R. R. 1976. *Symbiosis of Chlorella and Chlorhydra viridissima.* PhD thesis. Univ. California, Los Angeles, Calif. 122 pp.
84. Pool, R. R. 1979. The role of algal antigenic determinants in the recognition of potential algal symbionts by cells of *Chlorohydra. J. Cell Sci.* 35:367–79
85. Preer, J. R. 1975. The hereditary symbionts of *Paramecium aurelia. Symp. Soc. Exp. Biol.* 29:125–44
86. Provasoli, L., Yamasu, T., Manton, I. 1968. Experiments on the resynthesis of symbiosis in *Convoluta roscoffensis* with different flagellate cultures. *J. Mar. Biol. Assoc. UK* 48:465–79
87. Reimer, A. A. 1971. Observations on the relationships between several species of tropical zoanthids (Zoanthidea, Coelenterata) and their zooxanthellae. *J. Exp. Mar. Biol. Ecol.* 7:207–14
88. Ricard, M., Salvat, B. 1977. Faeces of *Tridacna maxima* (Mollusca, Bivalvia), composition and coral reef importance. *Int. Coral Reef Symp.* 3:495–501
89. Roth, S. 1973. A molecular model for cell interactions. *Q. Rev. Biol.* 48: 541–63
90. Sará, M. 1971. Ultrastructural aspects of the symbiosis between two species of the genus *Aphanocapsa* (Cyanophyceae) and *Ircinia variabilis* (Demospongiae). *Mar. Biol.* 11:214–21
91. Schmalljohann, R., Rottger, R. 1978. The ultrastructure and taxonomic identity of the symbiotic algae of *Heterostegina depressa* (Foraminifera, Nummulitidae). *J. Mar. Biol. Assoc. UK* 58:227–37
92. Schnepf, E., Brown, R. M. 1971. On relationships between endosymbiosis and the origins of plastids and mitochondria. In *Origin and Continuity of Cell Organelles,* ed. J. Reinert, H. Ursprung, pp. 299–322. New York, Heidelberg, Berlin: Springer-Verlag
92a. Schoenberg, D. A. 1976. *Genetic variation and host specificity in the zooxanthella Symbiodinium microadriaticum Freudenthal, an algal symbiont of marine invertebrates.* PhD thesis. Yale Univ., New Haven, Conn. pp. 183
93. Schoenberg, D. A., Trench, R. K. 1976. Specificity of symbioses between marine cnidarians and zooxanthellae. In *Coelenterate Ecology and Behavior,* ed. G. O. Mackie, pp. 423–32. New York: Plenum
94. Schoenberg, D. A., Trench, R. K. 1979. Genetic variation in *Symbiodinium* (=*Gymnodinium*) *microadriaticum* Freudenthal and specificity in its symbiosis with marine invertebrates. I. Isoenzyme and soluble protein patterns of axenic cultures of *S. microadriaticum. Biol. Bull.* In press
95. Schoenberg, D. A., Trench, R. K. 1979. Genetic variation in *Symbiodinium*

(=*Gymnodinium*) *microadriaticum* Freudenthal and specificity in its symbionts with marine invertebrates. II. Morphological variation in *S. microadriaticum*. *Biol. Bull.* In press
96. Schoenberg, D. A., Trench, R. K. 1979. Genetic variation in *Symbiodinium* (=*Gymnodinium*) *microadriaticum* Freudenthal and specificity in its symbiosis with marine invertebrates. III. Specificity and infectivity of *S. microadriaticum*. *Biol. Bull.* In press.
97. Sharon, N., Lis, H. 1975. Use of lectins for the study of membranes. *Methods Membr. Biol.* 3:147–200
98. Siebens, H. C., Trench, R. K. 1978. Aspects of the relation between *Cyanophora paradoxa* (Korschikoff) and its endosymbiotic cyanelles *Cyanocyta korschikoffiana* (Hall and Claus). III. Characterization of ribosomal ribonucleic acids. *Proc. R. Soc. London Ser. B* 202:463–72
99. Smith, B. N., Epstein, S. 1971. Two categories of $^{13}C/^{12}C$ ratios for higher plants. *Plant Physiol.* 47:380–84
100. Smith, D. C. 1974. Transport from symbiotic algae and chloroplasts to animal hosts. *Symp. Soc. Exp. Biol.* 28:473–508
101. Smith, D. C., Muscatine, L., Lewis, D. H. 1969. Carbohydrate movement from autotrophs to heterotrophs in parasitic and mutualistic symbiosis. *Biol. Rev.* 44:17–90
102. Soldo, A. T. 1974. Intracellular particles in *Paramecium aurelia.* In *Paramecium: A Current Survey,* ed. W. J. van Wagtendonk, pp. 375–442. Holland: Elsevier
103. Spangenberg, D. B. 1971. Thyroxin induced metamorphosis in *Aurelia. J. Exp. Zool.* 178:183–94
104. Stanier, R. Y., Cohen-Bazire, G. 1977. Phototrophic prokaryotes: The cyanobacteria. *Ann. Rev. Microbiol.* 31: 225–74
105. Steele, R. D. 1975. Stages in the life history of a symbiotic zooxanthella in pellets extruded by its host *Aiptasia tagetes* (Duch. and Mich.). (Coelenterata, Anthozoa). *Biol. Bull.* 149:590–600
106. Steele, R. D. 1977. The significance of zooxanthella containing pellets extruded by sea anemones. *Bull. Mar. Sci.* 27:591–94
107. Steele, R. D., Goreau, N. I. 1977. The breakdown of symbiotic zooxanthellae in the sea anemone *Phyllactis* (=*Oulactis*) *flos-culifera* (Actiniaria). *J. Zool. London* 181:421–37
108. Steinman, R. A., Cohn, Z. A. 1974. The metabolism and physiology of the mononuclear phagocytes. In *The Inflammatory Process,* ed. B. W. Zweifach, L. Grant, R. T. McCluskey, pp. 449–510. New York: Academic. 2nd ed.
109. Sugiura, Y. 1964. On the life history of rhizostome medusae. II. Indispensability of zooxanthellae for strobilation in *Mastigias papua. Embriologia* 8:223–33
110. Taylor, D. L. 1969. On the regulation and maintenance of algal numbers in zooxanthellae-coelenterate symbiosis, with a note on the nutritional relationship in *Anemone sulcata. J. Mar. Biol. Assoc. UK* 49:1057–65
111. Taylor, D. L. 1971. On the symbiosis between *Amphidinium klebsii* (Dinophyceae) and *Amphiscolops langerhansi* (Turkellaria:Acoela). *J. Mar. Biol. Assoc. UK* 51:301–13
112. Taylor, D. L. 1971. Ultrastructure of the "zooxanthella" *Endodinium chattoni in situ. J. Mar. Biol. Assoc. UK* 51:227–34
113. Taylor, D. L. 1973. Cellular interactions of algae-invertebrate symbiosis. *Adv. Mar. Biol.* 11:1–56
114. Taylor, D. L. 1973. Algal symbionts of invertebrates. *Ann. Rev. Microbiol.* 27: 171–87
115. Taylor, D. L. 1974. Symbiotic marine algae: taxonomy and biological fitness. In *Symbiosis in the Sea,* ed. W. B. Vernberg, pp. 245–62. Columbia: Univ. South Carolina Press
116. Taylor, D. L. 1974. Symbiosis between the chloroplasts of *Griffithsia flosculosa* (Rhodophyta) and *Hermea bifida* (Gastropoda, Opisthobranchia). *Publ. Stag. Zool. Napoli* 39:116–20
117. Terry, R. J., Smithers, S. R. 1975. Evasion of the immune response by parasites. *Symp. Soc. Exp. Biol.* 29:453–65
118. Theodor, J. 1969. Contribution a l'etude des gorgones. VIII. *Eunicella stricta aphyta,* sous espece novelle sans zooxanthelles, proche d'une espece normalment infesteè par ces algues. *Vie Milieu* 20:635–37
119. Ting, I. P. 1976. Malate dehydrogenase and other enzymes of C_4 acid metabolism in marine plants. *Aust. J. Plant Physiol.* 3:121–27
120. Trager, W. 1974. Some aspects of intracellular parasitism. *Science* 183: 269–73
121. Trench, R. K. 1971. The physiology and biochemistry of zooxanthellae symbiotic with marine coelenterates. I. Assimilation of photosynthetic products of zooxanthellae by two marine coelenter-

ates. *Proc. R. Soc. London Ser. B* 177:225–35
122. Trench, R. K. 1971. The physiology and biochemistry of zooxanthellae symbiotic with marine coelenterates. II. Liberation of fixed ^{14}C by zooxanthellae *in vitro*. *Proc. R. Soc. London Ser. B* 177:237–50
123. Trench, R. K. 1971. The physiology and biochemistry of zooxanthellae symbiotic with coelenterates. III. The effects of homogenates of host tissues on the excretion of photosynthetic products *in vitro* by zooxanthellae from two marine coelenterates. *Proc. R. Soc. London Ser. B* 177:251–64
124. Trench, R. K. 1974. Nutritional potentials in *Zoanthus sociatus* (Coelenterata, Anthozoa). *Helgol. Wiss. Meeresunters.* 26:174–216
125. Trench, R. K. 1975. Of "leaves that crawl." Functional chloroplasts in animal cells. *Symp. Soc. Exp. Biol.* 29: 229–65
126. Trench, R. K. 1979. *Integrative mechanisms in mutualistic endosymbioses.* Ann. Coll. Biol. Sci. Colloq., 5th, Ohio State Univ., Columbus. In press
127. Trench, R. K., Boyle, J. E., Smith, D. C. 1973. The association between chloroplasts of *Codium fragile* and the mollusc *Elysia viridis.* I. Characteristics of isolated *Codium* chloroplasts. *Proc. R. Soc. London Ser. B* 184:51–61
128. Trench, R. K., Boyle, J. E., Smith, D. C. 1973. The association between chloroplasts of *Codium fragile* and the mollusc *Elysia viridis.* II. Chloroplast ultrastructure and photosynthetic carbon fixation in *E. viridis. Proc. R. Soc. London Ser. B* 184:63–81
129. Trench, R. K., Gooday, G. W. 1972. Incorporation of [^{3}H]-leucine into protein by animal tissues and by endosymbiotic chloroplasts in *Elysia viridis* (Montagu). *Comp. Biochem. Physiol.* 44A:321–30
130. Trench, R. K., Ohlhorst, S. 1976. The stability of chloroplasts of siphonaceous algae in symbiosis with marine slugs. *New Phytol.* 76:99–109
131. Trench, R. K., Pool, R. R., Logan, M., Engelland, A. 1978. Aspects of the relation between *Cyanophora paradoxa* (Korschikoff) and its endosymbiotic cyanelles *Cyanocyta korschikoffiana* (Hall and Claus). I. Growth, ultrastructure, photosynthesis and the obligate nature of the association. *Proc. R. Soc. London Ser. B* 202:423–43
132. Trench, R. K., Ronzio, G. S. 1978. Aspects of the relation between *Cyanophora paradoxa* (Korschikoff) and its endosymbiotic cyanelles *Cyanocyta korschikoffiana* (Hall and Claus). II. The photosynthetic pigments. *Proc. R. Soc. London Ser. B* 202:445–62
133. Trench, R. K., Siebens, H. C. 1978. Aspects of the relation between *Cyanophora paradoxa* (Korschikoff) and its endosymbiotic cyanelles *Cyanocyta korschikoffiana* (Hall and Claus). IV. The effects of rifampicin, chloramphenicol and cycloheximide on the synthesis of ribosomal ribonucleic acids and chlorophyll. *Proc. R. Soc. London Ser. B* 202:473–82
134. Trench, R. K., Wethey, D. S., Porter, J. W. 1979. Some observations on the symbiosis with zooxanthellae in the Tridacnidae (Mollusca: Bivalvia). *Biol. Bull.* In press
135. Tripp, M. R. 1975. In *Invertebrate Immunity: Mechanisms of Invertebrate Vector-Parasite Relations,* Ed. K. Maramorosch, R. E. Shope, pp. 201–23. New York: Academic
136. Troughton, J. H., Hendy, C. H., Card, K. A. 1971. Carbon isotope fractionation in *Atriplex* spp. *Z. Pflanzenphysiol.* 65:461–64
137. von Holt, C. 1968. Uptake of glycine and release of nucleoside polyphosphates by zooxanthellae. *Comp. Biochem. Physiol.* 26:1071–79
138. Weis, D. S., Berry, J. L. 1974. Correlation of mating behavior with species of endosymbiotic algae in *Paramecium bursaria. J. Protozool.* 21:416
139. Weiss, P. 1953. Specificity in growth control. In *Biological Specificity and Growth,* ed. E. G. Butler, pp. 195–206. Princeton, NJ: Princeton Univ. Press
140. Whelan, T., Sackett, W. M., Benedict, C. R. 1970. Carbon isotope discrimination in a plant possessing the C_4 dicarboxylic acid pathway. *Biochem. Biophys. Res. Commun.* 41:1205–10
141. Yonge, C. M. 1936. Mode of life, feeding, digestion and symbiosis with zooxanthellae in the Tridacnidae. *Sci. Rep. Great Barrier Reef Exped.* 1:283–321
142. Yonge, C. M. 1944. Experimental analysis of the association between invertebrates and unicellular algae. *Biol. Rev.* 19:68–80
143. Yonge, C. M. 1963. The biology of coral reefs. *Adv. Mar. Biol.* 1:209–60

Ann. Rev. Plant Physiol. 1979. 30:533–91

BIOSYNTHESIS AND ACTION OF ETHYLENE ❖7680

Morris Lieberman

Post Harvest Plant Physiology Laboratory, Beltsville Agricultural Research Center, USDA, Beltsville, Maryland 20705

CONTENTS

0066-4294/79/0601-0533$01.00

INTRODUCTION

Ethylene, the simplest unsaturated carbon compound, which is a gas under physiological conditions of temperature and pressure, exerts a major influence on many if not all aspects of plant growth, development, and senescence apparently at regulatory levels of metabolism. Ethylene is considered a plant hormone because it is a natural product of metabolism, acts in trace amounts, in conjunction with or antagonistic to other plant hormones, and is neither a substrate or cofactor in reactions associated with major developmental plant processes. Whether or not such a natural product of metabolism fits all the definitions of a hormone, or more specifically a plant hormone, may be open to some argument, but such arguments can only dwell on semantics. The fact is that ethylene is a powerful natural regulating substance in plant metabolism, acting and interacting with other recognized plant hormones in trace amounts, and its effects are observed especially during critical periods in the life cycle of higher plants.

Biochemical and physiological studies of ethylene biosynthesis and mode of action have increased significantly in the past 20 years. This subject has been reviewed periodically in this series (4, 50, 221) and elsewhere (156, 198, 217 255, 288, 291) and in addition, two monographs have been published (5, 219) in the past decade. Interest and activity in this field has been increasing in recent years with the realization that the influence of ethylene is of considerable importance in understanding not only fruit ripening and senescence but also general hormonal activity in plants. Two developments in the past 15 years were of considerable importance in recognition and expansion of research on ethylene in plant physiology. One is the ubiquitous use of gas chromatography which allowed a rapid, sensitive, and simple assay of ethylene evolved by plant tissues. As a consequence, ethylene is currently the easiest hormone to assay, because it is a gas which evolves from the tissues and requires no extraction or purification prior to analysis. The second important development was the rediscovery of the relationship between ethylene production by plant tissues and auxins (56, 196, 201, 293). Rediscovery of this relationship has led to interesting studies which also show interactions between ethylene and cytokinins (98, 122, 151, 159), ethylene and GA (257), and ethylene and ABA (74, 165).

Influences of ethylene have been observed in practically all aspects of plant growth and development: in seed germination (140), seedling growth (59), root growth (64), growth of leaves (212), many kinds of stress phenomena (291), and in regulating ripening, aging, and senescence (41). Ethylene is therefore an important component in the mix of hormonal regulatory factors that control growth, development, and senescence.

Current research in the area of ethylene biochemistry and physiology has as its aim the establishment of pathways of biosynthesis, regulatory controls, and mechanisms of action of this hormone in plant metabolism. In this review I will consider and evaluate the state of current research, knowledge, and thoughts on ethylene biosynthesis and mode of action.

BIOSYNTHESIS OF ETHYLENE

Ethylene Production by Microorganisms

ETHYLENE PRODUCTION IN *PENICILLIUM DIGITATUM* Production of ethylene by microorganisms has been studied mostly in *Penicillium digitatum* Sacc., the green mold of citrus fruit, which produces large quantities of ethylene when grown in static culture. The biochemical origin of ethylene in *P. digitatum* was associated with the TCA cycle (129, 281) and specifically with the middle carbons of the dicarboxylic acids, particularly fumarate. Ketring et al (141) concluded that ethylene derives from the methylene carbons of citrate as they pass through the TCA cycle, because monofluoroacetate, which inhibits conversion of citric acid to isocitric acid, also inhibits ethylene production by *P. digitatum.* Addition of isocitrate restored ethylene production and respiration, but α-ketoglutarate, succinate, or malate had no effect in restoring ethylene production in fluoroacetate-treated mycelia. Uniformly labeled ^{14}C-methionine, which is the substrate for ethylene in higher plants, was not converted to labeled ethylene by *P. digitatum* (141). It appeared from these studies that ethylene production in *P. digitatum* is associated with metabolism of isocitrate in the TCA cycle.

Chou & Yang (69), following up on the work of Ketring et al (141), demonstrated that in *P. digitatum,* ethylene is derived from carbons 3 and 4 of 2-ketoglutarate or glutamic acid. Since (2,3^{14}C) succinate was very inefficient as a substrate for ethylene production in the fungus, as compared to (3,4–^{14}C) glutamate or (3,4–^{14}C) 2-ketoglutarate, 2-ketoglutarate must be the branching point leading from the TCA cycle to ethylene biosynthesis. It was not clear, however, whether 2-ketoglutarate or glutamate is the immediate precursor of ethylene biosynthesis in the fungus. It is interesting that glutamate, the precursor of ethylene in *P. digitatum,* has some structural similarity to methionine, the precursor of ethylene in higher plants, and that in both systems ethylene derives from carbons 3 and 4 of their respective substrates.

Significant production of ethylene by *P. digitatum* grown in liquid culture occurs only when the fungus is cultured under static conditions (without shaking), and it appears to be related to the surface development of a mycelial mat (254). Recently Chalutz et al (67) demonstrated ethylene production from methionine in shake cultures in which methionine both induced and provided the substrate for ethylene. Production of ethylene in shake cultures induced by methionine showed a lag period and was inhibited by actinomycin D and cycloheximide (66). The filtrate of induced shake culture containing no cells also evolved ethylene by both enzymic and nonenzymic reactions, presumably from components which leak from the fungal cells. In shake culture, therefore, both physiological and nonphysiological systems may produce ethylene from methionine or a substrate derived from methionine. The nonphysiological system develops about 12–24 hr after induction of the physiological system. Methionine inhibits static cultures of *P. digitatum* and aminoethoxyvinyl glycine (AVG) and methoxyvinyl glycine (MVG), inhibitors of methionine-ethylene-forming systems in higher plants, inhibited both the static system and the shake-culture system (66). Inorganic phosphate was shown to regulate ethylene production in shake cultures which produced 100-fold more ethylene in O.O1 mM as compared to 100 mM phosphate (68). However, phosphate levels had little influence on ethylene production in static cultures, although the same substrate, glutamate, is involved in phosphate-controlled ethylene production in shake cultures (A. K. Mattoo et al, unpublished). These experiments show that *P. digitatum* is capable of producing small or large quantities of ethylene from different substrates, depending on the type of medium and conditions of culture under which the fungus is grown.

ETHYLENE PRODUCTION BY OTHER MICROORGANISMS Thomas & Spencer (267) showed that shake cultures of yeast, *Saccharomyces cerevisiae,* were induced by L-methionine to produce ethylene, and in addition, as with *P. digitatum* in shake cultures, L-methionine was also a substrate for ethylene in this system. In further agreement with Chalutz et al (67), production of methionine-induced ethylene was increased by glucose in the medium.

Ethylene production from methionine has been reported in *E. coli* (222), from *Mucor hiemalis,* and other microorganisms isolated from soils (43). There are some doubts about the physiological or metabolic nature of ethylene production by these microorganisms, especially since they require light and are greatly stimulated by reduced iron (174), conditions which convert methionine to ethylene nonenzymatically (291).

PROPOSED ROLES OF ETHYLENE IN MICROORGANISMS A fungistatic function for ethylene in soils has been suggested (247), but Archer (20) found no impairment in spore germination after ethylene treatment among several soil-inhabiting fungi, suggesting that ethylene may not be capable of inducing fungistasis in all soil types. Currently there is considerable research on the production, metabolism, and function of ethylene in soils (247), and an oxygen-ethylene cycle was proposed which may have a bearing on the biological balance in soils (247).

It is intriguing that production of ethylene in soils is associated with anaerobiosis (247), and yet it is a process that requires oxygen in higher plants (24, 169). This research area has recently been reviewed (246).

An interesting observation relating to the function of ethylene in microorganisms was reported by Russo et al (232). They showed that ethylene mediates the avoidance response of the sporangiophores of the fungus *Phycomyces blakesleeanus*. This fungus has the property of growing away from a barrier which is a few millimeters from the growing zone of the sporangiophores. Ethylene appears to mediate this response in an interesting way, involving variations in its concentration on opposite sides of the sporangiophores. The implications of this observation may be worth investigating in root or stem tip growth in higher plants wherein ethylene may function similarly.

Biosynthesis of Ethylene in Higher Plants

Tissues from fruits such as apples and tomatoes, which produce relatively high levels of ethylene, produce no ethylene upon homogenization to a cell-free state (193). Despite numerous attempts to isolate cell-free ethylene-forming systems from many different plant tissues, utilizing various techniques for preserving cellular components and organelles, no one thus far has succeeded in obtaining physiological-like cell-free production of ethylene. Where such systems have been reported (6, 102, 148, 180, 268), there is considerable doubt about the physiological nature of these systems. All cell-free ethylene-forming systems reported are suspect because, as will be shown later, production of ethylene can occur in many nonphysiological reactions utilizing substances extracted from cells. Therefore, rigid criteria are necessary to eliminate all possibilities of nonphysiological ethylene production. These experiences strongly suggest that the natural ethylene-forming system is highly structured and very delicately poised in the intact cell (185).

MODEL SYSTEMS FOR ETHYLENE PRODUCTION Inasmuch as homogenates or cell-free systems were inactive in ethylene production, advances in elucidation of ethylene biosynthesis came from studies of model systems. These model systems, which operate at ambient temperatures, convert linolenic acid (171, 173), methionine (168), methional or α-keto-γ-methylthiobutyrate (KMB) (163, 290), or propanal (161) to ethylene in the presence of metal catalysts or free radical generating systems. The question arising from these studies of model systems was whether or not these systems relate to ethylene biosynthesis.

Peroxidized linolenate systems In the first of these systems, peroxidized linolenic acid is degraded by reduced copper or iron to the family of hydrocarbon gases from methane to pentane and ethylene to pentene, their isomers, and perhaps their higher homologues (162, 173). The reduced metal catalysts accelerated peroxidation and decomposition of linolenic acid, presumably by formation of peroxides and accompanying free radical chain reactions (164). This model system was considered to be analogous to the natural ethylene-forming system, which presumably derived from membrane-bound linolenate released during membrane disintegration association with senescence (84, 99, 171, 192). Propanal appeared to be the degradation product of peroxidized linolenate which was the immediate precursor of ethylene in this model system (161). However, ^{14}C-labeled proponal (25) or ^{14}C-labeled linolenate (178) were not incorporated into $^{14}C_2H_4$ by apple, tomato, or cauliflower tissue. These results appear to eliminate linolenate and propanal as precursors of ethylene in vivo. There is, nevertheless, a possibility that ethylene may be formed from fatty acids in ripening avocado fruit, where it appears that ethylene formed early in the climacteric rise stage had a $^{13}C/^{12}C$ ratio different from the ethylene formed by climacteric and postclimacteric fruit (T. Solomos and G. G. Laties, personal communication). These studies will be discussed further in the context of ethylene precursors other than methionine.

Ethane production by plant tissues Ethane production by plants is associated with tissue injury (145) which may relate to membrane destruction and peroxidation of linolenic acid. Cell-free homogenates of apple and *Phaseolus* seedlings, freed of dissolved ethylene, produce no ethylene but do evolve significant amounts of ethane which appears to be derived from linolenic acid (79, 135, 171). Production of ethane relative to ethylene increased significantly in sugar beet leaf disks as the area of injury by spot freezing increased. The percentage of ethylene formed by the leaf disks dropped sharply after 50% of the leaf area was frozen, while ethane production increased linearly through 100% of the frozen leaf area (88). In contrast to ethane formation which arose from the frozen (killed) cells, ethylene production appeared to arise from the leaf area surrounding the frozen cells of the leaf, representing a wound response. Ethane production was observed in potato tissue slices and mitochondria which were stimulated by linolenic acid and inhibited by diphenylamine, a free radical scavenger (145).

Bisulfite-injured leaves of cucurbits evolve ethane and ethylene and both are enhanced by light, which may activate free radicals (284). AVG, which inhibits ethylene biosynthesis from methionine, partially inhibited evolution of ethylene and simultaneously increased evolution of ethane in leaf tissues of cucurbits wounded with bisulfite. This can be interpreted to indicate that ethylene and ethane are derived from different sources. However, there may be dual precursors for ethylene, one of which could be peroxidized linolenic acid. Additional evidence that ethylene and ethane came from different sources was obtained from SO_2 wounded alfalfa seedlings which showed elevated ethylene and ethane production (218). Light had opposite effects on the production of the two gases; ethane production rates were higher in the light, whereas ethylene production was higher in the dark. The different reactions to light and different time courses of ethane and ethylene evolution suggest that these two gases are formed by different mechanisms or from different substrates. Although light appeared to accelerate both ethylene and ethane in experiments by Wilson et al (284), the data of Konze & Elstner (145), and Peiser & Yang (218) suggest that ethane and ethylene are derived from different precursors, but other interpretations are possible. Since peroxidized linolenate can serve as a precursor for both ethane and ethylene (173), depending on conditions, it is possible that a mixture of substrates may provide precursors for ethylene. Whether or not lipid peroxidation is a source of ethylene in living tissues, especially injured and senescent tissues, upon breakdown of membranes is a question that is not fully settled, although some current evidence appears to rule it out and suggests only a minimal quantity of ethylene can be derived from linolenate in vivo.

Methionine model systems In the course of experiments with model systems using peroxidized linolenate, ascorbate, and copper to catalyze the production of hydrocarbon gases, including ethylene, methionine was added to the system as a possible free radical quenching agent to inhibit the lipid peroxidation reaction (265). Instead of suppression of ethylene production in the peroxidized linolenate system, ethylene production increased significantly. It soon became obvious that methionine can act as a substrate for ethylene in a copper-ascorbate catalyzed model system in the absence of peroxidized linolenate (168). Unlike the linolenate system, which evolves a family of hydrocarbon gases from methane to hexane and ethylene to hexene and their higher homologues (81, 162, 164), ethylene was the only hydrocarbon evolved from methionine (168). This reaction was mediated by hydrogen peroxide which is produced in a reaction between cuprous ions, hydrogen ions, and oxygen. Methional (β-methylthiopropionaldehyde) was identified as an intermediate in this reaction and is a more efficient substrate, yielding three to four times as much ethylene as methionine (168). Tracer studies with ^{14}C methionine revealed that carbons 3 and 4 of methionine were the atoms converting to ethylene. Methionine and ethionine were the only amino acids which evolved significant amounts of ethylene in this system, whereas methionine sulfoxide and methionine sulfone were inactive, suggesting the necessity for an unencumbered sulfur atom that could be converted to a sulfonium ion.

Methionine also converted to ethylene in a model system in which light-activated FMN (flavin mononucleotide) catalyzed the degradation of methionine to ethylene (290). As in the copper-ascorbate model system, methional was an intermediate, but copper and hydrogen peroxide inhibited, and catalase had little influence on this ethylene-forming system. In both the light-FMN activated system and the Cu-ascorbate system, it appears that free radicals generated in the reaction provide the driving forces which degrade the methionine molecule. The sulfur atom in methionine is converted to a sulfonium ion which becomes the electron pulling function of the molecule. Thus conditions are set up for a concerted elimination reaction mechanism which, in conjunction with electron withdrawing effects of free radicals in the reaction medium, results in the elimination of the methylthio group of the methionine molecule. The carboxy and amino groups of methionine provide centers for nucleophilic attack which provide a flow of electrons to the sulfonium ion (290). In general, a molecule possessing a central ethylenic group, $-CH_2-CH_2-$, with one end attached to a possible electron withdrawing center, such as a sulfonium ion, and the other end attached to a possible electron donating center, should evolve ethylene in a degradation reaction initiated by a free radical attack on the nucleophile at one end of the ethylenic group. Methionine, methional, the α-keto analog of methionine (α-keto-γ-methylthiobutyric acid), and ethrel (2-chloroethanephosphonic acid) (72) all possess this general structure, and these molecules are readily fragmented in a concerted elimination reaction yielding ethylene. The efficiency of conversion varies somewhat between these molecules. Ethephon is readily converted to ethylene in either slightly acid or basic solutions (287), while the other molecules require various levels of attack by free radicals in order to release ethylene.

Indirect evidence for the free radical nature of these ethylene-forming reactions was provided by Beauchamp & Fridovich (28). Methional, but not methionine, was readily converted to ethylene in an aerobic system using xanthine oxidase and xanthine to generate superoxide radicals and hydrogen peroxide, which react to produce the hydroxy radical. The hydroxy radical was pinpointed as the chemical species which actually reacts with methional to produce ethylene. However, in a recent study, Bors et al (45) found the mechanisms for the oxidation of methional by hydroxy radicals to be more complex than a simple fragmentation of an intermediary thiyl radical cation, as suggested by Yang et al (290), and Bors et al suggest that the conversion of methional to ethylene is not necessarily evidence for the presence of the OH radical. A recent report by Pryor & Tang (224) also concludes that conversion of methional to ethylene is not specifically due to the OH radical. A variety of organic radicals are suggested as possible intermediates in production of ethylene from methional.

Peroxidase model systems for ethylene production The discovery of model systems for ethylene production provided clues to the substrate and reaction mechanisms that could be involved in the biosynthesis of ethylene. Methionine appeared to be a strong candidate for the ethylene precursor in plant tissues since ^{14}C L-methionine labeled in carbons 3 and 4 was incorporated into ethylene by apple tissues as in the model system (168, 169). Mapson and colleagues (179–181), in a series of experiments, demonstrated the conversion of methional or the α-keto analog of methionine, KMB, to ethylene by cell-free extracts of cauliflower florets. Three enzymes were reported to be involved in the conversion of methionine to ethylene in the cauliflower extract system: a transaminase which converts methionine to KMB; glucose oxidase which generates hydrogen peroxide; and a peroxidase which utilizes hydrogen peroxide to produce free radicals in a reaction with methional or KMB to form ethylene (177). Mapson identified *p*-coumaric acid and methane sulfinic acid as components of the cauliflower extract which were cofactors in the enzymatic system giving rise to phenoxy and sulfinic free radicals necessary for reaction with methional or KMB.

Mapson's system for ethylene production was similar to a model system developed by Yang (286), in which horseradish peroxidase, sulfite, specific phenols, and manganese, or catalytic amounts of hydrogen peroxide, converted methional or KMB to ethylene. Both these systems involve production of a series of free radical intermediates via oxidation of sulfite by peroxidase in the presence of phenols and hydrogen peroxide. Upon cell disruption, plant homogenates may contain a mixture of peroxidases and substances which may act as cofactors, similar to those compounds necessary for ethylene production from methional or KMB in the model peroxidase system. It is difficult to determine whether such ethylene production has physiological significance or a relationship to the natural in vivo system. Such systems appear to be nonphysiological artifacts. For example, whereas homogenates of apple tissue, a potent ethylene-forming tissue when intact, produce no ethylene, a peroxidase isolated from apple tissue extracts formed ethylene from methional when fortified with *p*-coumaric acid, sulfite, and catalytic amounts of hydrogen peroxide (264). This apple enzyme simply substituted for HRP in the ethylene-forming peroxidase model system. Actually, methional does not stimulate ethylene production in apple tissue which is stimulated to produce ethylene by methionine (169). Also, ^{14}C-methionine was converted to ethylene nearly twice as efficiently as ^{14}C KMB (27). Increased ethylene production and incorporation of ^{14}C-KMB to ^{14}C-ethylene observed by Mapson et al (178) could possibly have occurred in an extracellular ethylene-forming reaction catalyzed by peroxidase and cofactors which leak from the tissue to the surrounding incubation medium (163). Stimulation of ethylene production by KMB in incubation solution filtrates from cauliflower florets or tomato and apple tissue was related to the activity of peroxidase in these solutions. Cauliflower and tomato tissue filtrates from incubation media showed high peroxidase activity and large increases in ethylene production in response to KMB. In contrast, apple incubation media had virtually no peroxidase activity and showed no response to KMB (163).

If peroxidase is related to ethylene production, there should be a direct relationship between induction of ethylene production and increased peroxidase activity. However, there was no relationship at all between extractable peroxidase and endogenous ethylene production in IAA-induced subapical pea stem sections and in cotton (91, 136).

Although the evidence outlined above militates against the possibility that an ethylene-forming peroxidase system similar to the in vitro HRP system of Yang (286) operates in vivo, there is a likelihood that some unique kind of peroxidase system or a related free-radical generating system may be associated with ethylene production in vivo.

Two facts emerge from studies of model systems for ethylene production and the suggested enzymatic peroxidase systems. One is the requirement for oxygen in these ethylene-forming systems and the other is the involvement of free radical intermediates. Reaction of peroxidase with hydrogen peroxide forms free radicals which can react with oxygen nonenzymatically or give rise to a chain of free radical intermediates (110), some of which can react with methional or KMB to cause fission of the molecule yielding ethylene. Reaction with methionine seems to require other types of free radical species perhaps with greater energy or reactivity. The model copper ascorbate system and the model peroxidase system show these characteristics. The light-activated FMN system was not significantly inhibited by catalase (290), which suggests that H_2O_2 is not a major intermediate in this reaction system. This system, however, shows characteristics of a free radical reaction chain but apparently does not require oxygen. There may be a requirement for trace quantities of oxygen, while large amounts of oxygen may only serve to quench radicals or produce methionine sulfoxide or sulphone, which are not substrates for ethylene (168).

ETHYLENE PRODUCTION SYSTEMS IN VIVO In a number of tracer experiments with apple tissue (169), pea stem segments and bananas (60), avocado fruit (27), and morning glory flower tissue (115), ^{14}C methionine was converted to $^{14}C_2H_4$. Labeled ethylene was produced by the tissues only when carbons 3 and 4 of methionine were labeled and the relative specific activity of ethylene to methionine indicated that methionine goes rather directly to ethylene (27, 115, 169). Oxygen stimulates and is essential for the conversion of methionine to ethylene (169), and the L-form of methionine is preferentially converted to ethylene by the tissue, suggesting a stereospecific enzymatic reaction (24). The α-hydroxy and the α-keto analog of methionine (KMB) (27) and homoserine (26, 27) are also converted to ethylene but less efficiently than methionine.

In the conversion of methionine to ethylene, carbon 1 forms CO_2 (60), C-2 forms formic acid (244) and CO_2 (M. Lieberman unpublished), C-3 and 4 form ethylene (169), and the CH_3S is retained in the tissue (60). The fate of the S atom is of some significance since Yang & Baur (289) have determined that the level of free methionine and protein methionine are too low to provide all the ethylene evolved by apple fruit. Climacteric apple fruit contains about 60 nmoles/g of free methionine plus protein methionine and produces ethylene on the average at a rate of 4 nmoles/hr/g. At this rate, methionine, the ethylene substrate, must be resynthesized to provide sufficient substrate for continued ethylene production. The carbon skeleton of methionine can be supplied readily from glucose. However, the sulfur atom must be recycled to reform methionine. Tracer studies with ^{14}C and ^{35}S indicate that both the carbon and sulfur of the methyl mercapto moiety of methionine are incorporated as a unit into S-methyl cysteine (26). Baur & Yang (26) proposed that S-methyl cysteine is metabolized to methionine via homoserine, cystathionine, and homocysteine. More recently, with the possibility that S-adenosyl methionine (SAM) is an intermediate between methionine and ethylene, 5'-methylthioribose has been postulated as the S-methyl carrier which reacts with homoserine to regenerate methionine for ethylene production (8).

Hanson & Kende (115) noted that S-methylmethionine (SMM) was the major soluble metabolite formed from L-methionine-U-^{14}C by immature rib segments from morning glory (*Ipomea tricolor*) flower buds. As the rib tissue senesced, the methionine content increased about tenfold mainly through the transfer of the methyl group of SMM to homocysteine, thus providing two molecules of methionine. The specific radioactivity in ethylene produced during senescence of the rib segments was close to the specific activity of carbons 3 and 4 of the free methionine in the tissue and was about half that in SMM, indicating that ethylene derives from methionine in these tissues. The authors suggest that the control of ethylene production depends on the supply of homocysteine as the methyl group receptor, which may increase during aging. Alternatively, homocysteine may become accessible to the methionine synthesizing system as a result of the breakdown of intracellular compartmentation barriers. SMM has been reported as a metabolite of methionine-methyl-^{14}C in ripe apple tissue (26), and its relationship to methionine synthesis and ethylene production merits further investigation.

Recently selenomethionine and selenoethionine were shown to enhance ethylene production in senescing flower tissue of *Ipomea tricolor* and in auxin-treated pea-stem sections (146). The enhancement was greater than that exhibited by methionine and was strongly inhibited by AVG. Whereas ethionine inhibited ethylene production about 50% in pea stem sections, selenoethionine was the most effective stimulator of ethylene production, even more effective than selenomethionine. Since these selenoamino acids significantly lowered the specific radioactivity in labeled ethylene, from tissue previously incubated in methionine U-^{14}C, even more so than cold methionine, it appears that these selenoamino acids are precursors of ethylene via the methionine pathway and combine with the endogenous methionine pool. It is curious that these selenoamino acids are more efficiently converted to ethylene than methionine. The greater effectiveness of selenomethionine, as compared to methionine, in stimulating ethylene production was also observed in apple tissue slices (M. Lieberman and J. D. Anderson, unpublished data).

Inhibitors of ethylene production Metal chelators such as EDTA, DIECA, and KCN at 1 mM, and cuprizone at 0.1 mM inhibited ethylene production in apple tissues (169), suggesting the involvement of metal cofactors in the ethylene-forming system. Cobalt at 0.1 and 1.0 mM concentration strongly inhibited ethylene production by apple tissues and mung bean hypocotyl segments (155). This effect of cobalt may explain its influence in promoting elongation, leaf expansion, and hook opening in excised plant parts treated with auxins.

The most significant inhibitors of ethylene production in plant tissues are the enol ether amino acid analogs, the L-2 amino 4-alkoxy-*trans*-3-butenoic acid molecules which are natural products found in fermentation broths of some microorganisms (214, 223, 240). The structures of the three amino acid analogs and their microbial origins are shown in Table 1. The struc-

Table 1 Structure and origin of enol ether amino acid analog inhibitors of ethylene production

Compound	Structure	Origin	Ref.
Rhizobitoxine [L-2-amino-4-(2-amino-3-hydroxypropoxy)-*trans*-3-butenoic acid]	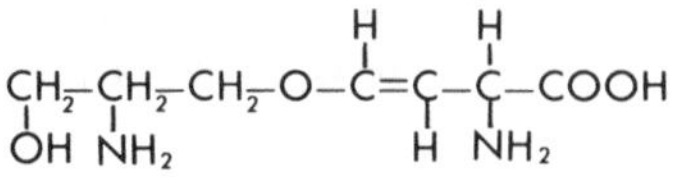	*Rhizobium japonicum*	214
Aminoethoxyvinyl glycine [L-2-amino-4-(2-aminoethoxy)-*trans*-3-butenoic acid]	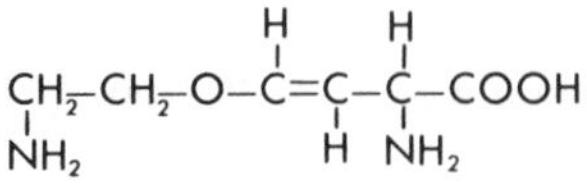	*Streptomyces* sp.	223
Methoxyvinyl glycine [L-2-amino-4-methoxy-*trans*-3-butenoic acid]	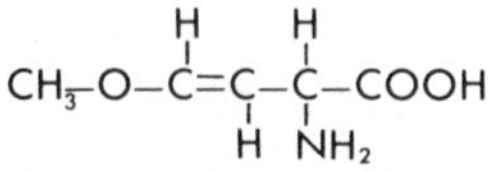	*Pseudomonas aeruginosa*	240

tural similarity of these amino acid analogs and methionine suggests that they may be competitive inhibitors for the substrate attachement site of the ethylene-forming enzyme system. Methionine could not reverse inhibition by rhizobitoxine, but it delayed inhibition by the amino acid analogs, depending on the relative concentration of the two compounds (170). The enol ether amino acid analogs are irreversible inhibitors of ethylene production in many tissues, and they appear to block the methionine binding site on an enzyme which may be involved in the initial step of the biochemical pathway from methionine to ethylene. Konze et al (146) found that neither selenomethionine nor methionine could protect the tissue against AVG. This raises the question as to whether the substrate for ethylene and the inhibitor (AVG) interact at the same site with the ethylene synthesizing system as reported by Lieberman et al (170). It is of some interest that the saturated analogs of the enol ether amino acids do not inhibit ethylene production and may act as substrates for ethylene (170). There are some tissues such as ripe tomatoes and avocado fruit which respond minimally to the enol ether amino acid analogs (21). This raises the possibility that substrates other than methionine may be precursors of ethylene in these tissues.

Alternatively, inhibition of ethylene production by the β-γ unsaturated enol ether amino acid analogs also suggests that a pyridoxal phosphate-linked enzyme may be involved in ethylene production since these β-γ-unsaturated amino acids have been shown to irreversibly inhibit β-cystathionase (104) and aspartate amino transferase (227), both of which are pyridoxal-linked enzymes. Canaline (α-amino-γ-aminoxybutyric acid), which is known to inhibit pyridoxal-dependent enzymes (226), inhibited the conversion of methionine to ethylene by apple tissue (203), which further implicates pyridoxal phosphate in ethylene biosynthesis. However, unlike inhibition with the unsaturated β-γ enol ether amino acid analogs, inhibition with canaline was partially reversed by methionine but not by pyridoxal phosphate (203). Since excess pyridoxal phosphate did not reverse the canaline-induced inhibition, as occurs in several aminotransferase and amino acid decarboxylase enzymes (226), these results suggest that the inhibition by canaline may be due to the structural similarity of canaline to methionine, involving stereospecific competitive inhibition on the enzyme at the substrate binding site. Experiments with canaline in the author's laboratory (unpublished) indicate that both methionine and pyridoxal phosphate reduce but do not eliminate the inhibition of ethylene production in apple tissue slices by canaline. It is possible that canaline may inhibit both by interfering with the binding of methionine and reaction with pyridoxal phosphate. Vinyl glycine, which irreversibly inhibits pyridoxal phosphate-linked amino transferase reactions (227) by specifically binding the pyridoxal phosphate moiety, inhibited ethylene production only about

20% as effectively as AVG. This suggests that in addition to inhibiting the α-amino binding site, AVG also inhibits by virtue of other structural characteristics of its molecule related to the ethoxy group. It is attractive to assume that the methionine-binding enzyme in the ethylene biosynthesis reaction is a pyridoxal phosphate-lined enzyme, since such enzymes are known to be involved in α-elimination reactions (80) and this could account for degradation of methionine to ethylene. However, more data are required to settle this question.

The enol ether amino acids have very little or no effect on ethylene production in the copper ascorbate or the FMN light model systems, which contain no enzymes. On the other hand, the enol ether amino acid analogs significantly stimulate ethylene production from KMB in the HRP model system for ethylene production (unpublished results of author).

Still another potent inhibitor of ethylene biosynthesis by apple tissue is 2,4-dinitrophenol (DNP) (203), the uncoupler of oxidative phosphorylation. The strong inhibition of ethylene production by DNP in both apples and mung bean hypocotyl sections suggests the involvement of ATP in the ethylene-forming system. Alternatively, however, DNP perturbs cellular membranes which may result in inhibition of ethylene biosynthesis (225).

Free radical inhibition of ethylene production Free radical quenching agents, such as *n*-propyl gallate and sodium benzoate, are another group of inhibitors which provide clues to the mechanism of ethylene production in plant tissues (21). These inhibitors, at 1 mM concentrations, significantly inhibit ethylene production even in tissues which are inhibited minimally by the unsaturated enol ether amino acid analogs (ripe tomato and avocado) (21).

Combinations of enol ether amino acid analogs and free radical quenching agents provide greater inhibition than either inhibitor alone (21). This further suggests that these inhibitors act at different sites in the pathway from methionine to ethylene. Since the model system for ethylene production all appear to involve free radicals (28, 45, 224), it is reasonable to assume that the in vivo system requires a free radical reaction. The fact that oxygen is required for ethylene production in vivo (27, 113, 169), supports the probability of a free radical intermediate step in the pathway from methionine to ethylene. The effective inhibition of ethylene production in cress roots by 3,5 diodo-4-hydroxybenzoic acid (DIHB) (230), a structure which suggests a free radical scavenger, also suggests the effectiveness of free radical scavengers as inhibitors of ethylene production.

Proteinaceous inhibitors of ethylene biosynthesis Sakai & Imaseki (234, 235) isolated a protein extracted from mung bean (*Phaseolus aureus* Roxb) hypocotyls which reversibly inhibited auxin-induced ethylene production from mung bean subapical hypocotyl segments. When the inhibited segments were washed to free them from added inhibitor, the rate of ethylene production reverted to that of auxin-stimulated control segments. The purified protein inhibitor did not degrade or bind IAA and had no peroxidase activity (235). The inhibitor binds to the surface of the cut segments, which suggests that the inhibitor acts on the epidermis and further suggests that ethylene stimulation by auxin also results from an interaction with epidermal cells. There is a possibility that this proteinaceous inhibitor of ethylene production may inactivate the ethylene-forming system during cell homogenization. On the other hand, it appears that this protein inhibitor only blocks auxin access to the active site with which auxin must react to stimulate the ethylene-forming system. There is a suggestion in these data that auxin-stimulated ethylene production occurs at the cell surface. The inhibitor had virtually no effect on ethylene production by apple slices. Thus there is some question as to whether this protein inhibitor is specific for the ethylene-forming enzyme or is uniquely effective in the mung bean system.

PROPOSED INTERMEDIATES IN THE PATHWAY FROM METHIONINE TO ETHYLENE Methionine can readily be degraded to ethylene in model systems, and methional has been identified as a major intermediate in this reaction, as already outlined above. However, there are currently no known enzymes which carry out this reaction in vivo. In the conversion of methionine to ethylene by tissues, C-1 is converted to CO_2, C-2 to formic acid, and C-3 and 4 to ethylene (60, 169, 244), much as in the model system. The sulfur atom and its related methyl group appear to be retained in the tissue (26, 60), thus differing from the model system. An additional important difference is that methional is not a substrate for ethylene in vivo (24, 169). The α-keto analog of methionine, which also serves as a substrate for ethylene in the HRP model system (286) and was suggested by Mapson et al (179) as an intermediate in vivo, was also shown not to be converted to ethylene as efficiently as methionine (24) and to produce ethylene by a nonphysiological extracellular reaction (163). Consequently, the two recognized intermediates between methionine and ethylene in model systems do not appear to be operative in the in vivo conversion of methionine to ethylene. Nevertheless, the products of the degradation of the first four carbons of methionine in vivo appear to be much the same as in the model systems. Metabolism of the methylthiol group of methionine appears to be different than in the model reaction systems. This result is expected, given the necessity to conserve the sulfur atom in tissues in which sulfur is limiting in the metabolic turnover of the methionine molecule (289).

Analysis of types of inhibitors of ethylene biosynthesis which do and do not inhibit reactions in the model system as compared to the in vivo system for ethylene production suggests possible mechanisms for ethylene biosynthesis. The ethoxy enol ether unsaturated amino acid AVG does not inhibit any of the model systems utilizing methionine as substrate for ethylene production (M. Lieberman, unpublished). The saturated analog of AVG is a substrate for ethylene production in the model system (M. Lieberman, unpublished). On the other hand, AVG is a potent inhibitor of ethylene production in tissue systems, and the saturated analog of AVG is either inactive or may even serve as a substrate for ethylene in vivo (170). Presumably the inhibition of the in vivo system by AVG occurs at an enzyme site, perhaps the substrate-binding site, and appears to act by virtue of its structural analogy to methionine. By contrast, the free radical quenching agents such as *n*-propyl gallate inhibit both the model system (M. Lieberman, unpublished) and the in vivo ethylene-forming systems (21). Apparently a free radical step is involved in both model and in vivo (enzyme-mediated) systems which may explain the requirement for oxygen in both types of systems.

Based on the effects of AVG and *n*-propyl gallate in model and tissue systems for ethylene production and the apparent absence of an intermediate between methionine and ethylene, a mechanism for ethylene production from methionine in vivo was proposed by Lieberman (157). It is suggested that methionine is degraded to ethylene directly while attached to an enzyme complex (Figure 1). If there is an intermediate, it is rapidly converted

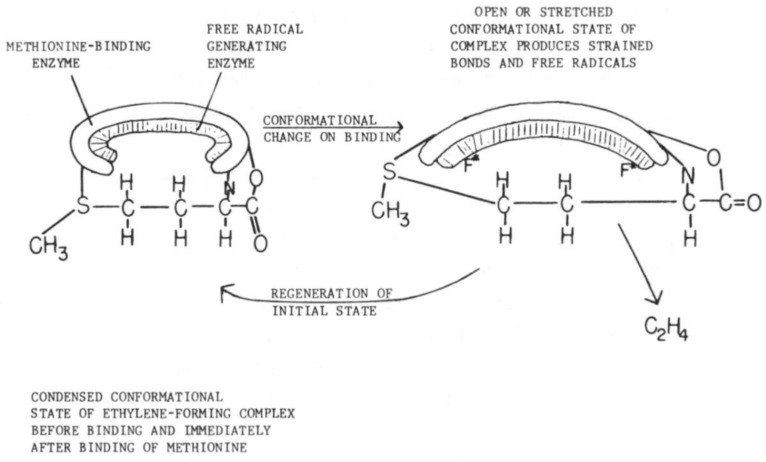

Figure 1 Hypothetical model of ethylene-forming enzyme system showing possible structural relationship between a methionine-binding ethylene-forming enzyme and a controlled free radical generating system.

to ethylene while still associated with the enzyme complex. In fact, it is conceivable that no free intermediate exists in the conversion of methionine to ethylene. The methionine molecule may break apart into its various products, releasing ethylene, CO_2, and other products in a rapid sequential disintegration of the molecule while still attached to the enzyme complex. Free radicals attack the methionine molecule, splitting it apart to form ethylene and other products of the reaction, including the methylthiol group which may be incorporated into the tissue for recycling to methionine. On binding of methionine to its substrate site on the ethylene-forming complex, the sulfur of methionine becomes a sulfonium ion which is then an electron-withdrawing center facilitating and directing attack by the generated free radicals, which are assumed to be in proper juxtapositions for reaction with the methionine molecule, bound and stretched on the binding enzyme.

In this scheme the free radicals formed are under rigid control of the ethylene-forming enzyme complex and closely associated with it both physically and biochemically. The entire complex is assumed to be delicately structured to provide for (*a*) triggering the free radical initiation reaction; (*b*) utilizing the free radicals to break the strained bonds of the methionine molecule; and (*c*) quenching the free radicals after they perform their function. The control of free radical production may reside in a conformational change in the complex which may either turn the generating system on or off. This enzyme complex, which must react with oxygen as well as methionine in order to produce ethylene, must be delicately poised and possibly localized in the plasma membrane and does not survive cell homogenization. No intermediate is theoretically necessary in this reaction scheme for conversion of methionine to ethylene.

There is no direct evidence for the existence of this hypothetical model for ethylene production. The rationale for the model rests on the following: (*a*) the ethylene-forming system does not survive cell homogenization, suggesting a delicate, highly organized structure; (*b*) inhibition by AVG and delay of inhibition in the presence of methionine suggests blocking of the binding site on the enzyme as the basis for inhibition; (*c*) inhibition by free radical scavengers and the requirement for oxygen suggests a free radical involvement in the reaction.

Adams & Yang (8) propose a different mechanism for in vivo ethylene production. Methionine is believed to be metabolized via S-adenosyl methionine (SAM) as an intermediate in the pathway to ethylene. The evidence for SAM as an intermediate was obtained after feeding ^{14}C methionine to apple tissue and identifying 5'methyl thioadenosine (MTA) and methylthioribose (MTR) which correlate with ethylene production. When ethylene production was inhibited by AVG, no MTA or MTR was detected in the tissue extracts. Preclimacteric apple tissue, which does not form ethylene, did not produce MTA or MTR from ^{14}C-labeled methionine. This provides indirect evidence that SAM may be an intermediate in the conversion of methionine to ethylene. Experiments in which the methylthiol group of methionine was labeled both with tritium and ^{35}S also showed that the methylthiol group was recycled as a unit into methionine after degradation of the molecule in the ethylene-forming reaction. The proposed pathway of methionine to ethylene and recycling of the methylthiol group is shown in Figure 2. In this scheme, conservation of the methylthiol group is provided for in the resynthesis of methionine.

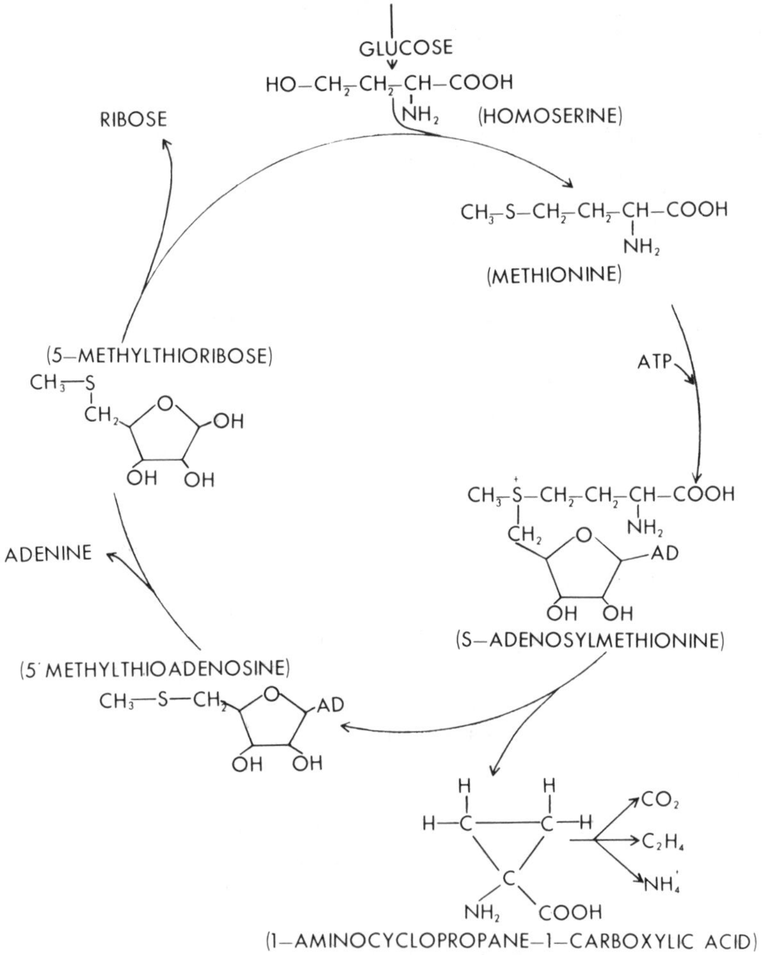

Figure 2 Proposed cycle for methionine metabolism related to ethylene biosynthesis according to Adams & Yang (8) and Yang (personal communications). AD = adenosine.

This hypothesis by Adams and Yang is supported by the fact that DNP, the uncoupler of oxidative phosphorylation, is a very potent inhibitor of ethylene production, implicating an ATP requirement for production of SAM in the ethylene-forming pathway. Also there is generally a lag period before incorporation of ^{14}C-methionine into ^{14}C-ethylene, which is consistent with the necessity of forming an intermediate (^{14}C-SAM) before ^{14}C-ethylene production can be observed. Additional indirect support for this hypothesis comes from the work of Kende (personal communication), who shows that selenomethionine is much more effective than methionine as a substrate for ethylene production. Since production of seleno-adenosyl methionine is kinetically more efficient than S-adenosyl methionine, the selenomethionine precursor would be expected to be a better substrate for ethylene. Teleologically, involvement of ATP in the reaction offers a mechanism for controlling ethylene production, and formation of SAM as an intermediate provides a mechanism for recycling the methylthiol group to methionine. While all these factors tend to support SAM as an intermediate, there is as yet no direct evidence for this pathway. It is difficult to introduce SAM into tissues because of its size and charge. Consequently no direct experiments have been possible with labeled SAM. The potent inhibition of ethylene production by DNP may be due mainly to disruption of membranes and not exclusively to its action as an uncoupler of oxidative phosphorylation (225). Nevertheless, the hypothesis of Adams and Yang is currently very attractive, but definitive proof of the intermediary nature of SAM in ethylene biosynthesis is still to be demonstrated conclusively.

Proposed immediate precursor of ethylene It has long been known that nitrogen atmospheres reduce and ultimately cause a cessation in ethylene production by pears and other fruits (113). Upon introduction of oxygen to the anaerobic fruit, there is an overshoot in ethylene production for a short time period and then a return to basal levels. There are a number of possible explanations for this overshoot in ethylene production after anaerobiosis. One explanation assumes total saturation of the methionine binding sites on the ethylene-synthesizing ethylene complex, which requires oxygen for free radical formation. Another explanation visualizes the buildup of an intermediate compound which needs to react with oxygen to produce ethylene.

Adams & Yang (9) studied the metabolism of ^{14}C methionine in apple plugs held in nitrogen and in air. At the end of a 6 hr incubation period the plugs were frozen and extracted in 80% ethanol. The plugs in N_2 contained a radioactive compound ("X") which when fed to apple plugs in air produced radioactive ethylene. When ^{14}C-methyl-labeled methionine was used in this type of experiment no compound "X" was observed, but labeled methylthioribose was formed. The "X" compound has recently been identified as 1-aminocyclopropane-1-carboxylic acid (ACC), which was effectively converted to ethylene in apple plugs under air but not in nitrogen atmospheres (S. F. Yang, personal communication). The reaction pathway from methionine to ethylene, according to Adams & Yang (9), may be depicted as in Figure 3.

AVG blocks conversion of SAM to compound ACC and nitrogen blocks conversion of ACC to C_2H_4. 1-Aminocyclopropane carboxylic acid was identified in pear and apple juice sometime ago by Burroughs (62). It does seem incongruous that ACC should build up in nitrogen atmospheres if it derives from SAM, a compound which is produced by reaction of methionine with ATP. Production of ATP is considerably curtailed by nitrogen atmospheres, but this may have little influence on production of SAM if there is sufficient ATP available from glycolysis and substrate level phosphorylation. Also, a sufficiently large SAM pool may exist to provide increased ACC.

LOCALIZATION OF THE ETHYLENE SYNTHESIZING SYSTEM IN APPLE TISSUES As already stated, the ethylene synthesizing system does not survive destruction of the cell. The very first step in isolating the enzyme system in vitro, tissue homogenization, destroys the system. Protoplasts prepared from apple tissue also did not exhibit ethylene production, but after culturing for several days (185), some ethylene production was obtained concomitant with regeneration of some of the cell wall. Production of ethylene by the protoplasts was dependent on addition of methionine, and the system was inhibited by AVG and *n*-propyl gallate. In more recent studies (J. D. Anderson, unpublished), protoplasts were obtained, using less drastic lytic methods which produce ethylene from ^{14}C-labeled methionine in complete absence of a cell wall. The experiments suggest that the ethylene-forming system is localized on the surface of the plasma membrane. Localization of the ethylene-forming system in the plasma membrane is

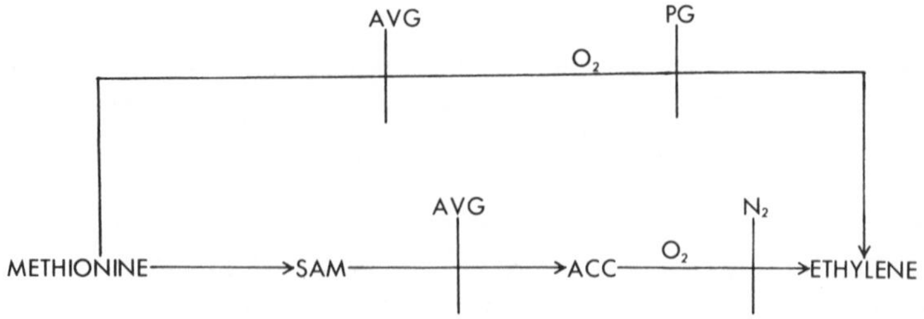

Figure 3 General scheme for metabolism of methionine to ethylene as proposed by Adams & Yang (9), with SAM and ACC as intermediates and inhibition by AVG and nitrogen where indicated. In the second pathway, proposed by Lieberman (157), methionine goes directly to ethylene while attached to enzyme complex. PG = propylgallate.

consistent with the localization of other hormonal activity in the plasma membrane (118).

The importance of proper osmolarity of the incubation solution surrounding apple slices (61, 126, 208) suggests the involvement of membrane stability for ethylene production. Stimulation of ethylene production by Ca^{2+}, shown by Lau & Yang (152), may also simply be related to membrane stabilization by high concentrations of calcium. The rapid decline of ethylene production shown by postclimacteric apple tissue slices incubated in 0.4 M sucrose or 0.6 M sorbitol can be stabilized considerably by 0.01 M calcium (M. Lieberman, unpublished data). This again implicates the importance of membrane stability to ethylene production. Additionally, Arrhenius plots of ethylene synthesizing systems of apple and tomato show discontinuities which suggest a lipid microenvironment for the ethylene synthesizing system in higher plants (184). These results are consistent with a lipo-protein matrix environment for the ethylene enzyme system as exists in membranes. Finally, the demonstration by Sakai & Imaseki (235) of the inhibition of ethylene production in mung bean seedlings by a protein which attaches to the cell surface of the tissue, and the reversal of the inhibition by washing off the inhibiting protein, further suggests that the ethylene synthesizing system is localized on the surface of the plasma membrane.

EVIDENCE FOR ETHYLENE PRECURSORS OTHER THAN METHIONINE Although much data indicate that methionine may be the sole precursor of ethylene in plant tissues, there are some suggestions that additional precursors and pathways may occur in some ripening fruits. Ethylene production in green tomato fruit is significantly inhibited by AVG, but as the fruit ripens, sensitivity to AVG declines considerably (21). However, ^{14}C-methionine is converted to $^{14}C_2H_4$ in ripe tomato fruit tissue, and AVG inhibits this conversion about the same as it does total ethylene. Avocado tissue slices are also relatively insensitive to AVG (21) and convert ^{14}C-methionine into $^{14}C_2H_4$ in climacteric rise and climacteric peak tissues, but not in preclimacteric tissues (27). These findings suggest other pathways from methionine to ethylene, and other precursors of ethylene production in ripening fruits. Solomos and Laties (personal communication) examined the ratio of the stable isotopes $^{13}C/^{12}C$ in ethylene collected from ripening avocados at various stages of ripeness. The prevalence of carbon 13 in this ratio indicates the type of endogenous substrate from which ethylene derives (130). The $^{13}C/^{12}C$ ratio in ethylene was found to vary in samples taken at different stages of the climacteric, suggesting that ethylene may derive from amino acids at the early stage of the climacteric rise and shift to fatty acids at the climacteric peak in the postclimacteric stage. These experiments suggest two physiological substrates for ethylene during ripening of the avocado fruit.

STRESS ETHYLENE PRODUCTION Many tissues which normally evolve little or no ethylene show a surge in ethylene production, three to ten times the basal levels, upon physical wounding (193, 236), bruising (116), chilling (75), freezing (88), contact with noxious chemicals (2), irradiation (1), contact with strong acid or base (M. Lieberman, unpublished), thigmotropism (131), attack by microorganisms (125), and other stresses.

Horizontally cut subapical pea segments, after a 26 min lag, exhibited a fourfold surge in ethylene production followed by a decline and a second rise to a peak to produce a characteristic double peak associated with wounding. This pattern of ethylene evolution following wounding lasted about 2.5 hr and suggests a dampened oscillation which may be related to negative feedback control in this tissue system (236, 237). Young Satsuma Mandarin fruit, which produced very little ethylene, provided albedo disks upon cutting which vigorously produced ethylene after a short lag period (121). Production of ethylene by albedo disks continued for 50 hr after excision, peaking in 30 hr. Intact immature green apple fruit, which do not produce ethylene, respond to chemical wounding by strong acid or base by producing ethylene in increasing amounts for many days following a 4 hr lag (M. Lieberman, unpublished data). Mature apple fruit, which produce high levels of ethylene, do not respond to such chemical wounding by increased ethylene production. There is, therefore, considerable variation between different tissues and tissues of different maturity in their response to wounding or stress as related to the length of the lag period, magnitude, and duration of increased stress ethylene production. In order to exhibit wound ethylene production, the wounding process must not destroy all the cells, because dead cells do not produce ethylene. It appears that the cells adjacent to the injured or dead cells are the ones that produce the surge in ethylene production. Ethylene production by wounded or stressed tissue was reviewed recently (291).

In response to water stress, leaves produce ethylene at increased rates (11–13, 111, 285). Water-stressed leaves which do not abscise produce only a transient rise in ethylene which lasts about 2.5 hr, in contrast to abscising leaves in which ethylene continues to increase under water stress conditions (13). Plants under water stress are known to show a rise in ABA and a decline in endogenous cytokinins, which taken together are believed to bring about rapid stomatal closure in leaves (272). Other plant hormones are also probably involved in the response to water stress. The surge in ethylene production by water-stressed leaves may represent a concerted hormonal response to water deficit which relates to stomatal closure.

The influence of ethylene on stomatal closure is brought into question by Pallaghy & Raschke (215), who clearly showed that ethylene has no effect on stomata of turgid *Zea* and *Pisum* leaves and did not act as a competitive inhibitor of stomatal response to carbon dioxide. On the other hand, Vitagliano & Hoad (279) showed that ethylene derived from ethephon always increased stomatal resistance in a number of different plant species. However, response of stomata to ethylene may be quite different in water-stressed leaves, especially in association with high levels of ABA. Aharoni et al (12) showed that while ABA applied to turgid lettuce leaves caused partial closure of stomata, GA and kinetin had no effect on stomatal aperture of those leaves. In desiccated leaves, however, GA and kinetin considerably retarded stomatal closure. Therefore, ethylene may similarly influence leaves under water stress despite inactivity in turgid leaves. Ethylene may also influence ABA levels which can then affect stomatal closure.

Waterlogging also appears to promote an increase in the concentration of ethylene which can be extracted from various parts of the plant (128, 138). Increased ethylene levels following waterlogging appears to be caused simply by the blockade of ethylene escape from the tissues by the water. Thus while the concentration of ethylene in the tissues of waterlogged plants is increased, the actual rate of ethylene production is decreased because of an oxygen deficit in the tissues. Increase in ethylene in waterlogged plants does not represent a surge of ethylene production as observed in other stress situations.

Injury to curcurbit leaf disks by SO_2 and bisulfite solutions was accompanied by evolution of both ethane and ethylene and was used to assay injury to these tissues (46). Light enhanced the evolution of both ethane and ethylene by cucurbit leaf tissue, and AVG, which inhibits ethylene production from methionine, caused increased evolution of ethane in response to bisulfite but only partially inhibited ethylene production. These results suggest that wounding leaf tissue by SO_2 or bisulfite may induce ethylene production from two sources, from methionine leading to ethylene and perhaps from peroxidized linolenate leading to ethane and some ethylene (78, 135, 171).

Peiser & Yang (218) treated alfalfa plants with SO_2 and also observed elevated ethylene and ethane production. The increase in ethylene production peaked in 6 hr and returned to control levels 18 hr following the end of the fumigation period. Ethane production peaked 30 hr after fumigation and was highest in plants held in the light after fumigation. In contrast, ethylene production was much higher in plants held in the dark. These experiments suggest two separate routes for ethylene and ethane production in SO_2-stressed plants. Ethylene production is assumed to derive from methionine, while ethane production, which was greatest in visibly injured plants, appeared to derive from lipid peroxidation stimulated by the chemical treatment.

The following three questions need to be answered regarding wound or stress ethylene production in plant tissues: (*a*) Does the production of wound or stress ethylene derive from the same source as "normal" physiological (nonwound or stress) ethylene production? (*b*) What are the controls which regulate wound-stress ethylene production? (*c*) What is the physiological role of wound-stress ethylene?

It does appear that most wound-stress ethylene production derives from methionine, perhaps in the same biochemical pathway that forms basal ethylene in the natural physiological system. However, as indicated above, upon extensive wounding, ethane may be produced by peroxidation of linolenic acid, which also results in some ethylene production (162). Abeles & Abeles (2) obtained large increases in ethylene production with bean leaves and tobacco leaves following application of toxic compounds such as $CuSO_4$, endothal, and ozone. The toxic compounds also increased conversion of ^{14}C-labeled methionine to ethylene. However, the relative specific activity of the conversion of ^{14}C-methionine to $^{14}C_2H_4$ was only 0.2% for controls and about 0.1% for stressed tissue in these experiments. This suggests a very extensive dilution of the label with endogenous methionine or with wound-induced endogenous unlabeled ethylene produced via another pathway.

Hanson & Kende (116) presented more convincing evidence implicating methionine as the precursor for wound ethylene production from rib segments excised from immature flower buds of morning glory flowers (*Ipomea tricolor*). Such immature segments evolve only a small quantity of ethylene, but upon wounding the rate of ethylene production rose more than tenfold within 1 hr and returned to a low rate after 3 hr. AVG virtually completely inhibited ethylene production in both control and wounded tissue, clearly implicating methionine as the precursor.

The lag in wound ethylene production suggests a control mechanism perhaps involving synthesis of enzymes for activating ethylene production. There is no direct evidence for such synthesis. However, Abeles & Abeles (2) and others (121) show effective inhibition of wound ethylene production by cycloheximide. Similar inhibition of ethylene production was also observed in the "normal" ethylene producing system in fruit (167) and in mung bean seedlings (122).

The physiological role of wound or stress ethylene production is unknown as is the precise general role or roles of ethylene in metabolism. However, some clues provide suggestions which invite simplistic speculations. For example, abscission of a damaged leaf or fruit could be accelerated by increased ethylene production, thereby ridding the plant of useless tissue. In young tissues ethylene production could be associated with wound healing and may play a role in defense against microorganisms which might invade an open wound. Increased activity of peroxidase in diseased or cut sweet potato roots was attributed to ethylene released from the tissues and associated with resistance of plant tissue to attack by microorganisms (183). The surge of ethylene production in water-stressed leaves appears to be correlated with drastic changes in other hormones. Perhaps stress upsets the entire hormone balance in the tissues, and depending on the magnitude of the stress, numerous metabolic changes can occur which may be related to restoration of proper hormonal balance in the stressed tissue. The damped oscillation curve for wound ethylene production in pea stem tissue (236) may reflect a return to a proper hormonal balance. Wound-stress ethylene may be symptomatic of the induced hormonal imbalance as a result of the metabolic stress. It may be useful to determine changes in auxins, gibberellins, cytokinins, and ABA upon wounding or stressing tissue to see if they relate in any way to the changes in ethylene levels.

MODE OF ACTION AT MOLECULAR LEVEL

General Hypotheses of Ethylene Action at Molecular Level

There is some question as to whether the metabolism of ethylene is involved in its action (5). Two possibilities are suggested to explain ethylene interaction with its receptor. One possibility envisages a dissociable ethylene-receptor complex, acting like a switch that could turn on a cascade of reactions when ethylene combines with its receptor. This complex is believed to be subject to rapid dissociation, perhaps after a conformational change in the receptor, which releases ethylene from the ethylene-receptor complex, and this in effect turns off the switch (7). In this view ethylene incorporation and metabolism is not associated with ethylene action. Such a mechanism is supported by the observations that: (*a*) many ethylene effects cease soon after ethylene removal; (*b*) there is generally an overshoot or overproduction of ethylene, presumably to prevent the dissociation of the ethylene-receptor complex; (*c*) there appears to be binding of ethylene to tissues and very little incorporation (49, 57); (*d*) hypobaric atmospheres delay or prevent ethylene action (83). Experiments with deuterated ethylene indicated that no exchange with H atoms occurred, which appeared to support the view that ready dissociation of the ethylene-receptor complex occurs (7, 29).

The alternative possibility envisages an actual reaction with ethylene at the receptor site which produces reactants from the ethylene molecule, resulting in physiological actions. Recent evidence supporting this hypothesis is reviewed in the following paragraphs. In both hypotheses ethylene may act to trigger a physiological event or events, but in the first the ethylene molecule is not altered whereas in the second the molecule must be metabolized in order to trigger physiological events.

Only very small quantities of exogenous ethylene are incorporated into tissues. Buhler et al (49) could obtain only 0.05% incorporation of ^{14}C-ethylene into ripe and green avocado fruit which appeared to be metabolized to organic acids, but they failed to observe incorporation of ^{14}C-ethylene into a number of other fruit. Jansen (132, 133) fed ^{14}C-ethylene to avocado fruit for 4 hr and obtained about 0.02–0.04% incorporation of the label into CO_2, toluene, and benzene. These experiments have been criticized (31) because of the absence of precautions to ensure that the label incorporated was actually derived from ethylene and not from impurities present in the ^{14}C-ethylene used. Radioactive ethylene derived from commercial sources contains many impurities presumably due to radiation decomposition. Impure commercial ^{14}C-ethylene trapped and released from mercuric perchlorate produced additional radioactive impurities which were preferentially taken up and metabolized by tissues (31, 132). Also, previous studies by Buhler et al (49) and Jansen (132) did not take precautions to avoid microbial contamination which could have produced the observed results (31). Work with ^{14}C-ethylene must assure the absence of radioactive impurities which are rapidly metabolized by tissues and microorganisms which may metabolize either ^{14}C-ethylene or the radioactive impurities. The incorporation of radioactivity from ^{14}C-ethylene, reported by various workers (112, 132, 133, 243), is probably due to artifacts resulting from impurities in the radioactive ethylene applied to the tissues. Possible microbial metabolism of applied ethylene is a second source of difficulty in evaluating the metabolism of ethylene by plant tissues. Even after purification, radioactive ethylene should be used immediately or stored over sodium hydroxide to avoid impurities due to self-irradiation (31).

EVIDENCE FOR METABOLISM OF ETHYLENE In a series of well-designed experiments in which use of ultrapure ethylene was combined with aseptic conditions, Beyer showed that ^{14}C-ethylene was metabolized in tissues which showed physiological responses to ethylene (30, 32, 35, 36, 40). ^{14}C-Ethylene was metabolized by etiolated pea seedlings (32), carnation flowers (35), and morning glory flowers (40), to $^{14}CO_2$, and also incorporated into tissue mostly in water-soluble metabolites (103). Metabolism of ^{14}C-ethylene to $^{14}CO_2$ appears to be greater than its incorporation into tissue components, and this metabolism requires physiological competence. Heat-treated tissue and severely wounded tissue are either incapable of metabolizing ^{14}C-ethylene or its metabolism of ethylene is severely reduced. The system metabolizing ^{14}C-ethylene to $^{14}CO_2$ is more unstable than the system which incorporates ^{14}C-ethylene into tissues, because upon homogenization only the tissue-incorporating system survives (32).

Metabolism of ^{14}C-ethylene by pea seedlings and other tissues was inhibited by Ag^+ ions, high CO_2 (7–10%), and low O_2 levels (33, 34, 38). These inhibitors also counteracted the physiological effects of ethylene in retarding epicotyl growth. The Ag^+ ion (100 mg/l) was the most effective whereas lowered O_2 concentration (5%) was the least effective or specific antagonist to ethylene. Ag^+ ions blocked the incorporation of $^{14}C_2H_4$ (0.2 μl/l) into tissues but had little effect on the oxidation of $^{14}C_2H_4$ to $^{14}CO_2$ in the same tissue. By contrast, 7% CO_2 inhibited $^{14}C_2H_4$ oxidation to $^{14}CO_2$ without affecting incorporation of $^{14}C_2H_4$ into the same tissue. This suggests two sites at which ethylene may be metabolized.

These data, along with other results in which Ag^+ ions antagonized such ethylene effects as sex expression in cucumbers, epinasty in tomatoes, and senescence and abscission in cotton (34), relate the metabolism of ethylene to its action. Two independent sites or pathways for ethylene metabolism are indicated: (*a*) a pathway leading to the oxidation of C_2H_4 to CO_2, which is inhibited by CO_2 at concentrations of 7% and higher; and (*b*) a pathway leading to incorporation of ethylene into a number of soluble tissue components (103) which is inhibited by Ag^+ ions. The inhibitory effects of the Ag^+ ions can be overcome by acetylene (37). Use of these inhibitors should be very revealing in probing further the mechanism of ethylene action at the molecular level.

Recent experiments by Aharoni, Anderson, and Lieberman (unpublished) examined the influence of Ag^+ ions and CO_2 on aging tobacco and bean leaf disks as determined by loss of chlorophyll and ethylene production. Ag^+ ions (10 ppm) reduced the rate of senescence of leaf disks as indicated by retention of chlorophyll, but the disks also showed a marked increase in ethylene production. The combination of CO_2 (10%) and Ag^+ ions (10 ppm) was even more effective in slowing down senescence and chlorophyll loss and induced an even larger production of ethylene from ^{14}C-methionine by the leaf disks. The large increase in ethylene production was not due to wounding by an excessive concentration of Ag^+ ions, since a dose response curve with Ag^+ ions indicated that 10 ppm Ag^+ ion does not produce a wounding effect. Results in these experiments were interpreted in terms of a possible negative feedback modulator of the ethylene-forming system which was linked to the metabolism of C_2H_4 at the Ag^+ sensitive receptor site for ethylene. Upon blockage by Ag^+ ion, the receptor site did not react with ethylene and no feedback modulation of ethylene production occurred. Consequently, six- to twentyfold increase in ethylene production was observed in the presence of Ag^+ ions.

Although Ag^+ and CO_2 respectively block $^{14}C_2H_4$ incorporation into tissues and its oxidation to $^{14}CO_2$ at a concentration of 0.2 μl/l to 1 μl/l of ethylene, higher ethylene concentrations of about 20 μl/l overcome to various degrees the Ag^+ and CO_2 blocks of both metabolism of ethylene and its physiological action (38). This suggests competition for sites or interaction between ethylene and Ag^+ or CO_2. The greatest problem in relating ethylene action to ethylene incorporation and metabolsim is that the specific activities for both tissue incorporation and oxidation to $^{14}CO_2$ do not saturate even at 140 μl/l of $^{14}C_2H_4$ (32). This is in contrast to most ethylene responses in etiolated peas which saturate around 10 μl/l (7). However, as suggested by Beyer (32), the rate-limiting step in the expression of an ethylene response may not occur at the initial step in which ethylene reacts at the receptor site. The metabolism of ethylene to CO_2 and incorporation into the tissue may represent the initial steps associated with the mode of action of this hormone or may involve a second pathway of metabolism of ethylene to CO_2 which is merely a disposal pathway of excess ethylene unrelated to ethylene action. The critical changes brought about by these initial events are still completely unknown.

Jerie & Hall (134) reported ethylene oxide as a major metabolite of $^{14}C_2H_4$ incorporated into developing cotyledons of broad bean. In excess of 95% of the $^{14}C_2H_4$ applied (3 μl/l) was metabolized in less than 25 hr with 85–95% of the ^{14}C appearing as ethylene oxide, which was clearly identified by GC mass spectrometry. It is rather surprising that in excess of 95% of the applied $^{14}C_2H_4$ was metabolized by this tissue when most previous experiments with $^{14}C_2H_4$ uptake and metabolism showed considerably less than 1% uptake and metabolism of the applied $^{14}C_2H_4$ (32). Although Jerie & Hall (134) emphasize that ethylene oxide is not an artifact in this system, it is possible that ethylene oxide is somehow produced artifactually within the living cell, perhaps catalyzed in the tissue in a nonphysiological reaction. Since the ^{14}C-ethylene was used without further purification, perhaps an impurity in the $^{14}C_2H_4$ may catalyze conversion of the $^{14}C_2H_4$ to ethylene oxide when in contact with cotyledon segments of *Vicia faba*. *Vicia faba* may be unique in this context. Further evidence is necessary to confirm the universal metabolism of ethylene to ethylene oxide in a variety of tissues and species. Ethylene oxide was found to inhibit ripening and senescence in some fruit and flowers (158, 172) but was never identified unequivocally as a natural product of metabolism. A reexamination of a possible role for ethylene oxide in ethylene metabolism may be warranted since metabolism of ethylene requires oxygen and ethylene oxide is a logical candidate for an intermediate, either in regulating the inhibition or suppression or acceleration of ethylene production and metabolism.

EVIDENCE FOR NONMETABOLISM OF ETHYLENE Sisler & Wylie (245) exposed tobacco leaves to $^{14}C_2H_4$ for 12 hr in the presence and absence of 1000 ppm of unlabeled ethylene, and studied binding and release of the labeled ethylene absorbed by the leaves. From such data a value of 3.5×10^{-9} M was calculated for the concentration of ethylene binding sites. Ethylene is assumed to be displaced easily from these sites. This value does not account for binding sites occupied by endogenous ethylene or possible nonphysiological nonspecific ethylene binding sites. The value obtained in Lineweaver-Burk plots for half-maximal response for the influence of ethylene on tobacco leaf respiration agrees closely with the values obtained in a Scatchard plot for displacement of $^{14}C_2H_4$ from the "attachment sites." This approach to a study of ethylene action makes the assumption that the binding of ethylene to its attachment site is weak and that it is readily displaced. Another assumption inherent in these experiments is that ethylene bound to its physiological receptor site does not react and is not metabolized. The action of ethylene upon attachment to its receptor site is assumed to release a switch, perhaps by inducing a conformational change in the protein receptor site, which sets in motion the physiological effects of ethylene. However, these experiments are subject to some criticism because unpurified ethylene was used, and the possibility that the ethylene absorbed or dissolved in the tissue and subsequently diffused out of the tissue may not all represent attachment to the specific physiological ethylene attachment sites. The use of respiration as an index of the physiological effect of ethylene may not be valid since there is not always close linkage between ethylene action and respiration. Lastly, the assumption that ethylene is not metabolized at its physiological attachment site is contrary to the evidence of Beyer (32), which clearly shows that ethylene is metabolized to CO_2 and is incorporated into tissue components (103). Recently Sisler (personal communication) also found that $^{14}C_2H_4$ was converted to $^{14}CO_2$ by leaves, but the same amount of labeled $^{14}CO_2$ was produced, from equal levels of $^{14}C_2H_4$, in the presence or absence of 1000 ppm of cold C_2H_4. These results may be due to labeled impurities or to a pathway which converts C_2H_4 to CO_2 which is not saturated by 1000 ppm C_2H_4 and may be unrelated to the physiological action of ethylene.

ETHYLENE IN GROWTH, DEVELOPMENT, AND SENESCENCE

Mechanism of Ethylene Action in Control and Development

The dramatic effect of ethylene on various aspects of plant growth has long been known (206), and its striking effect on growth of etiolated pea seedlings was standarized as the "triple response" assay for ethylene (77). This response consists of reduction in rate of stem elongation, increase in stem diameter, and absence of normal geotropic response (diageotropism), which at 1 ppm or less was presumably specific for ethylene. No other gas produced a similar response on all three factors, except perhaps at much higher concentrations. The triple response is associated with the subhook region of the pea epicotyl which is an active growing region (elongating region) of the seedling. In older plants ethylene induced epinasty (78), leaf abscission

(51), flower and fruit maturation and senescence, and other physiological effects associated with ripening and aging (57). These effects of ethylene appear to be associated with growth characteristics generally relevant to the action of auxin and other plant hormones, and therefore interaction of ethylene with other hormones is suspected (274).

The morphological changes associated with ethylene action may be related to the effects of ethylene on cell division, cell expansion, and auxin transport, among other of its influences. Ethylene at 50 ppm or 2,4-D at 10^{-4} M almost completely stops cell division and DNA synthesis in the apical hook region of etiolated pea seedlings (17). As the concentration of ethylene was reduced below 1 ppm the percent inhibition of cell division rapidly fell, but there was still about 20% inhibition at 50 ppb. This is about 10 times greater than the effective ethylene concentration for restoring normalcy to the diageotropica mutant (295) or the ethylene concentration level required for cytodifferentiation in lettuce pith cells (296). Nevertheless, it appears that ethylene at 0.1 ppm or higher significantly inhibits cell division in apical tips, which can explain the inhibitory effect of ethylene on growth in general. However, ethylene does not seem to inhibit cell division in tissues which are stimulated to divide by auxin (18) as in formation of adventitious roots (76). This points out that ethylene may effect specific tissues and cells quite differently (212, 213). Thus ethylene does not significantly effect RNA synthesis in the elongating subapex tissue (53), nor does it inhibit protein synthesis as indicated by the induction of cellulase (119) and phenylalanine ammonia lyase (229).

Inhibition of growth by ethylene in subhook regions of etiolated pea seedlings and other tissues is largely due to reduction in the capacity of the polar auxin transport system which supplies auxin to the cells (18, 199). The elongating subhook region of the etiolated pea epicotyl is especially sensitive to ethylene and shows pronounced swelling in ethylene at concentrations in excess of 100 ppb. Actually this does not fully represent inhibition of growth because, while there is a reduction in length in the presence of ethylene, there is no difference in the increase in fresh weight during the first 24 hr between the control and the ethylene-treated etiolated pea subhook region. However, during the next 72 hr in ethylene, the subhook region shows much greater increase in fresh weight than controls, though elongation of this region remains considerably reduced relative to the control (16, 18). In ethylene the tissue expands radially rather than longitudinally because of constraints on growth resulting from ethylene-altered orientation of new cell microfibrils from a radial to a longitudinal direction, resulting in isodiametric rather than rectangular shaped cells (18). Normal subapical tissue stops growing (elongating) after it reaches full length in 1 or 2 days, but ethylene-treated subhooks (cell enlarging regions) continue to grow in the presence of ethylene for a number of days with a linear increase in fresh weight and swelling (18). Therefore, ethylene does not inhibit growth in the subhook but alters it. In addition to altering the orientation of growth, ethylene in combination with high levels of 2,4-D may also influence cellular differentiation, as for example inhibiting xylogenesis and fiber lignification in the third internode of etiolated pea seedlings (19), or promoting xylogenesis in lettuce pith cells (297). Auxins in concentrations above 10^{-5} M, which induce relatively high levels of ethylene in subapical pea stem tissue, similarly influence swelling growth of the subapical region of etiolated pea stem.

Control of cell shape and size by the interaction of auxin and ethylene has been suggested (116, 211, 239), based on the opposing action of or interaction of these two hormones. Varying internal levels of auxin or ethylene by external application of hormones can produce cells of different shape and size (211). Auxin and ethylene regulate cell expansion in opposing ways, within given levels of physiological concentrations of these hormones. At certain physiological concentrations of auxin the cell is "buffered" against ethylene action, but the "buffering" effect can be overridden by high levels of ethylene. Thus, for example, concentrations of IAA (10^{-5} M) that enhance elongation growth of pea segments cause deposition of transverse microfibrils, restricting lateral growth (275). However, higher concentrations of IAA (10^{-3} M), which induce much greater ethylene production, lead to deposition of longitudinal microfibrils which result in considerable radial growth and swelling of pea stem segments (16, 211). At 10^{-3} M IAA the production of ethylene exceeds the buffering capacity of the auxin and the influence of ethylene dominates in determining growth.

The opposing action of auxin and ethylene can be observed not only in the disruption of polar auxin transport (39, 58) but also in the effects of ethylene on auxin concentration levels in tissues (160). IAA levels are diminished by 50% in epicotyls of 5-day-old etiolated pea seedlings treated with 10–36 ppm of ethylene for 18 to 24 hr. Similar reductions in tissue auxin content, due to ethylene, have also been attributed to reduction in synthesis, destruction, or conjugation (89, 196, 273). Goren et al (109) reported that ethylene did not induce formation of indole-acetylaspartate, which suggests that reduced auxin levels in the presence of ethylene is not due to conjugation of IAA. The suggestion of opposing actions of auxin and ethylene in many aspects of growth and development has been indicated in a number of studies, as mentioned above, and is not restricted only to the interaction between auxins and ethylene (166, 176, 220). Cytokinins, gibberellins, and ABA probably also play important roles in a network of feedback control mechanisms which serve to modulate normal growth and development, thus preventing odd overgrowths as observed when supraoptimal levels of ethylene, auxin, or other hormones are applied.

TARGET CELLS FOR ETHYLENE ACTION Osborne (213) identified certain cells, called target cells, based on their reactions to ethylene and auxin. Three cell types were identified: Type 1 cells are stimulated to elongate by auxin but not by ethylene, i.e. elongating cells of land-grown seedlings; Type 2 cells expand in response to ethylene but not to auxin, i.e. cells of abscission zones and developing fruits which show a dual sigmoidal growth pattern; Type 3 cells elongate and increase in size in response to both ethylene and auxin, i.e. rice seedlings and other water plants. There may be some question as to whether ethylene has no effect on cell size and growth in so-called Type 1 cells, since swelling in these cells, induced by ethylene, appears to be due to lateral growth of the cells (18). The final size and shape of the cell, as influenced by ethylene, is the result of interaction with not only auxin but also with GA (257) and perhaps cytokinin as well (98). At this stage of our knowledge it may be useful to classify different types of cell reactions with ethylene and auxin in terms of target cells. However, we must realize that this represents a tentative simplification of rather complex hormonal interactions which must involve gibberellins, cytokinins, and ABA, as well as auxin and ethylene. It seems clear that when ethylene acts to stimulate cell elongation, as in water plants, auxin and CO_2 enhance the ethylene effect (123, 147), which is exactly opposite to the effects shown by auxin and CO_2 versus ethylene in land plants, wherein they oppose the action of ethylene. Also, in contrast to the normal disruptive influence of ethylene on IAA transport in stems of land plants, transport of auxin is enhanced by ethylene through petiole sections of *Ranunculus scleratus,* a water plant (205).

AUXIN AND ETHYLENE Zimmerman & Wilcoxon (293) discovered that auxin stimulated ethylene evolution by tomato plants and suggested an hormonal role for ethylene in association with auxin. More recently the influence of supraoptimal levels of auxin (10^{-5}–10^{-3} M) on ethylene production was clearly shown by Morgan & Hall (201, 202), Abeles (3), and Burg (56). Continuous presence of auxin at high levels is necessary for stimulation of the ethylene-forming system in etiolated pea subhooks (136) and mung bean tissue segments (233), and it is considered to be the result of an induction process because of the necessity for a lag period of about 1 to 3 hr and the finding that inhibitors of RNA and protein synthesis prevent auxin-induced ethylene production (3, 167, 233). The decay of the ethylene-forming system in the presence of cycloheximide in mung bean hypocotyl segments (233) and the subhook segments of etiolated pea epicotyls (167) occurs very rapidly, which may explain the requirement for continuous presence of high levels of IAA for continuous stimulated ethylene production. Ethylene production in postclimacteric apple tissue (which already produces considerable ethylene) increases marginally in response to auxin (159), but fully expanded leaves producing a little ethylene showed large increases (five- to sevenfold) in response to auxin (11, 210). Ethylene production was rapidly and totally inhibited by cycloheximide in both auxin-induced seedling systems (167) and in the uninduced apple slice system (167), but the ethylene-forming system in the apple tissue had more than twice the half-life exhibited by the etiolated subhook stem system. These data indicate that the ethylene-forming enzyme system is in constant dynamic flux with both synthesis and degradation going on simultaneously.

Although a lag period of about 1 to 3 hr has been reported in IAA-induced ethylene production (56, 167, 233), wounding by excision of actively growing regions of a number of plant species, in the absence of auxin, rapidly (26 min) induces ethylene production (236). Actually a very low basal level of ethylene production may exist prior to wounding, and it is not clear whether wounding or added auxin acts on the basal system or induces a new one or perhaps does both (127). The lag period in auxin-induced ethylene production in mesocotyl segments of etiolated sorghum seedlings was recently found to last only 15–20 min (92). These results suggest that the initial effect of auxin on ethylene production is too rapid to involve synthesis of an ethylene-producing enzyme. The strongest argument for induction of an ethylene-forming enzyme system by auxin is its inhibition by inhibitors of RNA and protein synthesis (3, 167, 233). Although there have been reports that cycloheximide may not be specific in inhibiting protein synthesis (175), the use of other protein synthesis inhibitors and a variety of RNA synthesis inhibitors, all of which inhibit auxin-induced ethylene synthesis (167), supports the involvement of RNA and protein synthesis in auxin-induced ethylene production. Attempts to identify specific new proteins, extracted from IAA-treated seedling tissue, which may be associated with ethylene production has not been successful (14). However, this may require more precise techniques.

Steen & Chadwick (256) obtained inhibition of auxin-induced ethylene production in excised pea root tips with cycloheximide when high levels of IAA (10^{-5} M–10^{-4} M) were used, but no inhibition when 10^{-6} M IAA was used. Thus at low concentrations, IAA may increase ethylene production with no protein synthesis, but at higher concentrations IAA-increased ethylene production is associated with induced RNA and protein synthesis. Steen & Chadwick (256) proposed a basal ethylene-forming system, producing trace levels of ethylene, present at all times in actively growing tissues, which may be subject to regulatory control and stimulation by auxin without the necessity for de novo protein synthesis. This may be the system activated in 15 min or less by auxin (92) and observed early before wound ethylene appears (127). A second system subject to induction by high levels of auxin (above 10^{-6} M), appears to be associated with protein synthesis, and induces considerably larger quantities of ethylene as long as free auxin (above 10^{-6} M) is present in the tissues (136).

The rapid elongation response (with a 9–15 min latent period), observed when auxin is added to coleoptiles or stem segments (90) followed by sustained growth related to protein synthesis (with a latent period of 1 hr), is somewhat analogous to auxin-induced ethylene production, which may also represent a two-phase system, with a short and long time lag period. The short lag period, which related to a relatively small increase in ethylene production, may not be associated with RNA and protein synthesis (127), whereas the long lag period, which results in large increases in ethylene production, is associated with increased RNA and protein synthesis.

Auxin-induced ethylene has been shown to inhibit growth of pea root sections and has been associated with the general phenomenon of growth inhibition by supraoptimal concentrations of auxin (above 10^{-6}M) on rapidly growing tissues (63). Ethylene is produced in 15–30 min, if not sooner, after application of supraoptimal levels of auxin to roots, and roots cease growth almost immediately upon exposure to the gas. Thus there is clearly a correlation between supraoptimal auxin levels, ethylene production, and inhibition of root growth. Additional evidence linking the inhibitory action of supraoptimal auxin concentrations to its stimulation of ethylene production is provided by the action of CO_2, a competitive inhibitor of ethylene action (57). Carbon dioxide tends to reverse the inhibitory effect of supraoptimal levels of IAA on root growth. This relationship between auxin and ethylene may be considered part of a negative feedback system controlling rate of growth (166). Buds are also inhibited by supraoptimal levels of IAA, and this can be correlated to ethylene production resulting from the IAA treatment (59). Other evidence involving ethylene in bud growth was found in dormant buds of crabapple [*Malus hypehensis* (Pamp) Rehd.] released from dormancy by benzyladenine. Release from dormancy by benzyladenine was accompanied by increased ethylene production which could be inhibited by AVG (294). These results were interpreted to suggest involvement of ethylene in development of the bud after release from dormancy.

Andreae et al (15) pointed out a number of differences in the effects of IAA and ethylene on growth of excised pea root sections, thus raising doubts as to whether ethylene is a significant factor in mediating the inhibitory action of IAA. One major difference cited was the reversibility of IAA-inhibited growth in contrast to the nonreversibility of ethylene-inhibited growth. However, Andreae et al (15) compared relatively low concentrations of IAA (10^{-5} M) and relatively high concentrations of ethylene (100 ppm) in this study. Further work by Chadwick & Burg (64) presented additional data to support the hypothesis that most of IAA-induced inhibition of excised root tips and practically all such inhibition of intact roots result from IAA-induced ethylene production. They therefore proposed that root growth, including its geotropic response, is regulated by auxin-ethylene interactions.

Mer (194) also questioned whether the inhibitory effects of high levels of auxin on growth are simply due to the influence of auxin-induced ethylene. Mer suggests that the observed growth is due to independent action of auxin and ethylene which occur simultaneously. Mer (195) has also questioned the reported stimulation of mesocotyl growth by ethylene (258). In contrast to a number of reports, Mer (195) obtained no increased growth of oat seedling mesocotyls with ethylene treatment and attributes the observed increase in elongation growth of rice and oat seedlings (147, 261) to inadvertent exposure to excessive light which reduces the quantity of endogenous ethylene production (52) and consequently causes an increase in growth. This analysis by Mer is restrictive and narrow and does not consider that the influence of any hormone on growth must be integrated with that of other hormones and the growth observed is due to their interactions. Failure by Mer (195) to observe ethylene-induced elongation of oat mesocotyl is puzzling because there are a number of reports in addition to those of Suge (261) and Ku et al (147) in which stimulation of elongation growth by ethylene was also observed (123, 124, 205, 213, 260).

The evidence quoted above and additional literature (144, 262) indicate that effects of ethylene are not limited to depressing growth, elongation, and development, which is a common view of ethylene action (194). Ethylene can also promote elongation, growth, and development (292). The species, environment, stage of growth, type of cell and tissue, concentration level of ethylene, and interactions with other plant hormones all influence the effect of ethylene on plants. Smith & Robertson (248) demonstrated that root extension by some rice cultivars was stimulated by 0.1 ppm of ethylene, while roots of tomato and barley were inhibited at this concentration. Konings & Jackson (144) demonstrated that whereas root elongation was inhibited at 1 ppm ethylene in tomato, peas, and rice, elongation in all three species could be stimulated by concentrations below 1 ppm. Tomatoes required less than 0.02 ppm for stimulation, pea required less than 0.15 ppm, and rice required less than 1 ppm ethylene for stimulation. Thus extension of roots of all three species could be increased by ethylene, although different concentrations were required for each species.

Additional evidence linking ethylene metabolism and growth promotion or involvement in normal morphological development is suggested by the work of Zobel (295, 297) with the tomato mutant diageotropica. The diageotropica mutant is characterized by horizontal growth of shoots and roots, thin stems without large secondary xylem vessels, hyponastic leaf segments, and primary roots without lateral branching. Morphology of new growth in this mutant may be returned to normal by very low concentrations of ethylene (5 ppb) or high levels of auxin (10^{-4} M) (295). These studies suggest that morphological development in this mutant may be controlled by endogenous ethylene production resulting from an auxin-ethylene feedback mechanism. Apparently diageotropica lacks a normal auxin-induced ethylene forming system. Different effects of ethylene at high and low concentrations were observed in initiation of callus and tracheids in lettuce pith explants, in which tracheid initiation required ethylene concentrations below 100 ppb and was inhibited by ethylene above this concentration (296, 297). The conclusions reached in these studies are that 100 ppb of ethylene is a supraoptimal concentration which inhibits certain aspects of growth and development such as DNA synthesis (17) and cytodifferentiation (296). However, lower concentrations of ethylene perhaps in the low ppb range may be required for modulating normal plant growth and development in reaction to or in opposition to auxins in a feedback relationship (166). The conclusion by Burg et al (53) that nearly all responses of ethylene have the same dose response curve may not be valid.

Similar conclusions concerning ethylene and growth modulation were reached by Goldwin & Wain (108), using wheat coleoptile sections treated with various auxins. The faster a section elongated the faster it produced ethylene, and ethylene evolution followed elongation with a lag of about 3 hr. In all 34 compounds tested none caused large ethylene evolution without first promoting extension growth. Inactive isomers of auxins which had no influence on growth also did not increase ethylene production. These authors conclude that ethylene production occurs as a consequence of cellular growth and is not directly induced by auxin molecules. Ethylene may, in this context, be considered a natural plant growth modulator, which either promotes or inhibits growth and development in concert and reaction with other plant hormones (166).

CYTOKININS AND ETHYLENE Kinetin (10^{-4} M) stimulated ethylene production in 3- to 4-day-old etiolated pea seedlings and enhanced IAA (10^{-4} M)-induced ethylene production synergistically (98). Kinetin was more effective in enhancing ethylene production in 3-day-old seedlings than in 4- or 5-day-old seedling, but its synergistic interaction with IAA in increasing ethylene production was equally effective in 3- and 5-day-old seedlings. The synergistic effect of kinetin on auxin-induced ethylene production was attributed to enhanced IAA uptake and the suppression of IAA conjugation with aspartate, thus effectively increasing the concentration of free IAA (151). The role of kinetin was therefore assumed to be involved in maintaining the level of free IAA which is the active substance inducing ethylene. There is no doubt that the level of free IAA is important in inducing ethylene production in seedlings and that the different kinetics of ethylene production induced by IAA and 2,4-D illustrate the lability of IAA and its subsequent loss of effectiveness compared to 2,4-D (136). However, Fuchs & Lieberman (98) showed that exogenous kinetin (10^{-4} M) induced considerably greater ethylene production in 2- and 3-day-old seedlings than did IAA (10^{-4} M). Therefore it appears that kinetin can induce ethylene production independently of auxin, in addition to its ability to prevent conjugation of IAA. Lau & Yang (152) found that Ca^{2+} (10^{-2} M) and kinetin (10^{-4} M) produced a tenfold increase in ethylene production with mungbean hypocotyl segments. Synergistic interactions in ethylene production with these tissues was not obtained when Ca^{2+} (10^{-2} M) was added to IAA or GA_3 (10^{-5} M). It was further shown that Ca^{2+} increased both uptake and metabolism of kinetin and the kinetin also greatly increased the uptake of Ca^{2+} (153). Except for Sr^{2+} and to a lesser extent Fe^{2+}, none of the divalent ions tested (Ba^{2+}, Mg^{2+}, Ca^{2+}, Hg^{2+}, CO^{2+}, Ni^{2+}, Sn^{2+}, Zn^{2+}) showed synergism with kinetin in ethylene production. Although copper alone stimulated ethylene production after a 2 hr lag period, it showed no synergism with kinetin. However, when Cu^{2+} was applied with Ca^{2+} a synergism in ethylene production was obtained, and even further additive increase was obtained when kinetin was added to Cu^{2+} and Ca^{2+}. Copper enhanced the uptake of Ca^{2+}, which may explain the synergism observed (154). The stimulatory influence of Ca^{2+} on ethylene production may relate to stablilization of the ethylene-forming enzyme system which appears to be localized on the surface of the plasma membrane (185). Various other cytokinin molecules were shown to act synergistically with IAA and calcium to stimulate ethylene production in mung bean hypocotyls, much as kinetin does (149).

Imaseki et al (122) examined the synergistic effect of cytokinin on IAA-induced ethylene production and concluded that the influence of cytokinin cannot be explained solely on its suppressive effect on IAA conjugation. The rate of induced ethylene production doubled with a tenfold increase in exogenous IAA concentration or addition of 5×10^{-6} M benzyladenine, but the cytokinin increased free IAA in the tissues by only about 30%. The 30% increase in free IAA in the presence of cytokinin could not account for the doubling of ethylene production. Tissues preincubated with benzyladenine alone for 3 hr, followed by addition of auxin, showed a significantly higher rate of ethylene production than untreated or buffer preincubated tissues supplied with auxin and benzyladenine simultaneously. The stimulative effect of the cytokinin is formed in the absence of auxin and appears to require some metabolic process. Thus cytokinin seems to enhance ethylene production with IAA by an undetermined mechanism which is only marginally related to suppression of IAA conjugation.

Cytokinin-induced ethylene production was also observed in decapitated pea plants in which growth of cotylar buds were inhibited with 0.5% auxin on the epicotyl stump and cytokinin (BA) (0.5%) applied on the cotyledons. With this treatment more than a fourfold increase in ethylene production was observed in 72 hr, before growth of cotylars due to cytokinin application was manifested (120).

GIBBERELLINS AND ETHYLENE Scott & Leopold (242) observed opposing effects of GA_3 and ethylene in the lettuce hypocotyl elongation assay for GA, in the induction of invertase formation by GA in sugar beet tissue, and in the α-amylase induction assay for GA. There are also several growth and development functions in which ethylene and gibberellins act synergistically. These include reversal of induced dormancy in lettuce seed (85, 86), induced leaf abscission in cotton plants (200), and stem elongation in rice (259) and the fresh water plant *Callitriche platycarpa* (204). CO_2 which is generally known to oppose ethylene action (52) actually enhances the action of ethylene in elongating stems of water plants and in stimulating seed germination (140, 262). Where ethylene acts as a promoter of growth, CO_2 and GA enhance ethylene action, as does auxin (213), in contrast to systems in which ethylene acts as an inhibitor of growth wherein GA, CO_2, and auxins are antagonists of ethylene action (166).

GA_3 had no effect in stimulating ethylene production in etiolated pea seedlings except perhaps when combined with auxin and cytokinin (98). Ethylene could prevent the stem-elongating influence of GA_3 on etiolated pea seedlings stems, and, conversely, GA at the proper concentration could counteract the tendency of ethylene to cause swelling in the subapical region of these seedlings (257). GA has also been shown to oppose the action of ethylene in epinasty (216). These data suggest that interactions between GA and ethylene in the elongation region of pea seedlings may determine the ultimate size and shape of the stem cells.

ABA AND ETHYLENE Abscisic acid, like ethylene, inhibits growth of etiolated pea seedlings, but these seedlings do not show the triple response characteristic of ethylene-treated seedlings (165). ABA-inhibited seedlings also show depressed ethylene production (101, 166). The IAA and kinetin-induced stimulation of ethylene production by these seedlings is considerably suppressed by 10^{-4} M ABA. Addition of both ABA (10^{-4} M) and ethylene (4 ppm) to etiolated pea seedlings results in an increased inhibition of epicotyl growth above that obtained by application of either hormone alone (166). These data suggest that ABA and ethylene inhibit growth in etiolated pea seedlings by different mechanisms. On the other hand, the interrelationship between ABA and ethylene is less clear in aging of senescent tobacco leaf disks. Addition of ABA to senescent leaf disks caused a sharp rise in ethylene production, and it is not known whether this represents a primary or secondary effect, following an influence of ABA on aging (165). Gertman & Fuchs (101) obtained similar results showing inhibition of ethylene production by ABA in etiolated pea seedlings, both in endogenous and IAA-induced systems. They also showed that ABA stimulated ethylene production in mature orange peel plugs which already produced significant amounts of ethylene. Similar results showing inhibition by ABA of ethylene production were also observed in IAA-stimulated etiolated mung bean hypocotyls (143).

Interrelationship between ABA and ethylene in senescence was shown in carnation flowers, wherein ABA hastened senescence associated with stimulated ethylene production (186). Carbon dioxide, which inhibits ethylene action in many tissues, delayed the onset of ethylene production and senescence in these flowers, even after treatment with ABA. Hypobaric ventilation, which lowers endogenous ethylene to subnormal levels within tissues, extended flower longevity and negated enhancement of senescence by abscisic acid. These data suggest that ABA hastens senescence of carnations by advancing the onset of autocatalytic ethylene production (186). Conversely, application of 35 ppm ethylene to harvested citrus fruit resulted in accumulation of large amounts of free and bound ABA in the peel within 24 hr. Similar increases in ABA were observed in peel of fruit allowed to senesce on the tree (107). There appears to be a close correlation between senescence induced by ethylene and increased ABA levels and vice versa. The precise relationship between these two hormones associated with aging needs further clarification, but the evidence indicates that they behave differently, with respect to each other, in juvenile tissue and in mature or senescent tissue.

ETHYLENE AND SEED GERMINATION The influence of ethylene on germination of seeds, and especially dormant seeds, involves a complex interaction of the gas with light, gibberellins, cytokinins, and CO_2 in opposition to ABA and perhaps other inhibitors (85, 209). This subject has been reviewed recently (266) and will, therefore, be considered only briefly here.

A most interesting ethylene effect in promoting seed germination occurs with seed of the root parasite *Striga asiatica* Kuntze (witchweed), which has a dormant seed whose germination is enhanced by exudates of roots from susceptible plants. The active substance in the root exudates of cotton has been identified as strigol (72), which can induce germination at 10^{-15} M concentration. Ethylene is also a very effective stimulant of *Striga* seed germination, capable of inducing germination in preconditioned seed (kept moist and at 25° for at least 2 weeks) in a matter of 2 hr. Soils infested with

Striga seed have been treated with ethylene to induce premature germination. This results in subsequent death of the germinated seed which cannot survive in the absence of nourishment from the host (87).

ETHYLENE AND LIGHT Red light is one of the factors which suppresses ethylene production and may be involved in controlling its rate of synthesis (52). Formation of the seedling hook in the dark is attributed to ethylene production in the apex of the seedling, and when the seedling is exposed to light, ethylene production is suppressed and the hook opens, allowing its leaves to expand. The apical hook can also be opened in darkness if ethylene is removed by hypobaric pressure or by exposure to CO_2, which opposes ethylene action (137). Exposure of etiolated seedlings to a single dose of red light caused a transient decrease in ethylene production followed by an increase in plumlar expansion. Far-red irradiation following red light treatment virtually nullified the red effect, suggesting that ethylene production may also involve phytochrome action (105). Hypocotyl elongation was promoted by exposure of etiolated soybean seedlings to red light, which was related to a significant decrease in ethylene production. Far red irradiation following the red treatment reversed the red effect, again suggesting a role for phytochrome in ethylene synthesis and their interaction in controlling hypocotyl growth (238).

ETHYLENE AND SEX EXPRESSION Sex expression in plants can be modified by altering the hormonal complement of the plant, and ethylene appears to play a major role in determining the sex of flowers in monoecious plants. A shift of sex expression was observed in hemp following application of gibberellins, shifting to male flowers or to female or bisexual flowers by application of cytokinins (65). Ethylene was shown to be involved in sex expression in the *Annona* flower (*Annona hybrida*) which is protogynous. A rise in ethylene production preceded the male stage which could be hastened by addition of ethylene or delayed by hypobaric pressure, which removed endogenous tissue ethylene. The involvement of ethylene in sex expression in this flower is therefore indicated (44). In cucurbit plants (*Cucurbita pepo* L.) high levels of gibberellins are associated with maleness, and treatment with ethylene alters sex expression to femaleness (70). More ethylene was evolved from gynoecious than from monoecious floral buds of cucumber plants, and quantities of ethylene evolved from female buds were greater than from male buds. Also, plants grown under short days, which promotes femaleness, evolved more ethylene than those grown under long days. These data suggest that ethylene participates in controlling and regulating sex expression and is generally associated with promoting femaleness, at least in cucurbits (231).

Ethylene Action in Fruit Ripening and Senescence

Fruit ripening and senescence has for many years been associated with ethylene production and action, and in fact ethylene was, and still is, considered the triggering agent in ripening or the fruit ripening hormone (54) which sets in motion reactions associated with the ripening process (182). More recently, however, this concept has been questioned for some fruits (73). These doubts derive mostly from the lack of a correlation between ethylene production and ripening reactions in some fruits such as the grape berry (74) and tomato (189), or from experiments in which ripening is suppressed by auxin, despite increased ethylene production (93, 94, 276). In the grape berry, ABA is suggested as the "triggering agent" for the ripening process, and Frenkel et al (93, 96) propose that the products of oxidative degradation of auxins trigger ripening. It was demonstrated, however, that oxindoles, the oxidative products of IAA, do not induce ethylene production and cannot substitute for IAA in the mung bean hypocotyl ethylene-forming system (150). Despite questions about the triggering function, ethylene still is recognized as a major accelerating and integrating agent in senescence (139).

The seminal role of ethylene in fruit ripening and senescence is supported by the recent findings that AVG, the inhibitor of ethylene production, sprayed on apples before harvest delayed fruit ripening, reduced preharvest drop, and increased fruit removal force (23). Retardation of ripening in Anjou pears was also observed as a result of dipping stored fruit in AVG (282). Ripening and senescence of many fruits and flower varieties were also significantly retarded when ethylene was removed by hypobaric storage (22, 83). Therefore, ethylene appears to play a major role in ripening even at low temperatures and low oxygen tensions (22). In making a case for ethylene as a ripening hormone, one can invoke the well-known effect of exogenous ethylene in inducing ripening in green mature climacteric fruit and its remarkable stimulation of both growth and ripening in the fig (182). There is therefore considerable evidence to support a triggering role for ethylene in fruit ripening which may even justify considering ethylene a ripening hormone. On the other hand, additional evidence can be invoked to question this kind of role for ethylene, especially in specific fruits such as the grape berry. Specific fruits may show variable response to ethylene during their maturation and ripening which may not fit the classical typical climacteric fruit pattern shown by the apple or avocado.

HORMONES, ETHYLENE, RIPENING, AND SENESCENCE The concept of a ripening trigger or even a ripening hormone may be outmoded in light of new appreciations of hormonal interactions in plant metabolism. Since fruit ripening and senescence is a developmental process under hormonal control, some emphasis is now being placed on the integrated roles of hormones other than ethylene in ripening and aging metabolism (48, 166). Vendrell (276, 277) found that auxins delayed ripening in green banana slices and delayed but increased peaks in ethylene production and respiration. However, in whole green bananas, auxins accelerated ripening presumably as a result of increasing ethylene production. These results point out the differences in response to auxin of the peel and pulp. Presumably in the banana tissue slice, the added auxin has ready access to both peel and pulp, whereas in the whole fruit only the peel is in contact with auxin. Gibberellic acid had no effect on ethylene production on either whole fruit or on banana slices, but ripening was delayed in the whole fruit in contrast to the tissue slice in which it was somewhat accelerated. The ripening aspect delayed by GA_3 was retardation of yellowing of the peel in both the whole fruit and its slice wherein softening of the pulp was normal.

Banana slices treated with kinetin showed increased respiration, ethylene production, and pulp softening, but no degreening of the peel (280). A "green-ripe" condition was also obtained when slices were exposed to ethylene while in a solution of benzyladenine. These experiments suggest that the various ripening parameters such as degreening, pulp softening, respiration, and ethylene production are influenced by auxins, gibberellins, and cytokinins. Normally these changes probably occur in an integrated program for ripening and senescence modulated by endogenous hormones at a high level of metabolic control. However, when exogenous hormones are added in relatively large amounts, distorted ripening metabolism occurs, such as the "green ripe" condition in the banana.

Another example, which relates ripening to other hormones in addition to ethylene, was observed in premature ripening of Bartlett pears, a physiological disorder in which the fruits ripen prematurely on the tree and drop to the ground. This disorder, which involves a greatly accelerated rate of maturation and ripening, is associated with cool temperatures which may prevail 4–5 weeks prior to the normal harvest time. These cool temperatures correlate with an early increase in ethylene production that ages the fruit prematurely. Premature ethylene production and ripening could be counteracted by treating the fruit on the tree with GA_3 (283).

Still another suggestion of hormonal control of ethylene production and perhaps associated ripening reactions was observed in apple, tomato, and avocado fruit slices (159). In apple tissue, cytokinins suppressed ethylene production in both early climacteric rise and postclimacteric tissue, auxin suppressed it in early climacteric but stimulated it in postclimacteric slices, and GA_3 had little influence on ethylene production at any stage. Abscisic acid, often associated with senescence, stimulated ethylene production mainly in early climacteric-rise tissue slices. Cytokinins also suppressed ethylene production in both early climacteric rise and postclimacteric tomato and avocado tissue slices. However, auxin had little influence on early climacteric tomato slices, but stimulated ethylene production in postclimacteric slices. Response of avocado slices to auxin was exactly opposite to that of tomatoes, showing a stimulation in early climacteric rise tissue slices but virtually no effect on postclimacteric slices.

Intact avocado fruit, vacuum infiltrated with high concentration of IAA (100–1000 ppm), were stimulated in respiration and ethylene production, whereas kinetin, abscisic acid, and gibberellic acid had little or no effect on respiration, ethylene production, or ripening (10, 270). Low concentration of IAA (1–10 μM) delayed and reduced ethylene production and respiration in intact avocado fruit. This suggests a concentration-dependent multifaceted role for IAA in controlling ripening rates of fruit. On the one hand IAA tends to increase ripening rates by stimulating ethylene production, but IAA also acts to retard other aspects of ripening such as degreening and softening (94) and inhibition of abscission (10). These apparent opposing effects are also observed in other fruit with other hormones as well (see above). Interactions between all the hormones, acting in complex networks, may result in both synergism and antagonism to each other and thus may influence reactions in opposite directions at different stages during development of ripening (166).

It is of some interest that ethylene production is suppressed and stimulated by cytokinins and auxins, respectively, in old postclimacteric apple tissue (159). This suggests that metabolic control of ethylene production at the hormonal level continues even at a very late stage in senescence. The strong inhibitory effect of cycloheximide on ethylene production in apple tissue, pre- and postclimacteric, indicates that the enzyme system producing ethylene exists in a dynamic state and is constantly turning over (167). Thus auxin, cytokinins, and other hormones may play a role in regulating ethylene production at all stages of development. Does this imply a physiological function for ethylene in the very late stages of senescence?

In summary, the action of ethylene in inducing accelerating, and otherwise influencing maturation, ripening, and senescence of fruit can clearly be associated with interactions with the auxins, gibberellins, cytokinins, and ABA (47). The mechanisms involved in these interrelationships are quite vague, but there is evidence to suggest a general antagonism between ethylene and ABA on the one hand and auxins, gibberellins, and cytokinins on the other (48, 166, 188).

ETHYLENE AND THE NATURAL SENESCENCE INHIBITOR Some fruits do not ripen properly while attached to the parent plant (114), and some avocado fruit varieties do not ripen at all while attached to the tree (271). This inability to ripen on the tree has been attributed to an inhibitory substance (antiripening agent) presumably transmitted from the tree to the fruit as long as the fruit remains attached (55). Mature avocado fruit attached to the tree showed no abscission or ripening in response to 50 ppm ethylene applied for 48 hr. However, treatment with ethylene for 5–6 days caused the firm fruit to drop about 6 days after start of the treatment (100). Harvested mature avocado fruit also required some time period after harvest before acquiring sensitivity to ethylene (100). No evidence could be obtained to implicate an antiripening inhibitor in leaves (271), but low concentrations of auxin applied to avocado tissue could delay ripening (270). There are data which show that a time period is required, after detachment from the tree, before the fruit becomes sensitive to ethylene and its ripening effects. Presumably this time lag is necessary to dissipate some barrier, perhaps a substance or physiological-biochemical state, before the tissue is capable of entering an ethylene-induced or natural ripening state. It is possible that during this lag period there is a loss of auxins, gibberellins, and cytokinins or a shift in their dynamic functioning which allows a changeover from a mature-green nonripening state to a ripening state. Therefore, the so-called ripening inhibitor may be no more than the complement of growth hormones and their integrated activity with the whole plant which is maintained while the fruit is attached to the tree, but this is dissipated, lost, or altered with time after detachment from the tree. In varying degrees this situation may be comparable in other fruit but more subdued, i.e. Anjou pears ripen at 20° only after a period of cold storage after harvest (114).

TOMATO RIPENING MUTANTS AND ETHYLENE PRODUCTION The best example of mutants which show diversity in ripening capacity related to ethylene production is given by the tomato mutants rin and nor, which grow normally in most respects and reach full size at the same time as normal fruit but fail to undergo many of the changes associated with normal ripening (117, 189, 190). Endogenous ethylene levels in fully grown mature fruit of these mutants appear not to be very different from the normal Rutgers parent strain, but as they mature further toward ripening, virtually no ethylene is produced by the rin mutant and there is no climacteric in respiration (117, 189, 190). About 60–100 days from harvest the fruit turns yellow-green to yellow but undergoes very little softening. Exogenous ethylene did induce a stimulation of CO_2 production as long as ethylene was present (117). Application of ethephon (2-chloroethylphosphonic acid), which releases ethylene (287), or ethylene gas to attached rin tomato fruit induced additional ripening as measured by lycopene development, fruit softening, increased total soluble solids, and promotion of normal tomato flavor. However, the lycopene produced by the ethylene-treated attached rin fruit was only about 30% of the fully ripe Rutgers parent strain, and the "ripe" rin did not reach the fully ripe stage of Rutgers in all the other ripening parameters (197). Detached rin tomato fruit held in 10 ppm ethylene with increasing oxygen levels from 60–100% induced lycopene synthesis, whereas high concentrations of ethylene up to 1000 ppm in air had little or no effect on ripening (95). These fruit, however, produced only 10% of the red pigment present in normal tomatoes, and no mention was made of associated ripening parameters such as softening which apparently were not greatly influenced by the treatment. It is apparent that a barrier to ripening exists in this mutant which can be breached only to a small extent by exogenous ethylene treatment.

The other nonripening tomato mutant, designated nor, shows similar abnormal ripening characteristics. After reaching full size and maturity in the normal time period following anthesis, the fruit produces no ethylene, shows no respiration climacteric, develops very little color, and does not soften (207). Polygalacturonase activity and lycopene were virtually absent in the nor mutant, which may explain the lack of softening and red color formation. However, the fruit attained some lycopene pigment 90 days after anthesis and one month storage at room temperature (207). Heterozygote plants for the nor allele, produced by crossing normal and nor mutants, showed a delay in ripening, a delay and much reduced peak in respiration and ethylene production, intermediate softening, and much reduced polygalacturonase activity and lycopene content, suggesting dilution of the normal characteristics by the nor allele. Every characteristic of ripening was modulated by the nor allele in the heterozygous fruit.

The rin allele in heterozygous rin fruit (rin X normal) had no effect on the magnitude of the respiratory rise, but delayed the peak about 4 days. The peak in ethylene production was reduced about 60% in addition to the delay of the rise, and polygalacturonase activity was 35% of normal (269). Ethephon treatment of harvested fruit of both mutant heterozygotes hastened the onset of ripening but had no effect on the magnitude of the respiratory or ethylene peak or on color development.

These data taken as a whole indicate that the mutants rin and nor lack the capability to respond to ethylene in a normal manner. Increased levels of ethylene tend to increase some aspects of ripening responses, but there appears to be a deficiency in critical linked reactions necessary to produce the total integrated cascade of reactions necessary for normal ripening. Tigchelaar et al (269) consider ethylene production and action as a secondary event in tomato ripening, apparently because propylene (189) or ethylene application cannot overcome the ripening inability of these mutants.

Probably no one hormone acts alone to set in motion a major physiological event in the life cycle of a plant, although ethylene appears to be dominant in controlling the ripening process. However, its actions depend on controls at secondary and tertiary levels of metabolism. Anomalies may occur at the receptor sites for ethylene, subsequent reactions involving the receptor site, or in any of many of the second and third level reactions which precede and set in motion the various biochemical pathways leading to normal ripening. Perhaps the failure to observe activity of polygalacturonase is due to an error in its primary, secondary, or tertiary structure, causing its ineffectiveness as a softening enzyme. Alternatively, an endogenous inhibitor of polycalacturonase may be synthesized by the mutants which prevents softening. Studies of these mutants may provide insight not only to the ripening process but to the detailed chain of reactions linking hormonal action to metabolic events.

ETHYLENE AND CONTROL OF RESPIRATORY METABOLISM IN RIPENING FRUIT For many decades the action of ethylene has been associated with increased respiratory metabolism in ripening fruit, especially in the so-called climacteric fruit in which an upsurge in respiration occurs at what appears to be the onset of the ripening process (228). This so-called climacteric rise in respiration appears to be induced by a prior rise in endogenous ethylene, but low concentrations of exogenous ethylene can also trigger and accelerate the ripening process when applied to green mature unripe fruit. The applied ethylene appears to induce autocatalytic ethylene production and accelerates the ripening process. As a result of such observations, ethylene was and still is considered a natural ripening hormone, although there are some objections to this interpretation (nomenclature) because of the manifold effects of ethylene, especially as it relates to other plant hormones, and the apparent inactivity of ethylene in some mature green fruit, i.e. grape as discussed above. There is also some question as to whether the effects of applied ethylene always lead to autocatalytic production of ethylene (191, 278). The action of applied ethylene on induction of ripening appears to be more complex than was previously considered, but it is always associated with an upsurge of respiration in climacteric fruit and can induce a respiratory rise in nonclimacteric fruit (41, 42, 187). In nonclimacteric fruit, the magnitude of the respiratory response increases as the exogenous ethylene concentration increases and declines when the fruit is removed from the ethylene atmosphere. In contrast, exogenous ethylene applied to climacteric fruit sets in motion the ripening process, causing an upsurge of respiration to a peak level which is not significantly different over a wide range of ethylene concentrations. Furthermore, once production of endogenous ethylene is initiated, the respiratory rise and ripening continue, even after removal of the fruit from the atmosphere of exogenous ethylene. Apparently the climacteric fruit have a larger capacity for ethylene production, and their regulatory controls which govern ethylene production readily overshoot during ripening.

The ethylene-induced rise in respiration has been associated with a surge in protein synthesis related to onset of ripening which requires a new complement of enzymes (82). Alternatively, there were suggestions that the respiratory rise relates to a breakdown of "organizational resistance" or disorganization of the cell which may allow, for example, intermixing of cellular compartments and organelles normally separated (142). These are general explanations of ethylene-induced respiration. More recently, Solomos and Laties (106, 250–253) have presented data which indicate that ethylene stimulates respiration by shunting the flow of electrons in the respiratory election transfer chain from the conventional cytochrome system to an alternate cyanide-resistant oxidase. This results in an increased flow of electrons which relates to the increased respiration observed (249).

A common feature of both ethylene and cyanide-treated fruit tissue, in addition to a surge in respiration, is a sharp tenfold rise in fructose-1,6-diphosphate levels and a decline in glucose-6 phosphate and phosphoenolpyruvate (251). An increase in ATP production during the climacteric rise induced by ethylene or cyanide was also observed (253), presumably by increased electron flow through substrate level and site one phosphorylation sites.

Similar increases in respiration, glycolysis, and ATP synthesis, induced by ethylene and cyanide, were observed in potato tuber tissue in which ripening is not involved (252). In a survey of a number of different types of tissues, stimulation of respiration by ethylene was positively correlated with the presence of the cyanide-resistant alternate respiratory pathway (249). However, when *m*-chloro-benzhydroxamic acid (M-CLAM), the potent inhibitor of the cyanide-resistant alternate pathway (241), was added to tissues before ethylene treatment, the respiratory rise was still obtained as well as the increases in glycolysis (T. Solomos, personal communication). Consequently, the ethylene-induced surge in respiration is not due to induction of the alternate pathway, but to induced increase in glycolytic substrates mediated by a fivefold increase in phosphofructokinase and pyruvate kinase. The mechanisms by which ethylene activates these enzymes is probably not by direct modulation of the enzymes (97). This conclusion is based on comparative biochemistry of hormonal actions that generally appear to involve second messenger effects which derive from hormones acting on receptor sites on or in the plasma membrane (263).

CONCLUDING REMARKS

Although methionine was recognized as the major, if not sole, precursor of ethylene for more than a decade, up till now no clear evidence of the enzymes and intermediates in the pathway to ethylene could be demonstrated. Recently S-adenosyl methionine (SAM) and 1-aminocyclopropane-1-carboxylic acid (ACC) were suggested as possible intermediates between methionine and ethylene. Evidence for these intermediates is quite convincing and appears to provide the pathway from methionine to ethylene. The biggest gap in elucidating the pathway to ethylene remains the enzyme system which is apparently very delicate, highly structured, probably localized in the plasma membrane, perhaps on its surface, and does not survive destruction of the cell. This enzymatic system may be unique in that it includes a well-controlled free radical generating system and its activation may be closely regulated by ethylene utilization and by other hormones, particularly auxin. Clarification of the structure and action of this enzyme system should shed light not only on an intricate biochemical control mechanism in production and utilization of a plant hormone, but also provide insights on hormonal interactions and the importance of the structural integrity of the plasma membrane which appears to house the ethylene-forming system.

Some progress has also been made in determining whether or not ethylene metabolism is necessary in order to obtain its physiological effects. Beyer's work (30, 31) clearly relates physiological action of ethylene to its metabolism both to CO_2 and undetermined soluble products. However, there remain some questions raised by the influence of hypobaric conditions, which prevent ethylene action, and some experiments in which bound ethylene is displaced, suggesting that the hormone is not metabolized. The physical relationship between the ethylene-forming enzyme system and its receptor is of great interest in this connection. Are they in close proximity? Are they widely separated? These questions may relate to the general phenomenon of hormonal production and action in plants which may differ from that in animals in which sites of hormonal production and action are separated but linked by a circulating red blood cell system.

Some of the most intriguing recent findings on ethylene action relate to its interactions with other plant hormones at all levels of metabolism, especially in development of tissue systems and cytodifferentiation (296, 297). These findings reveal that extremely low levels of ethylene in the low ppb range may be required for growth and development, in contrast to higher levels in the low ppm range which may be inhibitory or antagonistic. One interpretation of these findings suggests that ethylene may be a modulator of the action of plant hormones in growth and development, and conversely other hormones may modulate the action of ethylene in ripening, aging, and senescence. The experiments concerning interaction of ethylene and other hormones are still at the level of description, for the most part. It is in this area where information on mechanisms needs to be developed. Such findings should reveal much about the control and regulation of plants at the hormonal level of metabolism.

ACKNOWLEDGMENTS

I wish to thank J. D. Anderson, A. Apelbaum, and J. E. Baker for reading the manuscript and providing helpful criticisms. I am also grateful to many colleagues who provided manuscripts of papers prior to publication or in press. My special thanks to Delores Sessions for her rapid and accurate typing of the manuscript.

Literature Cited

1. Abdel-Kader, A. S., Morris, L. M., Maxie, E. C. 1968. Physiological studies of gamma irradiation in tomato fruits. 1. Effect on respiratory rate, ethylene production and ripening. *Proc. Am. Soc. Hortic. Sci.* 92:553–67
2. Abeles, A. L., Abeles, F. B. 1972. Biochemical pathway for stress-induced ethylene. *Plant Physiol.* 50:496–98
3. Abeles, F. B. 1966. Auxin stimulation of ethylene evolution. *Plant Physiol.* 41:585–88
4. Abeles, F. B. 1972. Biosynthesis and mechanism of action of ethylene. *Ann. Rev. Plant Physiol.* 23:259–92
5. Abeles, F. B. 1973. *Ethylene in Plant Biology.* New York, London: Academic. 302 pp.
6. Abeles, F. B., Rubinstein, B. 1964. Cell-free ethylene evolution from etiolated pea seedlings. *Biochim. Biophys. Acta* 93:675–77
7. Abeles, F. B., Ruth, J. M., Forrence, L. E., Leather, G. R. 1972. Mechanism of hormone action. Use of deuterated ethylene to measure isotopic exchange with plant material and biological effects of deuterated ethylene. *Plant Physiol.* 49: 669–71
8. Adams, D. O., Yang, S. F. 1977. Methionine metabolism in apple tissue. *Plant Physiol.* 60:892–96
9. Adams, D. O., Yang, S. F. 1978. Effect of anaerobiosis on ethylene production and metabolism by apple tissue. *Plant Physiol. Suppl.* 61:90
10. Adato, I., Gazit, S. 1976. Response of harvested avocado fruits to supply of indole-3-acetic acid, gibberellic acid and abscisic acid. *J. Agric. Food Chem.* 24: 1165–67
11. Aharoni, N. 1975. *Hormonal regulation during senescence and water stress of detached lettuce leaves (Lactuca sativa L.)* PhD thesis. Hebrew Univ., Jerusalem
12. Aharoni, N., Blumenfeld, A., Richmond, A. E. 1977. Hormonal activity in detached lettuce leaves as affected by leaf water content. *Plant Physiol.* 59: 1169–73
13. Aharoni, N., Richmond, A. 1978. Relationship between leaf water status and endogenous ethylene in detached leaves. *Plant Physiol.* 61:658–62
14. Anderson, J. D., Lieberman, M. 1976. Relationship between IAA-stimulated ethylene evolution and protein synthesis in pea subhook sections. *Int. Conf. Plant Growth Subst., 9th Lausanne, Switzerland,* collected abstr., pp. 17–18
15. Andreae, W. A., Venis, M. A., Jursic, F., Dumas, F. 1968. Does ethylene mediate root growth inhibition by indole-3-acetic acid? *Plant Physiol.* 43:1375–79
16. Apelbaum, A., Burg, S. P. 1971. Altered cell microfibrillar orientation in ethylene-treated *Pisum sativum* stems. *Plant Physiol* 48:648–52
17. Apelbaum, A., Burg, S. P. 1972. Effect of ethylene on cell division and DNA synthesis in *Pisum sativum. Plant. Physiol.* 50:117–24
18. Apelbaum, A., Burg, S. P. 1972. Effects of ethylene and 2,4-dichlorophenoxyacetic acid on cellular expansion in *Pisum sativum. Plant Physiol.* 50: 125–31

19. Apelbaum, A., Fisher, J. B., Burg, S. P. 1972. Effect of ethylene on cellular differentiation in etiolated pea seedlings. *Am. J. Bot.* 59:697–705
20. Archer, S. A. 1976. Ethylene and fungal growth. *Trans. Br. Mycol. Soc.* 67: 325–26
21. Baker, J. E., Lieberman, M., Anderson, J. D. 1978. Inhibition of ethylene production in fruit slices by a rhizobitoxine analog and free radical scavengers. *Plant Physiol.* 61:886–88
22. Bangerth, F. 1975. The effect of ethylene on the physiology of ripening of apple fruits at hypobaric conditions. In *Facteurs et Regulation de la Maturation des Fruits, Colloques Int. C.N.R.S.* No. 238, Paris, 1974, pp. 183–88
23. Bangerth, F. 1978. The effect of substituted amino-acid on ethylene biosynthesis, respiration, ripening and preharvest drop of apple fruits. *J. Am. Soc. Hortic. Sci.* 103:401–4
24. Baur, A. H., Yang, S. F. 1969. Precursors of ethylene. *Plant Physiol.* 44: 1347–49
25. Baur, A. H., Yang, S. F. 1969. Ethylene production from propanal. *Plant Physiol.* 44:189–92
26. Baur, A. H., Yang, S. F. 1972. Methionine metabolism in apple tissue in relation to ethylene biosynthesis. *Phytochemistry* 11:3207–14
27. Baur, A. H., Yang, S. F., Pratt, H. K., Biale, J. B. 1971. Ethylene biosynthesis in fruit tissues. *Plant Physiol.* 47:696–99
28. Beauchamp, C., Fridovich, I. 1970. A mechanism for the production of ethylene from methional. *J. Biol. Chem.* 245:4641–46
29. Beyer, E. M. 1972. Mechanism of ethylene action. Biological activity of deuterated ethylene and evidence against isotopic exchange and cis-trans isomerization. *Plant Physiol.* 49:672–75
30. Beyer, E. M. 1975. ^{14}C-ethylene incorporation and metabolism in pea seedlings. *Nature* 255:144–47
31. Beyer, E. M. 1975. $^{14}C_2H_4$: Its purification for biological studies. *Plant Physiol.* 55:845–48
32. Beyer, E. M. 1975. $^{14}C_2H_4$: Its incorporation and metabolism by pea seedlings under aseptic conditions. *Plant Physiol.* 56:273–78
33. Beyer, E. M. 1976. A potent inhibitor of ethylene action in plants. *Plant Physiol.* 58:268–71
34. Beyer, E. M. 1976. Silver ion: a potent antiethylene agent in cucumber and tomato. *HortScience* 11:195–96
35. Beyer, E. M. 1977. $^{14}C_2H_4$: Its incorporation and oxidation to $^{14}CO_2$ by cut carnations. *Plant Physiol.* 60:203–6
36. Beyer, E. M. 1978. Rapid metabolism of propylene by pea seedlings. *Plant Physiol.* 61:893–95
37. Beyer, E. M. 1978. A method for overcoming the antiethylene effect of Ag^+. *Plant Physiol.* In press
38. Beyer, E. M. 1978. Effect of Ag^+, CO_2 and O_2 on ethylene action and metabolism. *Plant Physiol.* In press
39. Beyer, E. M., Morgan, P. W. 1971. Abscission. The role of ethylene modification of auxin transport. *Plant Physiol.* 48:208–12
40. Beyer, E. M., Sundin, O. 1978. $^{14}C_2H_4$ metabolism in morning glory flowers. *Plant Physiol.* 61:896–99
41. Biale, J. B. 1960. Respiration of fruits. Role of ethylene and plant emanation in fruit respiration. *Handb. Pflanzenphysiol.* 12:536–92
42. Biale, J. B. 1964. Growth, maturation, and senescence. *Science* 146:880–88
43. Bird, C. W., Lynch, J. M. 1974. Formation of hydrocarbons by microorganisms. *Chem. Soc. Rev.* 3:309–28
44. Blumenfeld, A. 1975. Ethylene and the *Annona* flower. *Plant Physiol.* 55: 265–69
45. Bors, W., Lengfelder, E., Saran, M., Fuchs, C., Michel, C. 1976. Reactions of oxygen radical species with methional: a pulse radiolysis study. *Biochem. Biophys. Res. Commun.* 70:81–87
46. Bressan, R. A., Wilson, L. G., LeCureux, L., Filner, P. 1978. Use of ethylene and ethane emissions to assay injury by SO_2. *Plant Physiol.* 61:93
47. Brisker, H. E., Goldschmidt, E. E., Goren, E. 1976. Ethylene-induced formation of ABA in citrus peel as related to chloroplast transformations. *Plant Physiol.* 58:377–79
48. Bruinsma, J., Knegt, E., Varga, A. 1975. The role of growth-regulating substances in fruit ripening. See Ref. 22, pp. 193–99
49. Buhler, D. R., Hansen, E., Wang, C. H. 1957. Incorporation of ethylene into fruits. *Nature* 174:48–49
50. Burg, S. P. 1962. The physiology of ethylene formation. *Ann. Rev. Plant Physiol.* 13:265–302
51. Burg, S. P. 1968. Ethylene, plant senescence and abscission. *Plant Physiol.* 43:1503–11
52. Burg, S. P. 1973. Ethylene in plant growth. *Proc. Natl. Acad. Sci. USA* 70:591–97

53. Burg, S. P., Apelbaum, A., Eisinger, W., Kang, B. G. 1971. Physiology and mode of action of ethylene. *HortScience* 6: 359–64
54. Burg, S. P., Burg, E. A. 1962. Role of ethylene in fruit ripening. *Plant Physiol.* 37:179–89
55. Burg, S. P., Burg, E. A. 1964. Evidence for a natural occurring inhibitor of fruit ripening. *Plant Physiol. Suppl.* 39:X
56. Burg, S. P., Burg, E. A. 1966. The interaction between auxin and ethylene and its role in plant growth. *Proc. Natl. Acad. Sci. USA* 55:262–69
57. Burg, S. P., Burg, E. A. 1967. Molecular requirements for the biological activity of ethylene. *Plant Physiol.* 42:144–52
58. Burg, S. P., Burg, E. A. 1967. Inhibition of polar auxin transport of ethylene. *Plant Physiol.* 42:1224–28
59. Burg, S. P., Burg, E. A. 1968. Ethylene formation in pea seedlings: Its relation to the inhibition of bud growth caused by IAA. *Plant Physiol.* 43:1069–74
60. Burg, S. P., Clagett, C. O. 1967. Conversion of methionine to ethylene in vegetative tissue and fruits. *Biochem. Biophys. Res. Commun.* 27:125–30
61. Burg, S. P., Thimann, K. V. 1960. Studies on the ethylene production of apple tissue. *Plant Physiol.* 35:24–35
62. Burroughs, L. F. 1957. 1-Aminocyclopropane-1-carboxylic acid: A new amino acid in perry pears and cider apples. *Nature* 179:360–61
63. Chadwick, A. V., Burg, S. P. 1967. An explanation of the inhibition of root growth caused by indole-3-acetic acid. *Plant Physiol.* 42:415–20
64. Chadwick, A. V., Burg, S. P. 1970. Regulation of root growth by auxin-ethylene interaction. *Plant Physiol.* 45:192–200
65. Chailakhyan, M. Kh., Khryanin, V. N. 1978. Influence of growth regulators absorbed by the root on sex expression in hemp plants. *Planta* 138:181–84
66. Chalutz, E., Lieberman, M. 1978. Inhibition of ethylene production in *Penicillium digitatum. Plant Physiol.* 61: 111–14
67. Chalutz, E., Lieberman, M., Sisler, H. D. 1977. Methionine-induced ethylene production by *Penicillium digitatum. Plant Physiol.* 60:402–6
68. Chalutz, E., Mattoo, A. K., Anderson, J. D., Lieberman, M. 1978. Regulation of ethylene production by phosphate in *Penicillium digitatum. Plant Cell Physiol.* 19:189–96
69. Chou, T. W., Yang, S. F. 1973. The biogenesis of ethylene in *Penicillium digitatum. Arch. Biochem. Biophys.* 157:73–82
70. Chrominski, A., Kopcewicz, J. 1972. Auxins and gibberellins in 2-chloroethylphosphonic acid-induced femaleness of *Cucurbita pepo* L. *Z. Pflanzenphysiol.* 68:184–89
71. Cook, C. E., Whichard, L. P., Wall, M. E. Egley, G. H., Coggon, P., Luhan, P. A., McPhail, A. T. 1972. Germination stimulants II. The structure of strigol. *J. Am. Chem. Soc.* 94:6198–99
72. Cooke, A. R., Randall, D. I. 1968. 2-Haloethanephosphonic acid as ethylene releasing agent for induction of flowering in pineapples. *Nature* 218:974–75
73. Coombe, B. G. 1976. The development of fleshy fruits. *Ann. Rev. Plant Physiol.* 27:207–28
74. Coombe, B. G., Hale, C. R. 1973. The hormone content of ripening grape berries and the effects of growth substance treatments. *Plant Physiol.* 51:629–34
75. Cooper, W. C., Rasmussen, G. K., Waldon, E. S. 1969. Ethylene evolution stimulated by chilling in citrus and *Persea* sp. *Plant Physiol.* 44:1194–96
76. Crocker, W., Hitchcock, A. E., Zimmerman, P. W. 1935. Similarities in the effect of ethylene and the plant auxins. *Contrib. Boyce Thompson Inst.* 7: 231–48
77. Crocker, W., Knight, L. I., Rose, R. C. 1913. A delicate seedling test. *Science* 37:380–81
78. Crocker, W., Zimmerman, P. W., Hitchcock, A. E. 1932. Ethylene-induced epinasty of leaves and the relation of gravity to it. *Contrib. Boyce Thompson Inst.* 4:177–218
79. Curtis, R. W. 1969. Oxygen requirement for ethane production *in vitro* by *Phaseolus vulgaris. Plant Physiol.* 44: 1368–70
80. Davis, L., Metzler, D. E. 1972. Pyridoxal-linked elimination and replacement reactions. *The Enzymes* 7:33–74
81. Dillard, C. J., Dumelin, E. E., Tappel, A. L. 1977. Effect of dietary Vit. E on expiration of pentane and ethane by the rat. *Lipids* 12:109–14
82. Dilley, D. R. 1970. Enzymes. In *The Biochemistry of Fruits and Their Products,* ed. A. C. Hulme, 1:200. London, New York: Academic. 620 pp.
83. Dilley, D. R. 1977. Hypobaric storage of perishable commodities—fruits, vegetables, flowers, and seedlings. *Acta Hortic.* 62:61–70
84. Dumelin, E. E., Dillard, C. J., Tappel, A. L. 1978. Breath ethane and pentane as measures of Vit. E protection of

Macaca radiata against 90 days exposure to ozone. *Environ. Res.* 15:38–43

85. Dunlap, J. R., Morgan, P. W. 1977. Reversal of induced dormancy in lettuce by ethylene, kinetin, and gibberellic acid. *Plant Physiol.* 60:222–24
86. Dunlap, J. R., Morgan, P. W. 1977. Characterization of ethylene/gibberellic acid control of germination in *Lactuca sativa. Plant Cell Physiol.* 18: 561–68
87. Egley, G. H., Dale, J. E. 1970. Ethylene, 2-chloroethylphosphonic acid and witchweed germination. *Weed Sci.* 18: 586–89
88. Elstner, E. F., Konze, J. R. 1976. Effect of point freezing on ethylene and ethane production by sugar beet leaf disks. *Nature* 263:351–52.
89. Ernest, L. C., Valdovinos, J. G. 1971. Regulation of auxin levels in *Coleus blumei* by ethylene. *Plant Physiol.* 48: 402–6
90. Evans, M. L. 1974. Rapid responses to plant hormones. *Ann. Rev. Plant Physiol.* 25:195–223
91. Fowler, J. L., Morgan, P. W. 1972. The relationship of the peroxidative indoleacetic acid oxidase system to *in vivo* ethylene synthesis in cotton. *Plant Physiol.* 49:555–59
92. Franklin, D., Morgan, P. W. 1978. Rapid production of auxin-induced ethylene. *Plant Physiol.* 62:161–62
93. Frenkel, C. 1975. Role of oxidative metabolism in the regulation of fruit ripening. See Ref. 22, pp. 201–9
94. Frenkel, C., Dyck, R. 1973. Auxin inhibition of ripening in Bartlett pears. *Plant Physiol.* 51:6–9
95. Frenkel, C., Garrison, S. A. 1976. Initiation of lycopene synthesis in the tomato mutant rin as influenced by oxygen and ethylene interactions. *HortScience* 11:20–21
96. Frenkel, C., Haddon, V. R., Smallheer, J. M. 1975. Promotion of softening and ethylene synthesis in Bartlett pears by 3-methylene oxindole. *Plant Physiol.* 56:647–49
97. Fuchs, Y., Gertman, E. 1973. Stabilization of enzyme activity by incubation in an ethylene atmosphere. *Plant Cell Physiol.* 14:197–99
98. Fuchs, Y., Lieberman, M. 1968. Effects of kinetin, IAA, and gibberellin on ethylene production and their interaction in growth of seedlings. *Plant Physiol.* 43:2029–36
99. Galliard, T., Hulme, A. C., Rhodes, M. J. C., Wooltorton, L. S. C. 1968. Enzymatic conversion of linolenic acid to ethylene by extracts of apple fruit. *FEBS Lett.* 1:283–86
100. Gazit, S., Blumenfeld, A. 1970. Response of mature avocado fruits to ethylene treatments before and after harvest. *J. Am. Soc. Hortic. Sci.* 95:229–31
101. Gertman, E., Fuchs, Y. 1972. Effect of abscisic acid and its interactions with other plant hormones on ethylene production in two plant systems. *Plant Physiol.* 50:194–95
102. Ghooprasert, P., Spencer, M. 1975. Preparation and purification of an enzyme system for ethylene synthesis from acrylate. *Physiol. Veg.* 13:579–89
103. Giaquinta, R., Beyer, E. M. 1977. $^{14}C_2H_4$: Distribution of ^{14}C-labeled tissue metabolites in pea seedlings. *Plant Cell Physiol.* 18:141–48
104. Giovanelli, J., Owens, L. D., Mudd, S. H. 1971. Mechanism of inhibition of spinach β-cystathionase by rhizobitoxine. *Biochim. Biophys. Acta* 227:671–84
105. Goeschl, J. D., Pratt, H. K., Bonner, B. A. 1967. An effect of light on the production of ethylene and the growth of the plumular portion of etiolated pea seedlings. *Plant Physiol.* 42:1077–80
106. Goldmann, D., Laties, G. G. 1976. The initiation of the climacteric rise in peeled banana fruit by HCN and CO under hypobaric conditions. *Plant Physiol. Suppl.* 57:27
107. Goldschmidt, E. E., Goren, R., Even-Chen, Z., Bittner, S. 1973. Increase in free and bound abscisic acid during natural and ethylene-induced senescence of citrus fruit peel. *Plant Physiol.* 51: 879–82
108. Goldwin, G. K., Wain, R. L. 1973. Studies on plant growth-regulating substances XXXV ethylene production by coleoptiles treated with auxin type chemicals. *Ann. Appl. Biol.* 75:71–81
109. Goren, R., Bukovac, M. J., Flore, J. A. 1974. Mechanism of indole-3-acetic acid conjugation on induction by ethylene. *Plant Physiol.* 53:164–66
110. Haber, F., Weiss, J. 1934. The catalytic decomposition of hydrogen peroxide by iron salts. *Proc. R. Soc. London Ser. A* 147:332–51
111. Hall, M. A., Kapuya, J. A., Sivakumaran, S., John, A. 1977. The role of ethylene in the response of plants to stress. *Pestic. Sci.* 8:217–23
112. Hall, W. C., Miller, C. S., Herrero, F. A. 1961. Studies with ^{14}C ethylene. *Proc. 4th Int. Conf. Plant Growth Regul.*, pp. 751–78. Ames: Iowa State Univ. Press

113. Hansen, E. 1942. Quantitative study of ethylene production in relation to respiration of pears. *Bot. Gaz.* 103:543–58
114. Hansen, E. 1966. Postharvest physiology of fruits. *Ann. Rev. Plant Physiol.* 17:459–80
115. Hanson, A. D., Kende, H. 1976. Methionine metabolism and ethylene biosynthesis in senescent flower tissue of morning-glory. *Plant Physiol.* 57: 528–37
116. Hanson, A. D., Kende, H. 1976. Biosynthesis of wound ethylene in morning-glory flower tissue. *Plant Physiol.* 57:538–41
117. Herner, R. C., Sink, K. C. 1973. Ethylene production and respiratory behavior of the rin tomato mutant. *Plant Physiol.* 52:38–42
118. Hertel, R., Thomson, K. St., Russo, V. E. A. 1972. *In vitro* auxin binding to particulate cell fractions from corn coleoptiles. *Planta* 107:325–40
119. Horton, R. F., Osborne, D. J. 1967. Senescence, abscission and cellulase activity in *Phaseolus vulgaris. Nature* 214:1–6
120. Hradilik, J. 1975. Estimation of ethylene produced by Pea (*Pisum sativum* L.) plants inhibited with auxin and treated with cytokinin. *Biochem. Physiol. Pflanz.* 167:459–61
121. Hyodo, H. 1977. Ethylene production by albedo tissue of Satsuma mandarin (*Citrus unshiu Marc.*) fruit. *Plant Physiol.* 59:111–13
122. Imaseki, H., Kondo, K., Watanabe, A. 1975. Mechanism of cytokinin action on auxin-induced ethylene production. *Plant Cell Physiol.* 16:777–87
123. Imaseki, H., Pjon, C. J. 1970. The effect of ethylene on auxin-induced growth of excised rice coleoptile segments. *Plant Cell Physiol.* 11:827–29
124. Imaseki, H., Pjon, C. J., Furuya, M. 1971. Phytochrome action in *Oryza sativa* L. IV. Red and far red reversible effect on the production of ethylene in excised coleoptiles. *Plant Physiol.* 48: 241–44
125. Imaseki, H., Uritani, I., Stahmann, M. A. 1968. Production of ethylene by injured sweet potato root tissue. *Plant Cell Physiol.* 9:757–68
126. Imaseki, H., Watanabe, A. 1978. Inhibition of ethylene production by osmotic shock. Further evidence for control of ethylene production by membrane. *Plant Cell Physiol.* 19:345–48
127. Jackson, M. B., Campbell, D. J. 1976. Production of ethylene by excised segments of plant tissue prior to the effect of wounding. *Planta* 129:273–74
128. Jackson, M. B., Gales, K., Campbell, D. J. 1978. Effect of waterlogged soil conditions on the production of ethylene and on water relationships in tomato plants. *J. Exp. Bot.* 29:183–93
129. Jacobsen, D. W., Wang, C. H. 1968. The biogenesis of ethylene in *Penicillium digitatum. Plant Physiol.* 43: 1959–66
130. Jacobson, B. S., Smith, B. N., Epstein, S., Laties, G. G. 1970. The prevalence of carbon-13 in respiratory carbon dioxide as an indicator of the type of endogenous substrate. *J. Gen. Physiol.* 55:1–17
131. Jaffe, M. J., Biro, R. 1977. Thigmomorphogenesis: Role of ethylene in wind-induced growth retardation. *Proc. 4th Ann. Meet. Plant Growth Regul. Work. Group,* pp. 118–24
132. Jansen, E. F. 1963. Metabolism of labeled ethylene in the avocado: appearance of tritium in the methyl group of toluene. *J. Biol. Chem.* 238:1552–55
133. Jansen, E. F. 1964. Metabolism of labeled ethylene in the avocado, benzene and toluene from ethylene ^{14}C, benzene from ethylene $-^{3}H$. *J. Biol. Chem.* 239:1664–67
134. Jerie, P. H., Hall, M. A. 1978. The identification of ethylene oxide as a major metabolite of ethylene in *Vicia faba* L. *Proc. R. Soc. London Ser. B* 200:87–94
135. John, W. W., Curtis, R. W. 1977. Isolation and identification of the precursor of ethane in *Phaseolus vulgaris* L. *Plant Physiol.* 59:521–22
136. Kang, B. G., Newcomb, W., Burg, S. P. 1971. Mechanism of auxin-induced ethylene production. *Plant Physiol.* 47: 504–9
137. Kang, B. G., Yocum, C. S., Burg, S. P., Ray, P. M. 1967. Ethylene and carbon dioxide, mediation of hypocotyl hook-opening response. *Science* 156:958–59
138. Kawase, M. 1976. Ethylene accumulation in flooded plants. *Physiol. Plant.* 36:236–41
139. Kende, H., Hanson, A. D. 1977. On the role of ethylene in aging. *Proc. 9th Int. Conf. Plant Growth Substances, Plant Growth Regulation,* ed. P. E. Pilet, pp. 172–80. Berlin, Heidelberg: Springer-Verlag
140. Ketring, D. L., Morgan, P. W. 1972. Physiology of oil seeds. IV. Role of endogenous ethylene and inhibitory regulators during natural and induced after-ripening of dormant Virginia-type peanut seeds. *Plant Physiol.* 50:382–87

141. Ketring, D. L., Young, R. E., Biale, J. B. 1968. Effects of monofluoroacetate on *Penicillium digitatum* metabolism and on ethylene biosynthesis. *Plant Cell Physiol.* 9:617–31
142. Kidd, F., West, C. 1930. Physiology of fruit. I. Changes in respiratory activity of apples during their senescence at different temperatures. *Proc. R. Soc. London Ser. B.* 106:93–109
143. Kondo, K., Watanabe, A., Imaseki, H. 1975. Relationships in actions of indoleacetic acid, benzyladenine and abscisic acid in ethylene production. *Plant Cell Physiol.* 16:1001–7
144. Konings, H., Jackson, M. B. 1974. Production of ethylene and the promoting and inhibiting effects of applied ethylene on root elongation in various species. *Ann. Rep. Agric. Res. Counc., Letcomb Lab.,* pp. 23–24
145. Konze, J. R., Elstner, E. F. 1976. Ethylene and ethane formation in leaf disks, plastids, and mitochondria. *Ber. Dtsch. Bot. Ges.* 89:547–53
146. Konze, J. R., Schilling, N., Kende, H. 1978. Enhancement of ethylene formation by selenoamino acids. *Plant Physiol.* 62:397–401
147. Ku, H. S., Suge, H., Rappaport, L., Pratt, H. K. 1970. Stimulation of rice coleoptile growth by ethylene. *Planta* 90:333–39
148. Ku, H. S., Yang, S. F., Pratt, H. K. 1967. Enzymic evolution of ethylene from methional by a pea seedling extract. *Arch. Biochem. Biophys.* 118: 756–58
149. Lau, O. L., John, W. W., Yang, S. F. 1977. Effect of different cytokinins on ethylene production by mung bean hypocotyls in the presence of IAA or calcium ions. *Physiol. Plant.* 39:1–3
150. Lau, O. L., John, W. W., Yang, S. F. 1978. Inactivity of oxidation products of indole-3-acetic acid on ethylene production in mung bean hypocotyls. *Plant Physiol.* 61:68–71
151. Lau, O. L., Yang, S. F. 1973. Mechanisms of a synergistic effect of kinetin on auxin-induced ethylene production: suppression of auxin conjugation. *Plant Physiol.* 51:1011–14
152. Lau, O. L., Yang, S. F. 1974. Synergistic effect of calcium and kinetin on ethylene production by the mung bean hypocotyl. *Planta* 118:1–6
153. Lau, O. L., Yang, S. F. 1975. Interaction of kinetin and calcium in relation to their effect on stimulation of ethylene production. *Plant Physiol.* 55:738–40
154. Lau, O. L., Yang, S. F. 1976. Stimulation of ethylene production in the mung bean hypocotyls by cupric ion, calcium ion and kinetin. *Plant Physiol.* 57:88–92
155. Lau, O. L., Yang, S. F. 1976. Inhibition of ethylene production by cobaltous ion. *Plant Physiol.* 114–17
156. Lieberman, M. 1975. Biosynthesis and regulatory control of ethylene in fruit ripening. A review. *Physiol. Veg.* 13:489–99
157. Lieberman, M. 1977. Biosynthesis and bioregulation of ethylene in fruit ripening. *Proc. Int. Soc. Citriculture,* ed. W. Grierson, Vol. 3
158. Lieberman, M., Asen, S., Mapson, L. W. 1964. Ethylene oxide an antagonist of ethylene in metabolism. *Nature* 204:756–58
159. Lieberman, M., Baker, J. E., Sloger, M. 1977. Influence of plant hormones on ethylene production in apple, tomato, and avocado slices during maturation and senescence. *Plant Physiol.* 60: 214–17
160. Lieberman, M., Knegt, E. 1977. Influence of ethylene on indole-3-acetic acid concentration in etiolated pea epicotyl tissue. *Plant Physiol.* 60:475–77
161. Lieberman, M., Kunishi, A. T. 1967. Propanal may be a precursor of ethylene in metabolism. *Science* 158:938
162. Lieberman, M., Kunishi, A. T. 1968. Origins of ethylene in plants. In *Biochemical Regulation in Diseased Plants or Injury,* T. Hirai, Z. Hidoka, I. Uritani, pp. 165–79. Phytopathol. Soc. Japan, Tokyo
163. Lieberman, M., Kunishi, A. T. 1971. An evaluation of 4-S-methyl-2-keto butyric acid as an intermediate in the biosynthesis of ethylene. *Plant Physiol.* 47:576–80
164. Lieberman, M., Kunishi, A. T. 1971. Synthesis and biosynthesis of ethylene. *HortScience* 6:355–58
165. Lieberman, M., Kunishi, A. T. 1971. Abscisic acid and ethylene production *Plant Physiol. Suppl.* 47:22
166. Lieberman, M., Kunishi, A. T. 1972. Thoughts on the role of ethylene in plant growth and development. In *Plant Growth Substances, 1970,* ed. D. J. Carr, pp. 549–60. Berlin, Heidelberg, New York: Springer-Verlag. 837 pp.
167. Lieberman, M., Kunishi, A. T. 1975. Ethylene-forming systems in etiolated pea seedlings and apple tissue. *Plant Physiol.* 55:1074–78
168. Lieberman, M., Kunishi, A. T., Mapson, L. W., Wardale, D. A. 1965. Ethyl-

ene production from methionine. *Biochem. J.* 97:449–59
169. Lieberman, M., Kunishi, A. T., Mapson, L. W., Wardale, D. A. 1966. Stimulation of ethylene production in apple tissue slices by methionine. *Plant Physiol.* 41:376–82
170. Lieberman, M., Kunishi, A. T., Owens, L. D. 1975. Specific inhibitors of ethylene production as retardants of the ripening process in fruits. See Ref. 22, pp. 161–70
171. Lieberman, M., Mapson, L. W. 1962. Fatty acid control of ethane production by subcellular particles and its possible relationship to ethylene biosynthesis. *Nature* 195:1016–17
172. Lieberman, M., Mapson, L. W. 1962. Inhibition of the evolution of ethylene and the ripening of fruit by ethylene oxide. *Nature* 196:660–61
173. Lieberman, M., Mapson, L. W. 1964. Genesis and biogenesis of ethylene. *Nature* 204:343–45
174. Lynch, J. M. 1974. Mode of ethylene formation by *Mucor hiemalis. J. Gen. Microbiol.* 83:307–11
175. MacDonald, I. R., Ellis, R. J. 1969. Does cycloheximide inhibit protein synthesis specifically in plant tissues? *Nature* 222:791–92
176. Malik, C. P., Mehan, M. 1975. Correlative effects of auxin, gibberellic acid and kinetin on the elongation of pollen tubes in *Calotropis procera. Biochem. Physiol. Pflanz.* 167:295–300
177. Mapson, L. W. 1970. Biosynthesis of ethylene and the ripening of fruit. *Endeavour* 39:29–33
178. Mapson, L. W., March, J. F., Rhodes, M. J. C., Wooltorton, L. S. C. 1970. A comparative study of the ability of methionine or linolenic acid to act as precursors of ethylene. *Biochem. J.* 117:473–79
179. Mapson, L. W., March, J. F., Wardale, D. A. 1969. Biosynthesis of ethylene 4-methylmercapto-2-oxobutyric acid: An intermediate in the formation from methionine. *Biochem. J.* 115:653–61
180. Mapson, L. W., Wardale, D. A. 1967. Biosynthesis of ethylene. Formation of ethylene from methional by a cell-free enzyme system from cauliflower florets. *Biochem. J.* 102:574–85
181. Mapson, L. W., Wardale, D. A. 1968. Biosynthesis of ethylene; enzymes involved in its formation from methional. *Biochem. J.* 107:433–42
182. Marei, N., Crane, J. C. 1971. Growth and respiratory response of fig (*Ficus carica* L. cv. Misson) fruits to ethylene. *Plant Physiol.* 48:249–54
183. Matsuno, H., Uritani, I. 1972. Physiological behavior of peroxidase isozymes in sweet potato root tissue injured by cutting or with black rot. *Plant Cell Physiol.* 13:1091–1101
184. Mattoo, A. K., Baker, J. E., Chalutz, E., Lieberman, M. 1977. Effect of temperature on the ethylene-synthesizing systems in apple, tomato and *Penicillium digitatum. Plant Cell Physiol.* 18:715–19
185. Mattoo, A. K., Lieberman, M. 1977. Localization of the ethylene-synthesizing system in apple tissue. *Plant Physiol.* 60:794–99
186. Mayak, S., Dilley, D. R. 1976. Regulation of senescence in carnation (*Dianthus caryophyllus*). Effect of abscisic acid and carbon dioxide on ethylene production. *Plant Physiol.* 58:663–65
187. McGlasson, W. B. 1970. The ethylene factor. See Ref. 82, pp. 475–519
188. McGlasson, W. B. 1978. Role of hormones in ripening and senescence. *IFT Symposium Post Harvest Biology and Bio-Technology Proceedings.* Basic Symp., Philadelphia, 1977, ed. H. O. Hultin, M. Milner, pp. 77–96. Food Nutrition Press
189. McGlasson, W. B., Dostal, H. C., Tigchelaar, E. C. 1975. Comparison of propylene-induced responses of immature fruit of normal and rin mutant tomatoes. *Plant Physiol.* 55:218–22
190. McGlasson, W. B., Poovaiah, B. W., Dostal, H. C. 1975. Ethylene production and respiration in aging leaf segments and in disks of fruit tissue of normal and mutant tomatoes. *Plant Physiol.* 56:547–49
191. McMurchie, E. J., McGlasson, W. B., Eaks, I. L. 1972. Treatment of fruit with propylene gives information about the biogenesis of ethylene. *Nature* 237:235–36
192. Meigh, D. F., Jones, J. D., Hulme, A. C. 1967. The respiration climacteric in the apple, production of ethylene and fatty acids in fruit attached to and detached from the tree. *Phytochemistry* 6: 1507–15
193. Meigh, D. F., Norris, K. H., Craft, C. C., Lieberman, M. 1960. Ethylene production by tomato and apple fruit. *Nature* 186:902–3
194. Mer, C. L. 1974. On the ethylene-auxin interaction in growth control. *New Phytol.* 73:653–55
195. Mer, C. L. 1974. On the inhibition of

mesocotyl growth by ethylene. *New Phytol.* 73:643–51
196. Michener, H. D. 1938. The action of ethylene on plant growth. *Am. J. Bot.* 25:711–20
197. Mizrahi, Y., Dostal, H. C., Cherry, J. H. 1975. Ethylene-induced ripening in attached rin fruits, a non-ripening mutant of tomato. *HortScience* 10:414–15
198. Morgan, P. W. 1976. Effects on ethylene physiology. In *Herbicides: Physiology, Biochemistry, Ecology,* ed. L. J. Audus, 1:255–80. New York: Academic. 608 pp.
199. Morgan, P. W., Beyer, E., Gausman, H. E. 1968. Ethylene effects on auxin physiology. In *Biochemistry and Physiology of Plant Growth Substances,* ed. F. Wightman, G. Setterfield, pp. 1255–73. Ottawa, Canada: Runge
200. Morgan, P. W., Durham, J. I. 1975. Ethylene-induced leaf abscission is promoted by gibberellic acid. *Plant Physiol.* 55:308–11
201. Morgan, P. W., Hall, W. C. 1962. Effect of 2,4D on the production of ethylene by cotton and grain sorghum. *Physiol. Plant* 15:420–27
202. Morgan, P. W., Hall, W. C. 1964. Accelerated release of ethylene by cotton following application of indolyl-3-acetic acid. *Nature* 201:91
203. Murr, D. P., Yang, S. F. 1975. Inhibition of *in vivo* conversion of methionine to ethylene by L-canaline and 2,4-dinitrophenol. *Plant Physiol.* 55:79–82
204. Musgrave, A., Jackson, M. B., Ling, E. 1972. *Callitriche* stem elongation is controlled by ethylene and gibberellin. *Nature New Biol.* 238:93–96
205. Musgrave, A., Walters, J. 1973. Ethylene-stimulated growth and auxin transport in *Ranunculus sceleratus* petioles. *New Phytol.* 72:783–89
206. Neljubow, D. 1911. Geotropismus in der Laboratoriumsluft. *Ber. Dtsch. Bot. Ges.* 29:97–112
207. Ng, T. J., Tigchelaar, E. C. 1977. Action of non-ripening (nor) mutant on fruit ripening of tomato. *J. Soc. Hortic. Sci.* 102:504–9
208. Odawara, S., Watanabe, A., Imaseki, H. 1977. Involvement of cellular membranes in regulation of ethylene production. *Plant Cell Physiol.* 18:569–75
209. Olatoye, S. T., Hall, M. A. 1972. Interaction of ethylene and light on dormant weed seeds. In *Seed Ecology,* ed. W. Heydecker, pp. 233–49. University Park: Penn. State Univ. 578 pp.
210. Osborne, D. J. 1968. Ethylene as a plant hormone. In *Plant Growth Regulators. Soc. Chem. Ind. London Monogr. 31,* pp. 236–49
211. Osborne, D. J. 1976. Control of cell shape and cell size by the dual regulation of auxin and ethylene. In *Perspectives in Experimental Biology,* ed. N. Sunderland, 2:89–102. Oxford: Pergamon
212. Osborne, D. J. 1977. Ethylene and target cells in the growth of plants. *Sci. Prog. Oxford* 64:51–63
213. Osborne, D. J. 1977. Auxin and ethylene and the control of cell growth. Identification of three classes of target cells. See Ref. 138, pp. 161–71
214. Owens, L. D., Lieberman, M., Kunishi, A. T. 1971. Inhibition of ethylene production by rhizobitoxine. *Plant Physiol.* 48:1–4
215. Pallaghy, C. K., Raschke, K. 1972. No stomatal response to ethylene. *Plant Physiol.* 49:275–76
216. Palmer, J. H., Halsall, D. M. 1969. Effect of transverse gravity stimulation, gibberellin and indoleacetic acid upon polar transport of IAA C-14 in the stem of *Helianthus annus. Physiol. Plant* 22:59–67
217. Pegg, G. F. 1976. The involvement of ethylene in plant pathogenesis. *Encycl. Plant Physiol.* (New Ser.) 4:582–91
218. Peiser, G. D., Yang, S. F. 1978. Ethylene and ethane production from SO_2 injured plants. *Plant Physiol. Suppl.* pp. 61–90
219. Phan, C. T. 1971. *L'Ethylene, Metabolisme et Activite Metabolique.* Paris: Masson et Cie. 130 pp.
220. Poapst, P. A., Durkee, A. B., McGugan, W. A., Johnston, F. B. 1968. Identification of ethylene in gibberellic-acid-treated potatoes. *J. Sci. Food Agric.* 19:325–27
221. Pratt, H. K., Goeschl, J. D. 1969. Physiological roles of ethylene in plants. *Ann. Rev. Plant Physiol.* 20:541–84
222. Primrose, S. B. 1976. Formation of ethylene by *Escherichia coli. J. Gen. Microbiol.* 95:159–65
223. Pruess, D. L., Scannell, J. P., Kellett, M., Ax, H. A., Janecek, J., Williams, T. H., Stempel, A., Berger, J. 1974. Antimetabolites produced by microorganisms X, L-2-amino-4-(2-aminoethoxy)-trans-3-butenoic acid. *J. Antibiot.* 27: 229–33
224. Pryor, W. A., Tang, R. H. 1978. Ethylene formation from methional. *Biochem. Biophys. Res. Commun.* 81:498–503
225. Racker, E. 1976. *A New Look at Mecha-*

nisms in Bioenergetics, p. 60. New York: Academic. 197 pp.
226. Rahiala, E. L., Kekomaki, M., Janne, J., Raina, A., Raiha, N. C. R. 1971. Inhibition of pyridoxal enzymes by L-canaline. *Biochim. Biophys. Acta* 227: 337–43
227. Rando, R. R. 1974. β γ-unsaturated amino acids as irreversible inhibitors. *Nature* 250:586–87
228. Rhodes, M. J. C. 1970. The climacteric and ripening in fruits. See Ref. 82, pp. 520–33
229. Riov, J. S., Monselise, S. P., Kahan, R. S. 1969. Ethylene controlled induction of phenylalanine ammonia-lyase in citrus fruit peel. *Plant Physiol.* 44:1371–77
230. Robert, M. L., Taylor, H. F., Wain, R. L. 1975. Ethylene production by cress roots and its inhibition by 3,5 diiodo-4-hydroxybenzoic acid. *Planta* 126:273–84
231. Rudich, J., Halevy, A. H., Kedar, N. 1972. Ethylene evolution from cucurbit plants as related to sex expression. *Plant Physiol.* 49:998–99
232. Russo, V. E. A., Halloran, B., Gallori, E. 1977. Ethylene is involved in the autochemotropism of *Phycomyces. Planta* 134:61–67
233. Sakai, S., Imaseki, H. 1971. Auxin-induced ethylene production by mung bean hypocotyl segments. *Plant Cell Physiol.* 12:349–59
234. Sakai, S., Imaseki, H. 1973. A proteinaceous inhibitor of ethylene biosynthesis by etiolated mung bean hypocotyl sections. *Planta* 113:115–28
235. Sakai, S., Imaseki, H. 1973. Properties of the proteinaceous inhibitors of ethylene synthesis: action on ethylene production and indoleacetyl aspartate formation. *Plant Cell Physiol.* 14:881–92
236. Saltveit, M. E., Dilley, D. R. 1978. Rapidly induced wound ethylene from excised segments of etiolated *Pisum sativum* L., cv. Alaska. I. Characterization of the response. *Plant Physiol.* 61: 447–50
237. Saltveit, M. E., Dilley, D. R. 1978. Rapidly induced wound ethylene from excised segments of etiolated *Pisum sativum* L., cv. Alaska II. Oxygen and temperature dependency. *Plant Physiol.* 61:675–79
238. Samimy, C. 1978. Effect of light on ethylene production and hypocotyl growth of soybean seedlings. *Plant Physiol.* 61:772–74
239. Sargent, J. A., Stack, A. V., Osborne, D. J. 1973. Orientation of cell growth in the etiolated pea stem. Effect of ethylene and auxin on cell wall deposition. *Planta* 109:185–92
240. Scannell, J. P., Pruess, D. L., Demny, T. C., Sello, L. H., Williams, T., Stempel, A. 1972. Antimetabolites produced by microorganisms V. L-2-amino-4-methoxy trans-3-butenoic acid. *J. Antibiot.* 25:122–27
241. Schonbaum, G. R., Bonner, W. D. Jr., Storey, B. T., Bahr, J. T. 1971 Specific inhibition of the cyanide-insensitive respiratory pathway in plant mitochondria by hydroxamic acids. *Plant Physiol.* 47:124–28
242. Scott, P. C., Leopold, A. C. 1967. Opposing effects of gibberellin and ethylene. *Plant Physiol.* 42:1021–22
243. Shimokawa, K., Kasai, Z. 1968. A possible incorporation of ethylene into RNA in Japanese morning glory seedlings. *Agric. Biol. Chem.* 32:680–82
244. Siebert, K. J., Clagett, C. O. 1969. Formic acid from carbon 2 of methionine in ethylene production in apple tissue. *Plant Physiol.* 44:(S)30
245. Sisler, E., Wylie, P. A. 1978. In vivo measurement of binding to the ethylene-binding site. *Plant Physiol. Suppl.* 61:91
246. Smith, A. M. 1976. Ethylene in soil biology. *Ann. Rev. Phytopathol.* 14: 53–73
247. Smith, A. M., Cook, R. J. 1974. Implications of ethylene production by bacteria for biological balance of soil. *Nature* 252:703–5
248. Smith, K. A., Robertson, P. D. 1971. Effect of C_2H_4 on root extension of cereals. *Nature* 234:148–49
249. Solomos, T., Biale, J. 1975. Respiration and fruit ripening. See Ref. 22, pp. 221–28
250. Solomos, T., Laties, G. G. 1973. Cellular organization and fruit ripening *Nature* 245:390–92
251. Solomos, T., Laties, G. G. 1974. Similarities between the actions of ethylene and cyanide in initiating the climacteric and ripening of avocados. *Plant Physiol.* 54:506–11
252. Solomos, T., Laties, G. G. 1975. The mechanism of ethylene and cyanide action in triggering the rise of respiration in potato tubers. *Plant Physiol.* 55: 73–78
253. Solomos, T., Laties, G. G. 1976. Effects of cyanide and ethylene on the respiration of cyanide-sensitive and cyanide-resistant plant tissues. *Plant Physiol.* 58:47–50
254. Spalding, D. H., Lieberman, M. 1965. Factors affecting the production of eth-

ylene by *Penicillium digitatum*. *Plant Physiol.* 40:645–48
255. Spencer, M. S. 1969. Ethylene in nature. In *Fortschr. Chem. Org. Naturst.* 27:31–80
256. Steen, D. A., Chadwick, A. V. 1973. Effects of cycloheximide on indoleacetic acid-induced ethylene production in pea root tips. *Plant Physiol.* 52:171–73
257. Stewart, R. N., Lieberman, M., Kunishi, A. T. 1974. Effects of ethylene and gibberellic acid on cellular growth and development in apical and subapical regions of etiolated pea seedlings. *Plant Physiol.* 54:1–5
258. Suge, H. 1971. Stimulation of oat and rice mesocotyl growth by ethylene. *Plant Cell Physiol.* 12:831–37
259. Suge, H. 1974. Synergistic action of ethylene with gibberellins in the growth of rice seedlings. *Proc. Crop. Sci. Soc. Jpn.* 43:83–87
260. Suge, H. 1976. Ethylene and carbon dioxide as factors regulating initial growth and development in several perennial aquatic weeds. *Proc. 5th Asian Pac. Weed Sci. Soc. Conf. Tokyo, Japan,* 1975, pp. 44–49. Published by Asian-Pacific Weed Sci. Soc.
261. Suge, H., Katsura, N., Inada, K. 1971. Ethylene-light relationships in the growth of the rice coleoptile. *Planta* 101:365–68
262. Suge, H., Kusanagi, T. 1975. Ethylene and carbon dioxide: regulation of growth in two perennial aquatic plants, arrowhead and pondweed. *Plant Cell Physiol.* 16:65–72
263. Sutherland, E. W. 1972. Studies on the mechanism of hormone action. *Science* 177:401–8
264. Takeo, T., Lieberman, M. 1969. 3-Methylthiopropionaldehyde peroxidase from apples: an ethylene-forming enzyme. *Biochim. Biophys. Acta* 178: 235–47
265. Tappel, A. L. 1965. Free radical lipid peroxidation damage and its inhibition by vitamin E and selenium. *Fed. Proc.* 24: (1) 73–78
266. Taylorson, R. B., Hendricks, S. B. 1977. Dormancy in seeds. *Ann. Rev. Plant Physiol.* 28:331–54
267. Thomas, K. C., Spencer, M. 1977. L-Methionine as an ethylene precursor in *Saccharomyces cerevisiae*. *Can. J. Microbiol.* 23:1669–74
268. Thompson, J. E., Spencer, M. S. 1966. Preparation and properties of and enzyme for ethylene production. *Nature* 210:595–97
269. Tigchelaar, E. C., McGlasson, W. B., Franklin, M. J. 1978. Natural and ethephon stimulated ripening of F_1 hybrids of the ripening inhibitor (rin) and non-ripening (nor) mutants of tomato (*Lycopersicon esculentum* Mill.) *Aust. J. Plant Physiol.* In press
270. Tingwa, P. O., Young, R. E. 1975. The effect of indole-3-acetic acid and other growth regulators on the ripening of avocado fruits. *Plant Physiol.* 55: 937–40
271. Tingwa, P. O., Young, R. E. 1975. Studies on the inhibition of ripening in attached avocado (*Persea Americana* Mill.) fruits. *J. Am. Soc. Hortic. Sci.* 100:447–49
272. Vaadia, Y. 1976. Plant hormones and water stress. *Philos. Trans. R. Soc. London Ser.* B 273:513–22
273. Valdovinos, J. G., Ernest, L. C., Henry, E. W. 1967. Effects of ethylene and gibberellic acid on auxin synthesis in plant tissues. *Plant Physiol.* 42:1803–6
274. Van der Laan, P. A. 1934. Der einfluss von Aethylen auf die Wachsstoffbildung bei *Avena* und *Vicia*. *Rec. Tran. Bot. Neerl.* 31:691–742
275. Veen, B. W. 1970. Orientation of microfibrils in parenchyma cells of pea stem before and after longitudinal growth. *K. Ned. Akad. Wet.* 73:113–17
276. Vendrell, M. 1969. Reversion of senescence: effects of 2,4-dichlorophenoxyacetic acid and indole acetic acid on respiration, ethylene production and ripening in banana fruit slices. *Aust. J. Biol. Sci.* 22:601–10
277. Vendrell, M. 1970. Acceleration and delay of ripening in banana fruit tissue by gibberellic acid. *Aust. J. Biol. Sci.* 23:553–59
278. Vendrell, M., McGlasson, W. B. 1971. Inhibition of ethylene production in banana fruit tissue by ethylene treatment. *Aust. J. Biol. Sci.* 24:885–95
279. Vitagliano, C., Hoad, G. V. 1978. Leaf stomatal resistance, ethylene evolution and ABA levels as influenced by (2-chloroethyl) phosphonic acid. *Sci. Hortic.* 8:101–6
280. Wade, N. L., Brady, C. J. 1971. Effects of kinetin on respiration, ethylene production, and ripening of banana fruit slices. *Aust. J. Biol. Sci.* 24:165–67
281. Wang, C. H., Persyn, A., Krackov, J. 1962. Role of the Krebs cycle in ethylene biosynthesis. *Nature* 195:1306–8
282. Wang, C. Y., Mellenthin, W. M. 1977. Effect of aminoethoxy analog of rhizobitoxine on ripening of pears. *Plant Physiol.* 59:546–49

283. Wang, C. Y., Mellenthin, W. M., Hansen, E. 1971. Effect of temperature on development of premature ripening of Bartlett pears. *J. Am. Soc. Hortic. Sci.* 96:122–26

284. Wilson, L. G., Bressan, R. A., Lecureux, L., Filner, P. 1978. Bisulfite-induced ethylene and ethane production dependence on light and photosynthetically competent tissue. *Plant Physiol. Suppl.* 61–93

285. Wright, M. 1974. The effect of chilling on ethylene production, membrane permeability, and water loss of leaves of *Phaseolus vulgaris. Planta* 120:63–69

286. Yang, S. F. 1967. Biosynthesis of ethylene, ethylene formation from methional by horse radish peroxidase. *Arch. Biochem. Biophys.* 122:481–87

287. Yang, S. F. 1969. Ethylene evolution from 2-chloroethylphosphonic acid. *Plant Physiol.* 1203–4

288. Yang, S. F. 1974. The biochemistry of ethylene: biogenesis and metabolism. *Recent Adv. Phytochem.* 7:131–64

289. Yang, S. F., Baur, A. H. 1972. Biosynthesis of ethylene in fruit tissues. See Ref. 166, pp. 510–17

290. Yang, S. F., Ku, H. S., Pratt, H. K. 1966. Photochemical production of ethylene from methionine and its analogues in the presence of flavin mononucleotide. *J. Biol. Chem.* 242:5274–80

291. Yang, S. F., Pratt, H. K. 1978. The physiology of ethylene in wounded plant tissue. In *Biochemistry of Wounded Plant Storage Tissues,* ed. G. Kahl. Berlin: de Gruyter. In press

292. Zimmerman, P. W., Hitchcock, A. E. 1933. Initiation and stimulation of adventitious roots caused by unsaturated hydrocarbon gases. *Contrib. Boyce Thompson Inst.* 5:351–59

293. Zimmerman, P. W., Wilcoxon, F. 1935. Several chemical growth substances which cause initiation of roots and other responses in plants. *Contrib. Boyce Thompson Inst.* 7:209–29

294. Zimmerman, R. H., Lieberman, M., Broome, O. C. 1977. Inhibitory effect of a rhizobitoxine analog on bud growth after release from dormancy. *Plant Physiol.* 59:158–60

295. Zobel, R. W. 1973. Some physiological characteristics of the ethylene requiring tomato mutant diageotropica. *Plant Physiol.* 52:385–89

296. Zobel, R. W. 1978. Effects of low concentration of ethylene on cell division and cytodifferentiation in lettuce pith explants. *Can. J. Bot.* 56:987–90

297. Zobel, R. W., Roberts, L. W. 1974. Control of morphogenesis in the ethylene requiring mutant diageotropica. *Can. J. Bot.* 32:735–41

Ann. Rev. Plant Physiol. 1979. 30:593–620

THE STRUCTURE OF CHLOROPLAST DNA

❖7681

John R. Bedbrook

Plant Breeding Institute, Trumpington, Cambridge CB2 2LQ, England

Richard Kolodner

Sidney Farber Cancer Institute and Department of Biological Chemistry, Harvard Medical School, Boston, Massachusetts 02115

CONTENTS

0066-4294/79/0601-0593$01.00

INTRODUCTION

The early observation of non-Mendelian inherited mutant phenotypes affecting chloroplasts in higher plants (3, 33) suggested the existence of genetic material in chloroplasts. Extensive research efforts over the years since have demonstrated a unique DNA species in the chloroplasts of higher plants and algae. Genetic experiments have made possible the isolation of chloroplast mutants, provided evidence that ctDNA codes for several chloroplast proteins, and have culminated in the construction of a chloroplast genetic map in *Chlamydomonas reinhardii.* These results have all been discussed recently in several reviews (43, 50, 81, 91, 129, 130, 132, 150) and books (59, 82, 128). In the last few years a considerable body of information on the physical structure of chloroplast DNA (ctDNA) has accumulated, and this information has stimulated workers in the field to apply newly developed techniques for gene mapping to the study of the genetic structure of the chloroplast genome. The purpose of this review is to discuss these most recent attempts to elucidate the physical and genetic structure of the chloroplast genome.

PHYSICAL STUDIES OF THE STRUCTURE OF ctDNA

Kinetic Complexity of ctDNA

The successful use of reassociation kinetics (20, 165) to study the complexity and organization of the nuclear DNA from eucaryotes is well known. A number of workers have used reassociation kinetics to study the structure of ctDNA from higher plants (83, 85, 109, 154, 163) and algae (1, 4, 73, 114, 136, 145, 148, 164). Optical methods have been mainly used to follow the renaturation, although some experiments have used chromatography on hydroxylapatite to distinguish between single-stranded and double-stranded DNA. With the exception of the ctDNA from *Acetabularia* (discussed below), it seems that all of the studied examples of ctDNA contain one major component which has a complexity of about 1×10^8 daltons, provided that the data of different workers are corrected to take into account the different standard DNAs that were used (85).

A rapidly renaturing component of ctDNA, having a complexity of $1\text{–}10 \times 10^6$ daltons and consisting of less than 10% of the ctDNA, has been reported in the ctDNA from *Chlamydomonas reinhardii* (164), *Chlorella* (4), *Polytoma obtusum* (136), and *Lactuca sativa* (163). Other workers who have failed to detect these rapidly renaturing sequences in some of these and other ctDNAs (1, 73, 83, 85, 145, 154) suggest that these rapidly renaturing sequences reflect incomplete denaturation of the ctDNA during the renaturation kinetics experiments (154). Rochaix (122) has shown, using the ring formation methods of Thomas and collaborators (155), that at least 20% of *C. reinhardii* ctDNA is located adjacent to interspersed reiterated sequences. However, ideally one would want to purify the rings that were formed in these experiments and demonstrate that they contain ctDNA and did not arise from contaminating nuclear DNA. Cross-hybridization of different restriction endonuclease fragments of ctDNA suggests that some reiterated sequences may exist in *Zea mays* and *C. reinhardii* ctDNA (6, 124). Regardless of the final resolution of this issue, it seems probable that only a small proportion of the ctDNA consists of highly reiterated sequences. Spontaneously renaturing sequences, which constitute another class of reiterated sequence, are discussed in a different section of this article.

Studies on the structure of *Acetabularia* ctDNA have demonstrated a kinetic complexity of 1.5×10^9 daltons for *A. cliftonii* ctDNA and 1.1×10^9 daltons for *A. mediterranea* ctDNA (56, 114). While these results are different from those obtained with other ctDNAs, they are difficult to explain on the grounds of contamination of the *Acetabularia* ctDNA with nuclear DNA or bacterial DNA. It is possible that the 1×10^8 dalton circular ctDNA has evolved by reduction of a larger form of ctDNA such as that found in *Acetabularia.* Alternatively, the 1×10^9 dalton ctDNA of *Acetabularia* may correspond to a component of high complexity which has not yet been found in higher plants or other algae, but which has been postulated to exist to explain certain genetic data (129, 130).

Circularity of ctDNA

The initial observation of intact ctDNA molecules was made by Manning et al (103), who used electron microscopy to detect circular ctDNA molecules having a contour length of 40 μm in lysates of *Euglena gracilis* chloroplasts. Large circular ctDNA molecules have since been observed in several species of higher plants (68, 83, 85, 102): the alga *C. reinhardii* (10) and liverwort *Sphaerocarpos castellanii* (10). In all cases the ctDNA molecules appeared to have a homogeneous contour length which differs in length from species to species (discussed below). The chloroplast preparations used in most experiments were extensively purified by floatation through sucrose solutions (103) or by incubation with deoxyribonuclease (68, 83, 85) to remove contaminating nuclear DNA, and they were sufficiently pure to contain only DNA having the buoyant density and renaturation behavior characteristic of ctDNA (81, 150, 154). In one case (89) purified circular ctDNA molecules were shown to have the same buoyant density as ctDNA. Furthermore, these large circular ctDNA molecules were found to constitute a large proportion of the total ctDNA. In *Pisum sativum* as much as 90% of the total ctDNA molecules could be demonstrated to be in the form of circular DNA molecules (83, 85). Relatively high proportions of large circular ctDNA molecules were observed in several other species of higher plants: 80% in *Antirrhinum majus* (68) and 70% in *Spinacia oleracea, Lactuca sativa, Zea mays,* and *Avena sativa* (85). Lower proportions of circular ctDNA molecules (10–50%) were observed in *Beta vulgaris* and *Oenothera hookeri* (68), *S. castellanii* and *C. reinhardii* (10), and in *E. gracilis* (100, 103), although these lower proportions may only reflect technical problems of DNA isolation. These data suggest that the large circular ctDNA molecules constitute the major, if not the only, component of ctDNA. This appreciation of the circularity of ctDNA and the observation of supertwisted forms (83, 100) led to the development of isolation procedures utilizing cesium chloride-ethidium bromide density gradients (85, 89), which made possible the isolation of large amounts of pure circular ctDNA.

An exception to these findings is that of *Acetabularia,* in which linear ctDNA molecules having lengths of up to 200 μm were observed (54, 55). The possibility that this organism has a larger chloroplast genome makes further experiments interesting.

Small circular DNA molecules having a length of 4.2 μm have been found associated with chloroplasts of *A. cliftonii* (54, 56). Similar circular DNA molecules were observed in *E. gracilis* (111), although there is no evidence for the association of these DNA molecules with the chloroplasts. There is a similar association between small heterogeneous circular DNA molecules and the mitochondria in some animal cells (137, 138); however, these were shown to be derived from nuclear DNA (137, 139). Alternately, these small circular DNA molecules may represent a "plasmid" DNA similar to the 2 μm plasmid DNA in yeast (60, 61, 94).

Size and Homogeneity of ctDNA

Electron microscopy is a highly accurate and reproducible method for determining the size of DNA molecules, provided standard DNA molecules are included in each experiment to correct for length variations that occur when DNA is mounted for electron microscopy under different conditions (35, 83). Since many of the experiments described above have used comparable internal length standards, a meaningful comparison of the different ctDNAs can be made and these data are presented in Table 1. It is clear that the sizes of the different ctDNAs vary considerably and range from a molecular weight of 85.2×10^6 for *S. castellanii* ctDNA to a molecular weight of 143×10^6 for *C. reinhardii* ctDNA. The significance of these size differences is unclear, and the size differences do not appear to parallel known evolutionary relationships (12). The size differences could imply that some parts of the ctDNA are not strictly required for chloroplast function.

Table 1 Molecular weights of ctDNA[a]

Organism	Ratio and standard	Molecular weight ($\times 10^{-6}$)	References
Sphaerocarpos castellanii	0.84 × *S. oleracea* ctDNA (±0.05)	85.2[b]	109
Zea mays	25.6 × ϕX174RF DNA (±0.2)	91.1	85
Avena sativa	25.9 × ϕX174RF DNA (±0.2)	92.2	85
Pisum sativum	26.7 × ϕX174RF DNA (±0.2)	95.0	83, 85
Euglena gracilis	27.3 × ϕX174RF DNA[c] (±0.2)	97.1	100
Beta vulgaris	0.98 × *S. oleracea* ctDNA (±0.07)	99.6[b, d]	68
Spinacia oleracea	28.5 × ϕX174RF DNA (±0.2)	101.4	85
Antirrhinum majus	1.0 × *S. oleracea* ctDNA (±0.09)	101.4[b, d]	68
Lactuca sativa	29 × ϕX174RF DNA (± 0.2)	103.2	85
Chlamydomonas reinhardii	1.41 × *S. oleracea* ctDNA (±0.11)	143.0[b]	10

[a] All molecular weights are ultimately standardized with respect to the molecular weight of ϕX174RF DNA which was determined by sequence analysis to be 3.558×10^6 (5375 base pairs × 662 daltons per average base pair).

[b] Calculated assuming a molecular weight of 101.4×10^6 for *S. oleracea* ctDNA (this table).

[c] Calculated in (85).

[d] These values are potentially the least accurate of the table since it is not clear if the contour lengths (as μm) reported by Herrmann et al (68) were all normalized with respect to the internal standard they used or if they were reported as raw values. For the purpose of this table we have assumed that the values were normalized.

The close agreement between the size of the circular ctDNA molecules as determined by electron microscopy and the complexity of the chloroplast genome determined by renaturation kinetics suggests that all of the circular ctDNA molecules in a given plant or alga have the same sequence. This suggestion was initially tested by denaturation mapping studies (75) on *P. sativum* ctDNA (84), which demonstrated that all of the *P. sativum* ctDNA molecules had the same denaturation pattern and therefore had an identical base sequence. Similar analyses have been carried out with the ctDNAs of *S. oleracea, L. sativa,* and *Z. mays* (88; R. Kolodner, unpublished results). Recent studies on the structure of the ctDNAs from *Z. mays* (6), *P. sativum* (26), *S. oleracea* (70, 167), *E. gracilis* (51), and *C. reinhardii* (124) have also provided convincing evidence for the homogeneity of the circular ctDNA molecules. In these experiments a ctDNA was digested with a restriction endonuclease, and a series of discrete DNA fragments were obtained whose molecular weights together approximate the size of the circular ctDNA molecules. Furthermore, mapping the restriction endonuclease fragments always yielded a circular map which was, within experimental error, the same size as that observed for the circular ctDNA molecules by electron microscopy. These methods are probably not capable of detecting a species of ctDNA consisting of less than 1–2% of the total ctDNA. Furthermore, any species of ctDNA that was removed during the purification of the ctDNA would be overlooked. However, most of the restriction endonuclease mapping studies have utilized either total ctDNA or ctDNA that has been fractionated to exclude only those DNA fragments having molecular weights of less than 2–4 $\times 10^7$ and have not used purified circular ctDNA molecules. Thus, the data from restriction mapping studies provide further evidence that all of the ctDNA molecules are identical and constitute the major species of ctDNA.

Physical Properties of Circular ctDNA Molecules

The buoyant densities of many different ctDNAs have been the subject of investigations by many workers and will not be reviewed here (see references for reviews). The covalently closed circular ctDNAs from *P. sativum, S. oleracea,* and *L. sativa* have been further characterized by analytical ultracentrifugation (89). These studies have shown that covalently closed ctDNA molecules share many properties with other covalently closed circular DNA molecules: each ctDNA was found to have a characteristic sedimentation coefficient in neutral cesium-chloride; when covalently closed circular ctDNA was treated with γ – irradiation to introduce single-strand breaks into the molecules, their sedimentation coefficients in neutral cesium-chloride decreased to a characteristic value; and when covalently closed circular ctDNA molecules were sedimented in alkaline cesium-chloride, their sedimentation coefficients increased dramatically. These properties are typical of covalently closed circular DNA molecules in which the two complementary single strands are topologically interwound (159). Sedimentation of these ctDNAs in the presence of different amounts of ethidium bromide (57) made it possible to calculate their superhelix densities which ranged from -8.7 to -8.9×10^{-2} superhelical turns per 10 base pairs and are intermediate among values that have been observed for other naturally occurring covalently closed circular DNAs (2, 57, 89). The sedimentation studies also confirmed the different sizes of these three ctDNAs that were detected by electron microscopy (85). The molecular weight of *P. sativum* ctDNA was determined by equilibrium sedimentation (133) and was found to be 89×10^6 (89), which is in reasonable agreement with the value of 95×10^6 determined by electron microscopy (see Table 1).

Ribonucleotides in ctDNA

Perhaps the most fascinating observation made during the physical studies on the structure of the covalently closed circular ctDNA molecules was the alkali lability of ctDNA (90). When the covalently closed circular ctDNA molecules were sedimented in alkaline cesium-chloride, they were gradually converted from the rapidly sedimenting closed circular form [240–264 S, depending on the ctDNA (90)] to more slowly sedimenting nicked and denatured single strands of ctDNA [55–58 S, depending on the ctDNA (89)]. Incubation with a mixture of pancreatic RNase and RNase TI also converted covalently closed circular ctDNA molecules to open circular molecules, indicating that the alkali labile sites in ctDNA could be due to the presence of covalently inserted ribonucleotides in the ctDNA. Kinetic studies on the rate of nicking in the presence of NaOH suggested that *P. sativum* and *S. oleracea* ctDNA each contained 18 ± 2 ribonucleotides, while *L. sativa* ctDNA contained 12 ± 2 ribonucleotides. Mapping of the positions of the alkali labile sites in *P. sativum* ctDNA by electron microscopy indicated that they existed at 19 unique sites in the ctDNA. The results were also consistent with the idea of each alkali labile site being caused by the presence of one ribonucleotide in the ctDNA. These findings have not been extended, and there is little data available to allow one to speculate about the significance of the ribonucleotides in ctDNA. It is interesting to note that in animal mitochondrial DNA, which also contains ribonucleotides (58, 110, 171), at least some of the ribonucleotides have been demonstrated to be at unique sites (16). At least one of these sites is near the origin of mitochondrial DNA replication, which implies that the ribonucleotides may have something to do with DNA replication.

Inverted Repeats in ctDNA

Studies in which nicked circular ctDNA molecules were denatured and then examined in the electron microscope for the presence of spontaneously renaturing ctDNA (88), together with restriction endonuclease mapping studies (6, 70, 124, 167), have demonstrated that the ctDNAs from *S. oleracea, L. sativa, Z. mays,* and *C. reinhardii* all contain a large DNA sequence repeated one time in an inverted orientation. The sizes of the inverted sequences and the segments between the inverted sequences (called spacer regions) of the ctDNAs from *S. oleracea, L. sativa,* and *Z. mays* were determined by electron microscopy and these data are presented in Table 2. The inverted sequences are all similar in size, and denaturation mapping

Table 2 Lengths of inverted sequences and spacer regions[a]

	Small spacer (Base pairs)	Inverted sequence (Base pairs)	Large spacer (Base pairs)
Spinacia oleracea	18,500 ± 800	24,500 ± 500	86,000 ± 4,500
Lactuca sativa	19,500 ± 800	24,500 ± 500	87,000 ± 2,500
Zea mays	12,600 ± 1,000	22,500 ± 400	78,500 ± 3,300

[a] Data from (90), assuming that ϕX174RF DNA is 5375 base pairs long.

studies (88) showed that they are similar in sequence with one notable difference—the *Z. mays* inverted sequences are shorter by 2000 nucleotide pairs in the region of the small spacer than in *S. oleracea* and *L. sativa* inverted sequences. Both spacer regions are also shorter in *Z. mays* ctDNA. The size of the inverted repeat in *C. reinhardii* ctDNA is about 19,000 base pairs long (124).

The inverted repeat in *Z. mays* (8), *S. oleracea* (149, 167) and *C. reinhardii* (124) codes for the chloroplast rRNA genes (discussed below), but this leaves about 80% of the sequence of the inverted repeat unaccounted for. The function of the inverted orientation is not understood, and we have proposed (8, 88) that it might improve the stability of a repeated sequence as well as help maintain the similarity of the two copies of the inverted repeat by promoting continual recombination and heteroduplex repair events (157) between the two copies. Intramolecular recombination between the two inverted sequences might also lead to "flip-flop" heterogeneity in the ctDNA, as is observed in the yeast 2 μm plasmid DNA (60). In any event, the inverted repeat is not a required structure for ctDNA because *P. sativum* and *E. gracilis* ctDNA lack this inverted repeat (51, 52, 88).

A second type of inverted sequence has been observed in *C. reinhardii* ctDNA (46). These sequences were isolated by denaturing sheared *C. reinhardii* ctDNA and isolating the spontaneously renaturing material by chromatography on hydroxylapatite. The isolated material consisted of 150 base pair long duplex fragments which made up less than 5% of the total ctDNA. The significance of these inverted sequences is not understood, but such sequences have been observed in DNA from many sources (134, 169).

Circular Oligomers of ctDNA

Circular oligomers appear to make up a small fraction of the total circular DNA isolated from virtually any source. Both circular dimers and catenated dimers have been observed in the ctDNAs from *P. sativum, S. oleracea, L. sativa, Z. mays,* and *A. sativa,* with circular dimers making up as much as 10% of the ctDNA molecules and catenated dimers making up as much as 2.5% of the ctDNA molecules (68, 83, 85). Larger oligomers may exist but would be hard to isolate due to their large size.

The structure of the circular dimers of ctDNA has been investigated by denaturation mapping (84) and by observing the pattern of inverted repeats in circular dimers (88). In *P. sativum* (84, 88), all of the circular dimers were found to consist of two monomer units linked in a "head to tail" arrangement as is the case with most circular dimers (29, 80), while in *S. oleracea* and *L. sativa,* 80% of the circular dimers were in a "head to head" configuration and 20% were in a "head to tail" configuration (88), which is similar to the observations made about circular dimers of the yeast 2 μm circular DNA (60, 94). These results suggest that circular dimers of ctDNA are formed by recombination between two monomers—a "head to head" dimer would be formed by two monomers recombining at a site in their inverted sequences, leading to the insertion of one monomer into the other in an inverted orientation. By inference, all circular monomers of ctDNA probably recombine with each other at some as yet unmeasured frequency. Studies on the formation of circular dimers in bacteria have provided considerable evidence that circular dimers are formed by recombination (5, 11, 118, 157), supporting the idea that circular dimers of ctDNA are also formed by recombination.

Replication of ctDNA

The earliest experiments on the replication of ctDNA were carried out by Chiang & Sueoka (25), who used density transfer experiments (107) to study the replication of *C. reinhardii* ct- and nuclear DNA. They showed that both the ct- and nuclear DNA were replicated at a distinct and different time during the cell cycle. *E. gracilis* ctDNA has also been shown to replicate semiconservatively (101, 119) and at a distinct period of the cell cycle which is also the period of nuclear replication (119). An intriguing fact about *E. gracilis* ctDNA replication is the finding that even though the ratio of ctDNA to nuclear DNA always is constant, every time the nuclear DNA replicates once the ctDNA replicates 1.5 times, suggesting that some ctDNA is degraded or that D-loops (see below) in the ctDNA may be turned over and resynthesized. Some of the factors affecting the timing of DNA replication during the cell cycle have also been investigated, and it has been observed that inhibitors of chloroplast protein synthesis inhibit nuclear DNA replication in *C. reinhardii* (15), suggesting that a chloroplast protein may be involved in the control of nuclear DNA replication. Cell cycle mutants of *C. reinhardii* (72) may also help to identify factors that control DNA synthesis (15, 72). These results must be interpreted carefully because any factor (mutant or inhibitor) that blocks the cell cycle prior to the replication of a given DNA species will block the DNA's replication but possibly only indirectly.

The replicative intermediates of *P. sativum, Z. mays,* and *E. gracilis* ctDNA have been studied by electron microscopy (86, 87, 119). Replication of *P. sativum* and *Z. mays* ctDNA appeared to initiate by the formation of two D-loops (78) which were located on opposite strands of the ctDNA at unique sites. The D-loops appeared to elongate toward each other and fused to form a "Cairns" replicative intermediate (22) which continued DNA synthesis bidirectionally to complete the round of ctDNA replication. D-loops were not observed in *E. gracilis* ctDNA (119), although it might not have been possible to distinguish between D-loops and the denaturation bubbles that were found in the covalently closed circular *E. gracilis* ctDNA (119). Such denaturation bubbles have been observed in other naturally occurring negatively supertwisted circular DNA molecules, including *P. sativum* and *Z. mays* ctDNA (86). *E. gracilis* ctDNA was also observed to replicate by "Cairns" replicative intermediates (119) as was the case for higher plant ctDNAs (87). Rolling circle replicative intermediates of higher plant ctDNA (48), which appear to result from a continuation of the Cairns round of replication were also observed (87). These rolling circle intermediates did not appear to be broken Cairns intermediates since they started at a unique site on the ctDNA and some of the tails were longer than the circular ctDNA molecules they were attached to. The significance of the rolling circle replication is unclear, but it could relate to the over-replication of *E. gracilis* ctDNA (101) since a rolling circle intermediate can lead to synthesis of many copies of a DNA molecule even though replication had only initiated once. Rolling circle replication may also have some important genetic consequences relating to the genetic diploidy of *C. reinhardii* chloroplasts (130), which is observed even though *C. reinhardii* should contain 20–60 circular ctDNA molecules per chloroplast (10, 130).

A number of workers have demonstrated ctDNA synthesis in isolated chloroplasts, indicating the existence of a chloroplast DNA polymerase (47, 69, 135, 141, 152). These studies have encouraged several attempts to purify and characterize chloroplast DNA polymerases (79; R. McKown and K. K. Tewari, unpublished data). However, to date the proteins that replicate ctDNA remain largely unexplored and may provide a fruitful area for future research.

Some Conclusions

Thus far we have presented the results of experiments that have attempted to define the physical structure of the chloroplast genome and to explore some of its properties. We feel that the available data support the view that the entire chloroplast genome is represented by the sequence of one circular ctDNA molecule. Since a chloroplast contains more DNA than is represented by one ctDNA molecule (10, 130), chloroplasts must be polyploid and could contain as many as 20–60 copies of ctDNA per chloroplast. According to this model, it should be possible to locate the genes coding for various chloroplast components on these circular ctDNA molecules. Another critical test of this model will involve the demonstration that genes coding for chloroplast proteins affected by genetically mapped chloroplast mutations in *C. reinhardii* (21, 50, 64, 108, 112, 129, 130) are also located on the circular ctDNA molecules.

In the next two sections of this review we discuss biophysical and biochemical methods for determining and mapping the products encoded by chloroplast DNA. In the absence of a system for rapid genetic analysis, we believe that physical mapping techniques provide the best means of defining the genetic structure and role of ctDNA and of determining the relationship between the physical order of DNA sequences and their regulation and control in development.

PHYSICAL MAPPING OF THE RECOGNITION SEQUENCES FOR RESTRICTION ENDONUCLEASES ON ctDNA

The discovery and use of class II restriction endonucleases has revolutionized sequence analysis of DNA (for review see 121). Recognition sequences for these restriction endonucleases provide ideal markers for the construction of physical maps of genomes. CtDNA is well within the size range of DNA molecules directly amenable to restriction enzyme mapping techniques.

Requirements for Physical Mapping of ctDNA

Since restriction enzyme mapping is a prerequisite to the complete physical analysis of the coding properties of ctDNA, we consider here some of the technical aspects of the subject. Along with the obvious requirement for purity of the ctDNA, it is of importance that the DNA used for restriction enzyme mapping be of high molecular weight. The DNA need not be full length but should be significantly larger than the largest fragments produced by the restriction enzymes used in the mapping. As the starting DNA size approaches the size of the largest DNA fragments produced by the restriction enzyme, the apparent stoichiometry of the larger fragments will decrease relative to smaller fragments. Methods for the preparation of intact supercoiled ctDNA from higher plants have been described (85, 89). In some species, for example *Zea mays,* the yield of supercoiled DNA was found to be low and represented only a small fraction of the total ctDNA present in lysates of DNAase treated chloroplasts. A useful method for preparing ctDNA is to prepare chloroplast lysates, extract with phenol, and precipitate the extract with ethanol as described (89), sediment the nucleic acids through 5 to 20 percent (w/v) sucrose gradients, and collect the fraction of DNA that is larger than linear lambda phage DNA. This nucleic acid fraction was found to be free of RNA and contained greater than 90% by weight ctDNA (J. Bedbrook, unpublished observations). Such DNA has proved to be highly suitable for restriction endonuclease mapping work and for DNA cloning experiments.

Several considerations should be borne in mind when restriction endonucleases are chosen for physical mapping of ctDNA. Initially the resolution of the map has proved to be less important than its completeness. These enzymes which cut the DNA infrequently provide the simplest tools for obtaining a map covering the entire genome. Those enzymes which recognize hexanucleotide sequences are most likely to cut the DNA into a workable number of fragments. For example, the enzymes Sal GI and Sma I have proved useful in mapping the higher plant chloroplast genomes of *Z. mays* (6), *S. oleracea* (70, 167) and *P. sativum* (26), while Sal GI, Bam HI, Pst I, and Xho I have been used in mapping ctDNA of *E. gracilis* (51, 52). Importantly, these enzymes cut the DNA in such a way that the entire genome can be displayed as fragments on a single agarose gel. For more detailed mapping, which is important for defining the exact location of a gene, combinations of hexanucleotide and tetranucleotide recognition sequence enzymes are required. Another prerequisite for restriction endonuclease mapping is the determination of the relative stoichiometry and complexity of the DNA fragments produced by various restriction endonucleases. Stoichiometry of DNA fragments in agarose gels that have been stained with ethidium bromide can be determined either by quantitating the fluorescence directly with a scanning fluorimeter, or by scanning photographic negatives of gels with a densitometer. Those gel bands containing the fluorescence equivalent of more than one copy per genome must then be further analyzed to determine their exact complexity. This is most simply determined by extracting the DNA fragment from the gel, digesting it with other enzymes, and fractionating the products by electrophoresis. If the sum of the molecular weights of the fragments produced by the second digestion is equal to the molecular weight of the starting fragment, then the starting fragment can be considered to be a repeated sequence in the genome. If the sum of the molecular weights of fragments produced by the second digestion is an integral multiple of the molecular weight of the starting fragment, then the extracted DNA band is likely to have consisted of two or more different sequences of equal size.

Methods of Physical Mapping

Many rapid methods for mapping restriction endonuclease fragments are available. CtDNA is sufficiently large and complex that the method of approach is important. It is possible to map recognition sites for restriction endonucleases on DNA by measuring the size of the DNA fragments produced by digestion with combinations of enzymes which when used alone would fragment the DNA into one or a small number of pieces. This method has been used successfully to map the ctDNA of *E. gracilis* (51, 52). This was possible because *Euglena* ctDNA contains only a few cleavage sites for Sal GI, Pst I, Bam HI, and Xho I. This type of approach is more complex and therefore less useful for higher plant ctDNA since these enzymes, with the exception of Sal I (6, 26, 70, 167), generally have more than 20 cleavage sites in higher plant ctDNA.

In work with *Zea mays* ctDNA, we found that for the initial work the simplest method for the construction of a map of the entire genome was to search directly for DNA fragments which overlapped the enzyme recognition sites we were interested in mapping. Thus, to map the Sal I sites on *Z. mays* ctDNA we determined by double digestions which Eco RI fragments or Bam HI fragments contained cleavage sites for Sal I. The Eco RI and Bam HI fragments containing sites for Sal I were prepared by electrophoresis in agarose gels. ^{32}P-labeled cRNA was made using these DNA fragments as template for *E. coli* RNA polymerase, and then the ^{32}P-labeled cRNA was hybridized to the DNA fragments produced by Sal I, using the method described by Southern (140). The hybridization results show which Sal I fragments are overlapped by the various Bam HI and Eco RI fragments, and therefore which Sal I fragments are adjacent.

A more rapid variation of this method has been described by Hutchison (74). In this method all the DNA fragments produced by one restriction endonuclease are simultaneously hybridized to all the fragments produced by another restriction endonuclease. The DNA fragments produced by digestion with a restriction endonuclease are fractionated on an agarose gel in such a way that the entire width of the gel is covered by the loaded sample, and after denaturation the DNA fragments are transferred to a nitrocellulose filter (140). DNA fragments produced by digestion of radioactively labeled DNA with a second enzyme are fractionated on an agarose gel in a similar fashion and then transferred to the same nitrocellulose filter at right angles to the first set of bands. The second transfer is performed under conditions which enable the radioactively labeled DNA fragments to hybridize to the DNA that had previously been transferred to the filter. Autoradiography reveals which DNA fragments from the first digest are complementary to those in the second digest. Thus, analysis of the hybridization pattern enables the regions of overlaps between the two sets of fragments to be determined. From this information the fragment order can be determined.

Detailed mapping of fractions of ctDNA can be performed by applying any of the above methods or by the analysis of partial digestion products produced with restriction endonucleases. Partial digestion products can be produced by digestion of DNA by a limited amount of enzyme or by carrying out the digestion in the presence of agents that can limit the action of a restriction endonuclease (116). The partial digestion products can then be purified by electrophoresis and then analyzed by digesting them to completion with the same restriction endonuclease and determining which DNA fragments a given partial product contains (34). Alternately, if a DNA fragment can be labeled with ^{32}P at only one end in a reaction with T4 polynucleotide kinase (120) and ^{32}P-ATP followed by digestion with an appropriate restriction endonuclease, then partial digestion products can be analyzed by the method of Smith & Birnsteil (139).

Clearly these are not the only methods that will be useful for mapping the positions of restriction endonuclease cleavage sites in ctDNA, and more methods will surely be developed. We have only tried to illustrate a few of the methods that have proved useful thus far.

Maps for the Recognition Sites of Restriction Endonucleases in ctDNA

To date, physical maps of ctDNA from the algae *E. gracilis* (51) and *C. reinhardii* (123, 124) and the higher plants *Z. mays* (6), *S. oleracea* (70, 144, 167), and *P. sativum* (26) have been determined using recognition sites for restriction endonucleases as physical markers. Importantly, this work has confirmed the view, already established by denaturation mapping analysis in the electron microscope (84), that the bulk of the circular ctDNA molecules in the cell are of identical sequence. The stoichoimetry and mapping properties of *Z. mays* and *C. reinhardii* ctDNA suggest that there may be a small amount of heterogeneity in ctDNA (6, 123, 124). It seems, however, that this apparent heterogeneity is unlikely to be related to the fact that *Z. mays* is a C_4 plant and contains two morphologically distinct plastid types (13), because in *Panicum* and *Z. mays*—both C_4 plants—the genomes of the two plastic types are identical (93, 161). Several other interesting features of these maps, such as the presence of a large sequence repeated in inverted orientation and the possible detection of a small number of dispersed repeats in the ctDNA, have been discussed above. In the next section we will consider the use of these maps for locating the positions of various chloroplast genes.

PHYSICAL MAPPING OF GENES IN ctDNA

The major incentive for the construction of physical maps of ctDNA is to provide a way of determining the nature and order of genes in the chloroplast genome. In this section we discuss the results of attempts to locate several genes on the ctDNA. We also discuss the experimental systems available for future research in this important area.

Genes for Chloroplast rRNAs

Lyttleton (97) first showed that chloroplasts contain a class of ribosomes distinct from those found in the plant cell cytoplasm. It was later demonstrated (27, 28) in *Brassica* that the cytoplasmic ribosomes were 86S and the chloroplast ribosomes were 68S. The presence of a smaller class of ribosomes in chloroplasts has proved to be a general feature throughout the plant kingdom. That chloroplast ribosomes contain RNA distinguishable from cytoplasmic ribosomal RNA was first demonstrated by Stutz & Noll (147). Loening & Ingle (95) showed that the major RNAs from chloroplast ribosomes were 23S and 16S. Chloroplast ribosomes also contain low molecular weight RNAs, a 5S RNA species (37, 38, 117), and the recently discovered 4.5S RNA (17, 37, 168). The 23S, 5S, and 4.5S RNAs are components of the 50S chloroplast ribosome subunit, and 16S is a component of the 30S subunit. RNA-DNA hybridization studies (153) demonstrated that ctDNA contains sequences complementary to the ribosomal RNAs. Thomas & Tewari (156), using quantitative RNA-DNA hybridization, showed that each ctDNA molecule from *P. sativum, S. oleracea, L. sativa, A. sativa, Phaseolus vulgaris,* and *Z. mays* contained two copies of sequences complementary to both the 23S and 16S RNAs. RNA-DNA hybridization of chloroplast ribosomal RNA to ctDNA fragments produced by digestion with various restriction endonucleases and electron microscopy of ribosomal RNA-DNA hybrids have been used to determine detailed physical maps of the chloroplast ribosomal RNA genes of *Z. mays* (6, 8), *S. oleracea* (70, 166, 167), *C. reinhardii* (126), *E. gracilis* (52), and *P. sativum* (26). These maps confirm the quantitative hybridization data showing two copies of the sequences for the 23 and 16S RNA molecules in *Z. mays, S. oleracea,* and *P. sativum* ctDNA, and show that *E. gracilis* ctDNA contains three copies of these genes in a tandem array. The physical mapping data also demonstrate that the ribosomal RNA genes are part of the inverted repeat sequence found in *Z. mays* (6, 8), *S. oleracea* (166), and *C. reinhardii* (126). The presence of the rRNA genes on the inverted repeat in *Z. mays* and *S. oleracea* ctDNA was confirmed by electron microscopy (8, 149). Denatured ctDNA was annealed to rRNA under conditions where RNA-DNA hybrids are more stable than DNA-DNA hybrids. The temperature of the hybridization reaction was then lowered to allow intramolecular renaturation of the inverted repeat sequences (8, 88, 149). Electron microscopy of these molecules showed that the rRNA hybridization resulted in the formation of displacement loops in the duplex region formed by the intramolecular hybridization of the inverted DNA sequences. The existence of such molecules confirms the presence of the ribosomal RNA genes in the inverted repeat. The inverted repeat arrangement for the rRNA genes is not strictly required in higher plants, as *P. sativum* ctDNA lacks this inverted repeat and has its two copies of the rRNA genes repeated in tandem, as determined by restriction endonuclease mapping and by electron microscopy (26, 88).

An inverted arrangement for ribosomal RNA genes has been found in other systems such as the extrachromosomal ribosomal RNA genes of *Tetrahymena pyriformis* (44, 77) and in *Physarum polycephalum* (160) and in the chromosomes of *Dictyostelium discoideum* (30). The frequent occurrence of an inverted arrangement for these genes suggests that this conformation may have some advantage over a tandem arrangement. We (8, 88) have proposed two possible advantages of an inverted arrangement for rDNA repeats, first as the basis of a mechanistically simple means of maintaining homogeneity and second in the interest of genetic stability.

The physical maps of chloroplast rRNA genes have shown that the order of the rRNA genes is 16S, 23S, and 5S RNAs, and is the same as that found in *E. coli* (96). The 16S and 23S genes are separated by a 2100 base pair spacer in *Z. mays* (8), by a 2200 base pair spacer in *S. oleracea* (149, 167), and by a 1680 base pair spacer in *C. reinhardii* (126). In *C. reinhardii* the DNA sequences for a 3S and 7S RNA map between the 23S and 16S RNA genes (126). The 23S, 16S, and 5S RNAs in *Z. mays* (8) and the 23S, 3S, 7S, and 5S RNAs (126) in *C. reinhardii* have been shown to be transcribed from the same strand. It is likely that they are synthesized as a common precursor (66), as is true for *E. coli* (76, 113).

Maizels (98) has pointed out that the linkage relationship between the 5S RNA genes and the genes for the large rRNAs is related to the genetic complexity of the organism. In *Xenopus* (115) and *Drosophila* (170) the 5S rRNA genes are not linked to the genes for the 18S and 28S RNAs. In yeast (127) and *Dictyostelium* (98) the 5S genes are linked to the genes for the larger rRNAs. In yeast (49), however, the 5S RNA is transcribed as a separate unit. In chloroplasts the relationship of the 5S rRNA gene to the larger rRNA genes is most like that in prokaryotes.

In *C. reinhardii* (126) all of the 23S RNA genes are interrupted by a 940 base pair intervening sequence at a site 270 base pairs from the 5' end of the 23S RNA coding strand. A similar situation has been found in *Drosophila melanogaster* (162), where about two-thirds of the rDNA repeats contain insertions in the 28S RNA gene. These insertions ranged in size from 500 to 6000 base pairs and occurred in distinct size classes which were multiples of 500 base pairs. Insertions were not detected in the rRNA genes of *Zea mays* (8), *P. sativum* (26), and *S. oleracea* (149) ctDNA.

It has recently been shown (17, 37, 168) that an additional low molecular weight RNA is present in the large subunit of the chloroplast ribosome.

This RNA, known as the 4.5S RNA, is present in all higher plant chloroplast ribosomes tested. Sequence analysis (57) showed that 4.5S RNA is not a tRNA or a variant of 5S RNA. Hartley (65) has demonstrated at least four copies of the sequence for 4.5S RNA per chloroplast genome, using quantitative hybridization experiments. *C. reinhardii* ctDNA does not appear to code for a 4.5S rRNA (126). In spinach ctDNA (168) the 4.5S RNA gene mapped adjacent to the distal end of the 23S RNA gene and adjacent to the 5S RNA gene. In *Zea mays* (T. A. Dyer and J. R. Bedbrook, unpublished) the 4.5S coding sequence was found to be located between the genes for the 23S RNA and the 5S RNA.

Genes for Chloroplast tRNAs

More than 20 tRNA genes (62, 104, 105, 151) have been identified on chloroplast DNA by filter hybridization of total 4S RNA to ctDNA and by hybridization of individually aminoacylated tRNA to ctDNA. In *S. oleracea* ctDNA (144), 23 tRNA species, representing 15 different amino acids, have been identified, while in *P. sativum* ctDNA, tRNAs representing 17 amino acids have been identified (104, 105). Failure to detect tRNAs for all amino acids could simply represent technical difficulties in charging all of the tRNAs. The physical arrangement of these genes on the ctDNA has been investigated in several species. In *S. oleracea* the tRNA species have been shown to map in the inverted repeats and in the single copy sequences flanking the inverted repeats (144). The isoaccepting tRNA species for serine, leucine, and methionine appear to represent distinct genes. In *E. gracilis* (63, 146), tRNA genes have been found to be distributed over the map, and in *C. reinhardii* (99, 125), 4S RNA was found to hybridize to at least 10 Eco RI and 7 Bam HI fragments. These fragments are distributed throughout the map. The spacer between the 16S and 23S rRNA genes codes for at least one 4S RNA in *C. reinhardii* (99, 125).

The Large Subunit of Ribulose Bisphosphate Carboxylase

Evidence that the large subunit of ribulose bisphosphate carboxylase (LS) is encoded by ctDNA is based on both genetic and biochemical experiments. Wildman and his colleagues have reported that the LS is maternally inherited in *Nicotiana* (23, 92). These workers took advantage of the fact that tryptic digests of LS from *N. gossei* show one more peptide than the tryptic digest of *N. tabacum.* The tryptic digest patterns were then used as a marker for the LS in reciprocal crosses. It was found that the extra peptide pattern characteristic of *N. gossei* was found in the F_1 progeny only when *N. gossei* was the maternal parent. More recently, isoelectric focusing variants of the LS have been used as a marker in demonstrating maternal inheritence of this protein (131). Maternal inheritance of the LS in a cross of *N. sylvestris* with *N. tomentosiforis,* using differences in amino acid composition as a genetic marker, has also been demonstrated by Kung and colleagues (cited in 91). Maternal inheritance of biochemical properties of the LS has also been shown in wheat (24) and *Avena sativa* (142).

Limitations in all these experiments can be discerned. First, it is possible that these analyses are following the maternal inheritance of a protein or proteins which modify the LS rather than the structural gene for the LS itself, and second, neither the F_2 generation of the reciprocal crosses nor backcrosses to the parents have been investigated. Nevertheless, these experiments provide the strongest evidence that LS is maternally inherited and therefore probably encoded by ctDNA.

Recently direct evidence that ctDNA contains the structural gene of the LS has been obtained for *C. reinhardii* (125), *Z. mays* (32) and *S. oleracea* (19). These experiments have involved the in vitro transcription and translation of total ctDNA (19) or of specific cloned fragments of ctDNA (32, 125). In *Z. mays* (32), using a linked transcription-translation system consisting of *E. coli* RNA polymerase and a translation system derived from rabbit reticulocytes, Coen et al (32) have shown that a 4000 base pair long cloned DNA fragment directed in vitro the synthesis of a polypeptide having a slightly larger size than the LS. This polypeptide had similar serological properties, limited proteolytic digestion products (31, 32), and the tryptic peptides (7) to the LS. This 4000 nucleotide long DNA fragment was found to be located (7) 30,000 base pairs from the 5' end of the closest of the two sets of rRNA genes and approximately 71,000 base pairs away from the other set of rRNA genes. A detailed physical map of the 4000 base pair long Bam HI fragment was made using various restriction endonucleases (7). In vitro transcription and translation of the 4000 base pair long DNA fragment digested with these various enzymes mapped the position of the structural gene for the LS within a 2500 base pair long sequence. Since the coding sequence required for the LS must be at least 1500 base pairs long, these results can be taken to mean that each copy of *Z. mays* ctDNA contains only one gene for the LS. In *S. oleracea* (19), LS has been found to be a product of in vitro transcriptions and translation of total ctDNA in a cell-free system derived from *E. coli.* As yet the precise location and number of copies per ctDNA molecule have not been determined.

In *C. reinhardii,* two approaches have been used in attempts to locate the gene for the LS. Howell and his colleagues (45, 71) used purified mRNA to identify restriction fragments and cloned DNA fragments containing sequences encoding the LS. They immunoprecipitated polyribosomes synthesizing LS and purified the RNA from these polyribosomes by electrophoresis in acrylamide gels. Two RNA bands from the gels directed incorporation of amino acids into polypeptides that were immunoprecipitated by antisera raised against LS. These RNA bands hybridized to a 4800 base pair long DNA fragment resulting from digestion of ctDNA with Eco RI, and to a lesser extent to other DNA fragments. When DNA clones were selected, which hybridized to LS mRNA, it was found that they also hybridized to ribosomal RNA. The hybridization was not competed out with purified rRNA. This result led to the fascinating suggestion (45) that the LS gene is directly linked to the rRNA genes and coordinately transcribed with the rRNA genes. However, no real estimate of the purity of the LS mRNA was given, and consequently there was no direct evidence that the RNA hybridizing to the 4800 base pair long DNA fragment was the same as the RNA that was translated in vitro. More recently, Rochaix & Malnoe (125) have reported that the gene for LS in *C. reinhardii* ctDNA is not associated with the rRNA genes. They found that a DNA clone containing a 5500 base pair long Eco RI fragment of *C. reinhardii* ctDNA directed in vitro synthesis of a polypeptide which was similar in size to the LS and was precipitated by LS antibody. This polypeptide had very similar tryptic peptides to the LS. The 5500 base pair fragment maps outside the inverted repeat and is present once per genome.

The 32,000 Dalton Membrane Protein

In *Z. mays,* evidence for the location of a gene for a chloroplast membrane protein has accumulated (9, 31; L. McIntosh and L. Bogorad, unpublished results). Steinback (143) found a protein of about 32,000 daltons which accumulates in chloroplast membranes during the light-dependent development of chloroplasts from etioplasts. A 34,500 polypeptide synthesized in isolated chloroplasts gave limited proteolytic peptides similar to those of the 32,000 dalton polypeptide (53). It was found (9, 31) that *Z. mays* ctRNA could direct the synthesis of a polypeptide which co-migrated with the 34,500 dalton polypeptide synthesized in isolated chloroplasts. The tryptic peptides (31) of the 34,500 material synthesized in isolated cloroplasts and in vitro were similar.

In a comparative analysis (9), RNA fractions extracted from etioplasts, "partically green" chloroplasts, and from "mature" chloroplasts were used to direct an in vitro translation system and for hybridization to the fragments of *Z. mays* ctDNA resulting from digestion of the ctDNA by Bam HI. It was found (9) that the amount of translation of the 34,500 polypeptide directed by the various plastid RNA franctions correlated with the amount of hybridization to Bam HI fragment 8 that was observed. The Bam HI fragment 8 is located within the Sal I fragment B of *Z. mays* ctDNA and, in several experiments (31; L. McIntosh and L. Bogorad, unpublished results) the Sal I fragment B and a derivative of the Bam HI fragment 8 was shown to direct the synthesis of a 34,500 dalton polypeptide in a linked transcription-translation system. The Bam HI fragment 8 maps (6) at the interface between one of the inverted repeat segments and the large spacer region and is approximately 68,000 base pairs away from the gene for the LS.

Regulation of Chloroplast Genes

In *Zea mays* the availability of a physical map for the ctDNA has made it possible to begin studying the regulation of several chloroplast genes. In one study (9), the changes in pools of plastid mRNA during the light-dependent development of chloroplasts from etioplasts were investigated. One class of mRNA was observed to increase tenfold during the development of chloroplasts; this mRNA mapped on the Bam HI fragment 8 of the *Z. mays* ctDNA (6) and probably codes for a chloroplast membrane protein (31, 53, 143). Similar developmental regulation of the mRNA coding for the photosynthetic lamellar P32,000 protein in *Spirodella* (40) has been observed, and an analogous protein may also have been observed in *Pisum sativum* (42) and *Lemna gibba* (158). In a second study in *Z. mays* (93), a DNA fragment containing the coding sequence for the LS was used to study the distribution of LS mRNA in mesophyll and bundle sheath cells. The gene for the LS is present in the ctDNA in both the mesophyll and bundle sheath cells (161) even though the LS is only found in bundle sheath cells. By preparing separate chloroplast RNA fractions from mesophyll and bundle sheath cells and hybridizing the RNA to the DNA fragment containing the LS gene [Bam HI fragment (9)], Link et al (93) determined that the LS mRNA, like the LS polypeptide, was only present in bundle sheath cells. The reason for this difference in the RNA populations of the two cell types is unclear but could involve the control of transcription or rapid degradation of LS mRNA in the mesophyll cells.

CONCLUSION

In this review we have discussed experiments which attempted to elucidate the physical and genetic structure of ctDNA. We feel that the combined results of the experiments on the physical structure of ctDNA support the idea that in most plants and algae the entire chloroplast genome is represented by the sequence of a single circular ctDNA molecule. The mapping experiments have demonstrated that the structural genes for several chloroplast components are located on the circular ctDNA molecules. This is expected if the circular ctDNA molecules represent the chloroplast genome. It should be noted that the chloroplast components that have been mapped so far represent major chloroplast components. Methodology is now available to map other chloroplast components which will lead to a more complete picture of the chloroplast genome.

A major problem encountered when considering mapping of the chloroplast genome is the identification of other components that are coded for by the chloroplast genome. This is even more complicated (and more interesting!) if one considers that a ctDNA molecule with a molecular weight of 1×10^8 could encode on the order of 100 average size proteins. A good indication of which proteins are likely to be encoded by the ctDNA comes from experiments involving protein synthesis by isolated chloroplasts (14, 42). This system takes advantage of the fact that only intact chloroplasts are capable of producing the photosynthetically derived energy necessary to drive the protein synthesis observed in the system. This strategy eliminates confusion caused by proteins synthesized by contaminating cytoplasmic and mitochondrial protein synthesis systems. Intact organelle systems have been found to synthesize the LS (14, 18), three of the five subunits of the coupling factor CFI (41, 106), various membrane proteins (18, 39, 67), and cytochrome *f* (36). Indeed, Ellis (42) found that two-dimensional electrophoretic analysis revealed about 80 soluble products and at least 12 membrane proteins that were synthesized in isolated chloroplasts, although the products of one gene may yield more than one spot on a two-dimensional gel. Another indication of proteins coded for by the chloroplast genome comes from the genetic analysis of chloroplasts in *C. reinhardii* (50, 129, 130). These studies have identified a number of chloroplast mutations (50, 135) and specific chloroplast proteins whose structure is altered by some of these mutations (21, 64, 108, 112). Physical mapping of the genes for proteins that have been identified genetically is certainly possible with the types of techniques available today and would, most importantly, be capable of unifying the results obtained from the studies on the transmission genetics of the chloroplast genome with the physical mapping studies of the chloroplast genome.

In addition to helping us better understand the physical and genetic structure of the chloroplast genome, several of the experiments discussed here have pointed a way toward future understanding of chloroplast gene expression and its regulation, as well as a better understanding of the genetic relationship between chloroplasts and the rest of the cell. One cannot help but feel that in the next few years we will see a number of very exciting results in these areas.

ACKNOWLEDGMENTS

We would like to thank Drs. L. Bogorad, D. Coen, P. Dunsmuir, T. Dyer, R. Sager, and K. K. Tewari for helpful discussions, and Ms. D. Bretz for her help in preparing the manuscript. R. K. was supported by N. I. H. grant GM26017-01 and J. R. B. was supported by an EMBO Fellowship.

Literature Cited

1. Bastia, D., Chiang, K. S., Swift, H., Siersma, P. 1971. Heterogeneity, complexity and repetition of the chloroplast DNA of *Chlamydomonas reinhardii*. *Proc. Natl. Acad. Sci. USA* 68:1157–61
2. Bauer, W., Vinograd, J. 1968. The interaction of closed circular DNA with intercalative dyes: The superhelix density of SV40 DNA in the presence and absence of dye. *J. Mol. Biol.* 33:141–71
3. Bauer, E. 1909. Das wesen und die erblichkeitsverhaltnisse der "Varietates albomarginateae hort" von *Pelargonium zonale*. *Z. Vererbungsl.* 1:333–51
4. Bayen, M., Rode, A. 1973. Heterogeneity and complexity of *Chlorella* chloroplastic DNA. *Eur. J. Biochem.* 39:413–20
5. Bedbrook, J. R., Ausubel, F. M. 1976. Recombination between bacterial plasmids leading to the formation of plasmid multimers. *Cell* 9:707–16
6. Bedbrook, J. R., Bogorad, L. 1976. Endonuclease recognition sites mapped on *Zea mays* chloroplast DNA. *Proc. Natl. Acad. Sci. USA* 73:4309–13
7. Bedbrook, J. R., Coen, D. M., Beaton, A. R., Bogorad, L., Rich, A. 1979. Location of the single gene for the large subunit of ribulose bisphosphate carboxylase on the maize chloroplast chromosome. *J. Biol. Chem.* In press
8. Bedbrook, J. R., Kolodner, R., Bogorad, L. 1977. *Zea mays* chloroplast ribosomal RNA genes are part of a 22,000 base pair inverted repeat. *Cell* 11:739–49
9. Bedbrook, J. R., Link, G., Coen, D. M., Bogorad, L., Rich, A. 1978. Maize plastid gene expressed during photoregulated development. *Proc. Natl. Acad. Sci. USA* 75:3060–64
10. Behn, W., Herrmann, R. G. 1977. Circular molecules in the β satellite DNA of *Chlamydomonas reinhardii*. *Mol. Gen. Genet.* 157:25–30
11. Benbow, R., Zuccarelli, A., Sinsheimer, R. 1975. Recombinant DNA molecules of bacteriaphage ϕX174. *Proc. Natl. Acad. Sci. USA* 72:235–39
12. Benson, L. 1957. *Plant Classification*. Boston: Heath. 355 pp.
13. Black, C. C., 1973. Photosynthetic carbon fixation in relation to net CO_2 uptake. *Ann. Rev. Plant Physiol.* 24:253–86
14. Blair, G. E., Ellis, R. J. 1973. Protein synthesis in chloroplasts I. Light-driven synthesis of the large subunit of Fraction I protein by isolated pea chloroplasts. *Biochim. Biophys. Acta* 319:223–34
15. Blamire, J., Flechtner, V. R., Sager, R., 1974. Regulation of nuclear DNA replication by the chloroplast in *Chlamydomonas*. *Proc. Natl. Acad. Sci. USA* 71:2867–71
16. Bogenhagen, D., Gillium, A. M., Martens, P. A., Clayton, D. A. 1978. Replication of mouse L-cell mitochondrial DNA. *Cold Spring Harbor Symp. Quant. Biol.* 43. In press
17. Bohnert, H. J., Driesel, A. J., Herrmann, R. G. 1976. Transcription and processing of transcripts in isolated unbroken chloroplasts. In *Acides Nucleiques et synthese des proteines chez les vegetaux,* ed. L. Bogorad, J. A. Weil, pp. 213–17. New York: Plenum
18. Bottomley, W., Spencer, D., Whitfeld, P. R. 1974. Protein synthesis in isolated spinach chloroplasts: comparison of light-driven and ATP-driven synthesis. *Arch. Biochem. Biophys.* 164:106–17
19. Bottomley, W., Whitfeld, P. R. 1979. The products of *in vitro* transcriptions and translation of spinach chloroplast DNA. In *Chloroplast Development,* ed. G. Akoyunoglou. Amsterdam: Elsevier. In press
20. Britten, R. J., Kohne, D. E. 1968. Repeated sequences in DNA. *Science* 161:529–40
21. Brugger, M., Boschetti, A. 1975. Two-dimensional gel electrophoresis of ribosomal proteins from streptomycin-sensitive and streptomycin-resistant mutants of *Chlamydomonas reinhardii*. *Eur. J. Biochem.* 58:603–10
22. Cairns, J., 1963. The chromosome of *Escherichia coli*. *Cold Spring Harbor Symp. Quant. Biol.* 28:43–46
23. Chan, P. H., Wildman, S. G. 1972. Chloroplast DNA codes for the primary structure of Fraction I protein. *Biochim. Biophys. Acta* 277:677–80
24. Chen, K., Gray, J. C., Wildman, S. G. 1975. Fraction I protein and the origin of polyploid wheats. *Science* 190:1304–6
25. Chiang, K. S., Sueoka, N. 1967. Replication of chloroplast DNA in *Chlamydomonas reinhardii* during the vegetative cell cycle: Its mode and regulation. *Proc. Natl. Acad. Sci. USA* 57:1506–13
26. Chu, N., Tewari, K. K. 1979. Arrangement of the ribosomal RNA genes in the restriction endonuclease map of pea chloroplast DNA. Submitted for publication

27. Clark, M. F. 1964. Polyribosomes from chloroplasts. *Biochim. Biophys. Acta* 91:671–74
28. Clark, M. F., Matthews, R. E. F., Ralph, R. K. 1964. Ribosomes and polyribosomes in *Brassica pekinensis*. *Biochim. Biophys. Acta* 91:289
29. Clayton, D. A., Davis, R. W., Vinograd, J. 1970. Homology and structural relationships between the dimeric and monomeric circular forms of mitochondrial DNA from human leukemic leukocytes. *J. Mol. Biol.* 47:137–53
30. Cockburn, A. F., Newkirk, M. J., Firtel, R. A. 1976. Organization of the ribosomal RNA genes of *Dictyostelium discoideum:* Mapping of the nontranscribed spacer regions. *Cell* 9:605–13
31. Coen, D. M. 1978. *Identification and mapping of protein coding sequences in maize chloroplast DNA.* PhD thesis. Massachusetts Inst. Technol., Cambridge, Mass. 220 pp.
32. Coen, D. M., Bedbrook, J. R., Bogorad, L., Rich, A. 1977. Maize chloroplast DNA fragment encoding the large subunit of ribulose bisphosphate carboxylase. *Proc. Natl. Acad. Sci. USA* 74:5487–91
33. Correns, C. 1909. Vererbungsversuche mit blass (gelb) grunen und bluntblattrigen Sippen bei *Mirabilis, Urtica, und Lunaria. Z. Vererbungsl.* 1:291–329
34. Danna, K. J., Sack, G. H., Nathans, D. 1973. Studies on simian virus 40 DNA. VII. A cleavage map of the SV40 genome, *J. Mol. Biol.* 78:363–75
35. Davis, R. W., Simon, M., Davidson, N. 1971. Electron microscope heteroduplex methods for mapping regions of base sequence homology in nucleic acids. *Methods Enzymol.* 21D:413–28
36. Doherty, A., Gray, J. C. 1979. Synthesis of a cytochrome F by isolated pea chloroplasts. Submitted for publication
37. Dyer, T. A., Bowman, C. M., Payne, P. I. 1976. The low-molecular-weight RNAs of plant ribosomes: their structure, function and evolution. In *Nucleic Acids and Protein Synthesis in Plants,* ed. L. Bogorad, J. H. Weil, pp. 135–54. New York: Plenum
38. Dyer, T. A., Leech, R. M. 1968. Chloroplast and cytoplasmic low-molecular-weight ribonucleic acid components of the leaf of *Vicia faba* L. *Biochem. J.* 106:689–98
39. Eaglesham, A. R. J., Ellis, R. J. 1974. Protein synthesis in chloroplasts II. Light-driven synthesis of membrane proteins by isolated pea chloroplasts. *Biochim. Biophys. Acta* 335:396–407
40. Edelman, M., Reisfeld, A. 1979. Identification, *in vitro* translation, and control of the main chloroplast membrane protein synthesized in *Spirodela*. See Ref. 19
41. Ellis, R. J. 1976. The synthesis of chloroplast proteins, nucleic acids and protein synthesis in plants. See Ref. 37, pp. 195–212
42. Ellis, R. J. 1979. Synthesis and transport of chloroplast proteins inside and outside the cell. See Ref. 19
43. Ellis, R. J., Hartley, M. R. 1974. Nucleic acids of chloroplasts. In *Biochemistry of Nucleic Acids,* ed. K. Burton. Vol. 6 of *MTP International Review of Science,* Biochemistry series. Baltimore: University Park Press
44. Engberg, J., Andersson, P., Leick, V., Collins, J. 1976. Free ribosomal DNA molecules from *Tetrahymena pyriformis* GL are giant palindormes. *J. Mol. Biol.* 104:455–70
45. Gelvin, S. R., Heizmann, P., Howell, S. H. 1977. Identification and cloning of the chloroplast gene coding for the large subunit of ribulose-1,5-bisphosphate carboxylase from *Chlamydomonas reinhardii. Proc. Natl. Acad. Sci. USA* 74:3193–97
46. Gelvin, S. R., Howell, S. H. 1979. Small repeated sequences in the chloroplast genome of *Chlamydomonas reinhardii. Mol. Gen. Genet.* In press
47. Gibor, A. 1967. DNA synthesis in chloroplasts. In *Biochemistry of Chloroplasts,* ed. T. A. goodwin, Vol. 2. New York: Academic
48. Gilbert, W., Dressler, D. 1968. DNA replication: the rolling circle model. *Cold Spring Harbor Symp. Quant. Biol.* 33:473–84
49. Gilbert, W., Maxam, A. M., Tizard, R., Skryabin, G. K. 1977. A promoter region for a yeast 5S rRNA gene. *Eucaryotic Gene Expression,* ed. J. Abelson, G. W. Wilcox. New York: Academic
50. Gilham, N. W. 1974. Genetic analysis of the chloroplast and mitochondrial genomes. *Ann. Rev. Genet.* 8:347–91
51. Gray, P. W., Hallick, R. B. 1977. Restriction endonuclease map of *Euglena gracilis* chloroplast DNA. *Biochemistry* 16:1665–71
52. Gray, P. W., Hallick, R. B. 1978. Physical mapping of the *Euglena gracilis* chloroplast DNA and ribosomal RNA gene region. *Biochemistry* 17:284–89
53. Grebanier, A. E., Coen, D. M., Rich, A., Bogorad, L. 1978. Membrane proteins synthesized but not processed by

isolated maize chloroplasts. *J. Cell Biol.* 78:734–46
54. Green, B. R. 1976. Covalently closed minicircular DNA associated with *Acetabularia* chloroplasts. *Biochim. Biophys. Acta* 447:156–66
55. Green, B. R., Burton, H. 1970. *Acetabularia* chloroplast DNA: electron microscope visualization. *Science* 168:981–82
56. Green, B. R., Muir, B. L., Padmanabhan, U. 1977. The *Acetabularia* genome: small circles and large kinetic complexity. *Progress in Acetabularia Research,* ed. C. L. F. Woodcock, pp. 107–22. New York: Academic
57. Grey, H. B., Upholt, W. B., Vinograd, J. 1971. A buoyant method for the determination of the superhelix density of closed circular DNA. *J. Mol. Biol.* 62:1–19
58. Grossman, L. I., Watson, R., Vinograd, J. 1973. The presence of ribonucleotides in mature closed-circular mitochondrial DNA. *Proc. Natl. Acad. Sci. USA* 70:3339–43
59. Grun, P. 1976. *Cytoplasmic Genetics and Evolution.* New York: Columbia Univ. Press. 435 pp.
60. Guerineau, M., Grandchamp, C., Slonimski, P. P. 1976. Circular DNA of a yeast episome with two inverted repeats: Structural analysis by restriction enzyme and electron microscopy. *Proc. Natl. Acad. Sci. USA* 73:3030–34
61. Guerineau, M., Slonimski, P. P., Avner, P. R. 1974. Yeast episome: oligomycin resistance associated with a small covalently closed non-mitochondrial circular DNA. *Biochem. Biophys. Res. Commun.* 61:462–69
62. Haff, L. A., Bogorad, L. 1976. Hybridization of maize chloroplast DNA with transfer RNA. *Biochemistry* 15:4105–9
63. Hallick, R. B., Gray, P. W., Chelm, B. K., Rushlow, K. E. 1979. *Euglena* chloroplast DNA structure, gene mapping, and RNA transcription. See Ref. 19
64. Hansen, M. R., Davidson, J. N., Mets, L. J., Bogorad, L. 1974. Characterization of chloroplast and cytoplasmic ribosomal proteins of *Chlamydomonas reinhardii* by two-dimensional gel electrophoresis. *Mol. Gen. Genet.* 132:105–18
65. Hartley, M. R. 1979. The synthesis and origin of chloroplast low-molecular-weight ribosomal ribonucleic acid in spinach. Submitted for publication
66. Hartley, M. R., Ellis, R. J. 1973. Ribonucleic acid synthesis in chloroplasts. *Biochem. J.* 134:249–62
67. Hearing, K. J. 1973. Protein synthesis in isolated etioplasts after light stimulation. *Phytochemistry* 12:277–82
68. Herrmann, R. G., Bohnert, H. J., Kowallik, K. V., Schmitt, J. M. 1975. Size, conformation and purity of chloroplast DNA from some higher plants. *Biochim. Biophys. Acta* 378:305–17
69. Ho, C., Lipsich, L., Fisher, G., Keller, S. J. 1974. DNA replication in isolated chloroplasts of *Chlamydomonas reinhardii. J. Cell. Biol.* 68:104a
70. Hobom, G., Bohnert, H. J., Driesel, A., Herrmann, R. G. 1977. Restriction fragment map of the circular plastid DNA from *Spinacia oleracea.* See Ref. 37, pp. 195–212
71. Howell, S. H., Heizmann, P., Gelvin, S., Walker, L. L. 1977. Identification and properties of the messenger RNA activity in *Chlamydomonas reinhardii* coding for the large subunit of D-ribulose-1,5-bisphosphate carboxylase. *Plant. Physiol.* 59:464–70
72. Howell, S. H., Naliboff, J. A. 1975. Conditional mutants of *Chlamydomonas reinhardii* blocked in the vegetative cell cycle. *J. Cell. Biol.* 57:760–72
73. Howell, S. H., Walker, L. L. 1976. Informational complexity of the nuclear and chloroplast genomes of *Chlamydomonas reinhardii. Biochim. Biophys. Acta* 418:249–56
74. Hutchison, C. A. 1979. Two-dimensional hybridization mapping of restriction fragments. Submitted for publication
75. Inman, R. B., Schnös, M. 1970. Partial denaturation of thymine and 5-bromouracil-containing λ DNA in alkali. *J. Mol. Biol.* 49:93–98
76. Jaskunas, S. R., Nomura, M., Davies, J. 1974. *Ribosomes,* ed. M. Nomura, P. Lengyel, pp. 333–68. New York: Cold Spring Harbor Lab.
77. Karrer, K. M., Gall, J. G. 1976. The macronuclear ribosomal DNA of *Tetrahymena pyriformis* is a palindrome. *J. Mol. Biol.* 104:421–53
78. Kasamatsu, H., Robberson, D. L., Vinograd, J. 1971. A novel closed-circular mitochondrial DNA with properties of a replicating intermediate. *Proc. Natl. Acad. Sci. USA* 68:2252–57
79. Keller, S. J., Biedenbach, S. A., Meyer, R. R. 1973. Partial purification of a chloroplast DNA polymerase from *Euglena gracilis. Biochem. Biophys. Res. Commun.* 50:620–28
80. Kim, J. S., Sharp, P. A., Davidson, N. 1972. Electron microscope studies of heteroduplex DNA from a deletion mu-

tant of bacteriophage φX174. *Proc. Natl. Acad. Sci. USA* 69:1948–52

81. Kirk, J. T. O. 1971. Will the real chloroplast DNA please stand up? *Autonomy and Biogenesis of Mitochondria and Chloroplasts,* ed. N. K. Boardman, A. W. Linnane, R. M. Smillie, pp. 267–76. Amsterdam: North-Holland
82. Kirk, J. T. O., Tilney-Bassett, R. A. E. 1967. *The Plastids: Their Chemistry, Structure, Growth and Inheritance.* London: Freeman. 608 pp.
83. Kolodner, R., Tewari, K. K. 1972. Molecular size and conformation of chloroplast deoxyribonucleic acid from pea leaves. *J. Biol. Chem.* 247:6355–64
84. Kolodner, R., Tewari, K. K. 1975. Denaturation mapping studies on the circular chloroplast deoxyribonucleic acid from pea leaves. *J. Biol. Chem.* 250:4888–95
85. Kolodner, R., Tewari, K. K. 1975. The molecular size and conformation of chloroplast DNA from higher plants. *Biochim. Biophys. Acta* 402:372–90
86. Kolodner, R., Tewari, K. K. 1975. The presence of displacement loops in the covalently closed circular chloroplast deoxyribonucleic acid from higher plants. *J. Biol. Chem.* 250:8840–47
87. Kolodner, R., Tewari, K. K. 1975. Chloroplast DNA from higher plants replicates by both the Cairns and rolling circle mechanism. *Nature* 256:708–11
88. Kolodner, R., Tewari, K. K. 1979. Inverted repeats in the chloroplast DNA from higher plants. *Proc. Natl. Acad. Sci. USA* In press
89. Kolodner, R., Tewari, K. K., Warner, R. C. 1976. Physical studies on the size and structure of the covalently closed circular chloroplast DNA from higher plants. *Biochim. Biophys. Acta* 447: 144–55
90. Kolodner, R., Warner, R. C., Tewari, K. K. 1975. The presence of covalently linked ribonucleotides in the closed circular deoxyribonucleic acid from higher plants. *J. Biol. Chem.* 250:7020–26
91. Kung, S. D. 1977. Expression of chloroplast genomes in higher plants. *Ann. Rev. Plant Physiol.* 28:401–37
92. Kung, S. D. 1976. Tobacco Fraction 1 protein: a unique genetic marker. *Science* 191:429–34
93. Link, G., Coen, D. M., Bogorad, L. 1978. Differential expression of the gene for the large subunit of riboulose bisphosphate carboxylase in maize leaf cell types. *Cell* 15:725–32
94. Livingston, D. M., Klein, H. L. 1977. Deoxyribonucleic acid sequence organization of a yeast plasmid. *J. Bacteriol.* 129:472–81
95. Loening, V. E., Ingle, J. 1967. Diversity of RNA components in green plant tissues. *Nature* 215:363–67
96. Lund, E., Dahlberg, J. E., Lindahl, L., Jaskunas, S. R., Dennis, P. P., Nomura, M. 1976. Transfer RNA genes between 16S and 23S rRNA genes in rRNA transcription units of *E. coli. Cell* 7:165–77
97. Lyttleton, J. W. 1962. Isolation of ribosomes from spinach chloroplasts. *Exp. Cell. Res.* 16:312–17
98. Maizels, N. 1976. Dictyostelium 17S, 25S and 5S rDNAs lie within a 38,000 base pair repeated unit. *Cell* 9:431–38
99. Malnoe, P., Rochaix, J. D. 1979. Localization of 4S RNA genes on the chloroplast genome of *Chlamydomonas reinhardii. Mol. Gen. Genet.* In press
100. Manning, J. E., Richards, O. C. 1972. Isolation and molecular weight of circular chloroplast DNA from *Euglena gracilis. Biochim. Biophys. Acta* 259: 285–96
101. Manning, J. E., Richards, O. C. 1972. Synthesis and turnover of *Euglena gracilis* nuclear and chloroplast deoxyribonucleic acid. *Biochemistry* 11: 2036–43
102. Manning, J. E., Wolstenholme, D. R., Richards, O. C. 1972. Circular DNA molecules associated with chloroplasts of spinach, *Spinacia oleracea. J. Cell Biol.* 53:594–601
103. Manning, J. E., Wolstenholme, D. R., Ryan, R. S., Hunter, J. A., Richards, O. C. 1971. Circular chloroplast DNA from *Euglena gracilis. Proc. Natl. Acad. Sci. USA* 68:1169–73
104. Meeker, R. R., Tewari, K. K. 1979. tRNA genes in pea chloroplast DNA. Submitted for publication
105. Meeker, R. R., Tewari, K. K. 1979. Divergence of tRNA genes in the chloroplast DNA from higher plants. Submitted for publication.
106. Mendiola-Morgenthahler, L. R., Morgenthaler, J. J., Price, C. A. 1976. Synthesis of coupling factor CF protein by isolated spinach chloroplasts. *FEBS Lett.* 62:96–100
107. Meselson, M., Stahl, F. W. 1958. The replication of DNA in *Escherichia coli. Proc. Natl. Acad. Sci. USA* 44:671–82
108. Mets, L., Bogorad, L. 1972. Altered chloroplast ribosomal proteins associated with erythromycin-resistant mutants in two genetic systems of *Chlamydomonas reinhardii. Proc. Natl. Acad. Sci. USA* 69:3779–83

109. Meyer, Y., Herrmann, R. G. 1973. Analysis of DNA of male and female *Sphaerocarpos donnellii* Aust. (Liverwort). *Mol. Gen. Genet.* 124:167–76
110. Miyaki, M., Koide, K., Ono, T. 1973. RNase and alkali sensitivity of closed-circular mitochondrial DNA from rat ascites hepatoma cells. *Biochem. Biophys. Res. Commun.* 50:252–58
111. Nass, M. M. K., Ben-Shaul, Y. 1972. A novel closed circular duplex DNA in bleached mutant and green strains of *Euglena gracilis. Biochim. Biophys. Acta* 272:130–36
112. Ohta, N., Sager, R., Inouye, M. 1974. Identification of a chloroplast ribosomal protein altered by a chloroplast mutation in *Chlamydomonas. J. Biol. Chem.* 250:3655–59
113. Pace, N. R. 1973. Structure and synthesis of the ribosomal ribonucleic acids of prokaryotes. *Bacteriol. Rev.* 37:562–603
114. Padmanabhan, U., Green, B. R. 1979. The kinetic complexity of *Acetabularia* chloroplast DNA. *Biochim. Biophys. Acta.* In press
115. Pardue, M. L., Brown, D. D., Birnstiel, M. L. 1973. Location of the genes for 5S ribosomal RNA in *Xenopus laevis. Chromosoma* 42:191–203
116. Parker, R. C., Watson, R. M., Vinograd, J. 1977. Mapping of closed-circular DNAs by cleavage with restriction endonucleases and calibration by agarose gel electrophoresis. *Proc. Natl. Acad. Sci. USA* 74:851–55
117. Payne, P. I., Dyer, T. A. 1971. Characterization of cytoplasmic and chloroplast 5S ribosomal ribonucleic acid from broad-bean leaves. *Biochem. J.* 124:83–89
118. Potter, H., Dressler, D. 1976. On the mechanism of genetic recombination: Electron microscopic observation of recombination intermediates. *Proc. Natl. Acad. Sci. USA* 73:3000–4
119. Richards, O. C., Manning, J. E. 1975. Replication of chloroplast DNA in *Euglena gracilis. Le Cycles Cellulaires et Leur Blocage,* pp. 213–24. C.N.R.S.
120. Richardson, C. C. 1965. Phosphorylation of nucleic acid by an enzyme from T4 bacteriophage infected *Escherichia coli. Proc. Natl. Acad. Sci. USA* 54:158–65
121. Roberts, R. J. 1976. Restriction endonucleases. *CRC Crit. Rev. Biochem.,* pp. 123–64
122. Rochaix, J. D. 1972. Cyclization of DNA fragments from *Chlamydomonas reinhardii. Nature New Biol.* 238:76–78
123. Rochaix, J. D. 1977. Restriction enzyme analysis of the chloroplast DNA of *Chlamydomonas reinhardii* and construction of chloroplast DNA plasmid hybrids. See Ref. 37, pp. 77–83
124. Rochaix, J. D. 1979. Restriction endonuclease map of the chloroplast DNA of *Chlamydomonas reinhardii. J. Mol. Biol.* In press
125. Rochaix, J. D., Malnoe, P. 1979. Gene localization on the chloroplast DNA of *Chlamydomonas reinhardii.* See Ref. 19
126. Rochaix, J. D., Malnoe, P. 1978. Anatomy of the chloroplast ribosomal DNA of *Chlamydomonas reinhardii. Cell* 15:681–70
127. Rubin, G. M., Sulston, J. E. 1973. Physical linkage of the 5S cistrons to the 18S and 28S ribosomal RNA cistrons in *Saccharomyces cerevisiae. J. Mol. Biol.* 79:521–30
128. Sager, R. 1972. *Cytoplasmic Genes and Organelles.* New York: Academic. 450 pp.
129. Sager, R. 1977. Genetic analysis of chloroplast DNA in *Chlamydomonas. Adv. Genet.* 19:287–340
130. Sager, R., Schlanger, G., 1976. Chloroplast DNA: Physical and genetic studies. *Handbook of Genetics* ed. R. C. King, 5:371–423. New York: Plenum
131. Sakano, K., Kung, S. D., Wildman, S. G. 1974. Identification of several chloroplast DNA genes which code for the large subunit of *Nicotiana* Fraction I protein. *Mol. Gen. Genet.* 130:91–97
132. Schiff, J. A. 1973. The development, inheritance, and origin of the plastid in *Euglena. Adv. Morphog.* 10:265–312
133. Schmid, C. W., Hearst, J. E. 1969. Molecular weights of homogeneous coliphage DNAs from density-gradient sedimentation equilibrium. *J. Mol. Biol.* 44:143–60
134. Schmid, C. W., Manning, J. E., Davidson, N. 1975. Inverted repeat sequences in the *Drosophila* genome. *Cell* 5:159–72
135. Scott, N. S., Shah, V. C., Smillie, D. M. 1968. Synthesis of chloroplast DNA in isolated chloroplasts. *J. Mol. Biol.* 38:151–52
136. Siu, C. H., Chiang, K. S., Swift, H. 1975. Characterization of cytoplasmic and nuclear genomes in the colorless algae Polytoma V. Molecular structure and heterogeneity of leucoplast DNA. *J. Mol. Biol.* 98:369–91
137. Smith, C. A. *Closed circular DNA in animal cells.* PhD thesis. California Inst. Technol., Pasadena, Calif. 216 pp.

138. Smith, C. A., Vinograd, J. 1972. Small polydisperse circular DNA of HeLa cells. *J. Mol. Biol.* 69:163–78
139. Smith, H. O., Birnstiel, M. L. 1976. A simple method for DNA restriction site mapping. *Nucleic Acids Res.* 3:2387–98
140. Southern, E. M. 1975. Detection of specific sequences among DNA fragments separated by gel electrophoresis. *J. Mol. Biol.* 98:503–17
141. Spencer, D., Whitfeld, P. R., Bottomley, W., Wheeler, A. M. 1971. The nature of the proteins and nucleic acids synthesized in isolated chloroplasts. See Ref. 81, pp. 373–82
142. Steer, M. W. 1975. Evolution in the genus *Avena:* inheritance of different forms of ribulose diphosphate carboxylase. *Can. J. Genet. Cytol.* 17:337–44
143. Steinback, K. 1977. *The organization and development of chloroplast thylakoid membranes in Zea mays.* PhD thesis. Harvard Univ., Boston, Mass. 189 pp.
144. Steinmetz, A., Mubumbila, V., Keller, M., Burkard, G., Weil, J. H., Driesel, A. J., Crouse, E. J., Gorgon, K., Bohnert, H. J., Herrmann, R. G. 1979. Mapping of the tRNA genes on the circular DNA molecule of *Spinacia oleracea.* See Ref. 19
145. Stutz, E. 1970. The kinetic complexity of *Euglena gracilis* chloroplast DNA. *FEBS Lett.* 8:25–28
146. Stutz, E. 1979. Gene mapping on *Euglena gracilis* chloroplast DNA. See Ref. 19
147. Stutz, E., Noll, H. 1967. Characterization of cytoplasmic and chloroplast polysomes in plants. Evidence for three classes of ribosomal RNA in nature. *Proc. Natl. Acad. Sci. USA* 57:744–81
148. Stutz, E., Vandrey, J. P. 1971. Ribosomal satellite of *Euglena gracilis* chloroplast DNA. *FEBS Lett.* 17:277–80
149. Swift, H., Whitfeld, P. R. 1979. Orientation of the duplicate ribosomal RNA cistrons in spinach chloroplast DNA as shown by R-loops. Submitted for publication
150. Tewari, K. K. 1971. Genetic autonomy of extracellular organelles. *Ann. Rev. Plant Physiol.* 22:141–68
151. Tewari, K. K., Kolodner, R., Chu, N. M., Meeker, R. R. 1977. The structure of chloroplast DNA. See Ref. 37, pp. 15–36
152. Tewari, K. K., Wildman, S. G. 1967. DNA polymerase in isolated tobacco chloroplasts and the nature of the polymerized product. *Proc. Natl. Acad. Sci. USA* 58:689–96
153. Tewari, K. K., Wildman, S. G. 1968. Function of chloroplast DNA. I. Hybridization studies involving nuclear and chloroplast DNA with RNA from cytoplasmic (80S) and chloroplast (70S) ribosomes. *Proc. Natl. Acad. Sci. USA* 59:569–76
154. Tewari, K. K., Wildman, S. G. 1970. Information content in the chloroplast DNA. *Control of Organelle Development,* pp. 147–79.
155. Thomas, C. A., Hamkalo, B. A., Misra, D. N., Lee, C. S. 1970. Cyclization of eucaryotic deoxyribonucleic acid fragments. *J. Mol. Biol.* 51:621–32
156. Thomas, J. R., Tewari, K. K. 1974. Conservation of 70S ribosomal RNA genes in the chloroplast DNAs of higher plants. *Proc. Natl. Acad. Sci. USA* 71:3147–51
157. Thompson, B. J., Escarmis, C., Parker, B., Slater, W. C., Doniger, J., Tessman, I., Warner, R. C. 1975. Figure-8 configuration of dimers of S13 and ϕX174 replicative form DNA. *J. Mol. Biol.* 91:409–19
158. Tobin, E. M. 1978. Light regulation of specific mRNA species in *Lemna gibba* L.G-3. *Proc. Natl. Acad. Sci. USA* 75:4749–53
159. Vinograd, J., Lebowitz, J., Radloff, R., Watson, R., Laipis, P. 1965. The twisted form of polyoma viral DNA. *Proc. Natl. Acad. Sci. USA* 53:1104–11
160. Vogt, V. M., Braun, R. 1976. Structure of ribosomal DNA in *Physarum polycephalum. J. Mol. Biol.* 106:567–87
161. Walbot, V. 1977. The dimorphic chloroplasts of the C_4 plant *Panicum maximum* contain identical genomes. *Cell* 11:729–37
162. Wellauer, P. K., Dawid, I. B. 1977. The structural organization of ribosomal DNA in *Drosophila melanogaster. Cell* 10:193–312
163. Wells, R., Birnstiel, M. 1969. Kinetic complexity of chloroplastal deoxyribonucleic acid and mitochondrial deoxyribonucleic acid from higher plants. *Biochem. J.* 122:777–86
164. Wells, R., Sager, R. 1971. Denaturation and renaturation kinetics of chloroplast DNA from *Chlamydomonas reinhardii. J. Mol. Biol.* 58:611–22
165. Wetmur, J. G., Davidson, N. 1968. Kinetics of renaturation of DNA. *J. Mol. Biol.* 31:349–70
166. Whitfeld, P. R., Atchison, B. A., Bottomley, W., Leaver, C. J. 1976. Analysis of the coding capacity of EcoRI restriction fragments of spinach chloroplast DNA. *Genetics and Biogenesis of Chlo-*

roplasts and Mitochondria, ed. T. Bucher. Amsterdam: Elsevier
167. Whitfeld, P. R., Herrmann, R.G., Bottomley, W. 1978. Mapping of the ribosomal RNA genes on spinach chloroplast DNA. *Nucleic Acid Res.* 5:1741–51
168. Whitfeld, P. R., Leaver, C. J., Bottomley, W., Atchison, B. A. 1979. On 4.5S RNA in higher plant chloroplasts. *Biochem. J.* In press
169. Wilson, D. A., Thomas, C. A. Jr. 1974. Palindromes in chromosomes. *J. Mol. Biol.* 84:115–44
170. Wimber, D. E., Steffensen, D. M. 1970. Localization of 5S RNA genes on *Drosophila* chromosomes by RNA-DNA hybridization. *Science* 170:639–41
171. Wong-Staal, F., Mendelsohn, J., Goulian, M. 1973. Ribonucleotides in closed-circular mitochondrial DNA from HELA cells. *Biochem. Biophys. Res. Commun.* 53:140–48

AUTHOR INDEX

A

B

H

N

Q

R

S

SUBJECT INDEX

B

C

D

N

CUMULATIVE INDEXES

CONTRIBUTING AUTHORS, VOLUMES 26–30

CHAPTER TITLES, VOLUMES 26–30

ORDER FORM ANNUAL REVIEWS INC.

Please list on the order blank on the reverse side the volumes you wish to order and whether you wish a standing order (the latest volume sent to you automatically upon publication each year). Volumes not yet published will be shipped in month and year indicated. Prices subject to change without notice. Out of print volumes subject to special order.

NEW.... to be published in 1980

ANNUAL REVIEW OF PUBLIC HEALTH — $17.00 per copy ($17.50 outside USA)
Volume 1 available May 1980

SPECIAL PUBLICATIONS

ANNUAL REVIEW REPRINTS: CELL MEMBRANES, 1975-1977 (published 1978)
A collection of articles reprinted from recent Annual Review series.
Soft cover — $12.00 per copy ($12.50 outside USA)

THE EXCITEMENT AND FASCINATION OF SCIENCE (published 1965)
A collection of autobiographical and philosophical articles by leading scientists.
Clothbound — $6.50 per copy ($7.00 outside USA)

THE EXCITEMENT AND FASCINATION OF SCIENCE, VOLUME 2:
Reflections by Eminent Scientists (published 1978)
Hard cover — $12.00 per copy ($12.50 outside USA)
Soft cover — $10.00 per copy ($10.50 outside USA)

HISTORY OF ENTOMOLOGY (published 1973)
A special supplement to the ANNUAL REVIEW OF ENTOMOLOGY series.
Clothbound — $10.00 per copy ($10.50 outside USA)

ANNUAL REVIEW SERIES

Annual Review of ANTHROPOLOGY	$17.00 per copy ($17.50 outside USA)
Volumes 1-7 (1972-1978) currently available	Volume 8 available October 1979
Annual Review of ASTRONOMY AND ASTROPHYSICS	$17.00 per copy ($17.50 outside USA)
Volumes 1-16 (1963-1978) currently available	Volume 17 available September 1979
Annual Review of BIOCHEMISTRY	$18.00 per copy ($18.50 outside USA)
Volumes 28-47 (1959-1978) currently available	Volume 48 available July 1979
Annual Review of BIOPHYSICS AND BIOENGINEERING	$17.00 per copy ($17.50 outside USA)
Volumes 1-7 (1972-1978) currently available	Volume 8 available June 1979
Annual Review of EARTH AND PLANETARY SCIENCES	$17.00 per copy ($17.50 outside USA)
Volumes 1-6 (1973-1978) currently available	Volume 7 available May 1979
Annual Review of ECOLOGY AND SYSTEMATICS	$17.00 per copy ($17.50 outside USA)
Volumes 1-9 (1970-1978) currently available	Volume 10 available November 1979
Annual Review of ENERGY	$17.00 per copy ($17.50 outside USA)
Volumes 1-3 (1976-1978) currently available	Volume 4 available October 1979
Annual Review of ENTOMOLOGY	$17.00 per copy ($17.50 outside USA)
Volumes 7-23 (1962-1978) currently available	Volume 24 available January 1979
Annual Review of FLUID MECHANICS	$17.00 per copy ($17.50 outside USA)
Volumes 1-10 (1969-1978) currently available	Volume 11 available January 1979

(continued on reverse side)

Annual Review of GENETICS	$17.00 per copy ($17.50 outside USA)
Volumes 1-12 (1967-1978) currently available	Volume 13 available December 1979
Annual Review of MATERIALS SCIENCE	$17.00 per copy ($17.50 outside USA)
Volumes 1-8 (1971-1978) currently available	Volume 9 available August 1979
Annual Review of MEDICINE: Selected Topics in the Clinical Sciences	$17.00 per copy ($17.50 outside USA)
Volumes 1-3, 5-15, 17-29 (1950-1952, 1954-1964, 1966-1978) currently available	Volume 30 available April 1979
Annual Review of MICROBIOLOGY	$17.00 per copy ($17.50 outside USA)
Volumes 14-32 (1960-1978) currently available	Volume 33 available October 1979
Annual Review of NEUROSCIENCE	$17.00 per copy ($17.50 outside USA)
Volume 1 currently available	Volume 2 available March 1979
Annual Review of NUCLEAR AND PARTICLE SCIENCE	$19.50 per copy ($20.00 outside USA)
Volumes 9-28 (1959-1978) currently available	Volume 29 available December 1979
Annual Review of PHARMACOLOGY AND TOXICOLOGY	$17.00 per copy ($17.50 outside USA)
Volumes 1-3, 5-18 (1961-1963, 1965-1978) currently available	Volume 19 available April 1979
Annual Review of PHYSICAL CHEMISTRY	$17.00 per copy ($17.50 outside USA)
Volumes 9-29 (1958-1978) currently available	Volume 30 available November 1979
Annual Review of PHYSIOLOGY	$17.00 per copy ($17.50 outside USA)
Volumes 19-40 (1957-1978) currently available	Volume 41 available March 1979
Annual Review of PHYTOPATHOLOGY	$17.00 per copy ($17.50 outside USA)
Volumes 1-16 (1963-1978) currently available	Volume 17 available September 1979
Annual Review of PLANT PHYSIOLOGY	$17.00 per copy ($17.50 outside USA)
Volumes 10-29 (1959-1978) currently available	Volume 30 available June 1979
Annual Review of PSYCHOLOGY	$17.00 per copy ($17.50 outside USA)
Volumes 4, 5, 8, 10-29 (1953, 1954, 1957, 1959-1978) currently available	Volume 30 available February 1979
Annual Review of SOCIOLOGY	$17.00 per copy ($17.50 outside USA)
Volumes 1-4 (1975-1978) currently available	Volume 5 available August 1979

To ANNUAL REVIEWS INC., 4139 El Camino Way, Palo Alto, CA 94306 USA (415-493-4400)
Please enter my order for the following publications:
(Standing orders: indicate which volume you wish order to begin with)

______________________, Vol(s). _____ Standing order _____

______________________, Vol(s). _____ Standing order _____

______________________, Vol(s). _____ Standing order _____

______________________, Vol(s). _____ Standing order _____

Amount of remittance enclosed $______ California residents please add sales tax.
Please bill me for the amount $______ Prices subject to change without notice.

SHIP TO (Include institutional purchase order if billing address is different)

Name ______________________

Address ______________________

____

______________________ Postal Code ________

Signed ______________________ Date ________

___ Send free copy of annual Prospectus for current year

___ Send free back contents brochure for Annual Review(s) of ________

PLACE
STAMP
HERE

ANNUAL REVIEWS INC.

4139 EL CAMINO WAY

PALO ALTO, CALIFORNIA 94306, USA